DYNAMIC
ANATOMY &
PHYSIOLOGY

Fourth Edition

DYNAMIC ANATOMY & PHYSIOLOGY

L. L. Langley, Ph. D., LL. B.
Associate Dean and Professor
University of Missouri—Kansas City
  School of Medicine

Ira R. Telford, Ph. D.
Professor of Anatomy
Georgetown University
  Schools of Medicine and Dentistry

John B. Christensen, Ph. D.
Director, Education Resources and Professor
University of Missouri—Kansas City
  School of Medicine

McGraw-Hill Book Company   A Blakiston Publication

New York St. Louis San Francisco Düsseldorf Johannesburg Kuala Lumpur London
Mexico Montreal New Delhi Panama Paris São Paulo Singapore Sydney
Tokyo Toronto

DYNAMIC ANATOMY & PHYSIOLOGY

Copyright © 1958, 1963, 1969, 1974 by McGraw-Hill, Inc.
All rights reserved. Printed in the United States of America.
No part of this publication may be reproduced,
stored in a retrieval system,
or transmitted,
in any form or by any means,
electronic, mechanical, photocopying, recording, or otherwise,
without the prior written permission of the publisher.

1 2 3 4 5 6 7 8 9 0   VHVH   7 9 8 7 6 5 4

This book was set in Elegante by York Graphic Services, Inc.
The editors were Cathy Dilworth and Sally Barhydt Mobley;
the designer was J. E. O'Connor;
and the production supervisor was Thomas J. LoPinto.
The drawings were done by Biagio Melloni, Peter Stone, and Jane Hurd.
Von Hoffmann Press, Inc., was printer and binder.

Library of Congress Cataloging in Publication Data

Langley, Leroy Lester.
    Dynamic anatomy and physiology.

    "A Blakiston publication."
    Bibliography: p.
        1. Human physiology. 2. Anatomy, Human.
I. Telford, Ira Rockwood, date   joint author.
II. Christensen, John B., date   joint author.
III. Title [DNLM: 1. Anatomy. 2. Physiology.
QS4 L283d 1974]
QP34.L348 1974   612   73-17153
ISBN 0-07-036274-2

CONTENTS

v

PREFACE

This textbook underwent a major revision in preparation for the third edition, which was published in 1969. Two new coauthors were added, the format was altered, and new art was prepared. Acceptance proved to be great, as measured by comments from users and a dramatic increase in sales. Accordingly, this fourth edition retains the most outstanding features of the third.

In addition to generally updating the text, especially Chapters 15, 18, 19, 25, 26, 29, 31, and 33, and making changes suggested by users and reviewers, we have paid special attention in this edition to the chemical transmission of the nervous impulse; acupuncture; the dynamics of circulation; cyclic AMP; the hypothalamic hormones; and sexual reproduction, including discussions of contraception, calcitonin, and the pineal gland. The very important concept of renal clearance, inexplicably left out of the third edition, has been reinstated.

This fourth edition has been given an elegant new design. Approximately seventeen new line illustrations have been prepared, and a number of the original drawings have been altered in response to user comments and in an effort to correct the seemingly inevitable errors. Several charts have also been added, and where feasible the metric system has been employed in place of the units inherited from Great Britain.

There is always a problem of selection. New material can easily be added; the harder task is to delete the old unless, of course, it is simply no longer tenable. As a result, the size of a textbook grows inexorably with time. But we have made an effort to keep the expansion of the book to a minimum.

Until very recently, the scientific research effort, especially in the United States, was expanding at a very vigorous rate. The result is a tremendous accretion of knowledge and an even greater blizzard of published material. Despite improved methods of communication, in the final analysis, one still must wade through that blizzard at a rate limited by one's ability to read and to assimilate. And, as everyone knows, it is a losing struggle. No matter how much time is spent, inundation is inevitable. All of this is by way of apologizing to those whose favorite concepts are omitted or not discussed to their satisfaction.

The authors gratefully acknowledge the contributions of the many people whose helpful suggestions are reflected in this edition. In particular, we wish to thank Mr. Peter Clason of Oakland Community College and Dr. Dean Schick of the State University of New York, College at Cortland, for their careful review of the third edition and recommendations for revision.

Once again, we solicit your comments and suggestions for future editions.

L. L. Langley
Ira R. Telford
John B. Christensen

DYNAMIC ANATOMY & PHYSIOLOGY

PART 1
ORIENTATION TO ANATOMY AND PHYSIOLOGY

1 INTRODUCTION TO ANATOMY AND PHYSIOLOGY

Since the beginning of recorded history man has been intrigued by the form and function of his body. Interest in the normal as well as the diseased human body is expressed in the earliest writings of the Egyptians and Babylonians, and in Chinese literature. It is likely that man's curiosity about himself is as old as man himself.

Persons who study the human body are primarily concerned with understanding man's relationship to his environment. Since external forces are constantly acting on man from conception to death, we should recognize that we are not static but constantly subject to change. The slowest of these forces is evolution, which affects very subtly and minutely, yet relentlessly, our physical form. Evolutionary forces require thousands of generations before any appreciable diversity of structure is discernible.

While the physical form of man may appear to be remarkably stable and uniform, we possess a body of almost unlimited adaptiveness, as evidenced by our ability to live under the most hostile conditions, such as extremes of temperature, atmosphere, and pressure, and famine and disease. Recent explorations from outer space to the ocean depths, from the frigid ice cap of Antarctica to the searing heat of the Sahara all attest to the ability of man not only to survive but to perform intricate, difficult tasks under adverse conditions.

Anatomy and physiology are branches of biology concerned with the form and functions of our body that enable us to live in our environment in reasonable health and comfort. Briefly defined, anatomy is the study of the structure and interrelationships of the parts of the body. Physiology is the science of the functions of the body. A basic understanding of the interaction and interdependence of structure and function in man, the most complex creation in nature, is the keystone to our knowledge of life processes.

ANATOMY

Human anatomy may be subdivided into a number of special fields, some of which will be described in the following sections.

Gross Anatomy

The parent study, from which the term anatomy (*ana-*, apart; *-tomy*, to cut) is derived, is gross anatomy. Ideally the gross anatomist studies the human body by dissection (*dis-*, apart; L., *secare*, to cut) of a cadaver. Gross anatomy may be studied either regionally or systemically. In regional anatomy the interrelationships of muscles, bones, and nerves in a particular region of the body, such as the head, are studied concurrently. In systemic anatomy all the structures concerned with a similar function, for example the muscles, are studied as a unit. Where dissection is a part of the curriculum, anatomy is usually studied regionally. The systemic approach is the more logical method to follow if dissection is not an integral part of the course.

Surface Anatomy

Surface, or topographic, anatomy involves body structures as they appear on, or are related to, the surface of the body. For instance, surface anatomy relates the jaw muscles to the place on the face or skull where they can be seen and felt as they close the mouth. In surface anatomy of the newborn baby, the margins of certain bones of the skull are readily observable at the anterior fontanel, or "soft spot," on the baby's head.

Applied Anatomy — *Specialized*

Surgical anatomy covers those special aspects of importance to the surgeon as he opens the body for observation, removal, or repair of structures.

Radiological anatomy considers basic relationships that enable the radiologist to interpret an x-ray film.

Kinesiology is the study of muscle action and the stress and pull of muscles on specific bones. Osteology is the detailed study of bones. Both of these special fields must be mastered by the orthopedic surgeon if he is to align fractures correctly or reconstruct parts of the body.

Microscopic Anatomy

During the Middle Ages the development of the microscope extended man's capability of studying the human body. This invention was a necessary prelude to the development of the specialized field of microscopic anatomy. Today the structure of the body is studied more extensively under the microscope than in the dissection laboratory. Cytology (*cyt-* or *cyto-*, cell; *-logy*, science) is the study of the cell, the basic unit of structure of any organism. Histology (*hist-* or *histo-*, tissue) is the study of the tissues of the body. Tissues are composed of similar cells which are organized to perform a specific function. Organology deals with the organization of primary tissues—the epithelial, connective, muscular, and nervous tissues—as they combine to form the organs of the body.

Important recent scientific advances enable man to understand better the interaction of the cellular components of the body. The development of the electron microscope has increased the magnification of the cell more than a hundredfold over that of the light microscope. The shape, characteristics, and relationships of cellular components are now as visible to the microscopist as are the muscles, tendons, and bones to the gross anatomist.

Developmental Anatomy

In developmental anatomy we study an organism from its beginning as a fertilized egg until birth. Embryology encompasses the growth of the embryo. In man this period lasts from conception through implantation and the initial development of the organ systems at about the eighth week of intrauterine life. In fetal anatomy the further elaboration and growth of the body systems, until the birth of the baby, is studied. Teratology is a subspecialty of developmental anatomy concerned with abnormal development.

Genetics, the study of the action of the genes and chromosomes, is an important adjunct to developmental anatomy. The genes are the agents responsible for the inheritance of individual characteristics from generation to generation.

Neuroanatomy

Neuroanatomy covers both gross and microscopic aspects of the nervous system. It is a specialized category because of its complexity and importance.

ANATOMICAL TERMINOLOGY

Specialized terminology develops with the unfolding of any science. In this course of study you will amass an entirely new vocabulary which will give you a basis for understanding the language of medicine and related sciences.

Some terms, many of which have been in use for over twenty-five centuries, tell something specific about a structure. A term may indicate function or position; for example, the flexor digitorum superficialis muscle is the superficial muscle which acts to bend (flex) the fingers (digits) in making a fist. A term may suggest shape or location; for example, the biceps brachii (L., *biceps,* two-headed; *brachii,* of the arm) and the biceps femoris (L., *femur,* thighbone) are muscles of the arm and the thigh, respectively, with two heads of origin.

As you encounter new terms, you should link them to their origin or meaning whenever possible. This will accomplish two purposes. First, knowing the definition of the root of a word will greatly simplify your mastery of the vocabulary in this textbook. Second, it will increase your appreciation, as well as your command, of much of your present vocabulary because of its Greek or Latin background.

The prefixes, suffixes, and roots of many anatomical terms are derived from Greek and Latin. Make friends of these words. They will help you change this course from one of rote memorization to one of living, dynamic study. For example, the definition of the stem *my-* or *myo-* (muscle) will give you a partial definition of the following terms, which are but a few of the words in which this stem occurs:

Myoblast (*-blast,* germ)—the embryonic cell which becomes a muscle cell
Myocardium (*-cardium,* heart)—muscle of the heart
Myocele (*-cele,* hernia)—herniation of a muscle
Myoenteron (*enteron,* intestine)—muscle of the intestine
Myography (*-graphy,* to record)—a technique used to record muscular activity
Myoma (*-oma,* tumor)—tumor of muscle tissue
Myonecrosis (*necrosis,* death)—death of muscle tissue
Myoplasty (*-plasty,* to form)—surgical repair of a muscle

Similar lists could be made of the second stem of each of the above terms.

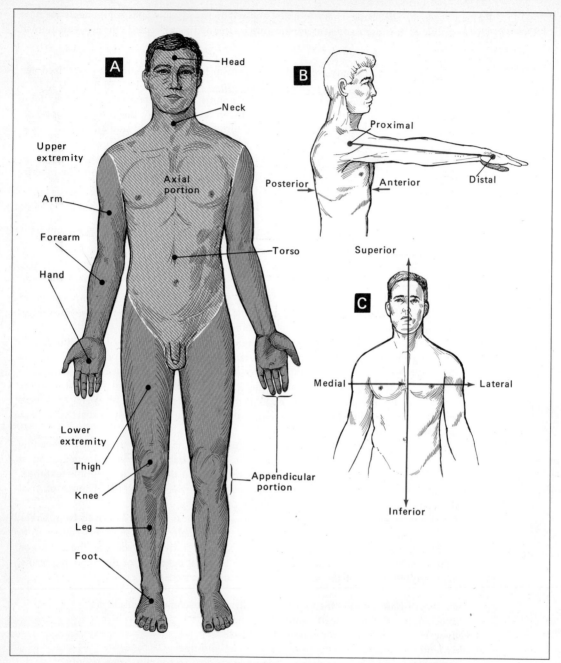

Figure 1-1. A. Regions of the body. The axial portion is shown in striped color; the appendicular portion, in solid color. The subject is in the anatomical position. **B, C.** Directional terms.

Terms of reference have been standardized to describe human anatomy. This is necessary in order to communicate coherently and clearly about various regions of the body, as well as to describe different structures within a given region. In order for you to read descriptive anatomy more easily, this general terminology should be mastered at the outset of your study.

In anatomy all terms of reference are relative to the anatomical position (Figure 1-1). This arbitrary position has the body standing erect, face forward, and the hands at the sides with the palms forward. Facing a person in the anatomical position, you see the front, or anterior, aspect of his body; the back, or posterior, aspect is hidden from view.

Many of these general terms are words which are already familiar to you but which now must assume specific anatomical meanings. For example, the word "superior" in your present vocabulary implies excellence, or qualitatively better. This word in anatomy means "toward the head," while its antonym, "inferior," means "toward the feet." Anatomical terms are always used to relate the part or area being described to the entire body. The clavicle (collarbone), for example, is described as being superior to the thorax but inferior to the head. The importance of mastering these general terms cannot be overemphasized. Words like "above" and "below," "over" and "under" may be misleading or even confusing if they are used to describe parts of the body. When one is lying on one's back, the umbilicus (navel) is obviously above, over, or on top of the abdominal organs. However, if one turns over, it could be described as being below or underneath the abdominal organs. In anatomical terminology the umbilicus is always anterior to the abdominal organs, irrespective of the position of the body.

Directional terms relating to the anatomical position and examples of their usage are given in Table 1-1.

Planes of Reference

In addition to directional terms, certain planes of reference of the body should also be understood (Figure 1-2).

TABLE 1-1. TERMS OF REFERENCE

TERM	SYNONYM	DEFINITION
Superior	Craniad, cephalad	Toward the head
Inferior	Caudad	Toward the feet
Anterior	Ventral, volar, palmar (latter two refer to hand)	Toward the front of the body
Posterior	Dorsal	Toward the back of the body
Medial		Toward the middle of the body
Lateral		Toward the side of the body
External	Superficial	Toward the surface of the body
Internal	Deep	Away from the surface of the body
Proximal		Toward the main mass of the body
Distal		Away from the main mass of the body
Central		Toward the center of the body
Peripheral		Away from the center of the body
Plantar		Undersurface of the foot

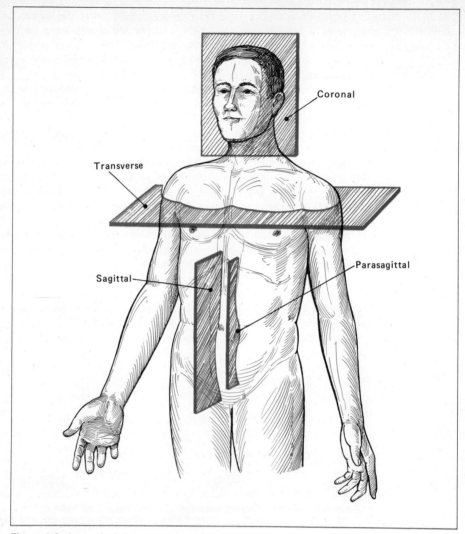

Figure 1-2. Anatomical planes of reference.

Sagittal This vertical plane extends in an anteroposterior direction. Slicing the body in the sagittal plane would split it into identical right and left halves. A parasagittal plane parallels the sagittal plane but would divide the body into unequal parts.

Coronal, or Frontal This vertical plane passes from side to side at right angles to the sagittal plane. Sectioning the body in the coronal plane would cut the front, or anterior, part of the body away from the back, or posterior, part.

Horizontal, Transverse, or Cross This plane extends across the long axis of the body. Cutting in the transverse plane would separate a superior portion of the body from the inferior part.

Planes of reference are also used in histology. Tissues may be sectioned in longitudinal, transverse or cross, and oblique planes.

REGIONS OF THE BODY

In a regional description of the body a general area may be subdivided into specific parts (Figure 1-1). Thus the body may be subdivided into an axial and an appendicular portion. The axial portion consists of the central axis of the body, namely the head, neck, and trunk; the appendicular portion is the superior and inferior limbs, or extremities.

Axial Portion

The head consists of the skull and its contents, such as the brain and portions of the respiratory and digestive systems. Specialized sensory organs associated with sight, smell, taste, and hearing are also in this region.

The neck is located between the head and thorax. All structures passing between these regions must traverse the neck. Additional structures, such as the thyroid and parathyroid glands, are entirely within the neck.

The trunk, or torso, is formed by the thorax superior to the diaphragm and the abdomen and pelvis inferiorly. Although the pelvis is a subdivision of the trunk, bones and muscles contributing to its walls are elements of the appendicular portion of the body. The area between the thighs at the inferior aspect of the trunk is the perineum. This region includes the external genital organs as well as the termination of the digestive and urinary systems.

Appendicular Portion

The superior extremity consists of the arm, forearm, hand, and associated joints. Certain bones and muscles of the shoulder, such as the superficial muscles of the chest and the clavicle and scapula, are located on the trunk, but they function with the appendicular portion of the body and are therefore considered components of the superior extremity. The arm extends between the shoulder and the elbow, the forearm extends between the elbow and the wrist, and the hand is distal to the wrist. An additional region of the superior extremity, the axilla (armpit), is located where the superior extremity attaches to the trunk. The axilla is bounded laterally by the arm and mediad by the thorax.

The inferior extremity consists of the thigh between the hip and the knee, the leg between the knee and the ankle, and the foot distal to the ankle. As in the upper extremity, bones and muscles topographically related to the trunk at the hip are functional components of the inferior extremity. Additional subdivisions of the inferior extremity include the inguinal region, or groin, at the anterior aspect of the junction of the inferior extremity and the trunk, and the popliteal region, the area at the posterior aspect of the knee joint.

BODY CAVITIES

The body cavities are lined with serous membrane and are located within the axial portion of the body (Figure 1-3). It should be emphasized that the body cavities are completely filled with viscera (organs) or fluid. Certain cavities are completely enclosed with no communication with other areas of the body.

The thoracic and abdominopelvic cavities extend from the base of the neck to the floor of the pelvis and are separated by the thoracic diaphragm.

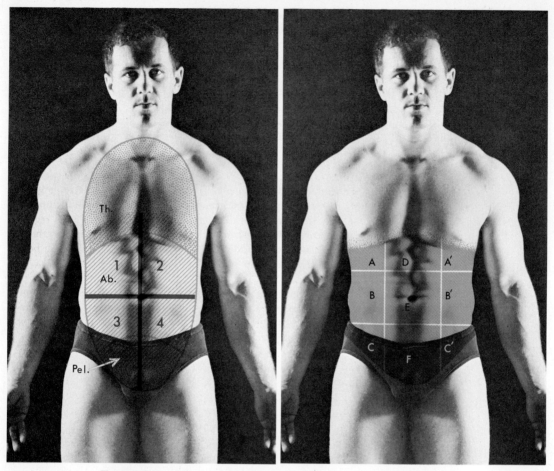

Figure 1-3. (*Left*) The major cavities of the trunk. The thoracic and pelvic cavities are shown stippled. The striped area indicates the abdominal cavity. The thoracic diaphragm separates the thoracic from the abdominal cavity. Numbers indicate the major quadrants of the abdomen. **Figure 1-4.** (*Right*) The nine regions of the abdomen. The verticals are the midclavicular lines. The upper horizontal is the transpyloric line, which crosses at the lower margin of the rib cage. The lower horizontal is the intertubercular line, which passes between the upper parts of the hipbones. In the midline are the epigastric (D), umbilical (E), and hypogastric (F) regions. At the subject's right are the right hypochondriac (A), right lumbar (B), and right iliac or inguinal (C) regions. The letters with primes denote the corresponding regions of the left side.

Thoracic Cavity

The thoracic cavity is enclosed by the rib cage, that is, the ribs, sternum, and thoracic vertebrae, and the diaphragm. It contains the pericardial cavity, which encloses the heart with its great vessels, the lungs, enveloped by the pleural cavity, and additional structures passing between the neck and the abdomen.

Abdominal Cavity

The abdominal cavity contains the major portion of the digestive system, i.e., the stomach, large and small intestine, liver, and pancreas. The spleen, kidneys, and part of the ureters are also contained within this body cavity.

The description of the locations of organs within the abdominal cavity is facilitated by subdividing this area topographically into regions. The midline and a horizontal line at the level of the umbilicus may be used to divide the abdomen into four regions (Figure 1-3): right and left upper quadrants and right and left lower quadrants. If more precision is needed, two vertical and two horizontal lines may be used to divide the abdominal cavity into nine regions (Figure 1-4). The vertical lines bisect the clavicles and are called the midclavicular lines. The upper horizontal line, the transpyloric line, is at the level of the lower margin of the rib cage; the lower intertubercular line passes between the uppermost part of the hipbones. The nine regions thus demarcated are, from superior to inferior, in the midline area, the epigastric, umbilical, and hypogastric regions and to either side the right and left hypochondriac, lumbar, and iliac, or inguinal, regions.

Inferiorly the abdominal cavity merges into the pelvic cavity; no definitive structure separates the two. At the lower end of the trunk the area internal to the pelvic bones is divided into the false and true pelvic cavities. The term false pelvic cavity is misleading. This portion of the trunk is actually the inferiormost extent of the abdominal cavity.

The true pelvic cavity houses the internal reproductive organs, the urinary bladder, and the lower parts of the digestive system. The demarcation between the abdominal cavity and true pelvic cavity is at the inlet of the pelvis, which is formed by the junction of the pubic bones anteriorly and the sacrum posteriorly.

Cavities in the Head Region

Several cavities are located in the head region (Figure 1-5). The largest, containing the brain, is the cranial cavity. At the foramen magnum this cavity is continuous with the space which encloses the spinal cord, the vertebral canal. The two spaces are sometimes referred to as the dorsal body cavity. The ventral body cavity is then described as consisting of the thoracic and abdominopelvic cavities.

The orbital cavity in the head contains the eye and its associated structures. The middle ear cavity, normally filled with air, is located between the external and internal subdivisions of the ear. It contains the three small bones of the middle ear and their associated muscles. The nasal and oral cavities open onto the face at the nostrils and the mouth. Posteriorly they communicate with respective nasal and oral portions of the pharynx.

PHYSIOLOGY

At the outset, the student should have in mind a clear understanding of the field encompassed by the term physiology. It is often stated that anatomy is concerned with the structure of an organism and physiology with function. This is an accurate statement, but perhaps it is not sufficiently descriptive. Physiology has also been said to be the analysis of what makes us tick, but although this expression may vividly convey the fundamental concept, it is somewhat broad because it includes the field of psychology. Quite simply, physiology is the study of activities characteristic of living organisms. This definition may also seem too broad; yet it does serve to underline the breadth of knowledge required for an understanding of the subject. Certainly familiarity with physics, anatomy, and biochemistry is essential. A comprehensive review of all the material essential to the study of physiology cannot, for obvious reasons, be included here, but

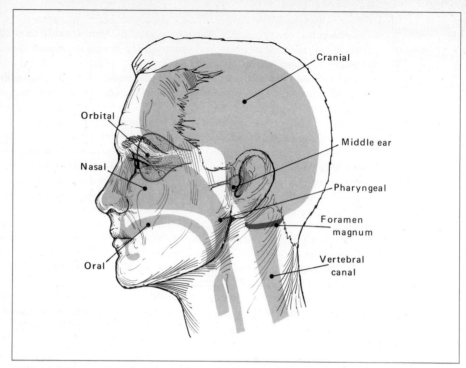

Figure 1-5. The cavities of the head. The cranial cavity is seen to be continuous with the vertebral canal. The nasal and oral cavities are continuous with the cavities of the pharynx. The orbital cavity and middle ear are shown stippled.

there does seem to be merit in discussing a few basic principles which derive from other disciplines as well as those which belong exclusively to physiology.

Homeostasis

By far the most important concept in the entire field of physiology now goes by the name homeostasis. The term was coined by the eminent United States physiologist, Walter Cannon. The concept is generally accredited to the great French scientist, Claude Bernard. Actually men before Bernard sensed the germ of the idea, but it was Bernard who stated so beautifully and succinctly, *La fixité du milieu intérieur est la condition de la vie libre.* He went on to explain that all the vital mechanisms, however varied they may be, have as their major object that of preserving constant the conditions of life in the inner environment. It is this constancy of the internal environment that Cannon called homeostasis. The word is derived from the Greek words *homoios,* meaning "like" or "similar," and *stasis,* "a standing still." Clearly the word is well chosen. But the student should not carry away the idea that this constancy represents a frozen, unchangeable, static state. The important point to understand and to remember is that the observed constancy is the result of marvelous dynamic mechanisms, processes properly referred to as homeostatic mechanisms.

There are many common, everyday illustrations of homeostasis. Everyone knows, for example, that the internal temperature of the body remains remarkably constant. In fact, we are so used to this that when the temperature varies, even a degree or two, we realize we are "sick," that something is wrong. Another example of homeostasis is the constancy of the blood-sugar level. Another is the blood pressure. As will be learned, there are so-called normal values for a host of factors. Clinically these values are determined because constancy is the norm, is expected. A major goal of the physiologist and of the student of physiology is to understand the various homeostatic mechanisms.

Merely to recognize that a certain physiological change maintains homeostasis is not to explain the sequence of events by which the change is brought about. For example, it is known that during exercise the contracting muscles require more oxygen than they do at rest and that more oxygen is taken in by the lungs. Thus the constancy of the blood oxygen content is maintained. But stating that the heart pumps more blood and the lungs take in more air because muscles need more oxygen does not explain *how* this change occurs. The student of physiology must ask himself: How is this change brought about? What are the mechanisms?

SOLUTIONS

A solution is a mixture of two or more substances. The substances may be in any of the three states of matter, solid, liquid, or gas. The type of solution most often dealt with in physiological considerations is a solid dissolved in a liquid. The liquid, that is, the dissolving agent, is called the solvent; the substance dissolved in the solvent is termed the solute.

Ionization

In Figure 1-6, the atomic structures of hydrogen, lithium, and oxygen are shown. Note that the number of protons in the nucleus of each atom is the same as the number of electrons surrounding each atom. These fundamental units of matter are charged, protons positively and electrons negatively. Thus, when the atom contains the same number of each, the entire atom is neutral; it has no charge. Atoms interreact to form compounds; in the case of sodium and chlorine, sodium chloride results. In solution the sodium and

Figure 1-6. Atomic structure of hydrogen, lithium, and oxygen.

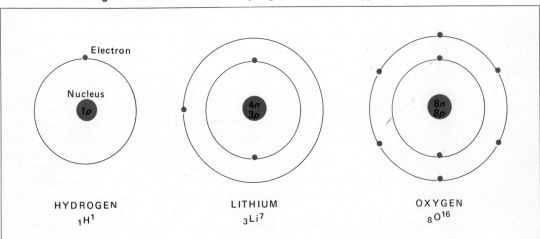

HYDROGEN
$_1H^1$

LITHIUM
$_3Li^7$

OXYGEN
$_8O^{16}$

chloride atoms separate, a process called dissociation. In so doing, the sodium atom gives up an electron to the chloride atom. This leaves the sodium with more protons than electrons, one more to be exact. It therefore has a positive charge, is termed a cation, and is written Na^+. The chloride atom has the extra electron, is termed an anion, and is written Cl^-. Because ions are formed in dissociation, the process is also termed ionization.

If an electrical current is applied to a solution containing ions, they will move to the electrodes. The anions are attracted to the positive pole, which is called the anode; the cations to the cathode. By virtue of this ionic movement the circuit between the electrodes is completed, and a current will flow. Because a solution containing ions conducts a current, it is called an electrolytic solution. The compound giving rise to the ions is termed an electrolyte. A compound that does not form ions in solution is a nonelectrolyte. Generally, inorganic compounds readily dissociate and form ions and thus are electrolytes. Organic compounds, on the other hand, usually do not form ions and are therefore nonelectrolytes. Formerly any compound derived from living tissue was considered organic. Now the term is reserved for compounds containing carbon, for example, glucose. All compounds without carbon are inorganic.

Concentration

The concentration of a solution may be expressed in several ways. It may be stated simply as weight of solute per unit volume of solvent. The most common form is milligrams percent, that is, the number of milligrams of the particular substance existing in one hundred milliliters of solution. This is often written as mg/100 ml. Also moles may be used. A mole of any substance is the gram-atomic, or gram-molecular, weight of that substance. For example, the molecular weight of NaCl is 58.5 g. Accordingly, 1 mole of NaCl is 58.5 g of the salt. A $1\,M$ solution would have 1 gram-molecular weight of the solute in 1 liter of final solution.

"Molar" should not be confused with "molal." In a molar solution 1 gram-molecular weight of the substance is dissolved in enough solvent to make 1 liter. But in the molal solution the gram-molecular weight of the substance is dissolved in 1,000 g of the solvent. The final solution in the latter case may be considerably more than 1 liter.

Because electrolytes ionize in solution and because these ions have charges which influence their behavior, it is often more meaningful to express concentration in terms that take the valence of the ion into consideration. In this way the electrical equivalents (Eq) of the various ions may be expressed. One equivalent of any ion is equal to its gram-atomic weight divided by its valence. For example, 1 Eq of Ca^{++} would be 40/2, or 20. This is usually expressed in terms of equivalents per liter. Thus if a solution has 60 g of Ca^{++} in 1 liter, it has a concentration of 3 Eq/liter. This is extremely concentrated. In living cells, the concentration of the various ions is so low that a milliequivalent (mEq), one one-thousandth of an equivalent, is used.

A similar concept is used in expressing normality (N). Actually valence is used in the calculation, but strictly speaking normality is based on the weight of a substance that is chemically equivalent to 1 Eq of oxygen, that is 8 g (16/2). If you wished to make a $1\,N$ solution of H_2SO_4, it would be necessary to divide the molecular weight by the valence of either ion multiplied by the number of those atoms in the molecule. For example, the molecular weight of H_2SO_4 is 98. The valence of H is 1, but there are two such ions, so it is multiplied by 2. The valence of SO_4 is 2, and there is only one in the molecule. This again equals 2. Thus, 98/2 is 49. If 49 g of H_2SO_4 is dissolved and the final solution made up to 1 liter, it will be a $1\,N$ solution.

Because some authors prefer to express concentrations in milligrams percent whereas others use milliequivalents per liter, the student should be able to convert one to the other. From the above discussion, the following should be clear:

$$\text{mg percent} = \frac{\text{mEq/liter} \times \text{gram-atomic weight}}{10 \times \text{valence}}$$

$$\text{mEq/liter} = \frac{\text{mg percent} \times 10 \times \text{valence}}{\text{gram-atomic weight}}$$

Finally, concentration is sometimes expressed in terms of osmolarity. This term refers to the osmotic pressure of the solution (see below), which depends upon the number of particles in solution. The osmole is the gram-molecular weight of a substance divided by the number of particles it produces in solution. A milliosmole is 1/1,000 of an osmole. It is difficult to determine exactly how many particles a substance produces in solution. For example, NaCl should produce two ions, or particles, per molecule. But this is true only if there is complete dissociation, and that is not invariably the case. Another way to determine the osmolarity of a solution is to take into consideration the temperature at which that solution freezes. The more particles in solution, the lower the freezing point of that solution. A 1-osmolar solution has a freezing point 1.86°C below that of pure water. Thus, to determine the osmolarity of any solution, divide the difference between its freezing temperature and that of water by 1.86.

Hydrogen-Ion Concentration

Hydrogen-ion concentration is generally expressed in terms of pH, which is defined as the negative logarithm of the hydrogen-ion concentration. This is simply a means of converting unwieldy numbers into a more convenient system. When pH units are used, a neutral solution has a value of 7. Values below pH 7 indicate degree of acidity; above pH 7, degree of alkalinity (Table 1-2).

TABLE 1-2. HYDROGEN-ION CONCENTRATION EXPRESSED AS pH

[H⁺]	LOG [H⁺]	pH (= −LOG [H⁺])	
10^{-0}	-0	0	
10^{-1}	-1	1	
10^{-2}	-2	2	
10^{-3}	-3	3	Acid
10^{-4}	-4	4	
10^{-5}	-5	5	
10^{-6}	-6	6	
10^{-7}	-7	7	Neutral
10^{-8}	-8	8	
10^{-9}	-9	9	
10^{-10}	-10	10	
10^{-11}	-11	11	Alkaline
10^{-12}	-12	12	
10^{-13}	-13	13	
10^{-14}	-14	14	

TRANSFER THROUGH MEMBRANES

The basic unit of all living organisms is the cell. The limiting part of the cell is the cell membrane. The cell membrane controls the passage of materials into and out of the cell. There are at least four ways in which transfer through membranes may occur: (1) diffusion, (2) filtration, (3) osmosis, and (4) active transport.

Diffusion

If common table salt, NaCl, is placed in the bottom of a glass and water added carefully so as not to stir up the salt, the salt nonetheless passes into solution. After a period of time, chemical analysis reveals that the salt is now evenly dispersed throughout the water. Actually, the top of the salt layer in contact with the water quickly dissolves, and once in solution the individual ions—that is, the sodium and chloride ions—shoot off in all directions. One of the basic properties of molecules and ions is this constant movement in solution. The rate of movement is a function of temperature. At absolute zero ($-273\,°C$), all ionic and molecular movement ceases. The warmer the solution, the more rapid the movement. Thus in water at room temperature, the sodium and chloride ions move in all directions at a relatively rapid rate. As each particle of salt dissolves, it shoots through the solvent, leaving another layer of salt exposed to the water. Eventually all the salt is in solution. The individual ions collide and bounce off one another until finally there is an even distribution of ions throughout the solution. This process of distribution is termed diffusion (Figure 1-7).

Diffusion of particles will take place across a permeable membrane as well. If such a membrane divides the container into two compartments and a salt solution is placed on one side and pure water on the other, after a few minutes the pure water will contain salt. Eventually the concentration of salt on both sides of the membrane will be identical. In other words, the salt has diffused through the membrane. This is explicable on the basis of ionic and molecular movement. The salt ions move at random, and when they strike the barrier, some of them pass through the minute openings which make the membrane permeable. In man diffusion takes place across many cell membranes. For example, some products of food digestion diffuse through the wall of the intestine and capillaries to enter the blood.

Figure 1-7. Diffusion in a solution. **A.** Molecular motion causes molecules to disperse. **B.** Even when the solution is divided by a permeable membrane, diffusion can occur.

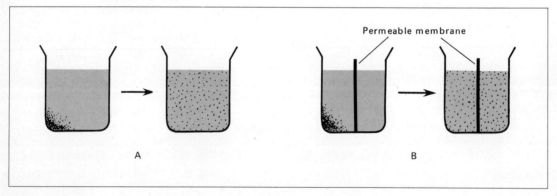

Permeable membrane

A   B

Filtration

In the chemistry laboratory, the student has probably had occasion to filter many substances. To do this he places a permeable piece of paper in a funnel and pours a solution into it. The solution slowly passes through, leaving behind molecules which are too large to escape through the small openings in the filter paper. This is filtration. If the chemist desires to speed the process, he reduces the pressure in the container below the filter (by applying suction). The lower the pressure, the faster the filtration process. Filtration then differs from diffusion. Filtration takes place not by virtue of molecular and ionic movement but rather as a result of an external force which actually pushes the substance through the filter. For example, when suction is applied, the atmospheric force on the top of the filter is greater than the pressure within the container. This pressure differential is the force which causes the fluid to pass through the filter. If no suction is employed, the only force is gravity. The rate of filtration depends on the external energy which tends to drive the substance through the filter. In the living organism the fluid component of blood, for example, filters through the capillary wall. The force in this instance is supplied by the blood pressure. The same is true in the kidneys, where substances filter through the glomerulus into the tubule.

Osmosis

The term osmosis is derived from a word which means "pushing." In the process of osmosis there is indeed a pushing, a forcing of substances through a membrane, but the force is different from that which is responsible for filtration. In order to appreciate osmosis, one must understand the principle involved in a semipermeable membrane. The term semipermeable means that some substances penetrate and others do not. If a membrane permits some substances to pass through but not others, there results a lack of balance on the two sides of the membrane, and this imbalance gives rise to a considerable force.

In discussing diffusion, it was mentioned that if a permeable membrane separates water from a salt solution, within a short period of time the salt and water diffuse through the membrane, and the concentration on both sides becomes identical. Thus, there results a balance. But if the membrane is semipermeable so that water molecules will pass but salt molecules cannot, a very different result is seen. The water will pass into the salt solution, but since there cannot be a simultaneous transfer of salt into the water solution, the column of salt solution rises because the volume progressively increases. It requires energy to support a column of fluid, since fluid possesses weight. So long as water molecules continue to enter the salt solution, the column will rise. However, as it does so it becomes progressively heavier. A point will ultimately be reached at which the weight of the column of fluid balances the force driving the water molecules into the salt compartment. Then movement will cease. In other words, the height of the salt solution above the water compartment is an index of the osmotic force or pressure (Figure 1-8).

In brief, whenever a semipermeable membrane separates two solutions one of which contains a substance which cannot pass, fluid from the other compartment will cross the membrane. This movement is osmosis, and the force created is called osmotic pressure.

If red blood cells are placed in a solution less concentrated than the solution within the cells, there will be movement of fluid into the cells because red blood cells contain molecules that do not readily pass through the membrane. In a much less concentrated solution the cells may swell until they burst. In a solution more concentrated than that of the cell interior there is movement of water out of the cells, with the result that the cells shrink. A set of terms has evolved to describe these different outcomes. If the cells

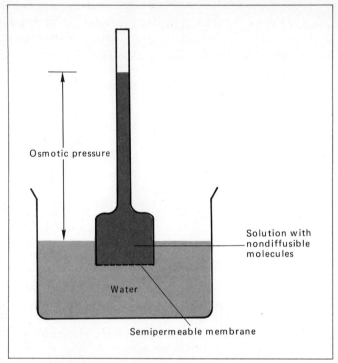

Osmotic pressure

Solution with
nondiffusible
molecules

Water

Semipermeable membrane

Figure 1-8. Osmosis. Because there are molecules in the bell of the tube which cannot diffuse through the membrane, water moves in and rises in the column until its weight balances the osmotic force.

swell, the solution is said to be hypotonic with respect to the cell interior. If the cells shrink, the solution is hypertonic. If no change in shape (in "tone") occurs, the solution is isotonic.

These facts have obvious practical importance. If an isotonic solution is injected into the bloodstream, as explained above, fluid will move into the cells, causing them to rupture, a process called hemolysis. Hypertonic solutions will also produce deleterious effects. Clearly, then, one must be able to calculate the osmolality of a solution to be used for injection, keeping in mind that the contents of the red cells have an osmolal concentration of about 310 mOsm/liter. Assume that a NaCl solution is to be used. The molecular weight is about 58.5, and dissociation gives rise to two particles per molecule; therefore $58.5/2 = 29.25$. Then

$$\frac{x}{29.25} = 310 \text{ mOsm/liter}$$

This works out to be about 9,000 mg, or 9 g, of NaCl per liter, which is a 0.9 percent solution. Such a solution is commonly referred to as physiological saline, or normal saline.

To give another example, glucose has a molecular weight of 180 and does not dissociate. Using the same calculation we see that a 5.5 percent glucose solution would be isotonic.

Another important consequence of osmotic imbalance is that seen following a decrease in the amount of protein in the blood, as occurs in malnutrition. The osmotic pressure of the blood decreases, and tissue fluid that would normally be attracted by osmotic pressure into the bloodstream accumulates in the tissue spaces of the body, thus causing swelling, a condition termed edema.

Active Transport

The term gradient refers to a difference in forces on the two sides of a membrane. Thus one speaks of concentration, osmotic, pressure, or electrical gradients. If a solute moves in accord with osmotic, concentration, or electrical gradients, it is descriptively said to be moving "downhill." This is a passive process and no energy is required, as everyone who has coasted down a hill realizes. Conversely, solutes may move in the opposite direction, against the gradients of osmotic pressure, concentration, or electrical attraction. In order for this to occur, energy must be expended. Thus the solute is said to be moving "uphill." Transfer uphill, that is, against a gradient, is termed active transport.

Simply because a solute moves in the direction of its electrical and chemical gradients, the conclusion that such movement is passive is not justified. A car may coast down a hill, but it may also be propelled down the hill by the energy of its engine. Some substances move faster than can be accounted for by a calculation of the gradients. There thus must be active transport added to the movement. Since this obviously requires difficult and precise measurements, active transport is generally studied when the solute is moving uphill.

There is no doubt that active transport occurs throughout the body. For example, in every cell the concentration of potassium ion is much greater inside the cell than in the fluid surrounding the cell; yet the cell wall is permeable to the ion, and thus there is movement in both directions. This can only mean that there is some active mechanism which moves more potassium ion into the cell in the face of the high concentration. Sodium ion, on the other hand, has a much higher concentration outside the cell than within. It is actively transported out of the cell. What is the mechanism? That question cannot yet be answered, but the current thinking is that the active process takes place *within* the cell membrane (Figure 1-9).

The basic mechanism is thought to involve a carrier which may be protein or a lipoprotein. At any rate, the substance to be transported diffuses into the membrane, where it is combined, in some manner, to the carrier which transports it across the cell membrane. At the inner surface of the membrane the substance is freed from the carrier and enters the cell (Figure 1-9). Energy is required. The energy comes from the mitochondria, but how the energy is utilized in this basic plan is not clear.

Probably most substances that undergo active transport have specific carriers. In some cases two or more substances may use the same carrier. For example, potassium and sodium ions are thought to do so. After the carrier has transported potassium ions into the cell, it combines with sodium ions to transport them out of the cell. In such cases the transport of the two ions is said to be coupled. The active transport of sodium ions out of the cell is referred to as a sodium pump.

ELECTRICAL PHENOMENA

One of the most used concepts in physiology is that of electrical potential. One speaks of resting potentials, action potentials, and injury potentials. Therefore a clear understanding of electrical potential is essential. Two closely placed charged particles exert a force. If the particles are of opposite charge and attract each other, in order to keep them apart or to separate them, work must be done. Conversely, if they have the same

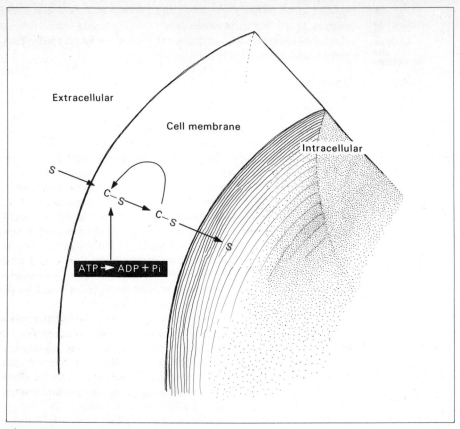

Figure 1-9. Active transport. First the substance (S) diffuses into the membrane. It then combines with the carrier (C) which transports it to the inner surface, where it is freed to diffuse into the cell.

charge, they will repel one another, and work must be expended to bring them together. The electrical potential is a measure of work done in moving such charges. By definition, the electrical potential at any distance between two charged particles is the work done to bring one of the particles from infinity closer to the other particle. If the two particles have the same charge, they will tend to repel one another, and so the work done is positive and the potential is positive. On the other hand, if the particles attract one another, the potential is negative.

Electrical Current

An electrical current is the quantity of electric charge transferred past a point per unit time. The unit of current is the ampere.

George S. Ohm showed that the current flowing in a conductor is proportional to the potential difference at the two ends of the conductor. If the potential difference remains constant, the current will be constant. When the potential difference is increased, the current increases. On the other hand, Ohm also showed that for any potential difference the current varied with the conductor used. A poor conductor may be thought of as resisting the flow of current. Thus various conductors may be classified according to their resistance to the flow. The unit called the ohm is a measure of resistance. The volt is a measure of potential difference.

Quite clearly, amperes, ohms, and volts are interrelated. That relationship is expressed as follows:

$$I = \frac{V}{R}$$

where I = current expressed in amperes
V = potential difference expressed in volts
R = resistance of the conductor expressed in ohms

A direct current is one in which the charged particles always flow in the same direction. Thus if the terminals of a battery are connected, there will be a flow in one direction, and it will be continuous for the life of the battery. But if instead of having the charged particles flow only in one direction they are made to flow first in one direction and then in the opposite direction, such a current is said to be alternating. The wave depicting the current will be biphasic (Figure 1-10). Each complete biphasic wave is said to constitute one cycle. In the usual house current there are 60 cycles per second (cps). Alternating current is rarely used to stimulate tissues, but it does have important usage in the clinic, namely, in diathermy treatments.

In many student laboratories, electrical stimuli are still produced by means of an inductorium. This type of stimulator consists of two coils of wire, one of which can be moved in relation to the other. A direct current flows through the so-called primary coil, and as a result there is an induced current in the secondary coil. In an induced current, the waveform is biphasic. The first wave results from the introduction of current into the primary coil. The second wave occurs when the current is stopped.

EXCITABILITY

The term excitability or, as it is often called, irritability, refers to the response of all living cells and organisms to sudden and significant change in their environment. If the lowly ameba is touched with a sharp object, it will usually retreat. Likewise if a person were to step on a nail, he would rapidly withdraw his foot. In both illustrations there has been a response. Beyond question, the reaction in the second instance is more complex, but fundamentally it is an example of excitability.

If the definition of the term excitability is analyzed, it will be found to contain two essential elements: (1) a change in the environment and (2) a response. For the former, the word stimulus may be substituted. This is perhaps the most used term in all of physiology. One speaks, for example, of the response of the nervous system, the heart, and the digestive glands to various stimuli. Physiology is the study not only of the basic

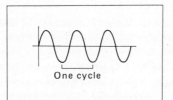

One cycle

Figure 1-10. Alternating current.

functions, but of the response of those mechanisms to stimuli. Yet, though the word is so commonly used, a precise definition is impossible. It will suffice to recognize that a stimulus is a change in the immediate environment of a living system, and to be familiar with the various types of stimuli.

Types of Stimuli

A change in many facets of the environment results in a response by the organism. For this reason it is recognized that there are (1) mechanical, (2) thermal, (3) electrical, (4) chemical, and (5) osmotic stimuli. This list is not complete, but it does include the more important stimuli.

In physiology, one grades stimuli according to the response of the part being excited. Thus the terms threshold, minimal (or liminal), and maximal have come into common usage. A threshold stimulus is one just strong enough to provoke a response. Threshold, minimal, and liminal are generally used interchangeably. A subliminal stimulus is one too weak to produce a response. As the strength of the stimulus is increased, a point is ultimately reached beyond which no augmentation of reaction is noted. At this point, the stimulus is said to be maximal.

Specific Energy

For the most part it is true that no matter what type of stimulus is used, if the part responds at all, it will always react in the same manner. This is the concept of specific energy. For example, a nerve may be activated mechanically, chemically, or electrically. In each instance an impulse is conveyed by that nerve in exactly the same way. The most striking illustration points out that if the optic nerve could be attached to the ear and the auditory nerve to the eye, the subject would see thunder and hear lightning! A more common example is the well-known phenomenon of "seeing stars" when a person receives a vigorous blow on the head.

To be sure, there are reactions which do not adhere to this concept of specific energy. For example, in response to mechanical stimulation, the ameba will retreat; but if it is stimulated with a receptive object such as food, it will surround and engulf it. This is a rather specialized instance. For the most part, especially in the higher organisms, the doctrine of specific energy is valid.

Anesthesia

Only mere mention can be given here to the important phenomenon of anesthesia, which is, in reality, the antithesis of excitability. Various agents are commonly used in medicine to raise the threshold of nerves in tissues so that they will not respond to stimuli. Thus a general anesthetic, which causes a person to lose consciousness, may be given, or the part to be operated on may be made less responsive by a local anesthetic. The precise mechanism by which anesthetics function is still open to question. Before this question can be answered, knowledge of excitability must be greatly enhanced.

ENZYMATIC ACTION

No matter what aspect of physiology is under consideration, the ultimate analysis lies within the myriad chemical reactions which take place within the cell. There are good reasons to believe that perhaps all these basic reactions are governed by substances known as catalysts. When the catalyst is of biologic origin, it is termed an enzyme.

Catalysts do not enter into the reaction in the sense that they are used up or converted into another form. Instead, a catalyst alters the rate of a chemical reaction and may be recovered intact when the reaction is finished. Most commonly the term is restricted to an agent that accelerates a chemical reaction, but there are substances, also termed catalysts, which inhibit chemical processes. Insofar as is known, all biocatalysts, that is, enzymes, are proteins and thus have a high molecular weight.

Classification

The main chemical compound which undergoes alteration in a chemical reaction upon which the enzyme exerts its influence is termed the substrate. Enzymes are often classified and named according to the substrate upon which they act. For example, if an enzyme acts on a phosphate compound, it may be termed a phosphatase. The suffix -*ase* is simply added to the specific name of the substrate. Another way to classify enzymes is according to the class of substances upon which they act. In this classification, the suffix -*lytic* is used to name the type of enzyme. For example, enzymes that act on lipids are termed lipolytic. These simple systems for naming enzymes are followed, for the most part, but there are exceptions. Some enzymes that have been long recognized were named prior to the general adoption of these suffixes and retain names that were given to them somewhat arbitrarily. One such example is ptyalin, an enzyme found in saliva. Under the more modern system of nomenclature this same enzyme is also known as amylase since *amyl-* is a combining form denoting "pertaining to starch."

In addition to classifying enzymes according to the substrate or class of substances upon which they act, there is a rather recent trend to classify them according to their chemical composition. But since so little is known concerning the chemical composition of the vast majority of enzymes, this system is still very limited.

Protein Nature of Enzymes

As previously mentioned, it is commonly believed that all enzymes are proteins. Until recently the thought persisted that the active agent itself was merely combined with, or in some way bound to, the protein molecule. However, enzymes have now been prepared in the crystalline state. These preparations retain full potency and lose their potency only when the protein is denatured. The evidence is therefore quite convincing that it is the protein itself which is responsible for the catalytic action. Enzymes exhibit all the classic properties of proteins, namely, high molecular weight, colloidal behavior, slow diffusion, inability to pass through most living membranes, and movement in response to an electrical current.

Factors That Alter Enzyme Action

Temperature, pH, substrate and enzyme concentration, and the presence of other chemical substances may all modify enzymatic activity. Usually, an elevation in temperature accelerates the action of any enzyme. It has been found that a rise of 10°C hastens the reaction two or three times. However, there is a limit to the acceleration of enzymes by heat, because above a certain temperature the enzyme is inactivated. Thus, if one plots the rate of reaction against increasing temperature, one finds that the rate at first increases and then decreases. The increase is due to two factors: (1) the influence of heat on the chemical reaction itself and (2) increased enzyme activity. There is thus an optimal temperature for each enzyme. There is also an optimum pH for each enzyme.

In most chemical reactions an increase in concentration of the reactants increases the rate of the reaction. The same is true for processes enhanced by enzymatic action. Thus the rate of reaction may be increased by raising the concentration of the substrate, of the enzyme, or of both.

Generally, enzymatic activity is restricted to a specific substrate. Thus lipase will hasten the digestion of lipids, but it exerts no influence on the catabolism of protein or carbohydrate. This property of enzyme action is referred to as enzyme specificity.

Many enzymes require the presence of certain organic substances in order to function. These substances are termed coenzymes and are nonproteins loosely bound to the enzyme. They usually function by accepting or donating the atoms or molecules which are added to or removed from the substrate by the catalytic action of the enzyme. Coenzymes are vital to the energy-building processes of the cell (see Chapter 2).

Mechanism of Enzyme Action

Enzymes may influence the rate of biochemical reactions in several ways. There is evidence to indicate that in some instances the enzyme reacts with the substrate to form a new compound. This intermediate compound then undergoes a further reaction which results in the liberation of the enzyme in its original form but the substrate in an altered form. Thus the overall change is in the substrate alone.

Another way in which enzymes function is by lowering the energy of activation of a reaction. Energy of activation is a term which has been applied to the minimal amount of energy required of a molecule to take part in a reaction. Not all molecules of the same substance have the same amount of kinetic energy. Thus the rate of a chemical reaction may well depend upon the concentration of high-energy molecules. A rise in temperature increases the concentration of high-energy molecules, and this is thought to be the explanation for the increase in the rate of chemical reactions at higher temperatures. Likewise, catalysts may exert their enhancing influence by somehow lowering the threshold, that is, the minimal amount of energy required for molecules to take part in the reaction. At this lower threshold more molecules can take part than at higher threshold, and therefore the reaction proceeds more rapidly.

This explanation of the mechanism of enzymatic action raises more questions than it answers. One wonders why some molecules have more energy than others and, more importantly, how an enzyme makes it possible for a reaction to proceed at a lower energy level. Whatever the answers to these questions may be, the fact remains that the energy of activation has been measured for various reactions, and in every case, in the presence of a catalyst that speeds the reaction, this energy has been found to be lowered.

QUESTIONS AND PROBLEMS

1 Describe the anatomical position of the human body.
2 Define the following terms: (*a*) systemic anatomy, (*b*) macroscopic anatomy, (*c*) histology, (*d*) cytology, (*e*) physiology, (*f*) kinesiology, (*g*) genetics, (*h*) embryology.
3 Briefly describe a specific situation or condition in which the following fields of knowledge would be especially pertinent for a person dedicated to maintaining or restoring the health of others: (*a*) fetal anatomy, (*b*) surface anatomy, (*c*) systemic anatomy.
4 Locate the following viscera in their normal body cavities and systems:

Organ	Body Cavity	System
Pancreas		
Ovary		
Cerebrum		
Eye		
Tongue		
Kidneys		
Large intestine		
Lungs		
Spinal cord		

5 Using anatomic terms how would you describe the location of the following organs:

a. The heart with respect to the (1) diaphragm, (2) stomach, (3) sternum?
b. The stomach with respect to the (1) diaphragm, (2) rectum, (3) anterior body wall?
c. The kidney with respect to the (1) diaphragm, (2) abdominal cavity?
d. Upper arm with respect to the (1) forearm, (2) shoulder?
e. Ears with respect to the nose?

6 Describe what happens to the charged particles of an electrolyte in an electric field.

7 Explain why certain molecules are capable of crossing a semipermeable membrane against a concentration gradient.

8 Why is an understanding of homeostasis important to those concerned with maintenance of health.

9 Differentiate between a molar, molal, and normal solution.

10 What is an enzyme?

2 THE CELL

In the first chapter the larger subdivisions of the body were introduced. We shall now examine the smallest structural and functional unit of the body, the cell. Then, in Chapter 3, the organization of the cells into tissues, of tissues into organs, and of organs into the systems of the body will be discussed.

Formerly, the physiologist was content to understand the functioning of the various organ systems. Now, as he analyzes the systems in greater detail, he is inevitably led to the cell. The ultimate answers he seeks are within the cell. Thus David Goddar's statement "When we truly understand the cell, we will understand life itself" can hardly be considered an exaggeration. Certainly if the student is to understand the way the living organism functions, if he is to understand himself, he must understand cell physiology.

Much of the biological knowledge explosion of the last few decades has centered around studies of cellular structure and function, which prove the minute so-called simple cell to be an extremely complicated structure. Expansion of knowledge of the cell was possible because of remarkable new tools for cellular investigation. For example, the electron microscope increases the magnifying power of the ordinary microscope by over 100 times (Figure 2-1). The phase microscope is used to study the fine detail of unstained living cells. With x-ray diffraction microscopes it is possible to analyze, with great accuracy, the architecture and physical details of crystalline structures of the cell.

In addition, new techniques of histochemistry are available to locate the sites of many enzymatic reactions as well as the organic and inorganic components of protoplasm.

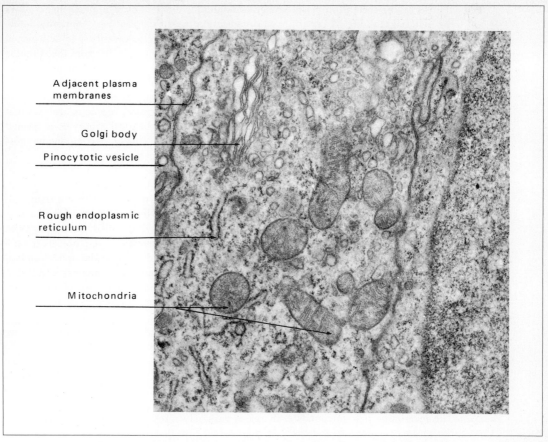

Adjacent plasma
membranes

Golgi body

Pinocytotic vesicle

Rough endoplasmic
reticulum

Mitochondria

Figure 2-1. Electron photomicrograph of a typical cell, showing some of the organelles ($\times$ 60,000).

Radioactive compounds, called isotopes, are used to study specific chemicals essential to cellular activities of proliferation, repair, and death. These and many other complicated, sophisticated tools of modern research are mobilized to explore more deeply and profoundly the great riddle of the universe—what is life?

ATTRIBUTES OF LIVING MATTER

Cells are composed of protoplasm, which has been defined succinctly by T. H. Huxley as "the physical basis of life." Protoplasm is composed of a number of different chemical substances organized in an aqueous solution. The concentration of inorganic compounds in protoplasm is similar to the concentration in sea water. There are also many organic compounds. The carbohydrates range from the simple sugars such as glucose and fructose up to the very complex molecules like glycogen. Some 20 amino acids and numerous fatty acids have been identified in protoplasm. In most animal cells, protoplasm contains about 80 percent water, 15 percent protein, 3 percent fat, and 1 percent carbohydrate, with electrolytes making up the remaining 1 percent.

As can be seen, protoplasm possesses no unique element or compound that endows the material with life. A certain fixed organization must exist. But we commonly define a living thing as one that shows attributes of life. One of these is response to an external stimulus (irritability). The response may be movement toward or away from the stimulus (contractility). Furthermore, notice of the stimulus can be transmitted from one part of the organism to another part or to other organisms (conductivity).

Living matter must be able to carry out continuous, complex biochemical processes that require oxygen and release carbon dioxide. These chemical reactions are essential to the life of the organism. The sum of these numerous chemical processes is termed metabolism. Living matter must be able to reproduce itself. Finally its components must grow in size and number in order to maintain and perpetuate itself.

Cellular Components

A typical cell consists of a delicate membrane enclosing a quantity of protoplasm, and near its center is a nucleus, itself enclosed by a membrane. The protoplasm of the nucleus is termed nucleoplasm. The remaining protoplasm of the cell is called cytoplasm. With an ordinary microscope we can observe in the cytoplasm several discrete objects (Figure 2-2). These are called cell organelles ("tiny organs") and include the mitochondria, lysosomes, endoplasmic reticulum, Golgi apparatus, centrioles (centrosomes), and fibers. The first four organelles mentioned are enclosed in delicate membranes of their own. These organelles are self-perpetuating; that is, when the cell multiplies, each daughter cell receives a full complement of them.

Inclusions are another group of constituents found in cells. Inclusions are nonliving substances such as proteins, carbohydrates, lipids, pigments, crystals, and secretory granules.

THE CELL MEMBRANE

The membrane that surrounds each cell is variously referred to as the cell membrane, the plasma membrane, and the unit membrane. The first is preferable and now most common. The membrane has been described as a fat sandwich with two layers of protein analogous to the bread. In the mid-1930s James Danielli and Hugh Davson suggested that the fat is phospholipid arranged in two parallel layers. The phospholipid provides the fundamental structure of the membrane and determines the "pore" size through which some molecules can squeeze and others cannot. The proteins serve as enzymes. As will be explained presently, they are essential for active transport to occur, and they are highly specific for a particular substance; they thus determine the character and function of the cell.

The important point to understand at this time is that the cell membrane is not simply a sac around the guts of the cell. The membrane is one of the most important, if not the most important, part of the cell. Its importance to virology, pharmacology, immunology, endocrinology, and basic physiology cannot be overstated. It is no exaggeration to insist that progress in many fields of medicine await better knowledge of the cell membrane.

Movement of Water through the Cell Membrane

There are several physical forces that move water through a membrane. First, of course, there is simple diffusion. If there are pores in a membrane, the simple molecular movement of water molecules will eventuate in some of them striking a pore and passing through. Also, if there is a difference in pressure on the two sides of the membrane, that is, a pressure gradient, water will be driven by that pressure through the membrane, a process which has been defined as filtration. The membrane is permeable to some molecules but not to others.

In addition, protoplasm contains protein, which does not readily move through the membrane. Thus, if a cell is placed in a hypotonic solution, it will swell because of the inward movement of water due to osmosis. The greater molecular concentration in the cytoplasm attracts water into the cell, and therefore the cell swells. Conversely, if the cell is placed in a hypertonic solution containing molecules which do not readily pass through the membrane, water from the cytoplasm will pass through the membrane, and the cell will shrink.

The movement of water into or out of the cell depends upon the permeability of

Figure 2-2. Sketch of a generalized cell, with the organelles emphasized.

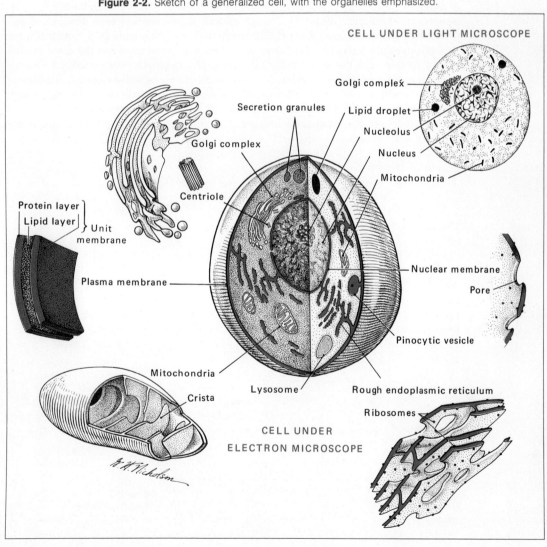

the membrane and the transmembrane osmotic pressure, that is, the difference between the osmotic pressure of the cytoplasm and of the surrounding fluid.

Movement of Solutes

The relationship between the size of the solute molecule and the diameter of the pore in the cell membrane is one factor which regulates the movement of solute through the membrane. Another important consideration is the degree of solubility of the solute in the substance of the membrane. Finally, there are various forces which influence movement of solutes through membranes. They are (1) the concentration gradient and (2), in the case of charged particles, electrostatic considerations.

Partition Coefficient The ratio of the solubility of a substance in a lipid solvent to its solubility in water is termed the partition coefficient. To put it another way, if a solute dissolves readily in lipid but not in water, it will have a high partition coefficient. The movement of various substances through the cell membrane has been determined and the rate of movement plotted against the partition coefficient (Figure 2-3). Note that there is a positive correlation between these two values, which indicates that the more soluble the solute is in lipid, the more readily it enters the cell. But this is not to say that molecular size plays no role. On the contrary, for example, if several substances with the same partition coefficient but with different molecular size are tested, the smaller molecules are found to enter the cell more readily than the larger ones. These results suggest that both factors, size and lipid solubility, are determinants and that lipid solubility is the primary determinant.

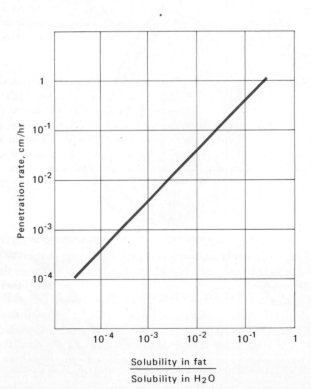

Figure 2-3. The effect of the partition coefficient on rate of movement through the cell membrane.

Concentration Gradient   If a molecule is capable of free movement through a membrane, the rate at which it will move is determined, if all other factors are constant, by the difference in concentration of that substance on the two sides of the membrane, that is, the concentration gradient. This follows from a simple consideration of diffusion. If the concentration on one side is greater than on the other, there will be more molecules of the solute in solution on that side. Accordingly, more molecules will pass from the concentrated solution to the less concentrated solution than vice versa.

Movement of Ions   The movement of ions through the cell membrane is determined not only by the size of the ion, its solubility in lipid, and the concentration gradient, but by still another factor, charge. A wide variety of experiments have led to the conclusion that the stronger the charge on the solute, the slower the penetration of the membrane. This means that weak electrolytes, that is, ones that do not readily ionize, enter more readily than do stronger ones. It also follows that monovalent ions penetrate the membrane more rapidly than do divalent or trivalent ions.

The movement of sodium and potassium ions has been studied perhaps more extensively than any other ionic movement. The studies show that potassium moves through the cell membrane more rapidly than sodium. Yet they both have the same charge. The explanation may lie in the fact that, in the hydrated form, the potassium ion is of considerably smaller diameter than the sodium ion.

The entire problem of the movement of ions through membranes into cells is complicated by the fact that in the normal living state there is a striking imbalance between the concentrations of various ions in the protoplasm and in the external environment. For example, in most cells there is very little sodium within the cell despite a high concentration in the surrounding fluid. Yet sodium ions can readily pass through the membrane. Clearly there must be an active process that pumps sodium out as fast as it enters.

Electrochemical Interrelationships   The movement of ions through membranes is determined by (1) the diameter of the hydrated ion, (2) concentration gradients, (3) osmotic pressures, and (4) electrical attraction by ions in the membrane and in the solutions on either side of the membrane. All these factors are included in the term electrochemical. If all the forces—that is, concentration, osmotic, and electrical—are known, then before movement takes place, the total force that will move that ion may be calculated. Such force is termed the electrochemical potential. The difference in the electrochemical potential between the two sides of the membrane determines the work required to move an ion from one side to the other.

Active Transport

Transfer of substances against a gradient has been termed active transport. There are many examples. Monosaccharides—the common sugars—penetrate most membranes and move in accord with concentration gradients. But in some instances they move in the opposite direction. The kidney is an excellent example. Normally there is no glucose in the urine. Yet, it has been clearly shown that glucose readily filters through the renal glomeruli. Obviously, all the glucose in the filtrate must be reabsorbed by the tubule cells (Figure 2-4). This can occur only by virtue of active transport. If the tubule cells are damaged or if the intracellular processes are chemically blocked, the glucose is not reabsorbed and appears in the urine. Another example involves sodium and potassium. As already mentioned, sodium can enter the cell, but most living cells rapidly extrude it so that the concentration on the outside is usually much higher than on the inside.

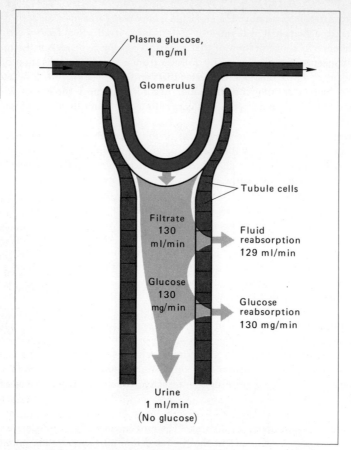

Figure 2-4. An example of active transport. The glucose is completely removed from the filtrate by the tubular cells.

Though there are many examples of active transport and though this physiological process has been extensively studied, as mentioned in Chapter 1, very little is known about the mechanism. It is known that interference with cell metabolism generally reduces or abolishes active transport. Likewise there are many so-called metabolic poisons which block certain metabolic processes and thus alter active transport. And there is evidence that certain enzymes or hormones control the active transport of specific substances. But just how is the question.

There is no convincing evidence that water is moved through a cell membrane by any active process other than pinocytosis, which means "a cell that drinks." In this process there is a discontinuous uptake of fluid, in the form of droplets, by the cell. The droplet seems to come in contact with the cell membrane, which then forms an invagination into which the droplet moves. Next the outer edges of the valley come together, forming a vesicle. The droplet is now completely surrounded by a membrane which lies within the cell (Figure 2-5). This membrane disintegrates, leaving the droplet within the cytoplasm.

A similar process involving solutes has been termed cytopemphis, which means "a cell pustule." In cytopemphis the particle is transported completely through the cell. In

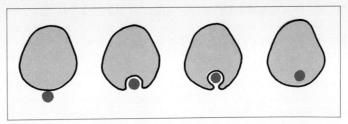

Figure 2-5. Pinocytosis. First an invagination appears. Next the droplet is surrounded as the invagination closes to form a vesicle. Finally the cell membrane forming the vesicle disappears, leaving the droplet within the cell.

other words, it is first engulfed and moved into a cell as outlined for pinocytosis. But then the reverse procedure occurs at the opposite membrane and the particle is discharged from the cell. This mechanism differs from phagocytosis, a process in which the substance is engulfed and then held or digested by the cell. The process by which a substance normally found within the cell is transported in this manner out of the cell is often referred to as reverse phagocytosis. Phagocytosis is illustrated by white blood cells, leukocytes, which exert a protective function by virtue of their ability to engulf foreign particles, such as bacteria, into the cell, where they are destroyed.

Whether or not the movement of solvent and solute by pinocytosis, cytopemphis, and phagocytosis is true active transport is not too important. The fact remains that by means of these processes there is movement against a gradient.

Bioelectrical Currents

In all living cells there is a potential difference between the inside of the cell and the outside. Consequently, if these two areas are connected by a conductor, a current will flow. The potential difference stems from the unequal distribution of ions between the inside and the outside of the cell. Because the ionic imbalance derives from the characteristics of the cell membrane and because the potential difference exists across the membrane, the resulting potential is termed the membrane potential. This potential exists while the cell is at rest and, therefore, is termed the resting potential. Membrane potential and resting potential are terms that are used synonymously.

Membrane Potential That an ionic imbalance exists has been known for some time. At first it was thought that the cell membrane was relatively impermeable to certain ions and thus some were kept out and others kept in. Radioisotope studies, however, have shown conclusively that, at least insofar as sodium, potassium, and chloride are concerned, they can diffuse through living membranes quite readily. Yet, by virtue of active transport of sodium and potassium ions, the ionic imbalance shown in Table 2-1 is maintained.

TABLE 2-1. INTRA- AND EXTRACELLULAR IONS*

ION	INTRACELLULAR IONS, mEq/LITER	EXTRACELLULAR IONS, mEq/LITER
Potassium	155	4
Sodium	12	145
Chloride	5	105

*Values in this table are merely representative. They vary considerably from cell to cell and from species to species.

A membrane potential can be produced simply by the active transport of ions or by the diffusion of ions after a concentration gradient has been established by active transport. For example, assume that a cell contains only sodium ions and an equal concentration of nondiffusible anions. Now turn on the sodium pump (see Chapter 1). As sodium ions are pumped out, the inside of the cell will become negative in relation to the outside because of the nondiffusible anions left within the cell. To understand how diffusion can give rise to a membrane potential, assume that potassium ions have been pumped into the cell so that they are in much greater concentration within than without. Again assume that there is an equal concentration of nondiffusible anions in the cell. Now, because of the concentration gradient, potassium ions will diffuse out of the cell, once again leaving the anions behind, and thus there is a membrane potential with the inside negative in relation to the outside.

The cations in greatest concentration are sodium and potassium, and they are transported actively. Chloride ions are not actively transported but rather distribute themselves on either side of the membrane in accord with the electrical potential. Since the inside of the cell is negative to the outside, the concentration of chloride ions is greater outside than inside.

The Nernst equation is used to calculate the membrane potential due to diffusion. It states that

$$\text{Membrane potential in millivolts (mv)} = 61 \log \frac{\text{concentration inside}}{\text{concentration outside}}$$

When we know the concentrations of potassium and sodium ions inside and outside a cell, the potential due to each ion can be calculated. Obviously, potassium causes a membrane potential that is negative inside, whereas sodium causes a potential that is positive inside the cell. Yet actual measurement of the potential reveals a membrane potential that is negative inside (Figure 2-6), and its magnitude is very close to that calculated for potassium using the Nernst equation. Thus, calculation may show a potential of -95 mv, while actual measurement reveals -85 or -90 mv. The difference is due to the diffusion of sodium. Because potassium diffuses through the membrane so much more easily than sodium, the actual membrane potential is almost wholly the result of potassium diffusion.

There is still another factor. Available evidence strongly suggests that the coupled

Figure 2-6. Membrane potential. Because the inside of the cell is negative with reference to the outside, a current will flow if they are connected as shown.

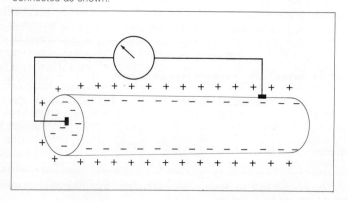

pump for sodium and potassium results in the extrusion of three sodium ions for each two potassium ions that are transported into the cell. If an equal number of ions were pumped in each direction, the pump would be said to be electrically neutral; if more ions move in one direction than the other, the pump is electrogenic. This in itself will create a membrane potential, since more cations are being pumped out than enter, leaving the inside of the cell negative in reference to the outside.

To summarize, a resting potential can exist only if the membrane permits the diffusion of an ion of one charge and is impermeable, or relatively so, to an ion of the opposite charge. In cells, potassium can diffuse but large anions cannot. Theoretically, the greater the potassium gradient across the membrane, the greater the membrane potential. Experimentally this has been confirmed.

Quite clearly there must be a mechanism that establishes the potassium concentration gradient. The potassium then diffuses down its concentration gradient until the electrical gradient balances it. At that point, equilibrium is reached and the membrane potential is maintained. But it can be maintained only so long as sodium ions are extruded as quickly as they diffuse into the cell in response to both concentration and electrical gradients. When the cell is killed, the pump becomes inoperable, sodium rushes in, and potassium out; all gradients are reduced to zero, and there is no longer a membrane potential.

Action Potential   In order to understand the discussion that follows, the student must appreciate the fact that there are several factors operating simultaneously which determine ion movement across the cell membrane. In the case of sodium ions there are at least four such factors: (1) the concentration gradient, (2) the electrical gradient, (3) the sodium pump, and (4) the membrane permeability, also termed conductance, for sodium ions. The first two tend to cause sodium ions to be driven into the cell; the last causes resistance in movement of the ions into the cell, and the pump drives them out. If it is assumed that the pump is working at a steady rate and the gradients remain unchanged, an increase in membrane permeability will permit more sodium to enter the cell. An increase in membrane permeability for sodium, therefore, results in a decrease in membrane potential.

The discussion thus far has been concerned with the resting potential. When the cell becomes active, there is a marked change in that potential. The altered potential is referred to as the action potential.

If electrodes are placed so that one is inside the cell and the other on the surface, a resting potential of about -70 mv will be observed. This, of course, varies for different types of cells. But for the sake of this discussion we will use a resting potential value of -70 mv. Now if the cell is activated, a marked alteration in the potential occurs (Figure 2-7). Note that there is a very rapid change in potential from the resting level of -70 mv and then to perhaps $+30$ mv or even higher. That is, there is a reversal of polarity called the overshoot. With almost equal suddenness the potential then returns to the resting level, and in most cases beyond, that is, to about -75 mv. Ultimately it returns to exactly the resting level. All these alterations together make up the action potential. The sharp rise and fall constitute the spike potential.

When a cell is at rest and the membrane potential is established and steady, the membrane is said to be polarized. Activation of the cell brings about the changes just described. When the membrane potential reaches zero, it is said to be depolarized.

Stimulation of the cell increases the membrane permeability for sodium. With less impairment to its entry and in response to the concentration and electrical gradients, this ion rushes into the cell. As it does so, the membrane potential changes from -70 mv

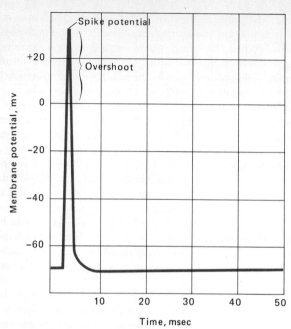

Figure 2-7. Action potential.

toward zero. As the potential changes in this direction, the permeability for sodium increases even more and therefore the influx accelerates. A self-sustaining sequence is thus set into motion. As sodium enters, it evokes an alteration that permits even more sodium to enter. For a brief period the influx is so rapid that it overwhelms the mechanism responsible for pumping it out.

If sodium diffused freely down its concentration gradient, it would create a potential of about +66 mv. However, the spike potential does not generally reach this peak for two reasons: (1) the rapid influx of sodium does not persist, and (2) there is increased diffusion of potassium out of the cell, which creates a more negative potential. In other words, during excitation of the cell the conductance for sodium and the conductance for potassium increase. But they do not increase simultaneously. The increase in sodium conductance occurs first; then, within a millisecond, the conductance for potassium increases. The initial inrush of sodium causes the rise in the spike potential. The slowing of sodium influx and speeding of potassium efflux cause the fall in the spike potential.

MITOCHONDRIA
Mitochondria are large, easy-to-identify organelles found in all cells except mature erythrocytes and bacteria. They have been referred to as the principal power plants of the cell.

Structure of Mitochondria
In the living cell mitochondria can be seen easily, even with only a light microscope, because they are relatively large cytoplasmic components and because they can be stained by the vital dye, Janus green. Mitochondria may also be visualized using a phase-contrast microscope.

In the living state mitochondria are seen to vary in number, shape, size, and distribution. The average cell contains a few hundred, but liver cells may contain one or two thousand. The typical mitochondrion is sausage-shaped, but in the living cell they have been described as granular or threadlike (Figure 2-2). The sausage-shaped mitochondria have an average diameter of 0.5 micron and a length of about 1.5 microns. However, mitochondria 6 or 7 microns long have been seen. They are not usually distributed evenly throughout the cell. Further, in any particular cell the distribution may change.

Mitochondrial Membranes Mitochondria are surrounded by two membranes. Each is a typical unit membrane about 75 Å thick and contains the usual protein and fat layers. Between the two membranes is a fluid rich in coenzymes.

The inner of the two membranes gives rise to structures that extend into the interior of the mitochondrion (Figure 2-2). They are elongated sacs which contain the same fluid that fills the space between the two membranes. The sacs are called cristae.

Mitochondrial Particles Very highly magnified electron micrographs have disclosed that the inside of the inner membrane and the outside of the outer membrane are covered with very small particles. They measure about 90 to 100 Å in diameter. The particles on the outer membrane differ from those on the inner membrane. The former generally appear as simple spheres packed closely together which give to the surface of the mitochondrion a rough or pimpled appearance. In some cells the particle on the inside of the inner membrane has a base, a stalk, and a spherical head (Figure 2-8). The particle, like a unit membrane, would seem to be composed of a protein cortex with a phospholipid core.

Function of Mitochondria

Energy is stored in the cell in a compound called adenosine triphosphate (ATP). The compound contains three phosphate molecules, two of which attach to the compound by means of so-called high-energy bonds; that is, unusual amounts of chemical energy are required to form the bonds and are released when the bonds are broken. The amount of energy released when 1 mole of ATP is decomposed is equal to 8000 calories.

The major function of mitochondria appears to be to produce ATP. ATP is manufactured by the addition of phosphate to adenosine diphosphate (ADP). This takes place in the mitochondrion in a sequence of events known as a respiratory chain. In a typical respiratory chain (Figure 2-9) the substrate is oxidized. Oxidation means loss of an electron. Reduction, on the other hand, means acquisition of an electron. For example, in the reaction

$$2H_2O \longrightarrow 4H^+ + 2O^{--}$$

note that each atom of hydrogen has lost an electron and therefore has acquired a positive charge, while each atom of oxygen has gained two electrons and thus two negative charges. According to the preceding definition, hydrogen has been oxidized and oxygen reduced.

Examination of Figure 2-9 will reveal that the electrons from the substrate are handed down the respiratory chain from compound to compound until ultimately molecular oxygen is reduced to an oxygen atom and this reacts with the two hydrogen ions to form water. In this process, the electron carrier is first oxidized and then, in the succeeding reaction, reduced.

So long as there is an adequate store of reduced substrate at one end of the chain and an adequate supply of oxygen or other electron acceptor at the other end, the wheels will turn, so to speak. For the wheels to turn, energy is required. The potential gradient

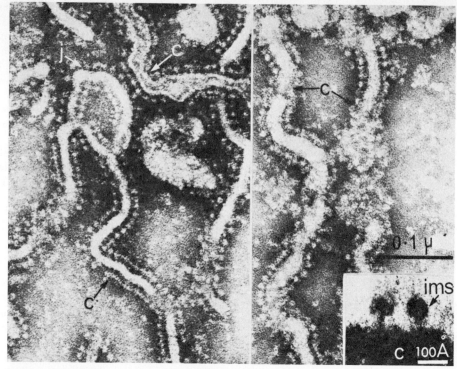

Figure 2-8. Electron micrographs of the particles associated with the mitochondrial cristae. (Left) The cristae (c) are narrowed into tubules which branch at some points (j). The surface of each crista is covered with knoblike subunits consisting of a head 90 Å in diameter and a stem 40 Å wide. (Right) Projecting subunits associated with the cristae (c) prepared from isolated mitochondria. (Inset) Higher-magnification view of two inner membrane subunits (ims). (*Courtesy of D. F. Parsons, from Science, vol. 140, fig. 1, May 31, 1963. Copyright 1963 by the American Association for the Advancement of Science.*)

represented in the system not only provides the energy to turn the wheels of the respiratory chain but also provides at three points in the system energy which is utilized to drive the ADP plus inorganic phosphate (P_i) reaction uphill, that is, to form high-energy ATP.

A word should be said about the coenzymes of the respiratory chain. The coenzyme most frequently acting as an acceptor of electrons from the substrate is nicotinamide adenine dinucleotide (NAD), commonly termed coenzyme I. (Changes in the terms used to designate two important coenzymes are summarized in Table 2-2.) Another coenzyme is flavin adenine dinucleotide (FAD), derived from the vitamin riboflavin. An important group of electron carriers are the cytochromes, which are substances containing iron, heme, and protein.

Role of the Particles   It is now known that the thousands of particles that line the outer surface of the outer membrane and the inner surface of the inner membrane are essential for the respiratory chain to function in the mitochondria. The outer particles serve a

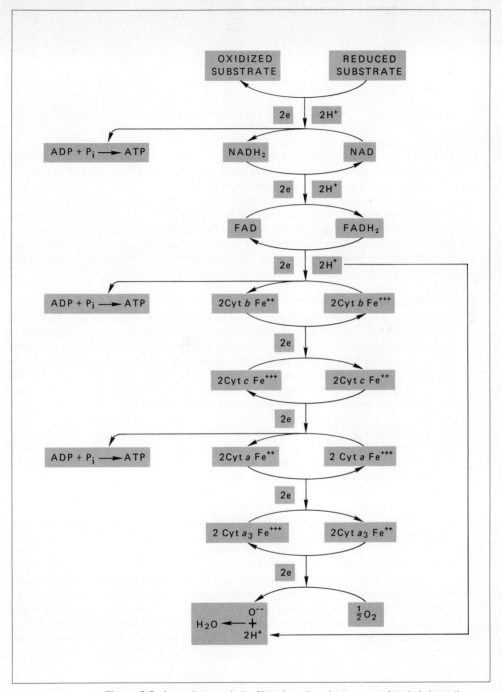

Figure 2-9. A respiratory chain. Note how the electrons are handed down the chain until molecular oxygen is reduced to an oxygen ion, which reacts with the two hydrogen ions to form water. In the process energy is liberated. This energy not only drives the system but also is used to form ATP, as shown.

TABLE 2-2. TERMINOLOGY OF COENZYMES

COENZYME	NEW TERM	OLD TERM
I	Nicotinamide adenine dinucleotide (NAD)	Diphosphopyridine nucleotide (DPN)
II	Nicotinamide adenine dinucleotide phosphate (NADP)	Triphosphopyridine nucleotide (TPN)

different function from those on the inner membrane. The outer particles are believed to carry out the oxidation reactions that supply electrons, and they are also the site for catalyzing of synthetic reactions powered by ATP. The transfer of electrons along a chain of complexes that synthesize ATP occurs in the inner-membrane particles. The overall picture as currently envisioned depicts the outer particles as providing electrons by the oxidation of various substances. The electrons from the outer particles are carried across the space between the two mitochondrial membranes and through the inner membrane to enter the inner particles. It is within the inner particles that the electrons move along a respiratory chain to produce three molecules of ATP.

Utilization of Energy To permit the process just outlined to "go," there must be an adequate input of the essential ingredients. There must be an adequate flow of electrons into the inner particles, and there must be sufficient oxygen to accept the electrons at the end of the chain. Oxygen, from exogenous sources, diffuses into the cell. Likewise, various substrates are supplied to the cell, where they are oxidized and thus provide electrons. The end result is the production of ATP, the ready source of energy. How the molecules of ATP move from the mitochondrial particles to the site where their energy is required and how this energy is utilized are questions that are only beginning to be answered.

Mitochondria move, and this movement is, apparently, purposeful. For example, mitochondria become concentrated at the site in the cell where energy demands are high. Some idea of how mitochondria move has been obtained, but what orients the direction of that movement is a complete mystery. Mitochondria move by virtue of change in the dimensions of their membranes. In this respect they function like independent cells, like amebas. The membrane can contract and relax. When the concentration of ATP in the mitochondrion is high, membrane contraction occurs; when low, the membranes relax. This may well be a self-regulating mechanism.

Control of Mitochondrial Function The suggestion has been made that mitochondrial function is self-regulated by the production of ATP. But this is apparently a basal regulation which can be altered, or reset, so to speak, by other influences. For example, a change in thyroid activity markedly alters the metabolic rate. The obvious explanation is that thyroid hormones, in some way, influence mitochondrial function. Evidence is accumulating that this is so. Probably other hormones and various vitamins have a similar influence.

LYSOSOMES

Only recently has the true nature and importance of the lysosomes come to light. They are organelles containing digestive enzymes capable of lysis, that is, the breakdown of substances. They were formerly identified as pericanalicular dense bodies because of their location. Now that their function is known, the term lysosome is vastly preferable.

Structure of Lysosomes

The shape and density of lysosomes vary broadly (Figure 2-2). Most of them are between 0.25 and 0.50 micron in size. In fact, their shape and appearance vary so greatly that they cannot be accurately identified solely on the basis of their appearance. Only an intracellular body which contains lytic enzymes is properly called a lysosome.

The exact cytoplasmic origin of lysosomes is not yet clear. Suggestions have been made that they are formed in association with the Golgi apparatus, or endoplasmic reticulum.

Lysosomes contain at least a dozen different enzymes. All of them are capable of splitting biological compounds in a slightly acid medium. They can digest proteins, nucleic acids, and polysaccharides.

Function of Lysosomes

Lysosomes apparently serve at least four functions: (1) the digestion of large particles that enter the cell, (2) the digestion of intracellular substances, (3) the digestion of the cell itself, and (4) the digestion of substances external to the cell.

Large particles and molecules are taken into the cell by phagocytosis. The cell engulfs the particle. The invagination is pinched off from the cell membrane and becomes an internal sac or body. This sac is referred to as a phagosome. The phagosome moves toward a lysosome. The two bodies come into contact, their membranes fuse, and somehow they then form one body (Figure 2-10). Now the enzymes from the lysosome can come into contact with the molecules brought into the cell in the phagosome, and digestion occurs. Once the molecules are degraded sufficiently, the smaller products can diffuse out of the lysosome or from what is now called by some the digestive vacuole.

Figure 2-10. Lysosomal digestion of external particles. The particles to the left of the cell are taken in by phagocytosis. In the cell the lysosome provides the enzymes for digestion of the particles. Various end products of digestion may diffuse out of the vacuole into the cytoplasm. The indigestible residue is eliminated by reverse phagocytosis.

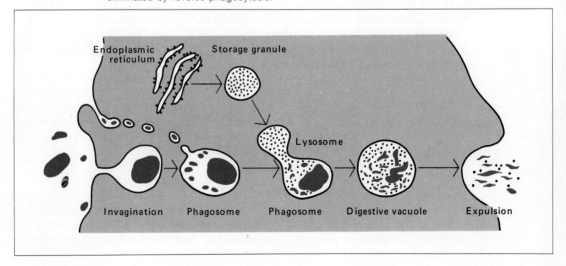

Just as in the digestive tract there remain residues that must be eliminated by defecation, so in this primitive digestive action provided by phagocytosis and the lysosomes, residues are left in the digestive vacuole. The vacuole now moves on to the cell membrane, where so-called reverse phagocytosis occurs. For this primitive intracellular digestive activity to continue to function, there must be, of course, lysosome production at a rate equal to that of the process of phagocytosis. Virtually nothing is yet known concerning the formation of lysosomes or the factors that control their rate of formation.

Portions of the cell somehow penetrate the cell's own lysosomes and are broken down, a process termed autophagy. How they get in is not clear, and what role autophagy plays in cell function can only be surmised.

When a cell dies, the lysosome membrane ruptures. The liberated enzymes now are free within the cell, which they quickly digest. One immediately wonders how the lysosome membrane keeps these enzymes from normally diffusing out or, even more bewildering, how the membranes are protected from the digestive action of the enzymes. But somehow the membrane does keep them in and is not digested; once death occurs, change is rapid.

A hypothesis has been advanced that this is a built-in mechanism for removing dead cells. In multicellular animals, many cells constantly form, live for a short period of time, and then die. Autodissolution is an efficient method of ridding the body of the debris.

There is also evidence that autodissolution may occur as a pathological mechanism. For example, if a cell is cut off from its oxygen supply or poisoned, the lysosome membrane may rupture, thereby permitting the enzymes to dissolve the cell. Also, when the lysosomes are overwhelmed by large amounts of foreign substances, they may rupture and spill their enzymes within the cell.

Quite clearly cellular activity, the very life or death of the cell, can be influenced by substances that act on the lysosome membrane. The suggestion has been made that the hormones of the adrenal cortex exhibit their well-known anti-inflammatory effect by virtue of their ability to stabilize the lysosome membrane. Conversely, unwanted cells could be eradicated by a substance which would cause the lysosome membrane to break down. The value of such a drug in the treatment of cancer is obvious.

Apparently cells can discharge lysosomal enzymes to destroy surrounding structures. Perhaps by reverse phagocytosis a packet of enzymes from a lysosome is released outside the cell, where it then digests contiguous structures. This is thought to explain how sperm penetrate the protective coating of the ovum. It may also explain how osteoclasts, cells that destroy bone, function. And this may also be the explanation for the well-known ability of white blood cells to pass quickly out of the blood vessels and into the tissue spaces at the site of an infection.

There is a very practical consequence of the release of lysosomal enzymes. For quite some time, it has been known that prolonged cardiopulmonary bypass, a procedure essential for heart surgery, has various deleterious effects. Large doses of corticosteroids prevent, to a great extent, these effects. Recent evidence strongly suggests that the corticosteroids function in this respect by preventing changes in the permeability of the lysosomal membrane and thus block the release of lysosomal enzymes.

OTHER CELLULAR STRUCTURES

There are several other organelles in the cell that have not yet been mentioned. These are described here only briefly, not because they are necessarily any less important than the ones already discussed, but rather because so much less is known concerning their function.

Golgi Apparatus

In 1898, Camillo Golgi, an Italian anatomist, described a network of threads in the cytoplasm of a cell that had been stained with silver nitrate or osmic acid. For years following this announcement, cell anatomists debated whether the so-called Golgi apparatus was really an organelle or merely an artifact caused by the way the cell had been handled in preparation for study. Today, however, there is general agreement that the Golgi apparatus, or complex, as it is also called, does exist and probably plays an important cellular role.

In cells stained with silver nitrate or osmic acid, the Golgi apparatus is seen to consist of several large, apparently empty vacuoles surrounded by a membrane (Figure 2-2). This organelle functions as an intracellular pump that regulates the movement of fluids in the cell and the expulsion of secretory products from the cell. It may well be necessary for the secretion of very large molecules, so-called macromolecules, by the cell. The Golgi region is the site where carbohydrates are linked to proteins.

Endoplasmic Reticulum

The endoplasmic reticulum is seen in Figure 2-2 to form a series of small canals through the cytoplasm. Various substances pass through these canals in moving from the cell membrane to the nucleus. Closely associated with the membranes which line the canals are tiny granules termed microsomes. Because they contain such a high concentration of ribonucleic acid (RNA), they are usually referred to as ribosomes. They are essential for protein synthesis.

The ribosomes, also referred to as Claude's particles, range in size from 100 to 150 Å in diameter and thus can be seen only by means of the electron microscope. They are so rich in RNA that they may contain as much as 60 percent of the total RNA in the entire cell.

Centrosome

The centrosome consists of two centrioles. Each appears as a short cylinder composed of bundles of parallel bars (Figure 2-2). The centrioles cannot be seen except during cell division, at which time they move away from one another until they lie on either side of the cell with the nucleus between them. They are thought to play an important role in cell division, but little definitive information is available. That the centrosome is not seen in certain plant cells, although these cells divide quite normally, is even more confusing.

Vacuoles

Vacuoles are cavities found in both plant and animal cells. They contain a variety of solid or fluid substances. Their size varies greatly. In some cells they can barely be discerned; in others they may occupy almost all the cell. They probably have several functions. In plants there are large vacuoles that contain sap and provide necessary rigidity for the support of the cell. Others probably participate in the processes of pinocytosis and phagocytosis. Vacuoles that contract probably move fluid within the cell.

Cilia

Some cells have hairlike processes called cilia capable of vibratory or lashing movements. Cilia serve to clean the surface. For example, the nasal mucosa is lined with ciliated epithelium. It is the cilia of these cells that move foreign particles into the pharynx, from where they may be expelled. Cilia are also present at the ovarian end of the fallopian tubes. Here they serve to conduct the ova into the tubes.

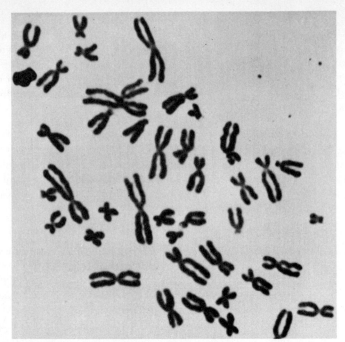

Figure 2-11. Photomicrograph of human chromosomes. (*Courtesy of Kyle W. Petersen.*)

THE NUCLEUS

Most mature cells contain a nucleus. If the nucleus is removed, the cell dies. This fact indicates that the cytoplasm is dependent upon, and is regulated by, the nucleus.

The nuclear membrane is a unit membrane. The nuclear fluid is termed nucleoplasm. Like cytoplasm, nucleoplasm contains several discrete structures. The most important of these are chromatin and the nucleolus.

Chromatin consists of dark-staining granules. During mitosis this material forms minute paired strands of protoplasm known as the chromosomes (Figure 2-11). In the interval between successive cell divisions the chromosomes unite to re-form the chromatin granules.

Structure of the Chromosomes

The predominant compound within the chromosome is deoxyribonucleic acid (DNA) (see p. 50). Through the functions of DNA, genetic features of the parents are transmitted to their developing offspring. The specific locus or spot on a chromosome carrying the genetic material or information for a specific characteristic or condition is called a gene, which is part of the DNA molecule.

Nucleolus

A tiny dark-staining body within the nucleus, the nucleolus, is involved in RNA synthesis and storage.

The two nucleic acids, DNA and RNA, are extremely important in the mechanism of inheritance. They will be discussed in detail presently (see p. 50).

CELLULAR REPRODUCTION

Two major types of cells exist: somatic cells, of which body tissues are composed, and sex cells, which are produced by the ovary and testis and which come together as a result of sexual union to create a new individual.

Cell multiplication occurs by division of a cell into two daughter cells. This process entails several intermediate steps and is called mitosis. Cell division of somatic cells is different in important respects from that of sex cells.

Mitosis

Somatic cells characteristically divide by mitosis, a procedure that can be divided into five stages or phases (Figure 2-12).

Figure 2-12. Mitosis. The stages of mitosis are depicted in the changes a pair of chromosomes undergoes during cell division. Note that in metaphase the chromosomes are aligned on the equatorial plate. Each has split into two chromatids, which are attached to spindle fibers preparatory to their movement toward the centrioles.

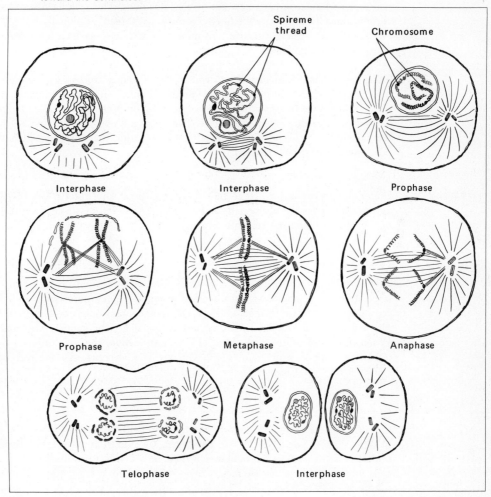

Prophase In the first stage, prophase, the chromatin differentiates into pairs of rod-shaped chromosomes. There is a characteristic number of chromosomes in each body cell for each species. In man this number is 46, divided into 23 pairs. During prophase each chromosome splits longitudinally, and each half is called a chromatid. At this point the nuclear membrane disappears, and the centrioles of the cytoplasm move to the opposite poles of the cell, trailing after them delicate, streaming fibers that eventually form a spindle between the poles.

Metaphase In the next stage, metaphase, the chromatids align themselves across the equator of the cell with each attached to spindle fibers at a special region, the centromere.

Anaphase During the third stage, or anaphase, the two chromatids of each chromosome become detached from each other and start moving to opposite poles along the spindle fibers.

Telophase In the telophase the chromatids have reached the poles. Nuclear membranes reappear and enclose each new set of chromosomes. The spindle fibers disappear, the chromosomes re-form into the long chromosomal thread, and finally the cell membrane between the new nuclei constricts, forming two new cells, each with the normal complement of 46 chromosomes.

Interphase These four phases proceed as a continuous process, each blending with the next phase. The whole process takes about $2\frac{1}{2}$ hours in rapidly dividing cells such as the cells lining the intestine. The interval between mitoses is often called the resting stage but is more correctly called interphase. The cell is not really resting but is growing while the chromosomes about double in size, thus assuring that the cell will have adequate DNA for the next cell division.

Amitosis

Under certain conditions, some cells divide by direct, or amitotic, cell division. During amitosis the cells simply elongate, and the nucleus and cytoplasm constrict near the center of the cell to divide the cell into two new cells.

Meiosis

Meiosis (see Figure 33-1) is a type of cell division restricted to the formation of reproductive cells (either spermatozoa or ova). One chromosome of each pair migrates to opposite poles in anaphase, creating two groups of 23 chromosomes each. Thus when reproductive cells are formed at telophase, they contain only one-half the normal complement of chromosomes; hence the alternative name for meiosis—reduction division. When the ovum unites with the sperm in fertilization, the resulting cell then has the normal number of chromosomes (46), which will be perpetuated in the succeeding mitotic cell division. (See Chapter 33 for more details.)

GENETICS

Cell division is essential for growth and development and for the generation of new cells to replace those which die or are destroyed. Cell division, as has just been seen, also underlies germ (sex) cell production. When cells divide to produce new skin, the new cells must have the same characteristics as did the old; that is, they must be skin cells, not nerve cells or liver cells. Likewise, as the embryo develops, a wide variety of cells, with many different functions, must be produced at the right place and at the right time. In short, cells are differentiated according to their configuration and function, factors which are under the control of the genes.

Gregor Mendel, the famous monk, began his studies of inheritance in 1854, using peas. As a result of his studies, laws of inheritance were formulated. These will be briefly outlined below. Mendel was an observer and an astute thinker, but he knew nothing about chromosomes or genes and nothing at all of cell physiology. This is not surprising, because it was only at about that time that the nucleus was discovered and not until 1880 that chromosomes were identified by the German biologist Walther Flemming. Because the threadlike structures in the nucleus of the cells that Flemming stained absorbed color, he called them chromosomes. He observed that these threads undergo orderly changes during cell division. He was first to describe mitosis.

Laws of Inheritance

Mendel made fundamental observations concerning patterns of inheritance. He formulated two basic laws:

1   Law of segregation. Mendel assumed that there were paired genetic determinants. From his observations he concluded that these factors must segregate from each other in the parent and come together again in the offspring. For example, as shown in Figure 2-13, there is a pair of black characteristics in one parent and a pair of white characteristics in the other. In the first generation, produced by mating of the parents, that is, F_1, both combinations are Ww; that is, both organisms appear white if white is the dominant factor. In the next generation, F_2, according to the law of segregation, there are three possibilities: (1) WW, (2) Ww, and (3) ww. There is a 3:1 ratio, 3 white and 1 black.

2   Law of independent assortment. Mendel also showed that where there is more than one pair of characteristics, or alleles, in each parent, the pairs segregate from each other independently. Figure 2-14 shows that if there are two pairs (size and color), there will be 16 possibilities in the F_2 generation. As will be explained later, the second law does not always hold because of what is known as linked genes.

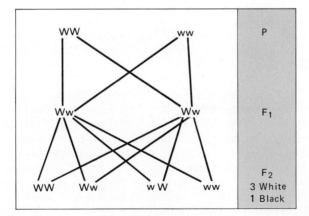

W = White dominant
w = Black recessive

Figure 2-13. Mendel's law of segregation. Note that because white (W) is dominant, the offspring in the F_2 generation have a ratio of 3 white to 1 black.

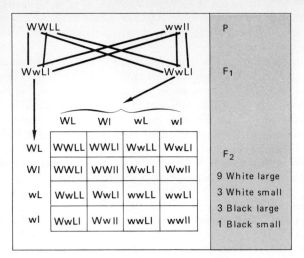

Figure 2-14. Mendel's law of independent assortment. Even when there is more than one pair of alleles, the pairs segregate independently. In this hypothetical case there are four characteristics, two dominant and two recessive. A ratio of 9:3:3:1 would be expected, and that is what Mendel reported.

With two characteristics that segregate independently, a ratio of 9:3:3:1 would be expected in the F_2 generation, and this is what Mendel found. He also studied peas with three characteristics. In the second generation a ratio of 27:9:9:9:3:3:3:1 would be anticipated, and these were exactly the figures reported by Mendel. As a matter of fact, the very exactitude of his findings caused more than a few eyebrows to be raised!

Linkage Exceptions to Mendel's second law gave rise to the concept of linked genes. If the chromosome is thought of as having a series of genes, unless the chromosome itself splits into fragments, the genes of any one chromosome must move together, that is, they are linked and cannot segregate independently.

Sex Determination The chromosomal pattern has also been shown to be responsible for sex determination. In addition to the usual chromosomes, the so-called autosomes, there are also special chromosomes that are primarily concerned with sex determination. These are the X and Y chromosomes. In some species there is only the X, but in others both X and Y chromosomes. In either case, as shown in Figure 2-15, the ratio of male to female in the offspring must be 1:1.

Sex Linkage The X chromosome, the one responsible for sex determination, carries many genes, and thus the characteristics resulting from these genes must be sex-linked, that is, related to the sex of the person exhibiting them (Figure 2-16). Examples of sex-linkage are hemophilia and color blindness. Hemophilia results from a mutation in the X chromosome, and it is a recessive characteristic. Because it is recessive, in order for hemophilia to afflict a female the mutation must be present on both X chromosomes. In the male, in contradistinction, since the Y chromosome carries no genes, the presence of the mutation on the one X chromosome suffices. To put it another way, to produce a hemophiliac female both parents would have to carry the mutation. The result of these facts is that women are often so-called carriers; that is, they have the mutation on one X chromosome, do not suffer from the disease, but can pass this chromosome to a male child, who will exhibit the disorder.

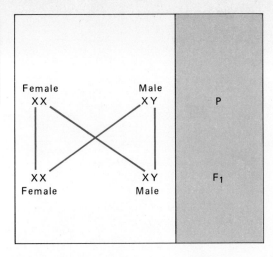

Figure 2-15. Determination of sex. Female cells have two X chromosomes. Male cells are XY. The ratio of male to female offspring is 1:1.

The Biochemistry of Gene Control

The laws of inheritance were enunciated before this century began, but only since about 1930 has there been any real understanding of why these laws hold. Since that date progress has been truly remarkable. An entire new discipline has developed around the biochemical mechanisms of gene control. The way in which the cell's characteristics are passed along to succeeding generations in line with Mendel's expectations has been elucidated. Furthermore, genes have been shown to direct a host of other activities through their control of the synthesis of protein. This aspect of gene control is of great importance for the cell. In the first place, proteins function as important structural units. For example, the unit membrane, which surrounds the cell, the organelles, and the nucleus, consists of protein combined with fat.

Figure 2-16. Sex linkage with respect to color. The Y chromosome carries no genes. Since the dominant characteristic (W) is on both X chromosomes in the female and the recessive characteristic (w) is on the X chromosome of the male, the female has no chance of being anything but white. The ratio of male to female is, of course, unaffected.

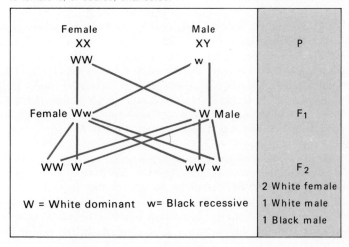

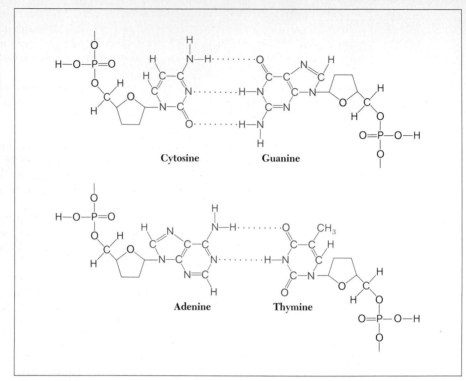

Figure 2-17. The structural formulas of two portions of a DNA molecule. The paired bases are shown in color, the hydrogen bonds by dotted lines.

Second, enzymes are proteins. Enzymes control the chemical reactions that take place within the cell. Thus, it can be said that genes, by directing synthesis of enzymes, determine whether the cell will behave like an epithelial cell, for instance, or a nerve cell.

Genes were quickly identified as part of the long DNA molecule, so it is time we examined this and another key molecule in the scheme of control, RNA.

Nucleic Acids About a hundred years ago a Swiss biochemist, Friedrich Miescher, treated cells with the enzyme pepsin and found that everything in the cell disintegrated except the nucleus. Because pepsin was known to digest protein, Miescher concluded that the nucleus must be made of some special nonprotein material. He called it nuclein and was able to show that it contains phosphorus. Later it became known as nucleic acid. Nucleic acid is now known to include nitrogen-containing compounds called bases.

There are two fundamental types of bases: those with a single-ring configuration and those with a double ring. Those with but a single ring are called pyrimidines; with a double-ring, purines.

Nucleic acids also contain carbohydrate. Two kinds have been identified. One is called ribose, and the other, which differs only in that it lacks an oxygen atom, is termed deoxyribose. Both have five carbon atoms and are thus termed pentoses.

A molecule of nucleic acid is made up of many nucleotides strung together to produce a very long molecule. A nucleotide consists of a base, a carbohydrate, and a phosphate. If the carbohydrate in the nucleotide subunit is ribose, the nucleic acid is referred to as ribonucleic acid; if deoxyribose, deoxyribonucleic acid. Everyone today refers to them as RNA and DNA. There is one other difference: in DNA the bases adenine, guanine, cytosine, and *thymine* are found; in RNA the bases are adenine, guanine, cytosine, and *uracil.* In other words, DNA has thymine; RNA, uracil. Figure 2-17 shows how nucleotides are linked to form nucleic acid.

The results of very brilliant research have established beyond reasonable doubt that chromosomes are molecules of DNA loosely bound to protein to form nucleoprotein, and the genes are specific parts of those DNA molecules.

Structure of DNA Knowing the composition of a compound is one thing; knowing its structure is something very different. The answer to the question of how to put the components together to form the molecule was supplied by James Watson, a young American scientist, and Francis Crick, a British chemist, in 1953. In 1962, they were awarded a Nobel Prize for their work.

From x-ray diffraction studies Watson and Crick learned that a molecule of DNA contains at least 200,000 nucleotide units. They also fathomed that the molecule must have a helical configuration. They concluded that DNA resembles a ladder that has been twisted into a spiral called a helix. The sides of this helical ladder, they showed, are formed by the deoxyribose-phosphate units. The two sides of the ladder are held together by rungs composed of pairs of bases, adenine and thymine forming one type of rung, cytosine and guanine forming another. The base pairs are joined through hydrogen bonding (Figure 2-18). Each half-turn of the helix contains five base pairs.

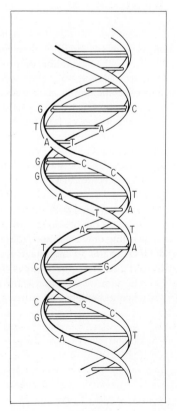

Figure 2-18. Schematic view of the DNA double helix.

DNA Replication   Having established the structure of DNA, Watson and Crick suggested that if the hydrogen bonds were broken, the two strands could separate and unwind. Each strand would then have naked bonds which would attract appropriate units. When complete, each original strand would be coupled with a new strand. Where there were at first two strands, there would now be four in the form of two DNA molecules. Thus DNA would make a replica of itself. This process is called replication.

Protein Synthesis

DNA replication is the crucial step in the inheritance of characteristics from cell to cell. To understand gene control of protein synthesis, however, we must examine the coordinated action of DNA and RNA.

Role of RNA   RNA is very similar to DNA, with two major differences: (1) RNA contains ribose instead of deoxyribose and (2) it generally exists in the form of a single- rather than a double-strand molecule. But DNA and RNA are similar enough so that a strand of DNA can not only replicate but also direct the sequence of nucleotides to form specific RNA molecules. Keep in mind that DNA is located in the nucleus. It is in the nucleus that the strands of DNA direct the formation of specific strands of RNA. These new strands of RNA then travel from the nucleus into the cytoplasm. Because they are molded by DNA and thus carry the DNA message from the nucleus to the cytoplasm, they are called messenger RNA (mRNA).

The strand of DNA, to make RNA instead of another DNA molecule, requires the pairing of the base adenine with uracil rather than thymine, and it requires the incorporation of ribose rather then deoxyribose. This is exactly what occurs, as radioactive isotope studies have now proved.

Admitting that DNA can rapidly unwind its two strands, that each strand can give rise to a specific strand of RNA, and that this messenger RNA then hurries out of the nucleus and into the cytoplasm where it directs the formation of protein, one is still bothered by the speed with which all this occurs. Thus it was suggested, again by Crick, that there may be a mechanism to gather up specific amino acids and to bring them to the RNA strand for proper incorporation into the protein molecule as directed by RNA. Such a mechanism does, in fact, seem to exist. Substances referred to as transfer RNA (tRNA) are visualized as scurrying about the cell, latching themselves to a certain amino acid, which they then transfer quickly back to the mother RNA strand to be incorporated into the finished protein molecule. As improbable as this may seem, there is excellent evidence that there is a separate transfer RNA molecule for each of the 20 amino acids.

Ribosomes   Thus far the sequence outlined by which DNA directs the formation of specific proteins has included the following steps: (1) the breaking of the hydrogen bonds that hold the two strands of DNA, (2) the unwinding of each strand from the characteristic helical configuration, (3) the formation of RNA chains in the image of the DNA chain, (4) the movement of these messenger RNA chains from the nucleus to the cytoplasm, and (5) the formation of polypeptides from amino acids brought to the messenger RNA by transfer RNA. A polypeptide is a compound containing two or more amino acids united through peptide linkage:

$$-\overset{\overset{\textstyle O}{\|}}{C}-\overset{}{\underset{\underset{\textstyle H}{|}}{N}}-$$

Proteins contain many polypeptides. The site of polypeptide synthesis in the cytoplasm is the ribosome.

Ribosomes can be seen only with the aid of the electron microscope. They are dense structures about 130 Å in diameter and are composed of protein and RNA in equal amounts. Ribosomes may be thought of as protein factories. They do not operate as single units but rather in clusters or groups termed polyribosomes, or simply polysomes (Figure 2-19). Ribosomes are thought to be held together in these clusters by strands of messenger RNA. It is envisioned that ribosomes move along the messenger RNA chain synthesizing protein as they go. As the ribosomes move along in this way, the appropriate amino acid, borne by transfer RNA, is brought to them to be added to the growing polypeptide chain. When synthesis is complete, the polypeptide is freed from the ribosome.

There is good evidence that each ribosome can produce a polypeptide. The advantage of having many ribosomes attached to a single strand of messenger RNA is one of speed. With many ribosomes moving as a unit, many polypeptides per unit time can be synthesized.

The Genetic Code   The word code has several meanings, one of which is "a system of signals for communicating." A pertinent example is the Morse code. Here we are dealing with a code, too, for messenger RNA has a means of communicating a message that directs the combination of amino acids in a specific sequence. Thus the expression "the genetic code" has become part of our language. There are 20 different amino acids

Figure 2-19. Electron micrograph of polyribosomes. The loop to the right is approximately 0.5 micron in diameter. (*Courtesy of A. Kleinschmidt and C. Vasquez.*)

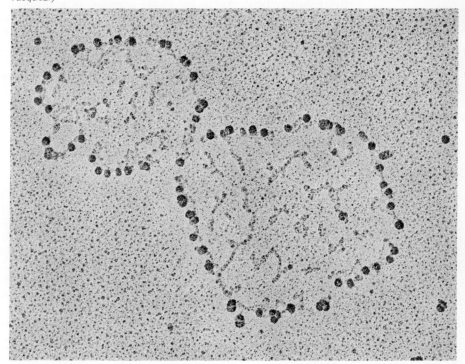

used by living organisms to synthesize protein. It is the sequence of the amino acids that differentiates proteins. This sequence is a code. Messenger RNA determines the amino acid sequence as the protein is being formed. But recall that RNA contains only four different bases. Thus the intriguing question is how can four bases control the sequence of 20 amino acids?

A moment of reflection and the use of simple mathematics suggest the answer. If each base has the responsibility of attracting a specific amino acid, that is a 1:1 relation. A molecule containing only four amino acids will result. Yet, there are 20 amino acids, and there are polypeptides with all 20 in a bewildering variety of combinations. If the relationship is assumed to be combinations of two bases for each amino acid, 16 combinations are possible, but that is still too few. If the units each contain three bases, 64 combinations may be formed, more than enough. It is possible to list all the combinations, using one-, two-, and three-base units, or codons, a codon being a group of bases that codes one amino acid. This exercise leads to one seemingly inescapable conclusion: to direct the sequence of 20 amino acids, codons containing at least three bases are required. This does not rule out, however, the possibility of larger codons. If four bases combine, the number of different arrangements becomes $4 \times 4 \times 4 \times 4$, or 256. Before going on, the mathematical-minded reader may wish to calculate how many different "words" may be formed using a 20-amino acid alphabet. The less mathematically oriented will get the same idea simply by recalling that the English alphabet contains only 26 letters.

Although the triplet, that is, the three-base codon, concept was generally accepted, investigators worried about the fact that there are 64 possible triplets but only 20 amino acids. Thus the suggestion was made that perhaps two or more triplets stand for each amino acid or that some triplets have nothing to do with amino acids but serve some other purpose. They could say, for example, in genetic code, "begin here" or "end here." Whatever their purpose, triplets that do not represent an amino acid are now referred to as nonsense triplets, a not altogether appropriate term.

The sequence, then, at least in its broadest outlines, is complete. The genes through RNA formation and subsequent protein formation control the rate of cell replication, the type of cell produced, and the function of that cell. Derangement of the genes can result in uncontrolled cell replication, termed cancer, or lead to grave abnormalities in reproduction.

QUESTIONS AND PROBLEMS

1 What are the qualifying attributes of living matter?

2 Differentiate between the following processes: (*a*) diffusion, (*b*) filtration, and (*c*) osmosis.

3 Describe the circumstances that could lead to: (*a*) rupture of a red blood cell, (*b*) shriveling of this cell. In both instances explain the role of cell factors and the environmental factors.

4 Discuss the properties of the plasma membrane which make it a semipermeable membrane.

5 Would a large molecule of a fatty substance which was polar (having an electrical charge) pass more quickly through a living membrane than a large molecule of a nonfatty, polar substance? Discuss your answer.

6 In your own words, explain what is meant by each of the following: (*a*) concentration gradient, (*b*) sodium pump, (*c*) electrochemical potential, (*d*) active transport, and (*e*) pinocytosis.

7 Consider the protective function of leukocytes, and explain how phagocytosis differs from cytopemphis and pinocytosis.

8 Under what circumstances would you expect to find no net change of ions across a cell membrane?

9 Why is the concept of a sodium pump necessary when considering the movement of sodium ions across the cell membrane?

10 Discuss why the electrical gradient is dependent upon the concentration gradient found on either side of the membrane.

11 How does the resting potential differ from the action potential?

12 Describe the events which can convert a polarized cell membrane to one which is depolarized. How can the membrane potential be restored to the resting state?

13 Why are mitochondria considered to be the power plant of the cell?

14 Define oxidation and reduction in terms of electron gain and loss. What is the mitochondrial site of the electron chain?

15 Why is ATP spoken of as a "high-energy" compound?

16 What is the source of electrons used in the production of ATP molecules within the mitochondrion?

17 Describe events illustrating the four functions of cellular lysosomes.

18 Match the organelle with the characteristic which best applies to it:

Characteristic	Organelle
1. Contains genetic material	a. Mitochondria
2. Appear during cell division	b. Ribosomes
3. Propulsion	c. Endoplasmic reticulum
4. Supply energy	d. Centrioles
5. Cytoplasmic canals	e. Cilia
6. Protein synthesis	f. Chromosome

19 Differentiate between cell division as seen in somatic cells and that seen in sex cells.

20 Explain what is meant by (a) law of segregation and (b) law of independent assortment.

21 Give an example of a sex-linked disorder and explain why it is said to be sex-linked.

22 Describe the events involved in protein synthesis.

3 TISSUES

A tissue is a group of similar cells joined together by their intercellular materials to perform a common function. Tissues are the building blocks for the organs of the body. There are four primary tissues in the body: epithelial, connective, muscular, and nervous tissue.

EPITHELIUM

Epithelial tissue, or epithelium, lines body cavities or organs and covers body surfaces. It forms the lining of all hollow organs, such as the intestine, bladder, trachea, and uterus. The external covering of the body, the epidermis of the skin, is also epithelium, as are the functional cells of glands.

Classification

Epithelial tissue is classified according to (1) the shape of the individual cell and (2) the arrangement of the cells into one or more layers (Figure 3-1). The shape of the cell may be either squamous (flattened), cuboidal (in which the dimensions of the cells are essentially equal), or columnar (if the cell is taller than it is wide). Epithelial cells may be arranged in a single layer (simple) or in many layers (stratified). By means of these two features, epithelium is classified as simple or stratified squamous, simple or stratified cuboidal, or simple or stratified columnar. Two additional types which do not fit the above

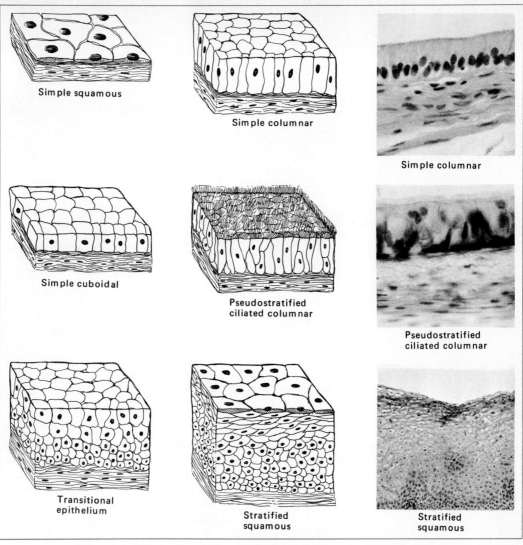

Figure 3-1. Epithelial tissue. Diagrams and photomicrographs illustrating the shape and arrangement of the various morphological classifications of epithelial tissue.

categories are pseudostratified (false stratification) and transitional. In pseudostratified epithelium a single layer of cells, extending from the basement membrane to the free surface, appears to be two or more layers because of the close packing of the cells. In transitional epithelium, for example, the cells lining the urinary bladder, cells undergo morphological change depending upon the physiological state, that is, distention of the bladder. All three kinds of cells—squamous, cuboidal, and columnar—may be present.

Modifications of Epithelial Cells

The free surfaces of some epithelial cells are modified. One modification, present in the respiratory system, consists of tiny hairlike projections, the cilia, that beat rhythmically

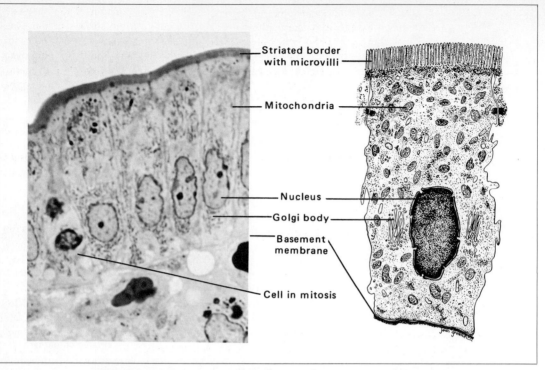

Figure 3-2. Photomicrograph (× 1200) of simple columnar epithelium of the intestine with a diagrammatic sketch of a single epithelial cell to show structure as revealed by electron microscope. (*Courtesy of Thomas G. Merrill.*)

to propel mucus and its entrapped foreign particles toward the nose and mouth. In certain tubules in the kidney very fine microscopic cellular projections, called the brush border, project into the lumen from the surface of the epithelial cells. Under the electron microscope, these irregular projections appear composed of tiny fingerlike processes, the microvilli. Another modification of the free surface of cells that occurs in intestinal epithelium is a thin covering, the striated border. Electron microscopic studies show this border also to be composed of microvilli (Figure 3-2).

Functions of Epithelium

Epithelial tissue has many functions. Some cells, such as the epithelium of the mouth or the skin, protect the body against mechanical trauma or damage. The entire body is covered with cornified layers of epithelial cells which protect tissues from dehydration (drying) and invasion of harmful bacteria.

Certain epithelial cells secrete products which are expelled directly into the bloodstream, or into ducts emptying into hollow organs, or onto the skin. (See ahead under Glands.)

Specialized cells filter waste products from the blood which are then excreted as urine (in the kidneys) or sweat (in the skin). We are nourished by the food we eat because of the absorptive function of the epithelial cells lining the digestive tract. Similarly the kidney recaptures sugar from the urine because of epithelial lining cells. We taste, smell, and hear by virtue of specialized epithelium. Even the perpetuation of the species is accomplished by epithelial cells. The outer layer of cells on the surface of the ovary consists of simple, cuboidal epithelial cells, the germinal epithelium. During embryonic development these cells proliferate and become incorporated into the developing ovary as oogonia, or primitive sex cells, some of which will eventually develop into mature ova. Similarly the stratified epithelium lining the tubules of the testis is germinal epithelium destined to become spermatozoa.

CONNECTIVE TISSUE

Connective or supportive tissue, including cartilage and bone, performs many mechanical functions in the body, such as supporting, anchoring, and binding various parts of the body to each other.

Connective Tissue Components

All connective tissues are composed of (1) intercellular materials (ground substance and fibers) and (2) cells, most of which are fibroblasts.

Ground substance is a homogeneous, amorphous material in which connective tissue fibers are embedded. It varies from a fluid to a semisolid (gel) state. Fibers present in connective tissue are collagenous (white), elastic (yellow), and reticular fibers.

Collagenous fibers, composed of the protein collagen, are the most common. They appear in loose connective tissue as long, slightly wavy, nonelastic bands. There are faint longitudinal striations due to the parallel arrangement of fine fibrils that make up a single fiber. Collagenous fibers are white in the fresh condition.

Elastic fibers are long, thin, wirelike threads that often branch to form a network. They contain elastin, a protein, which allows them to stretch about one and one-half times their original length and return to their former size. In the fresh state a mass of these fibers appears yellow.

Reticular fibers are relatively thin, short, branching threads that unite to form a delicate network (reticulum). Reticular fibers function principally to form an internal supportive meshlike lattice, the stroma, which suspends the functional cells of glands.

Connective tissue cells, though quite numerous and varied, are of two predominant cell types, the fibroblast and the macrophage. The fibroblast, as the term implies, gives origin to connective tissue fibers. Fibroblasts are found in all adult connective tissues, usually closely adjacent to collagenous fibers. They are typically stellate cells with flattened or oval nuclei. In dense connective tissue, such as tendon, the cells are so compressed that only the small, dark, flattened nuclei are visible.

Macrophages (histiocytes) are phagocytic cells that may be as numerous as fibroblasts in loose connective tissue but are usually absent in dense connective tissue. They resemble the stellate fibroblasts but may also be round or oval. Macrophages have small, uneven, dark nuclei, and the cytoplasm often contains clumps of ingested matter, such as carbon dust and pigment. The prime function of these cells is to engulf foreign matter and cellular debris by phagocytosis.

Other cells found in loose connective tissue include numerous fat cells, either singly or in groups; various white blood cells, especially monocytes, lymphocytes, and eosinophils; pigment cells containing mostly the pigment melanin; mast cells; and plasma cells. Mast cells are found chiefly along small blood vessels. Their cytoplasm is filled with coarse, dark-staining granules. They are known to release heparin, a substance that prevents coagulation of blood, and are believed to release histamine, which is thought to be responsible for the changes associated with an allergic reaction, and serotonin, which causes constriction of blood vessels. Plasma cells show characteristic small, dark chromatin masses on the periphery of their nuclei. They become common in connective tissue during chronic inflammation and are believed to manufacture antibody, which is the basis of the body's immune reactions (see Chapter 15).

Classification of Connective Tissue

Epithelium is classified by the type and arrangement of cells, but connective tissue, by the nature of the intercellular material and the fibers that predominate in the matrix, or ground substance. Connective tissue cells are distributed throughout the matrix, but in dense connective tissue they are often relatively few and inconspicuous.

Connective tissues may be classified as follows: (1) mesenchyme, the connective tissue of the embryo; (2) loose or areolar tissue, in subcutaneous tissue; (3) dense irregular tissue, in the dermis of the skin; (4) dense regular tissue, in tendons and ligaments; (5) compact (hard) tissue, in bone, cartilage, or teeth; (6) fluid tissue (blood and lymph); (7) special connective tissue, in adipose, reticular, and mucous tissues.

Mesenchyme   Mesenchyme is embryonic connective tissue. It is characterized by a network of elongated branching cells. The matrix is relatively free from fibers.

Loose, or Areolar, Tissue   Loose, or areolar, tissue is a rather unorganized arrangement of sheets of fibroelastic tissue (Figure 3-3). It occurs throughout the body as a "filler"

Figure 3-3. Loose areolar connective tissue. Note the variety of cells and fibers (× 335).

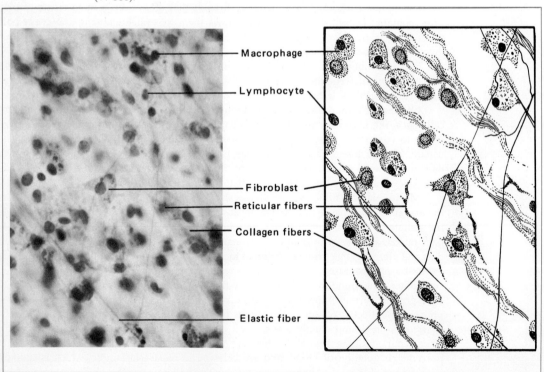

- Macrophage
- Lymphocyte
- Fibroblast
- Reticular fibers
- Collagen fibers
- Elastic fiber

tissue. It is present beneath the skin and mucous membranes, and it surrounds blood vessels, nerves, and ducts. Collagen fibers predominate, although many elastic elements are present. All connective tissue cells are found here; however, most of them are either fibroblasts or macrophages.

Dense Irregular Tissue Dense irregular connective tissue is simply a condensation of areolar tissue with compact layers of collagen and elastic fibers. The dermis of the skin is an example of this type of tissue.

Dense Regular Tissue Dense regular connective tissue contains mostly collagenous fibers arranged in parallel bundles. This tissue forms tendons of muscles and has the great tensile strength necessary for the constant pull of the muscle. Dense regular connective tissue is also found in ligaments. The fibers there are less regularly arranged, and some elastic fibers are present, which allow the ligaments to stretch slightly.

Cartilage Cartilage is a somewhat elastic, pliable, compact connective tissue. Most of the skeleton of the early fetus is composed of cartilage, which is later replaced by bone. Since cartilage has no blood vessels, the cartilage cells (chondrocytes) are dependent on the diffusion of nutrients from the capillaries of the perichondrium. The latter is a fibrous membrane which surrounds cartilage, having an outer, coarse, interwoven layer of collagenous fibers and an inner vascular layer containing chondroblasts. As cartilage grows or repairs itself, the inner layer of chondroblasts gives rise to the chondrocytes.

On the basis of the fibers embedded in the matrix, three types of cartilage are recognized: (1) hyaline, with very fine collagenous fibers (Figure 3-4); (2) elastic, with yellow elastic fibers; and (3) fibrous, with bundles of heavy collagenous fibers.

Hyaline cartilage is the most prevalent type. It covers all articulating surfaces of bone, makes up the skeleton of the early fetus, and forms the framework of respiratory passageways. The fine collagen fibers embedded in the matrix cannot be seen with the ordinary microscope. The matrix therefore appears clear or glassy; hence its designation hyaline cartilage. Large chondrocytes are entrapped singly or in groups of two or three, in small chambers, or lacunae, in the matrix.

Elastic cartilage has embedded in its matrix yellow elastic fibers surrounding many cells which are imprisoned in lacunae. Elastic cartilage is present in a structure or organ

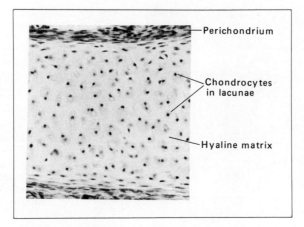

Figure 3-4. Hyaline cartilage from the trachea. Observe the large amount of ground substance between scattered chondrocytes in the lacunae (X 150).

subjected to constant bending, such as the external ear or some of the cartilages of the larynx. The presence of elastic fibers allows this type of cartilage to spring back to its former shape after it has been bent.

Fibrous cartilage has a large number of collagenous bundles in its matrix and very few cells. It is found in structures subjected to continuous heavy pressures, such as the intervertebral disks of the vertebral column, the internal cartilages of the knee joint, and the symphysis pubis.

Special Connective Tissue   Additional connective tissues include several unrelated varieties. Adipose tissue is composed largely of fat cells with a few fibroblasts, and of collagenous and reticular fibers. Mucous connective tissue is found in the umbilical cord in the fetus. Reticular tissue (Figure 3-5) is a network of reticular fibers and makes up the stroma of bone marrow, lymph nodes, spleen, and liver.

Bone and teeth, the compact connective tissues, are described with the skeletal and digestive systems, respectively. Blood and lymph are described in Chapter 15.

The remaining basic tissues, muscular and nervous tissues, are covered in chapters on the muscular and nervous systems, respectively.

RETICULOENDOTHELIAL (MACROPHAGE) SYSTEM

The reticuloendothelial system of cells is not a discrete anatomical entity. Rather, it consists of widely dispersed cells, arbitrarily grouped together because of a common function, namely, phagocytosis.

All reticuloendothelial cells are derived from mesodermal tissue but may differ considerably in morphology in different areas of the body. A typical macrophage is, however, usually described as a stellate or elongated cell with an oval, dark-staining nucleus. Its cytoplasm is heterogeneous, containing various granules and vacuoles, the latter often filled with carbon or dye particles.

These cells may engulf particulate matter (such as carbon particles) or dead or dying

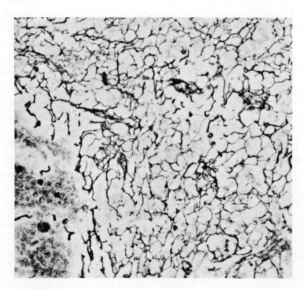

Figure 3-5. Reticular tissue. Note the branching of the short reticular fibers to form a network (reticulum) as the stroma of the spleen ($\times$ 270).

cells (such as senile red blood cells). As scavengers they rid the body of cellular debris and necrotic tissue that otherwise interfere with normal body processes.

Macrophages are found in large numbers in the following organs or tissues:

1 In loose connective tissue, such as in subcutaneous areas
2 Beneath the pleural and peritoneal membranes, especially in mesenteries
3 Lining blood sinuses in the bone marrow, spleen, and lymph nodes
4 Lining the blood sinusoids of the liver, as Kupffer cells
5 Lining the sinusoid blood vessels of the hypophysis and adrenal glands
6 In the alveoli and lymph nodes of the lung, as dust cells
7 Scattered throughout the brain and spinal cord, as microglia

Perhaps the largest accumulation of macrophages is in the spleen, where they function to ingest and remove worn-out erythrocytes from circulation.

EPITHELIAL MEMBRANES

Two kinds of epithelial membranes, serous and mucous, are present in the body. Each type contains both epithelial and connective tissue components. These membranes are kept moist by secretions of the epithelial cells. Because of their wet surfaces, these membranes are also called moist membranes.

Serous Membranes

The surface epithelium of a serous membrane consists of a sheet of simple squamous cells, the mesothelium, over a thin layer of loose connective tissue. The epithelial cells release a clear, watery secretion, the serous fluid. The serous membrane is a highly absorptive structure. It can rapidly transfer solutions (drugs, glucose, etc.) from its surface directly into the bloodstream.

Serous Cavities

Serous membranes line closed cavities in the body, form small closed sacs, or bursae, associated with certain muscles and joints of the body, and surround some muscle tendons as synovial sheathes. Examples of serous cavities are the peritoneal cavity in the abdomen and pelvis, the pleural cavity surrounding the lung, and the pericardial cavity around the heart. The serous membranes in these specific regions are called the peritoneum, pleura, and pericardium, respectively.

Embryologically these cavities form between developing organs and the body wall. The serous membrane in a fully developed serous cavity, therefore, usually has two layers. That portion of the membrane adjacent to an organ (viscus) becomes adherent to the surface of the organ as the visceral layer, while that portion lining the wall of the body cavity becomes the parietal layer.

Serous cavities surround organs that normally undergo considerable motion. For example, movement of the lungs occurs between the parietal and visceral layers of the serous membranes forming the pleural cavities. These layers are normally separated only by a thin film of serous fluid secreted by the membranes. This relationship of moistened surfaces in the serous cavities permits almost frictionless movement against the adjacent surfaces.

Mucous Membranes

Mucous membranes line many organs of the body, such as those of the respiratory, digestive, reproductive, and urinary systems. A typical mucous membrane (Figure 3-6)

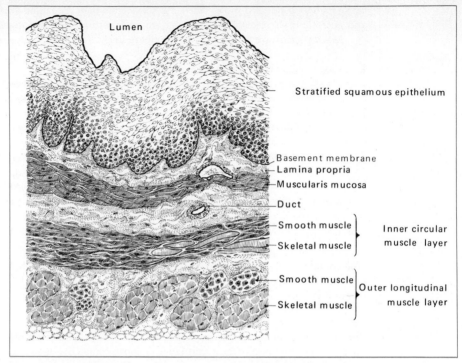

Figure 3-6. Cross section from the upper part of the esophagus. The four layers of the mucous membrane lining the esophagus are in the upper portion of the diagram.

consists of four layers: (1) a free surface or inner layer which may be formed by almost any type of epithelium (simple, stratified, squamous, or columnar), (2) a thin basement membrane composed of ground substance and reticular fibers, (3) a layer of areolar (loose) connective tissue, the lamina propria, and (4) a thin layer of smooth muscle (muscularis mucosa) forming the outer layer of the mucous membrane.

Mucous membranes are highly absorptive structures. For example, through the mucous membrane lining the digestive tract we receive nourishment from all the food we eat. Mucous membranes are also secretory. The secretions produced may be of a rather watery consistency, with an abundance of active enzymes, or more viscous when they are adapted to a lubricative function. In the stomach and intestines the enzyme-rich digestive juices aid in the breakdown of food into simple nutrients acceptable to the cells of the body. The rather chemically inert, thicker mucous secretion of the colon serves as an excellent lubricant for the undigested material that moves down the tract. In the respiratory system mucus entraps foreign particles from the inhaled air and may have some bactericidal properties as well.

GLANDS

Glands are composed of clusters of epithelial cells grouped together to perform a specific secretory or excretory function. The epithelial cells of glands synthesize or expel products peculiar to their individual metabolism; thus they secrete various products such as milk, sweat, hormones, enzymes, mucus, and oil. Glandular epithelium also specializes to remove waste products from the blood and provide for their elimination from the body. For example, urea is eliminated from the bloodstream by the kidney; carbon dioxide is separated from the blood by the lungs; bile pigments are excreted by liver cells into bile ducts which empty into the small intestine.

Individual cells may function as glands. The mucous (goblet) cells in the lining epithelium of the intestines or lining the air passageways of the lung act as unicellular glands. Mucus lubricates the passage of material along the intestinal tract. Secretions of the mucous membrane in the respiratory system moisten the inspired air as well as entrap inhaled foreign particles.

Most glands, however, are multicellular glands. There are two main functional types, exocrine and endocrine. Exocrine glands pour their products into a duct system which empties into the lumen of an organ or onto the surface of the body. Endocrine glands have no duct system—they are sometimes called "ductless glands"—and therefore deliver their secretions directly into the bloodstream or lymph stream. These glands secrete hormones which, by entering the bloodstream, become available to cells throughout the body.

Multicellular exocrine glands are also classified according to their shape and manner of secretion (Figure 3-7). Morphologically, glandular epithelium may be arranged as simple, branched, or coiled tubes, or it may end in small sacs (alveoli) or dilatations.

Figure 3-7. The morphological types of exocrine glands.

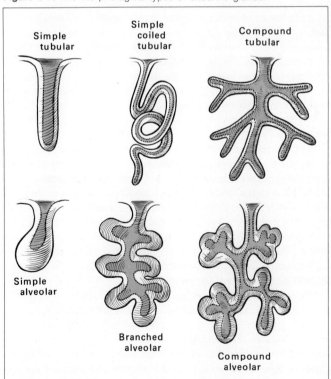

Some branched tubular glands have alveoli at their terminal ends and are called compound glands. The salivary glands are examples of compound exocrine glands.

According to their manner of secretion three different types of glands have been identified—merocrine, holocrine, and apocrine glands. In merocrine glands, for example, the sweat glands, the secretion passes through the cell membrane without damage or destruction to the cell. The secretion of holocrine glands accumulates in the cell and is released when the cell ruptures and dies. Sebaceous (oil) glands of the skin are of this type. Apocrine glands have secretory characteristics of both the merocrine and holocrine glands. In this type of gland, an example of which is the mammary gland, the end of the cell toward the lumen of the duct breaks off and is carried with the secretion while the remainder of the cell remains functional.

SKIN

The skin, or integument, is composed of two layers of epithelial and connective tissues. The outer or surface layer is the epidermis, which contains several layers of stratified epithelial cells. The thicker, deeper connective tissue layer is the dermis. These layers form an effective protective waterproof covering for the entire body. The skin contains many sensory nerve endings that keep us informed regarding our external environment, as well as myriad blood vessels that aid in temperature regulation. The skin is continuous with mucous membrane at the external openings of organs of the digestive, respiratory, and urogenital systems.

The epidermis develops from surface ectoderm as a single layer of cells, but by the second month of intrauterine life it is a double-layered covering. During the fourth month the fetus develops additional epidermal layers together with rudimentary hairs, nails, and sweat and oil glands. The thick dermis is derived from mesoderm except for nerves, glands, and hairs that invade this layer from their ectodermal origins.

Skin is modified in different areas of the body. For example, a thick, heavy epidermis covers the palms of the hands and the soles of the feet, in contrast to a thin layer over most of the rest of the body. The skin further adapts to environmental conditions, and considerable change occurs in aging.

The Epidermis

The epidermis is the outer, thinner layer of the integument. It consists of two or four zones of stratified squamous epithelium with increasing amounts of the protein keratin in the outermost layers. This arrangement decreases excessive water loss from the skin surface and renders the body relatively insensitive to minor abrasions or injuries. Furthermore the epidermis is devoid of blood vessels and has a limited distribution of nerve endings so that one can shave off several layers of cells without blood loss or pain.

In the epidermis of thick skin four distinct strata or zones of epithelial cells can be distinguished (Figure 3-8). From the surface inward they are stratum corneum, stratum lucidum, stratum granulosum, and stratum germinativum. In thin skin, only the stratum corneum and germinativum are present.

The stratum germinativum ("growing layer") contains several layers of cells undergoing mitosis. It is subdivided into basal layers and spiny layers. The basal, or deepest, layer is composed of columnar or cuboidal cells with dark-staining cytoplasm and indistinct cell outlines. The boundary between this layer of the epidermis and the dermis is uneven and undulating due to the large number of projections, the dermal papillae, that project into the basal area. Granules of melanin pigment are frequently present in

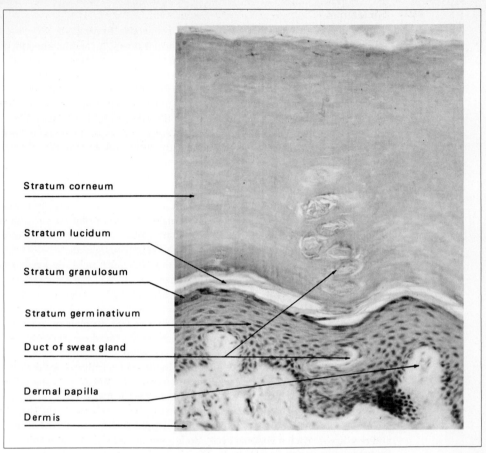

Stratum corneum

Stratum lucidum

Stratum granulosum

Stratum germinativum

Duct of sweat gland

Dermal papilla

Dermis

Figure 3-8. Cross section of thick skin. The four layers of the epidermis and the dermis are labeled (× 140).

the basal cells. The spiny layer consists of irregular, somewhat flattened cells with minute spinelike projections connecting them with others. Melanin granules may be present, and mitosis is frequent in this layer.

In the stratum granulosum the cells contain conspicuous granules (of keratohyalin) that stain darkly with basic stains. Only one or two layers of flattened spindle-shaped cells make up this zone.

Stratum lucidum is a narrow homogeneous layer, one or two cells thick. This zone has little affinity for stains but usually stains lightly with eosin. No nuclei or cell boundaries are present, and no mitosis is observed.

Stratum corneum is a broad zone of several layers of dead epithelial cells and makes up most of the epidermis. These flattened, dehydrated cells are constantly flaking off, often in irregular patches, as for instance after sunburn, and are replaced by cells migrating to the surface from the deeper epidermal layers. The dead cells provide an effective covering which protects against water loss and is also a poor conductor of heat. Thus brief contact with a hot object does not burn the skin.

The Dermis

The dermis, or corium, is a broad, dense connective tissue layer composed mostly of collagenous fibers with some elastic and reticular fibers. It contains the blood and lymph vessels, nerves, parts of the sweat and sebaceous glands, and hair roots.

The dermis consists of two zones: (1) an outer papillary layer which fits snugly against the epidermis and (2) an inner reticular layer that blends with the underlying subcutaneous tissue. The boundary between the zones is indistinct. The papillary zone is less compact, has more elastic and reticular fibers, and is thrown into folds or ridges (dermal papillae) that interdigitate with the epidermis. Dermal ridges are responsible for fingerprint patterns. The compact reticular zone is composed largely of collagenous fibers running in rows parallel to the skin surface.

Glands of the Skin

All the glands of the skin arise from the surface ectoderm as simple tubular downgrowths of solid cords of cells that later become canalized. They penetrate to variable depths in the dermis, where they either are thrown into coils, as in sweat glands, or terminate in grapelike clusters about hair shafts, as in the sebaceous glands.

Sweat glands are found over most of the body surface. Only small areas over the glans penis, the margins of the lips, the concave surface of the external ear, and the nail bed are free from sweat glands. They are heavily concentrated over the palms of the hands and the soles of the feet. The basal coiled portion of a gland is lined with cuboidal or columnar epithelium whose cytoplasm contains fine droplets or granules. Their thin, watery secretion is released into a narrow coiled duct and emerges on the skin surface at a tiny pore. This is a merocrine type of secretion.

Unusually large sweat glands are found in the axilla (armpit), anal region, and scrotum or labia majora. These glands arise embryonically from hair follicles, so most of their ducts open into hair follicles; others become separated and open independently of the hair follicle. Sweat from these cells contains parts of the secretory cells. Hence these glands are of the apocrine type. They undergo hypertrophy at puberty and, in the female, enlarge and recede with each menstrual cycle. Both types of sweat glands are stimulated to activity by heat, pain, or stress.

Sebaceous glands are associated with hair follicles. Their ducts empty an oily secretion (sebum) into the space between the hair follicle and hair root. The secretory portion of an oil gland is composed of a cluster of cuboidal cells with small, distinct central nuclei. Release of the sebum occurs when the central cells of the cluster disintegrate, a holocrine type of secretion. Cells thus eliminated are replaced by proliferating cells on the periphery of the cell mass.

Hair

Over most of the body the stratum germinativum dips down into the dermis to form epithelial pockets of cells, the hair follicles. Each follicle has two layers. The outer connective tissue zone loses its connection with the stratum germinativum. The inner epithelial layer retains its continuity with the epidermis of the skin. Upon proliferation of cells in the base of the inner layer (the hair bulb), a hair is formed. Continued growth of these basal cells forces the hair upward to finally emerge on the skin surface (Figure 3-9).

Each hair consists of a keratinized shaft, usually protruding beyond the skin surface, and a root lying within the follicle, whose enlarged termination is called the bulb. Under the microscope the hair proper, or shaft, can be seen to consist of the following epithelial layers, from without inward: (1) the transparent cuticle, which is composed of scales; (2)

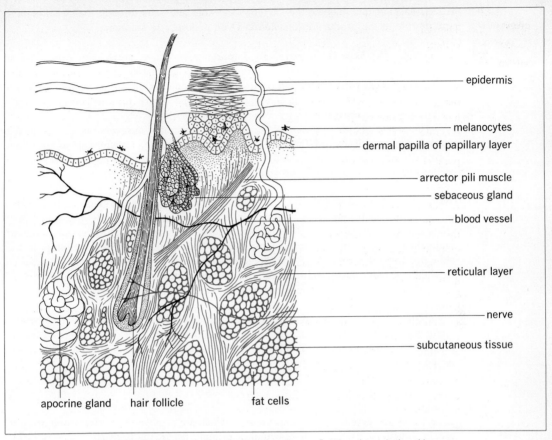

epidermis

melanocytes

dermal papilla of papillary layer

arrector pili muscle

sebaceous gland

blood vessel

reticular layer

nerve

subcutaneous tissue

apocrine gland   hair follicle   fat cells

Figure 3-9. Structures in subcutaneous tissue. Section through the skin.
Apocrine glands are found in greatest number in the skin of the axilla. (*From
R. M. DeCoursey, The Human Organism, 3d ed., McGraw-Hill Book Company,
1968.*)

the cortex, the larger layer, which consists of several flattened keratinized layers of cells
that contain pigments; and (3) the central medulla, which has one or more layers of
cuboidal cells interspersed with many air spaces.

Nails

On fingers and toes the stratum corneum of the skin is hardened into a transparent plate
of keratin, the body of the nail. It grows forward from the root of the nail. The latter
lies under a thin fold of skin, the cuticle or eponychium. The pink color of the body
of the nail results from the rich vascular plexus beneath the more or less transparent
nail.

Functions of Skin

The skin performs many functions in the body. As a covering of the entire body it offers
protection. Its numerous sensory receptors keep us aware of the external environment.
Its abundant blood supply aids in temperature regulation, as described in Chapter 19.

Additional functions include the excretion of water and salts through the sweat glands, and its role in the production of vitamin D on exposure to sunlight.

Protection   The skin is an extremely important protective organ. Not only does it successfully prevent the invasion of bacteria and foreign matter, but also it protects the underlying tissues from injury and drying. It is thickened at the palms of the hands and the soles of the feet, providing a tough covering necessary for the constant trauma occurring at these areas.

The protective role of the skin is dramatically illustrated in severe burn cases. When the integument is destroyed, the fluid loss from the underlying structures is so rapid that the total fluid balance is upset and, unless effective measures are promptly taken, the victim suffers circulatory collapse, goes into shock, and dies.

Sensation   Stimulation of the receptor endings of tiny nerves in the skin keeps us aware of the conditions in our immediate vicinity, for example, changes in temperature. Such changes set off alterations in the vasculature of the skin as well as the sweat glands to maintain a constant internal temperature of the body. Touch and pain receptors in the skin also provide information. Pain receptors are especially important. They enable us to withdraw from harmful situations and initiate additional adjustments necessary to survival in the ever-changing environment in which we live.

Excretion   Perspiration involving the sweat glands represents an avenue of excretion. Small quantities of nitrogenous waste products and sodium chloride are excreted by the skin. Both the volume and the composition of the sweat vary in accordance with the needs of the body.

ORGANS

An organ is an aggregate of different tissues which are arranged to perform one or more specific functions. There is one essential or primary tissue, the parenchyma, responsible for the specific function of the organ. An example is the hepatic cells of the liver. Other tissues function in secondary roles, such as supportive, nutritive, or conductive.

Most solid organs have an outer protective layer or capsule of irregular dense connective tissue. Strands of connective tissue from the capsule often penetrate the organ and serve as a stroma, which in turn supports the parenchymal cells.

SYSTEMS OF THE BODY

Organs which combine to perform a similar function form a system of the body. Organs of a given system are usually located in more than one region of the body. Some systems, such as the circulatory system, are dispersed throughout the body.

Skeletal System

Bones of the body are the principal components of the skeletal system. In certain areas this system protects the body by forming a hard exterior shield for internal vital organs. Bones are utilized as levers by the muscles in movement of the body and give support and form to the body. Joints or articulations allow the bones to move in many directions. Marrow cavities of certain bones are the sites of hemopoiesis, or production of blood cells. Bones also perform an important function in storing calcium and phosphorus, which may be released into the bloodstream if physiological processes require an increase of these substances.

Muscular System

Skeletal muscle forms the definitive muscles of the body. Thus, skeletal muscles are the organs of the muscular system. We usually think of muscles only in connection with body movements; however, they are needed for normal posture as well. For example, while a person is standing still, muscular activity is necessary to counteract the pull of gravity and maintain the body in an erect position.

Nervous System

This complex system is subdivided into (1) the central nervous system, formed by the brain and spinal cord, and (2) the peripheral nervous system, formed by the nerves of the body.

The nerves receive stimuli from the external environment or from the body itself and transmit impulses to centers in the spinal cord and brain. Here the information is correlated, interpreted, and integrated; impulses are then relayed peripherally to activate body structures. This total activity affects the body's response to a given stimulus. Divorced from external stimuli, the brain still functions in reasoning and other thought processes.

Circulatory System

This system includes the heart, arteries, veins, and capillaries. Lymphatic organs and vessels may also be included in this system.

The heart serves as a muscular pump to propel blood through the arteries to the tissues. Blood leaves the tissues through veins, the pumping force being largely the compression of veins by contracting muscles. Blood also carries the waste products to excretory organs, where such wastes are expelled from the body.

Lymph is fluid collected from tissue spaces once it has entered lymph vessels. The lymphatic system is principally concerned with filtering this fluid before returning it to the bloodstream. Phagocytic cells of the system engulf bacteria and foreign bodies and thus function in the defense of the body against disease.

Respiratory System

The respiratory organs consist of air passageways, namely, the nasal cavity, part of the pharynx, the larynx, the trachea, the bronchi, and the lungs. Within the lungs oxygen and carbon dioxide are exchanged across the wall of the smallest unit of the lung, the alveolus.

The larynx, in addition to being a segment of the air passageway, is modified for voice production.

Digestive System

This system extends from the mouth to the anus. It includes not only the alimentary canal, that is, the mouth, part of the pharynx, the esophagus, the stomach, and the small and large intestines, but also the liver, pancreas, and salivary glands.

These organs function in digestion—the breakdown of ingested proteins, fats, and carbohydrates into simpler substances for absorption across the epithelial lining of the intestine into the bloodstream. The alimentary canal transports the unused ingested food out of the body as semisolid wastes.

Glands associated with this system, as well as certain small glands within the walls of the alimentary canal, secrete specific enzymes necessary for digestion.

Urinary System

The urinary system comprises the kidneys, ureters, urinary bladder, and urethra. An elaborate filtration system in the kidney removes nitrogenous waste products from the blood and excretes them in the urine. The urine then flows through the ureters to the bladder, where it is stored until voided. Organs of this system are sometimes described as composing the excretory system; however, excretory functions also occur in the digestive, respiratory, and integumentary systems.

Reproductive System

Organs of the reproductive system are concerned with the propagation of the species. The major female organs, the ovaries, fallopian tubes, and uterus, are all located within the pelvic cavity. In the male the prostate, the seminal vesicle, and part of the ductus deferens are in the pelvic cavity, while the testis, the epididymus, and the remainder of the ductus deferens are outside the pelvic cavity.

Endocrine System

The endocrine system performs an integrating and controlling role over many metabolic activities of the body. Glands composing this system include the thyroid, parathyroid, hypophysis, and suprarenal, and portions of the pancreas, the ovary, the testis, and perhaps the thymus. The glands are ductless and empty their secretory products directly into the bloodstream, making the secretions immediately accessible to cells in all parts of the body. The secretions of the endocrine glands are called hormones.

QUESTIONS AND PROBLEMS

1 What are the primary tissues of the body?
2 Are all the primary tissues capable of mitotic division in the adult? Explain.
3 How do tissues relate to organs and to systems?
4 Where is epithelium found in the body?
5 Describe modifications found on epithelial cells, and name the organs where such cells are found. Give the function of the modifications in these organs.
6 How is pseudostratified epithelium different from stratified epithelium?
7 Describe a serous membrane. Where are serous membranes located, and what are their functions at these locations?
8 What is the function of epithelial tissue in the following structures: (a) epidermis, (b) germinal epithelium of the ovary, (c) columnar epithelium of the small intestine, (d) mammary gland?
9 In a laceration or incision of the skin, what type of tissue forms the scar?
10 Contrast the intercellular constituents of connective tissue in a young person with those in an elderly person.
11 What is the reticuloendothelial system? Describe its location, structure, and functions.
12 What are the two functional classifications of glands? What are the chief differences between them?
13 Describe the morphological types of glands. Describe the three types of glands distinguished by their manner of secretion.
14 Is the skin considered to be an organ? What tissues is it composed of? Name the six different layers. List the organs found in skin.
15 How does the epidermal epithelium differ from epithelium of the oral cavity?
16 What skin structures are involved in burns of the first, second, and third degrees? In skin pigmentation? In fingerprints?
17 Define (a) germinal epithelium, (b) stroma, (c) matrix, (d) serotonin, (e) dense irregular connective tissue, (f) lacunae, (g) macrophage, (h) corium.

4 EMBRYOLOGY

Embryology in its broadest sense is the study of the development of an individual from fertilization through intrauterine development and growth, until parturition, or birth.

Three stages may be recognized during intrauterine life. The first period, preimplantation, extends from conception, with the formation of the zygote (fertilized egg), until the conceptus is implanted into the uterine wall approximately 7 days later. Mitotic divisions of the zygote occur during this period as it passes through the fallopian tube into the uterine cavity. The second, or definitive embryonic, period begins at implantation and continues through the eighth week of life. During this time, all the organs and systems of the body are essentially established. The fetal period covers the last 7 months of pregnancy. This period is basically a period of growth, with elaboration and expansion of the body systems laid down during the embryonic stage.

FERTILIZATION

As a sperm penetrates the ovum, their nuclei unite, and at that instant fertilization has occurred. The resulting cell, the zygote, now has the normal 46 chromosomes, half maternal and half paternal. All subsequent cell divisions will result in cells with the full complement of chromosomes. Fertilization, or conception, usually occurs in the proximal third of the fallopian tube. As subsequent cell divisions in the zygote occur, the developing conceptus gradually moves through the tube into the cavity of the uterus.

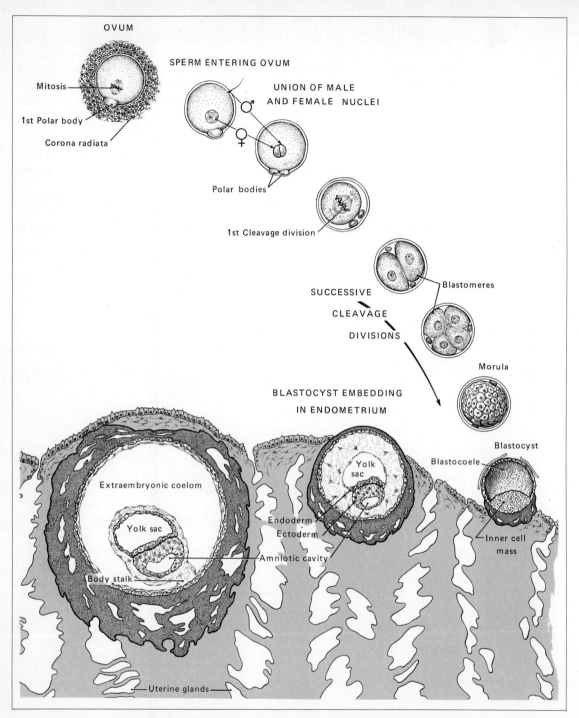

Figure 4-1. Early stages of embryogenesis. Schematic illustration showing fertilization, zygote formation, cleavage, blastocyst formation, implantation, and formation of germ cell layers. The endometrium is shaded.

Cleavage

Initially the zygote is a huge cell with a relatively large amount of cytoplasm and a very small nucleus. During the 3 or 4 days following fertilization, in a series of rapid mitoses called cleavage, there is synthesis of DNA with nuclear formation but no production of new cytoplasm. Thus many new cells are formed from the original cytoplasm of the ovum with no significant increase in the total mass. As a result, the nucleus-cytoplasm ratio returns to normal. The resulting solid ball of cells resembles a mulberry and is hence called the morula; the individual cells are termed blastomeres (Figure 4-1).

As the number of cells in the morula increases, the cells in the center of the ball undergo degeneration and form a fluid-filled cavity surrounded by an outer layer of cells. This sphere of cells is called the blastocyst, and the cells in the outer layer are the trophoblasts, or nutrient cells.

Subsequent uneven cell proliferation occurs and cells lining one area (pole) of the blastocyst divide much more rapidly than surrounding cells. This area buckles or in-vaginates into the cavity of the blastocyst. Such an infolding of tissue creates a mound

Figure 4-2. Forty-day-old embryo. The chorionic vesicle is cut to expose the embryo within the intact amnion. (*Chester F. Reather, photographer. Carnegie Institution of Washington.*)

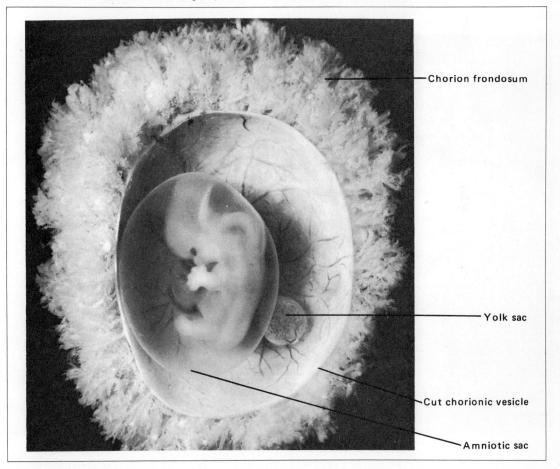

Chorion frondosum

Yolk sac

Cut chorionic vesicle

Amniotic sac

or mass of cells which projects into the blastocyst cavity and is thus called the inner cell mass. It is from the inner cell mass that the embryo will eventually develop. Some of the remaining blastomeres will later form the fetal membranes or sacs of the embryo and fetus, namely, the amnion, chorion, and yolk sac (Figure 4-2).

IMPLANTATION

Approximately the seventh day after fertilization the conceptus becomes implanted in the inner layer of the uterus, the endometrium (Figure 4-1). Implantation is accomplished by the action of special trophoblastic cells, the syncytial trophoblasts, which erode away the surface of the uterus and allow the blastocyst to sink into a highly vascular, uneven implantation cavity. By the eighth day the blastocyst is firmly implanted into the uterine wall and will remain embedded in uterine tissue until excessive growth forces the developing embryo into the uterine cavity. The attachment site of the blastocyst develops into the placenta. Both trophoblastic cells and tissues of the uterus contribute to the formation of the placenta (Figure 4-3).

Figure 4-3. Twenty-eight-day-old embryo. The chorionic vesicle at the left has been opened to show the embryo in situ at the right. Note the relatively large yolk sac. (*Chester F. Reather, photographer. Carnegie Institution of Washington.*)

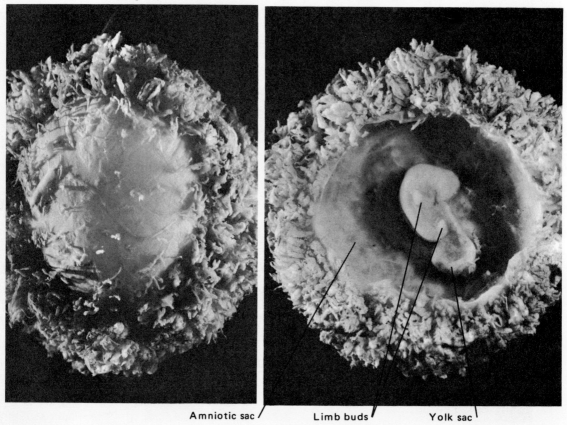

Amniotic sac   Limb buds   Yolk sac

TABLE 4-1. TISSUES DERIVED FROM THE PRIMARY GERM LAYERS

ECTODERM	MESODERM	ENDODERM
Nervous tissue	Muscle	
	Connective tissue, including bone, cartilage, blood, bone marrow	
Epithelium of the oral cavity and its glands; nasal cavity and paranasal sinuses; anal canal; special sensory organs	Epithelium of the kidney and ureter; blood and lymph vessels; gonads and their ducts; serous cavities; joint cavities and bursae; conjunctiva of the eye	Epithelium of the digestive system and related glands; respiratory system; urinary bladder; urethra and associated glands; middle-ear cavity and auditory tube
Epidermis of the skin and its related sweat and oil glands; hair; nails; mammary gland	Dermis of the skin	
Glands: hypophysis; epiphysis; adrenal medulla	Glands: adrenal cortex	Glands: thyroid; parathyroid; tonsils; thymus

GERM CELL LAYER FORMATION

As the blastocyst is settling into its nutrient nest in the endometrium of the uterus, the dividing cells of the inner cell mass form two layers in the initial phase, called gastrulation. The layer of cells nearest the uterine cavity becomes the endoderm, and those cells near the uterine wall form the ectoderm (Figure 4-1). As cells of the ectodermal layer multiply, a cavity develops which will become the definitive amniotic cavity surrounding the developing embryo. The thin roof of this cavity becomes the amniotic membrane.

The endodermal layer expands to form a second cavity, the blastocoele, or yolk sac. Cells in the area between these two cavities, where the endoderm and ectoderm are in contact, compose the embryonic disk and are the cells which will develop into the embryo. At the middle of the embryonic disk, surface cells multiply to form an oblong mass of cells which rapidly proliferate to extend between the ectoderm and endoderm and form a third germ layer, the mesoderm. Cells from each of these three germ layers give rise to specific tissues of the body (Table 4-1).

EMBRYONIC DEVELOPMENT

As the embryo begins to develop, the flat, elliptical embryonic disk grows more rapidly along its longitudinal axis than it does at its margins. It also grows more rapidly at the head end of the primitive streak. The center of the disk thus becomes elevated and bulges along the midline.

Eventually the head and tail regions and the sides fold under and the flat disk becomes a hollow tube. The tubular embryo is covered with an outer ectodermal layer, lined on the inside with endoderm, and between these two layers is mesoderm. This hollow cylinder is the beginning of the definitive embryo, and the endodermal cavity will become the primitive alimentary canal.

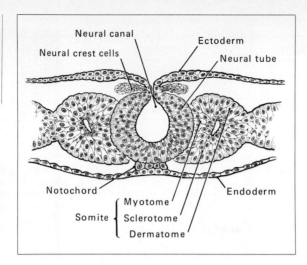

Figure 4-4. Diagrammatic cross section through the dorsal aspect of the embryo.

Brain and Spinal Cord

At about the same time that the digestive tract is developing, an overgrowth of ectodermal tissue on each side of the midline, on the dorsum of the embryo, forms the neural folds. The groove between these elevated, thickened ridges is the neural groove.

These folds continue to grow, thicken, and eventually fuse in the midline. This converts the neural groove into the neural tube (Figure 4-4). The head end of this tube enlarges markedly to develop into the brain. The less developed portion becomes the spinal cord.

Somite Formation

While the neural tube forms, discrete elevated blocks of mesoderm, the somites, appear on either side of the neural groove and tube. The somites give rise to muscular and bony structures in the body (Figure 4-5). About 40 to 44 somites develop in man. The cells of the somite form most of the voluntary muscles and skeleton of the body, and together with the overlying ectoderm give rise to the skin. As the somites develop, three regions are recognizable: the dermatome, myotome, and sclerotome. The lateral portion of the somite, the dermatome, develops into the dermis of the skin; the intermediate portion, the myotome, gives rise to the striated muscles of the body, and the medial sclerotome forms the vertebral column and ribs.

Limb Buds

During the fifth week of development limb buds, the rudiments of the extremities, appear on either side of the embryo. They are somewhat flattened mounds, or elevations, with the upper buds appearing first and differentiating ahead of the lower buds (Figure 4-6). These structures grow rapidly and elongate, and on each limb bud slight constrictions appear at the future site of the elbow and wrist, and at the knee and ankle joints. The distal ends of the limb buds flatten, and longitudinal furrows appear. These deepen and separate to form the fingers and toes.

Branchial Arches and Grooves

At about the fourth week of development a series of alternating ridges and depressions, the branchial arches and grooves, appear as external features along the side of the developing head and neck (Figure 4-6). These structures correspond to the gill slits in

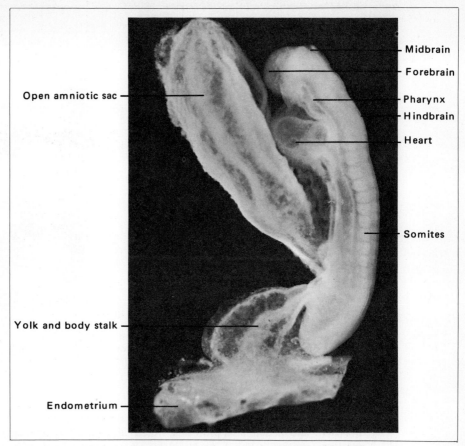

Open amniotic sac

Midbrain

Forebrain

Pharynx

Hindbrain

Heart

Somites

Yolk and body stalk

Endometrium

Figure 4-5. Lateral view of a 23-day-old embryo. The amniotic sac has been opened to show the somites more clearly. (*Chester F. Reather, photographer. Carnegie Institution of Washington.*)

a fish, but in man the grooves do not perforate into the pharynx as they do in a fish. At the upper end of the alimentary tube, in the region of the developing pharynx, the grooves project inwardly to form endodermal-lined pharyngeal pouches. Four of these outpocketings develop into glands of the neck region such as the thyroid, parathyroid, tonsil, and thymus (Figure 4-7).

The first branchial arch develops two protuberances, the maxillary and mandibular processes, which are destined to become the upper and lower jaws (maxilla and mandible). Between these structures a midline depression, the oral pit or stomodeum, develops into the mouth cavity.

The second arch forms the hyoid bone located at the base of the tongue. The remaining third and fourth arches contribute to the formation of the cartilages of the larynx.

Nose

Initially the nose appears as two thickened areas of depressed ectoderm, the olfactory pits. They are located on the sides of the head (frontonasal process) near the oral pit, adjacent to the medial ends of the developing maxillary processes. The olfactory pits

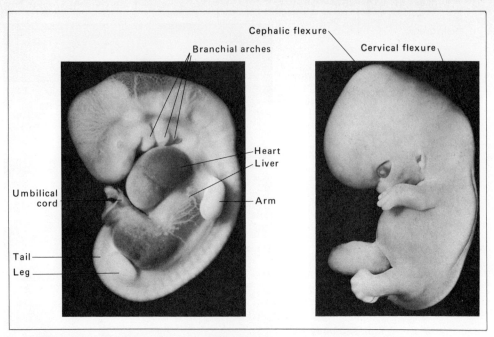

Figure 4-6. Photographs of a 31-day-old embryo on the left and a 40-day-old embryo on the right. The limb buds on the right show beginning finger and toe differentiation. Note the more advanced development of the upper limb. (*Chester F. Reather, photographer. Carnegie Institution of Washington.*)

elongate inwardly and develop into two grooves bounded on each side by two swellings, the lateral and medial nasal processes, which subsequently unite with each other to form the external nostrils. Rapid growth of this area results in the nose being elevated above the surface of the face. Internally the developing palate divides the upper nasal from the lower oral cavity with the nasal cavity being subsequently divided longitudinally by the development of the nasal septum.

Eye

During the fourth week the eyes begin to form between the frontonasal process and the first branchial arch. Development of the eye begins as two ectodermal depressions, the lens placodes, at either side of the head. Internally lateral outgrowths, the optic vesicles, bulge out from the head end of the neural tube and grow toward the lens placode. These evaginating optic vesicles soon differentiate into a large, spherical distal part, the bulb, and a slender proximal segment, the optic stalk. The latter retains its connection with the developing brain (neural tube) as a nerve tract of the brain. The hollow bulb becomes indented on its external surface until its two walls are in contact. This converts the bulb into a double-layered structure, the optic cup. The optic cup is destined to become the retina of the eye.

While the optic cup is forming, the lens placode thickens, migrates to a position opposite the open end of the optic cup, sinks into it, and ultimately develops into the lens of the eye.

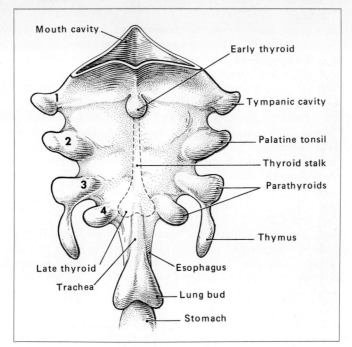

Figure 4-7. Primitive pharynx. The pharyngeal pouches are numbered, and structures arising from the pouches are indicated.

Loose embryonic connective tissue, the mesenchyme surrounding the optic cup and lens placode, forms additional layers of the eyeball. Mesenchyme immediately adjacent to the optic cup gives rise to the delicate, vascular choroid, or middle layer, of the eye. More peripheral mesenchyme condenses to form the tough, fibrous, outer capsule, or sclera, of the eyeball.

Ear

About the second month of development, a groove appears in the region of the first pharyngeal cleft. This groove deepens, and surface elevations or hillocks appear which fuse to form the primitive pinna, or external ear. The groove becomes the external auditory meatus.

The middle ear cavity arises from the first pharyngeal pouch and houses the three small auditory bones, the malleus, incus, and stapes, as well as two muscles and their tendons. These latter structures arise from mesenchyme associated with the developing middle ear cavity.

The internal ear first appears as a plate of thickened ectodermal cells, the otic placode, on either side of the head. The placodes sink into the head region, forming depressions called otic pits. The pits dilate to form hollow cavities called the otocysts. The lining epithelium of the otocysts gives origin either to the receptor neurons for equilibrium located in the semicircular canals or to the nerve cells associated with the cochlea, the organ of hearing.

TERATOLOGY

Teratology is the study of abnormal development. Some malformations occur in at least 10 percent of all live births. These may appear in any part of the body. The most frequent sites of occurrence are the face (for example, harelip and cleft palate), the heart (septal defects causing "blue" babies), the brain (hydrocephalus, or "water on the brain"), the digestive tract (pyloric stenosis, or narrowed stomach opening), the skin (birthmarks), and the skeletomuscular systems (talipes equinus, or "clubfoot").

Most congenital anomalies have their origin as organs are being formed during the first 8 weeks of development. During this critical period, organs are very sensitive to environmental stresses, such as lack of oxygen, poor nutrition, toxins, viruses, and other detrimental agents. Some malformations may occur after birth, such as failure of the foramen ovale to close between the right and left atria of the heart. This allows for the mixing of venous with arterial blood.

Chromosomal Defects

Hereditary factors may also cause anomalous development. If the chromosomes are altered, the developmental pattern is upset, and a baby may be born with extra digits, fused fingers, congenital cataract, color blindness, or other defects. Rearrangement of chromosomes during meiotic cell division may cause a sex cell to have too much or too little genetic material (Figure 4-8). For example, if a baby is born with a certain extra

Figure 4-8. A spread of human female chromosomes arranged in pairs. Abnormalities known to occur at pair 21 (against the blue panel) include trisomy (three chromosomes), which expresses itself in Down's syndrome. Abnormalities of the sex chromosomes (against the brown panel) include trisomy, leading to Klinefelter's syndrome, and monosomy (single chromosome), leading to Turner's syndrome. Drawings at the right show these abnormalities as well as the sex chromosomes of a normal male.

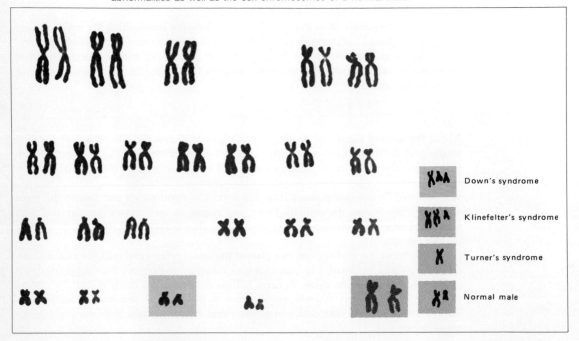

somatic (body) chromosome, the child will be mongoloid (Down's syndrome); if an X chromosome is lost in the female (44X), she will be sterile with imperfect development of primary and secondary sexual organs (Turner's syndrome); if an extra X chromosome is present in a male (44XXY), he will have underdeveloped sex organs and will usually be sterile (Klinefelter's syndrome).

QUESTIONS AND PROBLEMS

1 Describe the stages of development of the embryo from conception to parturition.
2 How and where is the zygote formed in the process of fertilization?
3 What is the diploid number of chromosomes in a normal human cell?
4 Describe the structure which implants itself in the endometrium of the uterus and the mechanism involved in implantation.
5 Differentiate between the cytotrophoblasts and syncytiotrophoblasts in structure and function. What role do they play in the formation of the placenta?
6 How does the blastocyst obtain its nutrition before and immediately after implantation?
7 How are the three germ layers formed in the young embryo? Give several examples of various tissues derived from each layer.

8 When does organogenesis take place in the embryo? Of what significance is this to the pregnant woman? Discuss the role of the nurse in counseling her.
9 Why does the head of the fetus develop much more rapidly than the rest of the body?
10 When and how can limb bud differentiation be affected during development?
11 When does the heart start to beat in the embryo?
12 How would you explain the formation of a harelip and cleft palate in an infant?
13 Of what significance are the pharyngeal pouches in man?
14 When do most congenital anomalies occur?
15 What kinds of known teratological agents produce congenital defects? How do these agents reach the developing embryo or fetus?

5 ANATOMY OF THE SKELETAL SYSTEM

Approximately 206 bones make up the skeleton. They are disposed in a midline axial segment, that is, the head and trunk, and in an appendicular portion, the upper and lower limbs.

Bony tissue consists of bone cells (osteocytes) and dense bundles of collagenous fibers entrapped in inorganic mineral salts. This intercellular matrix is responsible for the hardness of bone as well as for its specialized architecture.

Bones perform several functions. They contribute to the form of the body, as well as provide it with support and protection. For instance, the skull, thoracic cage, and pelvis are sturdy encasements for the brain, heart and lungs, and reproductive organs, respectively.

Another function of bone is to serve as a site of blood formation. In the adult, red blood cells are produced exclusively in the marrow cavities of certain bones.

Bones are essential for body movement. Most muscles attach to bone for leverage in moving the body.

Bone is a storehouse for inorganic minerals, such as calcium, phosphorus, and possibly magnesium and sodium. In pregnancy, for example, should the mother not include enough calcium in her diet, it will be removed from her bones and used in bone development of the baby.

Skeletal characteristics are useful in several respects. For example, the shape and size of bony components differ between races and so are helpful in identification. The sex of a specimen can be easily determined by examining the hipbone. Age can be measured in persons younger than 21 years by examining the secondary centers of ossification as they appear in radiographs of the extremities.

CLASSIFICATION OF BONES

Bones are long, short, flat, or irregular in shape. Most bones of the extremities are long bones, ranging from the short terminal phalanx of the little finger to the femur (thighbone). Long bones consist of a shaft and two extremities. The shaft, or diaphysis, is cylindrical and encloses an elongated marrow cavity. The extremities, or ends, of long bones are called epiphyses.

Short bones have approximately the same dimensions in all directions. The cubelike bones of the wrist and ankle are examples of short bones.

Flat bones are those with two surfaces roughly parallel and close to each other, as in the sternum, the scapula, and most skull bones.

Bones that fit none of the above categories are called irregular bones. The vertebrae, with projections extending from their bodies, are good examples.

DEVELOPMENT OF BONE

Bone begins to develop from mesoderm relatively late in embryonic life (sixth week), and many bones of the body continue to grow until a person reaches young adulthood. There are two methods by which bone develops. Most of the skeleton arises by endochondral bone formation. However, bones of the face and flat bones of the skull are formed directly in embryonic connective tissue membranes, that is, by intramembranous bone development.

Endochondral Bone Formation

In endochondral bone formation, a cartilage model, conforming to the shape of a future bone, is developed from mesenchyme (Figure 5-1). This cartilage model is subsequently

Figure 5-1. Stages of endochondral bone formation. Shading indicates extent of cartilage. Note the persistence of cartilage at the epiphyseal plate and on articular surfaces.

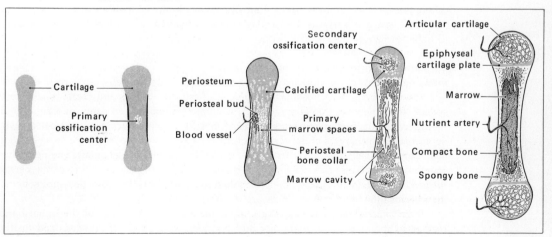

destroyed and reabsorbed as the bone is formed. The following describes the process in a typical long bone.

First a fibrous membrane grows around the cartilage model. This is the perichondrium. Then, midway along the shaft of the model, changes begin to occur at the center of the cartilage, which is termed the primary ossification center. Cartilage cells gradually enlarge, their boundaries break down, and small cavities form. Blood vessels from the perichondrium invade and occupy these cavities. They bring in and deposit calcium salts around the cartilage spicules—the spikelike remnants of the cartilage model. Osteoblasts (bone-forming cells) invade the area with the blood vessels and utilize the calcium to lay down bone on the cartilage spicules. The degenerating cartilaginous core of the spicule is eventually reabsorbed and withdrawn through the blood vessels, leaving bone spicules (also called trabeculae) that become loosely joined in a spongy network. This spongy, or cancellous, bone is characteristic of endochondral bone formation.

About the same time, marrow begins to form in the spaces of the spongy bone. Gradually the spaces toward the middle of the shaft coalesce and the marrow cavity, or canal, is created.

While spongy bone is being laid down at the center of the cartilage model, bone of the same constituents but different architecture, termed compact bone, is being created at the periphery of the shaft. It arises from cells lying between the cartilage and the enveloping perichondrium. As they lay down a sleeve of bone around the cartilage, the perichondrium becomes periosteum ("around bone"). As additional layers are deposited, the shaft grows in circumference. The bone is simultaneously being lengthened by proliferation of cartilage cells at the ends of the shaft. This zone of proliferation is termed the epiphyseal plate, or epiphyseal disk, and lies between the shaft and each epiphysis. Cells within this plate multiply, are pushed to the surfaces, and give way to ossification. Growth continues in this manner until early adulthood. Then the epiphyseal plate closes (ossifies) and growth in length ceases. Simultaneously the ossification centers—the primary center in the shaft and the secondary centers in the epiphyses—fuse.

Growth in diameter keeps pace with growth in length. As full stature is reached, compact-bone formation also halts, to be reactivated only in bone fracture when the quiescent osteoblasts in the periosteum are stimulated into bone-repair activity.

Within the epiphyses, in the secondary centers of ossification, bone formation follows essentially the same pattern as in the shaft. The center of the cartilage gives way to spongy bone, and the diameter is enlarged by layers of compact bone. However, there are two differences: (1) the epiphyses remain mostly spongy bone, without a marrow cavity, and (2) on the articular surfaces the cartilage remains hyaline cartilage rather than yielding to bone formation.

Intramembranous Bone Formation

Intramembranous ossification usually begins near the center of well-vascularized connective tissue membranes. Osteoblasts appear in these membranes and create small bone spicules in a network radiating in all directions. Osteoblasts in large numbers align themselves on the free margins or surfaces of these interlacing spicules. As the osteoblasts deposit successive layers of bone, some of the cells are entrapped in minute spaces called lacunae. These cells become osteocytes. Finally the spicules coalesce, and the definitive shape of the bone emerges, leaving osteoblasts crowded onto the periphery of the newly formed bone, where, limited by the periosteum that surrounds the entire developing bone, they enclose the spongy bone in layers of compact bone.

Bone is consistently undergoing remodeling and erosion. Much of the initial bone formed in the embryo is temporary and becomes reabsorbed, remodeled, and finally replaced by new bone as it grows into its final adult shape and size. Adult bone also undergoes changes of shape with reabsorption under stress. Osteoclasts, giant multinucleated cells, are responsible for bone reabsorption. These bone "destroyers" are usually embedded in shallow spaces called Howship's lacunae. Osteoclasts are not phagocytic cells but probably secrete an osteolytic enzyme to dissolve bone tissue.

Abnormalities of Development

Imbalance of the amount of growth hormone secreted by the anterior pituitary (see Chapter 31) affects the growth of the skeleton, especially the long bones of the body. If this imbalance occurs in someone prior to closure of the epiphyseal plates, it can decrease or increase his stature. For example, insufficient production of growth hormone causes an early closure of the epiphyseal plate and results in pituitary dwarfism in which the height is limited but body proportions are normal. With an excessive secretion of growth hormone, a person's height is increased due to a delayed closure of the epiphyses, resulting in gigantism.

Excessive secretion of growth hormone may occur after epiphyseal closure, that is, after normal growth in height is completed. The resultant condition, acromegaly ("large extremities"), affects bones and soft tissues of the hands, feet, and face, especially the lower jaw. These structures become excessively large and grotesque.

Achondroplasia (lack of cartilage formation) is a hereditary disturbance of endochondral bone formation. This results in the most common type of dwarf, the achondroplastic dwarf. In contrast to the pituitary dwarf the achondroplastic person is grossly deformed with a relatively large head, vertebral deformity (lordosis), and excessively short extremities.

HISTOLOGY OF BONE

We have seen that bone tissue arranged in compact layers forms the external portion of the bones of the body. Spongy bone fills the space between the cortical laminae of flat bones, the central areas of short bones, and the epiphyses of long bones, but is limited to the marrow cavity in the shafts of long bones.

Compact Bone

The structural unit of compact bone is the osteon, or Haversian system (Figure 5-2). An osteon consists mostly of inorganic material arranged in concentric rings, or lamellae, around a canal that runs parallel to the long axis of the bone. The compact layer of a given bone consists of many such osteons in arrays. The central canal within an osteon is called the Haversian canal and carries blood vessels. Haversian canals have right-angle branches, or channels, termed Volkmann's canals. Volkmann's canals originate in the periosteum and are perpendicular to the long axis of the bone.

Small spider-shaped osteocytes lie in lacunae between the lamellae of each osteon. These cells are simply osteoblasts trapped within the calcifying matrix during growth in circumference of the bone. The osteocytes extend their tentacle-like processes into minute canals, called canaliculi, which radiate in all directions and interconnect the lacunae. Canaliculi of the innermost lacunae communicate, in addition, with the Haversian canal. They are filled with tissue fluid and nutrients which enter through the blood vessels to diffuse throughout compact bone by way of these canaliculi.

Spongy Bone

Spongy bone is poorly organized tissue. It consists of irregular interlacing bone spicules and plates which are only a few layers thick. Nutrient materials reach the osteocytes of

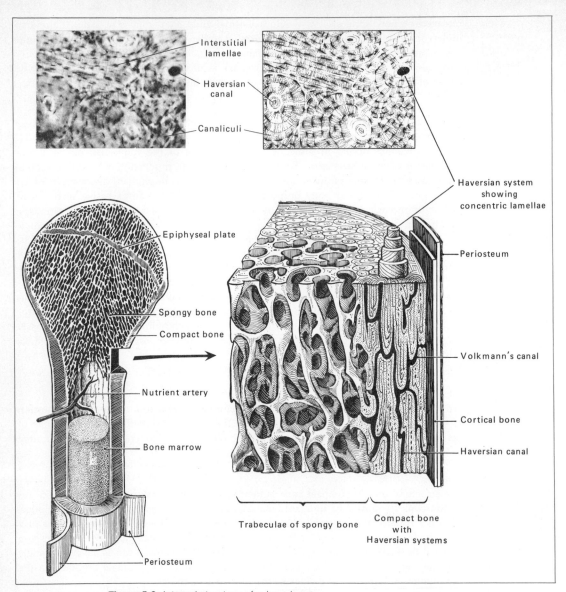

Figure 5-2. Internal structure of a long bone.

spongy bone from the marrow spaces between adjacent spicules, and therefore this type of bone tissue has no osteons.

Bone Marrow

Bone marrow is found between the spicules of the latticework of spongy bone and in the central marrow cavity of long bones. In the adult, the marrow of these two areas differs. Yellow marrow is present in the marrow cavity of long bones and is composed predominately of fat cells, or adipose tissue. Red marrow, associated with spongy bone, contains blood-forming tissue and a limited amount of adipose tissue.

SKULL

The skull is the most complex osseous structure in the body. For descriptive purposes it may be subdivided into a cranial portion (cranium), enclosing the brain, and a facial portion, forming the framework of the face. Bones in the junctional areas between these two subdivisions are modified to accommodate special sensory organs, namely, the orbital cavity for the eye, the nasal cavity for the olfactory apparatus, and within the temporal bone, the structures associated with hearing and equilibrium. Most of the bones of the skull are flat bones consisting of two thin layers of compact bone enclosing cancellous bone (the diploë) and a limited marrow cavity (Table 5-1).

TABLE 5-1. BONES OF THE SKULL AND NECK

PRIMARY LOCATION	NUMBER	NAME	SHORT DESCRIPTION
SKULL	28		
Calvaria	6	Frontal (1)	Forms forehead
		Parietal (2)	Form midportion of skull
		Occipital (1)	Forms back of skull
		Temporal (2)	Form sides, or temporal regions, of skull
Floor of the skull	1	Sphenoid (1)	Butterfly-shaped; forms bulk of floor of mid-portion of cranial cavity
Orbital cavities	2	Lacrimal (2)	On medial wall of cavity, additional skull bones contribute to this cavity
Nasal cavity	6	Ethmoid (1)	T-shaped; forms portion of nasal septum and roof of cavity; curled processes form superior and middle conchae
		Vomer (1)	Contributes to nasal septum
		Palatine (2)	L-shaped; contribute to lateral wall and posterior portion of hard palate
		Inferior concha (2)	Curled; on lateral wall of nasal cavity
Middle-ear cavities	6	Malleus (2)	Hammer-shaped
		Incus (2)	Shaped like an anvil
		Stapes (2)	Stirrup-shaped
Face	7	Nasal (2)	Form bridge of nose
		Maxilla (2)	Form upper jaw; contribute to oral, nasal, and orbital cavities
		Zygoma (2)	Form cheekbones
		Mandible (1)	Lower jaw
NECK	8		
		Vertebrae (7)	Form vertebral column between skull and thorax
		Hyoid (1)	U-shaped; at front of neck
Total	36		

Cranium

The calvaria, or skullcap, forms the rounded top portion of the cranium (Figure 5-3). It consists of curved portions of the frontal, temporal, occipital, and sphenoid bones, plus all the two parietal bones. Articulations between the bones forming the calvaria are nonmovable joints called sutures. The major sutures of the skull are the sagittal, coronal, and lambdoid sutures. The sagittal suture extends anteroposteriorly between the two parietal bones. The coronal and lambdoid sutures pass transversely across the skull, coursing at right angles to the sagittal suture. The anteriorly situated coronal suture is interposed between the frontal bone and the two parietal bones. Posteriorly, the lambdoid suture is between the two parietal bones and the occipital bone. Flanking the sagittal suture the parietal bones form the top of the skull. Anterior to the coronal suture the frontal bone forms the forehead, and behind the lambdoid suture the occipital bone forms the back of the skull.

The apposing edges of the flat bones of the calvaria are not ossified at birth. This condition allows the bones to override one another during birth, reducing the size of the head, which aids in its passage through the birth canal.

The membranous nature of the margins of the bones is readily observable at the

Figure 5-3. Lateral view of the skull. Note the three major sutures of the calvaria.

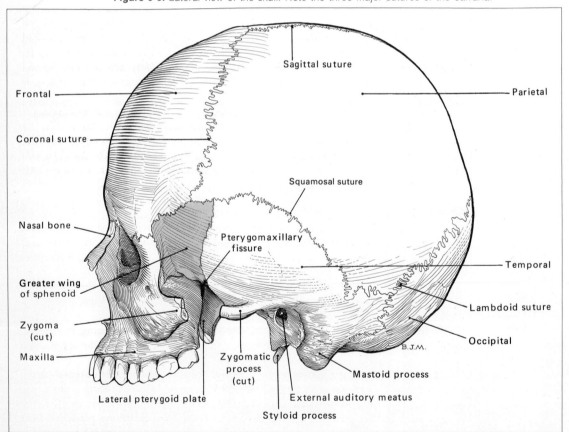

anterior fontanel, which is the junction of the parietal and frontal bones, or "soft spot" of the newborn baby's head. This region does not become entirely ossified until the infant is about 18 months of age. The flat bones of the skull continue to grow for several years before they become rigidly joined.

Premature or early closure of the sutures limits the further growth of the skull and prevents normal development of the brain. This results in an abnormally small head, a condition known as microcephaly. Persons with this abnormality are usually severely mentally retarded.

In hydrocephaly, an excessive amount of cerebrospinal fluid within the cavities (ventricles) of the brain results in a substantial increase in the intracranial pressure. In young children while the sutural joints are still slightly movable, this may cause a marked enlargement of the skull to accommodate the excessive fluid. Hydrocephalic children must have this excess fluid periodically drawn off to reduce pressure on the brain to prevent brain damage and mental retardation.

The midlateral portion of the skull is formed by the flattened, or squamosal, portion of the temporal bone (Figure 5-4).

The irregular inferomedial petrous portion of the temporal bone houses the middle and internal ear. Additional prominent features on the external surface of the temporal bone include a large opening, the external auditory meatus, which leads into the middle ear cavity. Just below and behind the opening is the mastoid process. The latter forms the large, rounded elevation behind the ear and contains the mastoid air cells, or sinuses. A long spinelike process, the styloid process, extends inferiorly from the temporal bone. Anteriorly the squamous portion of the temporal bone articulates with the sphenoid, superiorly with the parietal, and posteriorly with the occipital bones. Anteriorly, the zygomatic process joins the zygomatic bone to contribute to the zygomatic arch. Internal to the zygomatic arch the relatively deep infratemporal fossa contains the muscles of mastication.

In the back of the skull the basilar portion of the occipital bone curves forward to complete the base of the skull. Here the foramen magnum, the largest foramen in the skull, admits the spinal cord into the cranial cavity. Adjacent to the foramen magnum, two large occipital condyles articulate with the first cervical vertebra. The basal portion of the occipital bone forms a bed for the brainstem, the expanded superior continuation of the spinal cord.

Cranial Cavity

On the interior of the skull, the floor of the cranial cavity presents three shallow fossae (Figure 5-5). Anteriorly the greater wings of the sphenoid bone and posteriorly the petrous portions of the temporal bones limit the middle cranial fossa. The anterior and posterior fossae lie in front of and behind these boundaries, respectively.

The middle cranial fossa contains the temporal lobes of the brain. In the midportion of this fossa the body of the sphenoid bone is hollowed out in the shape of a saddle to form the sella turcica, which is the site of the hypophyseal gland.

The posterior cranial fossa contains the cerebellum. A prominent opening, the internal auditory meatus, in the petrous portion of the temporal bone leads into the internal ear cavity.

Anterior to the greater wings of the sphenoid bone, the floor of the anterior cranial fossa is formed by the orbital plate of the frontal bone and the cribriform plate of the ethmoid bone. This portion of the cranial cavity contains the frontal lobes of the brain. A prominent midline vertical process at the anterior end of the cribriform plate, the crista galli, provides the anterior attachment for a fold of dura mater known as the falx cerebri.

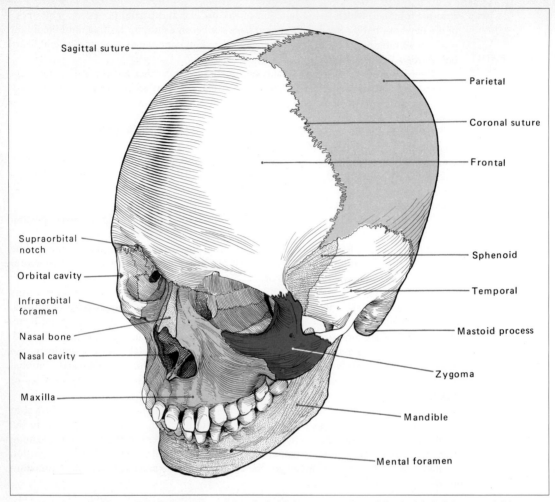

Figure 5-4. Oblique view of the skull. All the components of the facial skeleton are visible.

The inner surfaces of the bones of the calvaria are grooved to accommodate the meningeal arteries and veins and the dural (venous) sinuses.

A multitude of foramina are associated with the floor of the cranial cavity to permit the passage of blood vessels, nerves, and other structures.

Facial Skeleton

Four bones make up the bulk of the face (Figure 5-4). The frontal bone forms the forehead. Its most prominent features, the supraciliary, or brow, ridges above the eyes, are the sites of the frontal air sinuses. The frontal bone forms most of the roof of the orbital cavity.

The maxillae, the upper jaw bones, partially surround the nasal cavity. A maxillary air sinus within each bone is the largest of the paranasal sinuses. Teeth of the upper jaw are embedded in the alveolar process of each maxilla, and the horizontal palatine

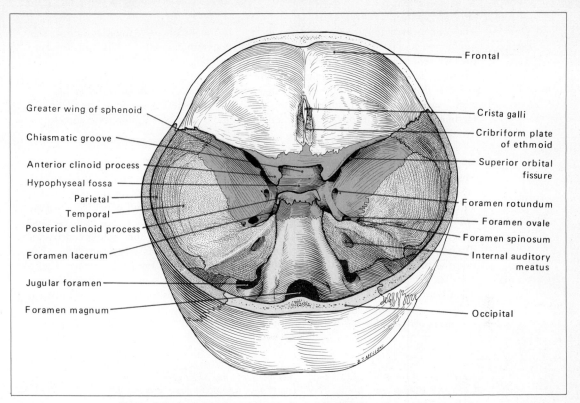

Figure 5-5. Cross section of the skull, showing the foramina and bones that contribute to the floor of the cranial cavity. The area shaded in color is the sphenoid bone. The area stippled in color is the temporal bone.

process of this bone forms the major portion of the hard palate. Laterally its zygomatic process articulates with the zygomatic bone to form the "cheekbone."

The nasal bones lie between the maxillary bones to form the bridge of the nose. Immediately below the nasal bones a large aperture bounded laterally by the maxillary bones opens into the nasal cavity.

The mandible, or lower jaw bone, completes the facial skeleton and, with the exception of the middle ear bones, forms the only movable articulation in the skull. The lower teeth are lodged in the alveolar process of the body of this bone. Just below and in front of the ear the body of the mandible makes a right angle bend and extends superiorly as the ramus. At the superior end of the ramus the anterior coronoid process is separated by a notch from the neck of the bone. At the top of the neck of the mandible the rounded condylar process articulates with the temporal bone in the mandibular fossa. On the internal surface of the mandible the large mandibular foramen opens into a mandibular canal which transmits the vessels and nerves supplying the lower teeth.

As the permanent teeth are lost, the alveolar sockets in which they are lodged are reabsorbed. In an edentulous (toothless) person, especially if prostheses (dental plates) are not used, the entire alveolar process will be reabsorbed. The loss of this bony process decreases the thickness (height) of the mandible by about one-half and makes the chin appear much more prominent.

93

Orbital Cavity

The orbital cavity (Figure 5-6) contains the eye, lacrimal gland, eye muscles, and vessels and nerves supplying the eye. The two cavities are pyramidal with the base opening onto the face, where it is formed about equally by the frontal, zygomatic, and maxillary bones. The optic foramen, through which the optic nerve and the ophthalmic artery pass, is at the apex of the cavity. Additional openings associated with the orbital cavity include the supra- and infraorbital fissures and the anterior and posterior ethmoid foramina. The lacrimal bone and portions of the sphenoid, ethmoid, and palatine bones complete the orbital cavity.

Nasal Cavity

Most of the bony structure of the nasal cavity (Figure 5-7) is formed by the complex ethmoid bone. The cribriform plate, which forms the narrow roof of the nasal cavity and admits filaments of the olfactory nerve, is a part of the ethmoid bone. Flanking the cribriform plate a series of small ethmoid air cells extend between the frontal and sphenoid bones. Curled processes of the ethmoid bone form the superior and middle conchae of the lateral wall of the nasal cavity, while the inferior concha is a separate bone. The perpendicular plate of the ethmoid is the largest bony part of the nasal septum (Figure 5-8). The septum is completed by the vomer inferiorly and the septal cartilage anteriorly. The palatine processes of the maxillary and the palatine bones form the floor of the nasal cavity as well as the roof of the oral cavity. Posterolaterally the palatine and pterygoid processes of the sphenoid bones complete the lateral walls of the nasal cavity.

Figure 5-6. The orbital cavity, frontal view. Note that the zygoma, maxilla (shaded light blue), and frontal bone (blue stipple) contribute the margin of the orbital cavity.

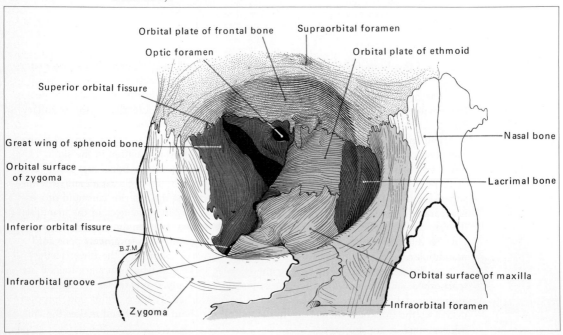

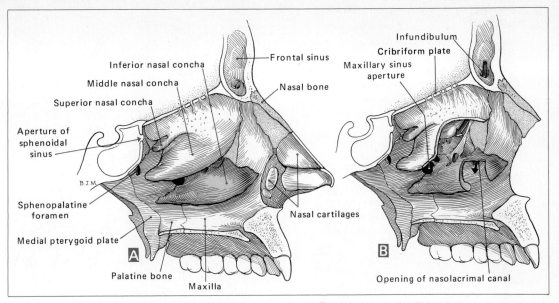

Figure 5-7. Lateral wall of the nasal cavity. **A.** The inferior concha (light blue) is separate from the other two conchae. **B.** Two conchae have been cut to expose the meatuses they overlie. The infundibulum is the opening from the frontal sinus.

Figure 5-8. The nasal septum. The septal cartilage, perpendicular plate of the ethmoid bone, and the vomer together form the medial wall of the nasal cavity.

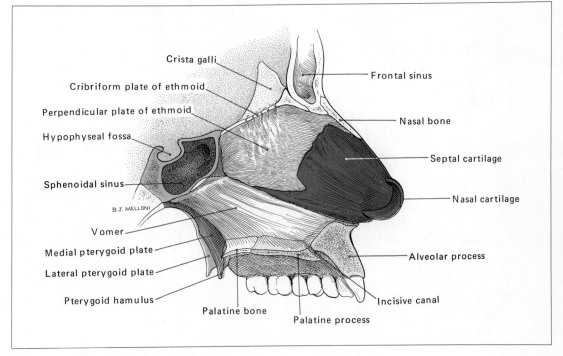

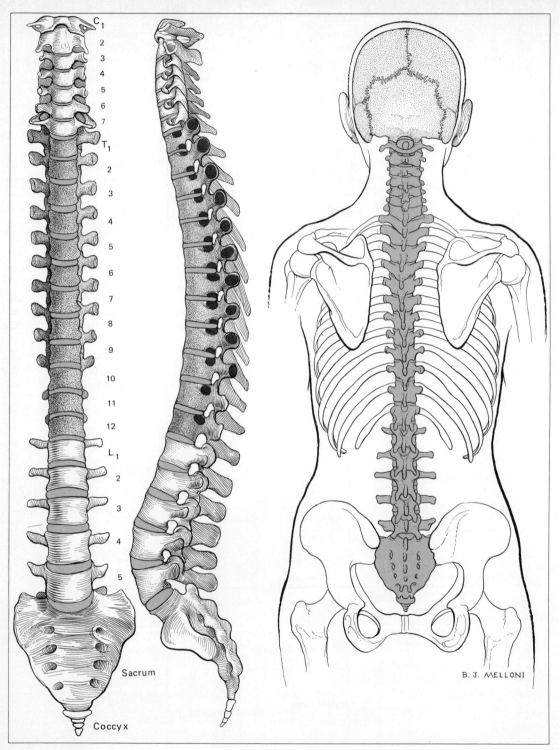

C 1
2
3
4
5
6
7
T 1
2
3
4
5
6
7
8
9
10
11
12
L 1
2
3
4
5

Sacrum

Coccyx

B. J. MELLONI

VERTEBRAL COLUMN

The vertebral column forms a slightly movable, rodlike, axial support for the body. It extends from the base of the skull to the coccyx and is divided into cervical, thoracic, lumbar, sacral, and coccygeal segments (Figure 5-9).

Seven vertebrae in the neck make up the cervical portion. The twelve thoracic vertebrae afford articulation for the ribs. Five lumbar vertebrae located in the small of the back contribute to the posterior boundary of the abdominal cavity. The wedge-shaped sacrum, formed by the fusion of five sacral vertebrae, makes up the posterior wall of the pelvic cavity. Terminally three or four small fused vertebrae form the coccyx, or tail bone (Table 5-2).

From an anterior, or front, view the vertebral column appears straight. From the lateral aspect, however, the vertebral column in the adult presents four curvatures. In the newborn infant only a single posteriorly convex curve is present, but as the baby develops, two secondary concave curvatures appear. One in the cervical region develops as the baby begins to raise his head and control his head movements, and the second curvature, in the lumbar region, develops as he stands and begins to walk. The primary convex curvatures persist in the thoracic and sacral regions.

Injury, disease, or poor posture may affect the normal alignment of the vertebral column. Kyphosis, or "hunchback," is an exaggeration of the curvature in the thoracic region; lordosis, or "swayback," is an increase in the curvature in the lumbar region. An imbalance or paralysis of the muscles on one side of the vertebral column will result in a lateral curvature, or scoliosis, of the normally straight column. Tuberculosis of the spine causes many of these deformities, especially in children.

A typical vertebra consists of an anterior body, a central vertebral foramen, and a posterior vertebral arch (Figures 5-10 and 5-11). The bodies of the vertebrae are essentially

TABLE 5-2. BONES OF THE TORSO

PRIMARY LOCATION	NUMBER	NAME	SHORT DESCRIPTION
Thorax	37	Vertebra (12)	Form vertebral column between neck and small of back
		Rib (24)	Slender shafts of bone which, with costal cartilages, form thoracic cage
		Sternum (1)	Breastbone
Abdomen	5	Vertebra (5)	Form vertebral column in small of back
Pelvis	2	Sacrum (1)	Wedge-shaped; formed from five fused vertebrae
		Coccyx (1)	Tail bone; formed from four fused vertebrae
Total	44		

Figure 5-9 (facing page). The vertebral column. The primary (convex) curvature in the thoracic and sacral regions is present at birth. Secondary (concave) curvatures develop in the cervical region when the baby begins to move his head, in the lumbar region when he begins to stand and walk.

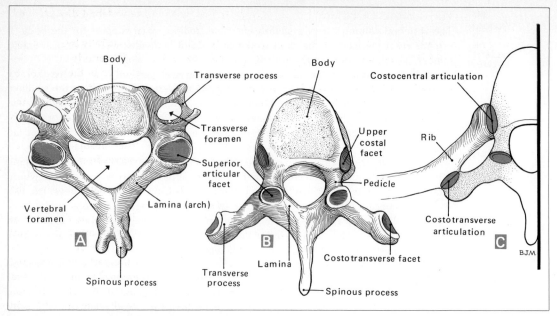

Figure 5-10. Typical vertebrae, seen from above. Articular surfaces are highlighted in color. **A.** A cervical vertebra. **B.** A thoracic vertebra. **C.** A thoracic vertebra in articulation with a rib.

Figure 5-11. Typical lumbar vertebra. Articular surfaces are highlighted in color. **A.** Superior aspect. **B.** Lateral aspect.

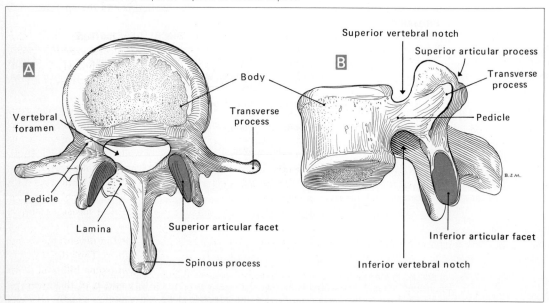

circular. They articulate with each other through an intervening vertebral disk of fibro-cartilage.

The vertebral foramen is bounded anteriorly by the body and laterally by cylindrical processes called pedicles, which project posteriorly from the body. Posteriorly the pedicles fuse with the flattened laminae, which, in turn, fuse with each other in the midline to complete the vertebral arch. The spinous processes project posteriorly from the midline junction of the laminae. Spinous processes may be felt along the entire length of the vertebral column. Adjacent vertebral foramina form the vertebral canal which surrounds the spinal cord.

Each pedicle has a superior and an inferior notch. These notches on adjacent pedicles form an intervertebral foramen which permits passage of spinal nerves from the vertebral canal.

The central portion of the intervertebral disk is formed by a gelatinous mass, the nucleus pulposus.

The fibrocartilaginous disk between adjacent vertebral bodies may be displaced or damaged. If this occurs, the nucleus pulposus may herniate and protrude into the vertebral canal or intervertebral foramen as a herniated disk ("slipped disk"). The disk may press upon the spinal nerve that leaves the vertebral canal through the partially obstructed foramen, thus causing pain or paralysis in the area of its distribution. The most frequent site of occurrence of a herniated disk is in the lower lumbar region. A disk herniation in this area involves the inferior extremity as attacks of sciatica. To alleviate this condition it may be necessary to remove the involved disk surgically and fuse the two adjacent vertebrae.

Vertebrae from each region are easily identifiable by the following distinctive features: in the transverse processes of cervical vertebrae there is a hole, the transverse foramen, through which the vertebral artery and vein course; vertebrae from the thoracic region have facets on their transverse processes and their bodies for articulation with the ribs; and lumbar vertebrae have neither of the above features.

The first cervical vertebra, the atlas (Figure 5-12), lacks all the typical vertebral features except foramina for the passage of the vertebral vessels and the spinal cord. It is essentially a ring of bone enlarged laterally.

The second cervical vertebra, the axis (Figure 5-12), differs markedly from other vertebrae in having a vertical projection, the dens or odontoid process, extending superiorly from its body. This is actually the body of the first cervical vertebra which, during evolutionary development, became detached and migrated inferiorly to fuse with the second vertebra.

The wedge-shaped sacrum lies between the two innominate bones (hipbones). Superiorly it articulates with the fifth lumbar vertebra, inferiorly with the coccyx (Figure 5-13). At either side the ear-shaped auricular surfaces of the sacrum are firmly attached to the iliac segment of the innominate bone to form the sacroiliac articulation. At the bottom of the posterior surface of the sacrum a small irregular aperture, the sacral hiatus, opens into the vertebral canal. The front of the first sacral segment, representing the body of the first sacral vertebra, bulges into the pelvic cavity at the inlet of the pelvis to form the sacral promontory. This is an important landmark for obstetrical measurements of the pelvis.

Spina bifida is a condition in which the vertebrae are arrested in development with failure of the vertebral lamina to fuse and spinous processes to form. Thus the vertebral canal is open posteriorly. This defect may be minimal or hidden (spina bifida occulta), or it may be extensive. In the latter case a meningocoele (herniation of the meninges and spinal cord) usually occurs. In the newborn infant this condition is usually observed

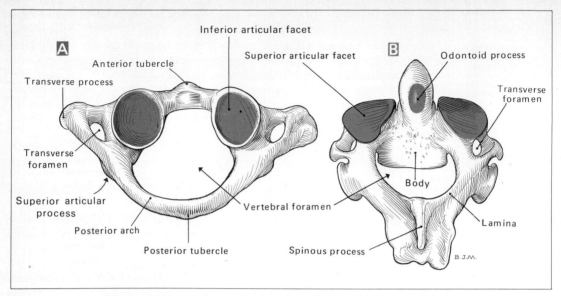

Figure 5-12. The first two cervical vertebrae. **A.** The atlas. Note that it lacks a body and spinous process. **B.** The axis. In addition to the typical features of a cervical vertebra the axis possesses an odontoid process.

Figure 5-13. The sacrum. **A.** Anterior aspect. Note that the sacral vertebrae fuse into a single bone at the transverse ridges. **B.** Posterior aspect.

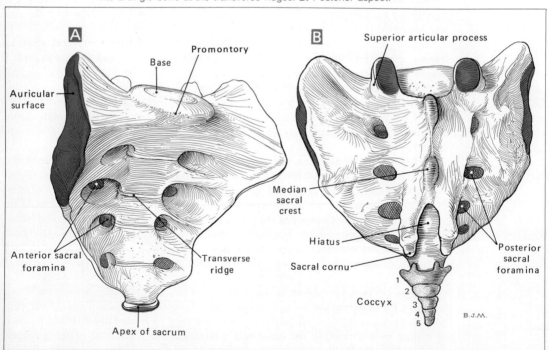

as a mass in the lumbar region of the back. Only skin covers the protruding meninges and spinal cord; repair of this defect is extremely difficult and infant mortality high.

THORACIC CAGE

Skeletal elements forming the thoracic cage include the sternum, ribs, costal cartilages, and thoracic vertebrae (Figure 5-14). The vertebrae have been described with the vertebral column on the preceding pages. Seen in cross section, the thoracic cavity is somewhat kidney-shaped with a greater lateral than anteroposterior dimension. Seen from the front, the thoracic cavity is funnel-shaped—narrow at the top and much wider at the junction of the thorax and abdomen.

The sternum, or breastbone, consists of three segments. The uppermost portion, the manubrium, can be felt at the base of the neck. The largest segment is the body. It forms the sternal angle (of Louis) as it articulates with the manubrium. The third, lowermost segment of the sternum is the small, moveable, midline, xiphoid process. At the sides of the sternum articular facets are present for the clavicles at the upper end of the manubrium and for the upper six ribs along the length of the sternum.

A sternal puncture is a frequently used procedure in the diagnosis of blood disorders. The cortex of the sternum is punctured by a trocar (large needle), and red bone marrow

Figure 5-14. The thoracic cage. **A.** Skeletal components. Cartilage is shown stippled. **B.** The sternum, lateral aspect. Articular facets for the costal cartilages are highlighted in color.

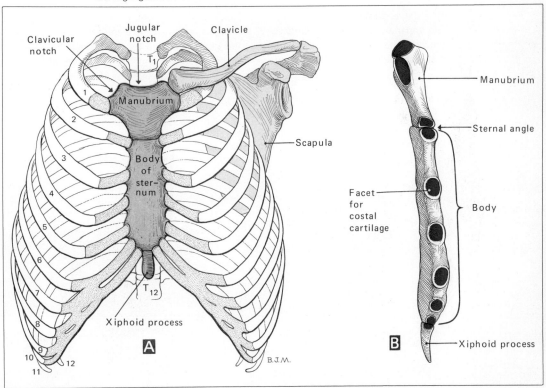

is withdrawn from the marrow cavity to be studied. This test is especially meaningful in following the progress of patients with leukemia, since red bone marrow is blood-forming tissue.

Ribs

The ribs consist of 12 pairs of elongated, curved, flattened shafts of bone which, except for the last 2 pairs, extend from the thoracic vertebrae to cartilage joining them to the sternum. Between ribs, in the intercostal spaces, are the intercostal muscles. Along the inferior border of each rib is a groove, called the costal groove, along which course the arteries, nerves, and veins to the intercostal muscles.

Each rib (Figure 5-15) consists of a head, neck, and shaft (body). The head of a rib articulates with the body of a thoracic vertebra. The short neck extends between the head and shaft of the rib. At the junction of the neck with the shaft, there is a small facet, the tubercle, for articulation with the transverse process of a vertebra. The shaft of the rib makes a sharp angle to continue anteriorly to its junction with the costal cartilage.

Costal cartilages, somewhat flattened bars of hyaline cartilage, join the ribs to the sternum. The cartilages of the upper seven ribs connect directly to the sternum. Cartilages of the eighth, ninth, and tenth ribs attach to the costal cartilages of the rib above, while the cartilages of the eleventh and twelfth ribs are rudimentary nubs that simply cap the ends of the ribs. The eleventh and twelfth ribs are embedded in the musculature of the abdominal wall and hence have no firm anterior attachment. They are sometimes referred to as floating ribs.

APPENDICULAR SKELETON

The appendicular skeleton consists of the upper and lower extremities. The extremities attach to the axial skeleton by the pectoral girdle (formed by the clavicle and scapula at the shoulder) and the pelvic girdle (consisting of the innominate bones at the hip).

Figure 5-15. A typical rib. Articular facets for thoracic vertebrae are highlighted in color. Compare this illustration with Figure 5-10**C.**

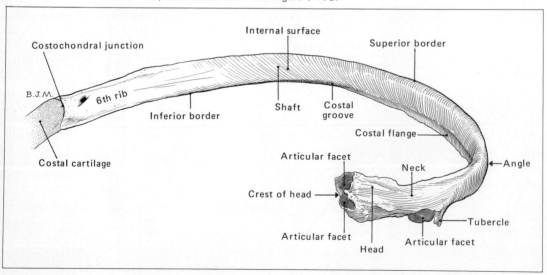

TABLE 5-3. BONES OF THE UPPER EXTREMITIES

PRIMARY LOCATION	NUMBER	NAME	SHORT DESCRIPTION
Shoulder	4	Clavicle (2)	Elongated S-shaped "collarbone"
		Scapula (2)	Flattened triangular "shoulder blade"
Arm	2	Humerus (2)	Long bone between shoulder and elbow
Forearm	4	Radius (2)	On thumb side of forearm
		Ulna (2)	On little-finger side of forearm
Wrist	16	Scaphoid (2)	Two rows of cube-shaped bones between hand and forearm
		Lunate (2)	
		Triangularis (2)	
		Pisiform (2)	
		Trapezium (2)	
		Trapezoid (2)	
		Capitate (2)	
		Hamate (2)	
Hand	38	Metacarpal (10)	Bones of palm
		Phalanx (28)	Bones of fingers—two in thumb, three in each finger
Total	64		

Superior Extremity (Table 5-3)

Scapula The scapula, or "shoulder blade" (Figures 5-16 and 5-17), is a flat triangular bone overlying the upper portion of the back. It is the site of attachment of superficial muscles of the back and certain muscles of the shoulder and arm. Its medial (vertebral) border lies parallel to the vertebral column, while its lateral (axillary) border follows, in general, the posterior wall of the axilla, or armpit. It has three prominent processes: the acromion, which forms the point of the shoulder and articulates with the lateral end of the clavicle; the spine, which extends along the posterior surface from the acromion toward the vertebral border and separates the supraspinous fossa from the infraspinous fossa, and the hooklike coracoid process on the anterior surface of the bone. Its glenoid fossa, at the lateral angle of the scapula, is modified to articulate with the head of the humerus.

Clavicle The clavicle, or "collarbone," shaped like an elongated S, extends from the sternum to the acromion. At its medial end it articulates with the sternum and the first rib. This articular surface, about 2.5 cm in diameter, is the only bony attachment between the trunk and the upper extremity.

Humerus The humerus, the bone of the arm, articulates with the scapula at the glenoid fossa and with the radius and ulna at the elbow joint (Figure 5-18). On the lateral

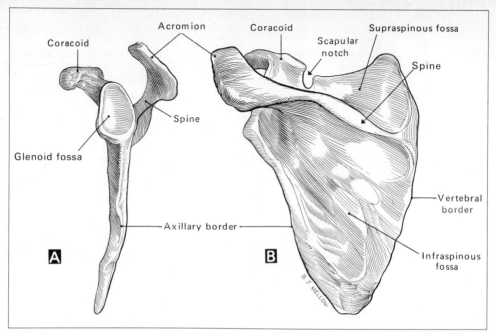

Figure 5-16. A. Lateral aspect of the scapula. **B.** Posterior aspect.

Figure 5-17. The scapula, anterior view, joined to the humerus and the clavicle. The clavicle articulates with the manubrium of the sternum. The head of the humerus articulates in the glenoid cavity.

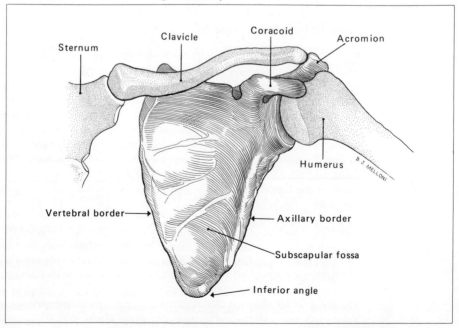

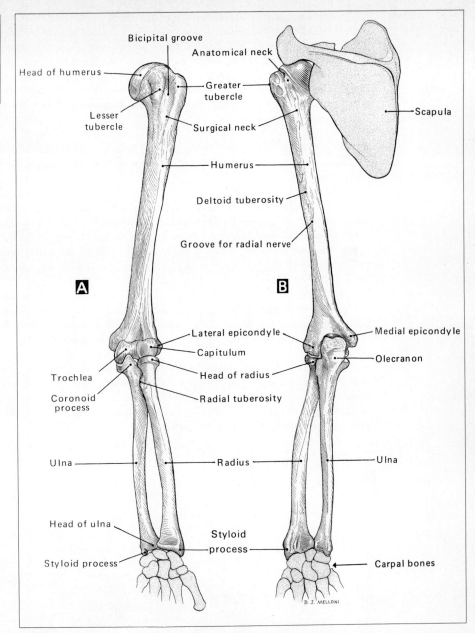

Bicipital groove

Anatomical neck

Head of humerus

Greater
tubercle

Scapula

Lesser
tubercle

Surgical neck

Humerus

Deltoid tuberosity

Groove for radial nerve

A

B

Lateral epicondyle

Medial epicondyle

Capitulum

Olecranon

Trochlea

Head of radius

Coronoid
process

Radial tuberosity

Ulna

Radius

Ulna

Head of ulna

Styloid
process

Styloid process

Carpal bones

B. J. MELLONI

Figure 5-18. Bones of the upper extremity. **A.** Anterior aspect. **B.** Posterior aspect.

aspect of the head of the humerus the bicipital groove is flanked by the greater and lesser
tubercles, which serve as sites of insertion for several shoulder muscles.

The constricted portion between the tubercles and the head is the anatomical neck
of the humerus. Distal to the tubercles a second slight constriction, a frequent site of
fractures, is referred to as the surgical neck of this bone. The deltoid tuberosity is a

roughened area on the lateral surface of the shaft of the humerus for insertion of the deltoid muscle. Opposite the deltoid tuberosity a faint groove or line, the radiospiral groove, marks the path of the radial nerve as it curves posteriorly around the humerus.

Distally the shaft of the humerus flares out into the medial and lateral epicondyles, which afford attachment for muscles of the forearm. At the distal extremity of the humerus the articulating surface is divided into two portions. The medial portion is the trochlea, which articulates with the ulna, while the lateral, rounded portion is the capitulum, which articulates with the radius.

Radius and Ulna   The bones of the forearm comprise the radius laterally and the ulna medially. They are joined together throughout most of their length by the interosseous membrane.

Salient features of the radius include the rounded proximal head which articulates with the capitulum of the humerus as well as with a facet on the olecranon process of the ulna. Near the head is the large radial tuberosity. Surrounding the neck of the radius is the annular ligament, which holds the head of the radius tightly against the ulna during supination and pronation. Distally the expanded end of the radius articulates with the proximal row of carpal (wrist) bones and at its lateral side with the distal end of the ulna.

At the proximal end of the ulna is the olecranon process, which articulates with the trochlea of the humerus. This process also forms the point of the elbow. The posterior border of the ulna can be felt along the entire length of the forearm. A fibrocartilaginous disk separates the ulna from the proximal row of carpal bones.

Wrist and Hand   Eight carpal bones are present in the wrist. Sliding joints between them permit only slight movement between individual bones but allow multidirectional movement of the entire wrist joint (Figure 5-19). Bones of the proximal row from the lateral to the medial side are the scaphoid, lunate, triangular, and pisiform. Bones of the distal row are the trapezium, trapezoid, capitate, and hamate.

Figure 5-19. Skeletal components of the hand. The carpal bones are highlighted in color.

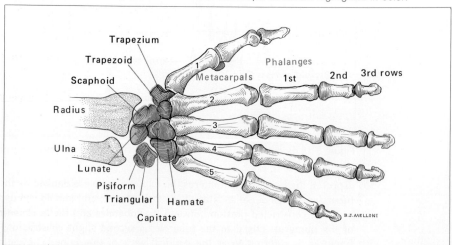

Metacarpal bones form the skeleton of the palm of the hand. Proximally they articulate with the distal row of carpal bones. Distally each metacarpal enlarges to form a prominent head for articulation with a proximal phalanx of a finger.

Fourteen phalanges form the finger bones, three for each finger and two for the thumb.

Inferior Extremity

The pelvic girdle is formed by the two innominate bones (hipbones), which attach the inferior extremity to the axial skeleton at the sacroiliac articulation. This very strong, stable joint transmits the weight of the torso onto the inferior extremity.

Innominate Bones The innominate bones articulate at either side with the sacrum posteriorly and with each other anteriorly to form the bony framework of the pelvic cavity (Figure 5-20). The internal aspect of the innominate bone and the promontory of the sacrum separate the false and true pelvis and, in the female, form the entry of the birth canal. The false pelvic cavity is the most inferior portion of the abdominal cavity. The true pelvic cavity extends from the brim of the pelvis to the outlet of the birth canal and is bounded inferiorly by the coccyx, the ischial tuberosities, the ischiopubic rami, and the symphysis pubis.

The dimensions of the true pelvic cavity are of paramount concern to the obstetrician in that during parturition the baby's head must pass through this area. Pelvimetry is the measurement of the size of the inlet and outlet of the birth canal. The anteroposterior inlet diameter is the distance between the top of the symphysis pubis and the sacral promontory, and the transverse inlet diameter is the greatest distance across the pelvis

TABLE 5-4. BONES OF THE LOWER EXTREMITIES

PRIMARY LOCATION	NUMBER	NAME	SHORT DESCRIPTION
Hip	2	Innominate (2)	Formed by fused ilium, ischium, and pubis. With sacrum and coccyx of vertebral column form skeleton of pelvic cavity
Thigh	2	Femur (2)	Long bone between hip and knee
Knee	2	Patella (2)	Flattened triangular "kneecap"
Leg	4	Tibia (2)	On medial side of leg, anterior border forms "shin"
		Fibula (2)	On lateral side of leg
Ankle	14	Calcaneus (2)	Short cube-shaped bones between leg and foot
		Talus (2)	
		Navicular (2)	
		Cuboid (2)	
		Cuneiform (6)	
Foot	38	Metatarsal (10)	Bones adjacent to instep of foot
		Phalanx (28)	Bones of toes—two in big toe, three in each of the others
Total	62		

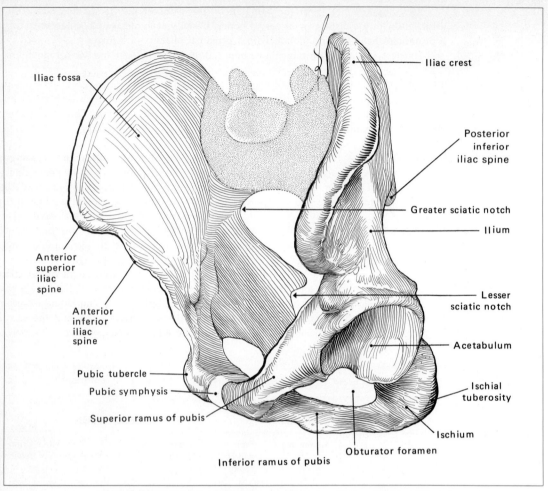

Iliac fossa

Iliac crest

Posterior
inferior
iliac spine

Greater sciatic notch

Ilium

Anterior
superior
iliac
spine

Anterior
inferior
iliac
spine

Lesser
sciatic notch

Acetabulum

Pubic tubercle

Ischial
tuberosity

Pubic symphysis

Superior ramus of pubis

Ischium

Obturator foramen

Inferior ramus of pubis

Figure 5-20. Oblique view of the innominate bones. They articulate with each other at the symphysis pubis anteriorly and with the sacrum (stippled area) posteriorly.

at the brim. The anteroposterior outlet diameter is the distance between the symphysis pubis (inferior aspect) and the coccyx, while the transverse outlet diameter is the distance between the two ischial tuberosities.

Diameters are usually determined by palpation through the vagina, but more accurate measurements can be made from an x-ray film of the pelvis of the patient. If the x-ray is taken near term, the size of the baby's head (cephalometry) may also be measured.

The innominate bone is composed of three fused portions: the ilium, ischium, and pubis (Figure 5-20). In early life these are represented by individual bones, but by late adolescence they have completely united. The area of fusion is in the acetabulum, a deep cup-shaped socket which articulates with the head of the femur. Each of the bones also contributes laterally to the margin of the obturator foramen, the largest foramen in the body.

Salient features of the ilium include the extensive superior iliac crest, easily palpable as the "hipbone," and at the anterior end of the crest, the anterior superior iliac spine. The wings, or ala, of the iliac bone form the extensive, shallow iliac fossa. Posteriorly the ilium articulates with the sacrum at the sacroiliac joint.

The ischium fuses to the posterior aspect of the ilium. It has greater and lesser sciatic notches, separated by the ischial spine. Inferiorly the large ischial tuberosities form the bony prominences upon which one sits.

The two pubic bones articulate with each other anteriorly at the symphysis pubis. Each has a lateral and an inferior extension. The lateral extension, the superior ramus, fuses with the ilium. The inferior extension, the inferior ramus, fuses with the ischium. The ischiopubic ramus of one side creates an angle, the subpubic angle, with the ischiopubic ramus of the other side. This angle is the skeletal characteristic that best distinguishes male from female. In the male, the usual angle is less than 90 degrees; in the female, greater than 90 degrees. Moreover, the female pelvis tends to be wider in all dimensions and made of lighter bones.

Femur The femur is the largest bone in the body. Proximally it articulates with the innominate bone at the acetabulum and distally with the tibia and patella (Figure 5-21). The neck of the femur forms an obtuse angle between the head and the shaft of the bone. At the junction of the neck and the shaft, two roughened prominences, the greater and lesser trochanters, are sites for muscular attachment. A prominent crest, the linea aspera, extends along the posterior aspect of the shaft of the femur. Inferiorly two large, rounded articular condyles, the medial and lateral condyles, separated by the intercondylar fossa, articulate with the flattened proximal head of the tibia.

The patella is at the anteroinferior aspect of the femur. This somewhat flattened triangular bone lies embedded in the tendon of the quadriceps femoris muscle and is the largest sesamoid bone of the body.

Tibia and Fibula The thick tibia and the slender fibula are the bones of the leg. The flat superior surface of the tibia has two facets for articulation with the condyles of the femur. The intercondylar eminence stands between the articular facets.

Just below the head of the tibia is the tibial tuberosity. Extending inferiorly from the tibial tuberosity the sharp anterior border of the tibia forms the "shinbone," which is subcutaneous throughout its length. Inferiorly the tibia is joined to the fibula by a strong ligament to form a mortise-type joint with the head of the talus, one of the tarsal bones. The bony prominence on the inside of the ankle, the medial malleolus, is formed by the distal end of the tibia; and the prominence on the outside of the ankle, the lateral malleolus, by the distal end of the fibula.

The entire shaft of the fibula is embedded in muscle with only its ends subcutaneous. Proximally it articulates with the side of the tibia near the tibial tuberosity.

Foot The tarsal bones of the foot (Figure 5-22) are homologues of the carpal bones of the hand. Posteriorly the roughened surface of the calcaneus, or heel bone, the largest of the tarsal bones, gives attachment to the Achilles, or calcaneal, tendon. Mounted on the superior surface of the calcaneus, the talus fits into the slot formed by the tibia and fibula as described previously.

The navicular, cuboid, and three cuneiform bones complete the tarsal bones. Distally the latter four bones articulate with the metatarsals. The metatarsals are larger than their homologues in the hand. Distally the metatarsals expand into heads which articulate with the proximal row of phalanges.

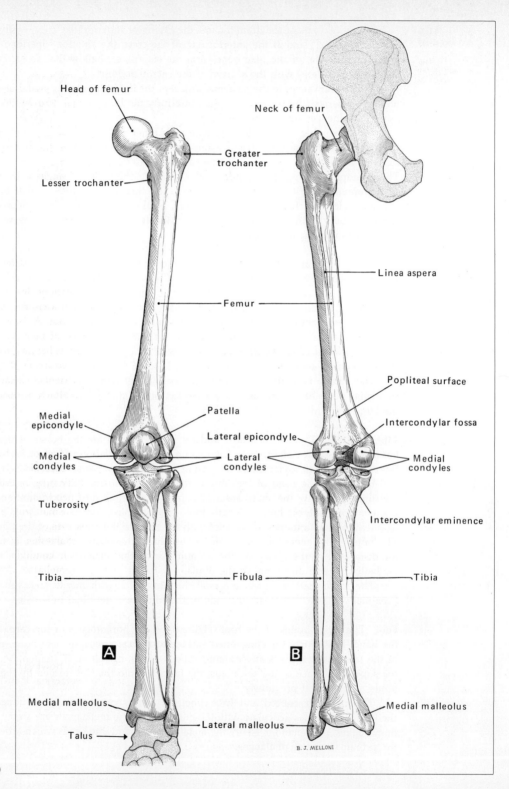

Head of femur

Neck of femur

Greater trochanter

Lesser trochanter

Linea aspera

Femur

Popliteal surface

Medial epicondyle

Patella

Lateral epicondyle

Intercondylar fossa

Medial condyles

Lateral condyles

Medial condyles

Tuberosity

Intercondylar eminence

Tibia

Fibula

Tibia

A

B

Medial malleolus

Medial malleolus

Talus

Lateral malleolus

B. J. MELLONI

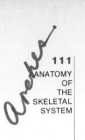

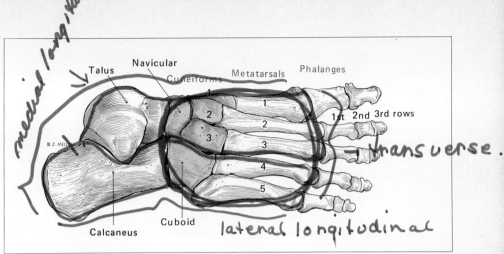

Figure 5-22. Skeletal components of the foot.

The tarsal and metatarsal bones together with their articulations form both the longitudinal arch and transverse arch of the foot. The longitudinal arch, located on the inside of the foot, extends from the calcaneus to the ball of the foot. At the distal end of the arch two small sesamoid bones are present in the tendons of muscles attaching to the plantar surface of the great toe. The transverse arch of the foot is basically associated with the distal ends of the metatarsal bones. The integrity of these two arches is dependent upon the shape of the bones of the foot, the strong intrinsic ligaments of the foot, and the tonus of the muscles of the leg which send their tendons to attach to the bones of the foot.

There are fourteen phalanges in the toes. They are similar to, but much shorter than, those of the fingers.

Several types of deformities may occur in the extremities. Most of them may be traced to under- or overdevelopment of normal differentiating processes. Perhaps the most frequent malformation of this type is a "clubfoot." The deformity may be minimal, as in talipes varus or talipes valgus, in which the foot is slightly turned in or out, or it may be extensive. An example of the latter is talipes equinus, in which the malformation causes such a shortening of the foot that the person must walk on his toes.

The widely publicized thalidomide deformities are examples of arrested development. Use of this drug during certain critical stages of pregnancy results in conditions in which the extremities, particularly the upper limbs, fail to develop (amelia) or are grossly underdeveloped (phocomelia).

ADDITIONAL CLINICAL ASPECTS
Rickets

Rickets is a nutritional disorder resulting from vitamin D deficiency in which there is a faulty calcification of bone that may cause skeletal deformities. Bowleggedness (genu varum) is often due to complications associated with the joint surfaces as a result of this

Figure 5-21 (facing page). Bones of the lower extremity. Because ligaments and cartilage are not shown, the knee joint appears disarticulated. **A.** Anterior aspect. **B.** Posterior aspect.

disease. Rickets that affects the thoracic cage causes a protrusion of the sternum ("pigeon breast") and localized enlargements at the costochondral joints. This latter complication looks like a string of beads beneath the skin of the anterior chest wall and has been named "the rosary of rickets."

Roentgenography

Roentgenography, the use of x-rays, is the common way of diagnosing bone disorders. In addition to enabling the physician to determine the position and extent of a fracture, it permits him to observe the degree of fracture healing or union. It is also useful as a diagnostic tool in several types of bone disease, such as osteosarcoma (bone cancer), osteoporosis (a loss of bone mineral with an increase of bone reabsorption cavities), and Paget's disease (an increased thickness of bone due to excessive deposition of bone minerals). In metabolic disorders increased or delayed bone growth may be diagnosed from the appearance of the epiphyses on an x-ray film.

Fractures

Fracture denotes the breaking or cracking of a bone, as opposed to dislocation, which refers to disruption of the normal alignment of bones as they form a joint, or to a sprain in which the ligaments supporting a joint are stretched or torn. The type of fracture varies with age. For example, in children the organic content of bone is greater than the mineral content, so that undue stress on a bone usually results in an incomplete fracture, or "greenstick" fracture, so named because the break resembles that of a green branch of a tree. In old age the organic content decreases, and bones become more fragile and break under much less stress. Bone repair is also then impaired, because changes in the circulatory system have diminished the blood supply.

Types of Fractures   Fractures are classified according to the manner in which the bones are broken.

Simple fracture—the skin of the fracture site is not broken.
Compound fracture—the broken end or ends of the bone protrude through the skin.
Depressed fracture—the broken bone is driven inward. This type is usually present in severe skull fractures.
Comminuted fracture—the bone is splintered into small fragments.
Avulsion fracture—a structure, for example the finger, is torn off.
Impaction fracture—the broken ends of a bone are driven into each other.
Pathological fracture—fracture as a result of destruction of bone by disease. There is usually minimal external stress.
Colles' and Pott's fractures—special types of breaks given additional consideration because of certain complications. Colles' fracture is a break of the lower end of the radius in which the hand is displaced posteriorly by the pull of forearm muscles. Pott's fracture is a break of the lower fibula. This is a serious injury because after it heals, movement at the ankle joint is frequently impaired.

Healing   Fractures also involve the soft tissue at the site of the break. The periosteum is always broken, vessels are usually torn, and nerves and muscles may be damaged. The ruptured blood vessels produce a massive blood clot (hematoma) at the fracture site which persists for 8 to 10 hr. The hematoma is then gradually infiltrated by fibroblasts, and the clot is organized into granulation tissue. The infiltrating fibroblasts produce collagenous fibers which, in turn, fill the space between the apposing broken ends of the bone with fibrous tissue (callus). New bone formed by the periosteum of the adjacent undamaged bone in time replaces the temporary fibrous callus.

Alignment Alignment of a broken bone may be performed as a closed or open reduction. In a closed reduction the bone is simply manipulated into its normal position and a cast applied. In more serious fractures it may be necessary to expose the bone surgically in order to align it properly. This procedure is called open reduction, and may involve providing additional support to assure proper fixation of the bone. A nail or screw may be inserted into the shaft of the bone, or a plate may be fastened to the broken portions of the bone to stabilize it as it heals.

Transplants Bone transplants are used in a fracture if a relatively large segment of bone is crushed or if a portion of a bone is diseased and must be removed. Transplants are of two types: The first is an autotransplant, in which a piece of living bone is transferred from one part of a patient's body to another. If this is not feasible, fresh bone may be taken from another person, or preserved bone from a bone bank may be used. This second type of transplant is a homotransplant. In either case the transplanted bone forms a framework utilized by the osteoblasts from adjacent bone to lay down new bone. The donor bone is slowly reabsorbed and replaced.

In certain conditions the damaged bone must be replaced with an artificial device called a prosthesis. An example of this is the replacement of the head of the femur with a metallic substitute when a fracture in this area fails to heal and the involved bone dies and must be removed.

JOINTS

A joint, or articulation, is the junction between two or more bones; it is surrounded by a fibrous joint capsule and held together by ligaments.

The joint capsule extends between the apposing ends of the bones forming the joint and is a sleeve of fibrous connective tissue. Reinforcing ligaments of the joint attach to the component bones some distance beyond their articular surfaces. Joints are further strengthened by muscle tendons that play over the joint, as well as by thickened fibrous bands called retinacula that embrace certain joints.

Types of Joints

Joints are classified according to their degree of motion as immovable, slightly movable, or freely movable.

Synarthrosis An example of an immovable joint, or synarthrosis, is a cranial suture. In the newborn child there may be some slight movement between the flat skull bones. However, these bones are later firmly united by dense connective tissue and finally become ossified and fused into a completely rigid structure.

Amphiarthrosis A slightly movable joint such as those between the bodies of adjacent vertebrae is termed amphiarthrosis. This type of articulation is designed for security and strength yet allows for limited motion.

Diarthrosis Most of the joints in the body are of the freely movable type, termed diarthroses. They all possess a joint capsule between the ends of the adjacent bones which is lined by a synovial membrane that secretes a lubricating fluid. The synovial membrane may send expansions into the joint cavity as synovial folds. Since all diarthrodial joints have this synovial membrane, they are also called synovial joints.

Hyaline cartilage lines the apposing ends of bones of a diarthrodial joint to provide a smooth, articular cartilage surface.

The range of movement, however, is not unlimited in a freely moving joint. Movement is limited by (1) the configuration or shape of the bones forming the joint, (2) the tautness or laxity of ligaments and the joint capsule, and (3) the position and action of the muscles which function on a particular joint.

Synovial joints are further classified by the type of movement possible at a given joint. The different types of diarthrodial joint are as follows:

1   A hinge, or ginglymus, joint has one concave and one convex articulating surface. Movement can take place only in one plane. An example is the articulation between the ulna and humerus at the elbow.
2   A pivot joint moves only in the vertical plane, similar to the turning of a key in a lock. The articulation at the elbow that allows us to turn the hand over and back is an example of a pivot joint.
3   A plane, or gliding, joint, which has essentially flattened or slightly curved articular surfaces, allows for sliding movements in all directions. The joints between the carpal bones of the wrist are of this type.
4   A ball-and-socket joint is present in the shoulder or hip joint. The rounded surface of the head (ball) of the femur fits into the acetabulum (socket) of the innominate bone. Such a joint has extensive movement in almost any direction or plane. Combinations of movements at such a joint would inscribe a circle.
5   A condyloid joint is a limited type of ball-and-socket joint, such as the joint between the metacarpals of the hand and the first phalanx of the fingers.
6   A saddle joint, as the term implies, is shaped like a saddle, as in the carpometacarpal joint of the thumb.

The Knee as a Joint

The knee joint has all the essential features of the freely movable joints of the body and therefore is described in some detail (Figure 5-23). This joint is essentially a hinge joint but also has slight pivotal as well as some gliding movement. It has three separate articulating surfaces, the femoropatellar and the two tibiofemoral joints, all of which are covered by hyaline cartilage. The articulating surfaces of the tibia are deepened by two C-shaped fibrocartilaginous disks called the menisci. These cover about two-thirds of the articular surface of the tibia and offer support and protection from the heavy and continuous pressure this joint must bear.

The fact that the rounded condyles of the femur articulate with the relatively flat surface of the tibia makes this joint naturally unstable. Internal and external structures help stabilize it. Initial support is provided by a tough fibrous capsule which is the strongest of all the joint capsules of the body. The capsule is strengthened by (1) a thickened band of deep fascia, the iliotibial tract, (2) expansions of the tendon of the quadriceps femoris into retinacula, (3) the lateral and medial collateral ligaments, (4) posterior reinforcements of the fibrous capsule by ligaments termed the oblique and arcuate popliteal ligaments, and (5) the attachment of thigh muscles to the tibia and fibula. The strong common tendon of the muscles of the front of the thigh forms the patellar tendon (ligament). The patellar bone is embedded in this tendon as a sesamoid bone.

In addition to these external structures there are two exceptionally strong internal

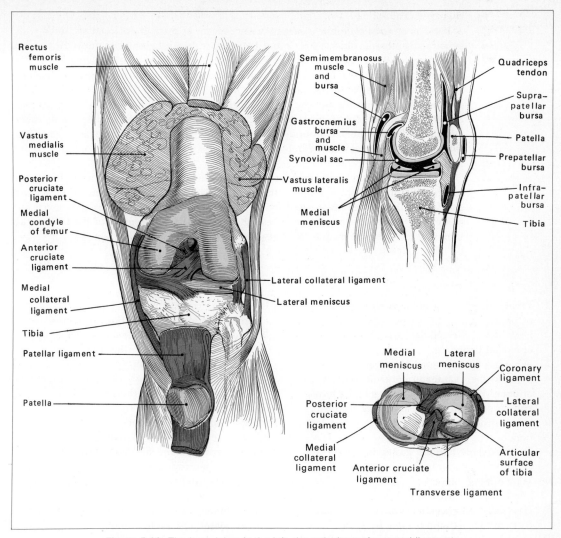

Rectus femoris muscle

Vastus medialis muscle

Posterior cruciate ligament

Medial condyle of femur

Anterior cruciate ligament

Medial collateral ligament

Tibia

Patellar ligament

Patella

Semimembranosus muscle and bursa

Gastrocnemius bursa and muscle

Synovial sac

Vastus lateralis muscle

Medial meniscus

Lateral collateral ligament

Lateral meniscus

Quadriceps tendon

Supra-patellar bursa

Patella

Prepatellar bursa

Infra-patellar bursa

Tibia

Medial meniscus

Lateral meniscus

Coronary ligament

Lateral collateral ligament

Articular surface of tibia

Transverse ligament

Anterior cruciate ligament

Medial collateral ligament

Posterior cruciate ligament

Figure 5-23. The knee joint. At the left, the articular surfaces and ligaments are shown in an anterior view. At the right the upper diagram shows a sagittal section through the joint, to expose muscles, tendons, ligaments, and bursae. At the lower right the articular surface of the tibia is depicted in cross section through the knee joint. In all three sketches articular surfaces are in light color and ligaments are in dark color.

ligaments, the anterior and posterior cruciate ligaments. They pass from the central portion of the head of the tibia to the internal aspect of the femoral condyles and are called cruciate ligaments because they cross one another like the limbs of an X.

The synovial membrane lining the joint capsule forms the most extensive synovial cavity in the body. It communicates with numerous bursae around the knee.

BURSAE

A bursa is a modification of a serous cavity present as a small enclosed sac lined with synovial membrane. Bursae are found between certain adjacent structures that move against each other. They function to reduce friction between these moving parts. For example, the subcutaneous prepatellar bursa is found between the skin over the kneecap and the underlying patella. A submuscular bursa lies between the lower portion of the gluteus maximus muscle and the bony ischial tuberosity. Additional bursae lie deep to or between the tendons of the thigh muscles which cross the knee to attach to the tibia and fibula. Several bursae around the knee communicate with its synovial cavity. The lining of these communicating bursae is continuous with the synovial membrane of the joint cavity (Figure 5-23).

Tendon sheaths are modifications of bursae which surround long tendons of certain muscles, particularly those crossing the wrist and ankle joint. These differ from bursae in that they have been invaginated by the muscle tendons so that a closed synovial sac completely surrounds the tendon. These sheaths facilitate movement of the tendons as they pass between the end of the forearm and leg to the fingers or toes.

QUESTIONS AND PROBLEMS

1 What is the structure of the intercellular matrix of bone?
2 How does intercellular matrix differ from epithelial tissue?
3 What are the functions of osseous tissue?
4 What are the organic and inorganic constituents of bone? Where are they located in the tissue?
5 How does one determine the sex of a skeleton?
6 What structural features characterize a long bone?
7 Describe how a bone grows in length and width.
8 Why doesn't a long bone grow in width proportional to its growth in length?
9 How does endochondral bone formation differ from intramembranous bone formation?
10 Describe the etiology of dwarfism and gigantism.
11 Where is spongy bone found in the adult?
12 Describe the Haversian system of compact bone and its significance in osseous tissue.
13 Discuss the location, function, and composition of bone marrow.
14 Describe the articulation of the bones of the calvaria in the newborn. In the adult.
15 Draw and describe a typical vertebra.
16 What are the dissimilarities of vertebrae of various regions of the vertebral column?
17 What structures traverse the intervertebral foramina?

18 What is meant by a herniated, or "slipped," disk? Why is this condition so painful?
19 Describe the skeletal elements, shape, articulations, and content of the thorax.
20 What are "floating ribs"?
21 How are the bones of the upper extremities attached to the bony thorax?
22 Differentiate between the:
    a Male and female pelvis
    b Acetabulum and glenoid cavity
    c Acromion and olecranon process
    d Carpus and tarsus
    e False and true pelvis
    f "Shinbone" and "kneecap"
    g Infant and adult vertebral column
    h Infant and adult cranium
    i Synovial cavity and bursa
23 Give an anatomical description of "fallen arches," or "flat feet."
24 How does a bone repair itself? Why does one immobilize a bone when a fracture occurs?
25 Describe the various articulations of the skeleton in relation to degree of range of motion permitted. Give specific examples.
26 When continual pressure is applied to bony prominences which are covered by skin only, what is the outcome? How can this condition be prevented?
27 How does bone receive its blood supply? Nerve supply?

28 What is the periosteum? What role does it play in the repair of a fracture?

29 State the bones and their parts involved in the following articulations. Identify the type of joint for each articulation.

a Skull with the vertebral column

b Distal end of the thighbone with the bone(s) of the leg

c Jaw with the skull

d Distal end(s) of the bones of the forearm with the bones of the wrist

e Sixth thoracic vertebra with the seventh thoracic vertebra

f Fourth rib with the sternum and the vertebral column

g Distal ends of the bones of the leg at the ankle

6 ANATOMY OF THE MUSCULAR SYSTEM

Primitive undifferentiated cells have the ability to contract. The cells of muscular tissue develop this property to a high degree and differentiate to become specialized as contractile units.

The fully developed muscle cell is termed a muscle fiber. Three types of muscle fibers are present in the body. Smooth muscle fibers are the predominant tissue in the walls of blood vessels, ducts, and hollow viscera (except the heart) and are usually arranged in sheetlike layers. Their contraction propels the contents of viscera along their course. Cardiac muscle fibers, found only in the heart and in the origins of the large vessels issuing from it, contract to pump the blood through the body. Skeletal muscle fibers make up the various muscles that move the body or its parts.

EMBRYONIC DEVELOPMENT OF MUSCLE

There are two stages in the development of muscle in the embryo. The first stage, histogenesis, involves the development of muscle fibers from primitive cells. The second stage, morphogenesis, entails the organization of muscle fibers into gross muscles.

Histogenesis

Almost all muscular tissue develops from primitive mesodermal cells called myoblasts. Only the intrinsic muscles of the eye arise from another source, namely, primitive ectodermal cells.

Smooth Muscle During the fifth week of embryonic life, changes occur around the epithelial-lined foregut (esophagus) that signal the appearance of smooth muscle cells. Certain mesodermal cells begin to elongate, and within them appear strands termed myofibrils (see page 120). These cells gradually become recognizable as myoblasts. They continue to differentiate and finally assume the typical spindle shape (with nucleus in the center) of the adult smooth muscle cell.

Cardiac Muscle The myoblasts that develop into cardiac muscle cells arise from the mesoderm that embraces the primitive tubular heart. The myofibrils that appear in the cytoplasm develop cross stripes, or striations, characteristic of adult cardiac tissue. Irregular, transverse markings, the intercalated disks, appear as the fiber matures. The nucleus remains in the center of the cell.

Skeletal Muscle Skeletal muscle cells may develop either from somites (see Chapter 4) or from mesodermal cells surrounding the branchial arches. Muscles of the trunk, neck, and limbs (hence most skeletal muscle) grow from somites, more precisely from the central part of the somite, the myotome. These muscles may be referred to as somatic muscles.

Muscles of the head and part of the neck grow from mesoderm surrounding branchial arches and therefore may be called branchiomeric muscles. Microscopically, somatic and branchiomeric muscles are identical. Their histogenesis is also the same. They develop from elongated myoblasts that join to form multinucleated fibers. The fibers are filled with cross-striped myofibrils. Their nuclei are flattened and located on the periphery of the cells.

Morphogenesis

Since it would be impossible to describe the individual development of the more than 650 named muscles in man, we shall restrict our discussion to a few comments on fundamental processes and forces that act upon the emerging muscles to determine their adult form.

Migration Muscles may desert their original site of development and migrate to some remote region. For example, both the diaphragm and the latissimus dorsi muscles arise from neck (cervical) myotomes but migrate inferiorly into the lower thoracic and lumbar regions. Muscles that migrate draw their nerve supply along with them; hence the muscles mentioned are innervated by cervical and not thoracic or lumbar nerves. With very few exceptions a nerve never abandons the muscle it initially innervates.

Change of Fiber Orientation Nearly all muscle fibers are originally oriented parallel to the long axis of the body. Few fibers retain this orientation.

Fusion Since myotomes are closely packed together, they often fuse to form a single muscle.

Splitting of Myotomes Myotomes may split in either the vertical or longitudinal plane to produce subdivisions of the myotomes.

Transformation of Myotomes All or portions of myotomes may be transformed into ligaments, fasciae, and aponeuroses. Most of these processes are completed in the embryo by the end of the eighth week.

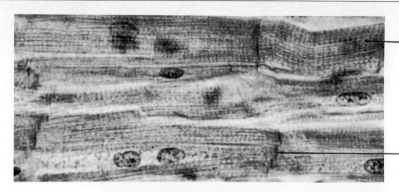

Cardiac fiber

Intercalated disk

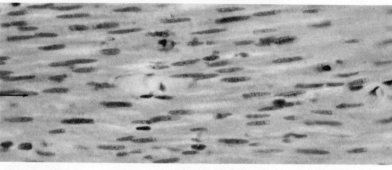

Muscle nuclei

Figure 6-1. Top photomicrograph shows typical features of cardiac muscle: cross striations, central nuclei, abundant sarcoplasm, and intercalated disks. Bottom photomicrograph is of smooth muscle. Note abundance of uniform, elongated nuclei, lack of cross striations, and indistinct cell boundaries.

HISTOLOGY OF MUSCLE
Smooth Muscle
Smooth muscle, also called involuntary or visceral muscle, is structurally the simplest of the muscle types. It is called smooth because it has no visible cross stripes; involuntary because it is not under conscious control; and visceral because it is predominately found in walls of hollow organs. The individual fibers are elongated, tapered, spindle-shaped cells with great range in length (20 to 200 microns), their length depending on the organ (Figure 6-1). In the pregnant uterus the length may be increased to 500 microns. The diameter of the smooth muscle cell may vary from 3 to 9 microns.

A cigar-shaped nucleus lies near the center of the fiber. The cytoplasm, which in muscle cells is called sarcoplasm, appears rather homogeneous even though it is filled with very fine myofibrils. The delicate cell membrane usually cannot be seen with the light microscope.

Skeletal Muscle
Skeletal muscle, also called voluntary or striated muscle, forms the muscles of the body that respond to conscious control.* The typical fiber of skeletal muscle is a giant multi-

*Neither "skeletal" nor "voluntary" is entirely appropriate as a descriptive term. Some skeletal muscles do not attach to the skeleton, while others are not always under the control of the will.

nucleated cell, cylindrical in shape with rounded ends, and enclosed in a distinct cell membrane, the sarcolemma (Figure 6-2). The fiber may attain relatively great length. Flattened, elongated nuclei are located at the periphery of the cell just internal to the sarcolemma. The fiber is filled with myofibrils that are prominently cross-striped. Sarcoplasm fills the limited spaces between the myofibrils.

Study with the electron microscope reveals that each myofibril is actually a complex structure. It contains many short, threadlike objects called myofilaments lying parallel to each other in overlapping arrays. The myofilaments are of two kinds. One is dark and thick and is composed of molecules of the protein myosin. The other is light and thin and is composed of another protein, actin. Dark and light filaments within a fibril lie in a regular pattern, and for that reason the fibril appears cross-striped. Furthermore, the dark and light bands of one fibril are accurately aligned with the dark and light bands of other fibrils in the muscle fiber. Hence the entire fiber appears cross-striped. The bands lie about 1 to 2 microns apart. It should be noted that these bands are actually three-dimensional cylinders and not simply surface markings.

The dark band is called the anisotropic, or A, band because it is doubly refractive

Figure 6-2. Microscopic structure of skeletal muscle. At the upper right-hand corner is a photomicrograph (× 750).

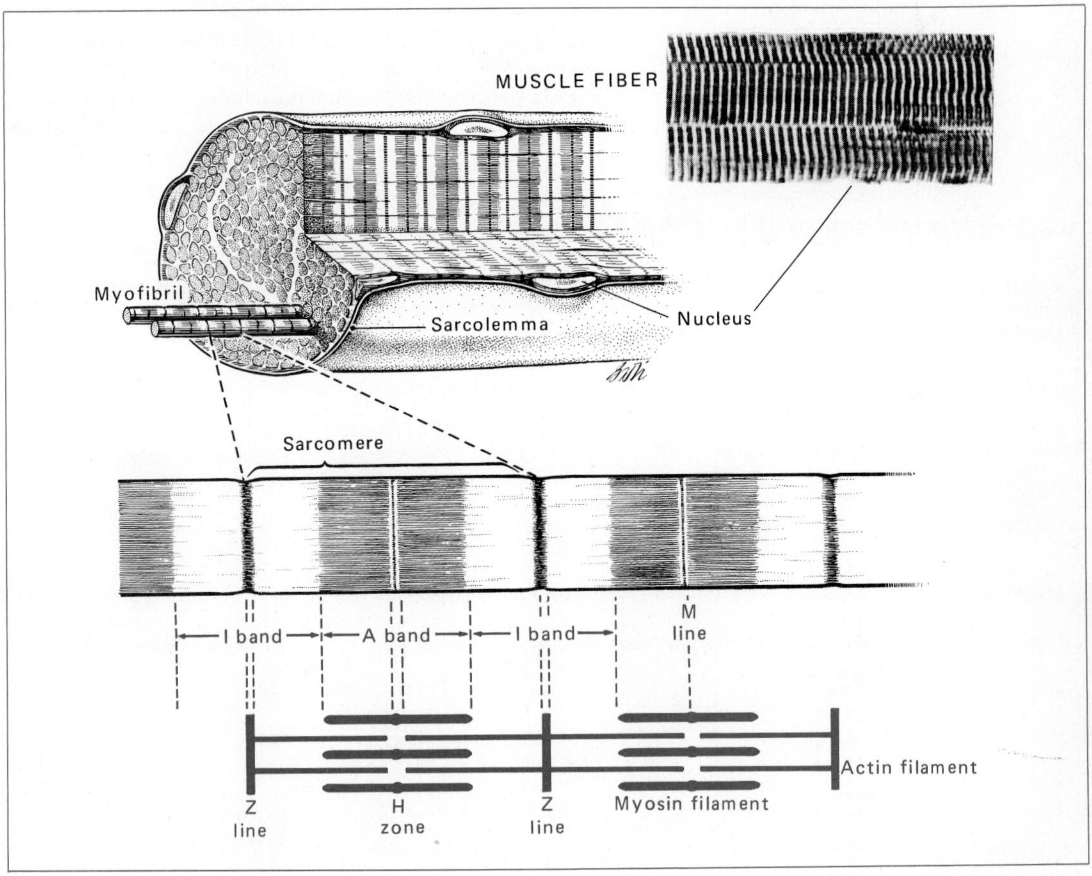

(anisotropic) to polarized light and appears dark in the fresh state. The light, or isotropic (hence I), band is singly refractive to polarized light and is pale in the living fiber.

Cardiac Muscle

Cardiac muscle, forming the bulk of the walls and septa of the heart as well as the origins of its great vessels, contracts to pump blood through the cardiovascular system. Like skeletal muscle the cardiac muscle fiber has distinct cross stripes; however, the myofibrils are more delicate, making the striations less prominent (Fig. 6-1). There is a greater amount of sarcoplasm surrounding the nuclei and the myofibrils than is present in the skeletal muscle fiber. The cardiac muscle fiber has a thin sarcolemma, and the oval, prominent nuclei are somewhat evenly spaced in the center, as they are in smooth muscle fibers.

Cardiac muscle differs from both of the other types in that its fibers are relatively short (50 to 100 microns) and branched in a complicated network.

Under the light microscope, dark lines, the intercalated disks, may be seen passing across the cardiac fiber at uneven intervals. Electron microscopy reveals these disks to be true cell membranes separating adjacent cardiac cells. The characteristics of the three types of muscle fibers are compared in Table 6-1.

Nerve and Blood Supply

All types of muscle are supplied with nerves. Nerve fibers that carry impulses to skeletal muscles terminate in motor end plates, which lie at the junction between nerve fiber and muscle fiber, the so-called neuromuscular junction (see Figure 8-4). In muscles that perform delicate movement, such as muscles of the eye, a neuron may supply just one muscle fiber. In certain other muscles, such as the postural muscles of the back, a neuron, by profuse branching, may supply over a hundred muscle fibers. In contraction all the

TABLE 6-1. COMPARISON OF MUSCLE TYPES

CHARACTERISTIC	SKELETAL MUSCLE	CARDIAC MUSCLE	SMOOTH MUSCLE
Location	Usually attached to skeleton	Heart wall	Walls of hollow organs
Cell shape	Long cylindrical or prismatic fibers	Short branching fibers	Long tapering cells
Sarcolemma	Definite membrane	Less definite	Replaced by delicate surface membrane
Sarcoplasm	Limited amount	Large amount—surrounds nucleus	Small amount
Myofibrils	Striated—fill cell	Striated—almost fill cell	Nonstriated—very fine—fill cell
Nuclei	Multiple at periphery of fiber	Several in center of fiber	Single—in middle of cell
Intercalated disks	None	Present	None
Blood supply	Very good	Profuse—double that of skeletal muscle	Fair
Type of contraction	Voluntary—often vigorous	Involuntary—rhythmic	Involuntary—sluggish—often rhythmic

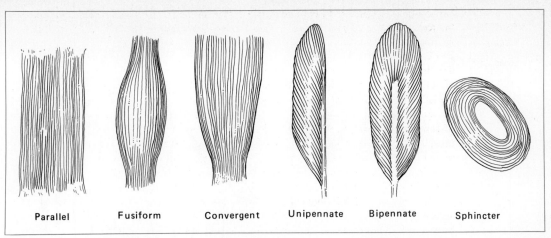

| Parallel | Fusiform | Convergent | Unipennate | Bipennate | Sphincter |

Figure 6-3. Types of muscle architecture, showing classification by orientation of the fibers.

muscle fibers supplied by a neuron act; that is, an impulse traveling over a single nerve fiber may cause contraction in anywhere from 1 to 100 muscle fibers, depending on the muscle involved. Each skeletal muscle of the body contains numerous muscle fibers and hence numerous end plates.

The blood supply to cardiac muscle is profuse, the richest of any muscle in the body. Skeletal muscle is also very vascular but receives only about half the blood flow per unit mass that cardiac muscle receives. Compared with other muscle types, smooth muscle receives only a moderate blood supply.

Muscle Patterns

Fiber directions in the skeletal muscles form patterns that are visible to the naked eye (Figure 6-3). Fibers of a flat muscle are aligned parallel to each other, as in the abdominal wall. The midportion of the muscle may be thickened into a belly, resulting in a fusiform, or cigar-shaped, muscle, as in the biceps brachii. Individual fibers of a muscle may converge from a broad attachment on one bone to a narrow attachment on another. Such a convergent pattern is evident in the deltoid muscle. A muscle may extend along one side of its tendon, passing fibers obliquely into the tendon to form a unipennate muscle. The tendon may be in the center with muscle fibers converging obliquely from both sides to form bipennate muscles. Palmar and plantar interossei muscles are unipennate, while the dorsal interossei are bipennate. Fibers of muscles adjacent to openings of the body may be circularly arranged around the opening to form sphincters.

Connective-Tissue Envelopes

Individual muscles are surrounded by a sheet of connective tissue, termed the epimysium, or fascial envelope. Within this envelope there is additional connective tissue called perimysium that surrounds groups of muscle fibers, each of which is termed a muscle bundle, or fascicle. Within a fascicle an additional microscopic connective tissue sheath called endomysium surrounds and separates individual muscle fibers.

In some areas of the body different groups of muscles are enclosed in envelopes, or septa, of connective tissue. In such compartments are usually found muscles which

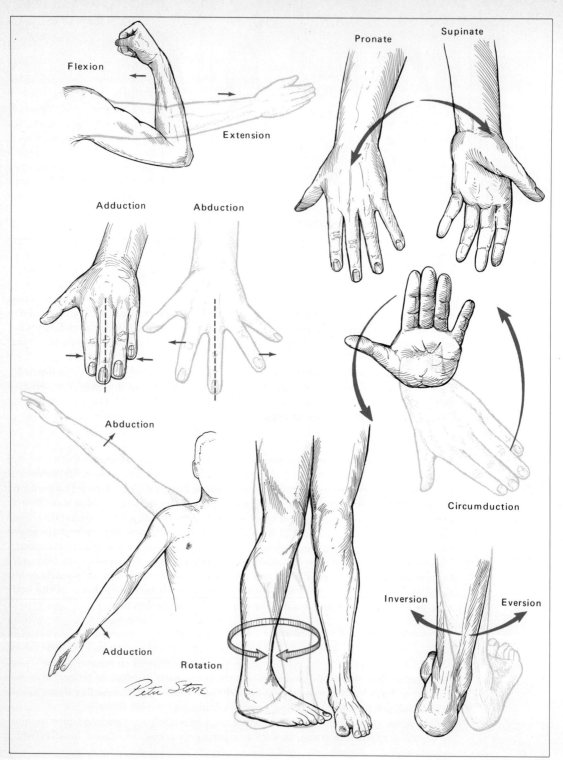

Flexion

Extension

Pronate

Supinate

Adduction

Abduction

Abduction

Adduction

Rotation

Circumduction

Inversion

Eversion

Peter Stone

124

perform a common function. In the thigh, for example, septa separate the extensor, flexor, and adductor groups of muscles.

Attachment of Muscles

Muscles are able to act because they are attached at their ends to movable parts of the body. The attachments are usually to bones, which act as levers in movement, but may instead be to soft tissue, to another muscle, to condensations of connective tissue, or even to skin. Muscles attach to bones by means of their connective tissue components. If the attachment is through the endomysium, the muscle may appear to arise directly from the surface of the bone. However, connective tissue elements usually coalesce at the ends of the muscle, increase in collagen content, and form a distinct cordlike structure called a tendon, which in turn attaches to the bone. In flattened, sheetlike muscle, the connective tissue elements do not converge at the end of the muscle but continue to their attachment in a broad, flat tendon termed an aponeurosis.

Origins and Insertions The attachments of muscles are called origins and insertions. During the primary action of a muscle, the part of the body that moves least is the attachment of origin; the part that moves the greatest distance is the insertion. The origin is usually proximal, or closer to the main mass of the body; the insertion is distal.

Muscles are referred to as being intrinsic or extrinsic. Intrinsic muscles are limited to a given region. In the hand, for example, they originate from, and insert on, bones of the hand. Extrinsic muscles are not so limited. In the hand, extrinsic muscles insert on the bones of the hand but originate from bones in the forearm or arm.

Kinds of Muscle Action Muscles seldom act independently. Any given movement of the body is almost always the result of coordinated action of several muscles. Nonetheless, a major or primary action is ascribed to each muscle of the body. These are the actions listed in the tables accompanying the descriptions of the various muscles. Muscles which act in concert to assist the principal muscle (prime mover) perform a given movement are synergists. Muscles which perform a movement opposite that of synergistic muscles are called antagonists. In order for a muscle to perform a given movement in any part of the body, an adjacent portion of the body must be stabilized to serve as a fixed point for one of the ends of the muscle. Fixating, or stabilizing, muscles (fixators) perform this function. In the various movements around a joint, a muscle at a given time may act as a prime mover, a synergist, an antagonist, or a fixator.

Movement occurring at joints will now be described, the hand being used as an example (Figure 6-4) and the anatomical position as the starting point. The movement of flexion decreases the angle between parts of the body by moving the adjacent parts closer together. The fingers are flexed in making a fist. The action of extension is the opposite of flexion. The extensor muscles of the fingers open or straighten out the closed fist. Abduction moves a part away from the body; adduction moves a part closer. In the hand, abduction spreads the fingers and adduction brings them back together. Rotation turns a part of the body around a pivotal point. Internal, or medial, rotation turns the body member toward the body; external, or lateral, rotation turns it away from the body. Circumduction is a coordinated action in which a cone is circumscribed by the movement of a part. If you stabilize your forearm and move your hand in a circle, you are circumducting your hand. Pronation and supination are actions limited to the forearm and hand. Pronation turns the palms downward; supination turns the palms upward. Inversion and eversion are limited to actions of the foot. Inversion turns the sole inward, or medially,

Figure 6-4 (facing page). Movements and the terms used to describe them.

toward the body; eversion turns the sole of the foot outward, or laterally, away from the body.

Muscle Nomenclature Muscles are usually named for their characteristics or locations. They may also be named according to their attachments, with the origin usually the first part of the name and the insertion the second part. As an example, the sternocleidomastoid originates from the sternum and clavicle (cleido) and inserts into the mastoid process.

Muscles may be named from their position or size. The pectoralis major is the large (major) muscle of the chest (pectoral). They are frequently named according to their shape. The trapezius muscles form a diamond-shaped trapezoid on the upper half of the back. Muscles are also named from their action. The flexor carpi ulnaris is the muscle located on the ulnar side of the forearm which flexes the wrist (carpus).

MUSCLES OF THE FACE AND SCALP
In the following section, general locations and functions of the skeletal muscles will be described. Specific origin, insertion, action, and nerve supply will be given in tables.

Muscles of the Face
The muscles of facial expression (Table 6-2) lie within the superficial fascia of the face and scalp, immediately beneath the skin. One of their attachments is to the deep layer of the skin, the dermis. In contraction the resultant movement of the skin produces a wide range of facial expressions. Individual muscles in this group frequently fuse with adjacent muscles. As a result, though individual actions are ascribed to each muscle, they usually act in concert. All the superficial muscles are innervated by the facial (seventh cranial) nerve. The muscles may be grouped by their association with the major regions of the face, namely, the scalp, eyes, nose, and mouth (Figure 6-5).

Muscles of the Scalp
The frontalis and occipitalis are sometimes listed as individual muscles but may be described as a single muscle, the occipitofrontalis. This broad musculomembranous sheet covers the entire top of the skull. The occipital belly attaches to the skin over the back of the head and the frontal belly to the skin over the forehead. The frontal portion extends to the supraorbital ridge, just above the eyebrow, where it interdigitates with the corrugator muscle. Between the frontal and occipital bellies there is a sheet of fascia, the epicranial aponeurosis. Both muscles act together to move the scalp. Acting by itself, the frontalis raises the eyebrows and forms transverse, or horizontal, wrinkles in the skin of the forehead, as in registering surprise (Figure 6-6).

Three small rudimentary muscles attach to the skin and cartilage of the external ear. Named from their position the superior, anterior, and posterior auricular muscles, they enable wiggling of the ear.

Muscles around the Eye
The sphincter muscle of the eyelids, the orbicularis oculi, encircles the margins of the orbital cavity. It consists of a thin central or palpebral portion and a thicker peripheral or orbital portion. It acts to close the eyelids tightly.

The corrugator muscle, lying deep to the eyebrows, interdigitates with the frontalis and the peripheral portion of the orbicularis oculi. It draws the eyebrows toward each other to furrow the brow in frowning.

TABLE 6-2. MUSCLES OF THE FACE

MUSCLE/NERVE*	ORIGIN	INSERTION	ACTION
Auricularis anterior, superior, and posterior	Temporal fascia, epicranial aponeurosis, and mastoid process	Front of helix, triangular fossa, and convexity of concha	Act in feeble movements of ear
Buccinator	Pterygomandibular raphe, alveolar processes of jaws	Orbicularis oris	Compresses cheek; accessory muscle of mastication
Corrugator	Brow ridge of frontal bone	Skin of eyebrow	Draws eyebrows together
Depressor anguli oris	Oblique line of mandible	Angle of mouth	Pulls corner of mouth downward
Depressor labii inferioris	Mandible adjacent to mental foramen	Lower lip	Draws lower lip downward
Depressor septi	Incisor fossa of maxilla	Septum and side of nose	Draws septum inferiorly, constricts nostrils
Dilator nares	Margin of piriform aperture of maxilla	Margin of nostril	Widens nostril
Frontalis	Epicranial aponeurosis	Skin of forehead	Raises eyebrows, wrinkles forehead
Levator anguli oris	Maxilla adjacent to canine fossa	Corner of mouth	Raises corner of mouth
Levator labii superioris	Maxilla and zygoma adjacent to floor of orbit	Upper lip and margin of nostril	Raises lip, dilates nostril
Mentalis	Incisor fossa of mandible	Skin of chin	Raises and protrudes lip
Nasalis	Maxilla adjacent to canine and incisor teeth	Side of nose above nostril	Draws margin of nostril toward septum
Occipitalis	Occipital bone	Epicranial aponeurosis	Moves scalp
Orbicularis oculi	Medial palpebral ligament, frontal, maxillary, and zygomatic bones	Skin encircling eye, lateral palpebral ligament	Closes eyelids, tightens skin of forehead
Orbicularis oris	Muscles adjacent to mouth	Muscles interlace to encircle mouth	Closes and purses lips
Platysma	Superficial fascia of upper chest	Skin over mandible, cheek, and mouth	Depresses lower jaw, ridges skin of neck
Procerus	Lower part of nasal bone, lateral nasal cartilage	Skin between eyebrows	Wrinkles skin over bridge of nose
Risorius	Fascia over masseter muscle	Angle of mouth	Retracts angle of mouth
Zygomaticus	Zygomatic arch	Corner of mouth	Raises corner of mouth

*All muscles are innervated by the facial (VII) cranial nerve.

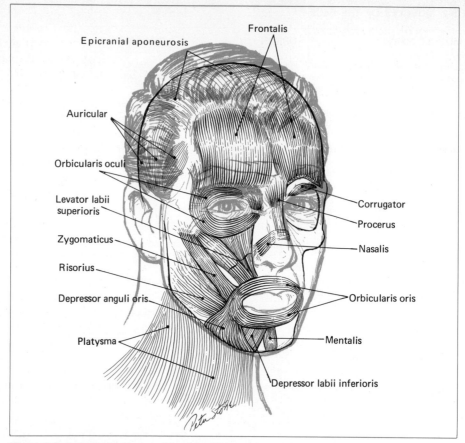

Figure 6-5. Superficial muscles of the face and neck.

Figure 6-6. (*Left*) The action of the frontalis muscle. **Figure 6-7.** (*Right*) The action of the platysma muscle.

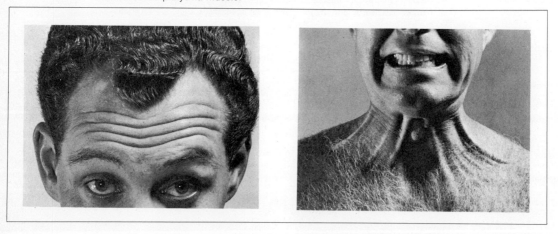

Muscles around the Nose

Four small slips of muscle are associated with the nose. The procerus covers the nasal bones at the bridge of the nose. Acting with the corrugator it produces transverse wrinkling of the skin between the eyebrows to register a fierce expression. The nasalis lies along the side of the nose. It acts to narrow the nostrils by drawing the lateral margin of the nostril toward the nasal septum. The dilator nares, adjacent to the nostril, has the opposite action of enlarging or flaring the nostril. As its name implies, the fourth muscle, the depressor septi, depresses the septum as the nostrils are constricted.

Muscles around the Mouth

The most prominent muscle of those around the mouth is the orbicularis oris. It is the thickest of all the facial muscles and makes up the bulk of the lips. In encircling the lips, its fibers interdigitate with muscles attaching to the lips. It is used to protrude the lips, as in kissing, and to compress the lips over the teeth.

Two quadrilaterally shaped muscles are associated with the lips. The more extensive levator labii superioris, assisted by two smaller muscles, raises the upper lip. The smaller depressor labii inferioris depresses and everts the lower lip to express terror or grief.

Four muscles attach to the angle of the mouth. The zygomaticus arises from the zygomatic bone and, in contraction, pulls the corner of the mouth upward and backward, as in smiling. The risorius retracts the angle of the mouth in grinning. When contracted, the levator anguli oris, arising from the maxilla and lying deep to the levator labii superioris, raises the corner of the mouth. The triangular depressor anguli oris extends inferiorly from the angle of the mouth to the mandible. It draws the angle of the mouth downward in expressing sadness.

Closely associated with the mouth is the largest of the oral muscles, the buccinator ("trumpeter"), which in contraction gives support to the cheek. This muscle draws the cheek inward, as in sucking and during mastication. Posteriorly the buccinator merges with the superior constrictor muscle of the pharynx.

The small, relatively thick mentalis muscle covers the front part of the chin. It protrudes the lower lip and wrinkles the skin of the chin, as in expressing doubt. If the muscles of both sides are well developed, a depression may appear between them, the dimple in the chin.

The most extensive, yet the thinnest, of the facial muscles is the platysma, which lies deep to the skin over the entire anterior half of the neck. Contraction of this muscle forms ridges in the skin of the neck in grimacing (Figure 6-7).

EXTRINSIC EYE MUSCLES

Two pairs of extrinsic eye muscles arise from a fibrous ring of connective tissue—the annular tendon, lying at the rear of the orbital cavity—and insert in the sclera of the eyeball (Figure 6-8, Table 6-3). These short straight (rectus) muscles are named according to the position of the muscles relative to the eyeball and the direction in which each moves it. The superior rectus crosses the upper portion of the orbital cavity and turns the eyeball upward. The inferior rectus crosses the lower portion of the cavity and moves the eyeball downward. The medial rectus extends medially through the cavity and turns the eyeball toward the nose, and the lateral rectus extends laterally and moves the eyeball laterally.

Arising just above the annulus, the superior oblique extends in the same direction as the recti muscles. Anteriorly, near the margin of the orbital cavity, its tendon passes

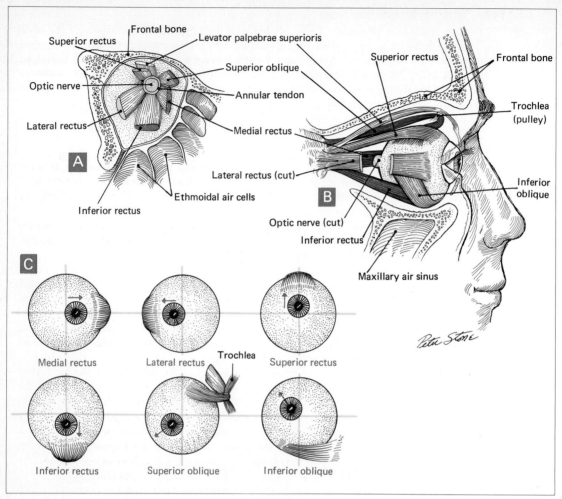

Figure 6-8. The extrinsic eye muscles. **A.** The eyeball has been removed to show the origin of the muscles in cross section. **B.** Lateral view. **C.** Action of the extrinsic eye muscles. Arrow indicates direction of movement of the eyeball.

through a small fascial pulley, the trochlea, reverses its direction, and inserts onto the upper aspect of the eyeball.

The inferior oblique is situated apart from the other eye muscles. It arises in the floor near the front of the orbital cavity and passes obliquely laterally, coursing below the lateral rectus, to insert close to the attachment of the superior oblique.

Only the medial and lateral rectus muscles have a straight pull on the eyeball and hence act in pure abduction and adduction. Most of the time the extrinsic muscles of the eye act synergistically, in the following combinations:

Adduction—medial, superior, and inferior recti
Abduction—lateral rectus and both obliques

TABLE 6-3. EXTRINSIC MUSCLES OF THE EYE

MUSCLE	NERVE	ORIGIN	INSERTION	ACTION
Inferior oblique	Oculomotor	Floor of orbital cavity at anterior margin	Between insertion of superior and lateral recti	Aids in elevation, abduction, and lateral rotation
Levator palpebrae	Oculomotor	Roof of orbital cavity	Tarsal plate of upper lid and superior fornix of conjunctivum	Raises upper lid
Rectus: inferior, lateral, medial, superior	Oculomotor (except lateral, innervated by abducens)	All recti muscles arise from tendinous cuff around optic foramen and insert, by aponeurosis into sclera just posterior to corneoscleral junction, on inferior, medial, lateral, and superior aspects, respectively		Medial and lateral recti act in pure abduction and adduction; others act in synergy (see text)
Superior oblique	Trochlear	Roof of orbital cavity anterior to optic foramen	Slender tendon passes through fibrous pulley and reverses direction to insert deep to superior rectus	Aids in abduction, depression, and medial rotation

Elevation—superior rectus and inferior oblique
Depression—inferior rectus and superior oblique
Medial rotation—superior rectus and superior oblique
Lateral rotation—inferior rectus and inferior oblique

The final extrinsic muscle of the eye, the levator palpebrae superioris, arises at the apex of the orbital cavity and extends forward above the superior rectus. As it reaches the eyelid, it widens into an aponeurotic sheet to insert into the tarsal plate and skin of the upper lid. Contraction of the levator palpebrae superioris raises the upper lid.

The trochlear (fourth cranial) nerve supplies the superior oblique muscle, the abducens (sixth cranial) nerve supplies the lateral rectus, and all other extrinsic muscles of the eye are supplied by the oculomotor (third cranial) nerve.

MUSCLES OF MASTICATION

Four strong muscles attach to the mandible to move the lower jaw in the action of chewing or mastication (Figure 6-9, Table 6-4). Two of the four muscles, the temporalis and masseter, act primarily to close the jaw, as biting muscles.

In forcibly clenching the jaw, the two biting muscles are easily palpable. The masseter can be felt over the angle of the mandible and the temporalis, above and in front of the ear, in the region of the temple.

The principal action of the other two, the medial and lateral pterygoids, is to move the jaw from side to side in grinding movements, as well as to open the mouth.

The lateral pterygoid lies superior to the medial pterygoid as it attaches to the neck of the mandible. Muscles of mastication are innervated by the trigeminal (fifth cranial) nerve.

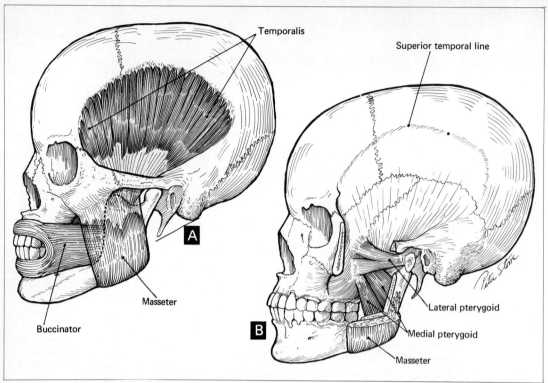

Figure 6-9. Muscles of mastication. **A.** The buccinator, an accessory muscle of mastication, has been included. **B.** The mandible has been sectioned to reveal the pterygoid muscles.

TABLE 6-4. MUSCLES OF MASTICATION

MUSCLE/NERVE*	ORIGIN	INSERTION	ACTION
Lateral pterygoid	Pterygoid plate and greater wing of sphenoid	Condyle of mandible	Opens and protrudes mandible and moves it side to side
Masseter	Zygomatic arch	Ramus of mandible	Raises and protracts mandible
Medial pterygoid	Maxillary tuberosity and lateral pterygoid plate	Medial surface of angle of mandible	Closes and protrudes mandible
Temporalis	Temporal fossa	Coronoid process of mandible	Raises and retracts mandible

*All the muscles of mastication are innervated by the trigeminal nerve.

MUSCLES OF THE TONGUE

The tongue is a muscular organ which functions in speaking, mastication, and swallowing. Paired groups of intrinsic and extrinsic muscles make up its musculature (Table 6-5). A median fibrous septum divides the tongue in the midline and is attached underneath to the hyoid bone.

The intrinsic muscles course in three different planes. They consist of vertical, horizontal, and longitudinal bundles which interdigitate with the extrinsic musculature. This group acts to alter the shape of the tongue.

The extrinsic muscles control the various movements of the tongue such as protrusion, retraction, and movement from side to side (Figure 6-10). The genioglossus arises from the inner surface of the mandible at the tip of the jaw and fans out backward to form the bulk of the substance of the tongue. The thin, vertical hyoglossus lies on the lateral aspect of the genioglossus. From the styloid process of the temporal bone the styloglossus passes anteriorly to insert into the lateral margin and tip of the tongue. An additional extrinsic muscle, the palatoglossus, is described below with the soft palate.

TABLE 6-5. MUSCLES OF THE TONGUE AND SOFT PALATE

MUSCLE	NERVE	ORIGIN	INSERTION	ACTION
Genioglossus	Hypoglossal	Genial tubercle of mandible	Ventral surface of tongue; body of hyoid	Protrudes, retracts, and depresses tongue
Hyoglossus	Hypoglossal	Body and greater horn of hyoid bone	Sides of tongue	Depresses and draws tongue laterally
Styloglossus	Hypoglossal	Styloid process	Sides of tongue	Retracts and elevates tongue
Palatoglossus	Pharyngeal plexus	Soft palate	Dorsum and sides of tongue	Elevates tongue and narrows fauces
Longitudinalis, transversus and verticalis linguae	Hypoglossal	Form intrinsic musculature of tongue, named according to relationship, and act to alter shape of tongue		
Tensor veli palatini	Trigeminal	Scaphoid fossa, cartilage of pharyngotympanic tube	Tendon hooks around hamulus to insert into soft palate	Tenses soft palate
Levator veli palatini	Pharyngeal plexus	Temporal bone, cartilage of pharyngotympanic tube	Midline of soft palate	Elevates palate
Palatopharyngeus	Pharyngeal plexus	Soft palate	Thyroid cartilage and wall of pharynx	Elevates pharynx, helps close naso-pharynx, aids in swallowing
Uvulus	Pharyngeal plexus	Palatine aponeurosis	Mucous membrane of uvula	Elevates uvula

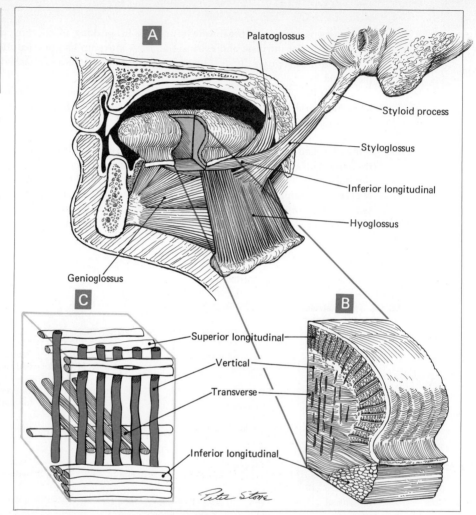

Figure 6-10. Muscles of the tongue. **A.** Extrinsic muscles. **B.** Intrinsic muscles shown in a section of the tongue. **C.** The arrangement of intrinsic muscles in planes.

MUSCLES OF THE SOFT PALATE

The levator veli palatini, arising in the region of the pharyngotympanic (eustachian) tube, descends obliquely to interdigitate with the muscle of the opposite side and forms the bulk of the soft palate (Figure 6-11). Arising from the scaphoid fossa the tensor veli palatini passes vertically downward as its rounded tendon hooks around the hamulus of the pterygoid plate to contribute to the formation of the soft palate. The tendon of the muscle fans out at right angles to the muscle belly to aid in the formation of the palatine aponeurosis. Paired intrinsic muscles of the soft palate, the uvulae, extend posteriorly at either side of the midline. At the posterior free margin of the soft palate, they hang down in a fingerlike projection, the uvula.

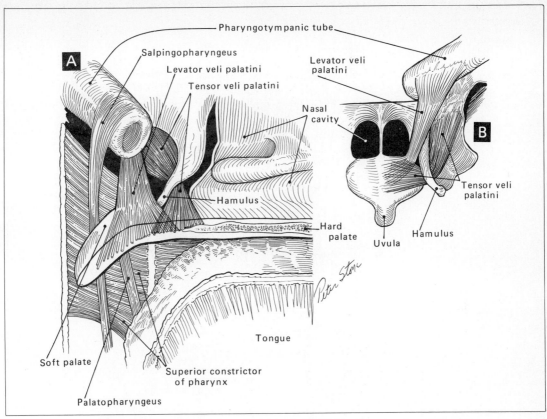

Figure 6-11. Muscles of the soft palate. **A.** Sagittal section, showing relation of the muscles to the pharynx. **B.** Posterior view.

In swallowing, these muscles elevate the soft palate to partially block the communication between the oral and nasal pharynx.

In the lateral wall of the oral cavity, at its junction with the oropharynx, there are two muscles. The palatoglossus extends vertically between the soft palate and the tongue; the palatopharyngeus, between the soft palate and the pharynx. A small area bounded anteriorly by the palatoglossus and posteriorly by the palatopharyngeus is the location of the palatine, or true, tonsil.

MUSCULATURE OF THE NECK

To facilitate the description of the neck, we shall let muscles serve as surface landmarks and subdivide the region into two triangles, an anterior triangle and a posterior triangle. The common boundary between the two is formed by the sternocleidomastoid muscle, which courses obliquely across the neck, from the sternum to the mastoid process. The other boundaries of the anterior triangle are the midline of the neck and the mandible. The sternum is at the apex of the triangle. The other boundaries of the posterior triangle are the clavicle and the anterior border of the trapezius, with its apex at the mastoid process. These larger triangles are further subdivided by their muscular contents into smaller triangles (Figure 6-12).

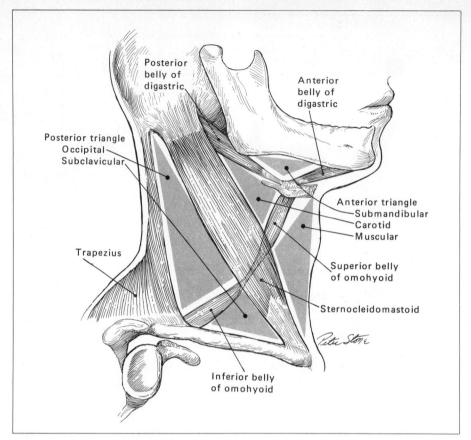

Figure 6-12. Triangles of the neck.

Muscles of the Anterior Triangle

The supra- and infrahyoid muscles within the anterior triangle are named for their relation to the hyoid bone (Figure 6-13, Table 6-6). They help form the floor of the oral cavity and attach to the hyoid bone and thereby aid in the functioning of the tongue as well as the larynx.

Suprahyoid Muscles The suprahyoid muscles are those muscles of the neck above the level of the hyoid bone, namely, the digastric, mylohyoid, and geniohyoid. The digastric has two bellies with an intermediate tendon which passes through a fascial pulley attached to the hyoid bone. Its anterior belly extends forward to the mandible; its posterior belly passes backward to attach close to the mastoid process. The bellies of this muscle together with the lower border of the mandible bound a subdivision of the anterior triangle, the submandibular triangle. Its anterior belly with the hyoid bone and the anterior belly of the muscle of the opposite side form the submental triangle.

From an extensive attachment along the internal surface of the mandible, the mylohyoid sweeps anteromedially to join its companion muscle of the opposite side in a midline raphe, or seam. The mylohyoid forms the greater part of the floor of the oral cavity.

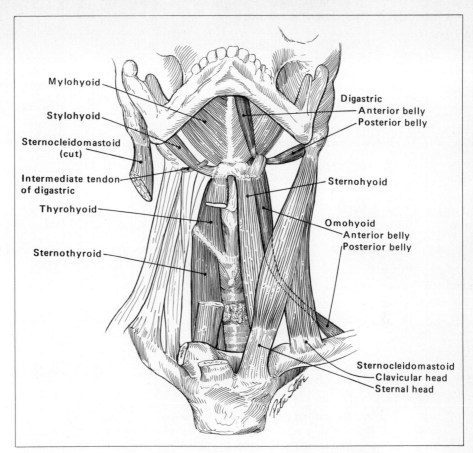

Mylohyoid

Stylohyoid

Sternocleidomastoid
(cut)

Intermediate tendon
of digastric

Thyrohyoid

Sternothyroid

Digastric
Anterior belly
Posterior belly

Sternohyoid

Omohyoid
Anterior belly
Posterior belly

Sternocleidomastoid
Clavicular head
Sternal head

Figure 6-13. The suprahyoid and infrahyoid muscles. Suprahyoid muscles are those above the hyoid bone; infrahyoid muscles are those below it. The hyoid is in the midline slightly above the center.

The small cylindrical geniohyoid lies immediately deep to the mylohyoid. These muscles act to depress the lower jaw and elevate the hyoid bone and thereby raise the larynx and lower portion of the pharynx as well. Thus they play a role in swallowing.

Infrahyoid Muscles   The infrahyoid muscles are four thin ribbon- or straplike muscles embracing the larynx. The sternohyoid is superficial to the sternothyroid as they extend downward from the hyoid bone and thyroid cartilage, respectively, to attach to the sternum. In the same plane as the sternothyroid the thyrohyoid extends superiorly from the thyroid cartilage to attach to the hyoid bone.

The superior belly of the omohyoid muscle diverges from the hyoid bone to pass deep to the sternocleidomastoid. Here it forms an intermediate tendon which passes through a fascial pulley formed on the deep surface of the sternocleidomastoid. The posterior belly continues across the posterior triangle to attach to the scapula. In its course the superior belly, together with the sternocleidomastoid and posterior belly of the digastric, forms the boundary of a superior carotid triangle. With the sternocleidomastoid

TABLE 6-6. MUSCLES OF THE ANTERIOR TRIANGLE

MUSCLE	NERVE	ORIGIN	INSERTION	ACTION
Digastric	Facial, trigeminal	Digastric notch at mastoid process	Mandible near symphysis	Raises hyoid and base of tongue, depresses mandible
Geniohyoid	Ansa cervicalis	Genial tubercle of mandible	Body of hyoid	Elevates hyoid
Mylohyoid	Trigeminal	Mylohyoid line on mandible	Median raphe and hyoid bone	Elevates hyoid and floor of mouth, depresses mandible
Omohyoid	Ansa cervicalis	Medial tip of supra-scapular notch	Lower border, body of hyoid bone	Depresses and retracts hyoid
Sternocleido-mastoid	C_2 and C_3, spinal accessory	Manubrium and medial third of clavicle	Mastoid process and superior nuchal line	Rotates and extends head, flexes vertebral column
Sternohyoid	Ansa cervicalis	Posterior surface of manubrium	Lower border, body of hyoid	Depresses hyoid and larynx
Sternothyroid	Ansa cervicalis	Posterior surface of manubrium	Thyroid cartilage (oblique line)	Depresses thyroid cartilage
Thyrohyoid	Ansa cervicalis	Oblique line of thyroid cartilage	Greater horn of hyoid	Depresses hyoid, elevates thyroid cartilage

and the midline, the superior belly describes the boundary of an inferior muscular triangle. Both of these small triangles are subdivisions of the anterior triangle (Figure 6-12).

The infrahyoid muscles stabilize the hyoid bone for action by the suprahyoid muscles. The sternothyroid depresses the thyroid cartilage, and the thyrohyoid brings the hyoid bone and thyroid cartilage closer together.

Muscles of the Posterior Triangle

Muscles of the posterior triangle lie between the sternocleidomastoid and the trapezius. Six muscles contribute to the floor of the posterior triangle (Table 6-7). They are, from anterior to posterior, the anterior, middle, and posterior scalenes; the levator scapulae; the splenius; and the semispinalis capitis. The scalene muscles extend obliquely from the vertebral column to the first or second rib. A major anatomical significance of the scalene muscles is their relation to the brachial plexus of nerves (see Chapter 9) and the subclavian vessels (see Chapter 17) as these structures pass from the neck into the upper extremity. The brachial plexus and subclavian artery pass over the first rib to course between the anterior and middle scalenes; the subclavian vein crosses the first rib anterior to the anterior scalene. The other muscles in the posterior triangle will be described with the muscles of the back.

Muscles in the posterior triangle attach to the rib cage, act as accessory muscles of respiration, and connect the pectoral girdle to the upper extremity. Through their attachment to the skull or vertebral column they assist the prevertebral and deep muscles of the back and neck in movements of the vertebral column and the head.

TABLE 6-7. MUSCLES OF THE POSTERIOR TRIANGLE

MUSCLE/NERVE*	ORIGIN	INSERTION	ACTION
Levator scapulae	Transverse processes, first four cervical vertebrae	Vertebral border of scapula	Elevates scapula
Scalenus anterior	Transverse processes, third to sixth cervical vertebrae	Scalene tubercle, first rib	Stabilizes or inclines neck to the side
Scalenus medius	Transverse processes, lower five cervical vertebrae	Upper surface, first and second ribs	Stabilizes or inclines neck to the side
Scalenus posterior	Transverse processes, fifth and sixth cervical vertebrae	Outer surface, second rib	Stabilizes or inclines neck to the side
Semispinalis capitis	Transverse processes, upper six thoracic and seventh cervical vertebrae	Skull between superior and inferior nuchal lines	Extends and inclines head
Splenius	Spinous processes, upper thoracic vertebrae	Mastoid process and superior nuchal line	Inclines and rotates head

*Muscles of the posterior triangle are segmentally innervated by cervical nerves.

Pharyngeal Muscles

Three paired muscles compose the bulk of the musculature of the pharyngeal wall (Figure 6-14, Table 6-8). These muscles, the superior, middle, and inferior constrictors, fit into one another like stacked flowerpots. Posteriorly the muscles on either side interdigitate with one another in a midline raphe. Anteriorly they extend to the margins of the larynx, and oral and nasal cavities.

The pharyngeal muscles initiate the voluntary phase of swallowing. When the bolus of food enters the pharynx, progressive circular constriction of these muscles passes it on to the esophagus. Once begun, however, swallowing becomes involuntary and is taken over by smooth muscle in the lower end of the pharynx and esophagus (Chapter 24).

A small muscle that also contributes to the pharyngeal wall is the salpingopharyngeus. It passes inferiorly from the cartilaginous portion of the pharyngotympanic (eustachian) tube (Figure 6-11). It opens the tube during swallowing to help equalize pressure in the middle ear cavity.

Muscles of the Larynx

The larynx has both intrinsic and extrinsic muscles. The extrinsic group, the sternothyroid, thyrohyoid, and inferior constrictor muscles, have been described previously. The small, sheetlike intrinsic muscles are enmeshed within the mucous membrane or ligaments which, with the cartilages, form the walls of this organ (Figure 6-15, Table 6-9). The laryngeal muscles are essentially named from their attachments to the cartilages of the larynx, which are described in Chapter 20.

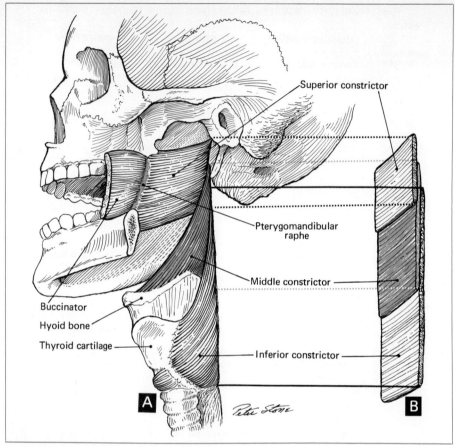

Figure 6-14. Muscles of the pharynx. These muscles initiate the voluntary phase of swallowing.

All but one of the intrinsic muscles of the larynx act by moving the arytenoid cartilages and in that way narrowing the size of the opening between the vocal cords. The one exception is the posterior cricoarytenoid, which widens the laryngeal opening, acting as a "safety muscle." Without the action of this muscle the vocal cords would move together, making it impossible to breathe. Another small muscle, the vocalis, lies parallel to and inserts into the vocal cords and in contraction produces fine modulation of the voice.

MUSCLES OF THE TRUNK

Muscles of the trunk include superficial and deep muscles of the back, superficial muscles of the chest, muscles of the thoracic and abdominal walls, and muscles which compose the floor of the pelvis.

Superficial Muscles of the Back

Although these muscles (Table 6-10) lie on the back, they function as muscles of the superior extremity. They are in two layers. The most superficial layer consists of the

TABLE 6-8. MUSCLES OF THE PHARYNX

MUSCLE/NERVE*	ORIGIN	INSERTION	ACTION
Constrictor			
Inferior	Cricoid and oblique line of thyroid cartilages	Median raphe of pharynx	Constricts pharynx in swallowing
Middle	Horns of hyoid and stylohyoid ligament	Median raphe of pharynx	Constricts pharynx in swallowing
Superior	Medial pterygoid plate, hamulus, pterygomandibular ligament, base of skull	Median raphe of pharynx	Constricts pharynx in swallowing
Palatopharyngeus	Soft palate	Wall of pharynx	Elevates pharynx, closes nasopharynx, aids in swallowing
Salpingopharyngeus	Cartilage of pharyngotympanic tube	Wall of pharynx	Opens pharyngotympanic tube
Stylopharyngeus	Styloid process	Borders of thyroid cartilage, wall of pharynx	Raises and opens pharynx

*Pharyngeal muscles are innervated by pharyngeal plexus formed by vagus and glossopharyngeal nerves.

Figure 6-15. Muscles of the larynx. **A.** Posterior view. **B.** Lateral view. The larynx is shown in sagittal section.

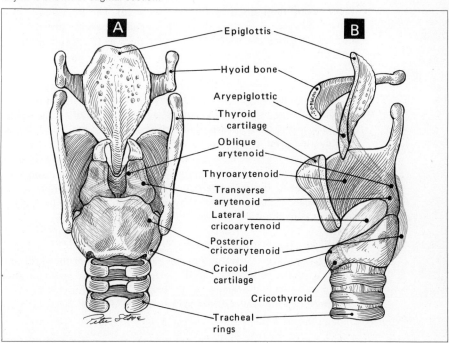

- Epiglottis
- Hyoid bone
- Aryepiglottic
- Thyroid cartilage
- Oblique arytenoid
- Thyroarytenoid
- Transverse arytenoid
- Lateral cricoarytenoid
- Posterior cricoarytenoid
- Cricoid cartilage
- Cricothyroid
- Tracheal rings

TABLE 6-9. MUSCLES OF THE LARYNX

MUSCLE/ NERVE*	ORIGIN	INSERTION	ACTION
Cricothyroid	Arch of cricoid cartilage	Lamina of thyroid cartilage	Chief tensor of vocal cords
Lateral cricoary-tenoid	Upper border of arch of cricoid cartilage	Muscular process of arytenoid cartilage	Closes vocal cords
Oblique arytenoid	Muscular process of arytenoid cartilage	Apex of opposite arytenoid cartilage, prolonged as aryepiglottic muscle	Closes vocal cords
Posterior cricoary-tenoid	Lamina of cricoid cartilage	Muscular process of arytenoid cartilage	Opens vocal cords
Thyroarytenoid	Lamina of thyroid cartilage	Muscular process of arytenoid cartilage	Aids in closure of laryngeal inlet
Transverse aryte-noid	Only unpaired muscle of the larynx. It passes between posterior aspects of arytenoid cartilages		Closes vocal cords
Vocalis	Lamina of thyroid cartilage	Vocal process of arytenoid cartilage	Adjusts tension of vocal cords

*Muscles of larynx except cricothyroid are innervated by inferior laryngeal nerve. Cricothyroid is innervated by external laryngeal nerve.

trapezius and the latissimus dorsi (Figure 6-16). The trapezius is so named because the two muscles join to form a diamond-shaped trapezoid covering the upper portions of the back. It forms the lateral taper of the neck as it sweeps inferiorly from the back of the skull to attach to both the clavicle and scapula. Its action is typified in shrugging the shoulders.

The second muscle of this layer, the latissimus dorsi, covers the lower half of the back. Its presence is responsible for the lateral taper of the chest. Inserting into the upper part of the humerus, it acts in synergy with the pectoralis major in drawing the upraised arm back to the chest, as when one climbs a rope.

Situated deep to the trapezius, the levator scapulae, rhomboideus major, and rhomboideus minor together make up the deeper layer of the superficial muscles of the back. All these muscles attach to the vertebral border of the scapula and act to draw the scapula toward the head and the vertebral column. An additional important action of the superficial muscles of the back is to stabilize the shoulder girdle (scapula and clavicle) during movements of the upper extremity.

Deep Muscles of the Back

The deep, or intrinsic, muscles of the back extend from the skull to the sacrum and act mainly upon the vertebral column. However, the upper segments attach to the skull and therefore also act in movements of the head. The deep muscles are arranged in two groups. The largest group, the erector spinae, forms a large mass of muscle which extends longitudinally along the entire back. The second, or transverse, group of muscles is located in the angle between the spinous and transverse processes of adjacent vertebrae.

The longitudinal erector spinae muscles lie at either side of the vertebral column in a groove between the spinous processes and the angles of the ribs (Figure 6-17, Table 6-11). Laterally they extend about the width of the hand from the midline. Subdivisions, or segments, of the erector spinae are the iliocostalis (the most lateral fiber bundles), the longissimus (intermediate), and the semispinalis (adjacent to the vertebral column). They are further subdivided according to the vertical position they occupy. The lumborum lies in the small of the back, the thoracis in the thoracic region, and the cervicis in the region of the neck. The capitis attaches to the head. For example, in the chest region the most lateral fasciculi would be the iliocostalis thoracis; the middle fasciculi, the longissimus thoracis; and the most medial fasciculi, the spinalis thoracis.

The longitudinal muscles act together to extend or straighten the vertebral column, or singly to twist the column on itself. They also serve to hold the trunk upright and counteract bending caused by extra weight at the front—as in pregnancy.

Rather than existing side by side, the transverse group of muscles are stacked on top of one another like layers of a sandwich (Figure 6-18, Table 6-12). The muscles are much less extensive than the longitudinal group, extending laterally only about a thumb's breadth from the midline. The transverse muscles include the multifidus, the rotatores, the interspinales, and the intertransversarii. The transverse group of muscles accounts for the minimal movement which occurs between adjacent vertebrae.

Superficial Muscles of the Chest

Three superficial muscles located on the chest wall insert into the appendicular skeleton (Figure 6-19, Table 6-13). The pectoralis major arises from abdominal, sternal, and clavicular heads and covers the entire anterior aspect of the chest. (In the female the breast attaches to the fascia covering this muscle.) Its tendon twists slightly upon itself as it inserts on the humerus adjacent to the attachment of the latissimus dorsi. This muscle adducts and flexes the arm and rotates it medially.

Immediately deep to the pectoralis major the much smaller pectoralis minor extends superiorly from the anterior aspect of the third, fourth, and fifth ribs to the coracoid process of the scapula. It functions in forced respiration.

On the side of the chest a broad sheet of muscle, the serratus anterior, covers the rib cage from the third to the eleventh rib. It hugs the side of the chest as it courses posteriorly to attach to the vertebral border of the scapula. It functions to stabilize and hold the scapula against the chest wall.

Muscles of the Thoracic Walls

The muscles which form the walls of both the thoracic and abdominal cavitites are arranged in three consecutively deeper layers.

Muscles of the thoracic wall fill the intervals between adjacent ribs, the intercostal spaces (Figure 6-20, Table 6-14). The external intercostals, the most superficial layer, are in intercostal spaces covering the area between the vertebrae and the costochondral junctions; the internal intercostals, between the sternum and the angle of the ribs.

The third layer of muscle lies on the internal surface of the thoracic cage. It is a discontinuous layer sometimes described as three muscles. The transverse thoracis is in the anterior portion of the rib cage, the innermost intercostal is located laterally, and the subcostalis is adjacent to the vertebral column. Fibers of the transverse thoracis pass more transversely than those of the other muscles of the thorax. The innermost intercostals

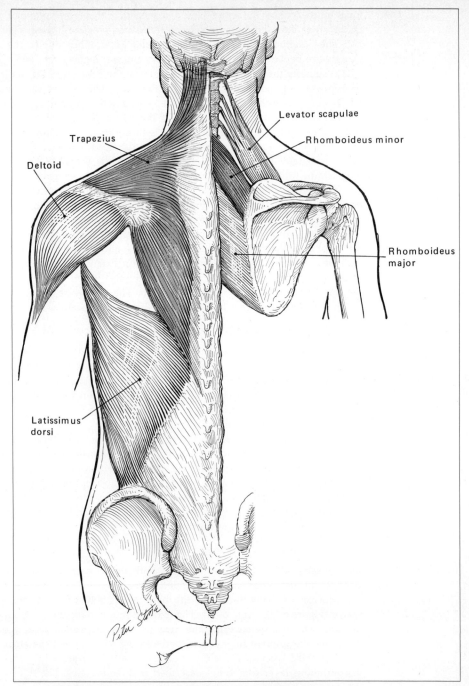

Levator scapulae

Rhomboideus minor

Trapezius

Deltoid

Rhomboideus major

Latissimus dorsi

Figure 6-16. Superficial muscles of the back. At the right, the trapezius has been removed. Compare with the photograph on the facing page. (*Photograph courtesy of R. D. Lockhart, from "Living Anatomy," 6th ed., Faber & Faber, Ltd., London, and Oxford University Press, New York.*)

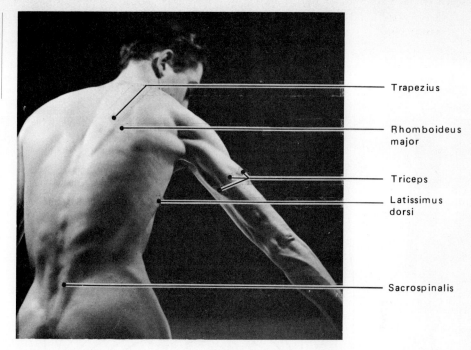

Trapezius

Rhomboideus major

Triceps

Latissimus dorsi

Sacrospinalis

TABLE 6-10. SUPERFICIAL MUSCLES OF THE BACK

MUSCLE	NERVE	ORIGIN	INSERTION	ACTION
Latissimus dorsi	Thoracodorsal	Spinous processes of lower six thoracic vertebrae, lumbodorsal fascia, crest of ilium	Floor of intertubercular groove of humerus	Adducts, extends, and rotates arm (medially)
Levator scapulae	Dorsal scapular	Transverse processes of first four cervical vertebrae	Vertebral border of scapula	Elevates scapula
Rhomboideus major	Dorsal scapular	Spinous processes of second through fifth thoracic vertebrae	Vertebral border of scapula	Adducts and laterally rotates scapula
Rhomboideus minor	Dorsal scapular	Spinous processes of seventh cervical, first thoracic vertebrae	Vertebral border at root of spine of scapula	Adducts and laterally rotates scapula
Serratus anterior	Long thoracic	Lateral surfaces of upper eight ribs	Vertebral border of scapula	Holds scapula to chest wall, rotates scapula in raising arm
Trapezius	Spinal accessory, C_3 and C_4	Superior nuchal line, ligamentum nuchae, spinous processes of seventh cervical and all thoracic vertebrae	Spine of scapula, acromion process, lateral third of clavicle	Adducts, rotates, and elevates scapula

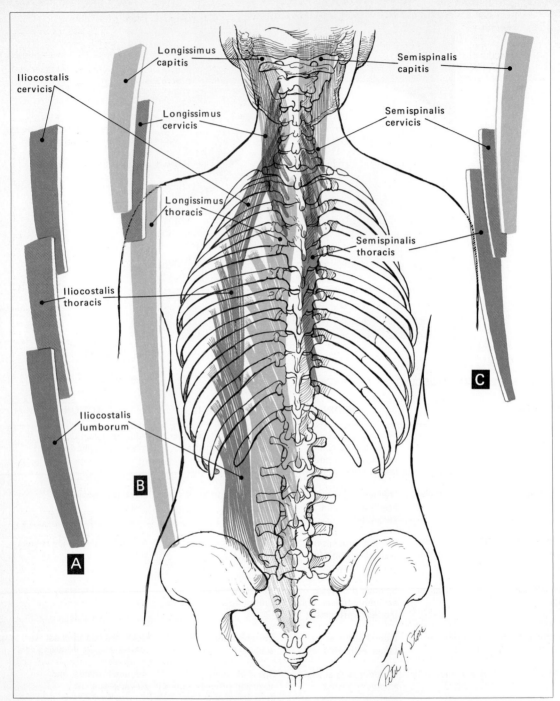

Figure 6-17. Intrinsic muscles of the back. **A.** The overlapping arrangement of the iliocostalis muscles. **B.** The arrangement of the longissimus muscles. **C.** The arrangement of the semispinalis muscles.

TABLE 6-11. DEEP MUSCLES OF THE BACK—LONGITUDINAL GROUP*

MUSCLE/NERVE†	ORIGIN	INSERTION
Iliocostalis thoracis	Lower seven ribs, medial to the angles of the ribs	Angles of upper seven ribs, transverse processes of seventh cervical vertebra
Iliocostalis lumborum	Iliac crest and sacrospinal aponeurosis	Lumbodorsal fascia, transverse processes of lumbar vertebrae, angles of lower six ribs
Longissimus capitis	Transverse processes of cervical vertebrae	Mastoid process of temporal bone
Longissimus cervicis	Transverse processes of upper six thoracic vertebrae	Transverse processes of second through sixth vertebrae
Longissimus thoracis	Sacrospinal aponeurosis, transverse processes of lower six thoracic and first two lumbar vertebrae	Transverse processes of lumbar and thoracic vertebrae, inferior borders of ribs
Semispinalis capitis	Transverse processes of upper six thoracic and seventh cervical vertebrae	Between superior and inferior nuchal lines
Semispinalis cervicis	Transverse processes of upper six thoracic vertebrae	Spinous processes of second through sixth cervical vertebrae
Semispinalis thoracis	Transverse processes of lower six thoracic vertebrae	Spinous processes of upper six thoracic and lower two cervical vertebrae
Spinalis cervicis	Transverse processes of upper six thoracic and seventh cervical vertebrae	Spinous processes of second, third, and fourth cervical vertebrae
Spinalis thoracis	Spinous processes of upper two lumbar and lower two thoracic vertebrae	Spinous processes of second through ninth thoracic vertebrae

*The longitudinal group is a series of muscles forming a mass, the erector spinae, which extends from the sacrum to the skull. Acting unilaterally it bends the vertebral column laterally; bilaterally it extends the vertebral column.

†All are segmentally innervated by posterior primary rami of spinal nerves.

are very thin and can be distinguished merely as fibers lying internal to the intercostal vessels and nerves. Posteriorly the fibers of the subcostalis pass in the same direction as do those of the internal intercostal. The subcostalis may be easily identified, however, because it extends over two or more intercostal spaces, whereas the internal intercostals span only a single space.

Action of the thoracic wall muscles is described in Chapter 21 in the discussion on the mechanism of breathing. During the major portion of their course, vessels and nerves supplying the muscles of the thoracic and abdominal walls pass between the second and third layers of muscles.

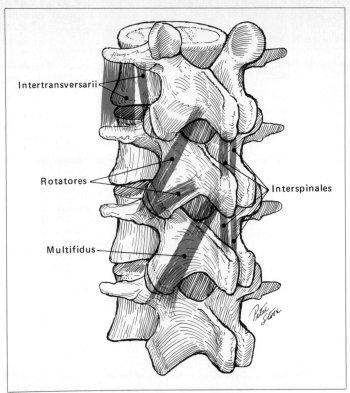

Figure 6-18. A section of the vertebral column showing relation of the transverse muscles of the back to the vertebrae. These muscles are intrinsic muscles of the back.

TABLE 6-12. DEEP MUSCLES OF THE BACK—TRANSVERSE GROUP*

MUSCLE/NERVE†	ORIGIN	INSERTION
Interspinales	Superior surface of spinous process of each vertebra	Inferior surface of spinous process of vertebra above vertebra of origin
Intertransversarii	Extend between transverse processes of cervical, lumbar, and lower thoracic vertebrae	
Multifidus	Sacrum and transverse processes of lumbar, thoracic, and lower cervical vertebrae	Spinous processes of lumbar, thoracic, and lower cervical vertebrae
Rotatores	Transverse processes of all vertebrae below second cervical	Lamina above vertebra of origin

*Transverse group is a series of muscles between adjacent vertebrae which act to slightly rotate, bend, or stabilize vertebral column.

† All are segmentally innervated by posterior primary rami of spinal nerves.

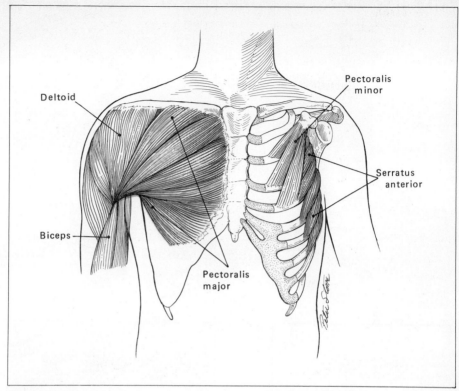

Figure 6-19. Muscles of the pectoral region. At the right the pectoralis major has been removed.

TABLE 6-13. MUSCLES OF THE PECTORAL REGION

MUSCLE	NERVE	ORIGIN	INSERTION	ACTION
Pectoralis major	Lateral and medial pectoral	Medial half of clavicle, sternum, and costal cartilages; aponeurosis of external abdominal oblique	Lateral lip of intertubercular groove on humerus	Flexes, adducts, and medially rotates arm
Pectoralis minor	Medial pectoral	Anterior aspect of second through fifth ribs	Coracoid process of scapula	Draws scapula inferiorly and anteriorly, elevates ribs
Subclavius	Nerve to subclavius	Junction of first rib and costal cartilage	Inferior surface of clavicle	Draws clavicle inferiorly and anteriorly

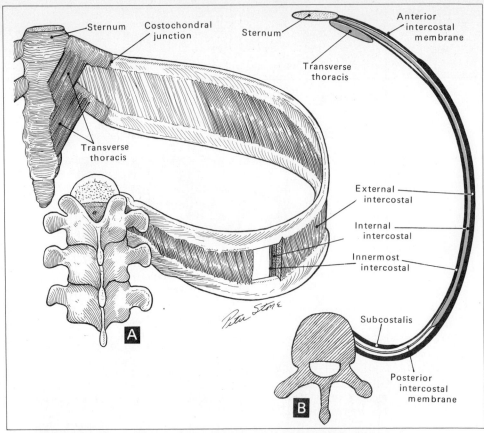

Figure 6-20. Intercostal muscles. **A.** Posterior aspect of the muscles between two ribs. **B.** Cross section, with ribs omitted, to show relationship of the muscles.

TABLE 6-14. MUSCLES OF THE THORACIC WALL

MUSCLE/NERVE*	ORIGIN	INSERTION	ACTION
Intercostals			
External	Inferior border of rib	Superior border of rib below origin	Draws ribs together
Internal	Superior border of rib	Inferior border of rib above origin	Draws ribs together
Innermost	Variable in extent, sometimes considered deep portion of internal intercostal separated by intercostal vessels. Origin and insertion same as internal intercostal		
Subcostalis	Inner surface of lower ribs near angle	Inner surface of second or third rib below rib of origin	Draws adjacent ribs together
Transversus thoracis	Xiphoid process, lower third of sternum	Costal cartilages of second to sixth ribs	Draws costal cartilages downwards, narrows chest

* All are segmentally innervated by intercostal nerves.

150

THE HUMAN BODY

a   If the vertebrae in the middle of the back are crushed, why may it be impossible to move the legs?

b   If the thyroid gland becomes massively enlarged, why may swallowing and breathing be affected?

c   Why is a nonpenetrating blow to the abdomen unlikely to injure the aorta —the large artery that carries blood from the heart through the chest and abdomen?

The answers to these questions, and many others, are anatomical: you can derive them for yourself by examining the "Trans-Vision" plates in this insert. The anatomy of the body is intimately associated with its function, and every educated individual should know something of the interrelationships of the internal organs and structures of the body with their functions.

As you study these "Trans-Vision" plates, keep in mind the *purpose* of each of the organs pictured. If necessary, look up the name of the organ in the index of the text and read about it in the appropriate section; try to remember the names and the positions of the other organs to which it is attached or related. The organization of the body into *systems* is of great help in this. Although the "Trans-Vision" plates do not show the continuity of organ systems completely, you should readily be able to reconstruct this continuity.

The principal systems are these:

1   **The gastrointestinal system:** the mouth and throat; the esophagus; the stomach; the small intestine, into which the pancreas and liver pour their secretions, including the first part, or duodenum; and the large bowel, or colon, which is divided into the ascending, transverse, and descending colons and the rectum.

2   **The cardiovascular system:** the peripheral veins, which empty into the venae cavae that returns blood to the heart; the heart itself; the pulmonary arteries and veins, which lead to and from the lungs; and the aorta, which leads to the peripheral arteries that take blood to the tissues.

3   **The urinary tract:** the kidneys, ureters, and bladder.

4   **The nervous system:** the brain; the spinal cord, which runs from the brain down through the vertebrae; and the peripheral nerves, which run out from the spinal cord to the entire body.

a   If the vertebrae are crushed, the spinal cord may also be damaged and the nerve impulses may be unable to travel from the brain to the legs.

b   If massively enlarged, the thyroid gland, which is just in front of the trachea and esophagus, may compress these structures, thus impeding their functions.

c   That part of the aorta that supplies blood to the abdomen is protected by the muscles of the anterior abdominal wall as well as by the other organs and structures which lie in the abdominal cavity.

(continued on back of insert)

PLATE I
Back view

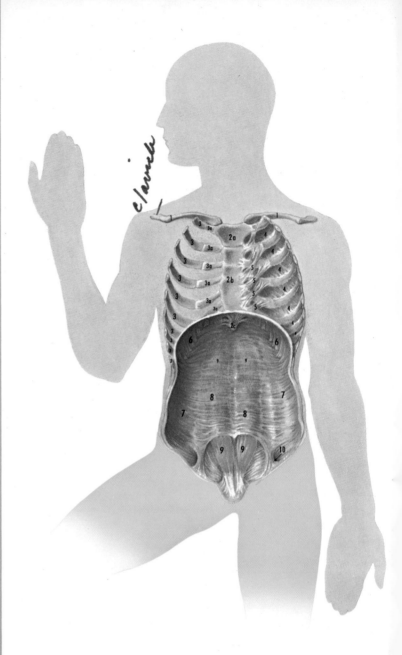

This plate shows the inside
surface of the front wall
of the chest cavity
and the abdominal cavity.

1. Clavicle
2a. Manubrium
2b. Body of the sternum
2c. Xiphoid process
3. Ribs
3a. Costal cartilage
4. Intercostal muscles
5. Transverse thoracic muscles
6. Diaphragm
7. Transverse abdominal
   muscles
8. Sheath of rectus abdominis
   muscles
9. Rectus abdominis muscles
10. Deep inguinal ring

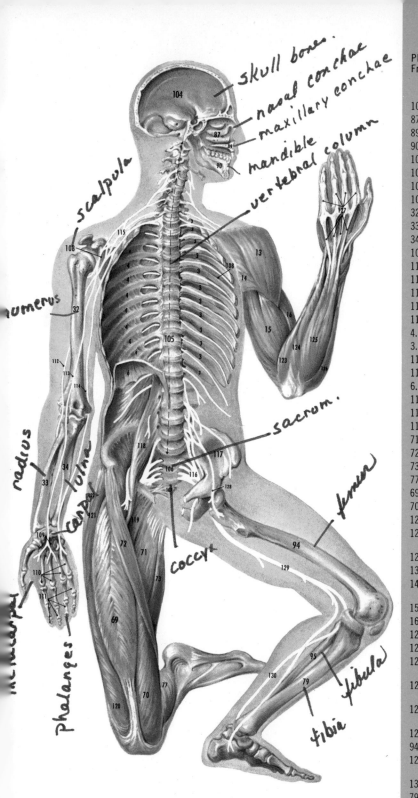

skull bones.
nasal conchae
maxillary conchae
mandible
vertebral column
scapula
humerus
radius
ulna
carpal
metacarpal
Phalanges
sacrum.
coccyx
femur
fibula
tibia

PLATE VI
Front view

104. Skull bones
87. Nasal conchae
89. Maxillary bone
90. Mandible
105. Vertebral column
106. Sacrum
107. Coccyx
108. Scapula
32. Humerus
33. Radius
34. Ulna
109. Carpal bones (wrist)
110. Metacarpal bones
111. Phalanges (finger bones)
112. Radial nerve
113. Median nerve
114. Ulnar nerve
4. Intercostal muscles
3. Ribs
115. Brachial plexus
116. Lumbo-sacral plexus
6. Diaphragm
117. Ilium (pelvic bone)
118. Psoas muscle
119. Pectineus muscle
71. Adductor longus muscle
72. Sartorius muscle
73. Gracilis muscle
77. Gastrocnemius muscle
69. Rectus femoris muscle
70. Vastus medialis muscle
120. Vastus lateralis muscle
121. Tensor fasciae latae
       muscle
122. Gluteus medius muscle
13. Deltoid muscle
14. Pectoralis major muscle
       (cut)
15. Biceps muscle (of arm)
16. Triceps muscle
123. Flexor carpi ulnaris m.
124. Extensor carpi ulnaris m.
125. Extensor digitorum
       communis m.
126. Extensor carpi radialis
       longus m.
127. Extensor tendons
       & sheaths
128. Femoral nerve
94. Femur
129. Peronaeus communis
       nerve
130. Tibial nerve
79. Tibia
95. Fibula

PLATE I shows the inside surface of the front wall of the chest and abdomen as it would be seen from within.

PLATE II shows the organs and structures of the chest and abdomen with the front chest and abdominal walls removed. It also shows the muscles of the arm and back of the neck.

PLATE III shows the organs of the chest and abdomen as seen from the back against the front wall of the body. It also shows the bones and arterial structure of the arm.

PLATE IV shows a cut section of the lungs, the heart, many blood vessels of the chest and abdomen, and the more posterior organs and structures of the abdomen, as well as the musculature of the leg.

PLATE V shows the contents of the chest and abdomen as seen when the structures of the back are removed. It also shows a cross section of the head and the bones and arterial structure of the leg.

PLATE VI shows the structures of the back, including the ribs, as seen with all the contents of the chest and abdomen removed. Some principal structures of the nervous system, the skeleton, the musculature of the arm and leg, and the interior portion of the skull are also shown.

Muscles of the Abdominal Wall (Table 6-15)

Muscles of the anterolateral wall of the abdominal cavity are the external abdominal oblique, internal abdominal oblique, and transversus abdominis. They form flat, wide sheets, lie in three layers, and have a relationship similar to that of the intercostal muscles (Figure 6-21). The fibers of the most superficial muscle, the external oblique, course downward and medially; fibers of the internal oblique course in an opposite direction; and fibers of the transversus abdominis, as its name implies, pass horizontally.

As these muscles pass toward the midline in front, their tendons form broad aponeuroses that insert into the linea alba. The linea alba is a depression that extends along the midline from the xiphoid process to the symphysis pubis.

The lower border of the external oblique aponeurosis folds back on itself to form the inguinal ligament, which extends between the anterior superior iliac spine and the pubic tubercle. This ligament can be felt in the groove at the front of the body between the abdomen and the thigh. Near the midline the ligament partly surrounds the inguinal

TABLE 6-15. MUSCLES OF THE ABDOMINAL WALL

MUSCLE	NERVE	ORIGIN	INSERTION	ACTION
External abdominal oblique	Lower intercostals, subcostal	External surface of lower eight ribs	Anterior half of iliac crest, linea alba	Rotates and flexes vertebral column, tenses abdominal wall
Internal abdominal oblique	Lower intercostals, iliohypogastric, ilioinguinal	Inguinal ligament, iliac crest, lumbodorsal fascia	Lower four ribs, linea alba by conjoint tendon to pubis	Rotates and flexes vertebral column, tenses abdominal wall
Iliopsoas—compound muscle formed from iliacus and psoas major, which join to form iliopsoas tendon				
Iliacus	Femoral	Iliac fossa, lateral aspect of sacrum	Lesser trochanter of femur with psoas major	Flexes and rotates thigh medially
Psoas major	L_2 and L_3*	Transverse processes of bodies of lumbar vertebrae	Lesser trochanter of femur with iliacus	Flexes and rotates thigh medially
Psoas minor	L_1*	Bodies of twelfth thoracic and first lumbar vertebrae	Pectineal line on ilium	Flexes vertebral column
Pyramidalis	Subcostal	Symphysis pubis	Linea alba	Tenses linea alba
Quadratus lumborum	Nerve to quadratus lumborum	Lumbar vertebrae, lumbodorsal fascia, iliac crest	Twelfth rib, transverse processes of upper lumbar vertebrae	Draws rib cage inferiorly, bends vertebral column laterally
Rectus abdominis	Lower intercostals, subcostal	Xiphoid process and fifth to seventh costal cartilages	Pubic crest and symphysis pubis	Flexes vertebral column, tenses abdominal wall
Transversus abdominis	Lower intercostals, iliohypogastric, ilioinguinal	Inguinal ligament, iliac crest, lumbodorsal fascia	Linea alba, conjoint tendon to pubis	Supports abdominal viscera

*Nerves named by the level at which twigs leave the spinal cord.

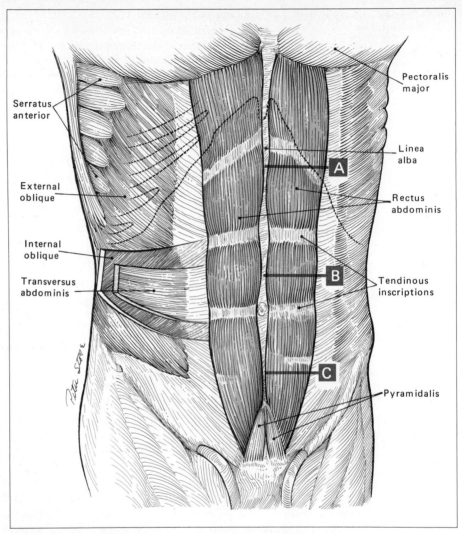

Serratus anterior

Pectoralis major

External oblique

Linea alba

Rectus abdominis

Internal oblique

Transversus abdominis

Tendinous inscriptions

Pyramidalis

Figure 6-21. Muscles of the anterolateral abdominal wall. The letters refer to sections in Figure 6-22. (*Photograph courtesy of R. D. Lockhart, from "Living Anatomy," 6th ed., Faber & Faber, Ltd., London, and Oxford University Press, New York.*)

canal (see Figure 32-4). The internal opening of the canal is the deep inguinal ring; the external opening, the superficial inguinal ring. The canal is only about 4 cm long. In the male it is traversed by the spermatic cord. In the female it transmits the round ligament of the uterus. This opening is a weak area and thus frequently the site of inguinal hernia. A hernia, or rupture, occurs usually during some act of straining, as in lifting a heavy weight, or during a difficult defecation, or in childbirth. In any case, some of the abdominal content is forced through the inguinal canal.

Anteriorly the rectus abdominis forms a vertical muscle mass about 10 cm wide which extends from the symphysis pubis to the sternum and lower margins of the rib

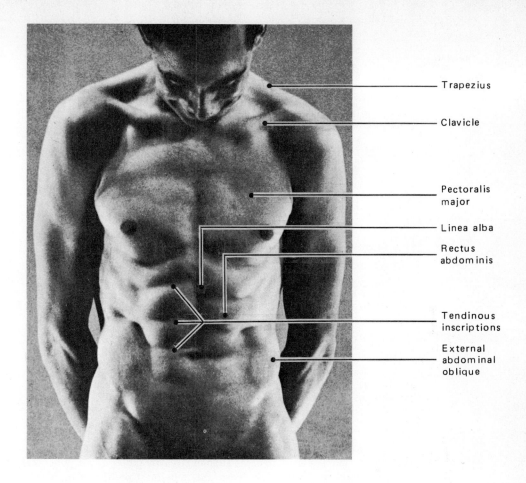

Trapezius

Clavicle

Pectoralis
major

Linea alba

Rectus
abdominis

Tendinous
inscriptions

External
abdominal
oblique

cage. It is enclosed in the rectus sheath formed by the aponeuroses of the anterolateral abdominal muscles (Figure 6-22).

In the midportion of the rectus sheath, that is, the area adjacent to the umbilicus, the aponeurosis of the internal abdominal oblique splits along its length. A portion of it, along with the external abdominal oblique, forms the anterior layer of the rectus sheath; the other portion, with the aponeurosis of the transversus abdominis, the posterior layer of the sheath. Midway between the umbilicus and the symphysis pubis, at the linea semicircularis, all three aponeuroses pass superficial to the rectus abdominis to form a strong anterior layer. Near the xiphoid process the transverse abdominis is absent, and both external and internal oblique aponeuroses form the anterior layer of the sheath.

The pyramidalis is a small triangular muscle which has its base attached to the pubic crest and its apex inserted into the linea alba. In some persons this muscle cannot be distinguished as a separate unit.

Since the abdominal cavity is closed, contraction of the muscles decreases the size of the cavity and thus elevates the intraabdominal pressure. Such an increase in pressure aids defecation, micturition, parturition, and vomiting. The intraabdominal pressure also forces the thoracic diaphragm up, which aids in expiration.

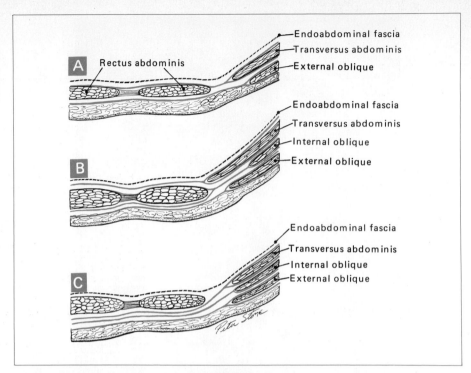

Figure 6-22. The rectus sheath in cross section. Note the contributions of the aponeuroses of the anterolateral abdominal muscles. Letters refer to levels indicated on Figure 6-21.

Additional muscles, the quadratus lumborum and the psoas major and minor, course vertically to complete the wall of the abdominal cavity posteriorly (Figure 6-23). The broad quadratus lumborum extends between the uppermost portion of the pelvic bone (iliac crest) and the last (twelfth) rib. It extends laterally about 10 cm, and at its lateral border a condensation of fascia gives attachment to the three anterolateral abdominal wall muscles.

Medial to the quadratus lumborum a long fusiform muscle, the psoas major, lying adjacent to the vertebral column, extends downward. Its tendon joins that of the iliacus to form the iliopsoas tendon, which exits from the abdominal cavity by passing deep to the inguinal ligament. A second small muscle, the psoas minor, absent in about 40 percent of all persons, lies anterior to the psoas major.

The Thoracic Diaphragm
The thoracic diaphragm (Figure 6-23), customarily called the diaphragm, is a flattened sheet of muscle and tendon separating the thoracic and abdominal cavities. It projects into the thoracic cavity as a double dome. The two lungs are set on the right and left domes of the diaphragm. The heart, lying between the lungs, rests on the central tendon of the diaphragm. In the abdominal cavity the liver, stomach, pancreas, spleen, and kidneys lie adjacent to the diaphragm.

Fibers of the diaphragm arise peripherally from the inner surface of the thoracic outlet and from the upper two lumbar vertebrae as the crura of the diaphragm. They

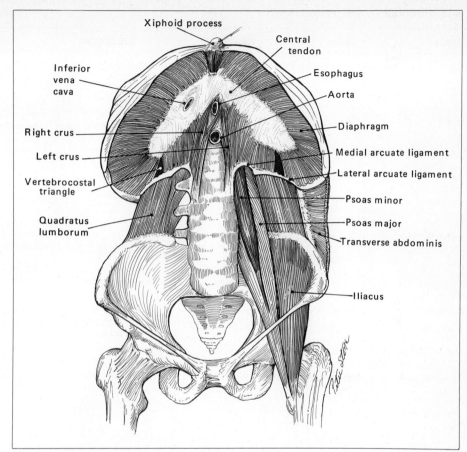

Figure 6-23. Muscles of the posterior abdominal wall and thoracic diaphragm (tipped upward).

extend centrally, where they insert into a flattened, leaf-shaped central tendon. All structures passing between the thoracic and abdominal cavities must pass through the diaphragm. The inferior vena cava passes through an aperture in the central membranous portion at the level of the eighth thoracic vertebra, while the aorta traverses the aortic hiatus at the level of the twelfth thoracic vertebra. Midway between the above openings the esophagus passes through the esophageal hiatus at the level of the tenth thoracic vertebra.

The diaphragm is the primary muscle of respiration. Its action is described in Chapter 21 in the discussion of the mechanism of respiration.

Muscles of the Pelvis (Table 6-16)

The musculature of the pelvis may be divided into three groups: (1) muscles applied to the internal surface of the innominate bone, or hipbone; (2) muscles forming the pelvic diaphragm or floor of the pelvic cavity, namely, the levator ani, the coccygeus, and their fasciae; and (3) the deep transverse perineus muscle and its fascia, which form the urogenital diaphragm.

TABLE 6-16. MUSCLES OF THE PELVIC WALL

MUSCLE	NERVE	ORIGIN	INSERTION	ACTION
Diaphragm, pelvic	Branches of sacral plexus	Composed of ischiococcygeus and levator ani (latter consists of iliococcygeus and pubococcygeus)		Forms floor to support pelvic viscera
Iliacus	Femoral	Iliac fossa, lateral aspect of sacrum	Lesser trochanter of femur with psoas major	Flexes and medially rotates thigh
Iliococcygeus	Branches of sacral plexus	Symphysis pubis and arching tendon over obturator internus	Coccyx and perineal body	Supports pelvic viscera
Ischiococcygeus	Branches of sacral plexus	Ischial spine and sacrospinous ligament	Coccyx	Supports pelvic viscera
Obturator internus	Nerve to obturator internus	Margins obturator foramen and obturator membrane	Greater trochanter of femur	Abducts and laterally rotates thigh
Pubococcygeus	Branches of sacral plexus	Symphysis pubis	Coccyx and perineal body	Supports pelvic viscera
Puborectalis	Branches of sacral plexus	Symphysis pubis	Interdigitates to form a sling which passes behind rectum	Holds anal canal at right angles to rectum
Levator prostatae	Branches of sacral plexus	Symphysis pubis	Prostate gland	Elevates prostate
Sphincter vaginae	Branches of sacral plexus	Symphysis pubis	Interdigitates around and interlaces into vaginal canal	Constricts vaginal orifice

The floor of the pelvic cavity supports the weight of both the pelvic and abdominal viscera. There is a median slitlike gap in the levator ani which permits passage of the urethra in the male and the urethra and vagina in the female. During childbirth muscles of the pelvic floor as well as the deep transverse perineus may be torn.

Muscles Lining the Innominate Bone Below the iliac crest two muscles arise from the internal surface of the innominate bone (Figure 6-24). The first one, the iliacus, contributes to the wall of the lowermost portion of the abdominal cavity, the false pelvic cavity. The iliacus lies in the iliac fossa. Joining the psoas major it forms the iliopsoas tendon, which leaves the abdominal cavity by passing deep to the inguinal ligament to attach subsequently to the upper end of the femur. This muscle acts to flex and rotate the thigh medially.

The second muscle, the obturator internus, covers the innominate bone below the brim of the pelvis, where it contributes to the wall of the true pelvic cavity. It arises from the margins of the obturator foramen, as well as from the obturator membrane which covers the foramen. Its tendon leaves the pelvic cavity through the lesser sciatic foramen to attach to the upper end of the femur. It extends and rotates the thigh laterally.

Muscles of the Floor of the Pelvis The major portion of the pelvic diaphragm or floor is formed by a sheetlike muscle, the levator ani (Table 6-16), which arises on the internal surface of the obturator internis (Figure 6-25). Two muscles, the pubococcygeus and

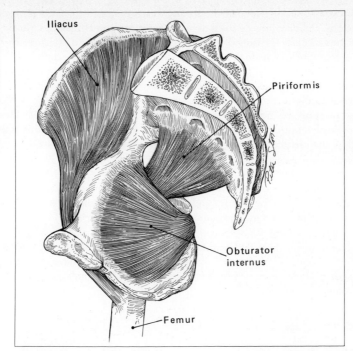

Figure 6-24. Muscles of the pelvic wall. The pelvis is in sagittal section.

Figure 6-25. Pelvic diaphragm. View from the interior of the pelvic cavity. The muscles that form the floor of the pelvic cavity of a male are shown in color.

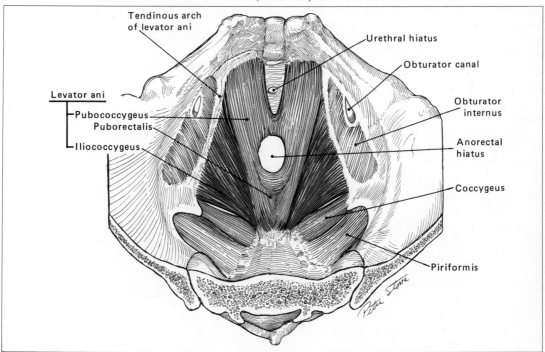

iliococcygeus, subdivisions of the levator ani, are named for their respective origins and insertions. Subdivisions of the pubococcygeus are sometimes described. In the male its most medial fascicles extend posteriorly from the pubis to attach to the prostate gland as the levator prostatae. In the female these fibers attach to the vagina as the pubovaginalis, while additional fibers of the pubococcygeus encircle the vagina to form the sphincter vaginae. A conspicuous thickened band of pubococcygeal fibers, the puborectalis, sweeps posterior to the rectum and interdigitates with fibers of the muscle of the opposite side to form a sling around the anorectal junction. The puborectalis holds these two portions of the large intestine at right angles to each other. In defecation this muscle must relax to allow the rectum and anal canal to straighten out before passage of fecal matter can occur.

The coccygeus (ischiococcygeus), the third muscle contributing to the pelvic floor, is a very thin muscular sheet which overlies the sacrospinous ligament. It extends between the lateral aspect of the sacrum and the spine of the ischium. At its anterolateral border it fuses to, and becomes continuous with, the iliococcygeus.

Muscles of the Urogenital Diaphragm   The urogenital diaphragm spans the gap between the pubic bones. This structure consists of two membranous sheets, the superficial and deep perineal membranes, which enclose muscles and glands associated with the perineal structures (Figure 6-26, Table 6-17).

Figure 6-26. Urogenital diaphragm of a male. The superficial layer has been reflected to expose the muscles of the diaphragm.

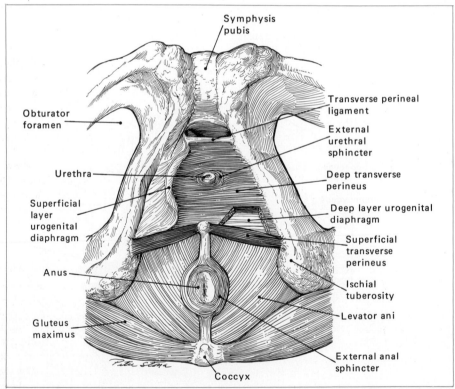

TABLE 6-17. MUSCLES OF THE PERINEUM

MUSCLE/NERVE*	ORIGIN	INSERTION	ACTION
Bulbocavernosus	Male, median raphe over bulb of penis, central tendon of perineum; female, central tendon of perineum	Male, corpus caverno-sus, root of penis; female, dorsum of clitoris, urogenital diaphragm	Male, compresses bulb; female, compresses vaginal orifice
Deep transverse perineus†	Inferior ramus of ischium	Central tendon, external anal sphincter	Fixes central tendon
External anal sphincter	Skin and fascia of anus, tip of coccyx	Central tendon of perineum	Closes anus
External urethral sphincter	Inferior ramus of ischium	Fibers interdigitate around urethra	Closes urethra
Ischiocavernosus	Ischium adjacent to crus of penis or clitoris	Crus near pubic symphysis	Maintains turgescence of penis or clitoris
Superficial trans-verse perineus	Ramus of ischium near tuberosity	Central tendon of perineum	Supports central tendon

*All are supplied by pudendal nerve, except that external anal sphincter is supplied by inferior rectal nerve.
†In females a portion of deep transverse perineal muscle is specialized to form constrictor vaginae, which acts to compress vaginal orifice and greater vestibular glands.

The deep transverse perineus spans the pubic arch between the ischiopubic rami as it lies between the superficial and deep perineal membranes. In both sexes muscle fibers of the central portion of this muscular sheet are modified to encircle the urethra as the external urethral sphincter. The internal urethral sphincter, formed by smooth muscle in the wall of the urinary bladder, is described with the urinary system. In the female, an additional modification of the deep transverse perineus muscle, the constrictor vaginae, encircles the vagina at the urogenital diaphragm.

The small superficial transverse perineus extends transversely along the posterior extent of the urogenital diaphragm to attach to the perineal body. The latter structure lies in the midline between the anterior urogenital triangle and the posterior anal triangle. It also serves as a site of attachment for the bulbocavernosus and external anal sphincter muscles of the perineum.

Muscles of the Perineum

Muscles of the perineum (Table 6-17) are associated with parts of the external genitalia and the anal canal. In the anterior, or urogenital, portion of the perineum thin sheets of muscle, the bulbocavernosus and the ischiocavernosus, form external coverings of the bulb and crura of the penis (Figure 6-27). In the female the bulb of the vestibule, the homologue of the bulb of the penis, lies on the urogenital diaphragm at either side of the vaginal opening. The bulbocavernosus muscle does not interdigitate with its opposite member as it does in the male but is present at either side of the vagina. The ischio-cavernosus in the female differs from the male only in its smaller size, which is to be expected since the crura of the clitoris are smaller than those of the penis.

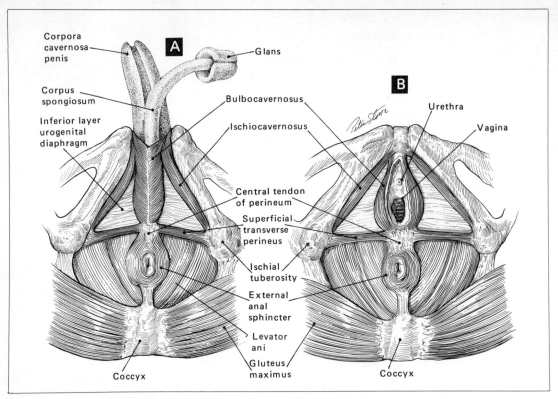

Figure 6-27. Muscles of the perineum. **A.** Male. **B.** Female.

In the anal triangle the external anal sphincter surrounds the anal canal. This muscle consists of three parts. One is the subcutaneous corrugator ani, which attaches to the skin of the anus. The second is an intermediate superficial portion, which fuses with the puborectalis as described with the pelvic musculature. The third is a deeper portion, which fuses with smooth muscle in the wall of the anal canal.

Smooth muscle in the wall of the anal canal forms the internal anal sphincter.

A relatively deep lateral excavation in the anal triangle, the ischiorectal fossa, lies alongside the termination of the large intestine. The fossa is bounded posteriorly by the gluteus maximus muscle and medially by the pelvic floor.

APPENDICULAR MUSCLES
Muscles of the Shoulder (Table 6-18)

Many muscles have been described that effect movement at the shoulder, either directly or indirectly. Yet these were described as muscles of the back or chest. Now we shall describe the definitive muscles of the shoulder: the supraspinatus, infraspinatus, sub-scapularis, teres major and minor, and deltoid (Figure 6-28). The scapula is almost completely covered by the first three of these muscles, the supraspinatus, infraspinatus, and subscapularis. All three muscles arise from the flattened surfaces of the scapula. Posteriorly the supraspinatus fills the supraspinous fossa, the infraspinatus fills the

TABLE 6-18. MUSCLES OF THE SHOULDER

MUSCLE	NERVE	ORIGIN	INSERTION	ACTION
Deltoid	Circumflex	Lateral third clavicle, acromion process, and spine of scapula	Deltoid tuberosity of humerus	Abductor of arm; aids in flexion, extension, and adduction
Infraspinatus	Suprascapular	Infraspinous fossa	Midportion of greater tubercle of humerus	Rotates arm laterally
Subscapularis	Upper and lower subscapulars	Subscapular fossa	Lesser tubercle of humerus	Rotates arm medially
Supraspinatus	Suprascapular	Supraspinous fossa	Superior aspect of greater tubercle of humerus	Initiates abduction of arm
Teres major	Lower subscapular	Axillary border of scapula	Medial lip of inter-tubercular groove of humerus	Adducts and medially rotates arm
Teres minor	Circumflex	Axillary border of scapula	Inferior aspect of greater tubercle of humerus	Laterally rotates and slightly adducts arm

infraspinous fossa, and the subscapularis covers the entire anterior aspect of the scapula as it fills the subscapular fossa. The tendons of these muscles together with the tendon of the teres minor attach to the tubercles on the head of the humerus. As they cross the shoulder joint, they blend with and strengthen the joint capsule.

The teres major and minor both attach to the axillary, or lateral, border of the scapula adjacent to the axilla. As the teres major passes to the humerus to insert into the bicipital groove opposite the attachment of the pectoralis major, it forms the posterior boundary of the axilla (armpit).

The deltoid muscle forms the large mass on the lateral aspect of the upper arm. Passing from an extensive origin on both the spine of the scapula and the clavicle, it inserts onto the deltoid tuberosity of the humerus. It covers the anterior, lateral, and posterior aspects of the shoulder joint. As a readily accessible large muscle, it is clinically important as a site of intramuscular injection.

The supraspinatus abducts the arm. The teres minor and infraspinatus act as lateral rotators and adduct the arm, while the subscapularis and the teres major adduct the arm and rotate it medially. By various combinations of muscular fascicles of the deltoid and because of its broad origin, this muscle can abduct, adduct, flex, extend, and rotate the arm, but its prime function is abduction.

Muscles of the Arm (Table 6-19)

Arm muscles are divided into compartments by intermuscular septa which extend inwardly from the deep fascia covering the arm to attach to the humerus. The anterior compartment contains flexor muscles—biceps, brachialis, and coracobrachialis—which act on both the elbow and shoulder joints. The posterior compartment contains a single muscle, the triceps, which extends the elbow and shoulder joints.

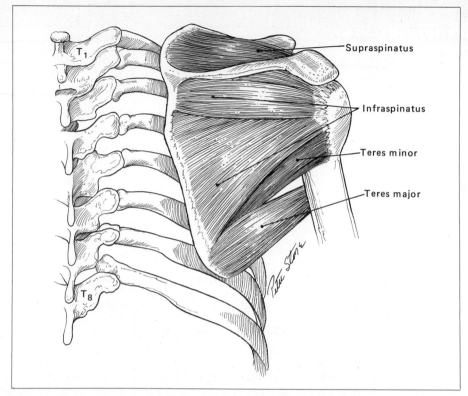

Supraspinatus

Infraspinatus

Teres minor

Teres major

Figure 6-28. Muscles of the shoulder, posterior aspect. The deltoid is visible in the photograph opposite. (*Photograph courtesy of R. D. Lockhart, "Living Anatomy," 6th ed., Faber & Faber, Ltd., London, and Oxford University Press, New York.*)

The biceps brachii has two points of origin, as "biceps" implies (Figure 6-29). The tendon of one head, the long head, arises from the supraglenoid tubercle and is surrounded by a synovial sheath. The long head passes through the shoulder joint, then between the greater and lesser tubercles on the head of the humerus to course along the bicipital groove. The second, or short, head arises from the coracoid process of the scapula and extends along the humerus to join the long head. The two together form the thick belly characteristic of this muscle. A long tendon extends the muscle to the radius. When the forearm is flexed, this tendon can be felt as it crosses the antecubital fossa to its insertion on the radial tuberosity. The main job of the biceps is to flex the forearm, but it assists in supination of the hand.

The brachialis arises from the humerus and inserts into the ulna. It lies deep to the biceps and so forms a bed for the belly of the biceps. It assists in flexion of the forearm.

The coracobrachialis extends from the coracoid process of the scapula to the middle of the medial surface of the humerus. It assists in flexion and adduction of the arm at the shoulder.

The triceps brachii (Figure 6-30) fills the posterior compartment of the arm. Its long head arises from the infraglenoid process at the margin of the glenoid fossa and combines with the lateral and medial heads arising from the shaft of the humerus. The three heads

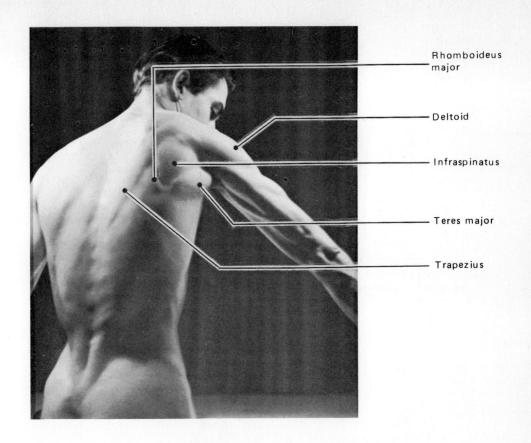

Rhomboideus major

Deltoid

Infraspinatus

Teres major

Trapezius

TABLE 6-19. MUSCLES OF THE ARM

MUSCLE	NERVE	ORIGIN	INSERTION	ACTION
Anconeus	Radial	Lateral epicondyle of humerus	Olecranon process, posterior surface of ulna	Weak extensor of arm
Biceps brachii	Musculocutaneous	Long head, supraglenoid tubercle; short head, coracoid process	Tuberosity of radius	Flexes forearm and arm, supinates hand
Brachialis	Musculocutaneous and radial	Distal two-thirds of humerus	Coronoid process of ulna	Flexes forearm
Coracobrachialis	Musculocutaneous	Coracoid process	Middle third of humerus	Flexes and adducts arm
Triceps brachii	Radial	Long head, infraglenoid tubercle; lateral head, proximal portion of humerus; medial head, distal half of humerus	Olecranon process of ulna	Extends arm and forearm

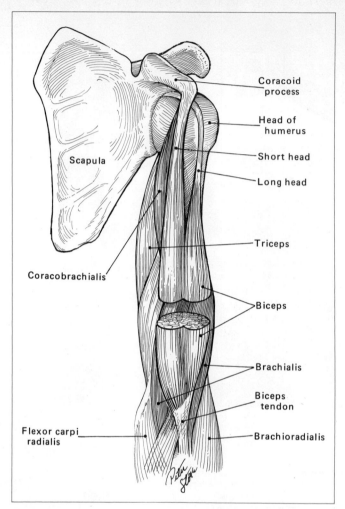

Coracoid
process

Head of
humerus

Short head

Long head

Scapula

Triceps

Coracobrachialis

Biceps

Brachialis

Biceps
tendon

Flexor carpi
radialis

Brachioradialis

Figure 6-29. Muscles of the arm, anterior view.

insert by a common tendon into the olecranon process of the ulna and function to extend both the arm and the forearm.

Muscles of the Forearm

Muscles of the forearm (Figures 6-31 and 6-32) act on both the fingers and the wrist. The bulk of the musculature is concentrated in two masses just below the elbow joint. The mass on the radial, or lateral, side is formed by the bellies of the extensor muscles (Table 6-20); that on the ulnar, or medial, side, by the bellies of the flexor muscles (Table 6-21). Tendons of the flexors pass essentially to the anterior, or volar, surface of the wrist or hand; those of the extensors, to the posterior, or dorsal, surface.

Movement of the Hand by Forearm Muscles   The principal extensors of the hand are two muscles. The extensor carpi ulnaris originates at the lateral epicondyle of the humerus

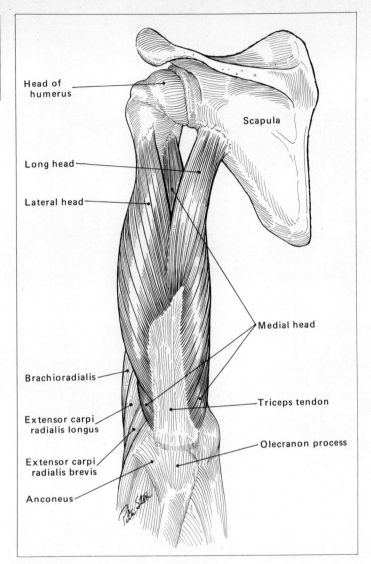

Head of
humerus

Scapula

Long head

Lateral head

Medial head

Brachioradialis

Triceps tendon

Extensor carpi
radialis longus

Olecranon process

Extensor carpi
radialis brevis

Anconeus

Figure 6-30. Muscles of the arm, posterior view.

and inserts on the fifth metacarpal bone. The extensor carpi radialis longus extends from the lateral supracondylar ridge of the humerus to the second metacarpal bone. They are assisted by a third muscle, the extensor carpi radialis brevis.

The principal flexors of the hand are three muscles, all of which arise from the medial epicondyle of the humerus. The flexor carpi radialis extends to the second and third metacarpal bones. The flexor carpi ulnaris extends to the pisiform, hamate, and fifth metacarpal bones. The palmaris longus attaches to the flexor retinaculum (a transverse ligament around the carpal bones) and the palmar aponeurosis.

For abduction of the hand the lateral flexor works with the lateral extensors—the

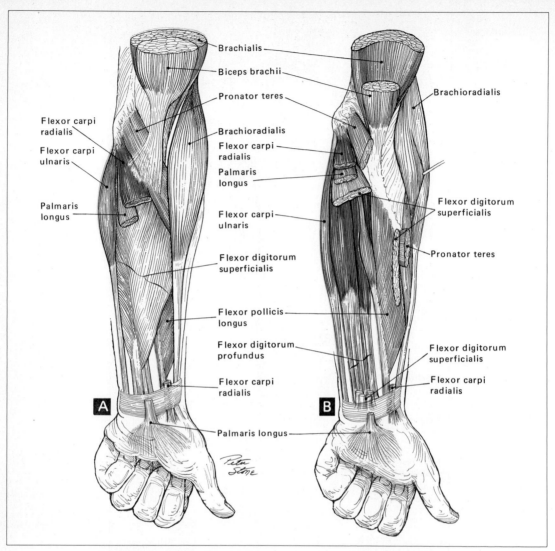

Brachialis

Biceps brachii

Pronator teres

Brachioradialis

Flexor carpi radialis

Flexor carpi ulnaris

Palmaris longus

Brachioradialis

Flexor carpi radialis

Palmaris longus

Flexor carpi ulnaris

Flexor digitorum superficialis

Flexor pollicis longus

Flexor digitorum profundus

Flexor carpi radialis

Palmaris longus

Flexor digitorum superficialis

Pronator teres

Flexor digitorum superficialis

Flexor carpi radialis

A

B

Figure 6-31. Muscles of the anterior aspect of the forearm. Compare with the photograph on the facing page. **A.** The more superficial muscles are shown here, and two are cut to expose the flexor digitorum superficialis. **B.** Deeper-lying muscles, as well as tendons at the wrist. (*Photograph courtesy of R. D. Lockhart, from "Living Anatomy," 6th ed., Faber & Faber, Ltd., London, and Oxford University Press, New York.*)

flexor carpi radialis with the extensor carpi radialis longus and brevis. For adduction, the medial flexor works with the medial extensor—the flexor carpi ulnaris with the extensor carpi ulnaris.

Muscles of the forearm also are responsible for supination and pronation of the hand. The muscle that supinates the hand (that is, turns the palm upward) is the supinator. It begins at the lateral epicondyle of the humerus and also at the supinator crest of the ulna and ends on the distal part of the radius. The biceps assists supination.

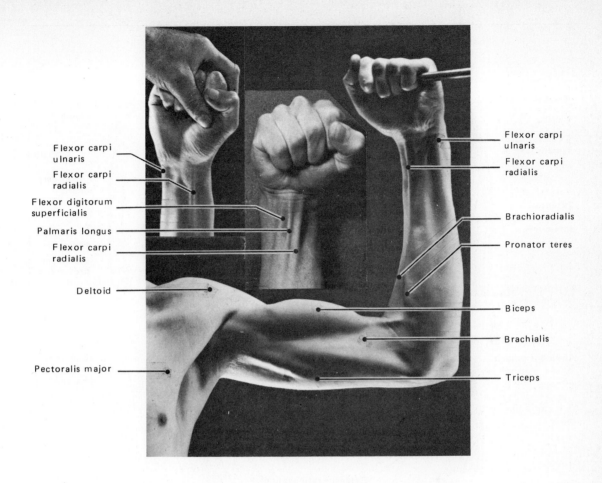

Flexor carpi
ulnaris

Flexor carpi
radialis

Flexor digitorum
superficialis

Palmaris longus

Flexor carpi
radialis

Deltoid

Pectoralis major

Flexor carpi
ulnaris

Flexor carpi
radialis

Brachioradialis

Pronator teres

Biceps

Brachialis

Triceps

Two muscles pronate the hand (turn the palm downward). The most important is the pronator teres. It has origins on both the humerus and the ulna and sweeps down and across the forearm to the radius. The second is the pronator quadratus, a square muscle that links the ulna and radius at the distal end of the forearm.

Movement of the Fingers and Thumb by Forearm Muscles Two muscles, in two layers, are the chief flexors of the fingers, excluding the thumb. Most superficial is the flexor digitorum superficialis, which arises on the radius and inserts on the middle phalanges. Deep to this is the flexor digitorum profundus, which arises on the ulna and inserts on the distal phalanges. In order to gain this insertion the four tendons of the more superficial muscle split to allow passage of the deeper tendons as they pass along the palmar surface of the digits.

The principal extensor of the fingers, again excluding the thumb, is the extensor digitorum communis. It arises on the lateral epicondyle of the humerus and sends tendons to the four fingers. With the hand extended these tendons can be seen as ridges along its back. They broaden into a hood (the extensor expansion) at the metacarpophalangeal joint and continue along the dorsum of the fingers to the distal phalanges.

167

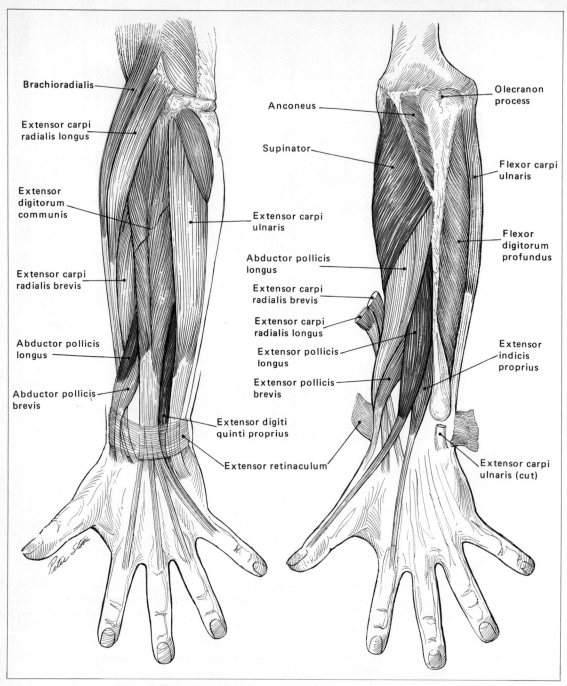

Figure 6-32. Muscles on the posterior aspect of the forearm. Superficial muscles are shown at the left, more deeply placed muscles on the right. Compare with the photograph on the facing page. (*Photograph courtesy of R. D. Lockhart, from "Living Anatomy," 6th ed., Faber & Faber, Ltd., London, and Oxford University Press, New York.*)

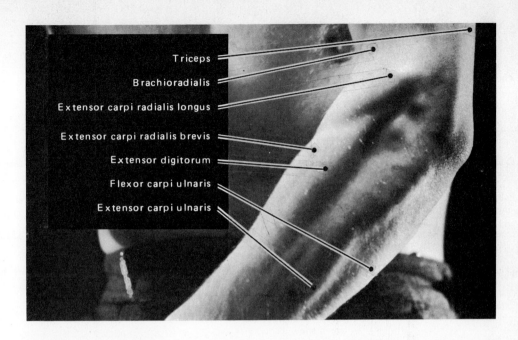

Triceps
Brachioradialis
Extensor carpi radialis longus
Extensor carpi radialis brevis
Extensor digitorum
Flexor carpi ulnaris
Extensor carpi ulnaris

TABLE 6-20. MUSCLES OF THE FOREARM (I)

MUSCLE	NERVE	ORIGIN	INSERTION	ACTION
Flexor carpi radialis	Median	Medial epicondyle of humerus	Second and third metacarpals	Flexes hand and forearm, aids in pronation and abduction of hand
Flexor carpi ulnaris	Ulnar	Medial epicondyle, olecranon process, posterior border of ulna	Pisiform, hamate, fifth metacarpal	Flexes and adducts hand
Flexor digitorum profundus	Median and ulnar	Proximal three-fourths of ulna and adjacent interosseous membrane	Distal phalanges of fingers	Flexes terminal phalanges and hand
Flexor digitorum superficialis	Median	Medial epicondyle, coronoid process, anterior border of radius	Middle phalanges of fingers	Flexes phalanges, hand, and forearm
Flexor pollicis longus	Median	Radius, adjacent interosseous membrane, coronoid process of ulna	Distal phalanx of thumb	Flexes thumb
Palmaris longus	Median	Medial epicondyle of humerus	Flexor retinaculum and palmar aponeurosis	Flexes hand
Pronator quadratus	Median	Distal fourth of ulna	Distal fourth of radius	Pronates hand
Pronator teres	Median	Medial epicondyle of humerus	Middle of radius	Pronates hand

TABLE 6-21. MUSCLES OF THE FOREARM (II)

MUSCLE	NERVE	ORIGIN	INSERTION	ACTION
Abductor pollicis longus	Radial	Posterior surface of ulna, middle third of radius	First metacarpal	Abducts thumb and hand
Brachioradialis	Radial	Lateral supracondylar ridge and intermuscular septum of humerus	Styloid process of radius	Flexes forearm
Extensor carpi radialis brevis	Radial	Lateral epicondyle of humerus	Third metacarpal	Extends and abducts hand
Extensor carpi radialis longus	Radial	Lateral supracondylar ridge of humerus	Second metacarpal	Extends and abducts hand
Extensor carpi ulnaris	Radial	Lateral epicondyle, posterior border of ulna	Fifth metacarpal	Extends and adducts hand
Extensor digitorum communis	Radial	Lateral epicondyle of humerus	Forms extensor expansion over fingers	Extends fingers, hand, and forearm
Extensor digiti minimi	Radial	Extensor digitorum communis interosseous membrane	Extensor expansion of little finger	Extends little finger
Extensor indicis	Radial	Posterior surface of ulna	Extensor expansion of index finger	Extends index finger and hand
Extensor pollicis brevis	Radial	Middle third of radius	Proximal phalanx of thumb	Extends and abducts hand
Extensor pollicis longus	Radial	Middle third of ulna, adjacent interosseous membrane	Distal phalanx of thumb	Extends distal phalanx of thumb, abducts hand
Supinator	Radial	Lateral epicondyle of humerus, supinator crest of ulna	Upper third of radius	Supinates hand

The index finger and the little finger each have an additional extensor muscle. For the index there is the extensor indicis proprius, which arises from the ulna and inserts into the extensor expansion. For the little finger there is the extensor digiti quinti proprius, which arises on the ulna and inserts into the extensor expansion over the little finger.

Four muscles act upon the thumb. For extension there are the extensor pollicis longus (between the ulna and the distal phalanx) and the extensor pollicis brevis (between the radius and the proximal phalanx). The third muscle is an abductor, the abductor pollicis longus, between the radius and the first metacarpal. The fourth muscle is a flexor, the flexor pollicis longus, linking the radius and ulna with the distal phalanx.

Muscles of the Hand (Table 6-22)

Muscles of the hand may be divided into three groups. Muscles which act on the thumb form the thenar eminence, the mass on the lateral side of the palm. Muscles which move

TABLE 6-22. MUSCLES OF THE HAND

MUSCLE	NERVE	ORIGIN	INSERTION	ACTION
Abductor digiti minimi	Ulnar	Pisiform, tendon of flexor carpi ulnaris	Proximal phalanx of fifth digit	Abducts fifth digit
Abductor pollicis brevis	Median	Flexor retinaculum, scaphoid and trapezium	Proximal phalanx of thumb	Abducts and aids in flexion of thumb
Adductor pollicis	Ulnar	Capitate, second, and third metacarpals	Proximal phalanx of thumb	Adducts and aids in apposition of thumb
Flexor digiti minimi	Ulnar	Flexor retinaculum, hook of hamate	Proximal phalanx of fifth digit	Flexes proximal phalanx of fifth digit
Flexor pollicis brevis	Median	Flexor retinaculum, trapezium	Proximal phalanx of thumb	Flexes and adducts thumb
Interossei Dorsal (4)	Ulnar	Adjacent sides of metacarpal bones	Proximal phalanges of index, middle, ring fingers	Abducts fingers to midline of hand (middle finger)
Palmar (3)		Medial side of second, lateral side of fourth and fifth metacarpals	Base of proximal phalanx in line with its origin	Adducts index, ring, little fingers; aids in extension of fingers
Lumbricales (4)	Lateral two by median, medial two by ulnar	Tendons flexor digitorum profundus	Extensor expansion distal to metacarpopha-langeal joint	Flexes metacarpopha-langeal, extends inter-phalangeal joints
Opponens digiti minimi	Ulnar	Flexor retinaculum, hook of hamate	Fifth metacarpal	Draws fifth metacarpal toward palm
Opponens pollicis	Median	Flexor retinaculum, trapezium	Lateral border of first metacarpal	Draws first metacarpal toward palm
Palmaris brevis	Ulnar	Flexor retinaculum	Skin of palm	Wrinkles skin of palm

the little finger form the hypothenar eminence, on the medial side of the palm (Figure 6-33). A third group, the midpalmar muscles, is situated between the metacarpal bones within the width of the palm.

Thenar muscles include the flexor pollicis brevis, the abductor pollicis brevis, the adductor pollicis, and the opponens pollicis. The abductor, flexor, and opponens arise from the lateral carpal bones, while the adductor arises from the second and third metacarpal bones. All the thenar muscles insert into the base of the proximal phalanx of the thumb. The opponens also attaches along the entire length of the metacarpal of the thumb.

Hypothenar muscles consist of the abductor digiti minimi, the flexor digiti minimi, and the opponens digiti minimi. Their origin and insertions on the medial side of the hand are similar to those of the thenar muscles on the lateral side of the hand.

Eleven muscles are found between the thenar and hypothenar mounds (Figure 6-33). Four small slips of muscle, the lumbricales, extend from the four tendons of the flexor

171

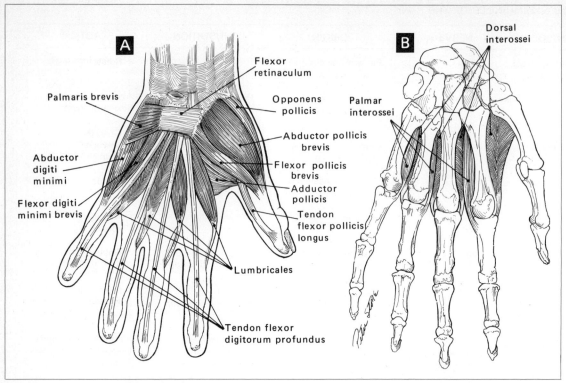

Figure 6-33. Intrinsic muscles of the hand, dorsal view. **A.** Superficial muscles. **B.** Superficial muscles have been omitted so as to reveal interossei.

digitorum profundus. They send delicate tendons into the extensor expansion over the dorsum of the digits and aid in the placing of the fingers at right angles to the palm. The palmar and dorsal interossei are in two layers between the metacarpal bones. In adduction of the fingers the palmar interossei move the index, ring, and little fingers toward the middle finger. In abduction the dorsal interossei spread the index, ring, and little fingers and move the middle finger from side to side.

MUSCLES OF THE INFERIOR EXTREMITY
Gluteal Musculature (Table 6-23)

The buttock, the largest mass of muscle in the body, is formed by three gluteal muscles (Figure 6-34). Like the deltoid of the upper extremity, this mass of muscle is of clinical significance as a site of intramuscular injection.

The gluteus maximus is the largest muscle in the body. It arises from the crest and adjacent external surface of the ilium and inserts into the iliotibial tract and the upper end of the shaft of the femur. It acts as the main extensor of the thigh and also rotates the thigh laterally.

The gluteus medius lies just beneath the gluteus maximus and completely covers the gluteus minimus. Both muscles arise from the external surface of the ilium, below the iliac crest. The origin of the gluteus medius is between the origins of the gluteus

TABLE 6-23. MUSCLES OF THE GLUTEAL REGION

MUSCLE	NERVE	ORIGIN	INSERTION	ACTION
Gemelli superior and inferior	Twigs from sacral plexus	Superior, upper margin of lesser sciatic notch; inferior, lower margin of lesser sciatic notch	Tendon of obturator internus, greater trochanter	Rotates thigh laterally
Gluteus maximus	Inferior gluteal	Upper portion of ilium, sacrum and coccyx	Gluteal tuberosity and iliotibial tract	Chief extensor, powerful lateral rotator of thigh
Gluteus medius	Superior gluteal	Midportion of ilium	Greater trochanter, oblique ridge of femur	Abducts, medially rotates thigh
Gluteus minimus	Superior gluteal	Lower portion of ilium	Greater trochanter, capsule of hip joint	Abducts, medially rotates thigh
Piriformis	S_1 and S_2*	Internal aspect of sacrum, sacrotuberous ligament	Upper portion of greater trochanter	Rotates thigh laterally
Quadratus femoris	L_4, L_5, and S_1*	Ischial tuberosity	Greater trochanter, adjoining shaft femur	Rotates thigh laterally
Tensor fascia lata	Superior gluteal	Iliac crest	Iliotibial tract	Tenses fascia lata

*Twigs leaving spinal cord at these levels supply innervation.

maximus and gluteus minimus. Both insert into the greater trochanter of the femur and act to abduct the thigh and rotate it medially.

In the same plane with the gluteus minimus a series of small muscles can be identified. All these small muscles rotate the thigh laterally. The piriformis, the most superior of these muscles, is located almost entirely within the pelvic cavity, being situated on the internal aspect of the sacrum. It forms a bed for the roots of the sacral plexus of nerves, originates from the anterior surface of the sacrum, and inserts into the greater trochanter of the femur. Lying immediately inferior to the piriformis, the gemellus superior arises adjacent to the spine of the ischium and fuses to the tendon of the obturator internis.

The latter exits from the pelvis by passing laterally through the lesser sciatic foramen to cross the posterior aspect of the hip joint. It inserts into the lesser tuberosity of the femur. The gemellus inferior, arising adjacent to the spine of the ischium, similarly fuses to the tendon of the obturator externis as the obturator externis crosses the posterior aspect of the hip joint. Inferiorly the quadratus femoris completes the lateral rotator group of muscles. Arising from the ischial tuberosity, it inserts into the upper end of the shaft of the femur.

Musculature of the Thigh

Thigh muscles may be conveniently divided into three groups: extensors, adductors, and flexors. They are housed in compartments formed by fascial sheets extending inward from the external investing fascia (fascia lata) of the thigh (see Figure 6-36B). The extensor

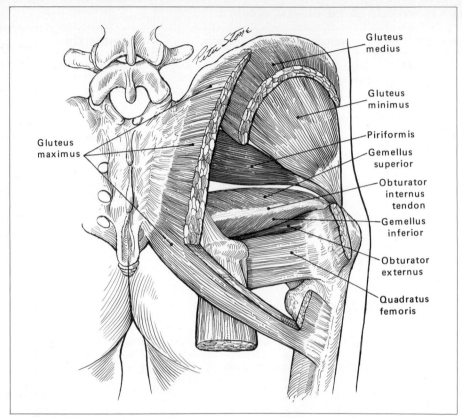

Gluteus
medius

Gluteus
minimus

Piriformis

Gemellus
superior

Obturator
internus
tendon

Gemellus
inferior

Obturator
externus

Quadratus
femoris

Gluteus
maximus

Figure 6-34. Gluteal muscles. The gluteus maximus and gluteus medius have been cut to expose the muscles lying in the plane of the gluteus minimus.

muscles are found in the anterior compartment, the adductors in the medial compartment, and the flexors in the posterior compartment.

Muscles located in each compartment are innervated by a single nerve, the femoral nerve to the anterior muscles, the obturator nerve to the medial muscles, and the sciatic nerve to the posterior muscles.

Anterior, or Extensor, Compartment   The quadriceps femoris muscle, forming the mass of the anterior group (Figure 6-35, Table 6-24), inserts by the strong patellar ligament into the tibial tuberosity. The largest sesamoid bone in the body, the patella, lies within the tendon of this muscle as it crosses the knee joint. Arising by four heads, the quadriceps femoris is usually described as four individual muscles. The vastus lateralis, vastus medialis, and vastus intermedius form a bed upon which the anteriorly placed rectus femoris lies.

The longest muscle in the body, the sartorius, crosses the anterior aspect of the thigh obliquely. It derives its name from its action of flexing the hip and the leg, enabling one to assume the cross-legged "tailor's position."

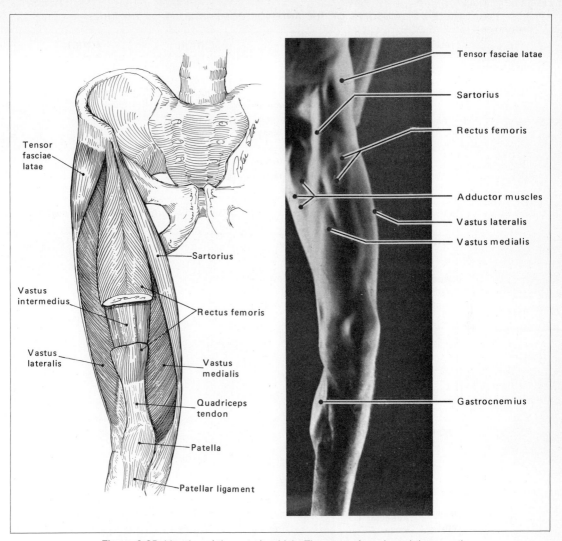

Figure 6-35. Muscles of the anterior thigh. The rectus femoris and three vasti muscles all insert into the tibia by means of the patellar ligament. (*Photograph courtesy of R. D. Lockhart, from "Living Anatomy," 6th ed., Faber & Faber, Ltd., London, and Oxford University Press, New York.*)

Medial, or Adductor, Compartment Three of the muscles occupying the medial compartment of the thigh are named adductors (Figure 6-36, Table 6-25). They originate from the pubic bone and the ischial tuberosity and insert into the linea aspera of the femur. As their names imply, they adduct the thigh to draw the legs together.

   The largest muscle of this group, the adductor magnus, is deeply situated and is innervated by both the tibial and obturator nerves. The small adductor brevis lies between the adductor magnus and the overlying adductor longus. The gracilis parallels the course of the adductor longus but extends farther distally to insert into the tibia.

TABLE 6-24. MUSCLES OF THE ANTERIOR COMPARTMENT OF THE THIGH

MUSCLE/NERVE*	ORIGIN	INSERTION	ACTION
Quadriceps femoris			
Rectus femoris	Inferior iliac spine and rim of acetabulum	Tibial tuberosity	Extends leg and flexes thigh
Vastus lateralis	Intertrochanteric line, linea aspera of femur	Tibial tuberosity	Extends leg
Vastus medialis	Intertrochanteric line, linea aspera of femur	Tibial tuberosity	Extends leg
Vastus intermedius	Upper shaft of femur	Tibial tuberosity	Extends leg
Sartorius	Anterior superior iliac spine	Upper medial surface of tibia	Flexes both thigh and leg

* All are innervated by the femoral nerve.

Figure 6-36. Adductor muscles of the thigh. **A.** Anterior view. **B.** Cross section showing compartments of the thigh.

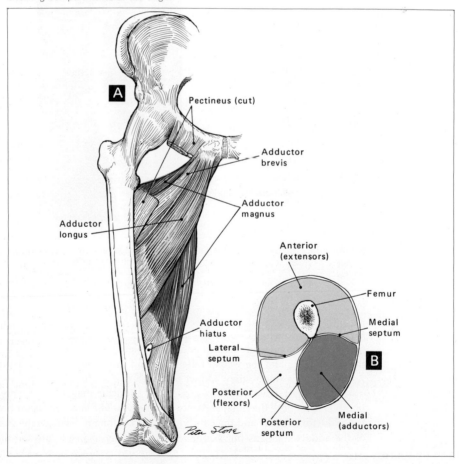

TABLE 6-25. MUSCLES OF THE MEDIAL AND POSTERIOR COMPARTMENTS OF THE THIGH

MUSCLE	NERVE	ORIGIN	INSERTION	ACTION
MEDIAL COMPARTMENT				
Adductor brevis	Obturator	Pubis, below origin of adductor longus	Upper part of linea aspera	Adducts, flexes, and laterally rotates thigh
Adductor longus	Obturator	Pubis, below pubic crest	Linea aspera of femur	Adducts, flexes, and laterally rotates thigh
Adductor magnus	Obturator and sciatic	Pubic arch, ischial tuberosity	Extensively into linea aspera, and adductor tubercle	Adducts, flexes, and laterally rotates thigh
Gracilis	Obturator	Lower half of pubis	Upper part of tibia	Adducts thigh, flexes and medially rotates leg
Obturator externus	Obturator	Margin of obturator foramen, obturator membrane	Intertrochanteric fossa of femur	Flexes and rotates thigh laterally
Pectineus	Obturator and femoral	Pectineal line of pubis	Femur between lesser trochanter and linea aspera	Adducts and aids in flexion of thigh
POSTERIOR COMPARTMENT				
Biceps femoris	Sciatic	Long head, with semitendinosus from ischial tuberosity; short head, supracondylar ridge of femur	Head of fibula and lateral condyle of tibia	Flexes knee, rotates leg laterally; long head extends thigh
Semimembranosus	Sciatic	Ischial tuberosity	Upper portion of tibia	Extends thigh, flexes and rotates leg medially
Semitendinosus	Sciatic	In common with biceps from ischial tuberosity	Upper portion of tibia	Flexes and rotates leg medially, extends thigh

Posterior, or Flexor, Compartment The posterior flexor group (Table 6-25) is also called the hamstring muscles because of cordlike tendons, which can be felt at the back of the knee (Figure 6-37). Names of these muscles reflect characteristics of their morphology. The semimembranosus extends from its origin as a membranous band for approximately half its extent before muscle fibers are grossly apparent. The semitendinosus begins as a fleshy belly in common with the long head of the biceps. About halfway along the thigh this muscle forms a long tendon which continues distally to insert into the medial surface of the proximal end of the tibia. The long head of the biceps femoris arises in common with the semitendinosus, the short head from the distal third of the femur. The biceps, like the adductor magnus, has a double nerve supply. The long head is supplied by the tibial division of the sciatic nerve, the short head by the peroneal division.

Musculature of the Leg

Like the thigh muscles, those of the leg may be studied by compartments. There are three compartments, but only two intermuscular septa are necessary since the medial

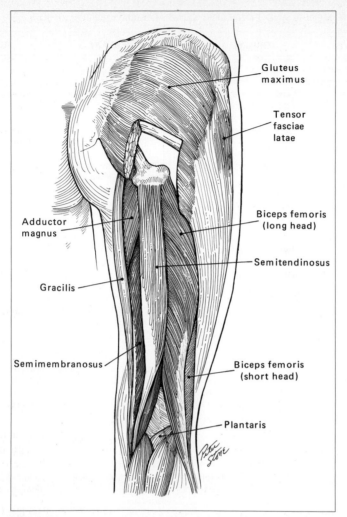

Figure 6-37. Posterior view of the thigh muscles. The ischial tuberosity (visible in the section cut from the gluteus maximus) is the origin of both the semitendinosus and the long head of the biceps femoris.

surface of the tibia forms a separation wall (Figure 6-38B). The three compartments are the anterior, or extensor, compartment; the lateral, or peroneal, compartment; and the posterior, or flexor, compartment. Muscles in the anterior compartment are innervated by the deep peroneal nerve, those in the posterior compartment by the tibial nerve, and those in the lateral compartment by the superficial peroneal nerve.

Lateral, or Peroneal, Compartment Two of the three peroneal muscles are located in this compartment (Figure 6-38, Table 6-26). Arising along the length of the fibula the tendons of the peroneus longus and brevis curve around the posterior aspect of the lateral

TABLE 6-26. MUSCLES OF THE LATERAL AND THE ANTERIOR COMPARTMENTS OF THE LEG

MUSCLE	NERVE	ORIGIN	INSERTION	ACTION
LATERAL COMPARTMENT				
Peroneus brevis	Superficial peroneal	Lower two-thirds of fibula	Fifth metatarsal	Everts, aids in flexion of foot
Peroneus longus	Superficial peroneal	Upper aspects of tibia and fibula	First metatarsal and first cuneiform	Everts, aids in flexion of foot
ANTERIOR COMPARTMENT				
Extensor digitorum longus	Deep peroneal	Tibia, upper three-fourths of fibula, interosseous membrane	By four slips forms extensor expansion over toes	Extends toes
Extensor hallucis longus	Deep peroneal	Middle of fibula and interosseous membrane	Distal phalanx of great toe	Extends great toe
Peroneus tertius	Deep peroneal	Distal fourth of fibula, interosseous membrane	Fifth metatarsal of deep fascia on dorsum of foot	Extends and everts foot
Tibialis anterior	Deep peroneal	Upper two-thirds of tibia, interosseous membrane	First cuneiform and first metatarsal	Extends and inverts foot

malleolus (ankle). Both tendons continue anteriorly along the lateral side of the foot, where the peroneus brevis inserts into the tarsal bones.

Lying adjacent to the tarsal bones, the tendon of the peroneus longus courses obliquely across the plantar surface of the foot to insert on the medial side of the foot adjacent to the attachment of the tibialis anterior (see below). The two muscles form a supportive sling for the longitudinal arch of the foot.

Anterior, or Extensor, Compartment The extensor digitorum longus (Figures 6-38 and 6-39) lies next to the peroneal compartment. It sends four tendons to the lateral four digits. The tendons spread out into extensor hoods over the dorsum of the toes.

The peroneus tertius is fused to the lateral side of the extensor digitorum longus. The large, fleshy belly of the tibialis anterior is easily palpable in the upper half of the leg as it lies immediately adjacent to the sharp anterior border of the tibia ("shinbone"). It sends its relatively long tendon to the inside of the foot, where its attachment to the tarsal bones aids in support of the longitudinal arch of the foot. The extensor hallucis longus lies deep in the anterior compartment. Near the ankle its large tendon is easily palpable as it emerges between the extensor digitorum longus and tibialis anterior to course toward its insertion into the distal phalanx of the great toe.

Posterior, or Flexor, Compartment The musculature of the posterior, or flexor, compartment (Table 6-27) is subdivided into a superficial group of three muscles and a deep group of four muscles (Figure 6-40).

The gastrocnemius and soleus form the bulging mass of the calf. They insert in

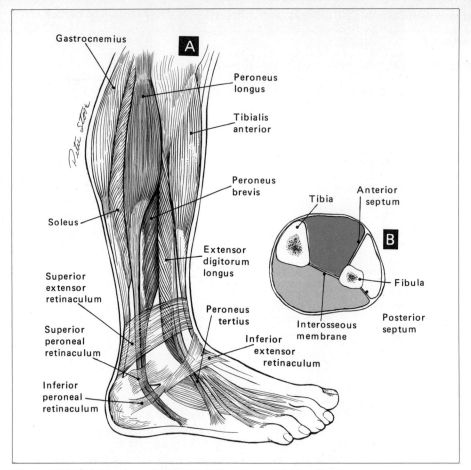

Figure 6-38. A. Muscles of the leg. **B.** Cross section. Extensors are between the tibia and anterior septum. Flexors are on the opposite side of the interosseous membrane. Peroneals are between the anterior and posterior septa. The photograph shows muscles of all compartments. (*Photograph courtesy of R. D. Lockhart, from "Living Anatomy," 6th ed., Faber & Faber, Ltd., London, and Oxford University Press, New York.*)

common by the calcaneal (Achilles) tendon into the back of the calcaneus, or heel bone. The more superficial gastrocnemius arises by two heads from the inner aspects of the lateral and medial epicondyles of the femur. The deeper soleus muscle has a more extensive origin from the upper portions of both the tibia and the fibula.

The third muscle of the superficial group is the small plantaris. Arising just above the origin of the lateral head of the gastrocnemius, its long slender tendon passes deep to the belly of the gastrocnemius and courses along the medial border of the calcaneal tendon. It may insert independently into the calcaneus but usually fuses with the calcaneal tendon at its lower end.

Four muscles in the deep group complete the musculature of the flexor compartment.

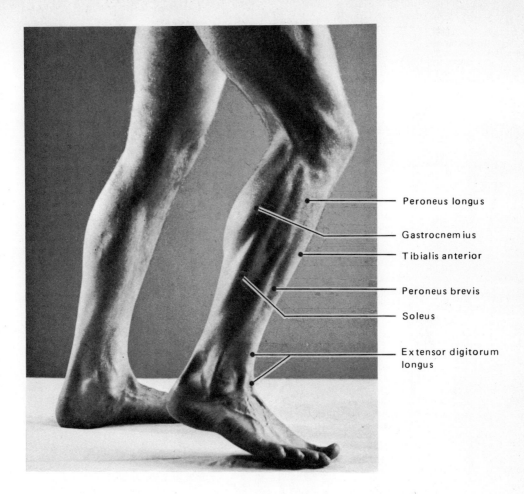

Peroneus longus

Gastrocnemius

Tibialis anterior

Peroneus brevis

Soleus

Extensor digitorum longus

The most superficial muscle of this group is the popliteus. It extends obliquely from the femur to the tibia just below the articular surfaces of the knee joint. It has an important function in the action of the knee. It rotates the femur medially, aligning the tibia and femur so that the posterior muscles of the thigh and leg can flex the joint. Thus, the popliteus "unlocks" the knee joint.

The flexor of the great toe, the flexor hallucis longus, arises from the lower two-thirds of the fibula to send its tendon behind the medial malleolus. From here it extends forward along the medial aspect of the plantar surface of the foot to pass with the third layer of muscles of the foot and attach to the great toe.

Arising from the tibia, the flexor digitorum longus hooks its tendon beneath the medial malleolus. The tendon crosses superficial to the long flexors of the great toe and separates into four tendons. Passing with the second layer of plantar muscles, a tendon from the flexor digitorum longus goes to each toe.

The tibialis posterior lying adjacent to the interosseous membrane is the deepest muscle of the posterior compartment of the leg. Its major attachment is to the navicular bone, but slips of its tendon fan out to insert into five of the seven tarsal bones.

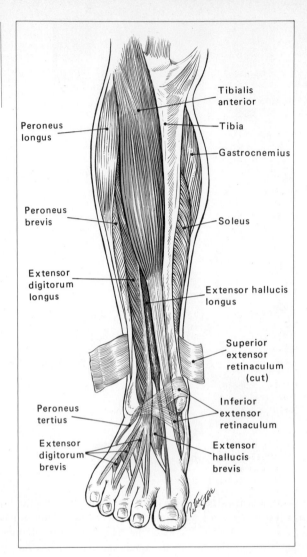

Peroneus
longus

Peroneus
brevis

Extensor
digitorum
longus

Peroneus
tertius

Extensor
digitorum
brevis

Tibialis
anterior

Tibia

Gastrocnemius

Soleus

Extensor hallucis
longus

Superior
extensor
retinaculum
(cut)

Inferior
extensor
retinaculum

Extensor
hallucis
brevis

Figure 6-39. Anterior view of muscles of the leg. Extensors are shaded in color.

Muscles of the Foot

Unlike the hand, the foot has an intrinsic muscle, the extensor digitorum brevis, on its dorsal aspect (Figure 6-39). Originating from the calcaneus it sends a tendon to each of the digits. The muscle fascicles of the tendon passing to the great toe are sometimes described as a separate muscle, the extensor hallucis brevis.

Muscles on the plantar surface of the foot (Table 6-28) parallel those of the hand in number and nomenclature. Functionally, however, they are quite different. The hand is adapted for grasping; the foot is essentially a compact, weight-bearing structure adapted for support.

TABLE 6-27. MUSCLES OF THE POSTERIOR COMPARTMENT OF THE LEG

MUSCLE/NERVE*	ORIGIN	INSERTION	ACTION
Flexor digitorum longus	Middle half of tibia	By four tendons into distal phalanges of lateral four toes	Flexes lateral four toes
Flexor hallucis longus	Lower two-thirds of fibula, intermuscular septum	Distal phalanx of great toe	Flexes great toe
Gastrocnemius	Condyles of femur	With soleus into calcaneus	Flexes foot and leg
Plantaris	Popliteal surface of femur	Medial side of calcaneal tendon	Weakly flexes foot and leg
Popliteus	Popliteal groove of lateral condyle of femur	Tibia above soleal line	From fixed tibia rotates femur laterally
Soleus	Upper third of fibula, soleal line of tibia	With gastrocnemius into calcaneus	Flexes foot
Tibialis posterior	Interosseous membrane adjoining tibia and fibula	Navicular, with slips to cuneiforms; cuboid; second, third, and fourth metatarsals	Principal inverter, aids in flexion of foot

* All are innervated by the tibial nerve.

The muscles of the foot are described either topographically or in functional groups. Topographically there are four layers (Figure 6-41). Functionally one group is associated with the great toe, one group with the small toe, and a third group with the lateral four digits.

Intrinsic muscles of the foot associated with the great toe include a flexor, an abductor, and an adductor. The belly of the flexor hallucis brevis lies deep to the tendon of the flexor hallucis longus and sends a tendon to each side of the great toe. The medial tendon joins the tendon of the abductor hallucis, and the lateral tendon unites with the oblique head of the adductor before they ultimately insert into respective sides of the proximal phalanx. The three intrinsic muscles perform the actions indicated by their names. In addition, the transverse head of the adductor hallucis helps maintain the transverse arch of the foot.

Intrinsic muscles of the foot associated only with the small toe are the abductor digiti minimi and the flexor digiti minimi brevis. The abductor lies along the lateral side of the foot in the first layer. The flexor, though located in the third layer, forms the fleshy portion of the distal half of the sole along the fifth metatarsal.

The remaining muscles act on the four lateral digits. The most superficial of the intrinsic muscles of the foot, the flexor digitorum brevis, differs from its counterpart in the hand. There the fleshy belly of the muscle is in the forearm, and therefore it is an extrinsic muscle. But the flexor digitorum brevis, like its homologue in the hand, sends

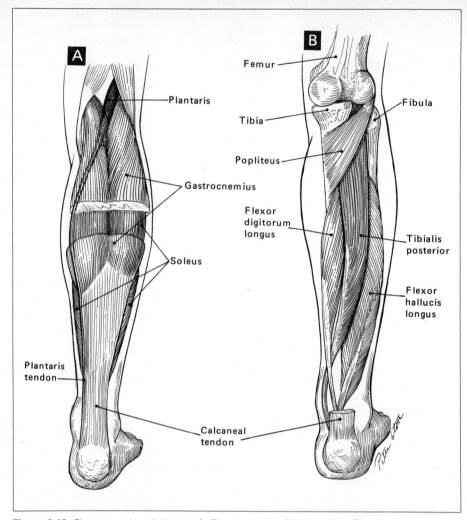

Figure 6-40. Flexor muscles of the leg. **A.** The more superficial muscles. **B.** The deep flexors.

four tendons to the digits. Furthermore, tendons of this muscle lie superficial to the tendons of the flexor digitorum longus and split to allow the deeper tendons to pass to a more distal insertion. The tendons of both flexors attach to the phalanges of the toes in the same manner as the two long flexors of the fingers.

In the posterior half of the foot the quadratus plantae lies immediately deep to the flexor digitorum brevis. It acts to straighten out or longitudinally align the pull of the flexor digitorum longus as the latter courses obliquely across the plantar aspect of the foot.

As in the hand, there are four lumbricales in the foot. Their origin from the flexor digitorum longus parallels their homologues in the hand. Their insertion into the extensor hood gives them a similar, though much weaker, function to the lumbricales in the hand, which is to flex the metacarpophalangeal joint and extend the interphalangeal joints.

TABLE 6-28. MUSCLES OF THE FOOT

MUSCLE	NERVE	ORIGIN	INSERTION	ACTION
Abductor digiti minimi	Lateral plantar	Calcaneus	Proximal phalanx of little toe	Abducts little toe
Abductor hallucis	Medial plantar	Calcaneus	With flexor hallucis into proximal phalanx of great toe	Abducts, aids in flexion of great toe
Adductor hallucis Oblique head	Lateral plantar	Long plantar ligament	With flexor hallucis into proximal phalanx of great toe	Adducts and flexes great toe
Transverse head		Capsules of metacarpo-phalangeal joints	With flexor hallucis into proximal phalanx of great toe	Supports transverse arch, adducts great toe
Extensor digitorum brevis	Deep peroneal	Dorsal surface of calcaneus	Four tendons into extensor expansion	Extends toes
Flexor digiti minimi	Lateral plantar	Fifth metatarsal and plantar fascia	Proximal phalanx of little toe	Flexes small toe
Flexor digitorum brevis	Medial plantar	Calcaneus and plantar fascia	Four tendons into middle phalanx of lateral four toes	Flexes lateral four toes
Flexor hallucis brevis	Medial and lateral plantar	Cuboid and third cuneiform	Two tendons to either side of proximal phalanx of great toe	Flexes great toe
Interossei Plantar (3)	Lateral plantar	Medial side of third, fourth, fifth metatarsals	Proximal phalanges of third, fourth, and fifth toes	Adducts lateral three toes toward second toe
Dorsal (4)		Adjacent metatarsal bones	Proximal phalanges of both sides of second toe, lateral side of third and fourth toes	Abducts lateral toes, moves second toe from side to side
Lumbricales	Medial and lateral plantar	Tendons of flexor digitorum longus	Extensor expansion over lateral four toes	Aids in flexion of toes
Quadratus plantae	Lateral plantar	Calcaneus and plantar fascia	Into tendons of flexor digitorum longus	Straightens pull of tendon of flexor digitorum longus

The deepest musculature of the foot consists of plantar and dorsal interossei. These muscles are again similar to those of the hand, the plantar being unipennate and the dorsal muscles being bipennate. But they abduct and adduct in relation to a line passing through the second toe, while in the hand action is relative to a line passing through the middle finger.

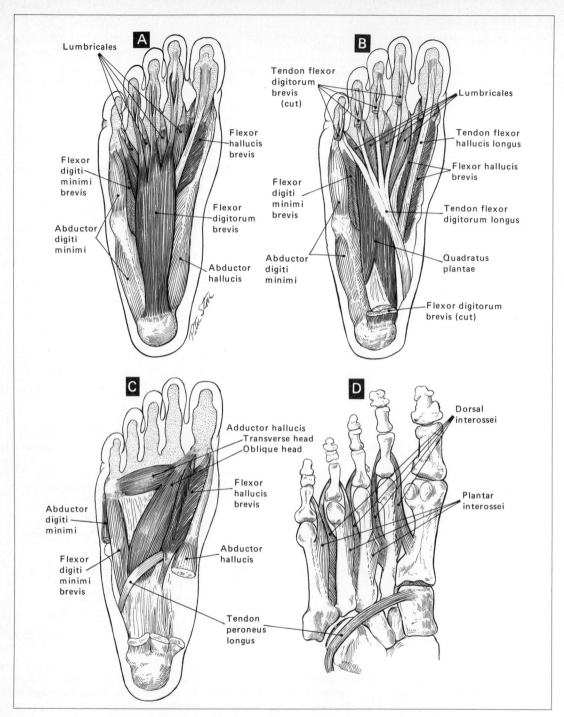

Figure 6-41. Muscles of the foot. In the four drawings successively deeper layers of muscles are shown.

QUESTIONS AND PROBLEMS

1 What type of muscular tissue is found in the following structures:
   a Heart
   b Rectus abdominis
   c Urinary bladder
   d Arteries
   e Bronchi
   f Sternocleidomastoid
   g Deltoid
   h Pancreatic duct
   i Ureter
   j Uterus
   k Pharynx

2 Compare muscular tissue with osseous tissue in structure and function.

3 Explain why the diaphragm is innervated by the phrenic nerve, which originates in the cervical region of the spinal cord.

4 What is meant by voluntary and involuntary control of muscles? Give examples of each type.

5 Do muscle fibers regenerate? Explain the consequences of this phenomenon when injury occurs in skeletal, smooth, and cardiac muscle.

6 Differentiate atrophy, hypertrophy, and hyperplasia, using muscle as an example. State several conditions which might cause one to occur.

7 Explain the means of attachment of skeletal muscles to the skeleton or other parts of the body.

8 What are sphincter muscles? Give several examples of them.

9 Describe and draw the pattern formed by muscles of the anterolateral abdominal wall.

10 When a patient is observed postoperatively, what measures must be taken to maintain a patent airway? Why?

11 In feeding a patient, especially a child, what precautionary measures would you employ? Why?

12 Describe the vascular supply of skeletal muscles. Of what significance is this in administering intramuscular injections?

13 Give the names of the muscles involved in an intramuscular injection at the buttock and upper arm.

14 What anatomical structures at each site can be injured if the injection is administered incorrectly?

15 What muscles constitute the pelvic floor in the female?

16 What is the perineum? What muscles are involved in an episiotomy?

17 Describe the inguinal canal and the structures traversing it.

18 What is a hernia? Name the anatomical sites where a hernia can occur.

19 For each muscle listed in column I, select the correct function from column II and place the corresponding letter in the space provided.

Column I	Column II
1. Hamstring muscles _____	a. Plantarflexes the foot
2. Biceps brachii _____	b. Adducts the arm
3. Sartorius _____	c. Extends the forearm
4. Trapezius _____	d. Flexes the leg
5. Triceps brachii _____	e. Extends the thigh
6. Gastrocnemius _____	f. Flexes the wrist
7. Latissimus dorsi _____	g. Flexes the thigh
8. Pectoralis major _____	h. Flexes the forearm
9. Gluteus maximus _____	i. Rotates the scapula
10. Gracilis _____	j. Adducts the thigh
	k. Extends the arm

7 PHYSIOLOGY OF MUSCLE

Movement is one of the fascinating properties of life. All body movement depends on the activity of muscles. As described in the previous chapter, there are three types of muscles: (1) skeletal, (2) smooth, and (3) cardiac. Because they differ not only structurally but also functionally, the physiology of each type is discussed separately in this chapter.

SKELETAL MUSCLE
The ability to move the body or parts of the body in response to the will is solely the function of skeletal muscle.

Structure of Skeletal Muscle
Under the microscope skeletal muscle myofibrils are seen to be divided into distinctive areas called bands, lines, and zones (see Figures 7-3 and 6-2). These bands, lines, and zones have all been designated by letters. H. E. Huxley has proposed an arrangement of the myosin and actin filaments that neatly explains each of the designated areas. According to him, the Z line is a dense structure to which actin filaments are attached. The I band consists only of the actin filaments, with no myosin overlap. In contradistinction, in the A band lie the thick myosin filaments which, at their ends, overlap the ends of the actin filaments. Thus the A band is seen to have thick, dark borders with a lighter central zone called the H zone. In the middle of the H zone is the dark M line. This

line is said to be caused by a bulge in the center of each myosin filament. The area between two Z lines is termed a sarcomere.

Huxley and his colleagues have succeeded in isolating individual myosin and actin filaments, which permits the structure of each to be studied in meticulous detail. Electron micrographs of myosin filaments (Figure 7-1) are characterized by a small zone at the center that differs from the two ends. This area is about 0.2 micron long and is responsible for the pseudo H zone, which for some time remained without satisfactory explanation. It differs from the ends of the myosin filaments in that so-called cross bridges are not present.

Most fascinating and impressive is the finding that filaments with the same configuration can be synthesized from purified solutions of myosin. The myosin molecule is about one-tenth the length and two-tenths the diameter of a myosin filament. This suggests that the myosin filament consists of several myosin molecules.

When a myosin molecule is enzymatically split, two fragments are obtained. They have been named light meremyosin and heavy meremyosin. The latter can split a phosphate group from ATP, and it can interact with actin. Light meremyosin can do neither. These two fragments have been viewed in the electron microscope and found to differ. The heavy form appears to have a globular head with a short tail, while the light form is but a simple strand. Thus it is proposed that the intact myosin molecule is arranged with the globular heavy meremyosin at one end with its short tail attached to the light meremyosin filament. By such an arrangement the intact filament would take on the appearance shown in Figure 7-1, that is, each end consisting of globular heads and the center thin and smooth. This arrangement explains the pseudo H zone and also provides a model for the sliding-filament hypothesis of muscle contraction.

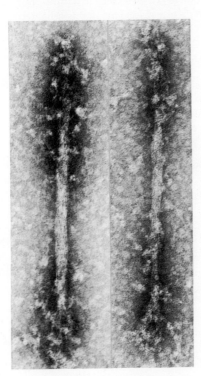

Figure 7-1. Myosin filaments. Note that the center appears bare. The bare area is thought to be devoid of cross bridges. (*Courtesy of H. E. Huxley.*)

Actin filaments are quite different. They consist of molecules of actin arranged in the form of a double helix. Two chains appear to be wound around each other, and the filament has much the appearance of two strands of pearls so wound. Not only does heavy meremyosin interact with the actin molecules, but it does so in such a manner as to cause an arrowhead pattern (Figure 7-2).

Mechanism of Contraction

The Z lines run at right angles to the filaments and provide a stable base to which the actin filaments are attached. According to the sliding-filament hypothesis, the two types of filaments produce contraction by sliding over each other. This brings the ends of the actin filaments progressively closer together, and since they are attached to the Z lines, these lines come closer together as the muscle shortens.

That this is exactly what occurs was shown beautifully in a series of micrographs taken during various stages of muscle contraction (Figure 7-3). Careful study of this series shows that definite changes occur in the I band and in the H zone. The I band is seen to become progressively more narrow. Figure 7-3 illustrates why this occurs. In the relaxed muscle the I band appears light because it contains only actin filaments. But during contraction, due to the sliding of the actin and myosin filaments over one another, the I band contains more and more myosin, so that the light area becomes narrower and narrower and ultimately disappears when the myosin filaments reach the Z lines.

The changes that take place in the H zone are a bit more complex. In relaxed muscle this part of the muscle also appears to be light—not as light as the I band but lighter than the borders of the A band. These relationships are explicable on the basis that in the H zone there is only myosin. At the borders of the A band there is overlap of myosin and actin. Thus, during contraction the H zone is seen to narrow until it disappears, leaving only the M line discernible. But then it reappears, not as a lighter area, but as a darker

Figure 7-2. Actin filaments. The actin molecules are arranged in the form of a double helix. Heavy meremyosin interacts with the actin molecules to give rise to the arrowhead pattern. (*Courtesy of H. E. Huxley.*)

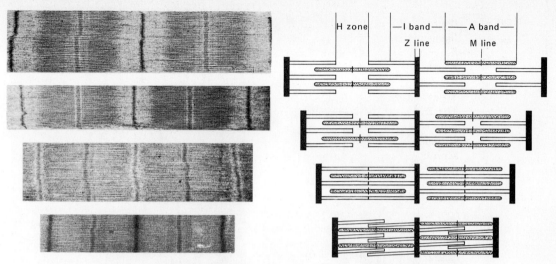

Figure 7-3. Muscle contraction. Note the change in the various zones on the electron micrograph longitudinal section (left) as the muscle contracts. These changes are well explained by the sliding-filament hypothesis shown schematically at the right. The stippled bars are myosin. (*Micrograph courtesy of H. E. Huxley.*)

one. The narrowing and disappearance of the H zone is consistent with the sliding-filament concept. As shortening occurs, the ends of the actin filaments approach each other until they completely overlap the myosin. Now all the A band is dark; there is no H zone. When this occurs, the muscle is only partially contracted. As it contracts even more, the new dense zone appears (Figure 7-3). According to Huxley, this new dense zone is caused by the ends of the actin filaments overlapping. The dense zone, then, consists of myosin and two filaments of actin. This is clearly evident in a section prepared by cutting transversely through the new dense region (Figure 7-4).

There is now general agreement concerning the fine structure of muscle and that during contraction the filaments do slide over one another as just described. What is not so well understood is how this sliding action is brought about.

Using special techniques and high magnification, electron microscopists have been able to reveal the so-called cross bridges which connect the myosin and actin filaments (Figure 7-5). These could serve to hold, like ratchets, the filaments. By so doing they would provide a mechanism for muscle to resist stretch. But in order for the muscle to shorten, these cross bridges must be broken and then new bridges formed. In other words, contraction is visualized as the sliding of the filaments over one another with bridges being formed, broken, and re-formed. The question is, of course, how?

The answer to this question is not yet available, but as mentioned previously, when isolated actin filaments are treated with heavy meremyosin, an arrowhead pattern results. The fascinating aspect is that there is perfect orientation of these arrowheads in opposite directions on the filaments which are attached to the Z line. They always point away from the line. This suggests that in the intact muscle the heavy meremyosin part of the myosin filament provides the active force for reacting with the actin filaments. If this is so and if the orientation is significant, it would be expected that when the actin filaments

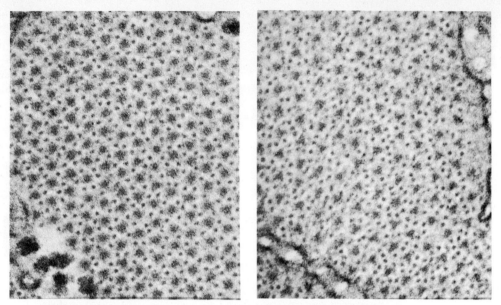

Figure 7-4. Overlap of actin filaments during contraction. In the electron micrograph cross section at the left, the muscle is not contracted; thus there are only a few small actin molecules around each large myosin molecule. At right is a section through muscle in the contracted state. Many more actin filaments can be seen for each myosin molecule, due to overlapping of the actin filaments. (*Courtesy of H. E. Huxley.*)

Figure 7-5. In this high-magnification electron micrograph cross bridges are seen between the filaments. (*Courtesy of H. E. Huxley.*)

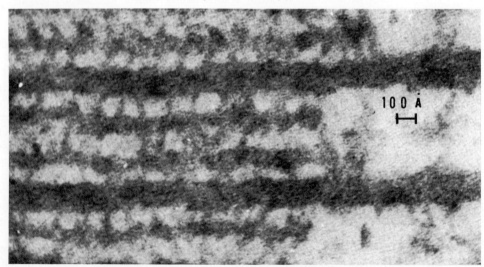

100 Å

overlap, generated tension would become progressively less. Physiologists have observed for years that there is an optimal length at which the muscle can generate maximum tension. As the muscle shortens from this length, the tension it can produce becomes progressively less. Actin overlap may well be the explanation.

The so-called ratchet theory is attractive but has several shortcomings; accordingly another hypothesis has been put forth. It is called the electrostatic solenoid theory. Instead of assuming chemical bonds between the cross bridges of myosin with the actin filaments, an electrostatic attraction is visualized. The hypothesis assumes that at rest both the myosin and actin filaments are electronegative and thus do not attract one another. Upon stimulation, calcium ions may bind with the myosin molecules, rendering them positive. Now there is an electrostatic attraction which causes the actin filaments to move, to slide. Again, when the ends of the actin filaments overlap, the electrostatic attraction would be decreased, thereby decreasing the force of contraction.

Models have been built out of metal plates which when suitably charged do exhibit typical sliding. The potential difference required is comparable to that occurring in muscle, and the force generated is comparable to that observed in muscle. These observations indicate that the electrostatic solenoid theory is feasible, but whether this actually occurs in muscle remains to be demonstrated.

Chemistry of Contraction

The chemical changes that occur during and after muscular contraction have been studied extensively, and the wondrously complex details have been fairly well elucidated.

ATP is essential for contraction, and large mitochondria are invariably present in muscle cells. There is general agreement that the conversion of ATP to ADP provides the immediate energy for contraction.

During maximal activity of a muscle the increase in rate of ATP utilization is far greater than the increase in rate of oxygen consumption by that muscle. Accordingly, unless there was a means other than oxidation, the supply of ATP would be quickly exhausted and the muscle would cease to contract. But ATP can be regenerated rapidly anaerobically, that is, in the absence of oxygen, by virtue of the interaction of ADP and phosphocreatine. Thus:

$$\text{Phosphocreatine} + \text{ADP} \longrightarrow \text{ATP} + \text{creatine}$$

Just as ATP must be regenerated for continued muscle action, so must phosphocreatine be re-formed from creatine and phosphate. This latter synthesis requires ATP. *This* ATP comes from the mitochondria, where it is generated by a sequence of events outlined in Chapter 2.

Mitochondria, to keep generating ATP, need an input of substrate, hydrogen ions, and oxygen. The oxygen comes from the atmosphere and is carried to the cells by the blood. Substrate and hydrogen ions come from the end products of food digestion. There are many complex interactions, but insofar as the chemistry of contraction is concerned the conversion of glycogen to lactic acid is of primary importance.

Glycogen is a polysaccharide represented by the formula $(C_6H_{10}O_5)_x$. In other words, it is a large molecule made up of many monosaccharides. The average molecular weight of glycogen in muscle is about 5 million. This permits the storage of a tremendous amount of potential monosaccharide. If muscle stored monosaccharide as a great number of individual small molecules rather than far fewer large molecules, the osmotic pressure within the cells would be insupportable.

Muscle glycogen is stored in muscle, where it releases monosaccharide as needed.

The monosaccharide provides fuel needed by the mitochondria for the generation of ATP. An important end product of the degradation of monosaccharide is lactic acid.

In the intact organism, the lactic acid formed during contraction diffuses out of the muscle and is carried to the liver, where it is converted once again to glycogen. The liver, like muscle, is capable of storing great quantities of glycogen. Liver glycogen constitutes an important pool of glycogen which can be drawn upon to maintain the level of blood glucose. Blood glucose is taken up by muscle cells, where it is either immediately used for the generation of energy (ATP) or is converted to glycogen and stored.

There is thus the cycle shown in Figure 7-6, which is known as the Cori cycle. The significance of this cycle, first described by the Coris, a husband-and-wife team of biochemists, is that some of the energy for muscle contraction is supplied by metabolic processes which take place in the liver.

To summarize, ATP is the central agent responsible for many interlocking relationships, as shown in Figure 7-7. ATP is required for actin and myosin to slide, thereby shortening the muscle. ATP also provides the energy needed to re-form phosphocreatine. The ATP for this re-formation comes from mitochondria. The conversion of glycogen to lactic acid provides fuel for the mitochondria to produce this ATP. Mitochondrial synthesis of ATP requires oxygen. Thus, though muscle may contract for a while in the absence of oxygen, oxygen must ultimately be supplied.

There is general agreement, as just outlined, that ATP is the key to contraction. But exactly how the energy from ATP is utilized for this purpose is only now becoming somewhat clearer. In addition, the sequence of events by which electrical stimulation evokes contraction is also being clarified. To understand this sequence one must be familiar with the sarcoplasmic reticulum of skeletal muscle. The sarcoplasmic reticulum consists of transverse, or T, tubules, and also longitudinal tubules. As the terms indicate,

Figure 7-6. (*Left*) The Cori cycle. During anaerobic muscle contraction muscle glycogen is converted to lactic acid which is carried by the circulation to the liver, where it is reconverted to glycogen. Liver glycogen is converted to glucose to be carried back to the muscle in the blood. **Figure 7-7.** (*Right*) Chemical interactions essential to contraction. Note that ATP is used for energy to cause muscle to contract, to convert creatine to phosphocreatine, and to convert lactic acid to glycogen.

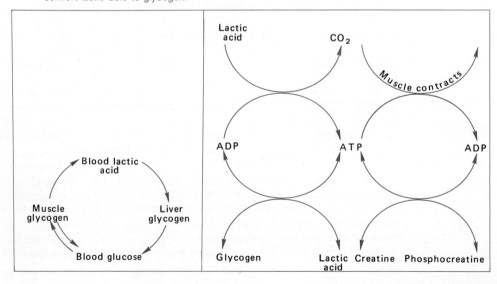

the longitudinal tubules run parallel to the myofibrils, whereas the transverse ones run at right angles to them. These T tubules pierce the muscle cell. To visualize them, we can think of a ball with a small tube running through it. The tube is open at both ends and thus is in contact with extracellular fluid which it contains. The longitudinal tubules come in contact with the T tubules. Note that the tubules are simply in contact with one another; they do not open into one another.

To start from the beginning and review the entire sequence of events, we recall from Chapter 6 that, in the intact organism, muscle fibers are innervated by nerve fibers which terminate in motor end plates at the neuromuscular junction. A single neuron may innervate very few muscle fibers. For example, muscles that must respond rapidly and with precision may have only a very few muscle fibers innervated by a single neuron, whereas large muscles, such as the ones that support the back, may have hundreds of muscle fibers innervated by a single neuron. These make up the motor unit, so termed because all the fibers contract as a unit.

At the neuromuscular junction, the nerve action potential causes the release of acetylcholine (ACh). ACh acts rapidly on the membrane of the muscle fiber to increase its permeability to sodium ions, which flow rapidly into the muscle cell. This influx of sodium ions changes the membrane potential, resulting in an action potential. There is thus an electrical current that flows in the T tubules, and at each contact with a longitudinal tubule the current crosses the walls of the tubules so that the current then flows in the longitudinal system as well as the T system. This results in stimulation of the entire muscle fiber by each action potential. It also assures that the fibers lying deep within the muscle will be stimulated.

When the current flows in the longitudinal tubules, calcium ions in the tubules are released to diffuse into the myofibrils. The calcium ions are thought to bind with ATP on the myosin cross bridges and also with ATP on the actin filaments. In some way this initiates sliding, that is, muscle contraction. As soon as this happens, an enzyme, ATPase, hastens the conversion of ATP to ADP, thereby breaking the bond between ATP and calcium. According to the ratchet theory, this forming and breaking of bonds are what cause the sliding to continue. Finally, when the electrical current in the tubules terminates, as it does in a few milliseconds, calcium ions move rapidly from the cell back into the longitudinal tubules. Now the muscle relaxes.

Clearly, the ACh liberated at the neuromuscular junction must be inactivated rapidly; otherwise it would continue to stimulate the muscle. At the neuromuscular junction there is a substance called cholinesterase which does just that.

An understanding of these relationships is essential for comprehension of the mechanism of drug action. For example, the drug curare causes paralysis. Curare has this ability because it blocks the action of ACh; thus the impulse from the nerve is not transmitted to the muscle. The drug eserine, on the other hand, inactivates cholinesterase and therefore potentiates the action of ACh.

Oxygen Debt

When one exercises vigorously and then stops, he nevertheless continues to breathe heavily for a period of time even though he is no longer active. To put it another way, oxygen is utilized at a rate greater than normal during the bout of exercise, and this high rate continues afterward for a period depending upon the severity and duration of exercise. An analogy with spending money shows that if one spends money faster than he earns it, a debt accumulates which must be paid off after expenditure ceases.

As already outlined, energy for muscle contraction comes from ATP, phosphocreatine, oxidative processes that go on continuously in the cells, and the conversion of glycogen to lactic acid. The oxidative processes can satisfy only limited needs. During bouts of heavy activity, the other sources are tapped. For the first few seconds after exercise begins the stores of ATP and phosphocreatine suffice, but thereafter glycolysis, that is, the breakdown of glycogen, is needed. As a result, lactic acid accumulates. The reconversion of lactic acid to glucose, which then re-forms glycogen, occurs, in part during the exercise but also afterward. The increased oxygen utilization following exercise over and above normal resting needs represents the oxygen debt (Figure 7-8).

Isotonic and Isometric Contraction

Contraction does not invariably involve shortening of the fibers and movement of the bone to which it is attached. This becomes apparent when one pushes vigorously against an immovable wall. Nothing moves, yet a great deal of energy is expended. This is an example of isometric contraction (*iso*, "the same"; *metric*, "measure"). Isometric, then, means that the length remains the same. In this type of contraction there is no change in length of the muscle, but there is a sharp augmentation of tension. In isotonic contraction, on the other hand, the tension remains constant while the length of the fiber shortens. In both cases the term contraction is properly used, and in both instances energy is utilized.

In our usual activities we sometimes contract muscles isotonically, sometimes isometrically, and often we use a combination of both. For example, in walking, the motion of the legs that provides forward movement of the body is the result of isotonic contraction. But as one leg swings forward and comes in contact with the ground, the muscles in that leg, both the flexors and extensors, contract isometrically, thereby converting the limb into a rigid pillar which momentarily supports the body before the leg becomes propulsive.

Electromyography

The study of action potentials produced by muscle is termed electromyography. The record of these action potentials is referred to as the electromyogram (EMG). Figure 7-9 shows an electromyogram recorded from the right masseter muscle of a man while he was chewing. Each burst of activity represents action potentials generated in the contracting muscle fibers. In order to obtain the record, very fine needle electrodes are inserted through the skin. The potential difference between the electrodes is amplified by suitable amplifiers and used to drive a pen recorder or displayed on a cathode-ray tube to be recorded photographically.

Electromyography is finding broad application both in the research laboratory and

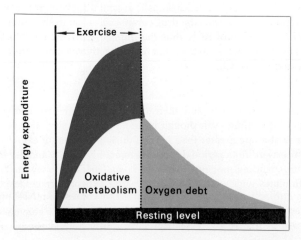

Figure 7-8. Oxygen debt. Note that during heavy exercise the rate of energy expenditure greatly exceeds the rate of oxidative metabolism. The oxygen debt results which is "paid back" after exercise ceases.

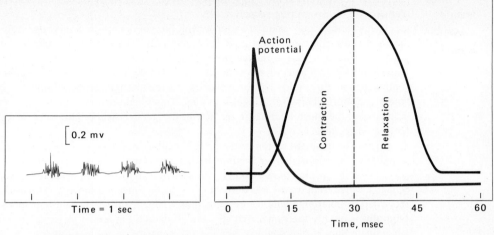

Figure 7-9. (*Left*) Electromyogram. These action potentials were recorded from the right masseter muscle while the subject was chewing. **Figure 7-10.** (*Right*) Response of skeletal muscle to a single stimulus. Note that the action potential is fired and the muscle is almost completely repolarized before it begins to contract.

in the clinic. For example, subjects quickly learn to alter the pattern of action-potential firing while observing the displayed record. In this way motor-unit training is possible and is used in physical therapy for teaching patients to relax specific muscles for the relief of painful spasms. Or physically handicapped patients can be taught to activate motor units connected electronically to electric motors which drive artificial limbs.

Response to a Single Stimulation

Under controlled laboratory conditions, the response of a muscle to stimulation can be studied. Figure 7-10 shows a typical response to a single stimulation. Two significant events occur: there is an action potential, and the muscle contracts. Note that the action potential occurs first, and only about 2 or 3 msec after the muscle is stimulated does it begin to contract. This delay is termed the latent period.

When sensitive recording methods are used, the record sometimes shows a brief period of muscle lengthening before it shortens. The lengthening phase is referred to as latency-relaxation. Once the muscle begins to shorten, it does so rapidly, quickly reaching a peak. This is the contraction phase, which is followed by the relaxation phase (Figure 7-9). The completed response to a single stimulation is called a twitch. The duration of the entire response varies from one muscle to another. Leg muscle fibers exhibit a duration of from 50 to 100 msec for a single twitch. In contrast, the twitch for an eye muscle may be little more than 10 msec.

All-or-Nothing Response

If a whole muscle is stimulated, the magnitude of response increases with stronger and stronger stimuli. If only one muscle fiber is tested, however, this relationship does not obtain. When a very weak stimulus is applied, there is no response. Such a stimulus is called subliminal, or subminimal. Then, as the stimulus is strengthened, a point is at last reached at which the muscle does respond. This is the liminal threshold, or minimal stimulus. But now, as the stimulus is further increased, no change in magnitude of response takes place. In other words, a single muscle fiber either responds wholeheartedly or it does not respond at all, a physiologic property of skeletal muscle fibers appropriately termed the all-or-nothing response.

This concept may seem at variance with human operation. Everyone knows that the force of contraction may be altered quite markedly. The explanation lies in the fact that each muscle fiber obeys the all-or-nothing law but different fibers possess different thresholds. Thus, if a very weak stimulus is applied to a muscle, only those fibers endowed with very low thresholds will be activated. As the strength of the stimulus is increased, more and more fibers are brought into play. This phenomenon is known as recruitment. Obviously, the more fibers that are active, the greater will be the force of contraction of the whole muscle.

There is an apparent contradiction to the all-or-nothing law which should be reconciled. If a single muscle fiber is stimulated at a moderate rate, the magnitude of response of the first few contractions steadily increases in a staircase-like manner. This ascending effect is called treppe, which is derived from the German and means "staircase." But with repeated stimulation a landing, or plateau, is attained, and after that the response to each stimulus is exactly the same. Actually, the all-or-nothing law in its complete form states that single muscle fibers will respond at full capacity or not at all as long as conditions are kept constant. When a muscle fiber begins to contract after a period of rest, there are definite temperature and acidity changes as well as shifts in ionic concentrations. Just as soon as these conditions reach a constant state, the response of the muscle fiber is truly all or nothing.

Of singular importance in determining the magnitude of response is the initial length of the muscle fiber. Figure 7-11 reveals that over a wide range the greater the initial length of the fiber, the more forceful will be its contraction. Again, this observation in no way violates the all-or-nothing principle, because this law emphasizes that conditions must be truly constant if the law is to hold.

Man unknowingly makes good use of this physiologic property of initial length of the muscle fiber. Greater weights can be lifted, for example, when the biceps muscles are partially extended (stretched) than when they are flexed. But it will be seen in Figure 7-11 that there is a critical point beyond which the efficiency of the muscle diminishes.

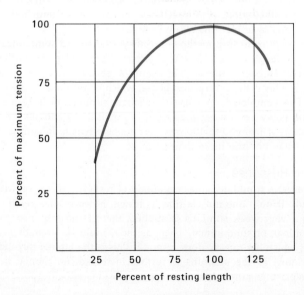

Figure 7-11. Effect of initial length on tension during contraction. By definition the length of the resting fiber that produces the greatest tension during contraction is the resting length. Lengths shorter or longer result in weaker contractions.

The sliding-filament theory of contraction provides an explanation for the relationship between initial length and force of contraction. Figure 7-3 suggests that if the muscle is stretched to the point where there is little or no overlap of the myosin and actin filaments, the ability of the muscle to contract will be minimal. At the other extreme, that is, when the muscle is approaching maximal contraction, the ends of the actin filaments are seen to overlap. This is, again, a very inefficient arrangement which would diminish the force of contraction. Between these two extremes the cross bridges are organized in a way that provides the power for the greatest force of contraction of which the muscle is capable.

Response to Repetitive Stimulation

Thus far only the response of a muscle to a single stimulus has been considered. After a variable latent period the muscle shortens, reaches a definite magnitude, and then relaxes. Other important physiological attributes of skeletal muscle are disclosed by repetitive stimulation. If a muscle is stimulated a second time and if that second shock quickly follows the first stimulus, there is no additional response. In other words, the wave so obtained is exactly the same as though only one stimulus had been given. The muscle at this time absolutely refuses to respond to a second stimulus. It is truly reluctant or refractive. The muscle now is said to be in the absolute refractory state. If the second shock is applied a fraction of a second later, that is, during the contraction phase, a second response can be elicited from the muscle, provided that a stronger-than-threshold stimulus is used. During this period the muscle is said to be in the relative refractory state. Finally, if the second stimulus is introduced at a still later time, a second contraction occurs in response to a threshold stimulus. The sensitivity of the muscle has, in fact, returned to normal.

Figure 7-12 demonstrates that the magnitude of the total contraction is greatest when the second stimulus is introduced during the contraction phase which results from the first stimulus. Presumably the second wave has been added to the first. This addition of contraction waves is known as summation. The phenomenon of summation might seem to violate the all-or-nothing law. This is not true. The all-or-nothing law states unequivocally that it is applicable only so long as conditions are kept constant. The length of the fiber at the time of stimulation determines the magnitude of response. Clearly, when the muscle is excited during the contraction phase, its length is less than it was at complete rest. It will be noted that the magnitude of the second response is really smaller than that of the first, but the total response is greater because the two contractions are added.

The maximal response a muscle can produce may be elicited by employing a volley

Figure 7-12. Response of skeletal muscle to repetitive stimulation. **A.** A second stimulus applied while the muscle is contracting causes summation. **B.** Several stimuli close together cause incomplete tetanus, or, **C,** complete tetanus. Note that although the contraction wave is smooth and continuous in complete tetanus, there are individual action potentials.

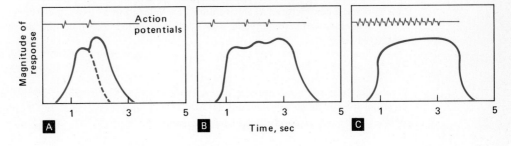

of stimuli so timed as to prevent the muscle from ever reaching the relaxation phase. Figure 7-12 shows that if the rate of stimulation is fast enough, the muscle contracts maximally and remains in that state as long as the stimulus continues or until the muscle fatigues. This smooth, sustained maximal contraction is termed tetanus. Tetanus may be complete or incomplete. These are relative terms. If the resulting wave is straight and smooth, the muscle is in complete tetanus. If the stimuli are so spaced as to allow the muscle some small degree of respite between the individual contractions, an undulating wave is produced. This is a demonstration of incomplete tetanus.

From this discussion of the physiological properties of muscle, one learns that man uses his muscles to produce their greatest work when the fibers are at their optimal length, when the stimulus applied to them is maximal (to excite all the fibers simultaneously), and when the rate of stimulation is rapid.

Muscle Tonus

Although the maximal muscular response is obtained when all the component fibers act in concert, a weaker but still smooth contraction is possible when only a fraction of the fibers are in complete tetanus. In man, at least during the waking hours, many muscles have some of their fibers in complete tetanus. Assume that only 10 percent of the fibers of a muscle are in this state. Then the muscle will be partially contracted. The muscle will feel firm; it will resist being stretched. The muscle, in this state, is said to display tonus. Tonus means "tension"; a muscle with tonus exerts a force, resists stretch. Muscle tonus may be defined as involuntary resistance to stretch.

In normal man all muscles will exhibit tonus. This condition is maintained without significant fatigue because there is a system of rotation. At any one moment some of the fibers are in a state of contraction; others are relaxed and rested so as to take up their work the next moment. This division of labor is indeed fortunate, for without it, it would be extremely difficult, if not impossible, to stand erect.

In the clinic, the physician routinely examines the tonus of the muscles. There are numerous conditions in which the muscles become completely devoid of tonus. These muscles appear soft and flabby and do not resist being stretched. The muscles in this state (and this is found in infantile paralysis) are described as being atonic. At the other extreme are those muscles which are tightly contracted. The muscles appear hard and hyperactive, and vigorously oppose stretch. These muscles (and this may occur in persons suffering with a stroke) are said to be hypertonic.

Muscle Fatigue

If a muscle is vigorously contracted, a point will ultimately be reached at which it cannot continue to contract or does so only weakly. The muscle is said to be fatigued. This state occurs when ATP is utilized more rapidly than it is resynthesized. There will be an abundance of lactic acid, and until ATP can be replenished, the muscle will not respond normally.

Interestingly, if the muscle is fatigued to the extent that ATP is completely depleted, the muscle not only will be fatigued but may go into a state of continuing contraction. This is termed contracture. ATP is required for relaxation as explained above. When it is not available, actin and myosin fibrils cling together and hold the muscle in the contracted state.

Abnormalities of Skeletal Muscle

Paralysis Paralysis, by definition, is a nervous affliction characterized by loss of motor function. It may be simply defined as loss of voluntary movement. In Chapter 10 the neurological basis of paralysis will be discussed.

Hypertrophy and Atrophy   Continuing, vigorous muscle activity causes an increase in size of the individual muscle fibers. This is termed hypertrophy. Not only does the diameter of the fibers increase, but so does the quantity of metabolic substances such as ATP and phosphocreatine. The hypertrophied muscle thus can do more work; it is stronger. The reverse occurs if a muscle is not used, as occurs in paralysis or if a limb is kept in a cast. The diameter of each fiber decreases. This is termed atrophy.

Rigor Mortis   After death all the muscles ultimately contract and remain in the contracted state for some time. This state of generalized, continuous contraction which occurs a few hours after death is called rigor mortis. The cause is thought to be the complete depletion of ATP. Recall that ATP not only is necessary to supply the energy for contraction but it also is needed to permit the myofibrils to separate during relaxation. Ultimately the muscle proteins undergo destruction after death, and when this occurs the muscles finally and permanently relax.

Regeneration of Skeletal Muscle

Skeletal muscle apparently has the ability to regenerate. For about two days following injury to the muscle there are no signs of regeneration. Following this period, however, there is considerable activity observable at the site of damage. Microscopic examination of the area reveals new protoplasmic shoots coming from the uninjured muscle into the injured region. Ultimately, new muscle fibers form. But in the case of areas of extensive damage the muscle is replaced with scar tissue.

SMOOTH MUSCLE

The physiology of smooth muscle is very inadequately understood. This is not due to a lack of interest in the problem but rather to the fact that smooth muscle rarely responds in a consistent manner even when conditions are rigidly controlled. The statement has often been made that smooth muscle accommodates itself rather than the investigator!

Structure of Smooth Muscle

Smooth muscle is differentiated, grossly, from skeletal and cardiac muscle by the absence of striations. However, under the electron microscope striations can sometimes be discerned. More generally the actin and myosin filaments are found to be poorly organized. This poor organization of the contractile elements probably accounts for the fact that smooth muscle contracts more slowly and with less force than does skeletal muscle. The chemistry of contraction is thought to be the same as in skeletal muscle, and in all probability the actin and myosin filaments slide over one another to bring about shortening. Yet there are definite differences in response that require further investigation.

Action Potential

The action potential associated with contraction of smooth muscle varies from moment to moment in the same muscle and from muscle to muscle. In some smooth muscles the action potential resembles that of skeletal muscle in that there is rapid repolarization; thus a characteristic spike is produced. In other smooth muscles the action-potential recording shows a long plateau during repolarization; thus the action potential resembles the characteristic one of cardiac muscle (see Figure 7-15).

The voltage of the resting potential also varies. In some cases it is only about −30 mv, whereas at other times it is as great as −75 mv. Quite often a progressive change in resting potential occurs between contractions. In Figure 7-13 note that following repolarization there is a slow depolarization which culminates in the action potential. This preliminary depolarization is termed the prepotential.

Smooth muscle contracts spontaneously. Prepotentials appear to be characteristic of all muscle that contracts spontaneously. This will be discussed when cardiac muscle is considered.

Contraction

The contraction of smooth muscle differs markedly from that of skeletal muscle. The response is considerably slower, and following a single stimulus the contraction may persist for several seconds. Smooth muscle is capable of spontaneous contraction (Figure 7-14). Even when it is completely denervated, it often continues to shorten and relax with variable rhythm.

As mentioned, smooth muscle contraction probably reflects the irregular arrangement of the contractile filaments. Another reason is the flow of current from one smooth muscle fiber to a contiguous one. Skeletal muscle fibers are so insulated that the activation of one fiber leaves surrounding ones unaffected. But in smooth muscle this is often not the case: there is a spreading of current, which thereby brings about progressive contraction of individual fibers.

Accommodation

The more tension placed upon skeletal and cardiac muscle, the greater will be the resistance to stretch. This is not true of smooth muscle. After smooth muscle is stretched, it seems to relax and lengthen so that the resistance remains about the same despite the stretching. For example, the urinary bladder, which is made up of smooth muscle, is stretched by accumulating urine. Yet the pressure within the bladder remains virtually unchanged. The same is true of the intestine and other smooth muscle viscera. This ability of smooth muscle to relax in response to greater stretch is termed accommodation. In other words, a larger volume is accommodated with little change in pressure or tension. Some authors prefer to label this response stress relaxation.

Regeneration

Smooth muscle has little or no ability to regenerate after injury. The damaged fibers usually die and are replaced by scar tissue.

Figure 7-13. (*Left*) Smooth muscle action potentials. Usually before each action potential there is a prepotential.   **Figure 7-14.** (*Right*) Spontaneous, rhythmic contractions of smooth muscle.

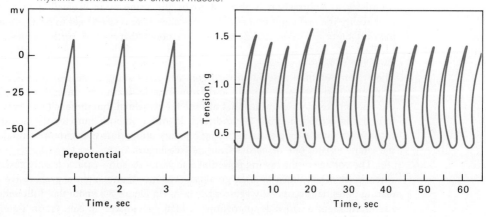

CARDIAC MUSCLE

A characteristic of cardiac muscle is that it possesses inherent rhythmicity. Skeletal muscle lies relaxed and quiet unless the innervating neurons are activated or unless the muscle is stimulated directly. And even under these conditions it will contract in response to the stimulus and then relax and refrain from further contractions in the absence of additional stimuli. Not so cardiac muscle. A heart may be removed from the body, thus being deprived of all innervation, and it will continue to beat rhythmically. Or it may be stopped and then later shocked to restart it. Thereafter it will beat normally, as shown dramatically by the transplantation operations.

Structure of Cardiac Muscle

Cardiac muscle is similar structurally, in some ways, to skeletal muscle. It is striated muscle, and prominent Z lines are present. For years cardiac muscle was described as being a syncytium; that is, protoplasmic continuity was thought to exist between the individual fibers. Electron microscopy, however, fails to support this conclusion. Virtually all the fibers are seen to be surrounded by a membrane and seemingly well-insulated from contiguous ones. Nonetheless, even though cardiac muscle may not be a syncytium, it contracts as one would expect a syncytial muscle to contract. For example, as emphasized previously, individual skeletal muscle fibers may contract while neighboring ones remain quiescent. In cardiac muscle this does not normally occur. Stimulation of any part of cardiac muscle generally causes all the fibers to contract. Thus, graded responses due to recruitment are not possible. In skeletal muscle the individual fibers exhibit an all-or-nothing response; in cardiac muscle the entire muscle responds this way.

Characteristic of cardiac muscle is the presence of intercalated disks (see Figure 6-1). These are thickened and tortuous membranes which usually cross the muscle in a stepwise manner. It is thought likely that current is transferred from one cardiac muscle cell to another by the intercalated disks. Thus, even though cardiac muscle is not now considered to be a syncytium, it probably acts like one because of the ease of the spread of current from cell to cell at the site of the disks.

Initial Length

The force of contraction of cardiac muscle, just like skeletal muscle, increases up to a critical point with the initial length of the fiber. Beyond that it decreases. In view of the fact that the structure and chemistry of contraction of skeletal and cardiac muscle are so similar, that they respond the same to changing lengths is not surprising.

In the heart, the initial length is determined by the volume of blood in the chambers of the heart just before contraction begins. Accordingly, the more blood that enters the heart, the greater will be the initial length of the fiber, the more forceful will be the contraction, and thus the greater will be the volume of ejected blood. This relationship is known as Starling's law of the heart. Its significance will be discussed in Chapter 16.

Action Potential

The action potential associated with the contraction of cardiac muscle differs considerably from those recorded from nerve and skeletal muscle. In Figure 7-15 the action potential recorded from cardiac muscle is seen to be very prolonged. As in nerve and skeletal muscle, the inside of the cardiac muscle cell is negative (about -80 mv) to the outside. During activation there is a typical rapid upswing of the spike, an overshoot, and then the process of repolarization begins. But instead of lasting a few milliseconds, in cardiac muscle the entire process takes at least 200 msec. When the contraction and action potential are recorded simultaneously, the muscle is seen to have finished its contraction phase and to be well into the relaxation phase before repolarization is complete. Why repolarization of cardiac muscle differs so strikingly from that of nerve and skeletal muscle is not known.

If action potentials are recorded from the so-called pacemaker part of the heart, the action potentials are seen to be different from those recorded elsewhere. The difference lies in the presence of prepotentials (Figure 7-16). The pacemaker is the part of the heart that initiates the contraction; the membrane of these cells permits sodium to seep into the cell, thus causing the prepotentials and the ultimate firing of the cell.

Refractory Periods

The refractory period is the time during activation when another stimulus has a different effect from that of the original stimulus. If the stimulus causes no response, the cell is in the so-called absolute refractory period. If the cell can be made to respond but a stronger-than-normal stimulus is required, the cell is in the relative refractory period.

In nerve the absolute refractory period extends throughout the upswing of the action potential and continues until repolarization is about one-third complete. The same is true in skeletal muscle. In cardiac muscle the absolute refractory period is even longer. It lasts through the entire contraction phase and, as can be seen in Figure 7-15, repolarization is about two-thirds complete at the peak of contraction. The relative refractory period then persists until repolarization is almost complete.

Because cardiac muscle will not respond to a stimulus during contraction, summation

Figure 7-15. (*Left*) Cardiac muscle contraction. Note that the action potential (black line) has a long duration and that the muscle remains depolarized well into the contraction phase. This accounts for the long refractory period of cardiac muscle. **Figure 7-16.** (*Right*) Cardiac muscle action potentials. The action potentials in the upper tracing were recorded from the sinus venosus of the frog heart; in the lower tracing, from its ventricle. Prepotentials are seen only in the pacemaker area (sinus venosus). (*Reproduced with permission from Hutter and Trautwein, Journal Gen. Physiol., 39:715, 1956. Reprinted by permission of the Rockefeller University Press.*)

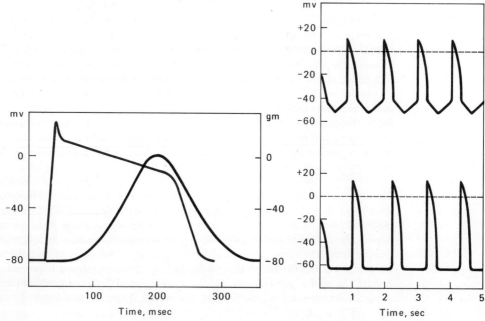

and complete tetanus cannot be evoked. If the heart were to go into complete tetanus, no blood would be pumped.

Regeneration

Cardiac muscle hypertrophies in response to an abnormal work load, but it does not regenerate when injured. Instead of new muscle, the injured area becomes a scar consisting of connective tissue. Thus, following a blockage of arteries supplying the heart muscle, that is, a coronary occlusion, if an area of cardiac muscle dies from oxygen deprivation, scar tissue will grow in its place. Scar tissue does not contribute to the force of contraction. If the injured area is large enough, there may not be enough heart muscle remaining to pump sufficient blood for survival (see Chapter 19).

QUESTIONS AND PROBLEMS

1 Draw a schematic representation of a skeletal muscle myofibril as seen under the electron microscope, labeling the light and dark areas as well as the filaments and cross bridges.

2 Differentiate between the structure of the two parallel filaments of protein—actin and myosin—found in skeletal muscle myofibrils.

3 Describe the mechanical and chemical events of Huxley's sliding-filament hypothesis for skeletal muscle contraction.

4 Describe the mechanical difference between isometric and isotonic contraction of muscle.

5 What difference does it make in the use of skeletal muscles that energy is derived from anaerobic as well as aerobic metabolism?

6 What is the advantage of the muscle cell's ability to convert glucose to glycogen and to store the polysaccharide?

7 Describe the Cori cycle and its relationship to skeletal muscle contraction.

8 In which is the ratio of muscle fiber to nerve fiber higher, the muscles of the finger or those of the legs? What is the significance of this?

9 Discuss the role of each of the following at the neuromuscular junction: acetylcholine, sodium, and cholinesterase.

10 Could a parenteral injection of curare affect respiration? How?

11 Explain the meaning of steady state and oxygen debt as they relate to exercise.

12 Explain why recovery heat production is greater in isolated muscle than that in the intact animal.

13 Of what clinical value is electromyography? Under what conditions would you expect to see it employed?

14 The all-or-nothing response refers to which of the following: (a) muscle fiber, (b) muscle bundle, or (c) myofilament?

15 What is the relationship between the initial length and force of contraction for skeletal muscle?

16 In the phenomenon of summation, why is the second contraction of the fiber less than the initial contraction?

17 What is the difference between tetanus and tone?

18 Under what conditions would you expect a patient to exhibit an absence of muscle tone? What are the implications for nursing care; that is, what will the nurse need to do for the patient that he can no longer do for himself?

19 Under what conditions would you expect that the leg muscles of an individual would atrophy?

20 What are the anatomical and physiological differences between the muscle fibers of skeletal and smooth muscle?

21 What is meant by the inherent rhythmicity of cardiac muscle?

22 Why is the structure of cardiac muscle no longer described as a syncytium?

23 How does myoglobin function in cardiac muscle?

24 What is meant by Starling's law of the heart?

25 In what area of the heart is contraction initiated?

26 Why is it not possible to demonstrate the properties of summation and complete tetanus in cardiac muscle?

27 Describe the events which could lead to the formation of an area of infarction in the myocardium.

PART 2
THE NERVOUS SYSTEM

8 THE NEURON

The nervous system and the hormones secreted by the endocrine system (see Chapter 31) have the task of maintaining balance among the many activities of the body and of preparing its responses to the external environment.

The nervous system is remarkably widespread in the body. A map of it would show the body and its organs in three dimensions. Furthermore, nerve impulses may be transmitted along this network virtually from one end to the other, so rich are the interconnections. Thus, the activities of one body region are quite likely to influence the events in others, and therefore the whole must be considered in any careful examination of activity within a part. Obviously the functioning of the nervous system can be extraordinarily complex, a fact that the student should remember during the necessarily simplified description that follows.

For descriptive purposes the nervous system is divided into central and peripheral portions. The central nervous system consists of the brain and spinal cord. The peripheral nervous system is made up of nerves that carry impulses between the central nervous system and muscles, glands, skin, and other organs.

ANATOMY OF THE NEURON

The basic functional and anatomical unit of the nervous system is the nerve cell, or neuron. The neuron is structurally the most complex cell of the body. It has a nucleus and

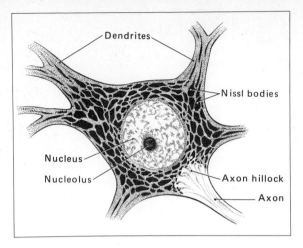

Figure 8-1 labels: Dendrites · Nissl bodies · Nucleus · Nucleolus · Axon hillock · Axon

Figure 8-1. A typical neuron. Also present but not shown in this drawing are neurofibrils, mitochondria, a Golgi body, and variable amounts of fat droplets and pigment granules.

cytoplasm like other cells, but the cytoplasm extends beyond the cell body into elongated processes.

Cell Body

Within the cell body is a large nucleus (with prominent nucleolus); many dark-staining granules called Nissl bodies; slender filaments termed neurofibrils; mitochondria; the Golgi body; and variable amounts of fat droplets and pigment granules (Figure 8-1). Almost always absent is a centrosome, which reflects the fact that nerve cells are incapable of mitosis and do not reproduce; once a cell body dies, the neuron is not replaced.

Only a small proportion of the billions of nerve cell bodies in the nervous system are outside the central nervous system. These are usually clustered in groups, called ganglia, surrounded by connective tissue. Within the brain and spinal cord cell bodies are found in clusters, but these are called nuclei. A nucleus or group of nuclei associated with a particular function is termed a center.

Processes

Nerve processes are threadlike extensions of the cell body. They contain cytoplasm and neurofibrils. The processes offer a way of classifying neurons (Figure 8-2). A neuron is said to be a unipolar neuron if a single process attaches to the cell body. During development these neurons are bipolar, a process growing from each pole of the cell body. Later the two processes fuse so that a single T-shaped extension projects from the cell surface, with one end directed toward the periphery and the other toward the central nervous system.

A bipolar neuron retains two separate processes. In the fully developed nervous system this type of nerve cell is found in the retina, olfactory mucosa, internal ear, and taste buds.

A neuron with several short processes and a single long one is termed a multipolar neuron. The cell body of a typical multipolar neuron lies in the ventral column of gray matter of the spinal cord and its long process extends out to skeletal muscle. Multipolar neurons whose cell bodies lie in the cerebral cortex are called pyramidal cells because of the pyramidlike shape of the cell bodies. Any process of a neuron may be called a nerve fiber.

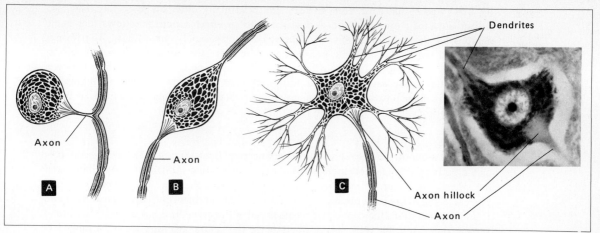

Figure 8-2. Types of neurons. **A.** Unipolar. **B.** Bipolar. **C.** Multipolar. Sensory neurons are typical unipolar neurons, except that in the retina, olfactory mucosa, taste buds, and internal ear the specialized sensory neuron is bipolar. Motor neurons are typical multipolar neurons.

Axon The long process of a neuron is termed the axon. It emerges from the cell body at an area free of Nissl granules known as the axon hillock. The axon may extend microscopic distances, as in some neurons within the spinal cord, or up to a meter in neurons that extend to skeletal muscles. It branches little until it reaches its termination, where it may divide into a spray of fibers. The axon contains neurofibrils but no Nissl granules. It is enclosed in a thin membrane, the axolemma. In addition, some fibers are sheathed in a fatty substance called myelin. Such fibers are known as myelinated, or medullated, fibers.

Outside the central nervous system the axon, whether myelinated or not, acquires another covering, the sheath of Schwann, or neurilemma. The neurilemma is actually the cytoplasmic membrane of a Schwann cell that enfolds the axon in more than one layer.

Along a myelinated axon at regular intervals are constrictions in the myelin where the neurilemma lies close to the axolemma. These constrictions are termed nodes of Ranvier, and it is here that a branch (collateral) may leave the axon. The area between two nodes is occupied by a single Schwann cell.

Dendrite The short processes of multipolar neurons are known as dendrites, and they are usually present in profusion. Dendrites contain Nissl substance and some neurofibrils. They are often densely covered by terminations of other neurons (see Figure 8-8).

Afferent, Efferent Neurons An impulse normally moves along a neuron in only one direction. One kind of process always carries an impulse toward the cell body, another always carries it away from the cell body. In the multipolar neuron the dendrites always bring impulses to the cell body, and the axon carries them away. In unipolar neurons the peripheral portion of the process brings the impulse to the cell body, and the central portion of the process carries the impulse away from the cell body. Yet two-way movement is necessary in the nervous system. Impulses containing sensory information pour into

211

the brain and spinal cord from muscle, skin, and many other outposts. Impulses also stream outward from the brain and spinal cord to effect action. It follows that some neurons handle only the inflow of impulses and others only the outflow. The former are termed sensory, or afferent, neurons, and the latter are motor, or efferent, neurons. Within the spinal cord and brain there are other neurons that act as bridges between sensory and motor neurons or that relay impulses to centers involved in reasoning, memory, or other functions. These neurons are named connection, association, or internuncial neurons.

Nerve

A nerve is a bundle of fibers enclosed in connective tissue (Figure 8-3). The large nerves of the body contain several bundles of fibers. Surrounding individual nerve fibers (and their myelin and neurilemma) there is a connective tissue sheath called the endoneurium. The enclosing tissue for a bundle of fibers is called the perineurium. Between and around a number of bundles the connective tissue is called the epineurium; it is laced with blood vessels and, often, fat cells. Nerve fibers leave the central nervous system arranged in nerves. There is frequent branching of nerves in the periphery, but the nerve fiber itself continues without interruption to its termination.

A motor nerve is one that contains only motor fibers, a sensory nerve one that contains only sensory fibers, and a mixed nerve one that contains both motor and sensory fibers. Most nerves of the body are mixed.

In the central nervous system fibers that transmit impulses associated with a specific modality, for instance, pain, lie together. Their common pathway is termed a tract, or

Figure 8-3. A peripheral nerve in cross section. The photomicrograph shows a nerve in longitudinal section.

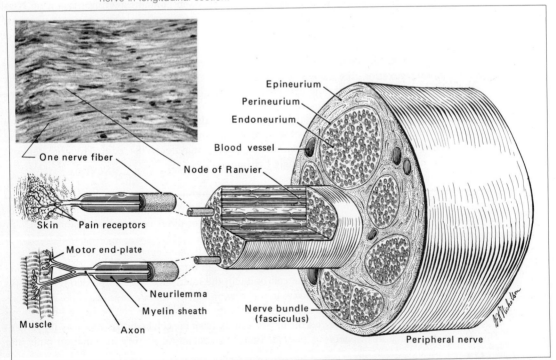

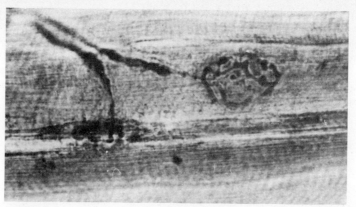

Figure 8-4. A motor end plate in skeletal muscle. A motor nerve fiber arborizes and terminates near muscle nuclei. (Muscle nuclei so situated are termed soleplate nuclei.) Collectively the terminations are a motor end plate.

fasciculus. The fibers of a tract are not enclosed in perineurium or epineurium as fibers are in the peripheral nervous system.

Specialized Nerve Endings

The motor nerve endings on skeletal muscle cells are well defined. The neuron arborizes and then terminates close to muscle nuclei. The terminations comprise a motor end plate (Figure 8-4). Motor nerve endings on cardiac muscle, smooth muscle, and glands are less well defined. These endings are known as visceromotor, or secretory, endings.

There are a number of modified sensory endings, called receptors, each of which is responsive to only a single type of stimulus (Figure 8-5). Special types of sensory

Figure 8-5. Sensory receptors.

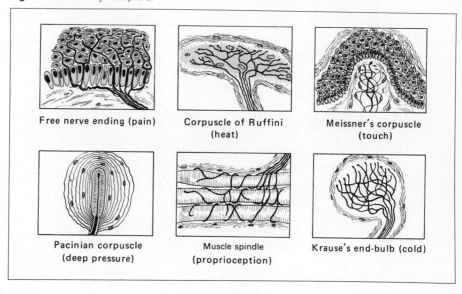

Free nerve ending (pain) Corpuscle of Ruffini (heat) Meissner's corpuscle (touch)

Pacinian corpuscle (deep pressure) Muscle spindle (proprioception) Krause's end-bulb (cold)

endings are found in nerves of the eye, nose, ear, and tongue and are described with vision, olfaction, hearing, and taste (Chapters 12 and 13). There are in addition general sensory receptors present in essentially all parts of the body that are responsive to pain, temperature change, touch, and pressure (see Chapter 10).

The simplest receptor is a bare, or free, nerve ending. It is bare in the sense that all external coverings are stripped away, leaving the axis cylinder exposed. These free nerve endings mediate painful stimuli.

Other nerve endings are encapsulated by connective tissue. Meissner's corpuscles, which are responsive to light touch, are composed of small swellings surrounded by a delicate connective tissue capsule. Ruffini bodies and the end bulbs of Krause are surrounded by heavier connective tissue sheaths. They are thermoreceptors. Pacinian corpuscles are sensitive to deep touch and pressure. These small bodies are composed of nerve endings surrounded by a series of concentric layers of connective tissue. Stretch receptors (muscle spindles) present in muscles and tendons mediate proprioception (awareness of the position of the body or its parts). This sensory modality is activated by pulling or stretching a muscle or tendon.

Nerve endings in the central nervous system are usually in the form of end feet or loops called terminal buttons, which terminate on other neurons.

Neuroglia The supportive tissue of the brain or spinal cord is composed of specialized connective tissue cells with many protoplasmic processes, the neuroglia. These cells are classified as astrocytes, oligodendroglia, and microglia.

Astrocytes are the largest of the glial cells (Figure 8-6). They have star-shaped cell bodies with many processes radiating outward. Some cells have slender, delicate, unbranched processes with fine fibrils and are called fibrous astrocytes. Other cells have

Figure 8-6. Neuroglial tissue with two types of astrocytes. The terminal processes of an astrocyte around a blood vessel are perivascular feet.

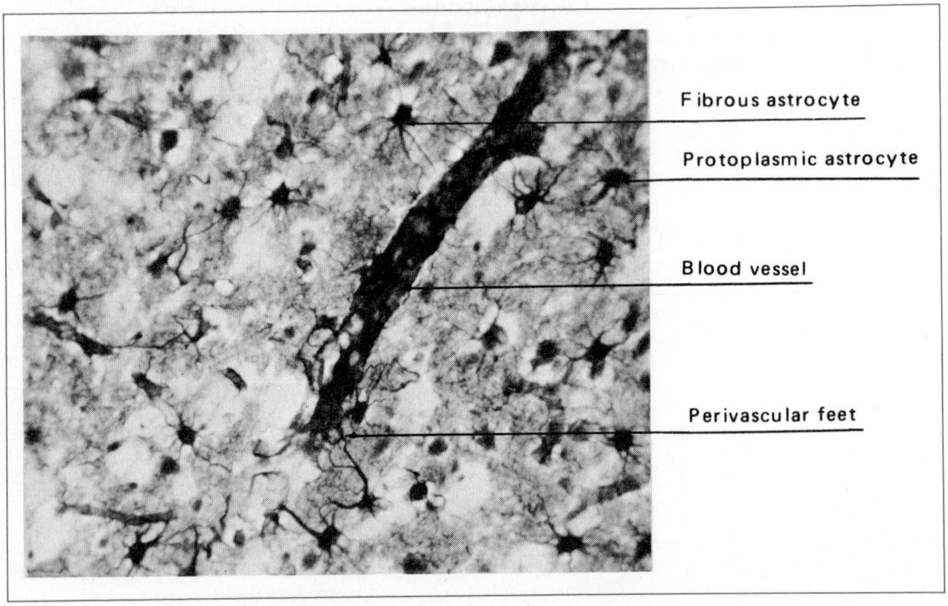

Fibrous astrocyte

Protoplasmic astrocyte

Blood vessel

Perivascular feet

shorter, thicker, nonfibrous processes and are designated as protoplasmic astrocytes. Many of the astrocytic processes end on blood vessels as perivascular feet.

Oligodendroglia are not as large as astrocytes. They have short, beaded processes with no fibrils. They are characterized by small, round, light-staining nuclei. These cells perhaps function in myelin formation in the central nervous system as do Schwann cells in the peripheral nervous system.

The microglia are the smallest of the glial cells. They have small, darkly stained, uneven nuclei, scanty cytoplasm, and slender branched processes. Microglia are phagocytic and function to remove cellular debris or metabolic breakdown products of the central nervous system.

PHYSIOLOGY OF THE NEURON

The neuron has been described as the structural and functional unit of the nervous system. Before we consider the nerves and tracts essential to perception and movement (Chapter 10), the mechanism by which neurons transmit information will be reviewed.

Neuronal Activation

In Chapter 2 resting and action potentials were explained. The point was made there that when a cell, such as a neuron, is at rest, the potential difference across the membrane is about 70 mv and does not change until the cell is stimulated. The neuron may be stimulated either directly or via the associated receptor. Some receptors are nothing more than free nerve endings, but others, as just discussed, are more complex. In any event, the appropriate stimulus will evoke a potential in the receptor. When the potential reaches threshold magnitude, the resting potential of the neuron is altered sufficiently to evoke an action potential. When this occurs, the neuron is said to have fired.

Action Potential Propagation

An action potential, at first, does not involve the entire cell, certainly not the typical neuron with long axon. The potential changes first occur at the point of stimulation. In that area, the outside of the membrane becomes negative while contiguous areas of the membrane are positive (Figure 8-7). The positive charges are therefore attracted to the negative ones. The area of negative charges is said to be a sink into which the positive charges flow. The removal of positive charges reduces the potential difference of the two sides of the membrane at that point. That is to say, the resting potential at the point becomes less negative; as a result the permeability increases and sodium quickly moves in. This part of the membrane then undergoes polarity reversal, and the outside becomes negative. The sequence is now ready to be repeated. Accordingly, there is a progressive series of action potentials which move in both directions along the nerve from the point of stimulation. The action potentials so generated constitute what is termed the nerve impulse. In the nervous system, the impulse begins at the dendrite or cell body and is then propagated throughout the length of the axon.

When myelin covers the neuron, impulse propagation is a bit different. Myelin acts as an insulator, but this insulation is broken at the nodes of Ranvier (Figure 8-3). Thus the positive charges must flow from one node to the next by passing over the myelin. Because the charges leap from node to node, this form of propagation is termed saltatory propagation, saltatory meaning "dancing" or "leaping."

Saltatory propagation provides high-velocity movement of the impulse. This results from the fact that the depolarization process leaps from node to node, a process that is far faster than the step-by-step depolarization process. Thus myelinated fibers propagate

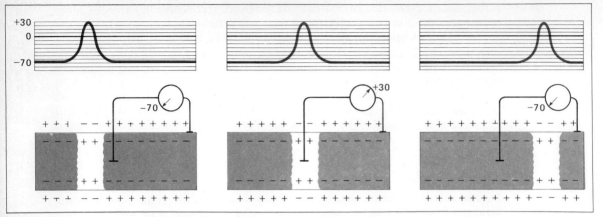

Figure 8-7. Propagation of an action potential. At the bottom an unmyelinated axon is represented in longitudinal section. A galvanometer measures the potential difference between the electrodes. During development of the action potential in a given area, the potential inside the cell changes as permeability of the cell membrane is altered. Positive ions from contiguous areas are then able to flow inward. The potential difference is first abolished and finally becomes positive. Further permeability changes lead to an outflow of positive ions, and the resting state is quickly restored.

impulses much faster than do unmyelinated ones. In addition, energy is conserved, because energy required for the ionic pumps is utilized only at the nodes rather than throughout the length of the fiber.

Frequency of Response

The absolute refractory period limits the number of impulses propagated per unit time, that is, the frequency of response. If a neuron has an absolute refractory period of 0.4 millisecond (msec), the maximum impulse frequency cannot exceed 2,500/sec. The absolute refractory period varies from fiber to fiber. The smaller the fiber diameter, the longer the refractory period, and therefore the maximal frequency of response must be lower in small fibers than it is in large ones. Why the length of the refractory period varies with fiber diameter is not known.

Velocity of Propagation

The velocity at which a neuron propagates an impulse is a function of its diameter. The fibers of greatest diameter have been found in the spinocerebellar tract (see Figure 10-4). They transmit impulses at a velocity of up to 130 m/sec. Large motor neurons propagate impulses at a velocity of 80 to 100 m/sec. Sensory nerves generally have a similar diameter and transmit impulses more slowly. In man, a velocity as low as 2 m/sec characterizes the very small sensory fibers. Even lower velocities occur in neurons of the autonomic nervous system (Table 8-1). As just explained, nonmedullated, that is, unmyelinated, neurons propagate impulses more slowly than do medullated ones.

All-or-Nothing Response

In the discussion of the physiology of muscle the point was made that the muscle cell responds maximally or not at all. The same is true of the nerve cell. Every neuron has

TABLE 8-1. RELATIONSHIP OF NEURON DIAMETER AND PROPAGATION VELOCITY

TYPE	DIAMETER, MICRONS	VELOCITY, M/SEC	FUNCTION
A (alpha)	12–20	80–130	Motor, proprioception
A (beta)	8–12	40–80	Touch
A (gamma and delta)	1–8	5–40	Pain, temperature
B	3	3–15	Autonomic preganglionic
C	1 or less	0.5–2.0	Autonomic postganglionic

a specific threshold of excitation, but once that threshold is exceeded, the response is unchanged by increasing the intensity of stimulus. A single threshold stimulus will evoke an action potential of a specific magnitude, configuration, and duration. The action potential and its velocity of propagation do not change when the stimulus intensity is increased. The number of impulses propagated by the neuron per unit time, however, is directly proportional to the rate of stimulation up to a maximum beyond which the neuron cannot respond because of the refractory period.

The threshold of excitation, the amplitude, and the duration of the action potential all seem to be related to the diameter of the neuron just as the velocity of propagation is. Thus, the greater the neuron diameter, the lower the excitation threshold, the greater the amplitude of the action potential, the longer its duration, and the greater the velocity of propagation.

Summation of Action Potentials

A nerve is composed of many neurons. The action potential recorded from a nerve depends upon the number of neurons firing. And because a nerve usually contains neurons with different activation thresholds, it is possible to alter the nerve action potential by increasing the intensity of stimulation. A weak stimulus activates only a few neurons; a strong one, more. The total nerve action potential represents the summation of the action potentials of all the active neurons. Ultimately, of course, a stimulus intensity is reached at which all the neurons respond. An increase in stimulus intensity beyond that point will not alter the summated, or compound, action potential.

Energy of Propagation

Propagation differs from conduction in that in the former there is a self-perpetuating process that requires energy. Conduction, in the sense that a wire conducts an electrical current, does not require the expenditure of energy on the part of the wire. The activity of the neuron depends upon an ionic imbalance. In order for ionic concentration differences to be created, maintained, or reestablished, work must be done. In the neuron the source of energy for this work is thought to result ultimately from oxidative processes, primarily the catabolism of glucose. The immediate source of energy is probably the high-energy phosphate bonds of ATP.

Degeneration

Many years ago Waller demonstrated that when an axon of a peripheral neuron is cut, the part separated from the cell body degenerates. This phenomenon has since been referred to as Wallerian degeneration. Not only does the severed segment of axon die, but the part still attached to the cell body also displays marked alterations. These changes are called retrograde degeneration.

Regeneration

If an axon is severed from its cell body, the neuron may regenerate a new axon. The presence of a neurilemma is essential for nerve regeneration (see Figure 8-3). The neurilemma is present only in peripheral neurons, that is, the ones which stream out from the brain and spinal cord. The neurons which run in the brain and spinal cord are not covered by a neurilemma, and they do not regenerate after being cut.

It must be understood that only the neuron process is capable of regeneration. If the cell body is destroyed, the entire neuron dies. Following section of a peripheral neuron, the distal portion of the axon degenerates so that an empty neurilemma remains. The axon grows from the proximal process into the empty neurilemma. It has been shown that an axon will regenerate at the rate of about 2.5 mm/day.

Many of the surgical techniques currently used are based on an appreciation of the property of degeneration and regeneration of neurons. For example, it is now feasible to cross-suture the regenerating end of one nerve with the degenerated segment of another. In this way it is possible to "order" a nerve to grow into a locale foreign to it and to innervate muscles which it has never innervated before.

THE SYNAPSE

The neuron propagates the impulse after being activated by the receptor. The pathways over which the impulse must pass are made up of a chain of neurons. Each link in the chain leads the impulse on to the next neuron. The junction between two neurons is called a synapse. The word synapse means "contact." A synapse, then, is a part of the nervous system in which two or more neurons are in functional association.

Actually, there is no protoplasmic continuity at the synapse. This is convincingly demonstrated when the axon of the primary neuron in a chain is severed from its cell body. As noted previously, this axon now degenerates. Histologic studies reveal that the process of degeneration occurs up to the synapse. The secondary neuron, with which the degenerating fiber is in synaptic union, is spared. It is clear, therefore, that a true anatomic gap exists at the synapse.

Organization of the Synapse

Figure 8-8 shows a diagrammatic representation of a synapse. Note that there are many terminal fibrils of the presynaptic neuron in contact with the dendrites or cell body, or both, of the postsynaptic neuron. As already mentioned, the terminal fibrils are named end feet, synaptic knobs, or terminal buttons.

The terminal fibrils may all come from a single neuron, or they may represent many neurons. The latter arrangement is termed convergence, because two or more presynaptic neurons impinge, that is, converge, upon a single postsynaptic neuron.

The opposite arrangement, in which a single presynaptic neuron subdivides and each branch forms a synaptic union with a separate postsynaptic neuron, is also common. Such an arrangement is referred to as divergence and is found in the motor system where a single large cell from the brain diverges upon neurons that innervate muscle. In this way, a single nerve cell controls many muscle fibers.

Synaptic Transmission

A closer examination of a terminal knob of a presynaptic neuron discloses that it contains many mitochondria and is loaded with vesicles (Figure 8-9). Note also that there is a

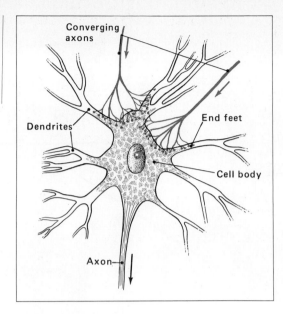

Converging axons

Dendrites

End feet

Cell body

Axon

Figure 8-8. Synapsis upon a neuron cell body. The arrangement shown is convergence: two presynaptic neurons impinge upon a single postsynaptic neuron.

space, the synaptic cleft, usually 200 to 300 Å wide between the knob and the postsynaptic neuron.

In 1936, Henry Dale and Otto Loewi received the Nobel Prize for their work which demonstrated that when the impulse reaches the synapse a chemical is liberated by the axon and it is this substance that stimulates the postganglionic nerve or, in the case of the neuromuscular junction, the muscle fiber. In 1971, the Nobel Prize went to B. Katz, Ulf von Euler, and Julius Axelrod for their work which further elucidated this mechanism.

The present concept envisions that depolarization opens so-called calcium gates in the terminal axon membrane. As a result, calcium ions move down their concentration gradient into the terminal knob of the axon. Here they initiate the quantal release reaction. The quantal release reaction involves a sequence of events by which the vesicles undergo frequent transient collisions with the axon membrane and ultimate fusion between the vesicular and axon membranes. Finally, the fused membrane opens up, thereby permitting the molecules of transmitter substance contained within the vesicles to be emptied into the synaptic cleft. This expulsion is termed exocytosis. Just how calcium ions bring about this sequence remains to be clarified. Current evidence suggests that the primary function of the calcium ions is to bring about fusion between the vesicular and axon membranes.

The most common chemical transmitters are acetylcholine (ACh) and norepinephrine, although other biogenic amines, such as dopamine and 5-hydroxytryptamine, may exist in certain junctions. ACh is the most ubiquitous. Quite clearly, whatever the transmitter, it must be quickly and abundantly synthesized in the terminal knob and then transported into the vesicles. Once evacuated into the synaptic cleft, it must affect the membrane of the postganglionic neuron. And, finally, there has to be a substance that inactivates the transmitter in order to prevent repetitive postganglionic responses.

The final step in the synthesis of ACh is catalyzed by cholineacetylase. The process involves the transfer of an acetyl group from acetyl coenzyme A to choline. Just how ACh is transported into the vesicles is not yet clear. Undoubtedly ATP plays a role. ATP is supplied by the abundance of mitochondria present in the terminal knob. Some

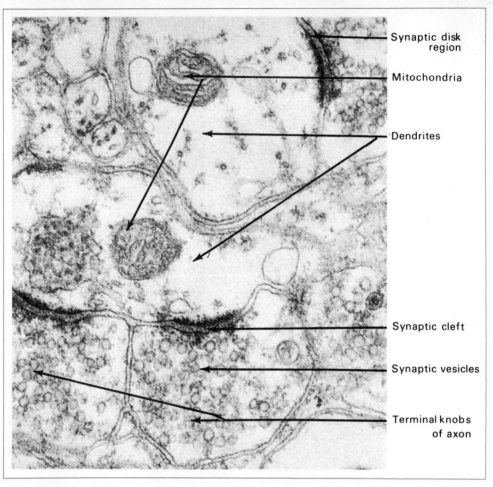

Synaptic disk region

Mitochondria

Dendrites

Synaptic cleft

Synaptic vesicles

Terminal knobs of axon

Figure 8-9. Electron micrograph of a synapse. The two rounded areas at the bottom are terminal knobs of axons. Their synaptic vesicles hold the transmitter substance. When the nerve impulse reaches the terminal knob, the vesicles release the substance into the cleft between the knob and an adjacent dendrite (center of micrograph). (*Courtesy of George D. Pappas.*)

authorities believe not only that the mitochondria provide ATP to supply the energy needed for ACh synthesis but that the mitochondria themselves may synthesize ACh.

In order for ACh to influence the postsynaptic neuronal membrane there must be so-called receptor sites, in this case termed the cholinoreceptors (ChR's). These receptors bind the transmitter to a specific site in the receptor and then produce the response of the postganglionic neuron. These receptor sites are embedded in the postsynaptic membrane.

In some manner, still to be discovered, these receptors, in response to the binding of ACh, increase the permeability of the postsynaptic membrane to sodium, which then flows into the cell, bringing about depolarization. Present in the synapse is acetylcholinesterase (AChE), which is an enzyme that is said to have the highest velocity of any known mammalian enzyme. This means that it inactivates ACh within a fraction of a millisecond.

An understanding of each step of the sequence of synaptic transmission is of interest not only in physiology but also for the treatment of several disorders. For example, the fatal outcome of poisoning with botulinum toxin is due to the block of the release of ACh which results in paralysis. In another disorder, called myasthenia gravis, there is a deficiency in the amount of ACh released at the terminals of the motor nerves. Treatment consists in the administration of drugs, such as neostigmine, that block cholinesterase.

Synaptic Delay

By use of the cathode-ray oscilloscope one can determine the time required for the impulse to be transmitted across the synapse. The synaptic delay, or period of time before the postsynaptic neuron fires after the impulse reaches the terminal knob, is about 0.5 msec. The cause of this delay is the time required for the liberation of the excitatory transmitter, the time it takes for the transmitter to diffuse across the synaptic gap, the time necessary for alteration of the permeability of the postsynaptic neuron membrane, and the time it takes for the influx of sodium to trigger the action potential.

Synaptic Fatigue

Because the transmission of the impulse depends upon the synthesis and liberation of a chemical transmitter, quite clearly, rapid and prolonged stimulation of the presynaptic neuron can exhaust the transmitter so that the postsynaptic neuron is no longer activated. Such a state is referred to as synaptic fatigue.

One-Way Transmission

A neuron is capable of propagating an impulse in either direction. In a chain of neurons, though, the impulse travels in only one direction because of the synapse. At the synapse, only the terminal knob liberates the transmitter and thus excites the postsynaptic neuron. If the latter neuron were first activated, say by stimulation of the axon, the impulse would be propagated back to the cell body and dendrite. But it could not activate the presynaptic neuron, because there is a gap and no transmitter is liberated by the cell body or dendrites. One-way transmission through synapses is known as the Bell-Magendie law.

Summation

In order for an impulse to be initiated in the postsynaptic neuron there must be sufficient alteration in the permeability of the membrane for the action potential to result. The activity of a single terminal knob is apparently inadequate. Yet, activation of a single knob does produce a change in the postsynaptic neuron. This change is a small decrease in the negativity of the resting potential and is referred to as the excitatory postsynaptic potential (EPSP). During the EPSP, which lasts only a few milliseconds, the postsynaptic neuron can be more easily depolarized. Accordingly, if another terminal knob becomes active at the same time, the membrane alterations resulting from its activity will be added to those resulting from the activity of the first knob. This addition, or summation, may then suffice to initiate an action potential. Because the additive factors come from two different terminal knobs that are separated by space, this type of addition is termed spatial summation.

The alterations evoked by the activity of a single terminal knob persist for a brief period of time. If the same terminal knob were to become active a second time, the two influences could be added. Because these additive factors are separated by time, this type of addition is termed temporal summation (Figure 8-10).

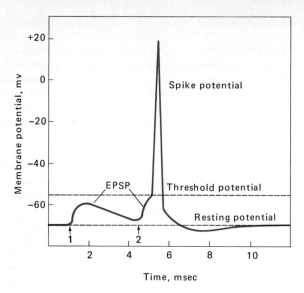

Figure 8-10. Excitatory postsynaptic potentials. The first EPSP depicted fails to reach threshold potential, and no action potential follows. The second EPSP, added to the first, does cause depolarization. This is termed temporal summation.

Facilitation

There are probably many presynaptic neurons which are incapable, by themselves, of activating postsynaptic neurons. Help is needed. Such help is provided in the form of another presynaptic neuron which also does not fire the postsynaptic neuron but which does evoke sufficient local alteration in membrane permeability, of which the first presynaptic neuron takes advantage. This aid is termed facilitation. In brief, one presynaptic neuron produces the EPSP which lowers the threshold of excitation of the postsynaptic neuron, thereby facilitating and aiding another neuron to fire it. Facilitation, then, depends upon spatial summation.

Inhibition

Synaptic transmission may be inhibited by the activity of specific presynaptic neurons. Apparently, such neurons liberate an inhibitory transmitter which raises the threshold of excitation of the postsynaptic neuron. Actual measurements of the membrane potential show that in response to the firing of the inhibitory presynaptic neuron there is hyperpolarization of the secondary neuron; that is, the membrane potential becomes more negative than it is at rest. This change is termed the inhibitory postsynaptic potential (IPSP).

The inhibitory transmitter has not yet been identified, although there is growing evidence that it is gamma aminobutyric acid (GABA). GABA is present in the central nervous sytem, and it does cause inhibition. GABA is thought to cause hyperpolarization by increasing the postsynaptic neuron membrane permeability for potassium and chloride, but not for the larger sodium ion. Thus, more potassium diffuses out of the neuron, and more chloride moves in. The net result is to make the inside more negative in reference to the outside. The resting potential, which is normally about −70 mv, becomes, under the influence of the inhibitory transmitter, about −80 mv and thus is more difficult to fire.

Another inhibitory transmitter may be dopamine; at least it has been found to function in this manner in the mammalian superior cervical sympathetic ganglion. It causes hyperpolarization of the postganglionic neurons but probably not as a direct action. A recent report shows that the dopamine causes an increase in the amount of cyclic AMP (adenosine 3′,5′-monophosphate) in the postganglionic neurons and that it is this increased cyclic AMP which is responsible for changing the membrane potential.

Inhibition depends upon the presence of specific short-axon neurons. Both excitatory and inhibitory neurons commonly impinge upon the same postsynaptic neuron. Thus a situation exists in which each influence may operate alone or they may operate in opposition to one another. This arrangement provides fine control for neuronal activity.

Function of the Dendrites

Neurons are said to obey the all-or-nothing law. This is true insofar as the axon is concerned, but recent evidence suggests that the dendrites have a graded response to the intensity of the stimulus. In this respect, the dendrites would be similar to receptors and, as a matter of fact, seem to have the same function. In other words, the postsynaptic potential in the dendrites varies according to the input. When the EPSP becomes great enough, it initiates an all-or-nothing response at the axon hillock. This local graded postsynaptic potential spreads along the dendrite to the axon electrotonically; it is not propagated as in the axon. Since there are usually many dendrites converging upon the cell body, there is spatial summation of the electrotonic effects of all active dendrites of that neuron.

Prolonged Neuron Activity

In the simple synaptic arrangement thus far considered, the presynaptic neuron fires the postsynaptic neuron, and once the impulse has been propagated the length of the chain, the activity terminates. There is in the central nervous system, however, activity of a more enduring nature. Circuits must exist in which the impulse of the presynaptic neuron evokes a continuing action. Many circuit patterns have been postulated. Figure 8-11 is a theoretical circuit in which the neurons would continue to fire until fatigue or inhibition brought it to an end. In all probability, such a circuit is always associated with an inhibitory fiber capable of quickly ending the oscillation of the impulse.

Figure 8-12 shows another circuit in which repetitive activity of the final neuron would occur. Since this neuron continues to fire after the primary neuron is no longer being activated, the response is termed after-discharge. Activity in the last neuron will cease after the impulse has traversed the longest pathway with the most synapses.

Figure 8-11. (*Left*) An arrangement of neurons for continuous firing. Once activated, the neurons in this circuit continue firing until fatigue or inhibition bring it to an end. **Figure 8-12.** (*Right*) An arrangement of neurons for producing after-discharge, that is, repetitive firing of the last neuron. The last neuron continues firing until neurons of the longest chain serving it have fired.

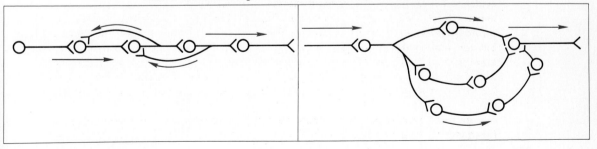

QUESTIONS AND PROBLEMS

1   Describe the function of (*a*) afferent neurons, (*b*) efferent neurons, and (*c*) internuncial neurons.

2   Compare the consequences of extensive damage to afferent neurons with the consequences of similar damage to efferent neurons.

3   Differentiate between the structure of a neuron, a nerve fiber, and a nerve.

4   If you place an ice cube against your skin, what receptors are stimulated?

5   When a patient is unable to change position in bed, he soon becomes uncomfortably aware of pressure and pain in areas of skin overlying the bony prominences. What specialized receptors make him aware of these sensations?

6   Comment on the following statement: Transmission of an impulse across the synapse is due to the physical contact of presynaptic membrane with postsynaptic membrane.

7   Compare impulse propagation in the axon of a nonmyelinated and a myelinated neuron.

8   Is nerve impulse transmission faster in a motor or a sensory neuron? What information supports your answer?

9   If a stimulus greater than threshold strength is applied to a neuron, does the configuration of the action potential change with increase in the strength of the stimulus? Give your reasons.

10   Explain the relationship between spatial summation and facilitation.

11   What is the relationship between summation and the all-or-nothing response?

12   What is meant by Wallerian degeneration? What evidences of such degeneration would you expect to see among sick and injured people?

13   If an axon is cut, will the distal portion degenerate, regenerate, or remain intact? Explain.

14   Define synaptic fatigue and synaptic delay, and explain how one could affect the other.

15   Of what advantage is a divergent neuronal circuit? A convergent neuronal circuit? Give an example of each.

16   What might happen at the synapse if the Bell-Magendie law was not observed?

17   Differentiate between EPSP and IPSP.

18   What effect does GABA have on the synaptic transmission of an impulse?

9 ANATOMY OF THE NERVOUS SYSTEM

As was outlined in the preceding chapter, the central nervous system is the term used to designate the brain and spinal cord. The peripheral nervous system refers to the nerves carrying impulses to and from the brain and spinal cord. A portion of the peripheral nervous system carries impulses to smooth muscle, cardiac muscle, and glands, and is termed the autonomic nervous system; this portion is reserved for special discussion, in Chapter 11.

EMBRYOLOGY

The nervous system develops from ectoderm on the dorsum, or posterior surface, of the embryo. At about the third week of development, the cells in this area increase in number and form a longitudinal depression, the primitive neural groove, which extends along the entire length of the embryo. Proliferation and overgrowth of ectodermal cells adjacent to the neural groove result in a closure of the groove, to form the neural tube (Figure 9-1). As the tube is formed, it separates from, and sinks below, the surface ectoderm. Clusters of ectodermal cells flanking the neural tube develop into the neural crest. Cells from the neural tube and neural crest form the entire nervous system including the brain, spinal cord, ganglia, and peripheral nerves.

The cells of the neural tube may develop into either spongioblasts, which become

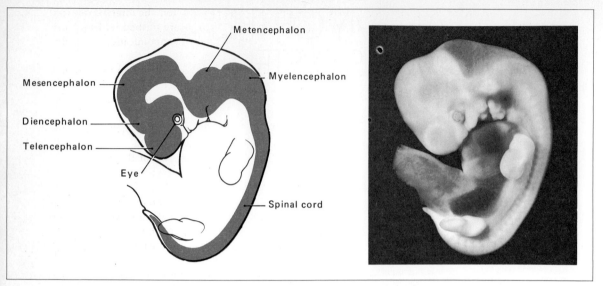

Figure 9-1. A. Development of the nervous system. The neural tube is shown in color. Note the expansion at the cranial end and the constrictions demarcating the subdivisions of the brain. **B.** Photograph of a 34-day-old embryo. The actual length is 11.6 mm. (*Chester F. Reather, photographer. Carnegie Institution of Washington.*)

the supporting, nonnervous glial cells of the adult nervous system, or neuroblasts, which develop into the functional nerve cells.

Development of the Brain

The brain develops from the expanded anterior portion of the neural tube. The growth is not uniform, and early in development (about 4 weeks) three unequal enlargements, the primary brain vesicles, appear. These are designated as the forebrain, or prosencephalon, consisting of the telencephalon and diencephalon; the midbrain, or mesencephalon; and the hindbrain, or rhombencephalon, subdivided into the metencephalon and myelencephalon (Figure 9-1). The uneven growth of the head end of the neural tube results further in three distinct bendings, or flexures. The first, the cephalic flexure, occurs in the region of the midbrain and thrusts the forebrain forward to curve under the embryo; the second, the pontine flexure, at the region which will develop into the pons, creates a sharp bend in the opposite direction; the third, the cervical flexure, curves the tube further in the direction of the cephalic flexure. These changes, accompanied by proliferation of cells, transform the head end of the tube into a greatly enlarged, convoluted portion from which the major subdivisions of the brain develop. The smaller, straight caudal end develops into the spinal cord. The lumen of the neural tube at the head end expands into the primitive vesicles which become the ventricles of the brain. In the caudal segment the lumen is retained as the central canal of the spinal cord.

At about the fifth week the forebrain develops prominent paired, lateral outgrowths, the cerebral hemispheres, which form the bulk of the adult brain. These lateral outgrowths further rapidly proliferate to envelop most of the rest of the brain as the telencephalon. The caudal part of the forebrain develops into the diencephalon, which differentiates into the thalamic portion of the brain.

In the region of the developing orbital cavity, outpouchings from the forebrain appear as the optic vesicles. The expanded ends of the optic vesicles are pushed in, becoming the optic cups, and develop into the retina of the eye. The communication, or stalk, between the optic cup and the brain becomes the optic nerve.

The midbrain, or mesencephalon, has no subdivision. It develops into the cerebral peduncles, which are massive fiber tracts connecting the lower centers of the brain with the cerebral hemispheres, and the corpora quadrigemina, four small elevations associated with auditory and visual functions.

The hindbrain is divided into two portions: The upper portion, or metencephalon, gives rise to the cerebellum, the pons, and part of the medulla oblongata. The lower part of the hindbrain, the myelencephalon, develops into the lower part of the medulla.

Development of the Spinal Cord

The caudal portion of the neural tube, which develops into the spinal cord, retains its tubular shape. It remains approximately the same diameter throughout except for slight enlargements in the cervical and lumbar regions. At these sites proliferation of cells results in formation of the brachial and lumbosacral nerve plexuses, nerve networks that supply the upper and lower limbs.

As the spinal cord develops, the nervous elements separate into three layers: (1) a central layer of cells, the ependymal layer, lining the central canal; (2) an extensive middle layer, the mantle layer; and (3) an outer marginal layer.

The mantle layer develops into the gray matter of the cord. Neuroblasts of this layer become grouped into a mass or column in the ventral part of the cord—the ventral gray columns. All the cells in this region develop into motor neurons. Smaller clusters of neurons locate in the dorsal part of the cord to become the dorsal sensory columns. Thus the cord in cross section appears as an H of central gray matter surrounded by white matter. The limbs of the H are formed by the dorsal and ventral columns on either side of the cord, the crossbar of the H by gray matter surrounding the central canal.

The peripheral white matter is derived from the marginal layer of the developing spinal cord. This layer is characterized by the presence of large numbers of spongioblasts, which give rise to glial (connective tissue) cells. The myelinated axons of the gray matter cell bodies extend into these developing glial fibers and are responsible for the white, glistening appearance of the peripheral portion of the cord.

Cells of the neural crest contribute to the formation of the ganglia of the peripheral nervous system.

ANATOMY OF THE CENTRAL NERVOUS SYSTEM

The bulk of the brain consists of the cerebrum, which almost completely fills the cranial cavity. The smaller cerebellum lies in the posterior fossa of the cranial cavity at the base of the skull under the cover of the cerebrum. The brainstem extends between the cerebrum and the spinal cord and, moving downward, consists of the midbrain, the pons (adjacent to the cerebellum), and the medulla oblongata, which is continuous with the spinal cord at the foramen magnum (Figure 9-2).

Cerebrum

The cerebrum is in the upper portion of the cranial cavity. The cortex, or surface, of the cerebrum is marked by a series of furrows which greatly increase its total area. A single furrow is termed a sulcus. The ridge between adjacent sulci is called a gyrus (Figure 9-3B).

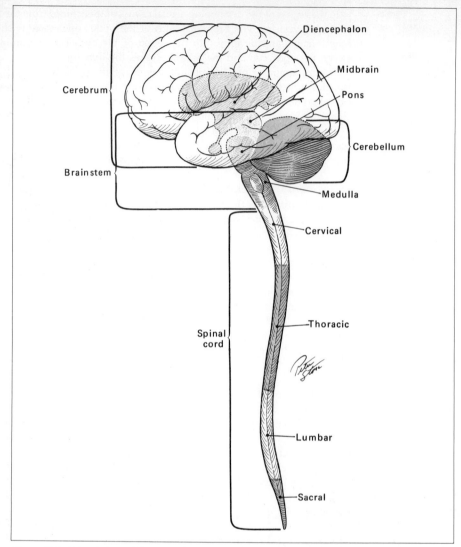

Figure 9-2. Lateral view of the central nervous system.

Seen from above, the cerebrum is oval. In the midline there is a deep groove, the longitudinal fissure, which divides it into right and left cerebral hemispheres. Additional fissures and sulci divide each hemisphere into functional lobes (Figure 9-3). A deep fissure in the side of each hemisphere, the lateral fissure (fissure of Sylvius), separates the temporal lobe from the remainder of the hemisphere. The temporal lobe is a tonguelike portion of the cerebrum lying deep to the temporal bone. The central sulcus extends laterally across the midportion of each hemisphere. It forms a dividing line between the frontal lobe and the parietal lobe. The frontal lobe lies beneath the frontal bone and the parietal lobe beneath the parietal bone. Farther back, the parietooccipital fissure forms the boundary between the parietal lobe and the occipital lobe.

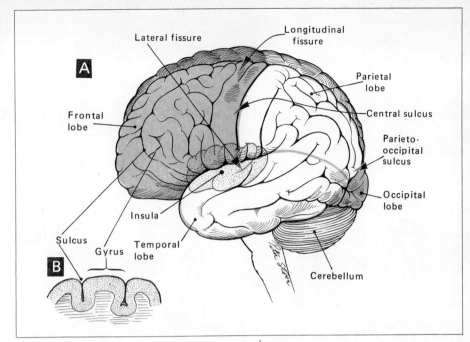

Figure 9-3. A. Lateral view of the cerebrum. Note the colored line that demarcates the parietal and temporal lobes. The insula is deep to the temporal lobe. **B.** A portion of the cortex in cross section.

An additional subdivision of the cerebrum, the insula, lies deep within the lateral fissure. The temporal lobe must be displaced away from the remainder of the cerebral hemisphere before the insula can be seen.

A specialized area on the inferior aspect of the frontal lobe, the olfactory trigone (triangle), gives superficial attachment to the olfactory tract of nerve fibers. The cerebral cortex consists essentially of gray matter, that is, neuron cell bodies. Beneath the cortex is white matter, that is, neuron fibers, which are arranged in tracts. Fibers that connect areas of the cortex within a single hemisphere are association fibers. Fibers that link the two hemispheres are commissures, the most prominent of which is a thick band of fibers known as the corpus callosum. Fibers that link the cortex to lower centers of the brain and to the spinal cord are projection fibers. It is only when stimuli from the body reach the cortex that full awareness of the stimuli occurs.

Functional Areas of the Cortex   From numerous research investigations, through observations of damaged areas of the brain and consequent loss of function in the body, and by the study of electrical stimulation to the brain, it has been determined that certain areas of the cerebral cortex are primarily concerned with specific functions (Figure 9-4). However, no single area functions by itself. Association and commissural fibers link various cortical areas so that several portions of the cortex are involved in any given action. For example, when a person hears a loud noise, the sound stimulus is recorded by his brain and in addition he usually turns around to see where the noise originated, blinks his eyes, and may decide to move away. Each sequence of this activity involves

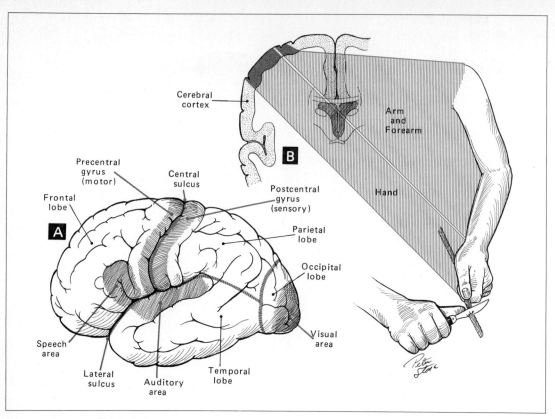

Figure 9-4. Lateral view of the cerebrum. **A.** Functional areas of the cortex. **B.** Representation of the left upper extremity on the primary motor (or sensory) areas of the cortex.

a specific area of the cortex; yet each is integrated with the whole of the activity by the association fibers.

Ascending and descending major tracts cross at some level in the brain or spinal cord from their side of origin to the opposite side of the body or brain. Thus the left side of the cortex receives stimuli and controls action from the right side of the body. In the following discussion only the major functional areas of the cortex will be described, though many additional areas have been identified.

Primary motor and sensory areas are located on the gyri immediately anterior and posterior to the central sulcus. Neurons controlling conscious, voluntary motor activity are located in the precentral gyrus, that is, in front of the central sulcus. The final neurons of sensory pathways are found in the postcentral gyrus, that is, behind the central sulcus, and thus it is here that general sensations reach consciousness. General sensations include heat, cold, touch, pressure, pain, and proprioception ("muscle sense").

The cortical area associated with vision is centered in the occipital lobe, with speech in the frontal lobe, and with hearing in the temporal lobe. The taste center is located deep in the lateral fissure. The prefrontal areas are those in front of the precentral gyrus. These areas, important in reasoning and memory, are discussed in Chapter 14.

Basal Ganglia The basal ganglia consist of collections of cell bodies located within the white matter of each cerebral hemisphere (Figure 9-5). The basal ganglia are the caudate, lenticular, and amygdaloid nuclei, and the claustrum. The basal ganglia are important links, or way stations, along various motor pathways of the central nervous system.

The caudate nucleus is the most medial of the basal ganglia. It is shaped like a comma with an extended tail. The head is the anterior portion. The body is arched, and the tail swings down and forward. The amygdaloid nucleus lies in a knob of gray matter at the tip of the tail, just internal to the cortex of the temporal lobe.

The claustrum is a thin plate of gray matter just beneath the cortex near the insula.

The lenticular nucleus is a lens-shaped body medial to the claustrum. It is actually composed of two groups of nuclei, the globus pallidus and the putamen.

In close association with the basal ganglia is the internal capsule. The internal capsule is a broad band of nerve fibers passing from the spinal cord and brainstem to the cortex, as well as fibers radiating from the cortex to lower centers. There is such a concentration of fibers that the internal capsule stands out as a discrete white structure. The lenticular and caudate nuclei, along with an adjacent portion of the internal capsule, are sometimes referred to as the corpus striatum.

Other centers, such as the substantia nigra, the nucleus of Luys, and red nucleus (see ahead under Midbrain), are related in function to the basal ganglia. The functions of basal ganglia are discussed in Chapter 14.

Figure 9-5. Basal ganglia. **A.** The relative positions of the chief basal ganglia. **B.** The location of these ganglia within the cerebrum.

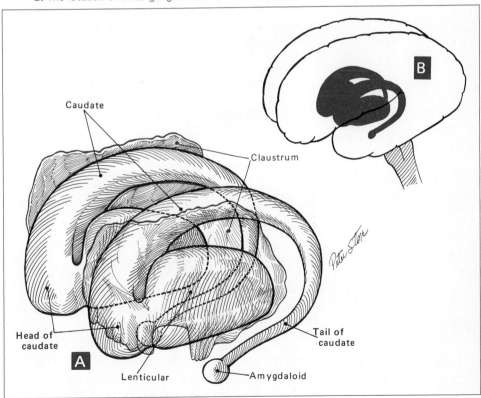

Caudate

Claustrum

B

Head of
caudate

Tail of
caudate

A

Lenticular

Amygdaloid

Diencephalon

Lying centrally within the brain between the cerebrum and the pons are the diencephalon and midbrain (Figure 9-6), covered at the sides by the cerebral hemispheres. The diencephalon lies to either side of the narrow, midline third ventricle (Figure 9-7). The cerebral peduncles are two rounded bands on the ventral surface of the diencephalon and midbrain. They consist of motor fibers that are part of the internal capsule.

Adjacent to the third ventricle the thalamic portion of the diencephalon is subdivided into an epithalamus, thalamus, hypothalamus, and subthalamus. The more important of these centers are the thalamus and hypothalamus.

The thalamus, an oval mass of nerve bodies, forms the bulk of the wall of the third ventricle. It is an important relay station between the cerebral cortex and the spinal cord. All incoming (sensory) impulses are projected to the thalamus and from there to the cortex. The thalamus is also under the control of the cortex so that impulses from the cerebral cortex to the spinal cord are relayed through this center. The thalamus acts as an integrating center through its connections with subcortical nuclei in the basal ganglia and the hypothalamus.

The hypothalamus forms the floor and a portion of the lateral wall of the third ventricle. Its functions are discussed in Chapter 11.

Figure 9-6. The diencephalon and brainstem. The letters refer to sections depicted in the next five illustrations.

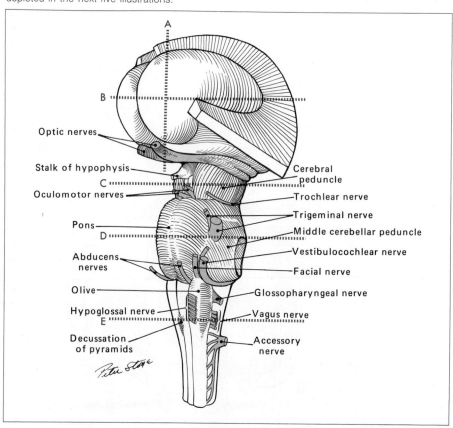

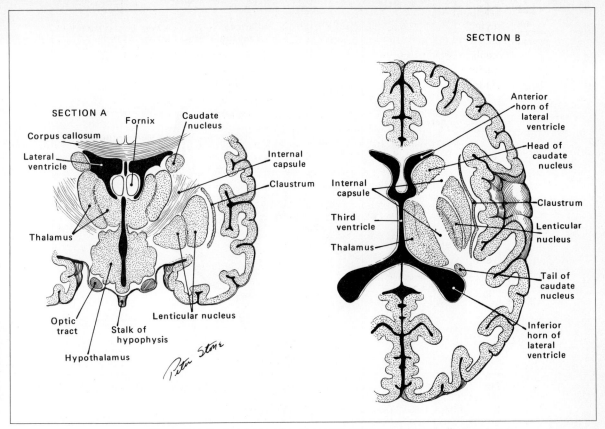

Figure 9-7. Section A is a coronal section of the diencephalon. Section B is in the transverse plane. See Figure 9-6 for the level of these sections.

Midbrain

The midbrain, or mesencephalon, consists of the cerebral peduncles and the corpora quadrigemina and surrounds the cerebral aqueduct, a channel between the third and fourth ventricles (Figures 9-8 and 9-11). As noted above, the cerebral peduncles are on the ventral surface. The corpora quadrigemina are four rounded prominences on the dorsal surface. Two of them are near the diencephalon and are termed superior colliculi. They are synaptic centers of the visual pathway, relaying impulses from the retina to the occipital lobes (see Chapter 12). The other two corpora are adjacent to the pons and are referred to as inferior colliculi. These are way stations for auditory impulses (see Chapter 13).

The red nucleus is located in the midbrain near the cerebral peduncles. It is part of the reticular formation and appears in fresh specimens as a large, pinkish yellow, oval of cells which extends from the level of the superior colliculus to the caudal margin of the diencephalon.

Pons

The pons, meaning "bridge," is a conspicuous bandlike segment that extends transversely across the brainstem; its basal portion is called the middle cerebellar peduncle (Figure

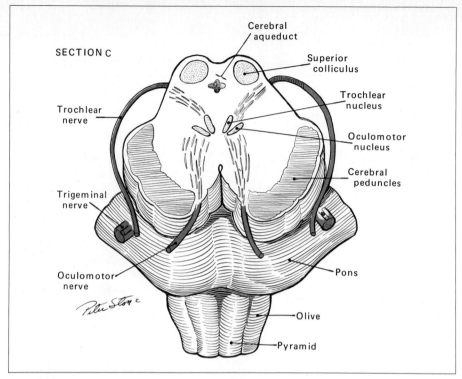

SECTION C

Cerebral
aqueduct

Superior
colliculus

Trochlear
nerve

Trochlear
nucleus

Oculomotor
nucleus

Cerebral
peduncles

Trigeminal
nerve

Pons

Oculomotor
nerve

Olive

Pyramid

Figure 9-8. Cross section of the midbrain (see Figure 9-6). The cerebral aqueduct is the communication between the third ventricle above and the fourth ventricle below. The superior colliculi are the two upper corpora quadrigemina.

9-9). Tracts in this portion of the brainstem link the cerebellum with the midbrain and the medulla oblongata.

Descending fiber tracts from the cortex and the thalamus, and ascending tracts from the spinal cord, course through the pons. On cross section scattered groups of cell bodies are evident near the ventral surface. These comprise the reticular formation (see ahead). Important nuclei of cranial nerves are located within the pons adjacent to the fourth ventricle.

Medulla Oblongata

The medulla oblongata is the upward extension of the spinal cord (Figure 9-10). It begins where the spinal cord passes through the foramen magnum and expands to become somewhat bulbar in shape. The central canal of the spinal cord greatly expands and dilates within the medulla, to become the fourth ventricle.

Fiber tracts that connect the spinal cord and cerebellum to higher centers of the brain become rearranged as they course through the medulla.

Grooves and ridges on the surface of the spinal cord are present also on the medulla. The fiber tracts of the cerebral peduncles appear as two rounded columns, the pyramids, on the anterior surface of the medulla. Near the junction of the medulla and the spinal cord these descending fiber tracts cross, and this crossing is named the decussation of

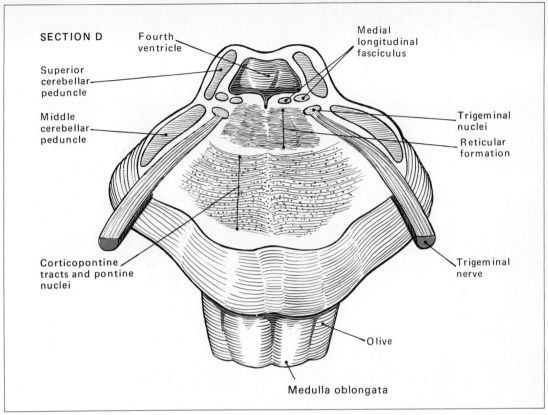

Figure 9-9. Cross section of the pons (see Figure 9-6). The cerebellum is adjacent to the pons, posterior to the fourth ventricle.

pyramids. The fibers then become the lateral corticospinal tracts of the spinal cord (see Chapter 10).

A flattened, oval mass, the olive, is situated between the anterior and posterolateral sulci at the upper end of the medulla.

On each side of the posterior aspect of the medulla there are two elevations, the fasciculus gracilis medially and the fasciculus cuneatus laterally, which are separated by the posterointermediate sulcus. These fasciculi are the upward extensions of the two large ascending tracts in the posterior funiculus, or posterior column, of the spinal cord.

In cross section the medulla is roughly square. Internal to the superficial structures already noted is the reticular formation, an area of scattered gray matter interspersed by white matter (myelinated fibers). In this region important centers regulating sleep, respiration, heartbeat, swallowing, and salivation are located (see Chapter 14).

Cerebellum

The cerebellum is the second largest part of the brain. It is located above the medulla at the back of the cranial cavity and is covered dorsally by the cerebral hemispheres (Figure 9-2). The cerebellum is separated from the cerebrum by the tentorium cerebelli, an extension of the dura mater (see ahead under Meninges).

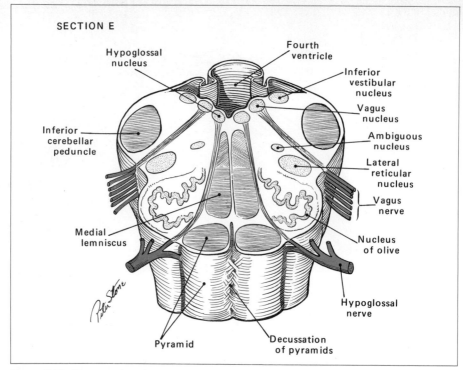

SECTION E

Hypoglossal
nucleus

Fourth
ventricle

Inferior
vestibular
nucleus

Vagus
nucleus

Inferior
cerebellar
peduncle

Ambiguous
nucleus

Lateral
reticular
nucleus

Vagus
nerve

Medial
lemniscus

Nucleus
of olive

Hypoglossal
nerve

Pyramid

Decussation
of pyramids

Figure 9-10. Cross section of the medulla oblongata (see Figure 9-6). The fourth ventricle becomes the central canal of the spinal cord below the medulla. The pyramids are the inferior extensions of the cerebral peduncles.

The cerebellum is composed of three lobes, the cerebellar hemispheres laterally and the vermis, which connects the hemispheres, in the midline.

As in the cerebrum the cortex of the cerebellum consists of gray matter covering deeply placed fiber tracts, which appear white. The cerebellar cortex is much thinner than the cerebral cortex. From the surface the cerebellum looks like a series of flattened plates or leaves, the folia, each having cortical gray matter covering a core of white matter (see Figure 14-2).

Cerebellar nuclei, deeply situated in the white matter, are connected to the midbrain via the superior cerebellar peduncles; to the pons via the middle cerebellar peduncles; and to lower centers of the brainstem and the spinal cord by way of the inferior cerebellar peduncles (see Figures 9-6 and 9-9).

Functionally the cerebellum acts to coordinate gross voluntary movement and is important in maintaining equilibrium (see Chapter 14).

Ventricles of the Brain

The lumen of the embryonic neural tube is retained in the fully developed central nervous system as the ventricles, or internal cavities, of the brain (Figure 9-11) and the central canal of the spinal cord.

The two lateral ventricles, by far the largest, are located in the cerebral hemispheres and extend into all their subdivisions. The main portion or body of the lateral ventricle

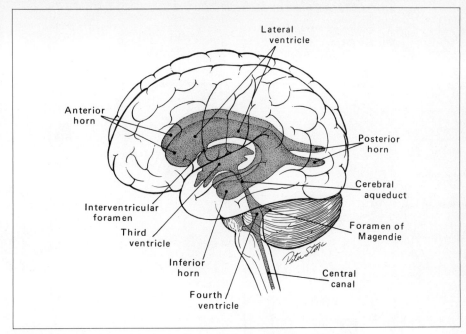

Figure 9-11. Ventricles of the brain. The ventricles are developed from the lumen of the neural tube of the embryo.

is located in the parietal lobe of each hemisphere, and from there projections extend into the frontal lobe as the anterior horn, into the occipital area as the posterior horn, and into the temporal lobe as the inferior horn.

Each lateral ventricle communicates with the third ventricle by way of an interventricular foramen (of Monro). The third ventricle stands between the right and left thalamus. The cerebral aqueduct (of Sylvius) connects the third ventricle with the fourth ventricle. The latter is a flattened pyramidal cavity between the pons and medulla anteriorly and the cerebellum posteriorly. At the lateral extent of the fourth ventricle there are two openings, the foramina of Luschka. In the midline, there is a single opening, the foramen of Magendie. These three openings allow communication between the ventricles and the subarachnoid space. At the termination of the medulla oblongata the fourth ventricle narrows sharply and continues as the central canal of the spinal cord.

The ventricles are associated with the production and circulation of the cerebrospinal fluid. In the lateral, third, and fourth ventricles, a delicate network of capillaries, the choroid plexus, is the site of production of cerebrospinal fluid (CSF). This fluid fills the ventricles and the central canal of the spinal cord. It flows through the foramina of the fourth ventricle and fills the subarachnoid space. CSF also extends for a short distance along the sheaths of cranial and spinal nerves as they leave the cranial cavity or the vertebral canal. (The production and function of CSF are discussed further in Chapter 19.)

Meninges

The central nervous system, in addition to being encased in bone and submerged in cerebrospinal fluid, is protected by fibrous membranes, the meninges (Figure 9-12). The

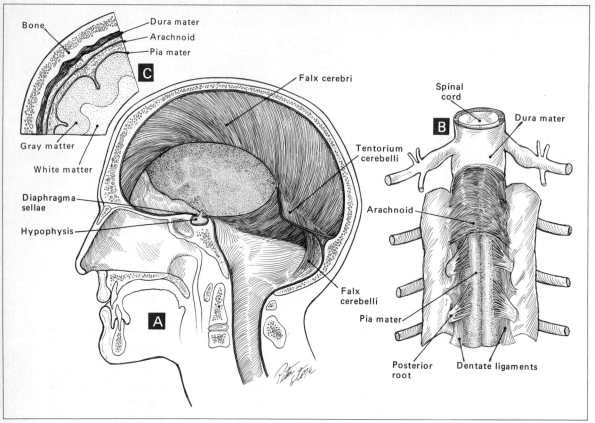

Figure 9-12. Meninges. **A.** Extensions of the dura mater in the cranial cavity. Sagittal view. **B.** The dura and arachnoid sheathe the spinal nerves at their origin. The dentate ligament separates dorsal from ventral roots and adheres to the dura. **C.** A vertical section through a portion of the calvaria and cortex. The meninges are depicted in greater detail in Figure 19-6.

meninges consist of three distinct membranes, namely, the dura mater, arachnoid, and pia mater, all of which completely enclose the brain and line the vertebral canal.

The dura mater forms a tough, outer covering. In the cranial cavity it is adherent to the endosteum of the skull, while in the vertebral canal a venous plexus lies between it and the vertebrae. In the cranial cavity extensions from the dura mater pass between various parts of the brain. The largest of these is the falx cerebri, which extends along the longitudinal fissure to partially separate the cerebral hemispheres. The falx cerebelli is its counterpart between the cerebellar hemispheres. Another dural extension, the tentorium cerebelli, passes transversely between the cerebrum and the cerebellum. At the anterior aspect of the tentorium cerebelli an oval gap, the tentorial notch, allows the brainstem to pass from the undersurface of the cerebrum into the posterior cerebral fossa. Dura mater also forms a roof, the diaphragma sellae, over the bony sella turcica, in which the hypophysis is situated. An aperture in the diaphragma sellae admits the stalk of the hypophysis to attach to the hypothalamus above.

The arachnoid is a filamentous, delicate covering immediately internal to the dura.

It is separated from the dura by the narrow (capillary) subdural space. Fine trabeculae extend from the arachnoid through the subarachnoid space to attach to the pia mater. Within the subarachnoid space there is cerebrospinal fluid, which acts as a liquid shock absorber for the brain and spinal cord.

The pia mater is a thin, highly vascular membrane, closely adherent to the brain and spinal cord. It follows the convolutions of the cerebral cortex and cannot be dissected away from the brain or spinal cord without damage to these structures.

Spinal Cord

The spinal cord extends from the foramen magnum to the level of the second or third lumbar vertebra (Figure 9-13). It is lodged within the vertebral column in the vertebral canal, submerged in CSF, and surrounded by the meninges. The cord is anchored in the dural sheath by paired longitudinal ligaments that pass laterally from the pia to the dura mater.

The vertebral canal continues below the termination of the spinal cord. Spinal nerves arising from the lower portion of the cord continue downward within the vertebral canal. They are called the cauda equina because of their resemblance to a horse's tail. The cauda equina is sheathed in dura and arachnoid membranes. The pia, too, extends below the spinal cord as a delicate glistening strand with a name of its own, the filum terminale. This strand attaches inferiorly to the coccyx.

The spinal cord, somewhat oval in cross section, has its greatest dimension from side to side. Anteriorly there is a deep groove, the anterior median fissure, along the entire length of the cord. Posteriorly there are less distinct grooves, one in the midline and two laterally. The lateral grooves demarcate a medial nerve tract, the fasciculus gracilis, from the fasciculus cuneatus laterally.

Thirty-one pairs of spinal nerves exit from successive levels of the spinal cord. A segment of the cord giving origin to a single pair of nerves is called a spinal cord segment, or level. From above downward the spinal nerves are designated as the cervical nerves (eight pairs), thoracic nerves (twelve pairs), lumbar nerves (five pairs), sacral nerves (five pairs), and coccygeal nerves (one pair). At levels where the nerves supplying the upper and lower extremities emerge, the diameter of the cord is markedly increased. The nerves of the upper extremities leave at the cervical enlargement. Those of the lower extremities emerge at the lumbosacral enlargement.

ANATOMY OF THE PERIPHERAL NERVOUS SYSTEM

The peripheral nervous system consists of 31 pairs of spinal nerves arising from the spinal cord and 12 pairs of cranial nerves which arise principally from the brainstem. Nerves of the autonomic portion of the peripheral nervous system (see Chapter 11) are associated with both spinal and cranial nerves.

All spinal nerves transmit both motor and sensory impulses and are distributed essentially to the body wall and limbs. Cranial nerves may transmit only sensory impulses, only motor impulses, or, like spinal nerves, both motor and sensory impulses. Cranial nerves are distributed to the head and neck region of the body. Exceptions to this rule are the spinal accessory nerve, which supplies the trapezius and sternocleidomastoid muscles, and the vagus nerve, which innervates viscera in the thorax and abdomen.

While the peripheral nervous system is primarily composed of nerve fibers, it also includes clusters of cell bodies which may be grouped as the cerebrospinal and autonomic ganglia. Cell bodies of all primary sensory neurons are found in the cerebrospinal ganglia, which are located on the dorsal roots of spinal nerves and interposed in certain of the

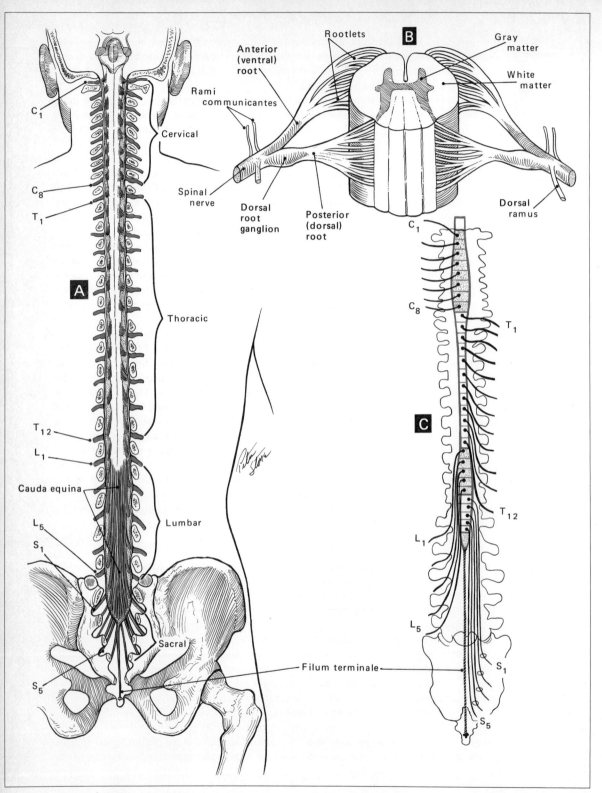

A

C₁

Cervical

C₈

T₁

Thoracic

T₁₂

L₁

Cauda equina

Lumbar

L₅

S₁

Sacral

S₅

B

Rootlets

Gray matter

White matter

Anterior (ventral) root

Rami communicantes

Spinal nerve

Dorsal root ganglion

Posterior (dorsal) root

Dorsal ramus

C

C₁

C₈

T₁

T₁₂

L₁

L₅

S₁

Filum terminale

S₅

240

cranial nerves. Autonomic ganglia are described in Chapter 11. Cell bodies of neurons associated with the organs of special sensation, namely the eye, ear, and nose, are found within these organs.

Spinal Nerves

Each spinal nerve of the pairs that arise from successive levels of the spinal cord begins as a series of fibers from both the dorsal aspect of the cord (the dorsal root) and the ventral aspect of the cord (the ventral root) (Figure 9-14). The dorsal root contains only sensory fibers, and the dorsal root ganglion contains cell bodies of primary sensory neurons. By contrast the ventral root transmits only motor fibers and acts as the final common pathway for all motor impulses leaving the spinal cord. Cell bodies of these motor fibers lie in the spinal cord. The dorsal and ventral roots merge as they leave the vertebral canal and become a spinal nerve. As the spinal nerve emerges, it is enclosed for a short distance by an extension of the meninges of the spinal cord.

The first cervical nerve leaves the vertebral canal by passing between the first cervical vertebra and the skull. Thereafter, successive cervical nerves pass between adjacent vertebrae. The eighth cervical nerve exits between the seventh cervical and the first thoracic vertebrae. Below the cervical region the spinal nerves leave the vertebral canal below their respective vertebrae; for instance, the first thoracic nerve leaves the canal below the first thoracic vertebra.

Shortly after the spinal nerve emerges from the vertebral canal, it divides into its first two branches, the dorsal ramus and the ventral ramus (Figure 9-14). The rami may also be referred to as the posterior primary ramus and the anterior primary ramus. The dorsal ramus passes posteriorly (dorsally) to supply a specific segment of the skin of the back, termed a dermatome, as well as the intrinsic back muscles. The ventral ramus behaves quite differently. In the cervical, lumbar, and sacral regions, ventral rami form an intricate network (plexus) of nerves. In the thoracic region the ventral rami course in the intercostal spaces to supply intercostal muscles as well as the skin overlying them. The sixth through twelfth thoracic nerves supply not only the intercostal space but also the muscles and skin of the lateral and anterior abdominal wall.

Somatic Nerve Plexuses

Ventral rami of spinal nerves in the cervical, lumbar, and sacral regions form plexuses named for these regions. Most nerves arising from these plexuses carry fibers of neurons from more than one segment of the spinal cord.

Cervical Plexus   The ventral rami of the first four cervical nerves form the cervical plexus (Figure 9-15 and Table 9-1). The nerves divide into ascending and descending branches. The ascending branch of C_2 (the second cervical spinal nerve) unites with the ventral ramus of C_1 to form a loop; a second and third loop are likewise formed by the ascending

Figure 9-13 (facing page). **A.** The vertebral column with the surface of the spinal cord exposed. Samples of cerebrospinal fluid (CSF) are usually taken in the lumbar region, where the meninges extend below the cord. **B.** A spinal cord segment. The meningeal covering of the spinal nerves is shown in Figure 9-12. **C.** Schematic view of the spinal cord. Note that, beginning in the thoracic region, spinal nerves emerge from the vertebral column at points progressively lower than their origin. The spinal nerves below the spinal cord are collectively termed the cauda equina.

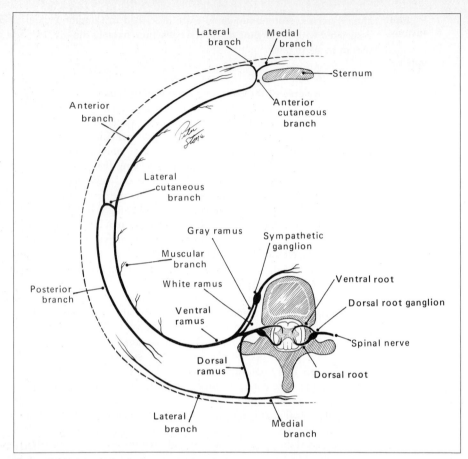

Figure 9-14. A typical spinal nerve and major rami, viewed from above. Note the communications with the sympathetic ganglion (see Figure 11-2).

and descending branches of C_2, C_3, and C_4. The descending branch of the ventral ramus of C_4 joins the brachial plexus.

The cervical plexus supplies segmental branches to the deep muscles of the neck, the infrahyoid muscles, the thoracic diaphragm, and the skin over the neck and upper chest. Nerves arising from the cervical plexus include the ansa cervicalis, which innervates the sternohyoid, sternothyroid, thyrohyoid, geniohyoid, and omohyoid muscles; the phrenic, which passes through the thoracic cavity to supply the diaphragm; and four cutaneous nerves. The four cutaneous nerves course as follows: The lesser occipital and greater auricular nerves innervate skin behind the ear, the transverse cervical nerve supplies skin over the front of the neck, and the supraclavicular nerves (medial, intermediate, and lateral) innervate the skin over the upper portion of the chest and over the shoulder.

Brachial Plexus The brachial plexus (Figure 9-16 and Table 9-2) is a complicated nerve network consisting of several component structures such as roots, trunks, divisions, cords, and branches. Ventral rami of the last four cervical nerves and the first thoracic nerve

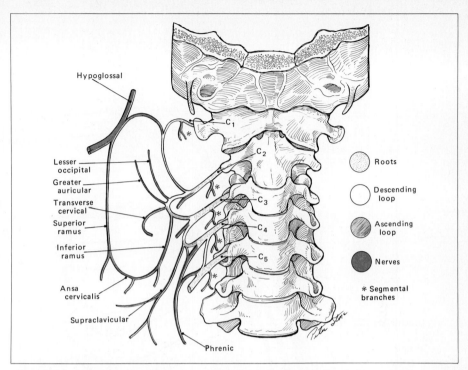

Figure 9-15. Cervical plexus. The major nerves which emerge are the ansa cervicalis, phrenic, and four cutaneous nerves.

TABLE 9-1. CERVICAL PLEXUS

NAME	ORIGIN FROM SPINAL CORD	DISTRIBUTION
Lesser occipital	$C_{2,3}$	Skin over lateral aspect of back of head
Greater auricular	$C_{2,3}$	Skin over lower portion of ear and in front of ear
Transverse cervical	$C_{2,3}$	Skin over front of neck
Supraclavicular	$C_{3,4}$	Skin over upper chest and shoulder by anterior, middle, and posterior branches
Ansa cervicalis Superior limb	$C_{1,2}$	Joins hypoglossal, then branches to join inferior limb and supply infrahyoid muscles; terminal thyrohyoid and geniohyoid branches supply their respective muscles
Inferior limb	$C_{3,4}$	Joins superior limb to form ansa and supply omohyoid, sternothyroid, and sternohyoid
Phrenic	$C_{3,4,5}$	Thoracic diaphragm
Segmental branches		Supply deep (prevertebral) muscles of neck and sternocleidomastoid and levator scapula

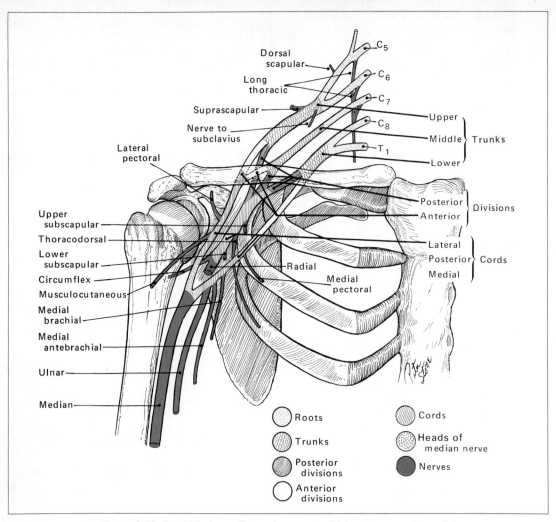

Figure 9-16. Brachial plexus. The major nerves which emerge are the radial, median, and ulnar.

exit from the vertebral column in the lower portion of the neck to form the roots of the brachial plexus. Passing between the anterior and middle scalene muscles, ventral rami of C_5 and C_6 unite to form the upper trunk, C_7 forms the middle trunk, and the ventral rami of C_8 and T_1 form the lower trunk of this plexus. The trunks pass deep to the clavicle to enter the axilla, where they divide into anterior and posterior divisions. Within the axilla the anterior divisions of the upper and middle trunk form the lateral cord, the anterior division of the lower trunk forms the medial cord, and the posterior divisions of all three trunks form the posterior cord. The cords are named from their relationship to the axillary artery.

Nerves arising from the brachial plexus supply all the musculature and skin of the upper extremity as well as muscles on the back and chest which move this extremity.

TABLE 9-2. BRACHIAL PLEXUS

NAME AND DERIVATION FROM PLEXUS	ORIGIN FROM SPINAL CORD	DISTRIBUTION
ROOTS		
Dorsal scapular	C_5	Levator scapulae, rhomboideus major and minor
Long thoracic	$C_{5,6,7}$	Serratus anterior
TRUNKS		
Suprascapular	$C_{5,6}$	Supraspinatus, infraspinatus
Nerve to subclavius	C_5	Subclavius
LATERAL CORD		
Lateral pectoral	$C_{5,6,7}$	Pectoralis major
Musculocutaneous	$C_{5,6,7}$	Biceps brachii, coracobrachialis, and brachialis
Lateral head of median nerve	(See below)	(See below)
MEDIAL CORD		
Median nerve (medial head)	$C_{5,6,7,8}$, T_1	Flexors of forearm except flexor carpi ulnaris and ulnar half of flexor digitorum profundus; flexor pollicis brevis, abductor pollicis brevis, opponens pollicis; skin of medial two-thirds of palm
Ulnar	C_8, T_1	Flexor carpi ulnaris, ulnar half of flexor digitorum profundus, intrinsic muscles of hand not supplied by median; skin of lateral third of hand
Medial pectoral	C_8, T_1	Pectoralis major and minor
Medial brachial cutaneous	C_8, T_1	Skin of medial aspect of lower portion of arm
Medial antebrachial cutaneous	C_8, T_1	Skin of medial surface of forearm
POSTERIOR CORD		
Radial	$C_{5,6,7,8}$, T_1	Extensor muscles of arm and forearm; skin of lateral two-thirds of dorsum of hand
Circumflex	$C_{5,6}$	Deltoid, teres minor
Thoracodorsal	$C_{6,7,8}$	Latissimus dorsi
Upper subscapular	$C_{5,6}$	Subscapularis
Lower subscapular	$C_{5,6}$	Subscapularis, teres major
ADDITIONAL CUTANEOUS BRANCHES		
Lateral brachial cutaneous (from axillary)		Skin of lateral aspect of arm
Posterior brachial cutaneous (from radial)		Skin of posterior aspect of arm
Intercostobrachial (from second intercostal)		Skin of upper portion of medial aspect of arm
Lateral antebrachial cutaneous (from musculocutaneous)		Skin of lateral aspect of forearm

The three largest nerves from this plexus are the radial, the median, and the ulnar. The radial supplies the extensor muscles in the arm and forearm and skin on the back of the arm, forearm, and hand; the median supplies most of the flexor muscles in the forearm, the thenar (thumb) muscles of the hand, and the skin of the lateral two-thirds of the palm; and the ulnar supplies two flexor muscles in the forearm, the intrinsic muscles of the hand not supplied by the median nerve, and the skin of the medial third of the hand.

Lumbar Plexus   Ventral rami of the upper four lumbar nerves divide into anterior and posterior divisions to form the lumbar plexus (Figure 9-17 and Table 9-3). Nerves from

Figure 9-17. Lumbar plexus. The major nerves which emerge are the femoral and obturator.

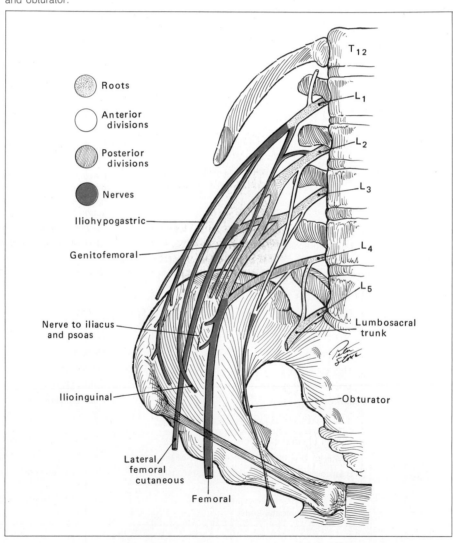

TABLE 9-3. LUMBAR PLEXUS

NAME	ORIGIN	DISTRIBUTION
Iliohypogastric	T_{12}, L_1	Muscles of anterolateral abdominal wall; skin over buttock and lower abdomen
Ilioinguinal	L_1	Muscles of anterolateral abdominal wall; skin over base of penis and scrotum
Genitofemoral	$L_{1,2}$	Skin over upper anterior aspect of thigh and base of penis and scrotum
Lateral femoral cutaneous	$L_{2,3,4}$	Skin over lateral aspect of thigh
Femoral	$L_{2,3,4}$	Extensor muscles in thigh
Obturator	$L_{2,3,4}$	Adductor muscles in thigh; skin over medial aspect of thigh
Additional cutaneous branches		
Medial femoral cutaneous		Skin over medial aspect of thigh
Intermediate femoral cutaneous		Skin over central portion of anterior aspect of thigh
Saphenous (from femoral)		Skin over medial side of leg and foot

this plexus contribute to the supply of muscles and skin of the anterolateral abdominal wall, skin of the upper part of the buttock, muscles and skin of the anterior and medial aspect of the thigh, and skin of the anterior aspect of the leg and foot.

The major nerves arising from this plexus are the femoral and obturator. The femoral supplies the iliacus muscle, all the muscles of the anterior, or extensor, compartment of the thigh, and skin over the anterior and medial aspect of the inferior extremity. The obturator supplies the muscles in the medial compartment of the thigh and contributes to the innervation of the skin over the medial aspect of the thigh.

Sacral Plexus   Ventral rami of L_4 and L_5 combined in the lumbosacral trunk plus ventral rami of S_1 to S_4 form the sacral plexus (Figure 9-18 and Table 9-4). The ventral rami divide into anterior and posterior divisions from which the nerves of the sacral plexus are derived.

The major nerve from this plexus is the sciatic, the largest nerve in the body. It supplies the flexor muscles in the posterior compartment of the thigh and divides in the popliteal region into the common peroneal and tibial nerves, which in turn supply all the muscles of the leg and foot. The common peroneal supplies muscles in the anterior and lateral compartments of the leg and skin on the anterior surface of the leg and on the dorsum of the foot. The tibial supplies flexor muscles on the posterior aspect of the leg, muscles on the plantar surface of the foot, and skin on the posterior aspect of the leg and plantar surface of the foot.

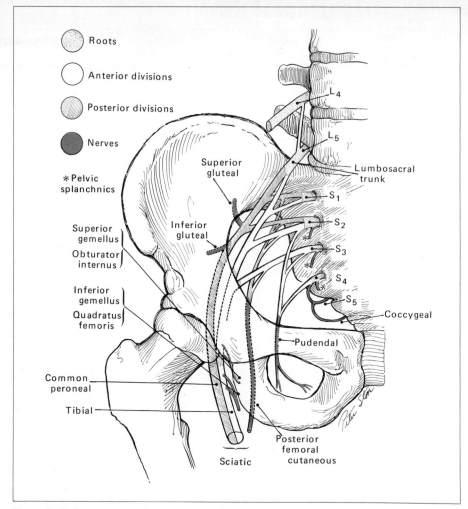

○ Roots

○ Anterior divisions

◐ Posterior divisions

● Nerves

✱ Pelvic
splanchnics

Superior
gluteal

Inferior
gluteal

Superior
gemellus
Obturator
internus

Inferior
gemellus
Quadratus
femoris

Common
peroneal

Tibial

Sciatic

Posterior
femoral
cutaneous

Pudendal

Coccygeal

L₄

L₅

Lumbosacral
trunk

S₁

S₂

S₃

S₄

S₅

Figure 9-18. Sacral plexus. The major nerve from this plexus is the sciatic.

Cranial Nerves

The twelve pairs of cranial nerves arise directly from the brain (Figures 9-19, 9-20). They supply structures in the head and neck and, in the case of the vagus nerve, structures of the trunk.

The olfactory (I), optic (II), and vestibulocochlear (VIII) nerves carry only sensory impulses to the brain. The oculomotor (III), trochlear (IV), abducens (VI), accessory (XI), and hypoglossal (XII) transmit only motor impulses from the brain. The trigeminal (V), facial (VII), glossopharyngeal (IX), and vagus (X) contain both motor and sensory fibers.

Olfactory (I) The olfactory nerve (Figure 9-21 and Table 9-5) receives and transmits impulses associated with the sense of smell. Cell bodies of the bipolar neurons of this nerve are located in the olfactory mucosa of the superiormost portion of the nasal cavity.

TABLE 9-4. SACRAL PLEXUS

NAME	ORIGIN	DISTRIBUTION
Superior gluteal	$L_{4,5}, S_1$	Gluteus medius and minimus
Inferior gluteal	$L_{4,5}, S_1$	Gluteus maximus
Nerve to gemellus superior and obturator internus	$L_5, S_{1,2}$	Gemellus superior and obturator internus
Nerve to gemellus inferior and obturator externus	$L_5, S_{1,2}$	Gemellus inferior and obturator externus
Nerve to quadratus femoris	$L_{4,5}, S_{1,2}$	Quadratus femoris
Posterior femoral cutaneous	$S_{1,2,3}$	Skin on posterior aspect of thigh
Sciatic Tibial division	$L_{4,5}, S_{1,2,3}$	Flexor muscles in leg
Medial plantar		Abductor hallucis, flexor hallucis brevis, medial two lumbricales in foot; skin over medial two-thirds of plantar surface of foot
Lateral plantar		Remaining intrinsic muscles of foot not supplied by medial plantar; skin over lateral third of plantar surface of foot
Common peroneal	$L_{4,5}, S_{1,2}$	
Superficial peroneal		Peroneus longus and brevis; skin over distal third of anterior aspect of leg and dorsum of foot
Deep peroneal		Muscles in extensor compartment of leg, extensor digitorum brevis of foot; skin between great and second toes
Additional cutaneous branches		
Inferior cluneal (from posterior cutaneous of thigh)		Skin over lower portion of buttock
Sural (from tibial and common peroneal)		Skin over posterior aspect of leg and lateral side of foot

Peripheral nerve processes receive the impulses, and central processes pass through the cribriform plate of the ethmoid bone in the roof of the nasal cavity. Here they synapse with the second neuron in the olfactory pathway, which is located in the olfactory bulb. The olfactory tract passes from the olfactory bulb to the olfactory trigone, on the inferior aspect of the frontal lobe of the brain.

Optic (II) The optic nerve (Figure 9-22 and Table 9-5) is formed by an evagination of the diencephalon. The primary sensory neurons are the rods and cones located in the retina. Rods and cones are responsive to light stimuli. They relay impulses to a second neuron in the retina, which in turn transmits the impulse to a third neuron, the fibers of which form the optic nerve. Shortly after the optic nerves leave the two orbital cavities,

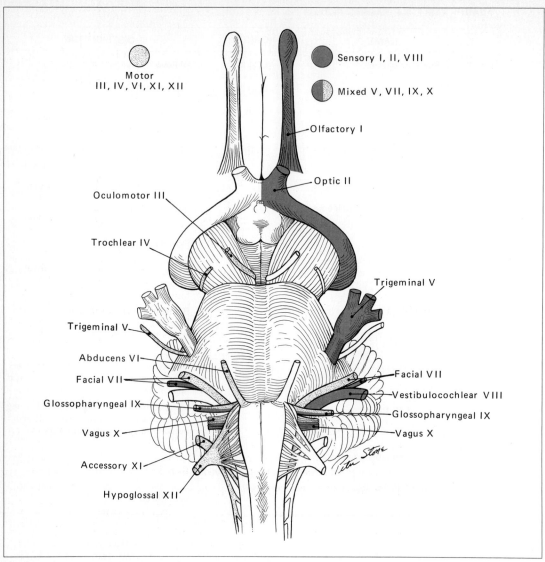

Motor
III, IV, VI, XI, XII

Sensory I, II, VIII

Mixed V, VII, IX, X

Olfactory I

Optic II

Oculomotor III

Trochlear IV

Trigeminal V

Trigeminal V

Abducens VI

Facial VII

Glossopharyngeal IX

Vagus X

Accessory XI

Hypoglossal XII

Facial VII

Vestibulocochlear VIII

Glossopharyngeal IX

Vagus X

Figure 9-19. Emergence of the cranial nerves from the ventral surface of the brainstem.

they fuse at the optic chiasma, then separate as the optic tracts, which encircle the midbrain and pass to the lateral geniculate bodies. A more detailed description of this visual pathway will be found in Chapter 12.

Oculomotor (III), Trochlear (IV), Abducens (VI) The oculomotor, trochlear, and abducens nerves (Figure 9-23 and Table 9-6) are motor in function. They supply the extrinsic muscles of the eye. The abducens supplies the lateral rectus, the trochlear innervates the superior oblique, and the oculomotor supplies the remaining extrinsic muscles of the eye.

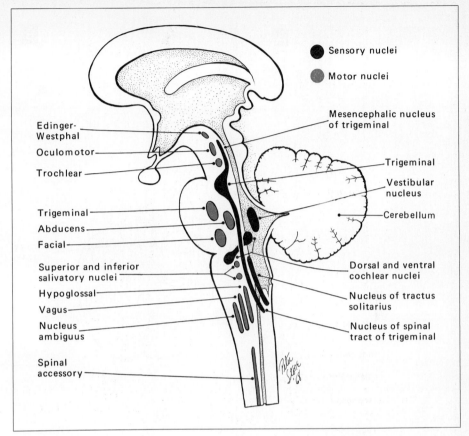

Sensory nuclei

Motor nuclei

Edinger-Westphal

Oculomotor

Trochlear

Trigeminal

Abducens

Facial

Superior and inferior salivatory nuclei

Hypoglossal

Vagus

Nucleus ambiguus

Spinal accessory

Mesencephalic nucleus of trigeminal

Trigeminal

Vestibular nucleus

Cerebellum

Dorsal and ventral cochlear nuclei

Nucleus of tractus solitarius

Nucleus of spinal tract of trigeminal

Figure 9-20. Cranial nerve nuclei. The vestibular nucleus is in the lateral wall of the fourth ventricle.

TABLE 9-5. FIRST AND SECOND CRANIAL NERVES

NAME	FUNCTION	DISTRIBUTION
Olfactory (I)	Sensory	Special (smell) fibers to olfactory mucosa; cell bodies of primary sensory bipolar neurons in olfactory mucosa; peripheral processes, distribution to upper portion of nasal septum and superior meatus; central processes, distribution to olfactory bulb
Olfactory bulb	Sensory	Contains cell bodies of secondary sensory neurons; fibers form olfactory tract
Optic (II)	Sensory	Primary and secondary neurons and cell bodies of tertiary neurons associated with vision located in retina of eye
Optic "nerve"	Sensory	Formed by axons of tertiary neurons which pass from eyeball to brainstem
Optic tract	Sensory	Optic nerves from each eye join in cranial cavity to form optic chiasma—optic tract extends from chiasma to brainstem

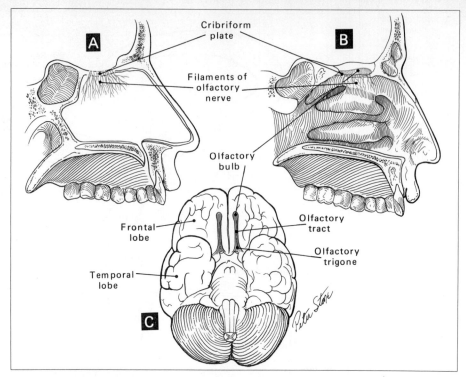

Figure 9-21. Olfactory nerve. **A, B.** Peripheral processes in relation to the nasal septum and conchae. The olfactory bulbs lie directly above the cribriform plate on either side of the crista galli. **C.** Relation of the olfactory bulb, tract, and trigone to the inferior surface of the brain.

Figure 9-22. Optic nerve, composed of tertiary neurons. Note that only the medial fibers cross at the optic chiasma.

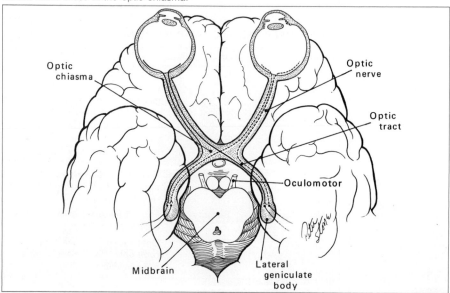

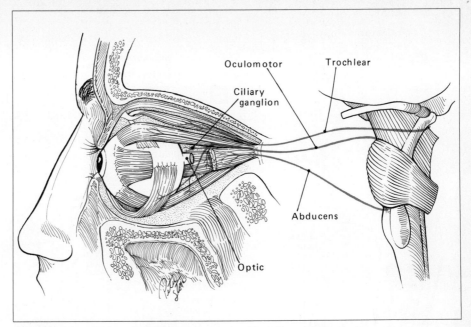

Figure 9-23. Innervation of extrinsic eye muscles. The ciliary ganglion is the origin of postganglionic parasympathetic fibers to the ciliary muscle and the constrictor muscle of the iris.

The oculomotor nerve also supplies autonomic fibers (see Chapter 11) to the intrinsic ciliary and sphincteric muscles of the iris.

Trigeminal (V)   The trigeminal nerve (Figure 9-24 and Table 9-7) is the largest of the cranial nerves. It has both motor and sensory fibers. Cell bodies of its primary sensory neurons are located in the large trigeminal ganglion located within the cranial cavity. Impulses reach the ganglion over three divisions of the trigeminal nerve. The ophthalmic is composed of sensory fibers which are supplied to the orbital cavity and skin above

TABLE 9-6. THIRD, FOURTH, AND SIXTH CRANIAL NERVES

NAME	FUNCTION	DISTRIBUTION
Oculomotor (III)	Motor	Medial, superior, and inferior rectus muscles; inferior oblique and levator palpebrae muscles; parasympathetic fibers
Ciliary	Parasympathetic	Preganglionic fibers to ciliary ganglion; postganglionic fibers to sphincter muscle of iris and ciliary muscle
Trochlear (IV)	Motor	Superior oblique muscle
Abducens (VI)	Motor	Lateral rectus muscle

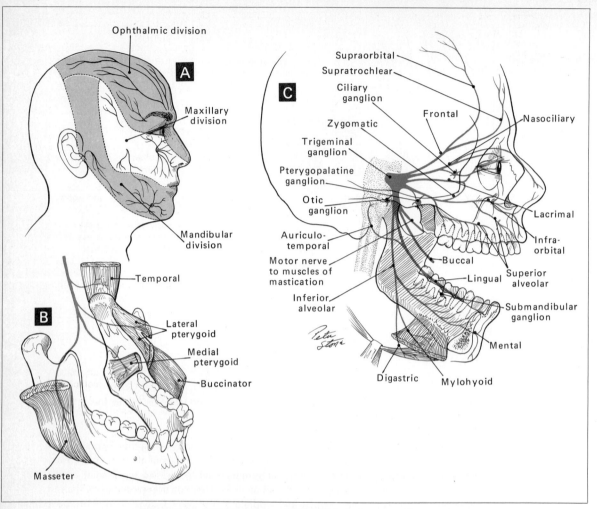

Figure 9-24. Trigeminal nerve. **A.** Distribution of sensory fibers to the skin by the three branches of the trigeminal. **B.** Distribution of the chief motor fibers to muscles of mastication. **C.** Distribution of terminal branches.

the level of the eye. The maxillary supplies sensory fibers to the nasal cavity, the skin of the midportion of the face between the mouth and the eye, and the upper teeth. The mandibular supplies sensory fibers to the lower jaw, floor of the mouth, anterior two-thirds of the tongue, and lower teeth.

The motor root of the trigeminal nerve passes independently out of the cranial cavity to join the mandibular division and supply eight muscles: the four muscles of mastication (lateral and medial pterygoids, masseter, and temporalis), the tensor tympani, the tensor veli palatini, the mylohyoid, and the anterior belly of the digastric.

Facial (VII) The facial nerve (Figure 9-25 and Table 9-8) carries both motor and sensory fibers. Leaving the brainstem, it accompanies the eighth nerve through the internal

TABLE 9-7. FIFTH CRANIAL NERVE

NAME	FUNCTION	DISTRIBUTION
Trigeminal (V)	Sensory and motor	Muscles of mastication; major sensory supply of face and head—cell bodies of primary sensory neurons in trigeminal ganglion
Ophthalmic	Sensory	Orbital cavity and area of face and scalp above level of eyes
Frontal	Sensory	Orbital structures and anterior two-thirds of scalp by terminal supraorbital and supratrochlear branches
Lacrimal	Sensory	Orbital structures, lacrimal gland, and upper eyelid
Nasociliary	Sensory	Orbital structures, nasal cavity, and skin of upper portion of nose and lower eyelid by terminal infratrochlear and external nasal branches
Maxillary	Sensory	Nasal cavity, area of head and face between mouth and eyes
Superior alveolar	Sensory	Upper teeth and gingiva by anterior and posterior branches
Descending palatine	Sensory	Palate
Sphenopalatine	Sensory	Nasal cavity
Zygomatic	Sensory	Skin of face and temporal regions adjacent to cheekbone by terminal zygomaticofacial and zygomaticotemporal branches
Infraorbital	Sensory	Skin of upper portion of cheek, upper lip, and lower eyelid
Mandibular	Sensory and motor	Muscles of mastication and face below mouth and area in front of ear
Muscular branches	Motor	Temporalis, masseter, internal and external pterygoids, tensor tympani, and tensor veli palatini
Lingual	Sensory	Anterior two-thirds of tongue
Buccal	Sensory	Skin of lower half of cheek
Auriculotemporal	Sensory	Skin in front of ear and region of temple
Inferior alveolar	Sensory	Lower teeth, gingiva, and skin of chin by terminal mental branch
Nerve to mylohyoid	Motor	Mylohyoid and anterior belly of digastric muscles

auditory meatus, then traverses the facial canal to emerge from the skull at the stylomastoid foramen, where it provides branches to the stylohyoid muscle, the posterior belly of the digastric, and the superficial muscles around the ear and back of the skull. It then courses through the substance of the parotid gland to divide into terminal branches supplying the remaining muscles of facial expression.

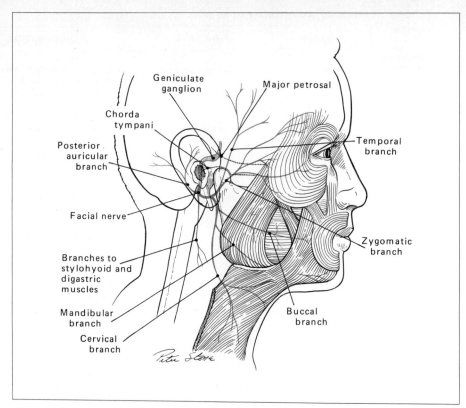

Figure 9-25. Facial nerve. Muscles of facial expression are innervated by the temporal, zygomatic, buccal, mandibular, and cervical branches.

The facial nerve distributes three branches as it courses through the facial canal within the temporal bone. The major petrosal leaves the canal and ultimately supplies fibers to the nasal cavity and lacrimal gland. The chorda tympani courses through the middle ear cavity and the infratemporal fossa, then joins the lingual nerve and supplies the sublingual and submandibular glands. Motor fibers of major petrosal and chorda tympani are autonomic, and their pathways are described in more detail in Chapter 11. The chorda tympani also provides special sensory fibers for taste to the anterior two-thirds of the tongue. The third branch arising from the facial nerve as it traverses the facial canal is the small nerve to the stapedius muscle, located in the middle ear cavity.

Vestibulocochlear (VIII)   The vestibulocochlear nerve (Figure 9-26 and Table 9-9) is entirely sensory and is distributed to the internal ear. Nerve cell bodies are located in two sites. The cell bodies of the cochlear division are in the spiral ganglion of the cochlea. Peripheral processes pass to the organ of Corti in the cochlea, where they pick up impulses related to sound. The central processes transmit these impulses to the brain. The cell bodies of neurons forming the vestibular portion are located in the vestibular ganglion. Peripheral processes pass to ampullae of semicircular canals, the utricle, and the saccule. The central processes transmit impulses associated with equilibrium from these organs to the brain.

TABLE 9-8. SEVENTH CRANIAL NERVE

NAME	FUNCTION	DISTRIBUTION
Facial (VII)	Sensory and motor	Facial muscles, tongue, salivary and lacrimal glands; cell bodies of primary sensory neurons in geniculate ganglion
Chorda tympani	Parasympathetic and special sensory	Preganglionic fibers to submandibular ganglion, postganglionic fibers to submandibular and sublingual glands; special sensory (taste) fibers to anterior two-thirds of tongue
Major petrosal	Parasympathetic	Preganglionic fibers to pterygopalatine ganglion, postganglionic fibers to lacrimal gland and nasal cavity
Posterior auricular	Motor	Posterior and superior auricular, occipitalis muscles
Nerve to stylohyoid	Motor	Stylohyoid and posterior belly of digastric
Temporal	Motor	Facial-expression muscles around eye, frontalis
Zygomatic	Motor	Facial muscles of upper cheek and around eye
Buccal	Motor	Buccinator and muscles around mouth
Mandibular	Motor	Muscles of facial expression overlying mandible
Cervical	Motor	Platysma (muscle of facial expression in neck)

Figure 9-26. Vestibulocochlear nerve. Cell bodies of the primary sensory neurons for equilibrium are in the vestibular ganglion; those of the primary sensory neurons for hearing are in the spiral ganglion of the cochlea.

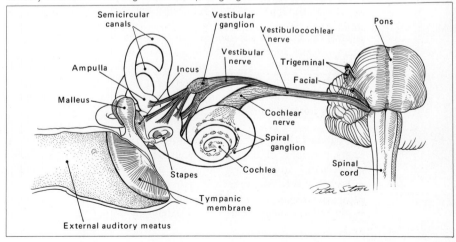

TABLE 9-9. EIGHTH CRANIAL NERVE

NAME	FUNCTION	DISTRIBUTION
Vestibulocochlear (VIII)	Sensory	Organs of equilibrium and hearing in internal ear
Cochlear	Sensory	Special (hearing) cell bodies located in spiral ganglion of cochlea; peripheral processes pass to organ of Corti; central processes form definitive cochlear portion of nerve
Vestibular	Sensory	Special (equilibrium) sensory cell bodies in vestibular ganglion; peripheral processes pass to cristae in semicircular ducts and utricle; central processes form vestibular portion of nerve

Glossopharyngeal (IX)   The glossopharyngeal nerve (Figure 9-27 and Table 9-10) is a mixed nerve carrying both sensory and motor fibers. As its name implies, it supplies the tongue and pharynx. Sensory fibers innervate the posterior third of the tongue. Some motor fibers supply the stylopharyngeus muscle directly, while others are distributed with branches of the vagus to form the pharyngeal plexus supplying both motor and sensory fibers to the pharynx. The glossopharyngeal nerve also contains autonomic fibers, which are distributed by way of the minor petrosal nerve to the parotid gland (see Chapter 11).

Vagus (X)   The vagus nerve (Figure 9-28 and Table 9-11) is the most widely distributed of the cranial nerves, hence its nickname, "the wanderer." It contains both motor and sensory fibers. The sensory fibers carry impulses from the lungs, heart and major vessels, larynx, pharynx, esophagus, stomach and small intestine, gallbladder, and some taste buds at the back of the mouth. As may be surmised, this broad distribution involves the vagus in the regulation of circulation, heart rate (see Chapter 16), respiration (see Chapter 23), and digestion (see Chapter 25). The motor component is equally important, since it innervates muscles of the larynx, pharynx, esophagus, bronchi, stomach and small intestine, heart, and glands which produce digestive juices.

Accessory (XI)   The accessory nerve (Figure 9-29 and Table 9-12) arises from both the brainstem and upper segments of the spinal cord. The spinal portion (root) passes upward along the side of the spinal cord and joins the cranial root from the brainstem. Fibers of the cranial root travel with the nerve for a short distance and then branch to join and be distributed with the vagus nerve. The spinal root descends as a separate nerve to supply the trapezius and sternocleidomastoid muscles.

Hypoglossal (XII)   The hypoglossal nerve (Figure 9-30 and Table 9-12) has only motor fibers. It supplies all the muscles, both extrinsic and intrinsic, of the tongue. A communicating branch from the cervical plexus travels for a short distance with the nerve before a portion of the fibers branch off as the anterior limb of the ansa cervicalis. A few fibers of the communicating branch remain with the hypoglossal nerve to innervate the thyrohyoid and geniohyoid muscles separately.

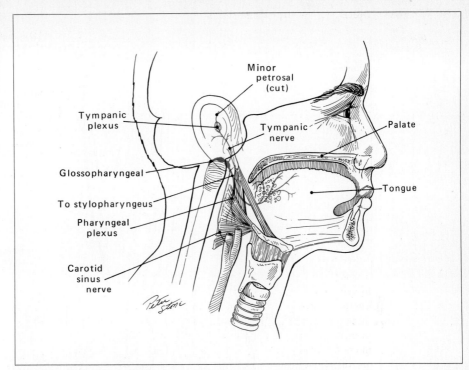

Figure 9-27. Glossopharyngeal nerve. The lingual branch is the chief conductor of impulses associated with taste.

TABLE 9-10. NINTH CRANIAL NERVE

NAME	FUNCTION	DISTRIBUTION
Glossopharyngeal (IX)	Sensory and motor	Pharynx, tongue, carotid sinus, and parotid gland; cell bodies of primary sensory neurons in superior and inferior ganglia
Tympanic nerve	Sensory and motor (parasympathetic)	Contributes to formation of tympanic plexus; supplies sensation to middle ear cavity
Lesser petrosal	Parasympathetic	Re-formed from tympanic plexus; pre-ganglionic fibers to otic ganglion, post-ganglionic fibers to parotid gland
Nerve to carotid sinus and body (nerve of Hering)	Sensory	To specialized pressure receptors in carotid sinus and carotid body
Lingual branches	Sensory	Posterior third of tongue with general and special (taste) fibers
Pharyngeal branches	Sensory and motor	Joins branches of vagus to form pharyngeal plexus

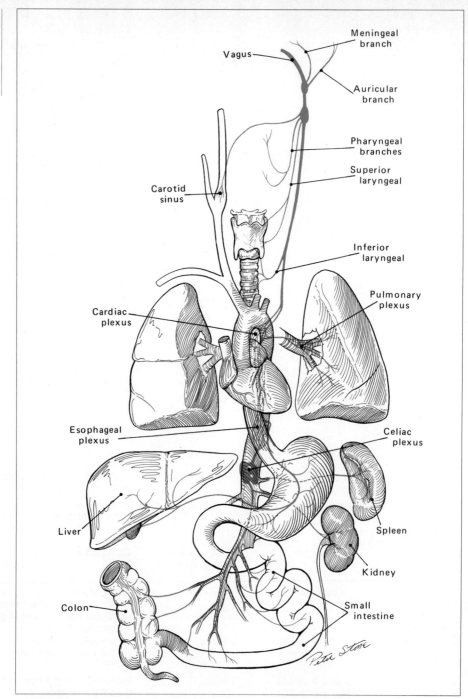

Figure 9-28. The vagus nerve and the major organs it supplies.

TABLE 9-11. TENTH CRANIAL NERVE

NAME	FUNCTION	DISTRIBUTION
Vagus (X)	Sensory and motor	Pharynx and larynx; viscera of neck, thorax, and abdomen; cell bodies of primary sensory neurons in superior and inferior ganglia
Cardiac branches	Parasympathetic	Preganglionic fibers to cardiac and pulmonary plexuses by superior and inferior branches
Pharyngeal branches	Motor and sensory	Join branches from glossopharyngeal nerve to form pharyngeal plexus
Superior laryngeal	Motor and sensory	Terminal internal laryngeal branch supplies sensation to larynx above level of vocal cords; terminal external laryngeal branch supplies cricothyroid muscle
Inferior laryngeal	Motor and sensory	Supplies sensation below level of vocal cords and all intrinsic muscles of larynx except cricothyroid
Esophageal plexus	Sensory and para-sympathetic	Vagus nerves from both sides of body form a plexus around esophagus as they traverse thorax
Gastric nerves	Sensory and para-sympathetic	Re-form from esophageal plexus as esophagus passes through diaphragm as anterior and posterior branches; distribute with autonomic plexuses of abdomen to viscera of abdomen and intestines to level of descending colon
Lingual branches	Sensory	Special (taste) fibers to root of tongue and epiglottis

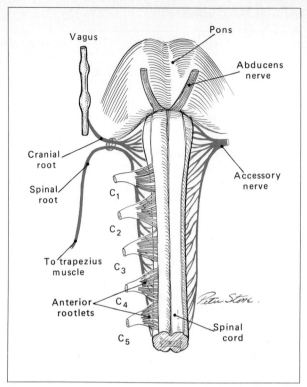

Figure 9-29. Accessory nerve. Its origin from both spinal cord and brainstem is shown at the left. The cranial portion is distributed with the vagus.

TABLE 9-12. ELEVENTH AND TWELFTH CRANIAL NERVES

NAME	FUNCTION	DISTRIBUTION
Accessory (XI)	Motor	Trapezius muscle and joins vagus to supply neck
Cranial root	Motor	Joins and is distributed with vagus fibers (probably supplies primarily larynx)
Spinal root	Motor	Trapezius
Hypoglossal (XII)	Motor	Supplies both intrinsic and extrinsic muscles of tongue

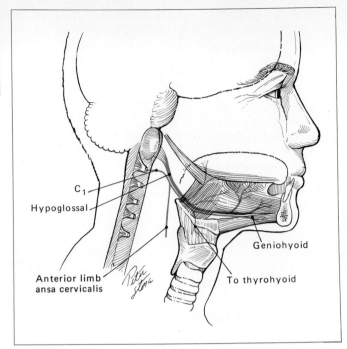

Figure 9-30. Hypoglossal nerve. Note how closely it lies to fibers of the first cervical nerve.

QUESTIONS AND PROBLEMS

1   What are the organs of the central nervous system? Where is the white and the gray matter located in these organs?

2   By what means are the brain and spinal cord protected?

3   Describe how the right cerebral hemisphere is connected to the left cerebral hemisphere.

4   By what means have scientists been able to determine and locate discrete functional areas of the brain?

5   Identify the voluntary motor and somesthetic areas of the cerebrum. Where would one apply an electrode in the cerebrum to evoke movement of the leg?

6   Define nuclei as referred to in the central nervous system. Locate the nucleus of the sixth cranial nerve—the abducens.

7   What is a dorsal root ganglion?

8   What is the internal capsule composed of?

9   What structures are included in the brainstem?

10   What is the anatomical relationship of the cerebellum to the spinal cord? Cerebrum? Pons?

11   Describe where the ventricles are located and name the connections between them.

12   Describe the components of a typical spinal nerve and how it is connected to the spinal cord.

13   What is the purpose of a nerve plexus? Where are the major plexuses located?

14   Is a neuron in a spinal nerve capable of regeneration?

15   Select from the left-hand column the appropriate response and place the corresponding

letter in the space provided in the right-hand column.

a. Facial

———innervates muscles of mastication

b. Glossopharyngeal

———responsible for movement of the tongue

c. Optic

———innervates muscles of facial expression

d. Trochlear

———innervates superior oblique muscle of eye

e. Abducens

f. Hypoglossal

———sensory to the face

———responsible for smile

g. Vestibulocochlear

h. Spinal accessory

———raises upper eyelid

———responsible for hearing

i. Vagus

———innervates trapezius muscle

j. Trigeminal

———responsible for equilibrium

k. Olfactory

l. None of these

10 PERCEPTION AND MOVEMENT

Not only is the nervous system the major integrative system of the body, but it is also responsible for permitting the organism to be aware of stimuli, that is, to sense. If the organism thus becomes aware of stimuli, it may or may not take action. If action is taken in the form of movement, that movement is said to be voluntary. On the other hand, stimuli may give rise to movement without awareness. Such movement is said to be reflex. In very broad terms, it is the function of the nervous system to permit awareness, to make possible voluntary and reflex movement, and to integrate all these activities.

In this chapter the role of the nervous system in perception and movement will be analyzed. The special senses, such as vision and audition, are considered in Chapters 12 and 13.

PERCEPTION

All knowledge depends on perception. Perception is knowledge gained through the senses of the existence and properties of matter in the external or internal world. Of fundamental import to perception are the senses. Normally more than one sense is used to gain knowledge. For example, even with the eyes closed, one can fondle an object and learn much about its properties. One knows if it is round or flat, long or short, hot or cold. One can ascertain whether it is metal, stone, wood, or rubber. From such information objects can be identified or described with amazing accuracy. Of course, the eyes are ordinarily kept open, and so when an object is handled, the sense of vision also contributes to our knowledge. The object may also be tapped and the resulting sound utilized to learn something of the object's composition. The point being made is that data are gained through several senses simultaneously. Through the use of all the senses, a steady growth of knowledge becomes possible. The senses, however, are not utilized exclusively for the acquisition of knowledge. They serve for protection, for pleasure, and for the integration of bodily functions as well.

Receptors

Everyone realizes that sound, in the form of electrical impulses, may be transmitted by telephone wires. But it should be clear that in order to have electrical impulses there must be some mechanism for the transformation of sound waves. In the telephone, the receiver carries out this transformation. In man, the receptors do this job. A receptor may be defined as a specialized nerve ending which is sensitive to a specific change in the environment. The environmental change constitutes the effective stimulus. For example, if a finger is placed in hot water, the change in the temperature (from the temperature of the air in the room to that of the hot water) activates an appropriate receptor in the skin of the finger. The person is thus able to perceive warmth.

Types of Receptors   A moment's reflection will convince the reader that there must be several types of receptors in the body. This fact is evident because we know that we can perceive not only warmth but pain, cold, touch, and pressure. Actually there are specific receptors for all modalities of sensation. The anatomical characteristics of various types were described in Chapter 8 (see also Figure 8-5). Table 10-1 contains a partial list of the receptors and the modality of sensation to which each responds.

Some of the receptors lie close to the body surface and, accordingly, serve to perceive changes in the external environment. They are therefore termed exteroceptors. Others, such as the chemoreceptors and pressoreceptors, are placed deep in the body in a position to respond to changes in the internal environment and thus are termed interoceptors.

Law of Specific Nerve Energies   The fact that there are specific receptors for each modality of sensation and therefore each sensory neuron is concerned with only that modality of sensation is termed the law of specific nerve energies. Were this not true, there would be no clear-cut recognition of the various changes in the environment. In other words, if a drop in external temperature stimulated the pain receptors as well as

TABLE 10-1. THE MAJOR RECEPTORS

RECEPTOR	SENSATION
Free nerve ending (nociceptor)	Pain
Corpuscle of Ruffini	Warmth
Krause's end bulb	Cold
Meissner's corpuscle	Touch
Pacinian corpuscle	Deep pressure
Golgi tendon organ	Proprioception
Muscle spindle	Proprioception
Pressoreceptor	Blood pressure
Chemoreceptor	Chemical change

Krause's end bulbs, one would perceive pain as well as cold. As a matter of fact, if the environment becomes very cold, pain may be experienced. This indicates that, under extreme or abnormal conditions, it is possible to fire some receptors with other stimuli. For example, when a person receives a blow on the eye, he often reports "seeing stars." The blow actually stimulates the receptors for vision and the person perceives light.

The law of specific nerve energies holds not only because there are specific receptors for the various modalities of sensation but also because the afferent chain of neurons fired by any one type of receptor ends in a specific region of the cerebral cortex.

Mechanism of Receptor Response   All the various types of receptors may be placed in six categories, depending upon their mechanism of response. Thus there are (1) mechanoreceptors, (2) thermoreceptors, (3) chemoreceptors, (4) electromagnetic receptors, (5) photoreceptors, and (6) osmoreceptors.

The basic mechanism of receptor response is alteration of the receptor membrane so that it becomes more permeable to sodium ions, which, as a result, move into the receptor cell, thereby altering the resting potential. The resulting alteration in the potential of the receptor is the receptor potential, also called the generator potential.

Figure 10-1 summarizes a highly revealing series of experiments. In the first place, the receptor core was found to respond whether the outer layers, in this case of a Pacinian corpuscle, were present or not. Still, it is the surrounding structure of the receptor which probably accounts for the specificity to a particular type of stimulus. But the mechanism is unclear. What is there about the Pacinian corpuscle's structure that makes it responsive to pressure while Ruffini corpuscles respond to heat? All that is known is that in some way the different modalities of sensation act on the receptor in a specific way to alter the permeability of the core.

Receptor Response to Increasing Stimulus Intensity

A weak stimulus can produce a receptor potential but not an action potential in the innervating neuron. Note in Figure 10-1 that potentials (a), (b), and (c) do not cause the neuron to fire. Only potential (d) does so. But also observe that progressively stronger stimuli evoke potentials of correspondingly greater magnitude. This is true, of course, only up to a certain point, after which there is no longer an increase in magnitude. With increasing stimulus strength, the magnitude of the potential increases, rapidly at first, then more slowly, and finally plateaus at about 100 mv (Figure 10-2). Most neurons can be fired with a generator potential much below 100 mv.

Not only does a stronger stimulus evoke a receptor potential of greater magnitude, but the number of impulses per unit time generated in the neuron also increases. According to the so-called Weber-Fechner law, the rate of firing is proportional to the logarithm of the stimulus strength. The significance of this relationship is that it provides the basis for the perception of a very great range of intensities. The best example is that of the eye, which can appreciate light of low intensity as well as light that is over 100,000 times more intense.

Adaptation of Receptors   The magnitude of the receptor potential and the frequency of response are functions of the stimulus intensity, but if a stimulus of constant intensity is prolonged, the magnitude of the potential and the frequency of response decrease (Figure 10-3). This decrease is termed adaptation. A common example is provided by placing a finger in warm water. After a short period, the sensation of warmth diminishes. Further, if the water temperature is very slowly increased, the person will not perceive any temperature change until the fluid becomes quite hot, that is, until the intensity of

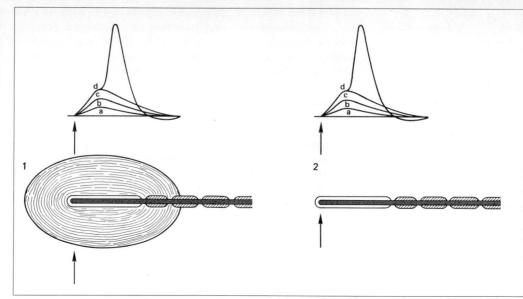

Figure 10-1. Receptor response to stimulation. Stimuli of increasing strength applied to a Pacinian corpuscle evoked generator potentials a through d. Generator potential d was able to depolarize the axon and create an action potential in the intact receptor (1), in the receptor with outer lamellae removed (2), and in the receptor after partial destruction of the core sheath (3). When the first node of Ranvier was blocked (4), no action potential occurred. After degeneration of the nerve ending (5), there was no response to stimulation. *(From W. Loewenstein, Scientific American, 203(2):98–108, 1960.)*

stimulus exceeds the ability of the receptor to adapt. From these observations it may be concluded that (1) receptors adapt to a stimulus and (2) in order for a receptor to be activated, the rate of change in the environment must be faster than the rate of adaptation. Not all receptors adapt at the same rate: touch and temperature adapt rapidly; pain receptors and proprioceptors adapt very slowly.

The Major Sensory Pathways

As a result of receptor activity, an afferent neuron is caused to propagate the impulse to the central nervous system. After the impulse reaches the spinal cord or brainstem, it may be conveyed by specific sensory pathways to higher centers for awareness.

Organization in the Spinal Cord The white matter (nerve fibers) of the spinal cord, which forms ascending and descending pathways, is divided into major regions—the anterior, posterior, and lateral funiculi (Figure 10-4). The posterior funiculus lies between the dorsal columns of gray matter, the anterior funiculus is between the ventral columns, and the two lateral funiculi are located between the dorsal and ventral columns on each side of the cord. A fasciculus is a bundle of fibers which forms a tract or pathway within a funiculus.

Major pathways are usually named from their point of origin to their point of destination; thus the lateral spinothalamic tract in the lateral funiculus originates at spinal levels and terminates in the thalamus.

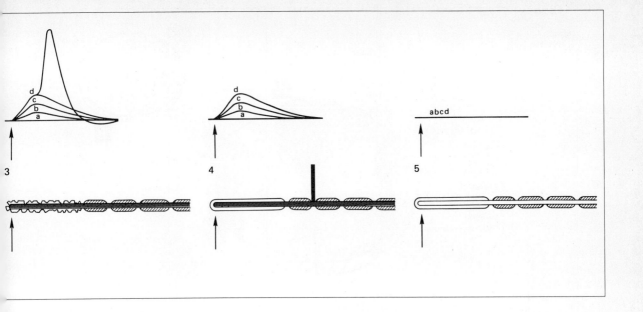

3 4 5

Figure 10-2. Relationship of the stimulus strength to the magnitude of the generator potential.

Figure 10-3. Adaptation of receptors under constant stimulus.

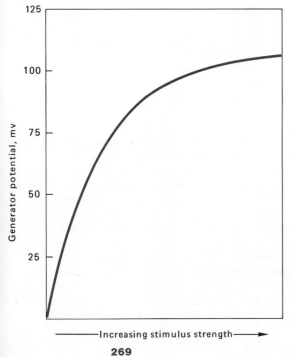

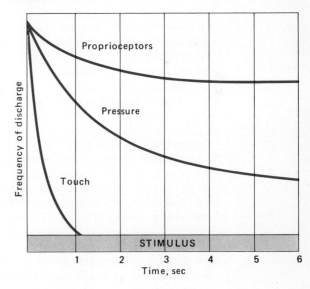

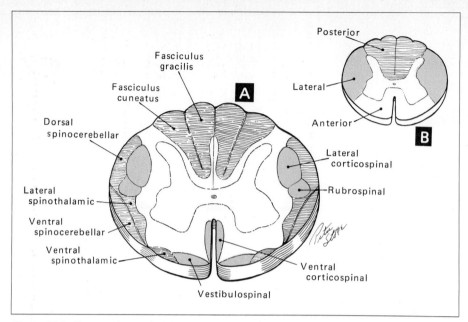

Figure 10-4. Nerve pathways in the spinal cord. **A.** Cross section at the cervical level. The major ascending tracts are shown in black, the major descending tracts in color. **B.** The funiculi.

The major ascending, or sensory, pathways are those for pain and temperature, touch and pressure, and proprioception.

Three neurons constitute the sensory chain. The primary sensory neuron, which conducts the impulse from the receptor to the spinal cord, is designated the first-order neuron. Its cell body lies in the dorsal root ganglion. The neuron that conveys the impulse within white matter up the cord to the thalamus is the second-order neuron, and its cell body lies at variable levels within the gray matter of the spinal cord or brainstem. The neuron that conducts impulses to the cerebral cortex is the third-order neuron, and its cell body lies in the thalamus.

Pain and Temperature   The first-order neuron in the pathway for pain and temperature synapses as it enters the spinal cord with a neuron in the dorsal column of gray matter. The axon of this second-order neuron crosses the gray matter to the opposite side of the spinal cord, where it courses upward as a component of the lateral spinothalamic tract (Figure 10-5). Passing to the thalamus the fiber synapses in a nucleus in this region with a third-order neuron, which relays the impulse to the sensory area of the cerebral cortex. Upon arrival of the impulse at the cortex, pain or temperature is consciously perceived.

President Nixon's trip to China in 1972 resulted in the introduction of acupuncture into the United States. Although Western scientists have long been aware of the widespread use of acupuncture in China, very little attention was paid to the technique, and few took it seriously. However, the visit of physicians, including the President's physician, to China, where they observed acupuncture being used during major surgery, generated

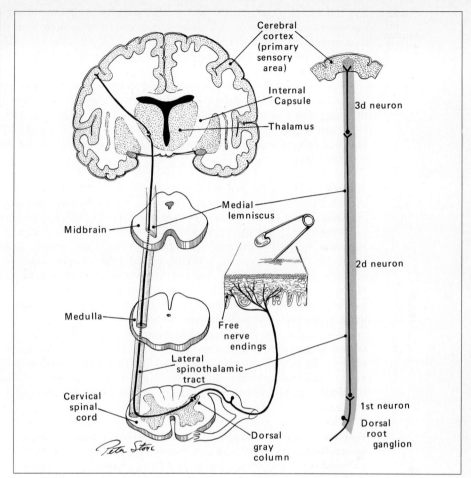

Figure 10-5. The central pathway for impulses perceived as pain, the lateral spinothalamic tract. Note that the fibers cross upon entering the spinal cord.

great interest and also a plethora of theories as to the mechanism by which acupuncture could block pain.

The term acupuncture refers to needles used generally to puncture the skin. Sometimes several needles are simply inserted, in other cases they are twirled after insertion, and in still other instances they are used as electrodes through which a weak current is passed. As practiced in China, these techniques are used not only to block pain but as treatment for many disorders. The question that interests physiologists is how acupuncture can block pain. The so-called spinal gating hypothesis has been proposed. This hypothesis requires the presence of afferent inhibitory neurons. Stimulation of these neurons would "close the spinal gate," thus preventing pain impulses from being propagated from the spinal cord to the brain. The gating mechanism is said to be embedded in the cells of the substantia gelatinosa, which would perform two vital functions for pain discrimination. First, it would sum up the net stimulus due to excitatory and inhibitory signals converging on the spinal cord from afferent fibers and would then transmit this net signal to brain centers. Second, it would coordinate pain on the one hand and its interpretation on the other. So by "closing the gate" either the pain impulses themselves would not reach the brain centers, or the coordination brought about by the gating mechanism would alter the interpretation, and pain would not be experienced.

Another hypothesis has been proposed which is similar to this one but assumes a "second gate" higher up in the central nervous system, either in the brainstem or the thalamus. The second gate would block impulses that bypass the first gate.

The entire field is now under investigation which it is hoped will not only clarify the underlying physiology but which may also put acupuncture and electroanalgesia on a rational basis.

Touch The first-order neuron transmitting impulses generated in response to light touch ascends in fasciculi of the posterior funiculus, that is, the fasciculus gracilis or fasciculus cuneatus (Figure 10-6). The former transmits nerve fibers from the lower portion of the body; the latter, from the upper portion of the body. Primary neurons of this pathway ascend to the level of the medulla oblongata, where they synapse, in either the nucleus gracilis or nucleus cuneatus, with the second-order neuron. The latter crosses to the opposite side of the medulla and ascends as a component of a tract called the medial lemniscus. Fibers of the second-order neuron pass to a nucleus of the thalamus and synapse with a third-order neuron which, in turn, transmits its impulse to the postcentral gyrus of the cortex to be interpreted as touch.

An additional pathway for crude touch lies in the anterior part of the cord. It is the ventral spinothalamic tract. The first-order neuron ends in the dorsal column of gray matter. The second-order neuron crosses the cord and ascends as a component of the ventral spinothalamic tract. At a higher level the fibers join the other touch fibers in the medial lemniscus and continue to the thalamus, where they thereafter follow the same course.

Proprioception Proprioceptive impulses reach consciousness, that is, the cerebral cortex, by way of the posterior funiculus and medial lemniscus (Figure 10-7).

Additional important proprioceptive impulses pass to the cerebellum. This portion of the brain acts in coordination and timing of muscular activity and is dependent upon proprioceptive impulses to function properly. Proprioceptive impulses destined for the cerebellum synapse in the dorsal column of gray matter with the second-order neuron. Fibers of the latter ascend through the cord as components of either the anterior or posterior spinocerebellar tracts and reach the cerebellum as components of the inferior cerebellar peduncles. The cerebellum, by means of additional pathways, in turn relays proprioceptive impulses to the cerebral cortex.

Function of the Thalamus

The thalamus is far more than merely a relay station on the way to the cortex. It is essential, at least in human beings, for many of the components of perception as well as for the emotional response to perception.

Awareness Perception demands awareness of stimuli. Even after complete removal of the cerebral cortex, a degree of awareness persists. But in the absence of the cerebral cortex, stimuli cannot be accurately located, and the intensity of the stimulation must be greater than normal in order to be appreciated.

All threshold stimuli result in impulses which are propagated by way of the thalamus

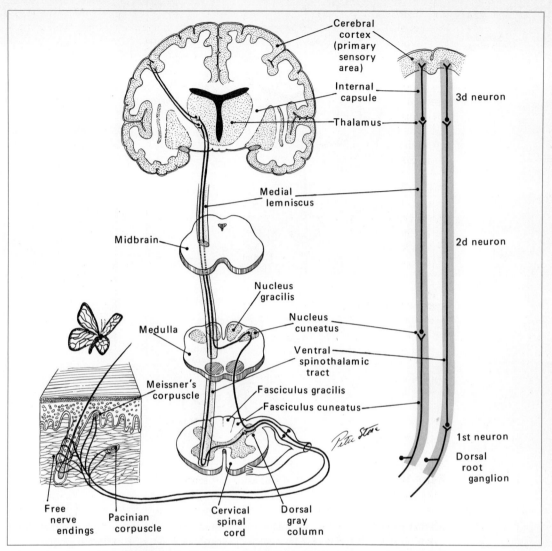

Figure 10-6. Central pathways for touch and light pressure perception, the ventral spinothalamic and the fasciculus cuneatus or gracilis. Some fibers cross as they enter the spinal cord, while others cross in the medulla at the nucleus gracilis or cuneatus.

to the cerebral cortex. There is awareness of the stimulation. But the fact remains that the direct stimulation of the cerebral cortex does not elicit such sensations. The person is aware of the cortical stimulation and localizes it as coming from a specific area of the body, but qualities of temperature, pain, pressure, or any other modalities of sensation are not present. This appreciation of sensation is thought to be a function of the thalamus.

Reactivity   The thalamus, in all probability, functions in some manner to determine the organism's reaction to stimuli, apparently by coordinating a wide variety of impulses.

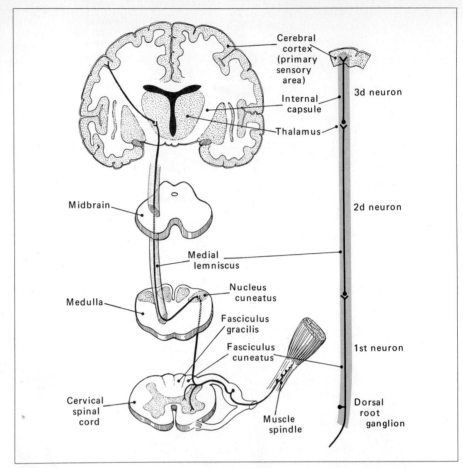

Figure 10-7. Central pathway for proprioception, the fasciculus cuneatus or gracilis.

All that precedes a specific stimulation determines the organism's reaction to that stimulation. In psychology, this is termed the psychological set, or conditioning.

Producing an emotional reaction to stimuli also seems to be a role played by the thalamus. One finds certain stimuli enjoyable, others unpleasant. Destruction of the thalamus exaggerates emotional reactivity: the patient may find that stimuli which were mildly unpleasant are now intolerable. Conversely, stimuli which were formerly enjoyable may now evoke ecstasy.

Sensory Functions of the Cerebral Cortex

Figure 10-8 shows the various functional areas of the cerebral cortex. Some of these areas are concerned primarily with perception, others with motor activities, and still others with integrating the two.

Sensory Areas Areas which have been designated areas 1, 2, and 3 are the primary areas for somatic sensation. Area 17 is the primary area for vision, and area 22, for hearing.

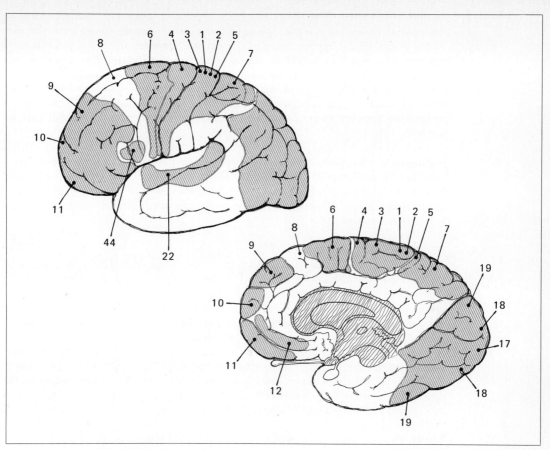

Figure 10-8. Important functional areas on the lateral and medial aspects of the cerebral cortex (see Table 10-2).

Surrounding the primary areas are the secondary sensory areas. These secondary regions are not necessary for naked perception but are utilized for interpretation of the impulses that are received by the primary areas. For example, damage to the secondary hearing areas does not cause deafness, but the person may no longer understand the meaning of the words he hears. Or if there is destruction of area 7, the secondary area for somatic sensation, dexterity which depends upon tactile sensation may be impaired.

Intensity of Sensation   The number of impulses which are propagated to the cerebral cortex per unit time is the clue essential to intensity discrimination. The impulse frequency depends upon (1) the number of receptors activated by the stimulus and (2) the response of each receptor. An intense stimulus not only activates the receptors in the immediate area but also receptors surrounding that area to a distance proportional to the intensity. Thus, the stronger the stimulus, the more receptors that respond and the faster they fire. The net result is a proportionately greater number of impulses propagated to the cerebral cortex.

Sensory Localization   The ability to localize a point of stimulation is independent of the sense of vision. For example, one can tell that pain is coming from a particular part of the body without seeing that part. Localization depends upon the point-to-point representation of all surface areas on the sensory cerebral cortex. The point-to-point representation plus the learning process enables one to associate stimuli with the particular part being stimulated. The importance of the learning process is underlined by the fact that a person has great difficulty in accurately localizing stimuli which emanate from regions not normally stimulated or which he cannot see, such as the internal parts of the body.

Stereognosis   The ability to recognize objects without visual assistance is termed stereognosis. The word implies knowledge of objects. By virtue of perception a person may handle an object and recognize many of its properties, such as size, shape, texture, and temperature. Lesions in the sensory pathways destroy this ability and create a condition known as astereognosis.

Abnormalities of Perception

Perception requires the anatomical integrity of the entire system from the receptor to the thalamus and the cerebral cortex. Disruption of the circuit, anywhere, results in impairment. For example, a lesion in the posterior funiculus causes loss of proprioception and impairment of the sense of touch and pressure. The senses of touch and pressure are not completely lost because these modalities are also carried by other pathways. In view of the anatomical fact that the afferent neurons subserving these senses turn upward as soon as they enter the spinal cord and ascend on the same side as the side they enter, spinal cord damage results in sensory loss on that side. In contradistinction, a lesion in the lateral spinothalamic tract causes loss of pain and temperature appreciation on the opposite side of the body, because the fibers that make up this tract enter on one side of the cord and cross to the other side before ascending.

Ataxia   Proprioception is the ability to appreciate changes in position and muscle tension. If there is loss of this sense, there will be difficulty in regulating the direction, force, rate, and extent of voluntary movements. Consequently, the usual high integration and smoothness of muscular activity is lost, a disorder termed ataxia.

Ataxia is commonly seen in advanced cases of syphilis, a disease which afflicts the posterior funiculus. The resulting loss of proprioception causes the syphilitic to utilize

**TABLE 10-2. AREAS OF THE
CEREBRAL CORTEX**

AREA	FUNCTION
1, 2, 3, 5, 7	General sensation
4, 6	Motor
8	Eye movements
9–12	Behavior
17–19	Vision
22	Audition
44	Speech

other senses to inform him of his position. Thus, unless he watches his feet, he will stagger as though intoxicated. He walks on a broad base, that is, with his feet wide apart, and he raises each leg unnecessarily high and then slaps it down, all because his awareness of the position of his legs is inadequate.

Abnormalities of Pain   A strong, and usually destructive, stimulus causes pain. Increased sensitivity to pain is termed hyperalgesia. If local conditions, such as a burned or diseased area, make the region more sensitive, there is primary hyperalgesia. An abnormality in the central nervous system may also increase pain sensitivity, a condition referred to as secondary hyperalgesia. Analgesia is the absence of pain.

In some instances, stimuli activating receptors in one area of the body will be inaccurately localized so that the person believes that an entirely different part is being stimulated. Because the pain is localized or referred to another area, this condition is termed referred pain. The explanation lies in the convergence of different afferent neurons onto the same secondary neuron. For example, one may have a neck pain when, in reality, the stimuli are arising in decayed teeth. Another common example is the referred pain of appendicitis in which the pain of the inflammatory process is referred to the surface of the abdomen (Figure 10-9). Afferent fibers from the appendix and from the surface enter the cord at the same level and then converge upon the same secondary neuron that then crosses the cord and ascends in the lateral spinothalamic tract. Since stimuli normally arise from the surface and not from the appendix, the impulses upon arriving in the cerebral cortex are interpreted as originating from the surface receptors.

Convergence of impulses onto a common secondary pathway also underlies referred hyperalgesia. In this condition, the action potentials from one primary pathway do not activate the postsynaptic neuron but simply lower its sensitivity, thus making it more responsive to impulses arriving from another area of the body. This is true of types of sensation other than pain and is then termed referred hyperesthesia.

If the elbow is struck, the fingers will seem to tingle, because no matter where a neuron is excited along its course, the person refers the stimulus to the usual site of activation, namely, the receptor. Thus, not infrequently after a limb has been lost, the patient complains of sensation, usually pain, emanating from the fingers or toes of the amputated limb, a condition termed phantom pain. The stimulus for the pain, of course, is at the point of amputation but is referred to the once-present receptor of that neuron in the extremity.

Faulty Perception   A false perception based on a stimulus is called an illusion. Figure 10-10 presents typical illusions. Perception in the complete absence of stimuli is termed hallucination. The drunk who gazes at the flowers on the wallpaper and insists that they are pink elephants is suffering an illusion. The drunk who gazes at the blank wall and sees elephants is having a hallucination.

MOVEMENT

All movement is dependent on the contraction of muscles. Skeletal muscle activity, in the intact organism, is controlled completely by the nervous system. Smooth and cardiac muscle, on the other hand, are capable of shortening even in the absence of all nerves. Still, despite this inherent ability, smooth and cardiac muscle are, for the most part, under nervous control. Thus, analysis of any movement must include a consideration of the role played by the nervous system.

Generally speaking, there are two types of movement: (1) reflex and (2) voluntary.

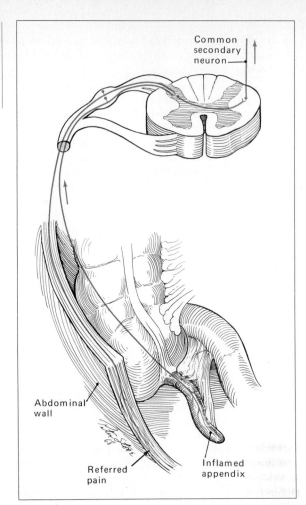

Figure 10-9. Referred pain. An inflamed appendix initiates impulses which are carried by an afferent neuron to the spinal cord. There the neuron synapses on a second-order neuron which ordinarily receives impulses from neurons in the abdominal wall adjacent to the appendix. Impulses from both sites are carried up the spinal cord by a common neuron and reach a single area of the cortex, where they are interpreted as arising from the abdominal wall.

For example, when one taps the patellar tendon of the leg, just below the knee, the limb shoots out. Much the same movement is carried out when kicking a football. But the first movement is reflex, the second voluntary. Specifically, a reflex is an involuntary response to a stimulus. It is an automatic act. The person may be aware of the movement; he may even be able to inhibit it; but when the activity occurs, it does so involuntarily and without conscious assistance. Although in this chapter only movement will be considered, it should be understood that a reflex may involve a gland instead of a muscle. Glandular tissue responds involuntarily to a stimulus. For example, when a foreign object strikes the eye, tears are secreted. The foreign body initiates mechanisms resulting in lacrimation. More will be said concerning glandular activity later.

Anatomy of Reflex Movement

The anatomy of most reflex patterns is extremely complex. However, all reflexes possess a basic design, the reflex arc. This consists of an afferent neuron which is sensory in nature; that is, it propagates the impulse from the periphery toward the central nervous

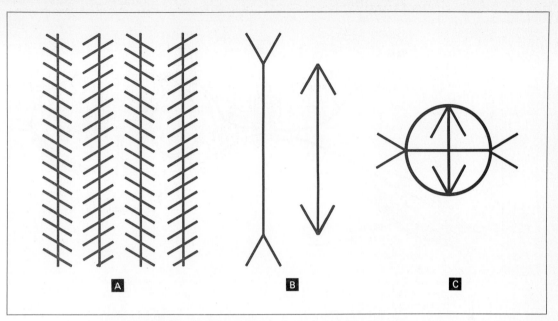

Figure 10-10. Classic illusions. **A.** The lines are actually parallel. **B.** The vertical lines are of equal length. **C.** The circular line is actually round.

system. It also includes an efferent, or motor, neuron which propagates the impulse from the central nervous system toward the periphery. In addition, there is ample evidence to suggest that normally all reflexes, even the most simple, are influenced by the higher brain centers. Thus it may be said that the pathways from these cerebral centers constitute another important component of the reflex arc.

The Afferent (Sensory) Neuron The afferent portion of the reflex arc consists of the receptor and its sensory pathway. A reflex may be elicited in response to variation in temperature, changes in pressure, alterations in position, or any other environmental disturbance adequate to stimulate the receptor mechanism.

The pathways by which impulses pass from the receptors to the sensory cortex, as already explained, embrace three components. The first link connects the receptor with the central nervous system. The second unit propagates the impulse to the thalamus. The third link connects the thalamus and the sensory cortex. In complicated reflex activity, the impulse may travel along all three components, but the more fundamental, simpler reflex patterns utilize only the first one.

In some reflexes the sensory neuron terminates in the ventral horn of the spinal cord in synaptic union with the efferent neuron. In others (Figure 10-11), the sensory neuron is seen to end in gray matter soon after it enters the cord. Then a short interconnecting, or internuncial, neuron conveys the impulse across the cord to the efferent neuron.

The Efferent (Motor) Neuron The efferent portion of the reflex consists of a motor pathway and an effector organ, which may be either muscle or gland. The motor tracts emanating from the higher brain centers consist of two links. One begins in the motor areas of the cerebral cortex and propagates the impulse to lower centers in the brainstem

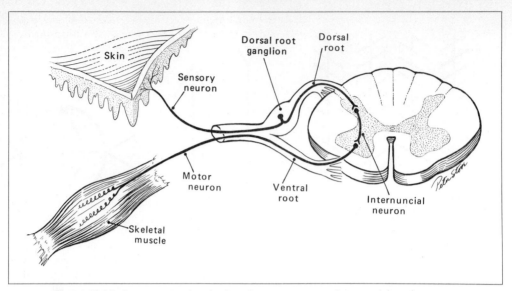

Figure 10-11. Components of a simple reflex: a sensory, an internuncial, and a motor neuron.

or spinal cord. The other then propagates the impulse from the central nervous system to the effector organ, that is, the muscle or gland.

The effector organ effects, or brings about, the reflex action. This activity may be one of muscular contraction or glandular secretion. In some cases the receptor which initiates the reflex lies within the effector organ. This arrangement is the basis of important self-regulatory mechanisms. For example, if a muscle is stretched, the proprioceptor within the muscle is activated, and as a result there is reflex shortening of that muscle which serves to oppose the stretching force. In other cases, the receptor is far removed from the effector organ.

Pathways from Higher Centers The reflex is transmitted from the sensory to the motor neuron across one or more synapses, depending on the complexity of the reflex. Synaptic transmission can be influenced by the activity of higher cerebral centers. These impulses are conveyed by pathways which descend the brainstem and spinal cord to end at the synapse between sensory and motor neurons. Some of these pathways are also utilized for impulses which originate volitionally. Other nerve routes serve no other function than to control reflex activity.

Under abnormal conditions the basic reflex continues to function even though the influence of the higher centers has been lost. For example, if the spinal cord is completely severed, reflex movement may still be elicited after a time. A person so afflicted cannot move his leg voluntarily, but if the patellar tendon is tapped below the knee, the typical patellar reflex occurs. Yet there is excellent evidence to indicate that in the normal person even the most simple reflexes are constantly under the guidance of these higher centers.

Physiology of Reflex Action

There are certain basic principles that govern all reflex action. In order that the student may appreciate the role played by reflex action in the total physiology of man, these must first be presented. Accordingly, three primary physiologic characteristics of reflex action will be considered, and then specific reflexes will be discussed to illustrate the part they play in the total pattern.

Reflex Latent Period The velocity of propagation of an impulse along a nerve fiber varies with the diameter of the fiber. A measurable interval of time is required for the impulse to cross a synapse as well as the neuromuscular junction. Time is also required for the muscle to contract after it has received the stimulus. Since even the simplest reflex involves all these components, the reflex reaction time, or reflex latent period, must be the sum of all the component latent periods. In the complicated reflex responses there are, in addition, various psychological factors.

Facilitation Reflex movement does not normally occur in the absence of higher-center influence. Impulses propagated by the sensory neuron of a reflex arc are thought to be subliminal, at least as they relate to excitation of the motor neuron. In other words, the impulses reach the synapse but fail to fire the motor neuron. However, at the same time impulses arrive at the synapse from higher centers. These impulses are also subliminal, but when they are summated with the subliminal impulses arriving over the sensory neuron, threshold strength is achieved and the motor neuron is fired. Put somewhat differently, the subliminal impulses from the higher centers, although they in themselves fail to activate the motor neuron, do alter the cell body in such a way that a central excitatory state is produced. This lowers the threshold of excitability of the motor neuron and allows it to be excited by the subliminal impulses arriving over the sensory neuron. Facilitation, by definition, means "making easier"; that is precisely what the impulses from higher centers do. They make it easier, at least according to this theory, for reflex movement to take place by aiding the impulses across the synapse.

The degree of facilitation may vary considerably. For example, if one is emotionally upset, a slight noise will cause one to jump, whereas under normal circumstances the same sound might pass unnoticed. Stimulation of various areas of the brain in experimental animals definitely influences reflex activity.

Inhibition Just as reflexes may be facilitated or aided, they may also be inhibited or impeded. Inhibition is a function of the central nervous system. It involves a short, specialized neuron which is a link between the neuron from higher centers and the lower motor neuron that innervates the muscle. This short neuron, when activated, frees a chemical that inhibits firing of the lower motor neuron by raising the threshold of activation. In effect, then, these specialized neurons create a central inhibitory state which makes it harder or even impossible to activate the lower motor neuron.

Spinal Shock In man, if the spinal cord is severed, all reflex responses below the level of transection are lost for a period of time. Such a person is said to be in spinal shock. After a period of time, reflex activity returns. Interestingly enough, if the cord is transected a second time just below the original site, reflex movement does not again disappear. This means that there is normally a preponderance of facilitatory impulses coursing down through the spinal cord. When the cord is cut, these impulses cannot reach the reflex synapse. Consequently, synaptic transmission does not occur, and there is no response. When reflex activity does return, it quite often becomes exaggerated. There results a state of hyperreflex activity termed hypertonicity. The mechanism of this increased reflex sensitivity is not clear.

Stretch Reflex When a muscle is stretched, it usually responds by contracting. In other words, it withstands, or opposes, the stretching force. The adequate stimulus is stretch, and thus the term stretch reflex. The receptors are proprioceptors in the muscles and tendons. As a result of proprioceptor firing, impulses are propagated to the spinal cord or brainstem, a motor neuron is activated, and the muscle that was stretched now contracts. By definition the muscle exhibits tone since tone implies involuntary resistance to passive stretch. If the reflex arc is destroyed or if it is inhibited or even not adequately facilitated, the muscle will not respond to stretch. Absence of response is termed atonia, diminished response hypotonia, and excessive response hypertonia.

Stretch reflexes are extremely important for the maintenance of the upright position. In the upright position the force of gravity will stretch certain muscles. As a result, they contract and thus oppose gravity. For this reason stretch reflexes are also called antigravity reflexes. They are also referred to as myotatic reflexes.

The familiar clinical test of eliciting the knee jerk, the patellar reflex, is an example of a stretch reflex. When the tendon below the kneecap is struck, the quadriceps muscle is stretched and responds by contracting. As a result, the leg swings forward.

The proprioceptor essential to the stretch reflex is the muscle spindle (Figure 10-12).

Figure 10-12. Muscle spindle. The spindle is activated by stretch. Contraction of the intrafusal fibers increases the sensitivity of the spindle. Contraction of the major muscle fibers releases the stretch on the spindle.

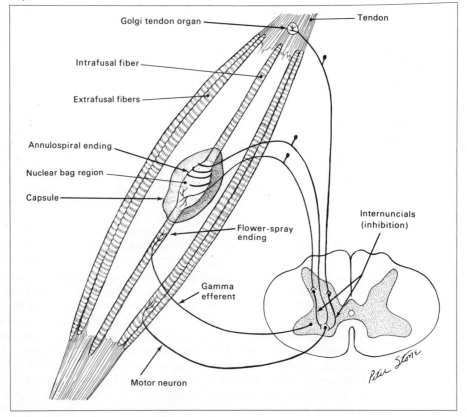

The muscle spindle is so termed because it is long and thin and has pointed ends; that is, it is spindle-shaped. It consists of modified muscle fibers called intrafusal muscle fibers. The sensory nerve fibers are wrapped around the intrafusal fibers near their middle, an area known as the equatorial region. Note in Figure 10-12 that there are motor neurons which innervate the intrafusal fibers of the spindle as well as the main muscle fibers.

Because the muscle spindle is in parallel with the muscle fibers, stretch of the muscle will also stretch the spindle, resulting in impulses which are propagated by the sensory nerve to the spinal cord, where synapse is made with the appropriate motor nerve. Consequently, the muscle contracts, and when it does, stretch of the spindle is relieved. Here then is an excellent servo control system which automatically regulates the length of the muscle in accord with the stretch placed upon it. Or to put it another way, as the muscle is loaded, the joint supported by that muscle is neither extended nor flexed, the angle of the two bones remaining fixed. For example, if one holds his hands out while someone places more and more weight on them, his hands will remain in just about the original position despite the increased load.

The fact that the intrafusal fibers can contract adds another dimension to this mechanism. Obviously, if the intrafusal fibers were completely relaxed and slack, considerable lengthening of the muscle would have to occur before the stretch reflex would be evoked. In contradistinction, were the intrafusal fibers contracted, the slightest stretch would bring about shortening. In this way the sensitivity of the reflex is regulated.

Autogenic Inhibition   Excessive stretch of a muscle causes it to relax reflexly. This response has been variously labeled autogenic inhibition, the inverse stretch reflex, and the lengthening reaction. If a joint is vigorously flexed, it will suddenly give way because of the reflex relaxation of the stretched muscle. Because of the similarity to the closing of a clasp knife, it is called the clasp-knife effect. The purpose is probably protective—to prevent muscle or tendon rupture from excessive stretch.

The receptor for this reflex is thought to be the tendon organ (Figure 10-12). Just as the muscle spindle is activated by stretch, so is the tendon organ. But when the muscle contracts, stretch on the muscle spindle is reduced whereas the tension on the tendon organ increases. Apparently, the tendon organ has a very high threshold and is activated only in cases of extreme tension. When it is activated, muscle relaxation, rather than contraction, occurs. There must, therefore, be an inhibitory neuron in the reflex circuit which inhibits the motor neuron.

Flexor Reflexes   If one inadvertently touches a hot stove, the arm is withdrawn. When one accidentally steps on a nail, the leg is lifted. These are everyday examples of the flexor reflexes. In all these instances the flexor muscles are involved, that is, muscle groups which are opposed to the extensor muscles utilized by the stretch reflexes.

The primary function of the flexor reflex is protection. The basic pattern is one of withdrawal from a painful or noxious stimulus. As in all reflexes, there must be a receptor. In these cases the receptor is called a nociceptor (*nocere,* "to hurt").

The flexor reflexes are somewhat more complicated than the stretch mechanisms in that they usually involve more muscle groups. A stretch reflex is often operative in a single muscle. For example, the quadriceps muscle plays the major role in the patellar reflex. But it is obvious that when one steps on a nail, one does not simply elevate a toe or flex the ankle. The entire limb is withdrawn. The efferent portion of this reflex clearly must involve several fibers innervating numerous muscles which contract to withdraw the entire lower extremity.

The flexor reflexes, like the stretch reflexes, are under higher-center control. During the period of spinal shock, these reflexes are also inactive. This evidence, along with other experiments, suggests that the flexor reflexes may be facilitated or inhibited.

Supporting Reflexes   Although the extensor muscles play the major role in supporting man against the force of gravity, the importance of the flexor musculature should not be overlooked. In the standing position the joints are fixed, thus converting the lower extremities into rigid pillars. In order to make possible such rigidity, both the flexor and extensor muscle groups must be partially contracted.

To avoid confusion at this point, some reiteration is necessary. Although the stretch reflexes are regarded as involving, for the most part, the extensor muscles, the flexors also contain proprioceptors. Hence, when they are stretched, these muscles respond by shortening. On the other hand, these very same muscles contract in response to painful stimuli. The muscle is one and the same; the reflex pathways are distinct and serve different purposes. The stretch-reflex component of the flexor muscles is essential in maintaining the limbs as rigid pillars which support the body against gravity. While a person is upright, if the knee flexes, the extensor muscles will be stretched and caused to contract. On the other hand, if the knee threatens to bend in the other direction (overextend), the flexor muscles are tensed and caused to shorten. The net result is that the knee joint is held immobile and the person remains upright.

Conditioned Reflex   A conditioned reflex involves a response to a stimulus that ordinarily would not evoke that response. In the classic experiments carried out in Russia by the famous physiologist Pavlov, dogs were conditioned so that when a bell was rung they would salivate. A reflex is conditioned by first using two stimuli simultaneously. In the Russian experiment, the dog would first be permitted to smell food each time a bell was sounded. The smell of the food caused salivation. After the period of conditioning, the sounding of the bell caused salivation even though no food was present. In short, a new afferent pathway was substituted for the old one.

Decerebrate Rigidity   The influence of higher-center control over reflex action is vividly demonstrated in a condition termed decerebrate rigidity. This state results when brainstem transection removes higher-center inhibition but leaves adequate facilitatory centers operative. As a consequence, all spinal cord reflexes are facilitated without the presence of balancing inhibition. The slightest stimulation, such as touching the feet, evokes a vigorous contraction of both extensor and flexor muscles. The limbs are converted into rigid pillars. An animal, in this condition, is paralyzed, but because of the great rigidity of the limbs it can be balanced in the upright position, and the rigid limbs will support the weight of the animal.

Voluntary Movement

Voluntary movement depends upon the integrity of the motor cortex and the motor pathways. Each area of the body is represented on the motor cortex. The same was noted to be true for the awareness of sensation on the sensory cortex. This point-to-point relation is inverted. The very top of the motor cortex, including the part folded into the central sulcus, controls the lower extremity. Stimulation of area 4 more laterally causes the muscles of the upper extremity to respond. Lower parts of this area control the muscles of the head and neck.

In addition to the point-to-point relation, there is also an unequal cortical representation (see Figure 9-4). A large part of area 4 is concerned with movement of the hand, for example. In contradistinction, the entire trunk is represented on the cortex by only a small region. The size of cortical representation is correlated with greater mobility and dexterity of the part.

When an axon reaches a muscle, it arborizes into terminal branches. If it arborizes profusely, as do axons supplying postural muscles of the back, for instance, many muscle fibers are innervated, and therefore controlled, by one neuron. On the other hand, an axon may arborize little, so that it innervates few muscle fibers, an example being the extrinsic muscles of the eye. There is far greater cortical representation for the muscle fibers in the second example than for those in the first example. As a consequence, while it is possible to contract small fascicles in an eye muscle, one can contract only large fascicles of the back muscles. In summary, muscles whose fibers have great cortical representation are more dextrous than muscles whose fibers do not.

Pyramidal Pathways   There are two important systems of motor pathways, pyramidal and extrapyramidal. Fibers of the pyramidal pathways originate in the pyramidal cells of the cortex. They terminate in the ventral horn of the gray matter within the spinal cord. Fibers from cells in the ventral horn then supply the final link to skeletal muscles.

The major tract for impulses that effect voluntary movement is the corticospinal pathway (Figure 10-13). "Cortico" signifies its origin in the cortex, "spinal" its destination in the spinal cord. Fibers of this pathway begin in pyramidal cells of the cortex, descend through the internal capsule, form the cerebral peduncles, and, once in the medulla, are visible on its anterior surface as the pyramids. At the lower end of the medulla most of these fibers cross to the opposite side of the spinal cord at the pyramidal decussation. They continue in the lateral funiculus as the lateral corticospinal tract. Fibers that fail to cross at the pyramidal decussation continue downward in the anterior funiculus as the ventral corticospinal tract. The majority of these finally cross to the other side of the cord at the spinal cord segment in which they terminate.

Another important pyramidal tract is the corticobulbar pathway. Fibers of this tract leave pyramidal cells in the cortex and descend to the brainstem, where they synapse with motor neurons in nuclei associated with the cranial nerves that supply muscles of the head and neck.

Most motor pathways involve two principal neurons. The neuron that has its cell body in the cerebral cortex is termed the upper motor neuron. The neuron that innervates the muscle is termed the lower motor neuron. Thus all corticospinal and corticobulbar neurons are upper motor neurons. All motor neurons of cranial and spinal nerves are lower motor neurons. Between upper and lower motor neurons there are often one or more internuncials. The upper motor neuron has at least two functions: (1) to initiate voluntary movement or (2) to influence reflex movement by facilitating or inhibiting the transmission of the impulse at the synapse. The lower motor neuron has but one function: to activate the muscle it innervates.

Extrapyramidal Pathways   The extrapyramidal pathways are less well defined than those of the pyramidal system. They are associated with phylogenetically older portions of the brain such as the basal ganglia and certain brainstem nuclei, namely, the subthalamic nucleus, red nucleus, and nuclei of the reticular formation.

Extrapyramidal pathways are involved in postural adjustment and gross movement patterns that are essentially reflex in character. Thus impulses traveling along extra-pyramidal pathways modify the operation of the pyramidal system. In any given movement, for example flexion of the elbow joint, contraction of the flexor muscles must be balanced by relaxation of the extensor muscles. Otherwise the action would be exaggerated and jerky. This balance is achieved in large part through the extrapyramidal system. In

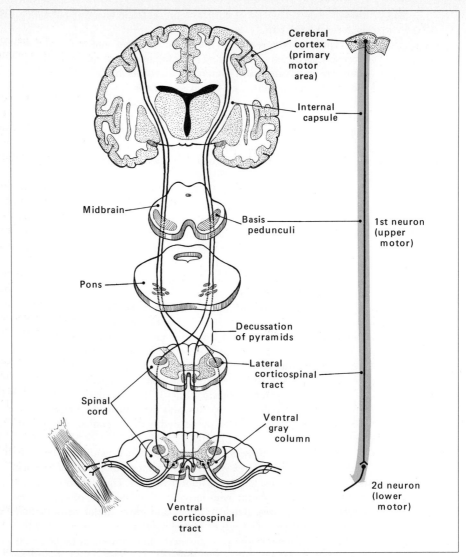

Figure 10-13. Pyramidal motor pathways, the corticospinal tracts. The tracts originate in pyramidal cells of the cortex. Fibers that cross at the medulla form the lateral corticospinal tracts, while the remaining fibers form the ventral corticospinal tracts. The bases pedunculi are part of the cerebral peduncles.

summary, then, the extrapyramidal system acts to enable coordinated, smooth movement of appropriate magnitude.

Fibers of the extrapyramidal system originating in the cerebrum synapse with neurons in certain basal ganglia and brainstem nuclei and then descend to terminate on lower motor neurons. Thus some extrapyramidal pathways involve two or even more upper motor neurons. Two extrapyramidal tracts are the rubrospinal and vestibulospinal tracts

(Figure 10-4). The former descends the cord from the red nucleus, the latter from the vestibular nucleus.

Because all impulses to skeletal muscles use a lower motor neuron to reach their destination, this neuron is frequently referred to as the final common pathway. Its significance will be discussed under spastic and flaccid paralysis.

Coordinated Movement Movement rarely results from the contraction of a single muscle or a single group of muscles. Most movement involves a pattern, a pattern that requires the coordination of several muscles. Area 4 has direct control over the individual muscles. Many other parts of the central nervous system are concerned with the coordination of these muscle movements into meaningful patterns. One such coordination center is area 6.

Exploratory stimulation of area 6 discloses that groups of muscles respond. In lower animals, for example, such stimulation may lead to the movement of an entire leg or a pattern that resembles walking, or chewing, or swallowing. Such experiments also indicate that there is cortical representation on area 6 like that on area 4. Stimulation of the upper regions of area 6 causes leg movement; the lower regions involve the head and neck.

The pathway by which impulses from area 6 are propagated to the muscles is apparently diverse and complex. Many of the impulses go by direct, short axons to area 4, but movement may still be elicited by area 6 stimulation after destruction of area 4. This indicates that there are other pathways, probably by way of the subcortical nuclei and the extrapyramidal tracts. More will be said of coordination in Chapter 14.

Decortication The removal of the cerebral cortex is termed decortication and can be done in experimental animals without fatal results. In man, a condition comparable in some ways to decortication is seen in cases of massive hemorrhage involving the internal capsule. In lower animals, decortication does not cause paralysis. The motor deficit becomes progressively more severe in higher forms. Yet, even in primates some movement is still possible following recovery from the operation. But this is very gross movement completely devoid of any influence of learning or conditioning. In addition, there is a degree of rigidity. Decorticate rigidity is not as marked as decerebrate rigidity.

Abnormalities of Movement

Lesions in many parts of the nervous system will cause abnormal movement. The more obvious disorder is paralysis. Disorders concerned with the centers of coordination will be discussed in Chapter 14.

Paralysis means the loss of voluntary movement. In many cases of paralysis, reflex movement of the part may still be elicited even though the same region cannot be moved volitionally.

Flaccid Paralysis Muscle tonus depends upon the activity of the stretch reflex. If the reflex arc is interrupted, the muscle becomes atonic. It is soft and unresponsive, that is, flaccid. In contradistinction, abnormal facilitation causes the muscle to be highly sensitive to stretch. It thus feels hard and tense, that is, spastic. If the lower motor neuron—the final common pathway—is severed, the muscle cannot be moved either voluntarily or by reflex action. There is paralysis. In addition, there is no longer any response to stretch; thus the muscle is flaccid. The combination is termed flaccid paralysis.

Flaccid paralysis may also be produced experimentally by discrete lesions involving only area 4 or only the pyramidal tract. Because area 4 is an important facilitatory center, such lesions result in overriding inhibition, and severe hypotonia results. The hypotonia combined with the loss of voluntary control causes flaccid paralysis. Clinically such discrete lesions rarely occur.

Spastic Paralysis   In spastic paralysis, the reflex arc remains functional but voluntary control is lost. This combination can result only from an upper-motor-neuron lesion. Immediately following this type of lesion there is areflexia, as in spinal shock, but after a variable period of time, reflex activity returns and is exaggerated. Thus muscle cannot be moved volitionally, but it readily responds to stretch.

QUESTIONS AND PROBLEMS

1   How does the human organism adapt to its environment?

2   What are the four types of receptors? Where are they located? What is the mechanism of receptor response?

3   Can a specific receptor respond to more than one modality of sensation? Explain.

4   How would you explain the fact that when you place your finger in warm water, the sensation of warmth diminishes after a short period?

5   How many neurons are involved in the major afferent pathways?

6   What is a fasciculus? Where is the lateral spinothalamic tract?

7   Trace the major pathways from the periphery to the higher centers in the brain for (1) pain and temperature, (2) touch, (3) proprioception.

8   Differentiate between first-, second-, and third-order neurons.

9   What role does the thalamus play in the afferent systems of the body?

10   What would be the effect of damage to the lateral spinothalamic tract on the left side of the cord at the level of the first lumbar vertebra? Of damage to the fasciculus gracilis on the same side?

11   Describe the physiologic events and anatomical structures involved in awareness that your big toe has touched an object.

12   What evidence would support the statement that pain receptors do not adapt as readily as temperature receptors?

13   Explain how it is possible for a blind person to know the exact site of a pinprick?

14   Explain why pain which originates in an inflamed appendix is felt on the surface of the abdomen.

15   How is it possible for a person to experience pain in his left foot after his left leg has been amputated at the knee?

16   How do you explain, physiologically, the heightened response to sensory stimuli observed in some persons who are emotionally upset? What does this imply for nursing care?

17   How could an astronaut maintain muscle tone when he is in a "weightless" condition?

18   What is a stretch reflex? Explain how it operates in a patient who demonstrates the knee jerk when the patellar tendon is struck.

19   Give an example of a conditioned reflex in your own behavior.

20   In what ways are the higher centers of the brain involved in reflex activity?

21   Why is sensory perception greater for the fingers and hands than for the entire trunk?

22   What symptoms would you expect in a patient following a cerebral vascular accident which resulted in damage to fibers of a pyramidal tract?

23   What is the pyramidal decussation?

24   What is meant by the term the final common pathway?

25   Differentiate between flaccid and spastic paralysis in terms of cause and effect.

11 THE AUTONOMIC NERVOUS SYSTEM

Thus far, perception and movement pertaining to the external regions of man have been considered. Within the body there are, of course, extremely important organs called the viscera. The viscera contain glands which secrete; in many cases, this secretion is under nervous control. The viscera also include smooth muscle which contracts; its contraction is usually governed by the nervous system. In each case it is the autonomic nervous system—a division of the peripheral nervous system—that is involved.

A reflex initiated by viscera travels along afferent pathways no different from those of a reflex initiated by skeletal muscle. But the efferent portion of the reflex follows a different course, one within the autonomic nervous system. Such reflexes are termed visceral reflexes because the resulting activity involves viscera, rather than skeletal muscle. The adjective "somatic" pertains to the framework of the body. Thus, in contradistinction to visceral reflexes, somatic reflexes involve skeletal muscles which move the skeleton. Visceral reflexes, like somatic reflexes, are under constant higher-center surveillance.

Before considering the autonomic nervous sytem in detail, a word of caution is required. This component is not something apart from the nervous system as a whole. It is but one part of the nervous mechanism and is, at all times, intimately integrated with all other components. An acute attack of appendicitis, for example, will give rise to pain. In addition, it may cause the legs to be drawn up to the abdomen. This illustrates the intimate interplay of a visceral afferent portion and a somatic efferent portion to produce a reflex act. Simultaneously, there may be marked alterations in blood pressure.

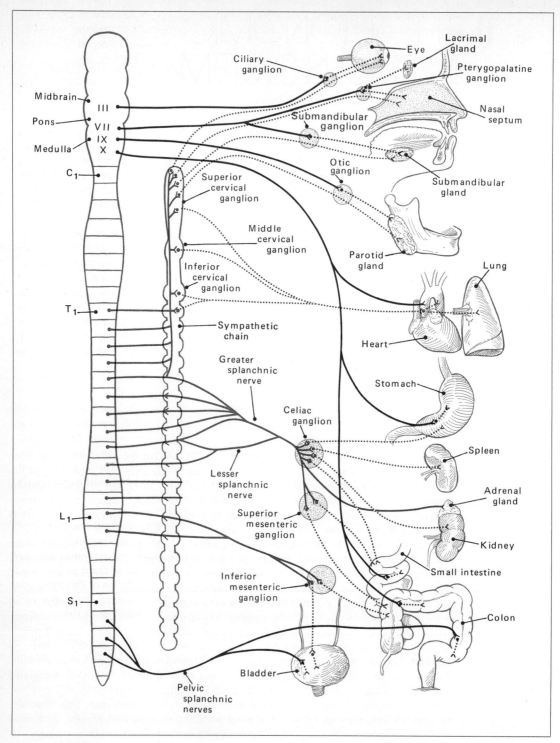

Ciliary ganglion

Eye

Lacrimal gland

Pterygopalatine ganglion

Midbrain

III

Pons

VII

IX

Medulla

X

Submandibular ganglion

Nasal septum

Otic ganglion

Submandibular gland

C₁

Superior cervical ganglion

Parotid gland

Middle cervical ganglion

Lung

Inferior cervical ganglion

Heart

T₁

Sympathetic chain

Greater splanchnic nerve

Stomach

Celiac ganglion

Spleen

Lesser splanchnic nerve

Adrenal gland

Superior mesenteric ganglion

Kidney

L₁

Small intestine

Inferior mesenteric ganglion

S₁

Colon

Bladder

Pelvic splanchnic nerves

This reflex involves the same afferent visceral link, but it is now in liaison with the visceral efferent unit of the autonomic nervous system. Numerous other illustrations could be cited to emphasize the fact that the autonomic nervous system is an integral constituent of the total nervous mechanism.

ANATOMY OF THE AUTONOMIC NERVOUS SYSTEM

The autonomic nervous system is a visceral motor system. It supplies the structures of the body not under voluntary control, namely, smooth muscle, cardiac muscle, and glands. Most organs innervated by fibers of this system have a dual nerve supply with opposing functions. One set of fibers stimulates the muscles to contract or glands to secrete, while the second group of fibers may inhibit such activities. These two divisions are named according to the location of their nerve cell bodies in the brain or spinal cord (Figure 11-1). Fibers arise from thoracic and lumbar segments of the spinal cord and form the thoracolumbar, or sympathetic, division. The other fibers come from cell bodies in the brain and sacral part of the spinal cord—the craniosacral, or parasympathetic, division.

The major anatomical characteristic of the autonomic nervous system is a two-motor-neuron pathway. That is, two motor neurons lie between the brain or spinal cord and the effector organ, in contrast to the innervation of skeletal muscle, where a single neuron with its cell body located in the brainstem or spinal cord sends a fiber directly to the organ to be innervated. Each neuron associated with the peripheral autonomic pathway is designated either preganglionic or postganglionic. Preganglionic nerve cell bodies are located in the brainstem or spinal cord. Postganglionic cell bodies are in ganglia outside the central nervous system.

Ganglia

A ganglion has been defined as a collection of nerve cell bodies. There are different types of ganglia within the body. For example, the dorsal root ganglion contains cell bodies of primary sensory neurons, but there are no synapses within the ganglion. On the other hand, in the ganglia of the autonomic nervous system most of the fibers that enter the ganglia synapse with one or more cell bodies of the postganglionic neurons found there. A second difference is that the dorsal root ganglion contains sensory neuron cell bodies, while the autonomic ganglia contain motor neuron cell bodies.

Sympathetic Ganglia   There are two groups of ganglia in the sympathetic portion of the autonomic nervous system. The first group is the vertebral ganglia which lie along both sides of the vertebral column. These ganglia are connected by filaments of nervous tissue and therefore are also referred to as vertebral, or sympathetic, chain ganglia. The vertebral chain of ganglia extends from the base of the skull to the tip of the coccyx. In the thoracic and lumbar regions there is usually a ganglion of the vertebral chain associated with each spinal nerve. But in the cervical and sacral regions, only two or three ganglia are found.

The second group of sympathetic ganglia is termed preaortic, because they lie in clusters close to the aorta and its chief branches. The arterial branches give these ganglia

Figure 11-1. (facing page) Schematic diagram of the autonomic nervous system. Parasympathetic (craniosacral) fibers are shown in black. Sympathetic (thoracolumbar) fibers are shown in color. Preganglionic fibers are depicted as solid lines, postganglionic fibers as dotted lines.

their names. The most prominent are the celiac, superior mesenteric, and inferior mesenteric ganglia. Like vertebral ganglia, preaortic ganglia contain exclusively postganglionic sympathetic nerve cell bodies.

Parasympathetic Ganglia There are two groups of ganglia associated with the parasympathetic portion of the autonomic nervous system. The first group consists of four small ganglia in the head in which the postganglionic cell bodies associated with the oculomotor, facial, and glossopharyngeal cranial nerves are located.

The second group of ganglia are those located near or within certain organs they supply. For example, postganglionic parasympathetic neurons that innervate the intestine are located in ganglia between the muscle layers or in the submucosal layer of the intestinal wall.

Autonomic Plexuses

At certain locations, slender nerve filaments extend from ganglia containing postganglionic nerve cell bodies and arrange themselves in a branching network. These concentrations are the autonomic nerve plexuses. For example, the celiac plexus is a meshwork of fibers originating in the postganglionic cell bodies of the celiac ganglion. The fibers surround the celiac artery and its branches. This plexus is sometimes referred to as the solar plexus.

The cardiac plexus, another network of autonomic fibers, lies on the surface of the left atrium and in the concavity of the aorta. Its fibers innervate heart muscle. This plexus in turn sends extensions along the pulmonary arteries to form the pulmonary plexuses and along the coronary arteries to form the coronary plexus.

In the neck region the cervical sympathetic ganglia (chiefly the superior cervical sympathetic ganglion) send fibers to form a skein around the carotid arteries and their branches. This is the carotid plexus. It provides the pathway for sympathetic fibers that innervate structures in the face and head.

Sympathetic Division

Preganglionic fibers of the sympathetic division have cell bodies in the lateral intermediate gray column of the spinal cord. The fibers emerge from the cord along the thoracic and first two lumbar segments, that is, from T_1 through L_2, 14 segments in all. The fibers, being motor in function, leave the cord in the motor root, that is, the ventral root.

The fibers then depart the spinal nerve in branches termed white rami communicantes which pass from the above spinal nerves to ganglia of the sympathetic chain (Figure 11-2). Thus, there are 14 pairs of white rami communicantes. They are white because such fibers are myelinated. Some preganglionic fibers terminate in ganglia at their level of origin. There they synapse with cell bodies of postganglionic neurons. Other fibers travel up or down the sympathetic chain to synapse with one or more postganglionic cell bodies in other ganglia of the chain. A third possibility for fibers entering by means of the white rami is to pass through the sympathetic chain without synapsing. Fibers that follow this course leave the vertebral ganglia in bundles termed splanchnic nerves.

Postganglionic fibers from nerve cell bodies of the sympathetic chain leave in bundles termed gray rami communicantes (gray because such fibers are unmyelinated). The gray rami pass to spinal nerves. These fibers course in the spinal nerve to terminate in smooth muscle of blood vessels, in hair follicles, and in sweat glands of the skin. In all spinal nerves there are some postganglionic fibers of the sympathetic division.

The largest of the splanchnic nerves pass from the vertebral chain to ganglia in the upper portion of the abdomen. The greater splanchnic nerve (T_5 to T_9) passes to the celiac ganglion, and the lesser splanchnic nerve (T_{10} to T_{11}) passes to the superior mesenteric

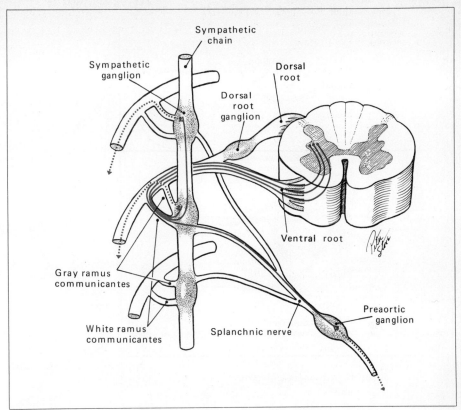

Figure 11-2. Relations between autonomic ganglia and the spinal cord. White rami are composed of preganglionic myelinated neurons (solid color). Gray rami are composed of postganglionic unmyelinated neurons (dotted color).

ganglion. The least splanchnic nerve, and nerves from the sacral portion of the trunk, pass to lower preaortic ganglia. Splanchnic nerves from the upper thoracic and cervical portions of the sympathetic trunk pass to the cardiac and pulmonary plexuses.

It should be understood that though preganglionic sympathetic nerve cell bodies are found only at the thoracic and upper lumbar levels of the spinal cord, postganglionic fibers ultimately fan out to cover the whole body, including the head and neck. Note that preganglionic fibers often synapse with more than one postganglionic cell body. Consequently an impulse traveling along a single such fiber can set off impulses in neurons leading to several effector organs. This is one more example of the divergence built into the nervous system. It leads to a diffuseness of response that is a leading characteristic of the sympathetic division of the autonomic nervous system.

Parasympathetic Division

Nerve cell bodies of the parasympathetic division are located in nuclei in the brainstem associated with the third, seventh, ninth, and tenth cranial nerves (Figure 11-3). They are also located in the lateral intermediate gray column of the second, third, and fourth sacral segments of the spinal cord.

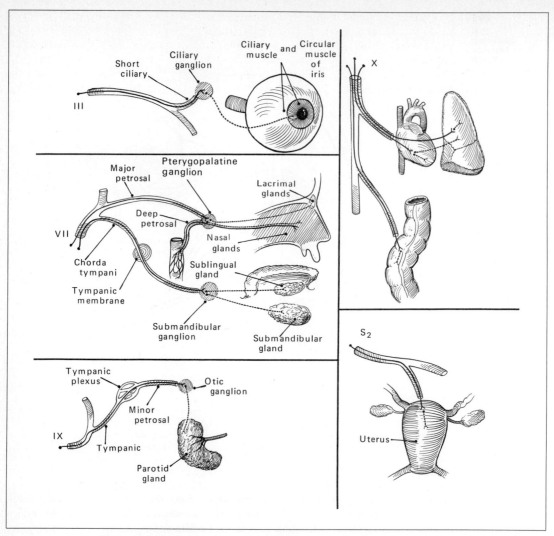

Figure 11-3. Pathways of parasympathetic neurons of the oculomotor, facial, glossopharyngeal, and vagus nerves, and a nerve of the sacral portion of the spinal cord.

Cranial Region Preganglionic fibers pass with the oculomotor nerve to the ciliary ganglion, located in the orbital cavity. Here they synapse with the postganglionic nerve cell bodies. Bundles of postganglionic fibers, leaving the ciliary ganglion as short ciliary nerves, pierce the sclera of the eyeball to innervate the ciliary and sphincter smooth muscles of the eye (see Chapter 12).

Preganglionic fibers in the facial nerve become components of the major petrosal and chorda tympani branches. The major petrosal nerve leaves the facial canal at the facial hiatus and courses along the floor of the middle cranial fossa. It is joined by the deep petrosal nerve from the carotid plexus (carrying postganglionic sympathetic fibers) to form the nerve of the pterygoid canal. This nerve courses through the pterygoid canal to enter the pterygopalatine fossa, where the preganglionic parasympathetic fibers contributed by the major petrosal synapse with postganglionic cell bodies located in the sphenopalatine ganglion. Postganglionic fibers from this ganglion reach the nasal cavity and the lacrimal glands. Fibers supplying the latter structure pass through communicating branches to the maxillary nerve, then course as components of the zygomatic branch of this nerve to the lacrimal gland located within the orbital cavity.

The chorda tympani nerve passes from the facial canal through the middle ear cavity and into the infratemporal fossa, where it joins the lingual branch of the mandibular nerve. The preganglionic fibers course through communications between the lingual nerve and the submandibular ganglion, where they synapse with the postganglionic cell bodies. Postganglionic fibers are distributed to both the submandibular and sublingual glands.

As the glossopharyngeal nerve passes through the jugular foramen, a small branch, the tympanic nerve, carrying preganglionic fibers, passes to the middle ear where it contributes to the formation of the tympanic plexus. Filaments from this plexus re-form as the minor petrosal nerve, which carries preganglionic fibers to the otic ganglion. Postganglionic fibers from this ganglion supply the parotid gland.

The most extensive distribution of parasympathetic fibers is transmitted by the vagus nerve. This nerve contributes to the cardiac plexus and the pulmonary plexus as it courses through the thorax on its way to the abdomen. In the abdomen the vagus nerve supplies branches to the celiac, superior mesenteric, aorticorenal, and inferior mesenteric plexuses. The preganglionic parasympathetic fibers intermingle with the postganglionic sympathetic fibers, forming the plexuses surrounding the distributing branches of these arteries. Parasympathetic fibers synapse on postganglionic cell bodies dispersed in the walls of the organ they supply. For example, postganglionic neurons supplying the gastrointestinal tract are located either in the submucosal area or between the smooth muscle layers forming the wall of the gut. These neurons form the submucosal (Meissner's) and myenteric (Auerbach's) plexuses, respectively.

Sacral Region   Parasympathetic fibers with their cell bodies in the sacral portion of the spinal cord are distributed as pelvic splanchnic nerves which pass directly from spinal nerves to the organs they innervate. Sacral parasympathetic fibers supply the distal portion of the gastrointestinal tract, the reproductive organs in the pelvic cavity, and genital structures in the perineal region.

PHYSIOLOGY OF THE AUTONOMIC NERVOUS SYSTEM

As has already been underscored, one of the most important concepts in the entire study of physiology is embraced by the word "homeostasis." The autonomic nervous system plays an indispensable role in the maintenance of the constancy of internal environment. It is, in short, a vital homeostatic mechanism.

Sympathetic and Parasympathetic Nerve Function

The two divisions of the autonomic nervous system differ not only anatomically but in their functions as well. Most viscera, but not all, are innervated by both types of fibers. Stimulation of fibers of one division usually produces effects just opposite to those noted on activation of the other member. Some representative examples of autonomic function are listed in Table 11-1. One reason for the antagonistic effects of the two divisions of the autonomic nervous system is the difference between the chemical substances freed

TABLE 11-1. AUTONOMIC FUNCTION

ORGAN	EFFECT OF SYMPATHETIC STIMULATION	EFFECT OF PARASYMPATHETIC STIMULATION
Heart	Increased rate	Slowed rate
Muscle	Increased force of beat	Decreased force of atrial beat
Arterioles	Dilation (?)	Constriction (?)
Systemic blood vessels		
Abdominal	Constriction	
Muscle	Constriction (adrenergic) Dilation (cholinergic)	
Skin	Constriction (adrenergic) Dilation (cholinergic)	Dilation
Blood		
Coagulation	Increased	
Glucose	Increased	
Lungs		
Bronchi	Dilation	Constriction
Blood vessels	Mild constriction	
Intestine		
Lumen	Decreased peristalsis and tone	Increased peristalsis and tone
Sphincter	Increased tone	Decreased tone
Eye		
Pupil	Dilation	Contraction
Ciliary muscle		Contraction
Glands		
Nasal	Vasoconstriction	Stimulation of secretion
Lacrimal	Vasoconstriction	Stimulation of secretion
Parotid	Vasoconstriction	Stimulation of secretion
Submaxillary	Vasoconstriction	Stimulation of secretion
Gastric	Vasoconstriction	Stimulation of secretion
Pancreatic	Vasoconstriction	Stimulation of secretion
Sweat	Copious secretion (cholinergic)	
Liver	Glucose released	
Kidney	Decreased output	
Ureter	Inhibition	Excitation
Bladder muscle	Relaxation	Contraction
Penis	Ejaculation	Erection
Basal metabolism	Increased	

by the postganglionic fibers when active. (The reader may wish to review the discussion of chemical transmission in Chapter 8.)

Adrenergic Fibers All except a very few of the postganglionic fibers of the sympathetic division when stimulated liberate norepinephrine. This chemical acts on the smooth muscle or gland which the nerve innervates, causing it to react. Neurons which liberate norepinephrine are termed adrenergic.

Cholinergic Fibers Insofar as is known, all postganglionic fibers of the parasympathetic division liberate acetylcholine (ACh) when activated. Such neurons are termed cholinergic.

All preganglionic fibers—whether sympathetic or parasympathetic—are cholinergic, as are the postganglionic neurons of the parasympathetic division. In short, only the postganglionic neurons of the sympathetic component are (with two exceptions) adrenergic. This breakdown is summarized as follows:

Adrenergic (liberate norepinephrine)	*Cholinergic (liberate acetylcholine)*
Sympathetic postganglionic fibers	Parasympathetic postganglionic fibers
	Parasympathetic preganglionic fibers
	Sympathetic preganglionic fibers
	Sympathetic postganglionic fibers (to sweat glands and those which cause vasodilatation in skeletal muscle)
	Skeletal nerve endings

Chemical Transmission

ACh and norepinephrine in this connotation are referred to as cholinergic and adrenergic mediators. Both are synthesized and stored in vesicles in the nerve endings. The mechanism of release of ACh was discussed in detail in Chapter 8. In brief, the nerve impulse results in an influx of calcium ions which initiate a sequence of events culminating in exocytosis, that is, the expulsion of the contents of the vesicles. Whether or not vesicles containing norepinephrine function in the same way is not certain. There is some evidence that adrenergic fibers also contain a small amount of ACh. The nerve impulse is thought to release the ACh, which then causes the vesicles containing norepinephrine to empty.

ACh is quickly inactivated by cholinesterase. There is also an enzyme present at adrenergic nerve endings, o-methyl transferase, which inactivates norepinephrine. In addition, norepinephrine, after being secreted by the nerve ending, is then reabsorbed back into the nerve. As a result of these two events, norepinephrine, like ACh, functions for only a short period of time.

Alpha and Beta Receptors

For years investigators have noted that norepinephrine and epinephrine have, in some cases, the same effect and in others quite the opposite. To explain these differences, so-called alpha- and beta-adrenergic receptors have been postulated. Recall that the discussion in Chapter 8 on synaptic transmission assumed the presence of ACh receptors on the postsynaptic neuron. This concept of specific binding sites on membranes to explain the mode of action of influencing substances is under intensive study.

Table 11-2 lists some functions of the supposed alpha and beta receptors. Evidence suggests that norepinephrine acts only on alpha receptors; epinephrine, on both alpha and beta. Although the presence of such receptors remains to be proved, the concept is so embedded in physiological and pharmacological jargon that one speaks of alpha-

TABLE 11-2. ALPHA- AND BETA-RECEPTOR FUNCTION

ALPHA RECEPTORS	BETA RECEPTORS
Cardioacceleration	Cardioacceleration Stronger cardiac contraction
Vasoconstriction Intestinal inhibition Pilomotor contraction	Vasodilation
Iris dilation	Bronchial inhibition

and beta-receptor blocking agents, which assumes that a substance that prevents a certain reaction does so by virtue of its influence on the receptors.

Visceral Reflexes

A complete study of visceral reflexes covers almost the entire range of physiology. In the next unit, devoted to circulation, it will be learned that heart action and blood pressure are controlled to a great extent by visceral reflexes. When respiration is studied, the rate of breathing will be seen to be regulated by similar nervous mechanisms. The ingestion and digestion of foodstuffs evoke numerous visceral reactions. These mechanisms will be analyzed in detail in the appropriate sections. For the present it is necessary to understand the anatomy and basic physiology of the visceral reflex arc.

Figure 11-4 shows a typical visceral reflex. This particular one controls the activity of the urinary bladder. The bladder wall is composed, for the most part, of smooth muscle. Within the muscle tissues are proprioceptors. As the urine accumulates, the bladder wall is stretched and the proprioceptors are activated. Consequently, impulses are propagated by the visceral afferent fibers to the spinal cord. Here they synapse with neurons of the parasympathetic division of the autonomic nervous system. These fibers propagate the impulse, via autonomic pathways, to the smooth muscle of the bladder wall and cause it to contract. Hence, urine is expelled and the stretch is relieved.

As with somatic reflexes, visceral ones are facilitated or inhibited by higher-center action. Thus, following spinal cord damage in man, visceral as well as somatic reflexes remain dormant for a period. Obviously, the individual in spinal shock is in a precarious state. The bladder then fills to an abnormal degree. There is no expulsion of fecal material. The blood pressure below the spinal level of damage drops precipitously. After spinal shock wears off, these reflexes reappear. They are then completely automatic, having been divorced from higher-center control. Thus, when the rectum fills to a critical point, it automatically evacuates; when urine accumulates to a threshold degree, the bladder empties spontaneously.

Visceral reflexes cannot be altered by volition. However, while it is quite possible to prevent urination or defecation, this inhibition is brought about by somatic sphincter control rather than by inhibition of the visceral reflexes. One cannot vary heart rate, raise blood pressure, or dictate the flow of digestive juices without voluntarily carrying on those activities which normally stimulate these responses, such as exercising, or eating, or entertaining arousing thoughts.

A variety of stimuli can evoke the visceral mechanisms. In the urinary bladder illustration, the effective stimulus is stretch. In the blood vessels, as will be seen later, there are receptors which respond only to pressure; others are responsive to changes in

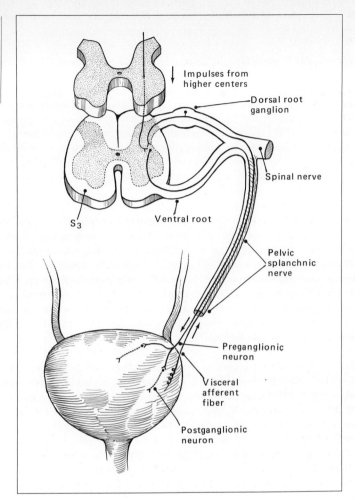

Impulses from
higher centers

Dorsal root
ganglion

Spinal nerve

Ventral root

S_3

Pelvic
splanchnic
nerve

Preganglionic
neuron

Visceral
afferent
fiber

Postganglionic
neuron

Figure 11-4. A typical visceral reflex, of the urinary bladder.

the chemical composition of the blood. Thus, through these highly specialized receptors, adjustments to a specific aspect of the internal environment can be made.

The Hypothalamus

Hypothalamus means "under the thalamus." This refers to its position at the base of the brain. The hypothalamus consists of a mass of nuclei which project fibers to all parts of the brain and into the spinal cord and which markedly influence autonomic function. Techniques have been developed which make it possible either to stimulate or to destroy specific nuclei in the hypothalamus. By these procedures, some hypothalamic nuclei have been shown to be concerned with sympathetic activity, and others with parasympathetic function. Although visceral reflexes are still operative after hypothalamic destruction, the experimental animal is far from normal. The major defect following hypothalamic destruction is the loss of rapid and effective adjustment to sudden environmental changes.

There are other centers of autonomic nervous system integration such as the brainstem nuclei which govern the circulatory and respiratory mechanisms. Cerebral and cerebellar cortices are also concerned with autonomic functions.

Cardiovascular Regulation   In Part 3, the circulatory system will be considered. Suffice it here to say that heart function and blood pressure are exquisitely controlled by the autonomic nervous system and that these activities are integrated by centers in the medulla and also in the hypothalamus. Again, stimulation experiments show that there are specific nuclei in the hypothalamus which exert neurogenic control of the cardiovascular system. For example, stimulation in the posterior and lateral nuclei increases heart rate and blood pressure; the preoptic area in the anterior hypothalamus decreases cardiac function and lowers the blood pressure. The control that the hypothalamus has over cardiovascular activity is exerted by impulses which impinge upon the cardiovascular centers in the reticular substance of the medulla.

Food Intake   The gastrointestinal tract consists of smooth muscle and glands. It is innervated by both divisions of the autonomic nervous system. Numerous visceral reflexes function effectively to digest and assimilate foodstuffs. Between meals the intestinal system is relatively quiescent, and if exercise is indulged in, the blood volume required for assimilation of digested foodstuffs is shunted away from the intestine to serve the actively contracting muscles. Sympathetic and parasympathetic activity produce opposite effects in the digestive activities. Obviously, therefore, to bring about effective alimentation, these two opposing forces must be coordinated. Under normal circumstances the hypothalamus serves this function (see Chapter 25).

Body Temperature Regulation   Normal internal body temperature is about 37°C (98.6°F). Unless the ambient temperature is extremely hot or cold or unless there is a disease process, this internal temperature does not vary much more than a degree above or below 37°C. Body temperature remains constant in a nude person subjected to an ambient temperature range from 15 to 55°C. One of the earliest experiments in this field was carried out by Charles Blagden, an English physician who subjected himself and a few colleagues to temperatures above the boiling point of water and found little variation in internal body temperature. To survive these extremes the humidity must be very low, but the body has remarkable homeostatic mechanisms which keep body temperature constant.

In order to regulate temperature there must be a source of heat and ways of ridding the body of excess heat. In addition there must be receptors sensitive to temperature changes and, finally, an integrating mechanism.

Body heat comes from cellular metabolic processes. The faster they proceed, the greater the heat produced. The body has several ways of increasing metabolism. A very effective one is muscle contraction. Shivering is an involuntary act that results in rhythmic muscle contraction with greater heat production. Voluntary exercise has the same effect. In addition, the metabolic processes themselves can be directly augmented by substances such as norepinephrine, epinephrine, and thyroxine, the hormone of the thyroid gland. All probably function by the uncoupling of oxidative phosphorylation. In the formation of ATP there is a definite ratio between phosphorus and oxygen. If oxidation proceeds at a faster rate, or if the formation of ATP is blocked, or if both events occur, the phosphorus-oxygen ratio will be lowered. This is referred to as uncoupling of oxidation phosphorylation. The net result is a greater utilization of oxygen and therefore an increased metabolic rate which produces more heat.

Not only is the rate of heat production regulated, but so is the rate of heat loss.

Clearly if the loss of heat from the body can be slowed, body temperature will be protected in cold environments. Heat loss is a function of radiation, conduction, convection, and evaporation.

A warm body loses heat to a cold body by means of the radiation of infrared heat rays. Usually the body is warmer than objects in the environment; thus heat is lost to them. The greater this difference, the greater the heat loss. The surface of the body functions as a radiator, and the rate of heat loss to cold objects is determined not only by the gradient of temperature difference but also by the rate of blood flow in the skin. This blood brings heat from the core of the body to the surface. The faster the flow, the greater the heat loss. Skin blood flow, as will be discussed later, is regulated by vasomotor activity. Vasoconstriction decreases flow, whereas vasodilation increases it. Skin blood flow can vary from 1 or 2 percent of the cardiac output (5,000 ml/min at rest) up to about 30 percent. This, then, is a very effective radiator system.

Heat is lost by conduction from parts of the body in direct contact with cold objects, for example, feet upon a cold bathroom floor. There is also conduction of heat to the cold air in contact with the body.

Convection refers to movement of air. If the body is enveloped in a stationary mass of air, the temperature of that air, by conduction from the body, will come to be the same as the temperature of the surface of the body. At that point heat loss by conduction to the air will cease. By convection, however, air continues to flow over the surface of the body. There is thus a steady stream of cool air to which heat is lost by conduction. The heat is carried away by convection currents. The rate of heat loss is a function of the temperature difference between the air and the body, and the rate of air movement. This is why one feels colder when the wind blows than when it is calm, at the same air temperature.

Heat is lost by the evaporation of fluids from the surface of the body. The faster the evaporation, the greater the heat loss. Normally the fluid is water which comes from perspiration. But if a highly volatile substance, such as ether, is put on the skin, the fast rate of evaporation removes heat rapidly and gives rise to a sensation of cold.

The evaporation of water from the skin depends upon the rate of sweating, the air temperature, and the percentage of humidity in the air. A high rate of sweating on a hot, dry day will result in very rapid evaporation, which removes heat quickly. One always perspires, and water is continuously lost from the lungs. This basal rate of sweating is referred to as insensible perspiration, because one is not conscious of it. But by means of insensible perspiration and evaporation from the lungs about 500 ml of water are lost per day. Since 0.58 Calorie (Cal) of heat is lost for each gram of water, approximately 300 Cal are lost per day in this way. But sweating can increase to well over 1 liter/hr in hot, humid weather, especially if vigorous exercise is attempted. This means the loss of over 600 Cal/hr.

To summarize, heat production and heat loss can be regulated. The center for this regulation is in the preoptic and adjacent anterior regions of the hypothalamus. For this reason the hypothalamus is often referred to as the body's thermostat. Quite obviously, there must be temperature input into the hypothalamus, and there must be means by which the hypothalamus can alter heat production and loss.

Input comes from the temperature receptors in the skin and from the blood which perfuses the hypothalamus. The latter is probably the more important and is certainly the most direct. If the temperature of the blood falls, sympathetic stimulation from the hypothalamus brings about secretion of norepinephrine, which increases metabolism. Also, the hypothalamus secretes thyrotropin-releasing factor, which stimulates the hypophysis to secrete thyrotropin, and thyrotropin acts on the thyroid gland to augment the output of thyroxine. Thyroxine increases metabolism. In addition, the hypothalamus brings about shivering, and sympathetic stimulation causes skin vasoconstriction which decreases heat loss by radiation. At the same time, sweating is reduced to minimal levels. In short, more heat is produced and less is lost. If blood temperature rises, all these mechanisms are reversed.

There is now good evidence that the setting of the thermostat at 37°C is a function of the balance between sodium and calcium ions within the posterior hypothalamus. If the ratio of these ions is altered so that the sodium concentration increases, body temperature rises; when calcium concentration increases, body temperature falls. These findings may well explain the mechanism of bacterial-induced fever. In such instances a decreased calcium blood level has been found. Clinically, hypernatremia has been reported in cases of prolonged fever. These observations are now being used to alter body temperature. For example, if body temperature can be lowered during surgery, tissue damage is reduced. Body temperature has been lowered by use of ice or cold water. It can also be lowered by perfusing the brain with solutions high in calcium.

A substance, such as bacteria, that elevates body temperature is called a pyrogen. Substances that lower body temperature are antipyrogens, or, more commonly, antipyretics; aspirin is the best known. How they work is still the question. As we have seen, the autonomic nervous system controls several mechanisms essential to body temperature regulation. Clearly, drugs that alter sympathetic or parasympathetic activity may alter body temperature. Undoubtedly some drugs act directly upon the hypothalamus.

Water Balance   The term water balance implies a constancy of the degree of hydration of the body. Here is an excellent example of homeostasis, and once again the autonomic nervous system and the hypothalamus are of singular importance. Just as in temperature control, the degree of hydration is a function of water intake or production on one hand and fluid loss on the other. Water is produced by many of the metabolic processes, but the prime source is, of course, ingestion. Obviously there must be some mechanism which controls the water traffic. Actually, the desire for fluid, or thirst, varies remarkably with the state of hydration. If there is excessive fluid loss, as there is with profuse sweating, the thirst is great, and large quantities of fluid are imbibed.

The major avenue of fluid loss is through the kidneys, although considerable amounts are lost by perspiration, through evaporation from the lung surfaces, and in the feces.

The problem of water balance cannot be discussed completely until kidney function is considered (Chapter 29). Suffice it to say here that not only does the hypothalamus, through the autonomic nervous system, control perspiration and possibly thirst, but it also sends a great parasympathetic tract into the posterior lobe of the hypophysis. Through this pathway the secretion of the posterior lobe is controlled. The secreted hormone is a potent antidiuretic; it influences the kidneys in such a way as to decrease the quantity of urine formed. This effectively conserves body water (see Chapter 31).

QUESTIONS AND PROBLEMS

1   Name the main types of tissues controlled by the autonomic nervous system and list 10 structures or organs which are under its influence.

2   Locate the cell bodies of the preganglionic neurons in the central nervous system. Where are cell bodies of postganglionic neurons located?

3   What kinds of fibers are transmitted in the gray rami communicantes? In the white rami communicantes? In the splanchnic nerves?

4   Explain what is meant by a nerve plexus, and

name at least one organ supplied by each of the following plexuses: (*a*) celiac plexus, (*b*) cardiac plexus, (*c*) coronary plexus, (*d*) carotid plexus, and (*e*) tympanic plexus.

5   If the vagus is stimulated, what is the effect on (*a*) heart rate, (*b*) bronchi, (*c*) gastric secretion, and (*d*) erection?

6   What is the primary difference between a visceral reflex and a somatic reflex?

7   How many neurons are involved in a simple visceral reflex arc? Where are these neurons located?

8   In which synapses would you expect to find norepinephrine released? In which would you expect to find acetylcholine?

9   Explain how the organs of the head and pelvis receive sympathetic innervation.

10   If you were responsible for nursing a person who had recently suffered spinal cord damage, what nursing activities would your plan of care have to include?

11   What evidence is there to support the statement that the hypothalamus functions as one of the integrative centers of the autonomic nervous system?

12   Describe the relation of the hypothalamus to the fever which accompanies some infections.

13   Explain why it is customary temporarily to leave a urinary catheter in the bladder of someone who has had spinal anesthesia.

12 THE SPECIAL SENSES: VISION

Vision is one of the most important of all the senses. Blind people can and do learn to depend on the other senses to a remarkable degree, but for the loss of vision there is never anything approaching complete compensation. One relies upon vision for protection, for equilibration, for coordination, for creation, and for pleasure.

Fundamentally, vision is similar to all other sensory modalities in that there are receptors (sensitive to light, in this instance), there is an afferent pathway which conveys the impulses to the cerebral cortex, and there is an area of the cerebrum responsible for the awareness of light. In addition, the afferent unit is connected by synaptic union with diverse efferent limbs for reflex responses to vision. But vision is far more complex than such relatively primitive sensory modalities as pain and temperature. Accordingly, a separate chapter is devoted to this important subject.

EMBRYOLOGY OF THE EYE

In embryos of about 4 weeks, the optic vesicles first appear as a pair of lateral outgrowths of the forebrain (Figure 12-1). Each vesicle grows rapidly and soon differentiates into the thin optic stalk, which becomes the optic nerve, and a larger distal part, the optic bulb. The latter becomes indented on its outer surface, transforming the bulb into the two-walled optic cup. The inner layer, which contains nervous tissue, thickens considerably more than the outer, pigmented layer. The two layers together form the retina.

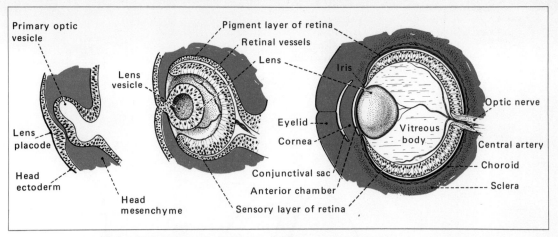

Figure 12-1. Embryonic development of the eye. At left the lens placode and primary optic vesicle are shown in formation. In the center the lens vesicle and retina have appeared. At right the principal components of the developed eye can be distinguished.

During this same time the lens placodes appear as bilateral thickenings of the surface ectoderm opposite the developing optic cups. Each placode becomes markedly depressed until it forms a closed sac or depression and then separates from the head ectoderm as the lens vesicle. Each detached vesicle comes to rest in the concavity of the optic cup, where it develops into the lens of the eye.

Loose mesenchyme of the head region surrounds the developing optic and lens vesicles. As these structures differentiate, the mesenchyme condenses about them, forming a two-layered fibrous capsule. The inner layer of the capsule becomes the highly vascular and delicate choroid coat of the eye. Anteriorly the choroid thickens and undergoes considerable development to form pigmented muscular structures, the ciliary body and the iris. The outer layer of mesenchyme completely surrounds the developing eye as a tough fibrous membrane, which becomes the cornea and the sclera.

ANATOMY OF THE EYE

The visual apparatus consists of the refractive media of the eyeball, the retina, the optic nerve, and the visual centers of the cerebral cortex. Also essential for normal vision are the focusing elements of the eye (intrinsic muscles) and the extrinsic muscles. In addition there are accessory, protective structures such as the eyelids and the lacrimal apparatus.

The eyeball rests in the orbital cavity, encased in the bulbar fascia over the posterior three-fourths. The fascia forms a socket in which the eyeball moves under the influence of extrinsic muscles (described on page 129). Between the bulbar fascia and the bony walls of the orbit is the bulbar fat pad. The fat pad infiltrates around all the structures within the orbital cavity.

The upper and lower eyelids (palpebra superior, palpebra inferior) form two protective curtains for the eye (Figure 12-2). The opening between the lids is the rima palpebrarum. The "corners of the eye" where the lids join are the medial (inner) canthus and lateral (outer) canthus. A dense plate of connective tissue, the tarsal plate, is present in each lid and provides additional protection. In each tarsal plate there are elongated

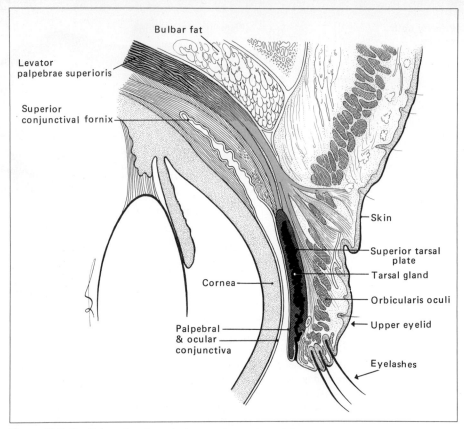

Figure 12-2. Sagittal section through the anterior portion of the eye.

sebaceous glands that secrete an oily substance onto the free margins of the eyelids. This secretion acts as an oily dam to prevent overflow of lacrimal fluid under normal conditions.

A number of short hairs, the eyelashes, emerge from the free margin of each lid. The internal surface of each lid is lined by the palpebral portion of a specialized mucous membrane, the conjunctiva. The conjunctiva also reflects onto the eyeball. This is termed the ocular portion. Reddened ("bloodshot") eyes are due to dilatation of the minute vessels in the conjunctiva. When the conjunctivae are infected, the condition is termed conjunctivitis, or pinkeye.

Lacrimal Apparatus

The lacrimal apparatus (Figure 12-3) provides additional protection in the form of a fluid film over the exposed part of the eyeball. The lacrimal gland is in the superolateral portion of the orbital cavity. Six to ten lacrimal ducts carry its secretions onto the conjunctiva. Involuntary blinking of the eyelids spreads a thin film of this fluid over the entire conjunctival surface and cornea. At the medial canthus, the fluid passes through minute openings, the puncta lacrimalia, into two small ducts, the lacrimal canaliculi, and from there into the lacrimal sac, the upper blind ending of the nasolacrimal duct. The duct courses through the maxillary bone to empty secretions onto the nasal mucosa.

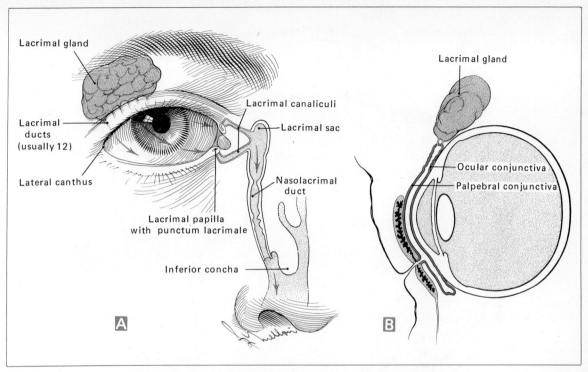

Figure 12-3. Lacrimal apparatus. **A.** Pathways of lacrimal fluid from the lacrimal sac to the nasal cavity. **B.** Relation of the lacrimal gland to the conjunctival surfaces with the eye closed. Parasagittal section.

Under emotional stress or following irritation of the conjunctiva, the lacrimal gland may so step up production of fluid that it overflows the oily dam of the eyelids and appears as tears.

Eyeball

The eyeball (Figure 12-4) (bulb of the eye) is housed in the orbital cavity and is spherical, with a slight anterior bulge. It is about 2.5 cm in diameter. The wall of the eyeball is composed of three coats, or layers. The outer layer consists of the sclera ("white of the eye") and cornea. The sclera is the tough and collagenous covering of the posterior five-sixths of the eyeball. The cornea is the transparent window over the anterior one-sixth. The cornea admits light rays and is the outermost refractive medium of the eye.

The middle layer of the eyeball consists of three structures. At the back is the choroid. Near the front is the ciliary body, made up of ciliary processes plus the ciliary muscle. Also at the front and suspended across the eye is the iris, with a central aperture, the pupil. Choroid means "skinlike" and, like the true skin, the choroid is extremely vascular. Its prime function is to supply blood to the other parts of the eyeball. Anteriorly, at the ora serrata, the choroid joins the ciliary processes, 60 to 80 small ridges that together form a ring around the inside of the eyeball. The anterior ends of the processes give attachment to the iris as well as to the suspensory ligaments, which insert into the lens (Figure 12-5). The ciliary processes also give attachment to the ciliary muscle, which

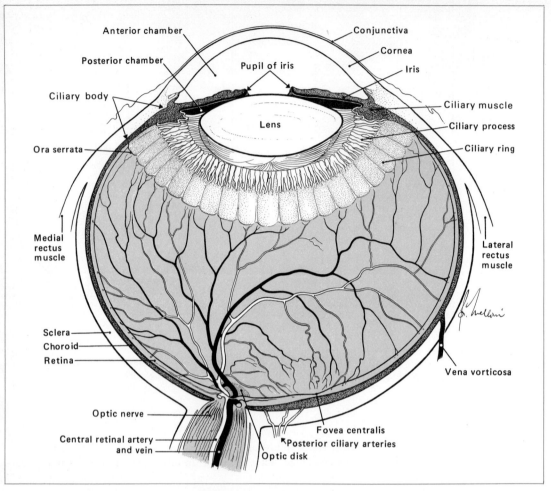

Figure 12-4. Blood vessels of the retina. The area in color represents the extent of visual receptors in the retina of the intact eye.

originates at the scleral spur. The ciliary muscle is responsible chiefly for change in the shape of the lens. Contraction of the muscle pulls the ciliary processes toward the scleral spur, releasing tension of the suspensory ligaments and allowing the lens to bulge slightly. The change in the shape of the lens is important in the eye's accommodation for near and far vision (see page 317).

The iris is a disklike body of connective tissue, muscle, nerves, and blood vessels. It is pigmented and gives the eye its color. At its center is a round hole known as the pupil. Two muscles of the iris control the size of the pupil and hence the amount of light passing to the interior. One encircles the pupil with its fibers and is the sphincteric muscle. The other lies with its fibers radiating outward from the pupil and is termed the dilator muscle. The iris is attached to the ciliary processes along its periphery. It is unattached around the pupil. As will be clear from later description, the iris is suspended in fluid of the chambers of the eye.

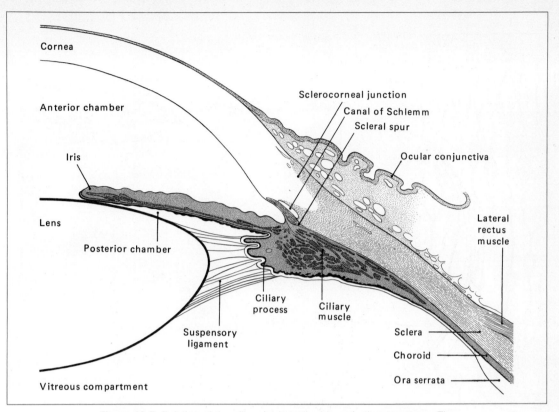

Figure 12-5. Relation of the ciliary body to the iris and other structures. The ciliary body consists of the ciliary muscle and ciliary processes.

The third, or internal, layer of the eyeball is the retina. It lines the entire eyeball but has three recognizable regions. The posterior portion of the eyeball forward to the ora serrata contains the visual receptors of the retina. The portion lining the ciliary body contains pigmented cells, as does the portion forming the internal lining of the iris.

Histology of the Retina The microscopic anatomy of the retina is quite complex (Figure 12-6). The neurons and their processes together with an inner and outer limiting membrane form 10 rather distinct layers. Only the more important of these layers will be described.

The layer farthest from the light, that is, adjacent to the choroid, consists of a single layer of pigmented epithelium securely bound to the choroid. Its cells absorb excess light that might reflect and interfere with the formation of the retinal image. The next layer anteriorly is the layer of rods and cones, which are the specialized photoreceptors that convert light energy to nervous impulses. In another layer the rods and cones synapse with bipolar neurons. These in turn synapse with large multipolar ganglionic cells. The axons of these cells converge toward the center of the retina and then turn and leave the retina together at its posterior aspect as the optic nerve.

It should be clear that light rays must penetrate many layers of the retina before falling on the photoreceptors, the rods and cones.

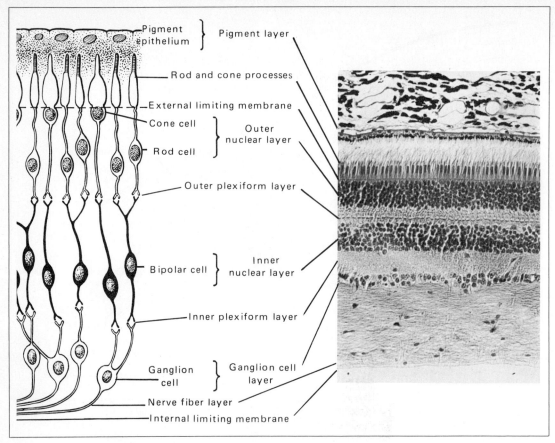

Figure 12-6. Diagrammatic sketch and photomicrograph of layers of the retina. The pigment epithelium at the top adjoins the choroid. Hence light, entering at the bottom, must pass through many layers to reach the receptors themselves, the rod and cone processes. (*Courtesy of Ronald G. Clark.*)

Rods are narrow and cylindrical. Cones are shorter, blunt, and flask-shaped. Rods respond to dim light; hence they enable discrimination of the outline of objects in poor light. Cones function in bright light, enabling perception of clear, sharp images. They are also responsible for color vision (see later).

In the exact center of the retina is a yellowish, poorly defined area known as the macula lutea, or "yellow spot." In its center is a small depression, the fovea centralis, in which there are only cones. The highest concentration of cones is in the fovea centralis. The retina is also thinner at the fovea. For both reasons it is the area of greatest visual acuity. Toward the periphery of the retina the proportion of cones decreases until, finally, they are entirely replaced by rods.

Optic Disk The axons composing the optic nerve leave the retina at a spot slightly medial to the fovea centralis. The convergence of these fibers displaces other cellular elements so that no rods or cones are present. This area is called the optic disk (Figure 12-7), or "blind spot." The reader may demonstrate the optic disk for himself by closing the left

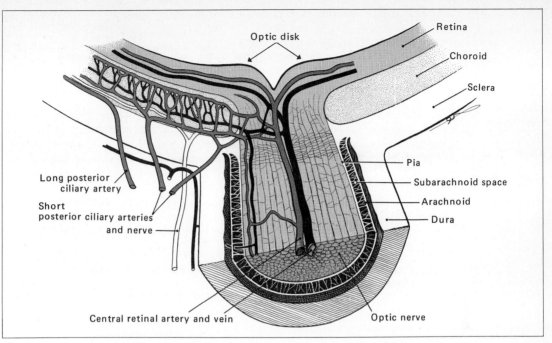

Figure 12-7. Section through the eye at the entrance of the optic nerve.

eye and focusing the right eye on the cross in Figure 12-8 from a distance of about 20 cm (about 8 in.). The circle may be seen, but if the book is moved a little closer or farther away from the eye, a point will be reached where light rays reflected from the circle converge on the optic disk and the circle disappears. The optic disk has clinical value in that it reveals abnormal fluid pressures within or in back of the eyeball to a physician using an ophthalmoscope. If intraocular pressure is high, the disk appears pushed backward. If intracranial pressure is high, the disk appears pushed forward.

Refractive Media   The refractive media of the eye consist of the cornea, the aqueous humor in the anterior chamber, the lens, and the vitreous humor. Their optical properties will be discussed later in the chapter.

The most important of these media is the lens, which is behind the iris and situated so that it separates the aqueous humor from the vitreous body. The lens is a transparent,

Figure 12-8. To demonstrate the optic disk (''blind spot'') the reader should cover the left eye and focus the right eye on the cross, then move the page back and forth until the black dot disappears. This occurs when its image falls on the optic disk, which is without rods and cones.

colorless body, consisting of a semisolid mass of specialized fibers and enveloped in a very thin capsule. It is held in position by the suspensory ligaments, which pass from the lens periphery to the ciliary processes.

The anterior chamber, lying in front of the lens, is filled with aqueous humor, a clear, watery fluid. Aqueous humor forms from the capillaries in the ciliary body and circulates through the pupil into the anterior chamber, from where it leaves the eyeball by way of the canal of Schlemm (see Figure 12-5).

The posterior compartment of the eyeball—the space behind the lens—is filled with vitreous humor. Vitreous humor is a refractive medium, aids nourishment of the retina, and helps to maintain the shape of the eyeball. Because of the gelatinous consistency of this fluid, the posterior compartment is sometimes called the vitreous body.

Optic Nerve

Nerve fibers leaving the eyeball form the optic nerve and pass through the optic foramen into the cranial cavity. The optic nerves from the two eyes converge at the optic chiasma, then separate and continue posteriorly as the optic tracts (Figure 12-9). The tracts pass

Figure 12-9. Visual pathways. Note the partial crossing of nerve fibers at the optic chiasma.

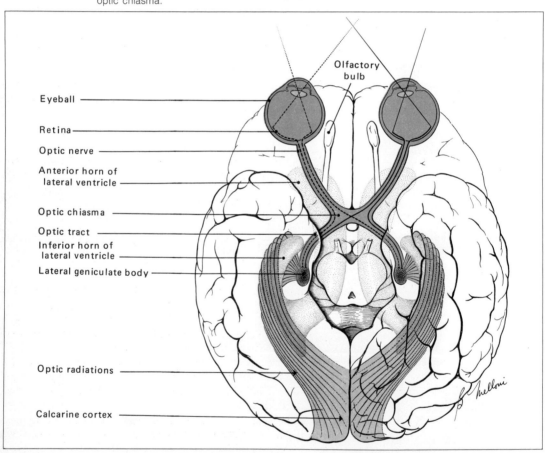

alongside the cerebral peduncles and terminate in synaptic union with neurons of the lateral geniculate bodies. From here fibers spread into optic radiations and pass to the cortex of the occipital lobe.

Fibers arising from the cell bodies in the nasal side of the retina cross in the optic chiasma to run in the opposite optic tract. For example, fibers from the nasal side of the left eyeball cross to the right side of the cortex. Those of the temporal side continue in the optic tract of the same side.

The optic nerve is actually a tract of the brain rather than a peripheral nerve. As an extension of the brain, it is therefore surrounded throughout its course in the orbital cavity by the meninges (Figure 12-7). Cerebrospinal fluid also extends along this ensheathment in the subarachnoid space between the pia and arachnoid. Where the optic nerve attaches to the eyeball, the dura mater fuses to the sclera.

Small vessels and nerves penetrate the wall of the eyeball at the area cribrosa, which surrounds the attachment of the optic nerve to the eyeball. The blood supply of the eyeball is derived from branches of the ophthalmic artery. The central artery of the retina, also a branch of the ophthalmic artery, courses in the center of the optic nerve to the retina. The remaining vessels course in the choroid.

Branches of the ophthalmic nerve supply sensation to the eyeball; autonomic fibers also are present. Parasympathetic fibers supply the circular sphincteric muscle of the iris and the ciliary muscle; sympathetic fibers innervate the radial dilator muscle of the iris. All these fibers enter at the area cribrosa.

PHYSICS OF VISION

Light is a form of radiant energy that travels in waves through air at a rate of about 300,000 km/sec. The wavelength of light visible to the human eye varies from about 3850 to 7200 Å.

Refraction

The word refraction means "bending" or "turning aside." This is precisely what occurs when a ray of light passes from one medium to another in which the speed of light is different. Light travels in glass more slowly than in air; thus as a light ray passes from one medium to the other, it is bent, or refracted.

The refractive index is a relative measure of the transmission of light—relative, that is, to the speed of light in air. For example, if light travels through a medium at a speed of 150,000 km/sec, that medium is said to have a refractive index of 2.0 (300,000/150,000). The lens of the human eye transmits light at a speed of about 214,285 km/sec and thus has a refractive index close to 1.4.

When light penetrates a transparent medium at a right angle to its surface, there is no bending. But as shown in Figure 12-10, refraction does occur at other angles. Assume that the light is passing from air into glass which has a refractive index of about 1.5. As the beam of light enters the glass at an angle other than 90 degrees, the rays on one side of the beam will still be traveling in the air while the rays on the other side are already moving more slowly in the glass. Consequently, the beam of light is bent toward a line drawn perpendicular to the surface. Conversely, when the beam emerges from the glass into air at an angle other than 90 degrees, there is again a bending of the beam, this time in the opposite direction.

In brief, light, upon passing into a medium of greater refractive index, is bent toward the perpendicular; upon passing into a medium of lesser refractive index, it is bent away from the perpendicular. The degree of refraction depends upon (1) the difference between

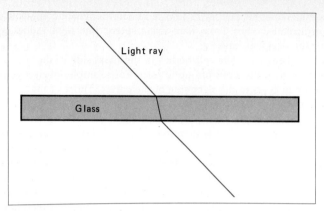

Figure 12-10. Refraction of light. Because the refractive index of glass differs from that of air, light rays are bent when they enter the glass at an angle other than 90 degrees and bent again when they reenter the air.

the refractive indices of the two media and (2) the angle at which the beam traverses the interface between the media. When light enters a different medium at right angles, all its rays change velocity simultaneously, and thus there is no refraction.

Lenses

A lens is a transparent substance which bends rays of light and causes the rays to converge or diverge. Man-made lenses are usually of glass. A biconvex lens is shown in Figure 12-11. Because each side of the lens is convex, parallel rays of light will converge on a

Figure 12-11. Refraction of light by lenses. A biconvex lens (middle) can bend diverging light rays to a point termed the conjugate focus. A biconcave lens (bottom) causes light rays to diverge; thus there is no focal point.

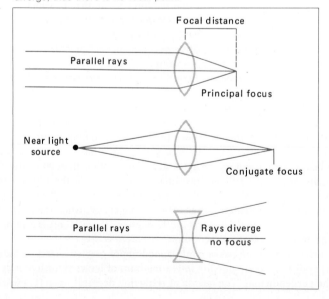

point, termed the principal focus of the lens. Rays of light are considered parallel when they emanate from a distant object, but as a pinpoint of light is brought closer to the lens, the rays of light will diverge as they approach the lens. If the lens has sufficient convexity, these diverging rays of light will be bent and ultimately converge on a point, termed the conjugate focus. Clearly, the distance to the conjugate focus is greater than to the principal focus for any one lens.

Figure 12-11 also shows that parallel rays of light are caused to diverge by a biconcave lens, and no focus is attained.

The power of a lens to refract light depends upon (1) the refractive index of the lens and (2) the curvature of the surfaces. Most lenses have the same refractive index; thus, lenses of different power are made by altering the curvature of the surfaces. The more nearly round the curvature, that is, the shorter the radius, the greater the refraction.

The refractive power of a lens is expressed in diopters. The distance between a biconvex lens and its principal focus is termed the focal distance. A lens that has a focal distance of 100 cm is said to have a power of 1 diopter (Figure 12-12). If the focal distance is but 20 cm, a lens has a power of 5 diopters (100/20). Because a biconcave lens causes rays of light to diverge and thus there is no focal point, its power must be expressed in terms of its ability to counteract the converging ability of a biconvex lens. Accordingly, by comparison with a biconvex lens of known power, the power of a biconcave lens may also be expressed in diopters. To differentiate the two, the power of a biconvex lens is expressed in plus diopters; that of a biconcave lens, in minus diopters.

Figure 12-12. The refractive power of a lens is expressed in diopters. The shorter the focal distance, the greater the refractive power of the lens. The refractive power of a biconcave lens is measured in terms of its ability to counteract the converging power of a biconvex lens and expressed in minus diopters.

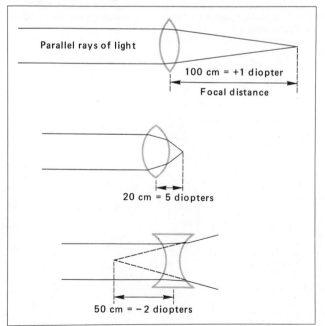

Optical System of the Eye

Rays of light first strike the cornea and then pass through the aqueous humor, the lens, and the vitreous humor, finally to strike the retina. The refractive indices of the media are cornea, 1.38; lens, 1.40; vitreous humor, 1.34. The outer surface of the cornea is convex and is in contact with air. There is a significant difference between the refractive indices of air and of the cornea; therefore this convex surface has considerable refractive power. It amounts to almost 40 diopters. To be sure, the inner surface of the cornea is concave and should diverge rays of light about as much as the anterior surface causes them to converge. However, the refractive indices of the cornea and the aqueous humor are so close together that this interface has little refractive power, certainly little more than -5 diopters. The lens is convex on both surfaces, but it is surrounded by media of very similar refractive indices; consequently, its total refractive power varies from 25 diopters to about 40 diopters at its most nearly round curvature. The cornea plus the lens, then, provide a system with a refractive power that ranges from about 60 to 75 diopters. A system with a refractive power of 60 diopters has a focal length of 16.7 mm (100/60), which is the approximate length of the normal adult eyeball.

PHYSIOLOGY OF VISION

Figure 12-13 is designed to illustrate the striking similarities between a camera and the human eye. They function in much the same manner. In the camera, the image is brought to focus upon the film by the lens. In the eye, the lens focuses the image upon the retina. In the camera, there is a mechanism for bringing the image into sharp focus; the eye is also so equipped. In the camera, there is a diaphragm to regulate the amount of light

Figure 12-13. Similarities between the eye and a camera. The iris of the eye and the diaphragm of a camera regulate the amount of light passing to the interior. In both, the lens causes the image to appear inverted on the light-sensitive surface. In the human being the image is righted in the process of interpretation by the brain.

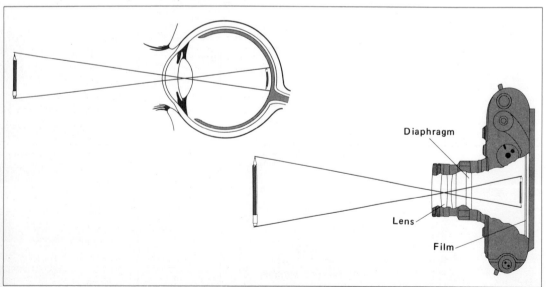

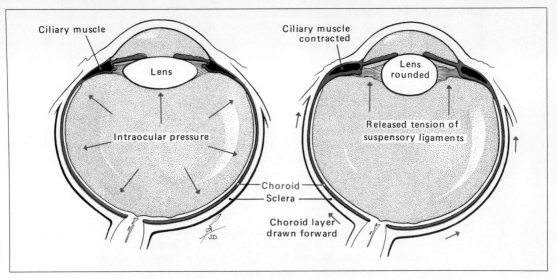

Figure 12-14. Mechanism of accommodation. Light rays from close objects diverge as they approach the lens. In order to focus the rays on the retina the lens must become rounder. The ciliary muscle therefore contracts and pulls the choroid forward, releasing the tension of the suspensory ligaments. The lens then becomes rounder because of its inherent elasticity.

which enters the lens. In the eye, the iris assumes this function. Finally, the image is inverted in both. Although one actually sees things upside down, the image is righted through a cerebral process. The student who is familiar with simple photographic techniques should have little difficulty in understanding the physiology of vision.

Formation of the Image

Accommodation The lens of the eye is biconvex. Light rays in passing through a biconvex lens are caused to converge to a focal point. Shifting the lens back and forth in a camera causes the image to pass in and out of focus. But in the human eye the distance between the lens and the retina cannot be varied. To effect a sharp image the curvature of the lens must change. Such change in lens curvature is termed accommodation.

The intraocular pressure created by the aqueous and vitreous humors tends to hold the eyeball round and firm. When the ciliary muscle is at rest, the suspensory ligaments are taut, which, with the intraocular pressure, holds the lens relatively flat. In this position, parallel light rays are brought to focus, in the normal eye, on the retina. As the object is brought closer, the rays diverge and must be refracted more. To do this the ciliary muscle contracts and draws the ciliary processes forward. This releases the tension on the suspensory ligaments, and the lens becomes more curved (Figure 12-14). Once again the light rays are focused on the retina. Thus the curvature of the lens is controlled by the ciliary muscles. A point must obviously be reached at which the ciliary muscles are completely contracted and the lens is as curved as it can possibly become. This point is appropriately termed the near point. If the object is now brought still closer, the eye will be unable to focus it clearly.

The range of accommodation thus depends upon the ability of the lens to become more curved. In the normal eye, the range of accommodation is in excess of 15 diopters. At the near point, the distance of the object from the normal eye is about 15 cm. To put it another way, the normal eye can see objects clearly from infinity to about 15 cm.

Helmholtz (circa 1885) contributed a considerable body of information to our knowledge of vision, and ever since that time the accepted teaching has been that at rest the ciliary muscle is relaxed and by virtue of the tension of the suspensory ligaments the lens is held relatively flat, a condition appropriate for distant vision, that is, to focus parallel rays of light. According to this concept, accommodation is the result of parasympathetic stimulation of the ciliary muscle, which pulls the insertions of the ligaments forward, thereby releasing the tension on the lens, and the lens, due to its inherent elasticity, assumes a more rounded configuration. Thus, accommodation would be an active process of ciliary muscle contraction which changes the power of the lens by some 15 diopters. Change in the opposite direction would occur when the muscle relaxes.

However, in 1937, D. G. Cogan stated:

> Unfortunately, an unwarranted terminology has been applied to accommodation which confuses the issue. Thus one speaks of accommodation as a property of changing one's focus from a distant object to a nearer object and calls the reverse relaxation of accommodation. If, as I believe, the radial fibres adapt one's focus for relatively distant objects, and the circular fibres adapt one's focus for near objects, the one is just as much an accommodation function as the other. They exist in a state of reciprocity similar to that of the dilator and sphincter muscles of the pupil, and it would be simpler if there were names for them similar to mydriasis and miosis, but since such names are lacking I am referring to the two as accommodation for distance and accommodation for near, respectively.

Since 1937 evidence has continued to accumulate which shows that at rest the lens is not at the point of least curvature but rather at a point about 1.5 diopters from there. This means that to see objects in focus that are brought closer to the eye, the parasympathetic system must contract the ciliary muscle, as already explained. But for focus of parallel rays of light, an active process is also required which is under the control of the sympathetic nervous system. Apparently these fibers, as Cogan says, contract the radial fibers, which results in greater tension placed upon the lens. A 1972 review by F. M. Toates presents the evidence for this concept. Clearly, the time to change our teaching in this respect is overdue.

Corresponding Points   In man, the object is brought into focus for clear vision, and both eyes must focus on similar areas so that one sees a single image instead of two. Fibers from the medial half of one retina cross in the optic chiasma and course to one of the lateral geniculate bodies with fibers from the lateral half of the other retina. Neurons from this body transmit the image to a specific place in area 17 of the occipital cortex. In other words, there are points on each retina which correspond to points on the other retina, so that the image is viewed as one. These are termed corresponding points.

Visual Acuity   In order for the eye to see an object clearly, the object must be brought into focus upon the retina. The region of the retina of greatest sensitivity is the fovea centralis. The retina at the fovea centralis is actually thinner than it is anywhere else, because many of the usual layers of the retina are not present here. For this reason the rays of light strike the sensitive cones almost directly, without having to traverse the other retinal layers. The factors which determine visual acuity are (1) the size of the object and (2) its distance from the eye. If two small dots are inscribed very close together on a piece of paper, they will appear as two distinct marks when the paper is held fairly close to the eye. However, at a distance they will be seen to fuse into one. In order for two objects to be differentiated as distinct entities, the resulting image of these objects must cover two or more retinal receptors.

In actual practice, visual acuity is tested with the Snellen chart (Figure 12-15). The usual distance from the chart to the eye is arbitrarily set at 20 ft. At this range, letters of a specific size can be identified by the normal eye. That line of letters is labeled 20. If the normal subject stands at 100 ft, the letters which he can identify are labeled 100, and so forth. If a myopic (nearsighted) person is now tested standing at 10 ft, he may possibly be able to read only the line marked 20. He will then be said to possess a visual acuity of 10/20. Normal vision is, of course, 20/20.

Visual Reflexes

When a distant object is brought progressively closer to the eye, three important changes in the visual mechanism take place: (1) convergence, (2) accommodation, and (3) constriction of the pupil. Accommodation is designed to focus the image sharply on the retinas. Convergence assures that the image strikes corresponding points. Constriction of the pupil decreases the amount of light which enters the eye. All these are essential to acuity. They are all involuntary; hence they are reflex mechanisms.

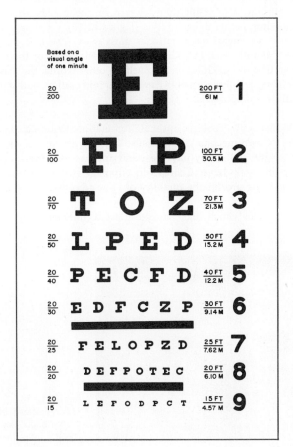

Figure 12-15. Snellen chart. The subject with normal visual acuity who stands 20 ft from the full-sized chart can read line 8, marked 20/20.

The Accommodation Reflex The accommodation reflex is complex and not completely understood. There are at least two stimuli that evoke the accommodation reflex: (1) convergence and (2) a defocused image. Either one seems to be adequate to cause accommodation. Thus, as an object is brought from afar up close, the eyes converge by virtue of the reflex just discussed. But exactly how convergence causes accommodation is not yet known, nor are the pathways. What is known is that convergence is not necessary to bring about accommodation. That is to say, if convergence is blocked, accommodation will still occur in response to a defocused image. Interestingly, if the object being viewed is put out of focus by a projector, accommodation does not occur.

The pathways involved in the accommodation reflex involve at least the optic nerve for the afferent limb and probably other neurons as well. Both parasympathetic and sympathetic neurons constitute the efferent limb. The central connections remain to be determined, but apparently the visual cortex plays a role.

The Convergence Reflex Here again the afferent limb is the optic nerve, but now the efferent limb consists of the motor fibers of the third, fourth, and sixth cranial nerves which innervate the extrinsic eye muscles. If the eyes did not converge as the object was brought closer, the image would not strike corresponding points. Apparently the formation of a double image is the effective stimulus which is transmitted to higher centers via the optic pathways and then back to the midbrain for activation of the nerves supplying the extrinsic eye muscles. As a result, the eyes turn synchronously, thus maintaining the image on corresponding points in each retina (Figure 12-16). Of course, in order for the eyes to turn inward, the lateral rectus muscles must relax. These, it should be recalled, are innervated by the abducens. In addition, there is relaxation of the superior oblique muscles, indicating that the trochlear is involved. In other words, the convergence reflex is extremely complex, involving superb reciprocal innervation and a high degree of coordination of all its components.

Light Reflexes Visual acuity requires optimal illumination. Within a considerable range, the eye possesses its own mechanism for controlling the amount of light which strikes the retina. On a bright and sunny day the pupil appears extremely small, whereas at dusk it is usually relatively large. Changes in pupillary size are a function of the iris. The larger the pupil, the greater the quantity of light admitted into the eyeball (Figure 12-17).

The smooth muscles which make up the iris are innervated by both divisions of

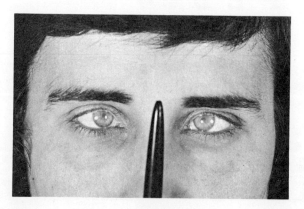

Figure 12-16. Convergence. This reflex maintains the image on corresponding points in each retina.

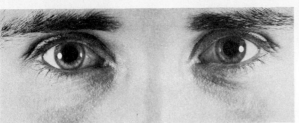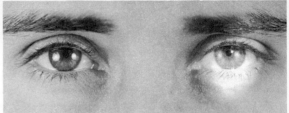

Figure 12-17. Light reflexes. At left the pupils are dilated, which is the reaction of the iris to dim light. At right the pupils of both eyes are contracted although a light has been focused on only one. This is the consensual light reflex.

the autonomic nervous system. The intensity of the light which strikes the retina is the effective stimulus for initiating the light reflexes. If the light proves too strong for optimal visual acuity, impulses are fired via appropriate nervous pathways. As a result, the pupil constricts, thereby decreasing the amount of light which enters the eye. Conversely, a decrease in illumination through a comparable reflex mechanism causes the pupils to enlarge. If a light is directed into only one eye, both pupils will rapidly constrict. This is termed the consensual light reflex (Figure 12-17, right).

Protective Reflexes   Blinking is a reflex mechanism of a protective nature. When the cornea is touched, as by a foreign object blown into the eye, receptors fire impulses to the brainstem, where they are transmitted across a synapse to motor neurons which effect the closing of the eyelids and the warding off of the foreign body.

Tears are copiously produced when any foreign body comes in contact with the cornea. Thus lacrimation, too, is a protective reflex. If the corneal reflex is unsuccessful in blocking the ingress of a foreign body, the lacrimal reflex supplies fluid to wash it out.

Vision under High and Low Intensities of Light

Under low illumination, the iris dilates, permitting all available light to strike the retina. Conversely, in bright light, the iris constricts, reducing the pupil to a very small size, thereby diminishing the amount of light entering the eye. In addition to this mechanism for vision under high and low intensities of light, the eye has another, namely, alteration of the sensitivity of the retina. Once again, the analogy with a camera is interesting. The so-called lens opening is varied according to the brightness of the illumination. But also, one may use film of different sensitivity. The eye makes excellent use of both possibilities.

Function of the Rods   A single neuron of the visual pathway arborizes to transmit the impulses from many rods. The exact number varies considerably, but synapses with 50 or more rods are not uncommon. For this reason, visual acuity using only rods is very poor, and peripheral vision (which employs only rods) lacks sharpness.

But the rods are far more sensitive to light than are the cones. The degree of sensitivity of the rods is a function of an elaborate chemical process that is still not completely understood. Rhodopsin, also known as visual purple, is the substance which is essential for the conversion of radiant energy of light into nervous impulses.

Bright light causes rhodopsin to bleach. Under these conditions, the rods are relatively inactive, and the eye depends chiefly upon the cones. When the light intensity is reduced, rhodopsin is re-formed and the rods take over. The time required for the resynthesis

321

of rhodopsin accounts for the delay before the eye adapts to so-called night vision. It explains why, for example, when one enters a darkened theater, he can see very little at first but after a few minutes is able to distinguish considerable detail. Figure 12-18 shows that light causes rhodopsin to be converted to lumirhodopsin, which is then rapidly converted to metarhodopsin. Next, metarhodopsin breaks down to scotopsin and retinene. These two substances are capable of slowly combining to re-form rhodopsin, thus completing the cycle.

Retinene is in equilibrium with vitamin A. If there is too little vitamin A, retinene is converted to the vitamin and thus is not available for rhodopsin formation. Accordingly, people with vitamin A deficiency have difficulty seeing under conditions of low illumination; that is, they are said to have night blindness.

Although the sequence of events of the rhodopsin cycle has been worked out in remarkable detail, the question of how this sequence alters the sensitivity of the rods to light remains unanswered. There is growing evidence, however, that the ubiquitous cyclic AMP may be involved.

Cyclic AMP results from the conversion of ATP, a reaction that is catalyzed by cyclase (adenylate cyclase). Cyclic AMP, acting through associated enzyme systems, mediates a number of specialized cellular functions such as melanosome dispersion, release of hormones, and transmission at certain synapses. The fact that cyclase is present in high concentration in the retinal photoreceptor structure suggests a role in light transduction. In addition, there is evidence that as rhodopsin is bleached by light, cyclase is inactivated. This may mean that under conditions of low illumination the presence of cyclase results in the production of cyclic AMP, which then increases conductance to sodium ion, rendering the receptor more sensitive.

Function of the Cones   The sequence of chemical events in the rods, just described, is fairly well established. Probably a similar sequence occurs in the cones, but instead of rhodopsin the cones contain so-called opsins. And instead of only one pigment the cones have at least three different ones, each sensitive to different colors: red, green, and blue. As will be discussed next, the cones are considerably less light-sensitive than are the rods. Under conditions of bright illumination, rhodopsin in the rods is so completely bleached that the rods are virtually insensitive and vision depends almost exclusively upon the cones.

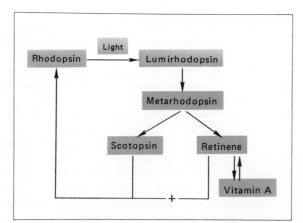

Figure 12-18. The sequence of chemical events in rod function. Rhodopsin, essential for rod activity, is converted by bright light to lumirhodopsin. The rods function chiefly at low light intensity, at which time rhodopsin is made available as shown.

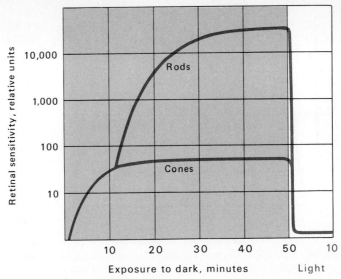

Figure 12-19. Dark adaptation. Both rods and cones become more sensitive to light under low illumination, but the major increase in retinal sensitivity is due to the rods. Note the rapid change in going from dark to light.

Dark Adaptation The retina is capable of changing its sensitivity to light as much as 500,000 times. That is to say, the rods can function over a remarkable range of light intensity. As indicated, the rods are responsible for most of this change, but there are alterations in the cones which also contribute. When one changes from a bright environment to darkness, cone adaptation takes place in approximately 10 min, whereas something like 90 percent of the rod adaptation occurs in 20 to 25 min (Figure 12-19). If the retina has been exposed to bright light for a prolonged period, complete dark adaptation may take several hours. In contradistinction, the alterations that occur when the eye is exposed to bright light after being dark-adapted take only a few minutes.

The terms photopic vision and scotopic vision are often used to refer to vision under conditions of high illumination and low illumination, respectively.

Color Vision

Color possesses three qualities: (1) hue, (2) brightness, and (3) saturation. The hue is the attribute commonly referred to simply as color. It depends upon the wavelength of the radiant energy. Brightness depends upon the intensity of the light rays. Accordingly, any particular hue may have an infinite degree of brightness. The saturation of a color is determined by the degree of purity of that particular hue. As white light is mixed with it, the degree of saturation decreases.

In 1802, Thomas Young, known as the English Leonardo da Vinci, published his theory of color vision. He theorized that the retina contains three types of cones which, when stimulated by light, elicit the sensations of red, green, and blue. He went on to state that color blindness must be due to a failure of one or more of these cones to respond. Helmholtz, the German physicist and physiologist, put the Young theories to experimental test some sixty years later and found them to be essentially valid. The so-called Young-Helmholtz theory of color vision, somewhat modified through the years, is generally accepted.

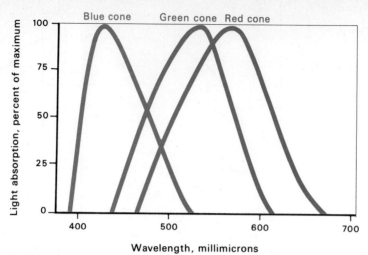

Figure 12-20. Light absorption by the cones. Different pigments have different peaks corresponding to the wavelengths of blue, green, and red.

As mentioned earlier, there are three different types of pigments in different cones, pigments that exhibit peak absorbencies at 430, 535, and 575 millimicrons (mμ) (Figure 12-20). These are in the blue, green, and red ranges, respectively.

Apparently, we see color only when at least two different kinds of cones are stimulated. This conclusion stems from the fact that a person with only one kind of cone pigment is totally color-blind; he does not perceive color even in the range of that sole cone type. This would seem to mean that the appreciation of color depends upon the relative rate of firing in neurons innervating the various types of cones. Examination of Figure 12-20 will show that light with a wavelength of 600 mμ stimulates the red cones far more than the green cones and the sensation is one of red. A wavelength at 500 mμ would stimulate the red and blue cones equally and the greens somewhat more, giving rise to a sensation that would be predominantly green. A sensation of yellow would result from light with a wavelength of about 560 mμ, which would stimulate the red and green cones equally. The same sensation results when red and green lights are shone into the eye simultaneously. Finally, if all three types of cones are stimulated equally, the sensation is of white light.

ABNORMALITIES OF VISION

Normal vision is a highly complex function involving many essential components. The more common disturbances concern the eye itself, the optic pathways, and the visual cones.

Refractive Errors

Myopia   Many people can see objects clearly only when they are held close to the eyes. Everything in the distance is blurred. These people have a tendency to squint, and so the term nearsightedness, or myopia, which means "closing the eye," is used to describe this condition. The difficulty in most cases is that the eyeball is abnormally long (Figure

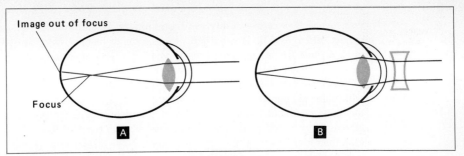

Figure 12-21. Myopia. **A.** Light rays from distant objects focus in front of the retina. **B.** A biconcave lens in front of the eye aids by causing these rays to diverge so that the lens of the eye can bring them to a focus on the retina.

12-21). As a result the focal point lies in front of the retina. The eye possesses no way in which it can accommodate itself to this condition. Only when objects are brought closer and the light rays have diverged will they be brought into focus upon the retina. Myopic persons, in order to see distant objects, must wear glasses with biconcave lenses. Figure 12-21 illustrates how such a lens causes parallel rays to diverge, thus permitting the lens of the myopic eye to focus them upon the retina.

Hyperopia The reverse of myopia, that is, farsightedness, or hyperopia, is also frequently encountered. The eyeball of the farsighted person is shorter than normal. Consequently, the curvature of the lens must increase even when he looks at a distant object. As the object is brought closer, greater accommodation is required. In such cases the near point will be farther away than it is normally. The term hyperopia means "beyond the eye." If the lens did not accommodate, even for parallel rays of light, the image would in fact be focused behind the eyeball (Figure 12-22). Most hyperopic young people possess enough accommodation to get along quite successfully without glasses. When they read, their eyes are maximally accommodated, and sometimes they report fatigue more readily than do others. There is a difference of opinion among experts as to whether or not glasses should be worn by persons with hyperopia. If the eyeball is so short that the person cannot do close work, glasses are definitely indicated. In this instance, biconvex lenses which cause rays of light to converge are used.

Figure 12-22. Hyperopia. **A.** Light rays from distant objects do not come into focus on the retina unless the lens accommodates. **B.** A biconvex lens in front of the eye aids by bending the rays so that the lens of the eye can bring them into focus on the retina without accommodating, as in the normal eye.

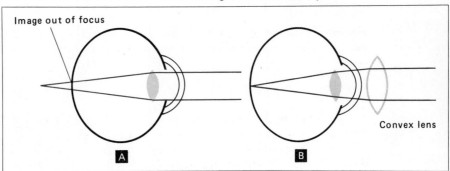

Presbyopia   As a person ages, various physical changes occur. One of these is a gradual loss by the lens of the ability to accommodate, a condition termed presbyopia, which means "old eye." The lens hardens (loses some of its elasticity), so that when the ciliary muscle pulls the ciliary processes forward, the lens does not become as curved as it formerly did. Therefore, the light rays cannot be bent as sharply. Thus there is a progressive recession of the near point as one grows older (Figure 12-23). Everyone is familiar with older people who must hold small print at arm's length in order to read it. When the aging processes finally cause almost complete loss of the power of accommodation, glasses are required at all times, and ones with two types of lenses are usually preferred. One then looks through the top of the lens to see distant objects and through the lower part (which is ground differently) to read or do close work. These glasses are called bifocals.

Astigmatism   The lens of the eye is rarely, if ever, part of a perfect sphere. Different parts of it will have varying powers of refraction. Therefore, when one looks at an object, some areas of it will be in focus whereas other parts will not be. The term astigmatism means "without a point." In other words, there is no point upon which the eye can focus all its rays. Everyone possesses a degree of astigmatism, but it is usually so slight as not to interfere with work or be noticeable. However, if the condition is severe, it proves very annoying and may promote fatigue and chronic headaches. Astigmatism is corrected by specially ground lenses which compensate for the inequalities of refraction.

Cataract

The lens may undergo changes which reduce its transparency. Ultimately, the lens becomes more and more opaque and is termed a cataract. If little or no light can be transmitted, the opaque lens is removed. Since the normal lens has a refractive power

Figure 12-23. Presbyopia. With age the lens loses its elasticity and ability to accommodate, and therefore the near point recedes.

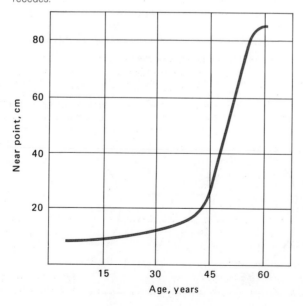

of about 25 diopters, glasses with comparable convex lenses must be substituted. And because removal of the lens removes all power of accommodation, bifocals, or sometimes trifocals, are used.

Glaucoma

The intraocular pressure is normally less than 20 or 25 mmHg. It is measured by means of a tonometer. The cornea of the eye is anesthetized with a local anesthetic, and then the tonometer is pressed against the cornea. The tonometer indicates the pressure required to displace the cornea inward. Intraocular pressure higher than 25 mmHg indicates a condition termed glaucoma. Actually, glaucoma is a misnomer since it means opacity of the crystalline lens, which may or may not occur. The primary problem is elevated intraocular pressure, which can lead to blindness due to pressure on the optic nerve.

Glaucoma can result from an imbalance between the rate of formation of aqueous humor and the loss of this humor through the canal of Schlemm. The disease is usually treated with a drug, Diamox, which reduces intraocular pressure by reducing the rate of aqueous humor formation. If this therapy fails, sometimes a small hole is made in the eyeball so as to permit the aqueous humor to leak out.

Strabismus

The term strabismus means "squint," but it is used to describe a condition in which the image does not fall upon corresponding points of the two retinas. Strabismus is generally due to an abnormality of the extrinsic eye muscles. As a result, two images result, a condition termed diplopia. Treatment consists of surgical alteration of the extrinsic muscles or of the lenses, in order to bring the image to bear upon corresponding points of the two retinas.

Argyll Robertson Pupil

If the pupil fails to change in size in response to light but still responds during accommodation, it is termed an Argyll Robertson pupil. This condition commonly results from syphilis. The explanation is that the pathway for the light reflex is different from that for pupillary response during accommodation.

Color Blindness

In order to have normal color vision, all three types of rods, each with a different pigment, must be present. The presence of these rods is controlled by sex-linked recessive genes. Thus it is that color blindness manifests itself in the male far more frequently than in the female. If all three rods are present, the individual is said to be a trichromat; if two are present, a dichromat; if there is only one, a monochromat. More specific terms are also used. One who does not have red cones is a protanope; he has protanopia. Lack of green cones causes deuteranopia. And absence of blue cones gives rise to tritanopia. In most cases there is not complete absence of cones, but rather greatly diminished numbers of one kind or more.

Tests for Color Blindness   The Ishihara test charts are now the most frequently used for the detection of color blindness. Most readers are already familiar with these charts, which consist of a series of colored dots arranged in such a way that one with normal color vision can discern a number in the array of dots whereas a person with a color deficit will see either no number or a different number.

Another test involves grouping tufts of colored wool which range over the entire color spectrum. If there is color blindness, tufts of different colors will be grouped together.

Lesions in the Visual Pathways

A review of the visual pathways will make it obvious that a lesion which interrupts these tracts, or a part of them, will cause visual deficits. Figure 12-24 summarizes the more common locations of such lesions. In each case there is partial, or half, blindness, termed hemianopsia. In designating the hemianopsia, reference is made to the visual field and not the afflicted retinal area. For example, if the left optic tract is destroyed so that no impulses are received from the left half of the left retina or from the left half of the right retina, the person is half blind, can see nothing to his right, and is said to have a right homonymous hemianopsia. The term homonymous refers to the fact that both right fields are involved. A careful study of Figure 12-24 will disclose the anatomical basis of the various types of disorders due to lesions of the visual pathways, along with the proper designation of each.

Lesions in the Visual Cortex

Area 17 Complete destruction of area 17 (see Figure 10-8) on both sides results in total blindness. Figure 12-24 shows that if this area is destroyed on only one side, there is

Figure 12-24. Lesions in the visual pathways are indicated by colored bars. The resulting deficit in the visual field is shown at left.

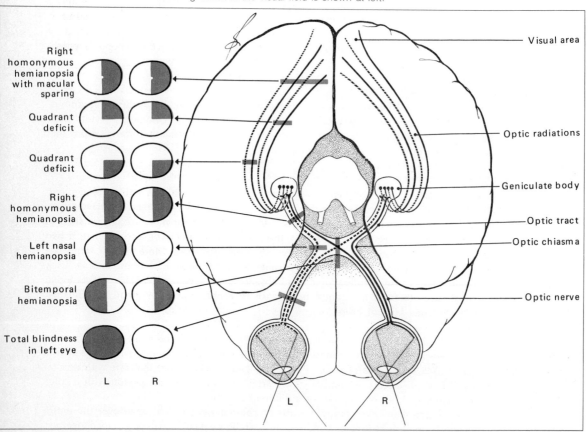

homonymous hemianopsia, as is to be expected, but the central area of the blinded visual field is spared. The explanation for this observation is not yet known, but it seems probable that some of the fibers which come from the central part of the retina cross over before they reach the visual cortex, so that there is actually a dual cortical representation for this central region. Thus destruction of only one side of the visual cortex leaves central vision relatively unaffected.

Areas 18 and 19 These regions serve to coordinate visual sensation with other bodily functions. However, since it is extremely difficult to remove these parts without interfering with area 16 or the optic radiations to area 17, the precise function of areas 18 and 19 cannot be stated. Patients with tumors involving these regions have been studied. Such neoplasms are often associated with visual hallucinations. These patients often report seeing objects, usually highly developed patterns, which do not in fact exist. There may also be an inability to interpret the written word. The patient can see the word but cannot summon its meaning. Yet in speech he will use the word in its correct context. In addition, there may be impairment of coordination between vision and muscular activity. Fortunately, such disorders in the visual cortex are not common.

QUESTIONS AND PROBLEMS

1 How soon after conception is it possible to identify optic vesicles in the embryo? What is the point of origin of these vesicles, and what structures develop from these vesicles?
2 Briefly describe the embryonic development of the following structures of the eye: (*a*) optic cup, (*b*) lens, (*c*) choroid layer, and (*d*) cornea.
3 Describe the three layers of the eyeball.
4 What structures protect the eyeball from injury?
5 If the lacrimal canaliculi in a newborn infant are not patent, what symptoms would be present? Can you suggest how the condition might be corrected?
6 Describe the process and the structure whereby the conjunctiva is kept moist.
7 Draw a midsagittal section of the eye. Label and describe the function of the following structures: (*a*) macula lutea, (*b*) fovea centralis, (*c*) optic disk, (*d*) optic nerve, and (*e*) area cribrosa.
8 How is the normal intraocular pressure created and maintained? Describe the pathology of glaucoma.
9 Myotics are drugs which reduce the size of the pupil and are used in glaucoma to reduce intraocular pressure. What structures of the eye are affected by myotics?

10 Trace the path of a beam of light as it enters the eye and finally stimulates the visual receptors.
11 What is the role of the autonomic nervous system in the functioning of the eye?
12 What is meant by refractive index?
13 Explain why a biconvex—but not a biconcave—lens will bring light rays to a focal point.
14 What is meant by accommodation? When is it necessary and how is it achieved?
15 What symptoms would you expect in a person whose oculomotor nerve has been destroyed?
16 Explain how 10/20 vision differs from 20/20. How do artificial lenses correct the type of visual deficit represented by 10/20 vision?
17 Why is it more difficult for an older person to see near objects as clearly as objects at a distance?
18 How could a physician test for your light reflex? Your convergence reflex?
19 What is the function of rods? Of cones?
20 What is the function of rhodopsin?
21 Explain the Young-Helmholtz theory of color vision.
22 Explain why vitamin A is important to vision.
23 Describe the pathology of cataracts.

13 THE SPECIAL SENSES: AUDITION, GUSTATION, AND OLFACTION

The perception of sound and the perception of various substances that come into contact with taste or olfactory receptors depend upon mechanisms similar to those already outlined. In every case there is a specific receptor, a chain of neurons, and higher centers essential for awareness of the stimulus. But, as in the case of vision, there are differences in the anatomy and physiology of these special senses which require additional elaboration.

AUDITION

The ear, concerned with the special sense of hearing and with equilibrium, is contained mostly within the temporal bone. For descriptive purposes, the ear is subdivided into external, middle, and internal portions (Figure 13-1). Sound waves directed by the external ear (auricle) travel through the external auditory meatus, strike the tympanic membrane, and cause it to vibrate. These vibrations move a series of small bones, the otic ossicles, which extend from the tympanic membrane to the internal ear and set in motion fluid in the labyrinth of the internal ear. Fluid movement in the internal ear stimulates primary sensory neurons there, which transmit impulses to acoustic and equilibrium centers in the brain.

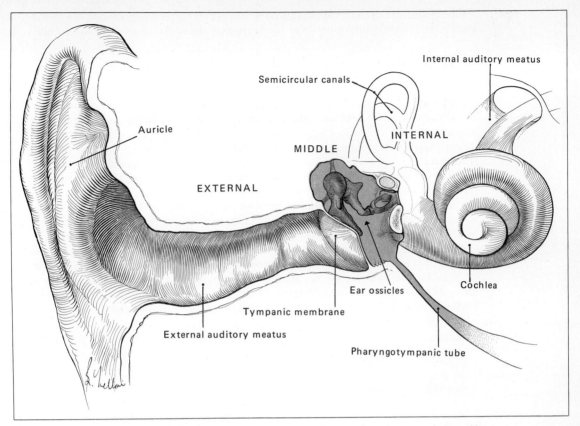

Figure 13-1. The ear and its subdivisions. The middle ear communicates with the nasopharynx by means of the pharyngotympanic tube.

Embryology of the Ear

The external ear arises from ectoderm surrounding the first (mandibular) and second (hyoid) arches adjacent to the first branchial groove. At about the sixth week of development a number of elevations, or hillocks, appear on the arches. These remodel and coalesce to form the auricle and other parts of the external ear. The branchial cleft, or groove, between the two arches deepens to form the external auditory meatus.

The cavity of the middle ear and the pharyngotympanic (eustachian) tube arise from an expansion of the first pharyngeal pouch. A distal dilatation of this pouch becomes the middle ear cavity, while a proximal constriction forms the cylindrical pharyngotympanic tube. Mesenchyme associated with the cavity differentiates into the three ear bones and two muscles.

The internal ear appears initially at about 3 weeks as a thickening of ectoderm of the hindbrain, the auditory or otic placode. Because of uneven growth, the placode develops a depression, the otic pit, near its center. The placode then becomes detached from the brain and develops a central cavity that becomes the otocyst. During the fifth week the otocyst differentiates into uneven cavities which form the semicircular canals, utricle, saccule, and cochlea. During the differentiation of internal ear structures, fibers from the eighth nerve extend into the epithelium of certain regions in the cavities. The epithelium thickens and modifies to become receptor cells for equilibration in the utricle, saccule, and semicircular canals and for hearing in the organ of Corti of the cochlea.

Anatomy of the Ear

External Ear The external ear is an appendage adapted for directing sound waves toward the eardrum. It consists of an expanded projecting portion, the auricle, or pinna (Figure 13-2), and a tube, the external auditory meatus, that ends blindly at the eardrum. The framework of the auricle is elastic cartilage, which gives it resiliency, gives it its characteristic shape, and minimizes damage when the ear is roughly handled. If the cartilage is ruptured, it often heals by an overgrowth of tissue, as seen in "cauliflower ear." The skin lining the external auditory meatus is supplied with modified sweat glands, the ceruminous glands, which secrete earwax, or cerumen.

The outermost rim of the auricle is called the helix; the semicircular ridge internal to the helix, the anthelix; the shallow depression near the top, the triangular fossa; and the deep well adjacent to the external auditory meatus, the concha. The earlobe, a vascular structure which may be used to obtain small blood samples, hangs down from the auricle. The tragus is a small projection anterior to the entrance to the external auditory meatus.

The external auditory meatus is a 2.5-cm-long canal that leads from the auricle to the middle ear cavity. The outer framework of the canal is cartilaginous, while the inner portion is formed by the temporal bone. Internally it is closed by the tympanic membrane.

Tympanic Membrane A flattened-cone-shaped fibrous sheet, the tympanic membrane, or eardrum, separates the external auditory meatus from the middle ear cavity. The

Figure 13-2. The auricle.

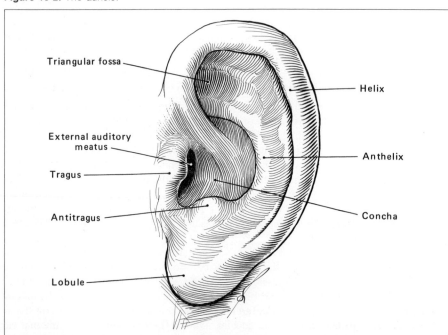

333

THE
SPECIAL
SENSES:
AUDITION,
GUSTATION,
AND
OLFACTION

concavity of this membrane is directed laterally. The point of maximal concavity is called the umbo. Throughout most of its extent the margin of the tympanic membrane is embedded in the tympanic groove. This groove lies at the periphery of the inner end of the external auditory meatus.

Middle Ear The tympanum, or middle ear cavity, is filled with air and contains the three ear ossicles, two small muscles, and branches of the facial and glossopharyngeal nerves. It has four walls, a roof, and a floor, all lined with mucous membrane. Thin plates of bone form the roof and floor separating the tympanum from the middle cranial fossa above and vessels in the neck below. The posterior wall has an opening, the aditus, that leads into the mastoid air cells, as well as a small conical mass of bone, the pyramid, which surrounds the stapedius muscle. The tendon of this muscle passes through an opening at the summit of the pyramid to attach to the stapes.

Also within the middle ear is the pharyngotympanic (eustachian) tube, which forms a communication between the nasopharynx and the middle ear. Mucous membrane lining the middle ear is continuous with the pharyngeal mucosa through the pharyngotympanic tube. The middle ear also communicates with the mastoid air cells. Throat infections often travel via the tube to the middle ear and may pass into the mastoid cells, where abscesses may occur. Infections of this type were common before the advent of antibiotics.

The lateral wall of the middle ear is formed by the tympanic membrane (Figure 13-3).

Figure 13-3. The lateral wall of the middle ear viewed from the interior. The stapes, omitted here, articulates with the lenticular process of the incus.

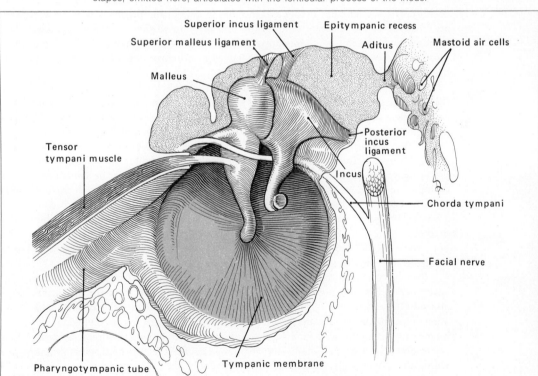

The handle of the malleus is embedded in the membrane, and this bone, with the incus and stapes, forms an osseous bridge to the medial wall. The rounded head of the malleus ("hammer") articulates with the body of the incus ("anvil") (Figure 13-4). The long process (crus) of the incus is directed inferiorly to articulate with the head of the stapes ("stirrup"). The foot plate of the stapes closes the oval window in the medial wall of the middle ear. The articulations of the ear ossicles are freely movable synovial joints. Slender ligaments attached to the walls of the middle ear cavity hold these bones in position.

The medial wall of the middle ear separates it from the internal ear (Figure 13-5). It has two openings, the oval window (fenestra vestibuli) and the round window (fenestra cochlea). The round window is closed by mucous membrane of the middle ear.

A mound, the promontory, lies between the round and oval windows. It is formed by the basal turn of the cochlea. A plexus of nerves, the tympanic plexus, formed by branches of the facial and glossopharyngeal nerves, is located on the promontory. This plexus supplies sensation to the middle ear and gives rise to the minor petrosal nerve.

Two small, longitudinal ridges are visible above the oval window. The smaller is formed by the lateral semicircular canal; the larger, by the facial canal containing the facial nerve as it courses through the temporal bone.

Internal Ear   The internal ear consists of a network of tubes hollowed out of the temporal bone, the osseous labyrinth, within which is a continuous membranous sac, the membranous labyrinth (Figure 13-6). The membranous labyrinth is joined to the wall of the osseous labyrinth by fibrous bands. It is filled with a fluid known as endolymph. The space between the two labyrinths is filled with tissue fluid, the perilymph.

The osseous labyrinth is divided into the semicircular canals, the vestibule, and the cochlea. The semicircular canals and the utricle, one of the organs within the vestibule, are involved in the sense of equilibrium and are termed the vestibular apparatus. The anatomy and physiology of this apparatus are discussed in Chapter 14. In this chapter these organs will be described in a general way and closer attention given the cochlea, the organ of hearing.

In relative position, the cochlea is anterior, the semicircular canals are posterior, and the vestibule lies in between. The oval window, on the medial wall of the middle ear cavity, opens into the vestibule. Vibrations induced at the oval window by the stapes (through its foot plate) are thus communicated to the perilymph of the internal ear.

Within the vestibule are two small membranous sacs, the utricle and saccule, which do not conform to the shape of the vestibule itself. The utricle is the chamber into which the semicircular canals open at both ends of their loop. The saccule connects directly with the membranous sac of the cochlea, the cochlear duct, through the ductus reuniens.

The cochlea is shaped like the shell of a common snail (Figure 13-7). It consists of a bony tube that spirals about $2\frac{1}{2}$ times around a central conical structure of bone, the modiolus. A thin shelf of bone, the spiral lamina, projects from the modiolus into the bony tube like the thread of a screw. The spiral ganglion of the eighth cranial nerve is located in the modiolus at the base of the spiral lamina.

The space within the bony canal of the cochlea is divided into three canals by the vestibular and basilar membranes. These membranes extend from the spiral lamina to attach to the inner surface of the bony canal. The scala media (cochlear duct), a continuation of the membranous labyrinth, lies between the two membranes. Lying on the basilar membrane within the scala media is a complex structure termed the organ of Corti, which is the organ of hearing. Finally, the scala media, like the rest of the membranous sac within the osseous labyrinth, is filled with endolymph.

On each side of the scala media, and sharing one of its limiting walls, are the scala

335

THE
SPECIAL
SENSES:
AUDITION,
GUSTATION,
AND
OLFACTION

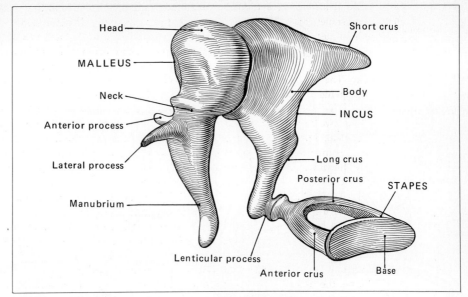

Figure 13-4. Ossicles of the ear. The foot plate of the stapes closes the oval window. The malleus is embedded in the tympanic membrane.

Figure 13-5. Diagrammatic representation of the middle ear with the lateral wall removed and the ossicles omitted. (*Modified from J. Maisonnet and R. Coudane, Anatomie clinque et operatoire, Doin, Paris, 1950.*)

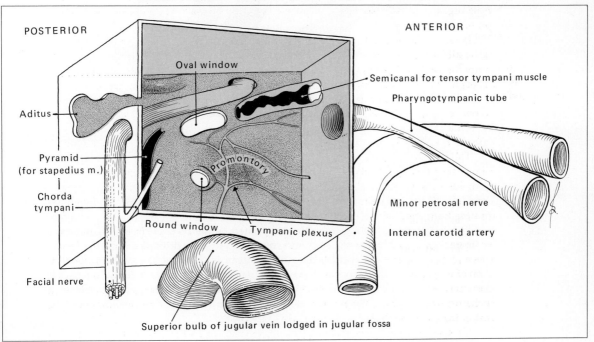

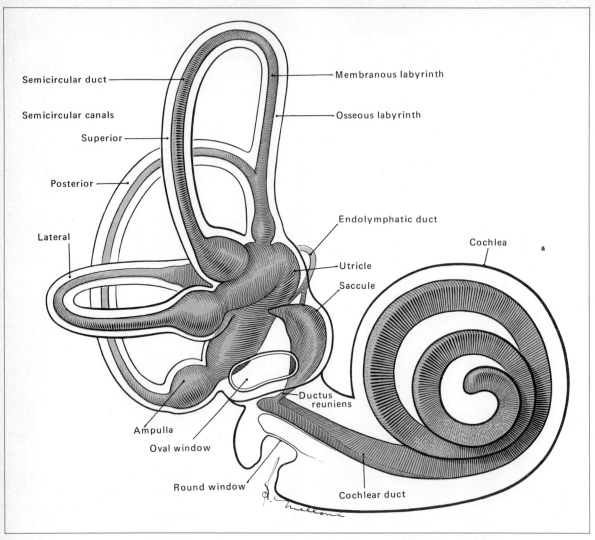

Semicircular duct

Semicircular canals

Superior

Posterior

Lateral

Ampulla

Oval window

Round window

Membranous labyrinth

Osseous labyrinth

Endolymphatic duct

Utricle

Saccule

Cochlea

Ductus reuniens

Cochlear duct

Figure 13-6. The internal ear. The membranous labyrinth (in color) is filled with endolymph; the osseous labyrinth, with perilymph.

vestibuli and the scala tympani. The scala vestibuli is between the vestibular membrane and the bony wall of the cochlea; the scala tympani is between the basilar membrane and the wall of the cochlea. These canals separated by the scala media follow the spiral of the cochlea in parallel to the apex and communicate there through a hole, the helicotrema. At the base of the cochlea the scala tympani is closed at the round window by a membranous flap, the secondary tympanic membrane. The scala vestibuli opens into the vestibule at a point close to the oval window. Thus movement of the foot plate of the stapes which closes the oval window sets the perilymph within the scala vestibuli in motion (see page 340).

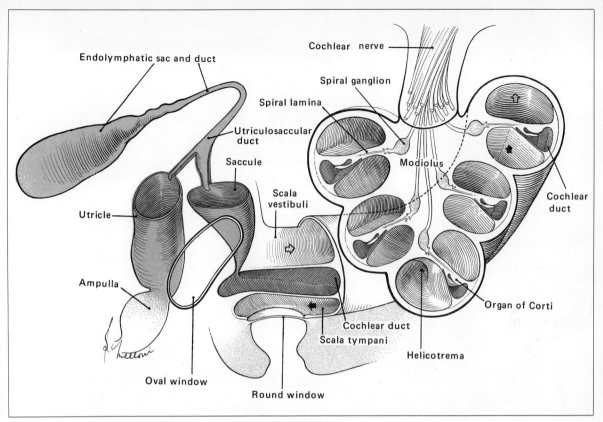

Figure 13-7. Section through the cochlea and vestibule. Sound waves pass through the perilymph of the cochlea in the direction indicated by arrows.

Organ of Corti This specialized receptor organ for hearing (Figure 13-8) is a complicated structure consisting of sensory hair cells, groups of supporting cells, and fibers from the bipolar neurons of the spiral ganglion. The bases of the hair cells are anchored, by means of supporting cells, to the basilar membrane. The tips of the hair cells are embedded in a delicate gelatinous structure termed the tectorial membrane.

Bipolar neurons are the link between the organ of Corti and the central nervous system (Figure 13-9). A peripheral process from each hair cell extends to a cell body in the spiral ganglion. Each cell body sends a central fiber to cochlear nuclei in the medulla. The central fibers course together as the cochlear portion of the eighth cranial nerve. Other neurons link cochlear nuclei in the medulla with the medial geniculate body of the brainstem. Neurons associated with the medial geniculate body in turn transmit impulses to the auditory area of the cortex, in the temporal lobe (area 22).

Properties of Sound

Sound has three important properties: (1) pitch, (2) intensity, and (3) quality. The pitch is a function of the number of vibrations of the sound waves per second. Figure 13-10 shows representative sound waves. Note that the greater the number of vibrations per

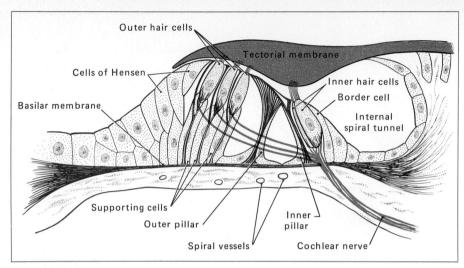

Figure 13-8. Section through the organ of Corti. Movement of the basilar membrane by sound waves in the perilymph generates action potentials in the hair cells which travel along fibers of the cochlear nerve.

unit time, the higher the pitch, or what is often referred to as the tone. The number of vibrations, or cycles per second (cps), is called the frequency. The normal human ear can usually detect sound with a frequency from 30 up to 20,000 cps.

The intensity of sound depends upon the amplitude of the waves. The greater the wave amplitude, the more intense the sound. The amplitude, or energy, of the sound wave can be measured directly but more conveniently on a relative scale. A bel is defined as an increase in sound intensity of 10 times. An increase of 100 times would be 2 bels, 1,000 times 3 bels, and so forth. These are such large changes that the decibel is more appropriate for expressing the range of audition; 10 decibels equals 1 bel.

For convenience the decibel has been given an absolute value. The starting point is 0 on the decibel scale, and this has been defined as 1 microwatt of sound energy per square centimeter. In more meaningful terms, a whisper at a distance of approximately 1 m has an intensity of 20 decibels; normal conversation at 3 m is about 50 decibels; an intensity of 120 decibels borders on the pain threshold. The difference between 20 and 120 decibels equals a 10-billionfold increase in intensity!

The quality, or timbre, of sound depends upon overtones, that is, so-called higher constituents. An example will make this clear. If a violin string is set in motion, it vibrates as a whole and then rapidly divides itself into segments which oscillate at their own frequency, which is inversely related to the length of each segment. Thus, middle C may be produced, but with it are vibrations with twice and three times the frequency. The latter are the overtones. Together they produce the quality of the sound. Thus, although middle C always has the same frequency and can be produced with the same intensity by a variety of instruments, one can certainly determine whether a tuba or a zither, for example, is giving rise to the sound.

Transmission of Sound Waves in the Ear

Sound waves enter the external auditory meatus, strike the tympanic membrane, and cause it to move in and out with the frequency of the sound waves. The tympanic membrane

339

THE
SPECIAL
SENSES:
AUDITION,
GUSTATION,
AND
OLFACTION

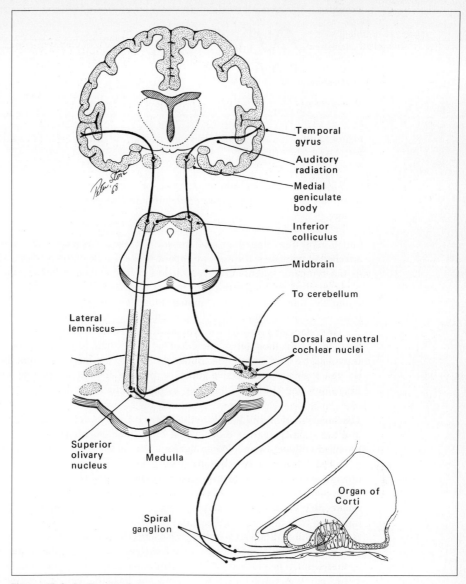

Figure 13-9. Auditory pathways.

responds accurately to most frequencies, and its movements stop almost instantly upon cessation of the sound.

The movements of the tympanic membrane cause the auditory ossicles to move. As a result, the foot plate of the stapes attached to the oval window moves in response to the motion of the incus and malleus. The tympanic membrane has an area about 20 times larger than that of the foot plate of the stapes in the oval window. Thus, the force exerted by the foot plate is some 20 times greater than the force of the sound waves acting on the tympanic membrane. Beyond the oval window, in the internal ear, there is perilymph.

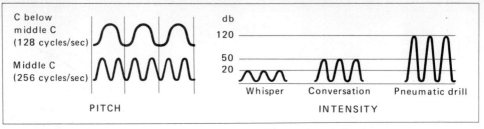

Figure 13-10. Pitch and intensity of sound waves. The number of vibrations per second determines pitch. The amplitude of the wave determines intensity.

Considerable energy is required to overcome its inertia. The force advantage due to the difference in size of the tympanic membrane and oval window provides the necessary energy.

Function of the Stapedius and Tensor Tympani Muscles The stapedius muscle is attached to the stapes; the tensor tympani, to the handle of the malleus, which is embedded in the tympanic membrane. Contraction of the stapedius moves the stapes outward; contraction of the tensor tympani moves it inward. If both contract, the stapes is made immobile, thus blocking, or damping, movement of all three ossicles and the tympanic membrane. Such contraction, then, attenuates sound transmission.

The stapedius and tensor tympani muscles are under reflex control, a reflex with the remarkably short latent period of no more than 10 milliseconds (msec). The reflex is evoked by loud sound; the louder the sound, the more vigorous the response. Thus, the reflex serves a protective function, permits the ear to tolerate sounds of different intensities, and is often referred to as the attenuation reflex.

Transmission of Sound Waves in the Internal Ear As described earlier, the scala vestibuli and the scala tympani are connected by a small opening, the helicotrema. These scali are filled with perilymph. When the foot plate of the stapes moves inward, the perilymph is moved toward the helicotrema (Figure 13-11). The movement continues through this opening and then back in the scala tympani, ultimately causing the round window to bulge into the middle ear. Without the round window there would be little fluid movement. Theoretically, if the stapes moves very slowly, the perilymph will move as just described with little or no change in the basilar membrane. Normally, though, the stapes moves rapidly. When this occurs, the increased pressure in the perilymph close to the oval window, that is, at the base of the cochlea, causes bulging of the basilar membrane at that point. The elasticity of the membrane causes it to rebound, and thus a wave is produced. This wave now travels along the basilar membrane toward the apex of the cochlea.

The amplitude of the wave increases as it travels along the basilar membrane. Note, however, that when the wave reaches the part of the basilar membrane that has a natural resonant frequency equal to the frequency of the sound that initiated the wave, the amplitude of the basilar membrane wave rapidly decreases. This is called the cutoff point (Figure 13-11).

Activation of the Auditory Receptors

As can be seen in Figure 13-8, showing a section through the organ of Corti, the hairs of the hair cells are embedded in the tectorial membrane. The organ of Corti is constructed

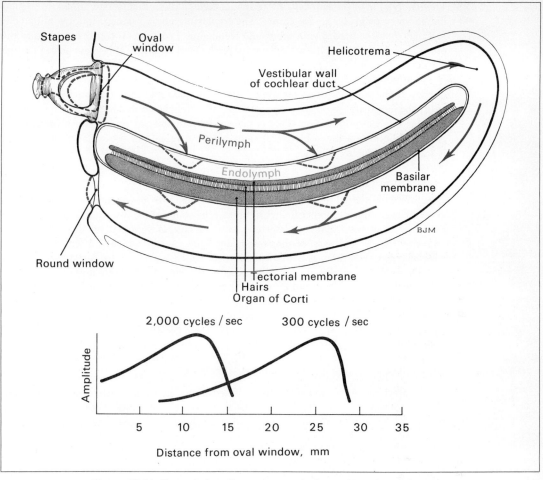

Figure 13-11. Transmission of sound waves in the cochlea. A sound wave traveling through the perilymph reaches its maximum amplitude at the place where there is resonance in the basilar membrane. This location is closer to the oval window for sound waves of high frequency than for those of low frequency.

in a way that results in the hairs being bent when the basilar membrane moves. When it moves up, the hairs will be bent in one direction; downward movement causes bending in the opposite direction. This bending, probably by altering the permeability of the hair cells, results in the receptor potential which activates neurons of the cochlear nerve.

Pitch Discrimination

The basilar membrane is about 0.04 mm wide at the base of the cochlea and about 0.5 mm wide at the helicotrema. The shorter fibers vibrate at a higher frequency than do the longer ones. As explained above, all sound frequencies initiate the traveling wave at the base of the cochlea, but each has a specific cutoff point; the higher the frequency, the closer to the oval window; the lower the frequency, the closer to the helicotrema (Figure 13-11).

The innervation of the hair cells has a spatial configuration that is maintained throughout the auditory pathways to the cortex. The basilar membrane is maximally stimulated at a specific place in accord with the sound frequency. The cerebral cortex can identify this place, thereby providing the major clue for pitch discrimination. This is the place principle for determination of pitch.

The number of impulses propagated by the auditory neurons varies with the frequency of the sound up to a maximum of about 1,000 cps. Thus, for pitches below this figure the number of impulses propagated per unit time is thought to provide an important clue for pitch discrimination. This is the volley principle, or periodicity principle. The current thinking on the subject embraces both the place and the volley principles and is termed the duplex theory.

Localization of Sound

A person with normal hearing can localize the direction from which waves emanate. The essential clue is the difference in arrival time of the sound at the two ears. A time difference as small as 10 msec may be detected. Another clue is supplied by the difference in intensity, that is, loudness, of the sound impinging on the two ears. Intensity decreases with distance; thus there is a slight difference, which sensitive ears can detect.

There is now evidence which indicates that when sound enters one ear just before it enters the other ear, inhibition occurs at the level of the superior olivary nucleus (see Figure 13-9). Apparently the neurons in the contralateral nucleus are excited while those in the ipsilateral superior olivary nucleus are inhibited. This inhibition persists for only a fraction of a millisecond, but it suffices to reinforce the difference in perception of sound from each ear, thus contributing to the ability to localize sound.

Intensity Discrimination

The appreciation of sound intensity is based upon the same mechanisms as those for all types of sensory intensity determination: (1) each receptor fires more rapidly, (2) more receptors fire, and (3) receptors with higher-intensity thresholds become active. In the case of audition, the basilar membrane undergoes greater amplitude excursions, which stimulate the hair cells to fire more rapidly. Second, hair cells surrounding the cutoff point fire. Third, high-threshold hair cells become active in response to the vigorous basilar membrane movements. The final result is that more impulses per unit time reach the auditory cortex. The activity of the high-threshold receptors may provide a special clue to loudness.

Retrograde Auditory Impulses

Pathways have been identified which carry impulses from the auditory cortex down to each level of the auditory system and ultimately to the cochlea. These impulses are inhibitory, and since they maintain a spatial organization, specific areas of the basilar membrane are suppressed. The function of these retrograde impulses is thought to be to aid in focusing attention on specific components of sound, for example, to listen to a conversation against background sound or to focus upon a single instrument in an orchestra.

Tests of Audition

Two types of auditory tests have proved useful: one evaluates sound-wave transmission; the other measures auditory acuity.

343

THE
SPECIAL
SENSES:
AUDITION,
GUSTATION,
AND
OLFACTION

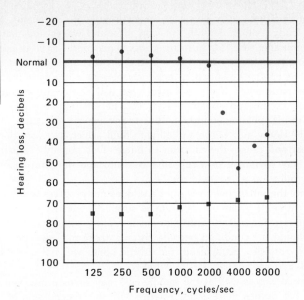

Figure 13-12. Audiogram. The audiometer is calibrated so that the normal ear will just perceive sounds of each pitch at the 0-decibel level. If the intensity must be increased for perception of the sound, the person is said to have a hearing loss of that many decibels. The dots indicate hearing loss only in the upper frequencies. The squares represent hearing loss at all frequencies.

Sound Transmission If a tuning fork is made to vibrate and then is pressed against the forehead, a person with normal hearing will localize the sound as being directly in front of him. But if there is transmission deafness in one ear, he will report the sound as coming from the impaired side. This is the so-called Weber test. The explanation lies in the fact that there are usually other sounds besides those of the tuning fork. The patient does not detect these extraneous sounds with the defective ear. With that ear he recognizes only the tuning-fork vibrations, because these are transmitted by bone conduction. The normal ear hears extrinsic sound plus the tuning fork; thus the tuning-fork sounds seem louder on the defective side, because there is no interference from extraneous sounds.

In the Rinne test the tuning fork is set to vibrating and then is held in front of the ear until it is no longer heard. Next it is placed on the mastoid process. If there is transmission deafness, the subject will perceive the sound when the fork is placed on the bone. The person with a normal ear will not.

Auditory Acuity The hearing threshold, or auditory acuity, may best be ascertained by use of the audiometer, which is an electronic oscillator capable of emitting pure tones over a wide range. The subject is placed in a soundproof room and fitted with a set of earphones. Various frequencies are selected and the intensities varied until the subject reports he can hear the sound. The threshold intensity for each frequency is recorded. The entire range of audible pitches is tested and plotted on an audiogram (Figure 13-12). The audiogram discloses deviations from normal values in the entire range of audible frequencies.

Abnormalities of Audition

Disturbances of any part of the auditory mechanism may eventuate in partial or complete deafness. This may be caused by a failure of transmission of sound waves, it may be associated with occlusion of the pharyngotympanic tube, or it may be an interruption of the central auditory pathways.

Transmission Deafness Any object in the external auditory meatus, as excessive wax or a foreign body, will buffer transmission of sound waves and may also damp the tympanic membrane, resulting in hearing loss. This condition is usually transient, and normal hearing is resumed just as soon as the obstruction is eliminated.

Acute otitis media is a condition in which there is, as the term indicates, an inflammation of the middle ear. It is often associated with the common cold, influenza, and other disorders of the upper respiratory tract. The inflammatory process spreads from the pharynx up the pharyngotympanic tube to involve the structures in the middle ear. In this way the cavity may become filled with fluid, which increases the pressure within the middle ear and causes the eardrum to bulge. This not only impairs hearing but is painful and may eventuate in a ruptured eardrum and the pouring out of a purulent exudate, referred to as a "running ear."

Ordinarily the inflammation is mild and does not progress to the point where fluid accumulates. In some cases the eardrum is surgically incised to allow the exudate to escape. In these instances the membrane heals rapidly without any subsequent hearing deficit. But should the condition become chronic, there may develop fibrous adhesions of the auditory ossicles. As a result, these little bones will not move so freely as they normally did, and hearing will be impaired.

The internal ear, as well as the middle ear, may become involved in inflammatory processes. If this occurs, there is interference with the transmission of the sound waves through the perilymph, as well as damage to the sensitive hair cells of the organ of Corti. A few cases of partial deafness have been reported in which a tumor within the cochlea has blocked sound-wave transmission. Finally, there are instances of a congenital defect in the formation of the inner ear resulting in varying degrees of hearing loss.

Central Deafness The auditory pathways undergo such extensive crossing at various levels that a unilateral lesion in these pathways does not cause deafness unless the lesion is in the cochlear nerve.

Bilateral destruction of the auditory centers in the cerebral cortex causes almost total deafness in man. In lower animals, nuclei in the brainstem and thalamus continue to function and to provide the animal with considerable auditory perception.

Occlusion of the Pharyngotympanic Tube Were it not for the pharyngotympanic tube, the middle ear would be a closed, air-filled chamber. The middle ear is separated from the outer ear by the tympanic membrane. Changes in the atmospheric pressure would cause the membrane to bulge in one direction or the other and could cause rupture. Fortunately, the middle ear is not a closed chamber but rather communicates to the pharynx via the pharyngotympanic tube. So long as this tube is patent, the air pressure in the middle ear will be the same as in the atmosphere. For example, when one ascends a mountain or in an airplane, atmospheric pressure decreases; therefore the middle ear pressure is relatively higher. This pressure differential suffices to force the tube open, and air leaves the middle ear until the pressure on either side of the tympanic membrane is the same. When descending, however, atmospheric pressure becomes greater than middle ear pressure. This differential collapses the tube. In order to bring about equalization, one must swallow, an act which opens the tube. This procedure is especially effective if one first builds up pressure in the pharynx and then swallows. As the tube is momentarily opened, air will be forced into the middle ear.

Occlusion of the pharyngotympanic tube, as generally occurs during a "cold," may make it impossible to equalize the pressures, and if the differential becomes great enough,

345

THE
SPECIAL
SENSES:
AUDITION,
GUSTATION,
AND
OLFACTION

rupture of the membrane occurs. In any event, the pressure differential damps the tympanic membrane, thereby diminishing auditory sensitivity.

GUSTATION

While taste, or gustation, is generally considered associated with the tongue, much of the discrimination we associate with taste is a function of olfaction. Gustatory discrimination associated directly with the tongue is limited to differentiation between sweet, sour, bitter, and salt tastes. These are referred to as the primary tastes.

Organ of Taste

The gustatory structures, or taste buds, are ovoid bodies of neuroepithelium embedded in the epithelium of the tongue (Figure 13-13). They are most numerous on the large vallate papillae at the back of the tongue. Taste buds are somewhat barrel-shaped with a narrow aperture (taste pore) opening onto the tongue surface. A taste bud contains two cell types, the spindle-shaped narrow taste cells (taste receptors) with modified free ends, the taste hairs; and the broader, crescent-shaped sustentacular, or supportive, cells. Each taste bud has several taste hairs, which are about 4 microns long. They lie in the taste pore of the bud, and their ends protrude above the surface. Nerve fibers enter the taste bud to terminate on the taste cells.

Figure 13-13. Photomicrograph of a longitudinal section through a papilla of the tongue ($\times$ 255).

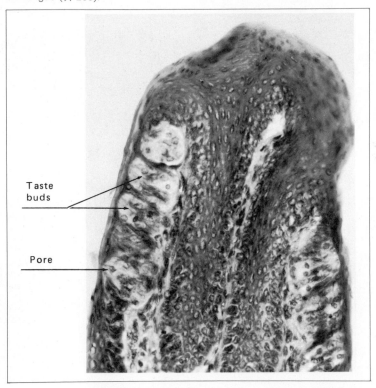

Taste buds

Pore

Tongue

The tongue is a fibromuscular organ located partially in the oral cavity and partially in the pharynx. It is shaped like a short boot turned upside down. The sole of the boot-shaped structure, forming the dorsum of the tongue, is the portion you can see with the mouth widely opened. The upper part of the "boot" is anchored to the hyoid bone. In addition to subserving taste functions, the tongue assists in speech, mastication, and swallowing. The bulk of the tongue is formed by muscles which are described on page 133.

The dorsum of the tongue is covered by stratified squamous epithelium. Its roughened appearance is due to numerous minute projections, the papillae, which vary in shape (Figure 13-14). Small, slender, threadlike filiform papillae, by far the most numerous, are dispersed over the entire dorsal surface; larger mushroom-shaped fungiform papillae, less numerous, are scattered throughout the dorsum of the anterior two-thirds of the tongue, while the 10 to 12 large vallate papillae, circular and dome-shaped, converge to form a wide V at the posterior part of the tongue. The foramen cecum is a small pit in the midline at the point of convergence of the two lines of vallate papillae.

Taste buds are the receptor organs of special sensory neurons carried in the facial, glossopharyngeal, and vagus nerves. The taste fibers in the facial nerve are distributed to the anterior two-thirds of the tongue; those carried with the glossopharyngeal nerve, to the posterior third. The vagus nerve supplies the area of the tongue adjacent to the epiglottis.

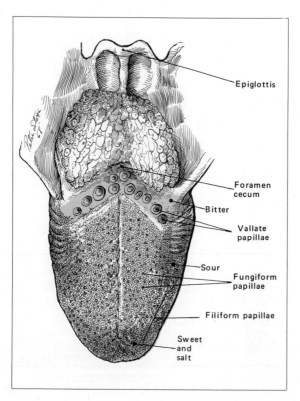

Figure 13-14. The dorsum of the tongue with the primary taste areas in color. Filiform papillae are found over most of the dorsum.

347

THE
SPECIAL
SENSES:
AUDITION,
GUSTATION,
AND
OLFACTION

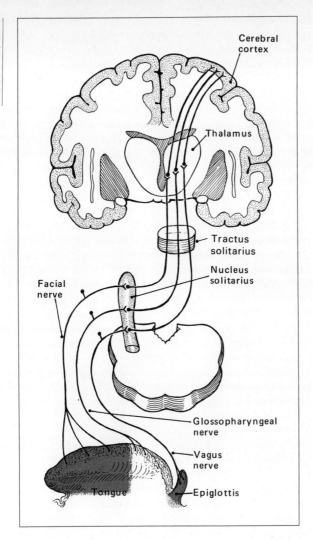

Figure 13-15. Taste pathways.

Taste Pathways

Taste impulses are conveyed in the seventh, ninth, and tenth cranial nerves to the medulla and pons, where they synapse with sensory nuclei associated with these nerves (Figure 13-15). Secondary fibers cross to the opposite side in the brainstem to course in the tractus solitarius to the thalamus, where they synapse with third-order neurons which pass to the cortical area of taste located in the parietal lobe. There must also be pathways ending in the hypothalamus to relate taste to feeding behavior.

Activation of the Taste Buds

The taste buds are chemoreceptors. In order to activate a taste bud, the substance must be in solution. Dry solids are not effective. Just how a solution of certain chemicals stimulates the taste buds is not understood. There is general agreement, however, that the taste chemical, in some way, alters the membrane of the hairs of the taste bud. Once the membrane is altered, a generator potential ensues by the usual mechanism. The unanswered question is how certain chemicals alter the membrane. Saliva or water that is placed in the mouth serves to clean the taste hair and thereby remove the taste stimuli.

Taste-Bud Adaptation

Action-potential studies show that when an effective solution is placed on a taste bud, there is at first a burst of impulses, but then within a second or two the rate of firing becomes much less, to reach a slow but steady pattern of firing. In other words, the taste buds quickly, but only partially, adapt. Yet despite the continued flow of impulses from the receptors, the sensation of taste is almost completely lost during prolonged exposure to most substances. In other words, of the total adaptation, the taste buds account for most, but not all. Somehow the central nervous system is responsible for the rest.

Taste Discrimination

Four basic modalities of taste are recognized: (1) salt, (2) sour, (3) sweet, and (4) bitter. The tongue is not uniformly sensitive to each modality. Sweet and salt sensitivity are greatest at the tip; sour, at the sides; and bitter, at the back (see Figure 13-14).

According to the classic concept of taste discrimination, there are taste buds specific for each of the primary tastes of salt, sour, sweet, and bitter. But no histological difference has been shown to differentiate the buds and, more importantly, action-potential studies disclose that any one bud will respond to all four of the primary tastes. However, the quantitative responses differ with different buds. That is to say, one bud, while responding to all tastes, responds more vigorously perhaps to salt, another to sweet, and so forth. The basic clue, then, for taste discrimination would appear to be the ratio of impulses a particular taste evokes. Additional clues are derived from the activation of extragustatory receptors such as those of temperature, pressure, pain, and olfaction.

Taste Preference

The selection of food by human beings is related to so many factors that the basic selection exerted by the gustatory mechanism is often dominated and obscured. In lower animals there is a more direct relationship. Studies have shown that when the taste mechanism is the controlling factor, the organism selects a diet that is remarkably attuned to his physiological needs. For example, an adrenalectomized animal will, if it has the opportunity, drink saltwater in preference to pure water. As a matter of fact, adrenalectomized rats can survive without administered hormone if they are allowed to drink saltwater freely.

The mechanism of food preference in response to bodily needs is not understood, but it is believed to be a negative response. That is, a physiological alteration in some way makes certain foods unattractive either from the standpoint of taste or because the consumption of those particular types of food gives rise to bodily discomfort or sickness. Thus those foods are avoided. But how the specific needed food, as the saltwater mentioned above, is elected, other than by trial and error, is not understood.

Taste Reflexes

Impulses evoked by various substances are not only propagated to the cerebral cortex for awareness but also synapse with the motor fibers that innervate the salivary glands. Again, by a little-understood mechanism, certain foods call forth a very dilute, watery saliva, whereas others stimulate a thick, mucous secretion.

349

THE
SPECIAL
SENSES:
AUDITION,
GUSTATION,
AND
OLFACTION

OLFACTION

Olfaction, or the sense of smell, is associated with the mucosal epithelium lining the uppermost portion of the nasal cavity. The olfactory epithelium, or membrane, develops from the olfactory placode, a derivative of the neural ectoderm in the head region of the embryo. In the adult it occupies a small area about the size of a dime on the midline nasal septum and adjacent lateral nasal wall above the superior meatus (see Figure 9-21). Olfactory epithelium resembles closely the pseudostratified ciliated columnar epithelium that lines the air passageways (Figure 13-16). It has three cell types: (1) supporting, or sustentacular, cells, (2) basal cells, and (3) small bipolar receptor neurons. The first two types play no role in olfaction but serve only to support and possibly nourish the olfactory cells.

The bipolar primary neurons of olfaction are scattered among the other epithelial components of the mucous membrane in the olfactory area. Slender peripheral processes project outward from the epithelium and are responsive to various odoriferous substances. Central processes of the olfactory neurons coalesce to form the olfactory nerve, 15 to 20 filaments or bundles which pass through the cribriform plate of the ethmoid bone to synapse in the olfactory bulb with second-order neurons called mitral cells. Processes of the mitral cells passing toward the brain form the olfactory tract (Figure 13-17). This band of fibers, situated on the inferior surface of the frontal lobe of the brain, courses posteriorly and spreads out into the olfactory trigone, near the optic chiasma. Some of the olfactory cells terminate here, others cross to the opposite hemisphere through the

Figure 13-16. Section through the olfactory area of the nasal cavity.

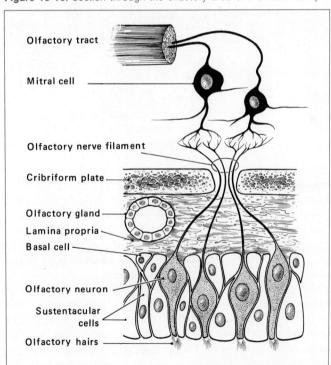

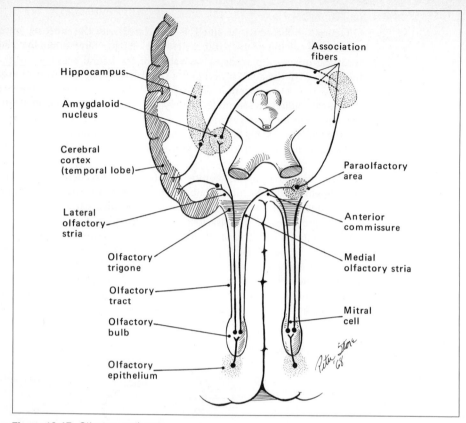

Figure 13-17. Olfactory pathways.

anterior commissure, but most of the cells pass to the olfactory area in the cortex of the temporal lobe (uncus and hippocampus).

Association fibers from the olfactory trigone course to many parts of the brain.

Olfactory Discrimination

Of all the senses, the sense of smell is the least understood. Undoubtedly one can detect, by olfaction, various substances and one can discriminate between them, in some cases, with remarkable accuracy. Unfortunately there are more theories than facts.

One theory is that there is a balance of catalysts in the olfactory receptors. Odoriferous substances then initiate reactions which require specific catalysts, and thus the balance is altered. This imbalance is said to be adequate to provide the clue for olfactory discrimination. The paucity of evidence accounts for the vagueness of this hypothesis.

Another theory suggests that there are certain basic odors and receptors which are relatively more responsive to each basic odor. This concept is clearly patterned on the current theory of taste discrimination. The primary odors are said to be camphoraceous, musky, floral, pepperminty, ethereal, pungent, and putrid—seven in all. A variation on this theme forms the basis for the so-called stereochemical theory, which envisions each of the seven primary substances as having a specific molecular shape. There are then thought to be seven different types of olfactory receptors, each of which will accept a

351

THE
SPECIAL
SENSES:
AUDITION,
GUSTATION,
AND
OLFACTION

molecule of appropriate configuration the way a socket takes a plug. Such receptors have not yet been identified, but action-potential studies have shown that different odors do selectively activate specific receptors and not others.

Whatever the mechanism, the substance to be perceived by the olfactory system must be volatile, at least slightly soluble in water, and also soluble in lipids in order to come into contact with the olfactory receptors in the olfactory epithelium. In all probability, once in contact with the receptor, the membrane is altered so as to give rise to an action potential.

The imprecise nature of odor discrimination is taken advantage of to mask undesirable odors. Households, hospitals, and other areas are often sprayed with odoriferous substances which either overwhelm the undesirable odor or, when one smells both of them, create an entirely new and desirable sensation. Perfume, of course, is used for the same purpose.

Olfactory Adaptation

The olfactory mechanism adapts even more rapidly and completely than does taste. Even a repugnant odor can soon be tolerated. Yet, action potentials recorded from the olfactory nerves show that the receptors continue to fire for prolonged periods when exposed to odoriferous substances. To be sure, there is a high-frequency initial burst of activity, but that quickly subsides to a steady output of about half the initial frequency. Thus, just as with taste, there must be adaptation at a higher level. The mechanism of higher-level adaptation is completely unknown.

Olfaction and Mating

The sense of smell is obviously related to mating, certainly in lower animals. Male dogs are attracted, via olfaction, to bitches in heat; male hamsters are sexually stimulated by an abundant, highly odorous substance exuded by the genital region of the female. The role of olfaction in human mating remains conjectural, although perfume manufacturers would have one believe that it is highly important. Data are sparse. A recent study, however, demonstrated that if the olfactory system in male hamsters is destroyed, they will not mate.

QUESTIONS AND PROBLEMS

1 What is the embryonic origin of the following: (a) pinna, or auricle, (b) middle ear cavity, (c) pharyngotympanic tube, (d) malleolus, and (e) cochlea?

2 In what way would the absence of either pinna or both affect hearing?

3 How is it possible for a neglected throat infection to lead to a middle ear infection?

4 Describe the transmission of sound waves produced by a sudden noise beginning with the external auditory meatus; conclude your description with the cerebral cortex.

5 How are each of the following involved in audition: (a) spiral ganglion, (b) organ of Corti, (c) oval window, (d) perilymph, and (e) stapedius muscle?

6 How does the pitch of sound differ from its quality and intensity?

7 In what way does the attenuation reflex protect your hearing?

8 Differentiate between central and transmission deafness. How could a tuning fork be used to make this diagnosis?

9 Describe the events involving the taste buds, taste pathways, and salivary glands when a cube of sugar is placed in your mouth.

10 Which of the three olfactory epithelial cell types is sensitive to odors?

11 What evidence is there to support the statement that different odors selectively activate specific olfactory receptors?

14 CENTERS OF COORDINATION

The nervous system does far more than make possible perception and movement. It is a system that brings about the remarkable and essential coordination characteristic of living organisms, especially the more complex ones such as man. The high degree of this coordination and its importance become dramatically apparent when there is an abnormality of the nervous system which interferes with or destroys that coordination. In this chapter the more important centers of coordination will be discussed and illustrative abnormalities briefly mentioned.

VESTIBULAR APPARATUS

The vestibular apparatus, located in the bony labyrinth of the temporal bone, consists of the semicircular canals and the utricle. The cochlea, located near the vestibular apparatus, is a part of the hearing mechanism and was described in Chapter 13. The function of the saccule is still uncertain, but there is a growing belief that it probably contributes proprioceptive information.

Anatomy of the Vestibular Apparatus

The three semicircular canals open into the expanded vestibule located between the canals and the cochlea. A dilatation, the ampulla, is present at one end of each canal as it enters the vestibule. (See Figures 13-6 and 13-7.)

The bony semicircular canals are lined with membranous ducts of the same shape. The ducts open into the utricle, and the latter communicates, by means of a duct, with the saccule. The utricle and saccule are membranous sacs within the vestibule. The entire membranous system of the internal ear is filled with endolymph.

Localized areas of neuroepithelium are found in the utricle and saccule. These areas consist of two kinds of columnar cells: (1) supporting (sustentacular), nonnervous cells and (2) sensory hair cells. The otolithic membrane covers these cells. It is a layer of gelatinous material with crystals of calcium carbonate, the otoliths, embedded in it. The free ends of the hair cells lie in this jellylike mass. As the head moves, the endolymph shifts, carrying with it the crystalline particles, which bend the hairs. This bending gives rise to the receptor potential.

Similar areas of neuroepithelium are found in the ampullae of the semicircular canals. Within each ampulla is a mound of hair cells termed the crista ampullaris. This mound also has a gelatinous membrane covering it, but it lacks otoliths. The hair cells, however, are stimulated by movement of the endolymph in the canal.

The primary sensory neurons have their cell bodies in the vestibular ganglion located on the eighth cranial nerve at the inner end of the internal auditory meatus. Impulses from the peripheral process of these bipolar neurons are transmitted via its central process to vestibular nuclei in the medulla (Figure 14-1).

Figure 14-1. Vestibular pathways.

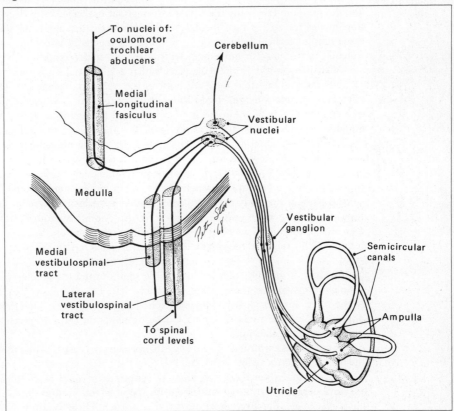

Impulses associated with equilibrium utilize a series of pathways in the central nervous system. Secondary neurons convey impulses to the spinal cord for reflex-type adjustments of posture and movement (locomotion); to nuclei of the third, fourth, and sixth cranial nerves for reflex eye movements which accompany movements of the body; and to salivatory and vomiting centers in the brainstem, associated with salivation and nausea in response to dizziness. Others go to the cerebellum. Some fibers undoubtedly wend their way, probably via the thalamus, to the sensory cortex so as to make one aware of the position and changes of velocity of the body.

Function of the Vestibular Apparatus

The vestibular apparatus serves a dual role: (1) to initiate appropriate reactions to movement and (2) to aid in the maintenance of the upright position. The semicircular canals are sensitive to movement, while the utricle is fundamentally concerned with position. The exact function of the saccule is uncertain, but it probably operates in concert with the utricle.

Reaction to Movement   Figure 13-6 shows that the semicircular canals are arranged at right angles to each other. Thus, no matter in what plane the movement takes place, at least one pair of canals (that is, one canal from the apparatus in each ear) is involved.

If a tube filled with fluid is moved rapidly through space, during acceleration there is a backward movement of the fluid, thus increasing the pressure against the back of the tube. When the movement becomes constant, the fluid regains its resting position, and the pressure is equally distributed throughout the tube. With deceleration, the fluid now moves forward, which increases the pressure against the front wall of the tube. If these simple physical facts are borne in mind, the means by which the semicircular canals function in response to movement will be easy to understand.

Visualize a pair of semicircular canals, that is, one canal from the vestibular apparatus of each ear. Assume that they are both in the horizontal plane and will, therefore, be stimulated by rotation of the body about its long axis, such as when the subject sits upright in a spinning chair. If the rotation is to the right, the endolymph will move toward the ampulla of the right canal and away from the ampulla of the left canal. The hairs of the hair cells which compose the crista are bent by the movement of the fluid. In this case the hairs of the right canal will be bent forward while the hairs of the left canal will be deflected backward. The movement of the hairs initiates impulses which are conducted by the vestibular branch of the eighth cranial nerve to the brainstem and to various parts of the body for reflex coordination. As a result, there are reflex movements of the subject's eyes as well as changes in the autonomic nervous system, plus the usual dizziness or giddiness which is associated with rapid rotation. In addition, following the termination of rotation, the person demonstrates characteristic errors in voluntary motion.

Nystagmus   If the subject is rotated to the right, during the acceleration phase there is a slow conjugate movement of the eyes to the left. This is followed by a rapid jerk back to the right, another slow movement to the left, and so forth. This rhythmic oscillation of the eyeballs is termed nystagmus. During acceleration, the slow phase is in the opposite direction to the rotation.

If rotation is continued at a steady rate, nystagmus ceases. When rotation decelerates, the nystagmus once again occurs, but this time the quick phase is in the opposite direction. These facts underline the mechanism of semicircular canal reaction, not to movement as movement, but rather to movement during acceleration and deceleration. Change in velocity activates these receptors, not velocity as such. This is because the crista is activated by being bent by the movement of the endolymph. The endolymph moves only during acceleration and deceleration.

Postrotational Errors of Movement   If a subject who has just been rotated attempts to stand up, he will invariably fall to one side. This is not a reflex act but an error in volitional movement. The subject's equilibratory mechanism has been disturbed by the rotation. Consequently, when he stands erect, he is under the impression that he is falling to one side when actually he is not. Therefore, he throws himself in the opposite direction in an attempt to compensate for the imagined movement.

It is precisely for this reason that pilots are instructed not to trust the "seat of their pants" when flying, especially in combat maneuvers which involve rapid changes in direction and acceleration. The resulting disturbances to the equilibratory mechanisms may prove fatal unless the pilot relies on his instruments more than his sensations.

Reactions to Change in Position   The utricle is stimulated by any alteration in the position of the body. The receptor in the utricle is the neuroepithelium covered by an otolithic membrane. The otoliths give weight to the hair cells so that the entire mass is influenced by gravity and, accordingly, moves with any deviation of the body. This movement of the receptor in the utricle initiates impulses which are conducted to the medulla and then down the spinal cord to synapse with the anterior horn cells. By virtue of this arrangement, volitional and reflex movements are modified. In this way a deviation of the body from the upright position will give rise to appropriate responses which oppose that particular movement.

Tests of Vestibular Function

The normal individual can maintain his upright position with his eyes closed. If a writing instrument is attached vertically to his head and a horizontal paper pad lowered until it comes into contact with the writer, a record is easily obtained of the small movements one makes in maintaining equilibrium with the eyes closed. If the utricles are not functioning, the patient may still be able to prevent falling to one side, but wavering from one side to another will be markedly increased. Of course, the myostatic reflexes play a major role in maintaining equilibrium; thus only in rare cases does the patient fall.

Robert Bárány, a Viennese ear specialist, received the Nobel prize in 1914 for his work on the physiology of the vestibular apparatus. He developed a test which involves rotating the subject in what is now referred to as a "Bárány chair." As already indicated, such rotation stimulates the semicircular canals, producing nystagmus and postrotary errors in movement. If the canals are not functioning, these results will not be seen. By placing the head in different positions during rotation, each of the three pairs of canals can be tested.

Bárány also showed that similar semicircular canal responses can be elicited by injection of warm or cold water into the external auditory meatus. The temperature differential causes currents in the endolymph, currents which activate the hair cells.

RETICULAR FORMATION

The reticular formation is a mass of gray matter that extends from the caudal end of the brainstem on up to the thalamus. Pathways from the basal ganglia, the cerebellum, various brainstem nuclei, and the spinal cord feed impulses into the reticular formation. The reticular formation, in addition, initiates pathways that go to these various structures.

The reticular formation, then, is a complex network of afferent and efferent neurons. In the reticular formation there are important centers for the regulation of circulation, respiration, and gastrointestinal activities. These functions will be outlined in their respective sections.

Motor Functions of the Reticular Formation

When the cephalic part of the reticular formation is stimulated, both discrete and more generalized movements are elicited. This does not mean that the cephalic area initiates movement but rather that it participates in movement which is initiated at higher levels. More importantly, stimulation experiments disclose that certain parts of the reticular formation inhibit, and other areas facilitate, movement. Thus, the reticular formation probably coordinates at least the grosser types of movement which involve large groups of muscles.

Reticular Activating System

All the sensory pathways, as they traverse the brainstem, send collaterals into the reticular formation. This input is utilized by the reticular formation in its role of coordinating motor activity. From the reticular formation there is a diffuse projection of impulses up to the thalamus, to the basal ganglia, and to many parts of the cerebral cortex. As a result, these higher centers are activated. In the absence of this reticular function, wakefulness is not maintained. Clinically this is seen in patients with tumors or hemorrhage involving the mesencephalic portion of the reticular formation. If there is severe damage to the reticular formation, the patient remains in a state of coma and cannot be awakened. Apparently, under normal circumstances, all sensory activity is funneled into the reticular formation, which then fires nonspecifically into higher centers to activate them.

CEREBELLUM

The cerebellum is a relatively large structure which lies just behind the larger cerebral hemispheres (Figure 9-2). In lower animals it is small and comparatively inconspicuous and apparently not extremely important to the well-being of the organism. But in man it is very prominent and highly essential for normal movement.

Anatomy of the Cerebellum

The cerebellum consists of a central, unpaired structure called the vermis and two lateral masses referred to as hemispheres. If a sagittal section of the cerebellum is prepared, the vermis is seen to be divided, by deep fissures, into 10 lobules. These lobules, along with subdivisions of the lateral hemispheres, may be grouped into the anterior, posterior, and flocculonodular lobes (Figure 14-2).

The cerebellum receives a barrage of impulses from many parts of the nervous system via afferent pathways. In turn, it fires impulses to various areas of the nervous system by means of efferent pathways. The fiber bundles linking the cerebellum with the brainstem are the cerebellar peduncles, of which there are three (superior, middle, and inferior) on each side of the cerebellum.

Afferent Pathways   The corticopontocerebellar system consists of neurons from areas of the cortex to nuclei of the pons, where they synapse on neurons extending to the cortex of the cerebellum (Figure 14-3).

Impulses from the brainstem pass to the cerebellum via the olivocerebellar tract, the vestibulocerebellar tract, and the reticulocerebellar tract. The fibers in the olivocerebellar

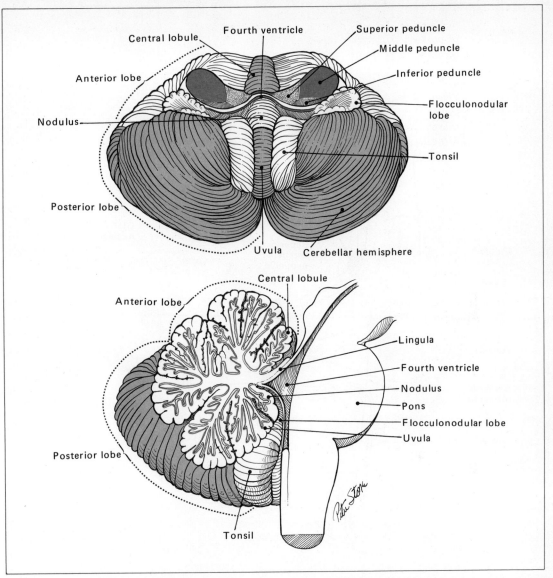

Figure 14-2. Cerebellum. At the top is the anterior aspect, normally hidden from view behind the brainstem. The cerebellar peduncles are shown severed. Below is a sagittal section.

tract originate in the inferior olive and radiate to many parts of the cerebellum. The vestibulocerebellar neurons come from the vestibular nuclei and from the vestibular apparatus and terminate in the flocculonodular lobe. The reticulocerebellar fibers carry impulses from the reticular formation to the cerebellum.

Although the cerebellum does not function for the awareness of sensation as do the sensory areas of the cerebral cortex, it does, nevertheless, receive impulses initiated by

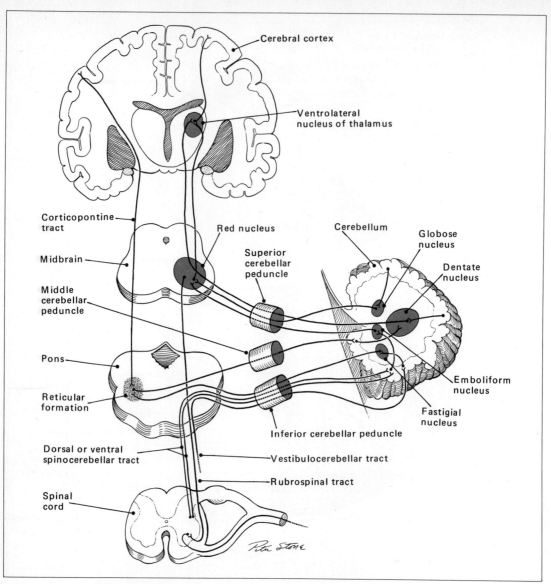

Figure 14-3. Schematic diagram of the chief cerebellar pathways. In addition to those shown there is the reticulospinal tract extending from the reticular formation into the spinal cord, where its fibers terminate on the anterior horn cell bodies which are the origin of the peripheral motor neurons.

peripheral receptors, namely, the muscle spindles and the tendon stretch receptors. The afferent pathways which propagate the impulses from these receptors to the cerebellum are the ventral and dorsal spinocerebellar tracts. The neurons in these tracts are very large and propagate impulses more rapidly than do any other neurons in the body. The pathways terminate in the cerebellum in an orderly arrangement representing various areas of the body, just as the sensory fibers have a point-to-point relationship on the sensory cortex of the cerebral cortex.

In Chapter 10 mention was made of proprioceptive impulses being transmitted along the posterior funiculi by fibers that terminate in the medulla. From these termination points, secondary fibers ascend to the cerebellum. Finally, there are afferent fibers in the spinothalamic tract which terminate in the reticular formation of the brainstem, from where secondary fibers pass to the cerebellum.

In short, the cerebellum is so "wired" that it receives information from the motor cortex, various nuclei in the brainstem, and peripheral receptors.

Efferent Pathways   Impulses originating in the cerebellar cortex pass via short fibers to one of four nuclei in the cerebellum which lie just below the cortex. They are the dentate, globose, emboliform, and fastigial nuclei.

Fibers from the dentate, globose, and emboliform nuclei pass to the red nucleus and the ventrolateral nucleus of the thalamus, from which there are connections to the cerebral cortex (Figure 14-3). Fibers from the dentate and fastigial nuclei go to the reticular formation. Neurons from the reticular formation pass in the reticulospinal tract to terminate on the anterior horn cell bodies which give rise to peripheral motor neurons.

Function of the Cerebellum

The flocculonodular lobe functions in connection with the vestibular mechanism. Accordingly, this lobe is essential to equilibration. The vestibular apparatus feeds information concerning the position of the head in space and its rotation into the flocculonodular lobe. This lobe, in turn, through its connections with descending motor pathways, is able to coordinate the information with movement and thus maintain equilibrium even during highly complex and violent movements. The importance of this lobe is emphasized when it is destroyed. Persons whose flocculonodular lobe has been damaged or destroyed have difficulty standing erect. When they walk, they do so with their legs wide apart (broad base), and still they tend to stagger and to fall.

The anterior and posterior lobes are "wired," as already described, in a manner that permits them to receive impulses from the motor cortex and, at the same time, from the contracting muscles and from the joints. Impulses from the visual and auditory receptors also reach these cerebellar lobes. On the efferent side, the cerebellum can send impulses to the motor cortex and to the lower motor neurons. It is, therefore, in a superb position to coordinate motor function in the light of sensory data, and that is exactly what it does. Each time the motor cortex sends out a stream of impulses to initiate movement, the cerebellum is immediately apprised of this activity. At that same instant it has information from visual, auditory, and proprioceptive receptors. It can now modulate the motor command of the cortex by a direct action on the lower motor neurons as well as by firing back to the motor cortex. In addition, as the movement proceeds, there is a continuous feedback from vision, from the proprioceptors, and from the auditory receptors. All this information is utilized to adjust, refine, and perfect the movement. In short, the cerebellum functions as a high-speed computer.

One need consider only a few everyday activities to understand why this high-speed computer is essential. Analyze the act of reaching for a water glass. By vision it is located. Then the arm must move forward, the hand open and, at the precise moment, the hand close around the glass, with just the appropriate strength to hold it. The sense of vision and the motor cortex in the absence of the cerebellum could never accomplish this act. Many muscles are involved, flexors and extensors. Clearly they must be coordinated. As the arm comes forward, it must start, accelerate, decelerate, and then stop. During this time the hand is opening just the right amount and then closing around the glass at the appropriate time. Engineers have designed equipment that can duplicate this act and far more complex ones, and all such equipment operates by virtue of a mechanism that evaluates and feeds back information to minimize error and to make the act smooth and precise. The cerebellum does that for human beings. Little wonder that the spinocerebellar tract contains fibers that can propagate impulses at velocities up to 130 m/sec. Such high-speed transmission is essential if the feedback of information is to arrive in time to effect error control.

Finally, to emphasize the function and importance of the cerebellum still further, a more complex act will be briefly mentioned. Consider the complexity of driving a car in a situation in which another car is to be passed on a road while a third car is some distance ahead, coming toward one. Information input must be rapid and it must be accurate. One needs to evaluate his own speed, that of the car he is about to pass, the speed of the car coming toward him, and the ability of his own car to accelerate under the particular set of circumstances existent at the moment. Then he must act on that information, information that comes from vision, from experience, from various movements. Fortunately, the cerebellum is indeed a very high-speed computer.

Abnormalities of Cerebellar Function

The loss of equilibration due to involvement of the flocculonodular lobe has already been mentioned. Lesions that disrupt other areas of the cerebellum cause ataxia, that is, lack of muscular coordination. More specifically, ataxia is movement characterized by abnormal rate, force, direction, or range.

Dysmetria In this disorder there is difficulty in gauging the extent of muscular movements. If the examiner holds his finger pointed to the patient and instructs him to touch it with his fingertip, the patient is unable to accomplish this seemingly simple task. He will point all about it or past it.

Intention Tremor Dysmetria is usually accompanied by a so-called intention, or terminal, tremor. Thus, not only does the patient find it difficult to point precisely, but as his finger approaches the target, it begins to shake. The tremor occurs only when there is great intention on hitting the mark, that is, at the end or termination of the act. Thus, if a seamstress attempts to thread a needle, as the thread is brought closer to the eye of the needle, her hand begins to shake, making threading difficult or impossible.

Muscle Tonus In view of the fact that the cerebellum can both inhibit and facilitate the final common pathway, that is, the neuron that innervates muscle, some cases of cerebellar disorder result in hypertonicity and others in hypotonicity.

Adiadochokinesis This term means "loss of successive movements." It thus aptly describes the deficiency noted in many patients with cerebellar damage. A normal person can pronate and supinate his wrist, that is, rotate it one way and then the other, with great rapidity. Cerebellar dysfunction often makes such rapid change of motion impossible. This is simply a manifestation of inadequate feedback essential for braking action in one direction in order to start action in the opposite direction.

Rebound Phenomenon If a normal person pushes against an examiner's hand and the hand is suddenly withdrawn, the subject's arm will jump forward just a few inches before he can successfully inhibit the motion. If a patient with cerebellar dysfunction is tested in the same way, his arm will move forward unchecked. In severe cases the forward movement of the arm, combined with loss of general coordination, suffices to throw the patient off balance.

Decomposition of Movement   The cerebellar patient executes movements like a robot, a poorly engineered robot. Every movement is broken down into its individual units. There is decomposition of the usually smooth and efficient pattern of movement. This result of cerebellar dysfunction dramatically illustrates the important role of the cerebellum in coordination.

BASAL GANGLIA

The basal ganglia, also called the subcortical nuclei, are extremely important centers of coordination. The basal ganglia include the caudate, putamen, and globus pallidus nuclei (Figures 9-5 and 9-7). Other centers, such as the substantia nigra, the red nuclei, and the nuclei of Luys, are closely associated with the basal ganglia. The putamen and globus pallidus together are referred to as the lenticular nucleus. The caudate and lenticular nuclei together with the internal capsule which separates them constitute the corpus striatum.

Function of Basal Ganglia

The clue to the function of the basal ganglia is provided by experiments in which the various nuclei are stimulated. If no movement is taking place, stimulation of the ganglia is usually without demonstrable effect. However, if movement is first elicited, for example, by stimulating the motor cortex, activation of the basal ganglia promptly inhibits that movement. From this and other lines of evidence, the concept has developed that the major function of these ganglia is inhibition. The general activation or facilitation resulting from motor and premotor cortical activation can thus be appropriately inhibited by the basal ganglia so as to bring about orderly, smooth, and purposeful muscular movement. Disorders of the cerebellum result in ataxia; disorders of the basal ganglia, in involuntary, uncontrollable, purposeless movements. In both instances, the movement is probably initiated by the motor cortex.

Abnormalities of Basal Ganglia

Lesions involving the basal ganglia produce several disorders, all characterized by involuntary, uncontrollable, purposeless movements.

Tremor   Tremor consists of involuntary rhythmic movements. The tremor which results from basal ganglion disorder is different from that resulting from a cerebellar abnormality. In the latter instance the tremor occurs only when a specific movement is attempted. In the former, the tremor persists at all times during the waking state. Interestingly, during voluntary movement it may be lessened or even disappear.

Parkinsonism   This disorder, also known as Parkinson's disease, is characterized by a persistent tremor, considerable hypertonia, and so-called pill-rolling movements. The patient is usually bent, he moves with great difficulty, his hands and arms shake, and his fingers carry out a continuous pill-rolling motion. All these findings are explicable on the failure of the basal ganglia, because of a lesion, to damp or suppress facilitation from the cortex and other areas. The globus pallidus and substantia nigra are most often involved.

Some success has been experienced in recent years in the treatment of Parkinsonism.

One procedure involves the surgical destruction of portions of the basal ganglia or the ventrolateral nucleus of the thalamus. As a result the characteristic rigidity and tremor disappear. Just why is not clear. In many cases, the administration of L-dopa ameliorates many of the signs and symptoms of the disease. Here too the rationale for the treatment remains to be established, but the thinking is that a derivative of L-dopa, dopamine, is a transmitter required by certain basal ganglia.

Chorea   In this disorder uncontrolled movements occur at random. The patient appears to be undertaking a peculiar dance; hence the sometimes-used term "St. Vitus' dance." In fact, the word "chorea" is derived from the Greek word meaning "dance." Upon autopsy, patients who have had this disorder are found to have a widespread lesion in the corpus striatum.

Athetosis   The term athetosis means "without position or place" and accurately connotes the disorder, for in fact the limbs and even the trunk of persons with athetosis are seldom at rest. There are constant, slow, sinuous, and tortuous gyrations. These movements cannot be controlled, and only during sleep do they subside. Lesions in the globus pallidus appear to be responsible.

Hemiballismus   This disorder is characterized by sudden movements affecting one side of the body and generally involving an arm or leg. These movements are irregular, occur without warning, and are usually quite forceful, even violent. Autopsy of patients who have had this disorder generally reveals a lesion involving the nuclei of Luys.

CEREBRAL CORTEX

The areas of the cerebral cortex thus far studied, that is, the sensory and motor regions and the areas essential to the special senses, occupy a relatively small portion of the total cerebral mantle. Much of the remainder constitutes what are known as association areas. These areas surround all the primary sensory areas, and a large association area extends forward from the motor cortex. They are apparently essential to coordinate incoming sensory data, and they probably play a role in influencing motor function in the light of sensory data.

Prefrontal Areas

In front of area 6 and extending forward the entire remaining distance of the cerebral cortex are the prefrontal areas. These areas of the cerebral cortex have been designated by numbers from 9 through 13 (see Figure 10-8). A lesion in this part of the brain produces remarkable personality changes. There does not seem to be any loss of intelligence, but the person responds to various situations without inhibition and with almost complete euphoria. There is usually loss of drive or ambition, and one who formerly could evaluate his environment and then act with good judgment now responds apparently without thinking, quite rapidly, and with little or no evaluation of the consequences. Because of these changes the prefrontal areas are supposed to have somewhat of an inhibitory function over other areas of the brain. In other words, the prefrontal areas somehow cause one to pause before responding, to evaluate all the factors and consequences, to plan his response, his course of action. In some people this inhibitory function is so exaggerated that they cannot reach a decision at all. They become incapable of responding. They are, in other words, abnormally depressed. The sum total of one's experience seems to be stored in the prefrontal areas so that all action is undertaken in the light of that experience.

When the areas are destroyed, the person responds, as a child would, quickly, happily, directly, but without judgment, without what is generally termed foresight. This is not to say that the prefrontal areas are the seat of intelligence, but to manifest whatever intelligence may exist is certainly a function of the region. The word "personality" better expresses the consequence of prefrontal-area activity.

Before the development of tranquilizers and other personality-altering drugs, an operation known as prefrontal lobotomy gained prominence for the treatment of intractable depression. The results were dramatic, but the new personality often had as many undesirable features as did the condition for which surgery was performed. This, plus the irreversible nature of the treatment and the development of various drugs, brought an end to this form of surgery.

Speech

Speech is an act that requires incredible coordination of a vast number of sensory and motor functions. Broca's area, area 44 (see Figure 10-8) coordinates the muscles that take part in articulate speech. If this area is destroyed, the muscles used for speech are not paralyzed. They can still be volitionally contracted, but they cannot be contracted with sufficient coordination to produce speech. Another part of the cerebral cortex, known as Wernicke's area, is also involved in speech. Wernicke's area lies between the auditory area and the angular gyrus which acts as a way station between the auditory and visual regions. Connecting Wernicke's and Broca's areas is the arcuate fasciculus. There is thus a large part of the cerebral cortex which coordinates visual, sensory, and motor function for speech. Damage to Broca's area results in slow and labored speech, but comprehension of language remains intact. Damage to Wernicke's area produces speech that has little content or comprehension. The patient speaks very rapidly and has good articulation, but the sentences come out sounding much like "double-talk." In short, Broca's area coordinates articulation of speech, whereas Wernicke's area is concerned with the intellectual content of the speech.

Although there are similar areas in both the left and right cerebral hemispheres, the primary language areas of the human brain are, in most individuals, in the left hemisphere, as shown by the fact that damage to the right side of the brain rarely causes language disorders as does damage to the left side. This one-sided dominance probably accounts for the speech problems that sometimes develop when a person who normally writes with one hand is forced to learn to write with the other hand.

The term aphasia means "loss of speech." This loss or inability may be due to a wide variety of conditions. If it is due to a failure of coordination, as results from disruption of Broca's area, the term motor aphasia is used. A lesion in the sensory association areas may produce inability to interpret a word that is seen or heard. This may result in an inability to speak, that is, sensory aphasia. There is also a condition in which there is no impairment on the motor or sensory side, but the patient simply cannot find words to express his thoughts. He understands the words, he can usually write them, but he cannot use them in speech. It is as though he had forgotten what he wanted to say; thus the term amnesic ("to forget") aphasia is used for this condition.

Limbic System

The limbic system consists of a part of the cerebral cortex called the limbic cortex and several subcortical structures, most importantly the hypothalamus. In addition, this system includes the hippocampus, portions of the basal ganglia, the anterior nuclei of the thalamus, the amygdala nucleus, and the paraolfactory area (Figure 14-4).

These structures would appear to be primarily concerned with the affective nature

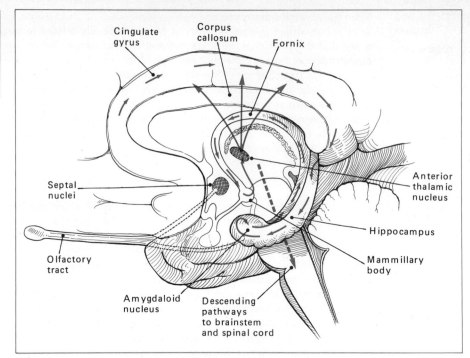

Figure 14-4. Limbic system. The direction in which impulses flow is indicated by arrows.

of sensory sensations, that is, whether such sensations are pleasant or unpleasant. These reactions to sensation are referred to as reward and punishment.

The experimental procedure that has contributed most to knowledge of the limbic system involves self-stimulation of individual parts of the system. Electrodes are precisely placed in the area to be studied and then wired in such a way that the experimental animal, by pressing a lever, can cause stimulation. Clearly if such stimulation gives rise to a pleasant sensation, the animal will continue to press the lever, whereas if stimulation produces an unpleasant sensation or pain, the animal will exhibit signs of pain and displeasure and will not repeat the experience.

This procedure strongly suggests that the hypothalamus is the most important part of the limbic system and the responses to stimulation of other parts, in many instances, are funneled through it. Stimulation of the ventromedial nuclei of the hypothalamus causes a pleasurable response, whereas stimulation of the perifornical nucleus of the hypothalamus gives rise to intense pain.

Brain Waves

So-called brain waves represent electrical manifestations of the activity of the cerebral cortex. This electrical activity is of very low voltage. It can be sufficiently amplified to be recorded by an instrument termed the electroencephalograph. The record so obtained is called the electroencephalogram, or EEG. Figure 14-5 shows a normal EEG that was recorded from the indicated areas of the cranium. Usually brain waves are classified into three categories: (1) alpha waves, which predominate and which have frequencies of about

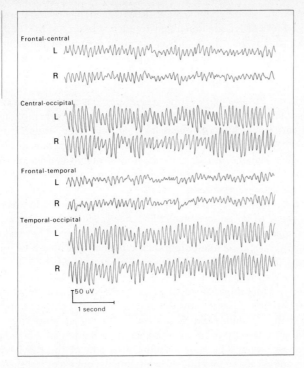

Frontal-central

L

R

Central-occipital

L

R

Frontal-temporal

L

R

Temporal-occipital

L

R

50 uV

1 second

Figure 14-5.
Normal electroencephalogram
recorded from different areas of
the cranium. These are alpha
waves.

6 to 13 cycles per second (cps) and an amplitude of about 50 microvolts; (2) beta waves with frequencies higher than 13 cps and a lower voltage; and (3) delta waves with frequencies lower than 6 cps but with high amplitude.

Brain-wave patterns vary during various activities. During sleep, for example, the frequency becomes very slow, but the amplitude—that is, the voltage—increases. On the other hand, low-amplitude, high-frequency waves are characteristic of intense mental activity.

Cortical electrical activity is generated by the most superficial layers. The waves do not come from the deeper layers that contain the larger motor cell bodies. If they did, a regular rhythm would not be expected. There is good evidence that the EEG has its origin in the mass of dendrites that form a dense network in the superficial cortical layers. Such a mass of dendrites also exists in the superficial layers of the cerebellar cortex, from which a similar pattern can be recorded. In order for a regular pattern to exist, there must be synchrony of firing. Just what causes this synchrony is the question. The regular, synchronous input from the thalamus is thought to be at least partially responsible.

The EEG is used clinically to aid in the diagnosis of brain disorders. There are characteristic changes in the EEG associated with epileptic attacks, with lesions, and with subdural hematomas.

Sleep

In both sleep and unconsciousness the individual is unaware of his surroundings, but sleep is a normal, recurring state from which one can be aroused with relative ease.

There are two states of sleep: (1) light sleep, or slow-wave sleep characterized by reduced electrical activity in the cerebral cortex and subcortical structures, and (2) deep sleep, fast-wave sleep, rapid-eye movement (REM) sleep, or dreaming sleep. Deep sleep is characterized by loss of muscle tone, high-frequency brain waves, and REMs. The fact that there are REMs and increased cortical electrical activity at the same time that the muscles become completely relaxed would appear to be a paradox, and thus deep sleep is often referred to as paradoxical sleep.

Mechanism of Sleep Many parts of the central nervous system are involved in sleep and wakefulness. The thalamus is thought to project a basic, synchronous rate of firing to the cerebral cortex which causes or at least permits sleep. When there is adequate sensory input, collaterals going to the reticular formation activate this area, which, in turn, fires to higher centers, bringing about arousal and wakefulness. Although these observations have been confirmed, there is now additional evidence to suggest that various chemical substances play major roles.

Serotonin is thought to be involved in the production of light sleep, whereas the catecholamines seem to be important to REM sleep. Most of the serotonin-containing neurons appear to be located in the so-called raphe system. The catecholamines are produced in the lateral part of the bulbopontine tegmentum. Just how these substances function and how they are controlled remain to be elucidated.

Duration of Sleep How much sleep an individual requires has never been accurately determined. It apparently varies greatly. All that can be said is that infants seem to require far more sleep than do adults, and their sleep has a higher proportion of REM sleep to light sleep. Infants start out in life sleeping some 18 hr/day, of which half is the REM type. Adults average about 7 hr with no more than 2 hr of the REM type. Inadequate sleep causes physiological and psychological alterations, but excessive sleep would appear to have no benefit.

Circadian Rhythms

The term circadian is derived from *circa* meaning "about" and *dies* meaning "day." The word thus implies a rhythm that repeats approximately each 24 hr. One such rhythm has just been discussed, namely, sleep-wakefulness. In certain animals there is a yearly rhythm of an aroused state and hibernation. In women the menstrual cycle has a 28-day cycle. But most biological rhythms have a 24 hr basis (Figure 14-6). Living organisms, because of these rhythms, are said to possess "physiological clocks."

Control of Circadian Rhythms The most obvious explanation for circadian rhythms would seem to be a rhythm in the environment such as light and dark. There are experiments which demonstrate the influence of light and darkness on various physiological functions, but the weight of evidence shows that the basic circadian rhythms are not imposed on the organism by the environment but are truly endogenous. The strongest evidence is derived from experiments in which animals are kept either in continuous light or continuous dark. In both instances the rhythms continue; however, the periodicity of the cycle is rarely exactly 24 hr even when the light in the environment is kept at 12 hr light and 12 hr dark. These observations suggest an internal "clock" that regulates the rhythms.

But under normal circumstances, the environment does influence the rhythms; that is, it alters the basic endogenous clock. The term entrained is used to indicate that a particular rhythm has been synchronized with some environmental factor. An environmental factor that entrains a physiological rhythm is called a *Zeitgeber,* a German term implying a giver of time. A rhythm that is not under the influence of a *Zeitgeber* is said to be free-running.

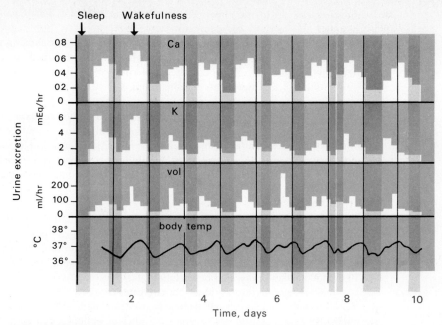

Figure 14-6. Circadian rhythms. Note how the physiological functions depicted vary in rhythm with the 24-hr schedule. (*Reproduced with the kind permission of Professor Jürgen Aschoff.*)

Practical Significance of Circadian Rhythms   The recognition of circadian rhythms is becoming an ever greater factor in the planning of human activities. The jet age has made it possible to move through time zones with great rapidity, thereby altering the influence of environment on endogenous rhythms. Entrainment to local time takes at least 2 or 3 days, and during this time mental and physical performance is below par.

Learning and Memory

Learning and memory are inseparably related. Learning is the acquisition of knowledge; memory involves the retention of that knowledge. In the ensuing discussion, the term memory will be used to imply both factors.

Types of Memory   There are two types of memory: (1) short-term and (2) long-term. If one is confronted with information for but a brief time span, he can generally immediately recall that information, but that recall fades rapidly. The actual length of time depends upon the nature of information, distraction by other types of new information, interest, and so forth. Long-term memory refers to the ability to recall information days, weeks, and years later. Short-term memory gives rise to neuronal activity, whereas long-term memory involves structural changes in the brain.

Chemical Basis of Memory   The claim has been made that transfer of learning has been accomplished through the extraction of nucleic acid from the brains of trained animals and the subsequent injection into so-called naïve animals. The authors suggest that there is a specific nucleic acid associated with learning, that it can be extracted, and that when injected into an untrained animal, that animal then responds as though it had learned the particular information involved. This report is still being debated and is not generally accepted.

The chemical basis of memory postulates that memory consists of changes in the base sequence of neural RNA, these changes being induced directly by electrical impulses impinging on the neuron. The altered RNA molecules then, in some unspecified way, modify the functioning of the neuron, giving rise to the phenomenon of memory.

Localization of Memory The site of memory continues to baffle investigators. The prefrontal areas seem to be involved in short-term memory, and certainly other parts of the cortex are essential for the intake of information, for example, the visual, auditory, gustatory, and somatic sensory areas. In addition, a recent report implicates the ventral reticular formation, the posterior thalamus, and part of the hippocampus.

The Commissures

As we have seen, the brain consists of two hemispheres bound together and bridged by the commissures. The largest, by far, is the great cerebral commissure, termed the corpus callosum (see Figure 9-7). Others are the hippocampal commissure, the habenular commissure, the posterior commissure, the anterior commissure, and the massa intermedia. The optic chiasma may also be included.

Interestingly, if these bridges are cut, after recovery the animal appears to function quite normally unless special testing is used. In man, the corpus callosum and other commissures are sometimes severed for relief of epileptic convulsions. Again, without special test procedures, it is impossible to note any change in the patient other than dramatic relief from the convulsions. With various testing methods, however, it can be demonstrated that the nerve fibers which make up the interhemispheral bridges transmit information from one-half of the brain to the other. As already noted, the left half of the brain is concerned primarily with the right side of the body, and the right half with the left side. If the commissures are cut, the experimental animal can be taught to respond, say with its left hand. The normal animal could then give the same response with the right hand, but following section of the commissures this is impossible. In short, the information learned by one side of the brain is available to the other side only if the commissures are intact.

In man, one hemisphere is dominant. In right-handed individuals it is the left hemisphere. By virtue of the intact commissures the right hemisphere can take over. If the commissures are cut, each half functions independently. It is as though there were two independent brains. Clearly, this could lead to interesting conflicts, and the suggestion has been made that such conflict between the two hemispheres may underlie certain mental illnesses.

QUESTIONS AND PROBLEMS

1 Locate, describe, and give the function of the otolithic membrane and the crista ampullaris.
2 Where do impulses conducted by the primary and secondary sensory neurons of the vestibular apparatus originate and terminate, and what related centers are ultimately stimulated?
3 Explain what is meant by the dual role of the vestibular apparatus.
4 Why are there three semicircular canals asso-

ciated with each vestibular apparatus? Why not two or one?
5 Explain why pilots are instructed *not* to trust the "seat of their pants" when flying combat maneuvers.
6 Describe the events that occur in your nervous and muscular systems when you are first thrown off balance and then quickly "right" yourself.

7 What is the reticular formation, and what is its relationship to discrete and generalized movements of the body?

8 List the afferent pathways that terminate in the cerebellum, and name their points of origin.

9 Select a simple activity which you perform daily, and describe how your cerebellum functions as a coordinating center and a high-speed computer.

10 What symptoms are evidenced when ataxia is present? Dysmetria? Intention tremor?

11 What are the basal ganglia, and why are they classified as important centers of coordination?

12 How does area 44 of the cerebral cortex function as a center of coordination?

13 What is the relationship of the hypothalamus to emotions?

14 Explain the clinical use and value of electro-encephalography.

15 What is the effect of sectioning the cerebral commissures?

PART 3
THE CIRCULATORY SYSTEM

15 THE BLOOD

In an average-sized person there are about 5,000 ml of blood in the circulation. This blood volume is propelled through the vessels by the pumping action of the heart. The blood receives its essential supply of oxygen in the lungs. It collects foodstuffs from the alimentary tract. Other vital substances are manufactured by specially designed cells and secreted into the circulating blood for transportation to other parts of the human body. Finally, undesirable materials are extracted from the blood by the kidneys, the skin, the lungs, and the liver. All these mechanisms will be considered in detail in subsequent chapters, but first the blood will be discussed.

THE COMPOSITION OF THE BLOOD

Blood may be considered an atypical, specialized type of connective tissue. The fluid portion, or plasma, of blood represents the ground substance of this tissue; the fibrous portion appears only during clotting, and the cellular components are either red blood cells or white blood cells. In addition, two types of irregularly shaped particles are present: platelets, involved in blood clotting, and chylomicrons, which are tiny fat droplets.

Fresh blood appears brilliant red, thick, opaque, and to the naked eye homogeneous. It is about three to four times more viscid than water and has a specific gravity of about 1.055.

When whole blood is allowed to stand or is centrifuged, it separates into two distinct

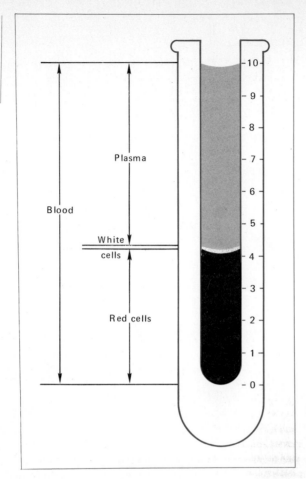

Figure 15-1. Determination of hematocrit. Whole blood is centrifuged in a test tube and allowed to settle. The hematocrit, that is, the volume of red cells per 100 ml of whole blood, is obtained by multiplying the number at the top of the red cell fraction by 10.

fractions. Figure 15-1 shows that after centrifugation less than half the tube is packed with the so-called formed elements, consisting of red blood cells, white blood cells, and platelets. The upper fraction is a clear straw-colored fluid called the plasma. The percentage of red cells, by volume, in whole blood after centrifugation is usually referred to as the hematocrit. In the normal male the red cells constitute about 45 percent of the whole-blood volume. In the normal female the hematocrit is approximately 42 percent.

If fibrinogen (see page 389) is removed from plasma either by clotting or by some other method, the remaining fluid is termed serum.

Red Blood Cells

The red blood cells, also referred to as erythrocytes, are shown in Figure 15-2. They are biconcave disks about 8 microns in diameter. The rim of the disk is thicker than the center. The thickness at the center is about 1 micron; at the rim, approximately 2 microns.

Erythrocytes are very flexible and elastic. They customarily dance through the blood vessels at great speed under high pressure and, as confirmed by slow-motion picture studies, are bent and twisted to a remarkable degree. On a slide, viewed through a light

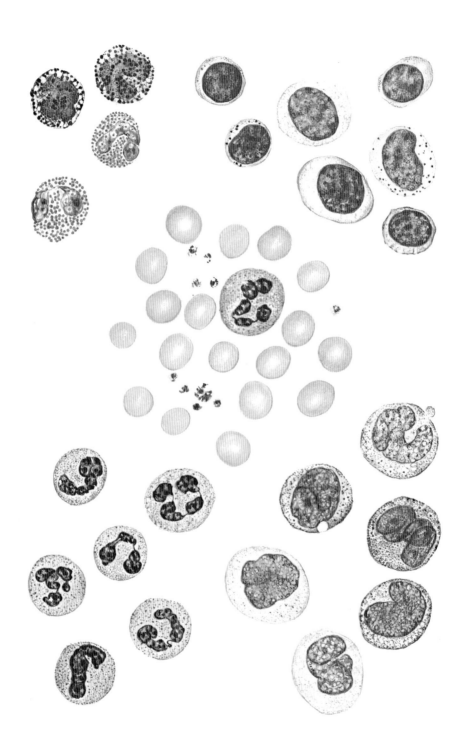

microscope, they are customarily seen sticking together, resembling stacks of coins. This is called rouleau formation. The red cell contains a nucleus only during its early formative period. When it is functioning in the bloodstream, the cell is without a nucleus.

Red Blood Cell Count Each cubic millimeter of blood contains, in man, about 5 million erythrocytes. As already mentioned, the normal blood volume is 5 liters. Thus in the total circulation there are approximately 25×10^{12} red blood cells (5,000,000 × 5,000 × 1,000).

Since the number of erythrocytes in the blood bears a direct relationship to health or disease, the physician very often needs to know that number. What he really obtains is the number per cubic millimeter. As just stated, the normal level is about 5 million. In some conditions it will be significantly less; in others, greatly increased. One might think that the determination of the number of red blood cells would be an extremely laborious task. Actually, it is quite simple. The older method utilizes an instrument called a hemocytometer. A hemocytometer contains a well that holds exactly 1 cu mm of diluted blood. The bottom of the well (Figure 15-3) is divided into squares by lines etched on the glass. The well is then viewed under a microscope, the cells counted in several squares,

Figure 15-2. Cells in circulating blood are shown on the color plate and identified below. (*R. O. Greep, "Histology," McGraw-Hill Book Company, New York, 1966.*)

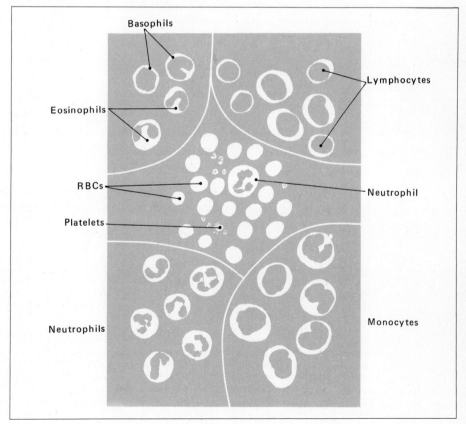

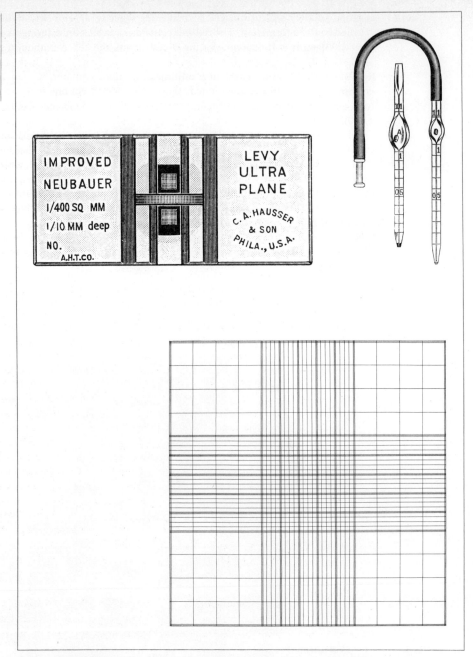

Figure 15-3. Equipment for blood cell counts. For red cell counts, blood is drawn into the center pipet, and a fluid added to achieve a standard dilution. A drop is placed in a hemocytometer counting chamber (upper left; enlarged at lower right). The center area, composed of 25 secondary squares, is examined under a microscope. Red cells in five of the secondary squares are counted, and a simple calculation gives the number of red cells per cubic millimeter of blood. The corner areas are for white cell counts.

and an average calculated, which is then multiplied by a factor that takes the degree of dilution into consideration. The more modern way of obtaining the red blood cell count is by means of an electronic instrument that forces the blood through a small opening, during which passage the red cells are automatically counted at high speed and great accuracy.

Function of Red Blood Cells The main function of the erythrocytes is to transport oxygen from the lungs to the cells. These cells have this unique ability because of their high hemoglobin content. The term hemoglobin is derived from *hemo,* which means "blood," and *globin,* which has reference to the protein nature of the substance. Hemoglobin is a complex protein-iron pigment. It is this blood protein, or pigment, which gives blood its characteristic red appearance. Hemoglobin readily unites with oxygen, and as will be learned when respiration is studied, it is essential for the transport of oxygen by the blood. When hemoglobin is carrying its full complement of oxygen, it is red. In the absence of oxygen, it turns somewhat blue. This explains why arterial blood is bright red and venous blood has a blue tinge.

Normal blood contains about 15 g of hemoglobin/100 ml. Thus the concentration of hemoglobin is given as the actual grams present per 100 ml of blood, or it may be reported as a percent of normal. For example, a patient may be found to have but 12 g of hemoglobin/100 ml of blood. This fact may be recorded that way, or it may be expressed as a hemoglobin of 80 percent ($12/15 \times 100$). Another way of expressing hemoglobin concentration is in terms of the red blood cells. The results are given as a color index. If a patient has 12 g of hemoglobin/100 ml of blood and 4.5 million erythrocytes/cu mm of blood, his color index would be 0.89 (80/90). The 80 represents the percentage of hemoglobin in relation to the norm ($12/15 \times 100$). And the 90 represents the percentage of erythrocytes in relation to the norm ($4.5/5 \times 100$). It should be clear, then, that when the color index is less than 1.0, each red blood cell must contain less than the normal amount of hemoglobin. On the other hand, a color index greater than 1.0 is possible. This means that each erythrocyte is carrying more than its share of hemoglobin. The color index is highly valuable in the analysis of anemia.

Bear in mind that it is the hemoglobin that carries the oxygen, not the red blood cell. One may then wonder why the red cells were brought into the design. Why not simply have the hemoglobin in the plasma? The answer may be that hemoglobin is a relatively small molecule, for a protein, and can pass through the wall of the capillary and thus be lost from the circulation. In other words, the red blood cells trap the hemoglobin and hold it in the circulation. Oxygen can easily diffuse through the plasma and the red cell membrane to come in contact with the hemoglobin.

Hemoglobin plays an additional role to its oxygen-carrying one. It is a good buffer and is primarily responsible for maintaining the acid-base balance in the blood (see Chapter 30).

When respiration is studied (Chapter 22), the transport of carbon dioxide will be seen to involve the reaction of this gas with water, a reaction which is catalyzed by the enzyme carbonic anhydrase. Carbonic anhydrase is concentrated in the red blood cells.

Red Blood Cell Formation In the embryo and fetus blood cells develop at sites throughout the organism, such as the mesenchyme, yolk sac, liver, spleen, thymus, lymph nodes, and bone marrow. After birth, blood cell formation normally occurs only in the bone marrow and lymph nodes. In certain blood diseases and following massive hemorrhage, however, blood formation may revert back to the liver and spleen as in the prenatal condition. At birth the cavities of all bones are filled with red bone marrow, which constitutes the main hemopoietic tissue. The red marrow of the long bones of the extremities is gradually replaced by fatty or yellow marrow, so that in the adult, the blood-forming red bone marrow is essentially limited to the bones of the skull, clavicles, vertebrae, sternum, ribs, and pelvis.

The most widely accepted hemopoietic theory states that all blood cell types are derived from a common primitive parent cell, the hemocytoblast. Structurally this cell resembles a large lymphocyte (see later), with considerable basophilic cytoplasm and a large nucleus containing one or more nucleoli.

Postnatally two types of hemopoietic tissue are recognized: (1) myeloid tissue (red bone marrow), which gives rise to red blood cells, granular leukocytes, and megakaryocytes (giant cells), and (2) lymphoid tissue, from which nongranular leukocytes (lymphocytes and monocytes) are formed.

In red blood cell formation the hemocytoblast undergoes a series of changes to develop into the mature erythrocyte found in the circulating blood. The initial cell of this series of transformations has deep blue (basophilic) cytoplasm and is hence called a basophilic erythroblast. This cell becomes reduced in size to form the polychromatophilic erythroblast, so named because its cytoplasm stains erratically, ranging from blue to purple-lilac to gray, due to the presence of hemoglobin, which begins to be produced in the cell at this stage. The developing red blood cell, at this point, is about half the size of the preceding cells and undergoes frequent mitosis. In further development the polychromatophilic erythroblast decreases further in size, the cytoplasm becomes more acidophilic as the hemoglobin content increases, and at this stage the cell is called a normoblast. As the hemoglobin content continues to increase, the nucleus becomes smaller, stains more darkly, and finally is extruded from the cell. The resulting nonnucleated cell is the functional, fully developed, highly acidophilic erythrocyte, or circulating red blood cell, which contains a maximal amount of hemoglobin in its cytoplasm.

Hemoglobin Formation   Hemoglobin is synthesized, starting in the erythroblasts, from acetic acid and glycine. The acetic acid is first converted to alpha-ketoglutaric acid, after which two molecules of this acid combine with glycine to form a pyrrole compound:

$$
\begin{array}{c}
\overset{\displaystyle |}{C}\text{——}\overset{\displaystyle |}{C} \\
\| \quad \| \\
H\text{—}C \qquad C\text{—}H \\
\diagdown \quad \diagup \\
N \\
| \\
H
\end{array}
$$

The pyrroles constitute the basic units of the hemoglobin molecule. Four pyrroles combine to form the heme molecule. This molecule contains one atom of iron. Four heme molecules combine with one molecule of globin to give rise to hemoglobin, which has a molecular weight of about 68,000. The important point to remember is that each molecule of hemoglobin contains four iron atoms. Oxygen is loosely bound to the iron; thus each molecule of hemoglobin can carry four molecules of oxygen. This relationship is considered in detail in Chapter 22.

Red Blood Cell Life-Span   In the bloodstream the red cells function for a period of time and then disintegrate. The average life-span of an erythrocyte is 120 days. The remains are removed from the blood by the liver and spleen. Just what causes the demise of the red blood cell is not known. The cell simply seems to grow old and become fragile, and then, probably while whirling through the circulation or filtering through sinusoids, it breaks apart. The load of hemoglobin is, of course, set free when the red cell breaks up. It then diffuses out of the plasma to be picked up by the reticuloendothelial cells (see Chapter 3), which reduce it first to a pigment called biliverdin and then to another pigment termed bilirubin. Bilirubin leaves the reticuloendothelial cells to be carried by the plasma to the liver, where it is removed to be excreted in the bile. It is this pigment that colors bile green-yellow. If bilirubin is not excreted adequately in the bile, it accumulates in the body, giving one a yellow color, a condition called jaundice.

The iron that is freed from hemoglobin in this process is available for the formation of new hemoglobin.

Obviously, for a steady concentration of red blood cells to be maintained, the rate of formation must keep up with the rate of disintegration. A simple calculation will disclose just how rapidly red cells must be produced. Red blood cells are produced at a rate sufficient to form 1,250 ml of blood per month. In 1,250 ml of blood there are normally 7.25×10^{12} erythrocytes. There are 2.6×10^6 sec in a month. This means that about 3 million red cells must be produced each second! As a matter of fact, this rate of production can be increased considerably, perhaps up to 10 times as fast. Certainly, after loss of blood, the rate of red cell production is stepped up markedly.

Control of Red Blood Cell Production The fact that the red blood cell count remains so constant and the fact that the loss of blood increases production suggest that there is a control mechanism sensitively attuned to the red blood cell concentration. In short, the constancy of the red cell count is a fine example of homeostasis. How does this mechanism work?

Many observations and a number of carefully controlled experiments have disclosed that whenever the quantity of oxygen being delivered to the tissues falls below normal, the rate of red blood cell production increases. A wide variety of conditions can decrease the delivery of oxygen. For example, heart failure, or clogging of blood vessels, will decrease the flow of blood, and thus the amount of oxygen delivered per unit time. If one goes to a high altitude where the partial pressure of oxygen in the atmosphere is diminished, the concentration of oxygen in the blood will fall, and the amount delivered to the tissues per unit time will decrease. The same end result follows various respiratory abnormalities. Finally, and significantly, the athlete undergoing prolonged, vigorous training increases his red blood cell count. In other words, any time the relationship between tissue cell oxygen demand and the supply of oxygen to tissue cells changes so that the demand exceeds the supply, red blood cell production is increased.

The homeostatic mechanism involved is a negative-feedback circuit (Figure 15-4). The diagram shows that when there is relative hypoxia, that is, the supply of oxygen is less than the demand, a substance called erythropoietin is produced by the kidneys. Erythropoietin is secreted into the blood, which carries it to the bone marrow. There it stimulates red blood cell production. As a result, the concentration of erythrocytes in the blood increases, more oxygen per unit time is transported, and the oxygen demand is satisfied; therefore, the secretion of erythropoietin decreases. The sequence is termed negative feedback because the feedback, in this case increased red blood cells, has a negative effect, the decrease in erythropoietin production.

The student can easily appreciate the simplicity, beauty, and effectiveness of this mechanism. It provides a means of constantly monitoring red blood cell concentration and a mechanism to regulate the rate of red blood cell production. The end result is a constancy of the red blood cell count, an increase when there is a need for such, and greater production when red blood cells have been lost.

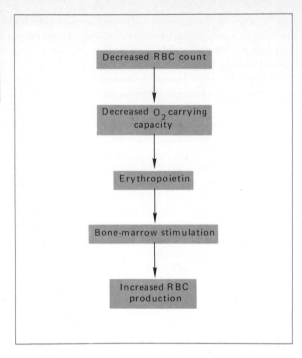

Figure 15-4. Homeostatic control of red cell production.

Red Blood Cell Abnormalities In some cases there are too few red blood cells, in others too many, and sometimes the count is normal but the cells themselves display abnormalities.

The term anemia means, literally, "a lack of blood." This is, of course, a misnomer. In most cases of anemia, there is no true lack of blood. The term is used to describe any condition in which the oxygen-carrying capacity of the blood is reduced, such as following hemorrhage, the inadequate production of hemoglobin or red blood cells, or the abnormal disintegration of erythrocytes. In each instance there is a decrease in the available hemoglobin. Since this is the agent responsible for the transport of oxygen, it logically follows that in all such cases there will result a reduced oxygen-carrying capacity of the blood.

Following loss of blood, by the mechanisms already outlined, red blood cells are rapidly formed; thus in a day or two following hemorrhage the red blood cell count returns to normal. If there is chronic loss of blood, for example in heavy, prolonged menstruation, there may be inadequate iron absorption from the intestine to form needed hemoglobin. Red cells will be formed, but they will contain inadequate hemoglobin. Thus the red blood cell count will be normal, but the hemoglobin concentration will be low. This condition is referred to as hypochromic anemia.

Aplastic anemia results from a malfunctioning bone marrow. The result is inadequate production of erythrocytes. As will be discussed later, the formation of red blood cells requires the so-called intrinsic factor which is produced by the stomach mucosa. If this factor is not produced or if the stomach is removed, so-called pernicious anemia results. In this respect, vitamin B_{12}, folic acid, is essential. A shortage of this vitamin also leads to anemia.

Sometimes the red blood cells are abnormal. They may be unusually fragile or assume odd shapes. In sickle-cell anemia, a disease that occurs almost exclusively in blacks, the red cells are sickle-shaped rather than biconcave disks, and they contain an abnormal type of hemoglobin. Instead of combining with oxygen, this hemoglobin precipitates, causing cell damage that leads to severe anemia.

The opposite condition, that is, a higher-than-normal red blood cell count, is termed polycythemia. A high red blood cell count does not necessarily indicate disease. As already mentioned, sojourn at high altitude or regular, vigorous physical activity results in increased erythrocyte production—a completely physiological response. On the other hand, excessive red blood cell production over and above the oxygen demands of the body is a serious condition called polycythemia vera. It generally results from a malignancy of the red bone marrow.

Anemia not only reduces the oxygen-carrying capacity but, by greatly reducing the viscosity of the blood, augments circulation, thereby increasing the work of the heart. The reason the number of cells in the blood affects blood viscosity is that the greater the number of cells, the more friction there will be between the layers of blood, and it is this friction that determines viscosity. As will be discussed later, blood does not flow as a single mass, but rather in layers. At the normal hematocrit of 42, the viscosity of blood as compared with water, to which a value of 1 is assigned, is about 3.5. At a hematocrit of 20 it is 2, and at a hematocrit of 60 it is about 5.5. In polycythemia, then, the blood is very viscous and therefore flows through blood vessels with great difficulty.

White Blood Cells

The white blood cells are also termed leukocytes (*leuko-*, "white"; *-cyte*, "cell"). There are various types of leukocytes which can be easily identified under the microscope. Examination of representative types of white cells shown in Figure 15-2 reveals that they differ in the shape of the nucleus and in the character and staining qualities of the cytoplasm. Leukocytes are all nucleated cells, void of hemoglobin in their cytoplasm, and may be classified into two main groups, the granular and nongranular cells. This classification is based upon routine methods in which the cells are stained to study their morphological characteristics. Three different granulocytes can be distinguished by the presence of characteristic cytoplasmic granules and the shape of the nucleus. The most prevalent granulocyte is the neutrophil, which makes up 65 to 75 percent of the circulating white blood cells. As its name implies, the cytoplasmic granules stain "neutral" to dyes, appearing a pale lilac instead of a definitive eosinophilic (red) or basophilic (purple) color. The cytoplasm of the neutrophil is filled with fine, somewhat obscure granules that take up the stain very poorly. The nucleus has three to five lobules connected together by delicate chromatin strands. The lobulated nucleus accounts for these cells also being referred to as polymorphonuclear leukocytes, or "polys."

The cytoplasm of the less numerous eosinophil (2 to 5 percent of circulating white cells) is filled with coarse, uniform, definitive eosinophilic (red-orange) granules. The nucleus is usually bilobed. The cytoplasm of the rare basophil (less than 0.5 percent of circulating cells) is packed with irregular basophilic (blue-purple) granules that overlie and partially obscure the bilobed nucleus of this cell. All granulocytes are about twice the diameter of a red blood cell, ranging from 12 to 15 microns.

The nongranular white blood cells, called agranulocytes, are lymphocytes and monocytes. Lymphocytes make up 25 to 30 percent of the circulating white blood cells; monocytes, 3 to 8 percent.

The lymphocyte has a relatively large, round, dark-staining nucleus encircled by a thin rim of clear blue cytoplasm. Lymphocytes are usually slightly larger than erythrocytes (8 microns) but may be considerably larger (12 microns).

The monocyte resembles a large lymphocyte except that it is usually larger (12 to 20 microns). It has a lighter-staining, often bean-shaped nucleus surrounded by considerable grayish-blue cytoplasm.

In the healthy adult the normal number of circulating white blood cells is about 7,000/cu mm of blood. This number is somewhat higher in children, and in the newborn is about 16,000.

Resistance to Infection   Because the major function of the white blood cells is to combat infection, a brief overview of the mechanisms available to the body for resistance to infection is presented first.

The first line of defense is by blocking bacteria, viruses, and other foreign agents from entering the body. There are so many possible entrances, however, that this is not a very effective defense. Infectious agents can enter any of the body orifices or through breaks in the skin, or they can be introduced in other ways, for example, by injection. The second line of defense is phagocytosis. Certain cells engulf the foreign agent and then destroy it. The third defense mechanism is provided by antibodies.

The so-called reticuloendothelial system consists of phagocytic cells as well as cells that give rise to antibodies. Many different types of cells are involved. Reticulum cells are basic cells that can become differentiated into hemocytoblasts, lymphocytes, plasma cells, and other types essential to combating infection.

Sometimes the infectious agent enters directly into the bloodstream. But under other circumstances it first gets into the tissue spaces. Most of these agents are too large to pass through the blood capillary wall; instead either they are destroyed by tissue histiocytes, or they are swept up by the lymph to enter the lymphatic system (see Chapter 18). Once in this system, they are filtered out by the lymph nodes, which contain many reticulum cells. These cells phagocytize the invading agent and thus prevent its entrance into the blood. If the agents do get into the blood, then the various phagocytic cells that line the vascular vessels or circulate in the blood take over. In addition the sinuses of the liver and of the spleen serve to filter these large molecules out of the blood. In the liver there are Kupffer cells, which are also reticulum cells. Kupffer cells and other types of reticulum cells in the spleen function to destroy the invader. The blood also passes through the bone marrow, which not only produces blood cells but also phagocytizes foreign agents.

The body contains several substances that combat infectious agents in general, and, in addition, specific substances can be produced to battle a particular invading agent. For example, the blood contains interferon, which inactivates certain viruses, and lysozyme, an enzyme that destroys bacteria. But the so-called immune reaction involves the production of a specific substance as a result of the invasion of the body by a particular agent. The terminology is as follows: The foreign agent is called the antigen; it stimulates the production of the combating substance, the antibody. When the antigen enters the body, in some manner it stimulates plasmoblasts to form plasma cells. As these cells are developing, they produce gamma globulin antibodies. These antibodies then combine with the antigens, rendering them innocuous.

One is made immune to a particular antigen by virtue of an initial exposure to that antigen. In many instances the one exposure assures lifelong protection. This is the basis of vaccination in which a small amount of the antigen is introduced so as to generate antibodies. Just how the antigen initiates a specific reaction in the plasma cells so that they thenceforth continue to produce the antibody is the intriguing but unanswered question. One hypothesis states that the antigen influences the genetic mechanism of the cell to produce a specific antibody as though the antigen acted as a template for the design of the antibody. Another hypothesis envisions the antigen as simply turning on the proper cell that already has the appropriate genetic mechanism to produce the necessary antibody.

The thymus gland obviously plays a major role in the immune reaction, especially in the fetus and newborn. It seems to be essential for the development of lymphoid tissue. If the thymus fails to form or is removed at birth, normal immunity does not ensue. And recent evidence suggests that it plays a role throughout life. For years a thymic hormone has been predicted, but only now has it been extracted. It is called thymosin. Its function is to stimulate the thymic production of lymphocytes. The importance of thymosin lies in its possible utilization to bolster defenses against infections and to treat allergies and perhaps cancer.

Function of White Blood Cells   The best-known role of the white cell is to aid in combating infectious processes. The neutrophils and monocytes are phagocytes; that is, they are capable of engulfing, ingesting, and destroying foreign agents. Phagocytes have the unique and even uncanny ability to move to the site of the foreign particles. The attraction is thought to be chemical. The process by which a chemical substance attracts leukocytes is termed chemotaxis. Such substances are generally degenerative products of the tissues which have been invaded by the foreign material. Probably the key chemical substance is generated at the site of infection and then diffuses to a nearby blood vessel, where it attracts the phagocytes. The phagocytes, by a poorly understood mechanism, pass through the tiny pores of the capillary wall to enter the tissue spaces. This process is called diapedesis. In the tissue spaces they move by self-propulsion in a manner similar to the movement of a primitive one-cell organism called an ameba. The behavior is thus termed ameboid movement.

Once the neutrophils and monocytes reach the site of the invasion, they rapidly engulf the offending particles. Within the phagocytes there are proteolytic enzymes which digest proteinaceous matter. Some of them also contain lipases that digest the lipid membranes of certain bacteria. Other enzymes are also present. As a consequence, most invading agents are digested and thus inactivated within the phagocyte. But this is accomplished at a price, and the price is engorgement of the phagocyte with the digestive end products, resulting in its death. The disintegration of the phagocyte probably provides more chemotaxic material to attract an ever greater number of phagocytes.

At the site of invasion, the battle between the phagocytes and the invading agents causes death of some local tissue as well as death of the invading agent and the engulfing phagocytes. The debris gives rise to a necrotic mixture termed pus.

Lymphocytes are capable of being converted to other cell types. For example, they can become monocytes and thus take on a phagocytic function. They may even enter the bone marrow and be converted into red blood cells. Thus it would appear that lymphocytes are rather undifferentiated cells that can be rapidly transformed into needed blood cell types. In the immune reaction, small lymphocytes become, in effect, antibodies. Normally, as stated above, the antibody is a specific gamma globulin that circulates in the blood to inactivate a particular antigen. But, in addition, lymphocytes are capable of becoming sensitized by an antigen in such a way that they attach themselves to the antigen. As a result the foreign agent undergoes lysis.

The role of the eosinophil is uncertain, although the available evidence strongly suggests that it too has an important defense function. The presence of foreign protein in the body greatly increases the eosinophil count. Thus, by implication, it is concluded that eosinophils function to digest or in some way to inactivate such protein. They may also release plasminogen, or profibrinolysin, which is converted to plasmin, or fibrinolysin, essential to clot dissolution.

Basophils are capable of secreting an anticoagulant called heparin. As will be discussed later in this chapter, heparin is a normal constituent of blood and plays a key role in preventing intravascular clotting. Basophils also contain other substances that influence the circulatory system, such as histamine and serotonin, but heparin secretion is thought to be their primary role.

White Blood Cell Formation   Granulocytes arise from hemocytoblasts in red bone marrow through a series of progressive cytoplasmic and nuclear changes similar to those described above for the erythrocyte. Early in their differentiation they acquire characteristically staining cytoplasmic granules which foreshadow their eventual fate, either as eosinophils with pink-staining granules, basophils with blue-staining granules, or neutrophils with neutral-staining granules.

Agranulocytes develop from mitotic division of hemocytoblasts in lymphoid tissues and, to a much lesser extent, in the bone marrow. Most lymphocytes arise in lymph nodes, while monocytes differentiate in the venous sinuses of the liver, spleen, and bone marrow.

Life-span of White Blood Cells   Because the white cells enter and leave the blood and, in essence, are present in the blood primarily for the purpose of transportation, one may speak of the total life-span or their life-span in the blood. Generally the latter is determined. Granulocytes, in the absence of infection, spend about 12 hr in the blood. Lymphocytes probably spend only a few hours, and since the monocytes seem to come and go, their life-span is impossible to calculate.

White Blood Cell Abnormalities   If the white blood cell count is too high, that is, over about 10,000, the condition is referred to as leukocytosis; if too low, below about 5,000, leukopenia.

Leukocytosis is the more commonly observed condition. Virtually all acute infectious processes evoke a leukocytosis. Thus, even in the absence of any other findings, a high white cell count causes the physician to suspect the presence of infection. On the other hand, there are conditions in which the leukocyte count soars to as high as 500,000/cu mm. This state is termed leukemia, which means "white blood." Leukemia is a form of cancer and usually proves fatal. In this condition the white cells are formed very rapidly in the bone marrow, so rapidly that the immature red cells are "starved out." This, of course, leads to a severe anemia, which predisposes the patient to many other diseases which, in turn, may lead to death. In addition, the tremendous number of circulating white cells may actually plug up important blood vessels in the brain, heart, lungs, and kidneys. The sequence of events may create alterations which are incompatible with life.

Leukemic cells are usually abnormal and often have little phagocytic ability. This probably stimulates an unfortunate vicious cycle in which more and more such cells are produced. At any rate they invade the bones, lymph nodes, liver, spleen, and other organs of the body. In very high concentration they interfere with the normal function of these organs or may bring about tissue destruction. In the bone marrow their destructive influence results in anemia. Finally, the material and energy required to produce this huge number of cells debilitate the body. The energy drain, the anemia, and the destruction of tissue all contribute as causes of death.

Leukopenia is less common. Mild cases are often associated with viral diseases such as measles, mumps, chickenpox, and poliomyelitis. More severe leukopenia, termed agranulocytosis, is usually the result of poisoning or excessive irradiation. The result of agranulocytosis is to deprive the organism of its main defense against infection.

Platelets

Blood platelets are also called thrombocytes. *Thrombo-* means "clot," and *-cyte* refers to "cell." This designation could stem from the fact that platelets are essential to the clotting mechanism or because platelets cling together in groups, or clots.

Platelets come from very large marrow cells, the megakaryocytes. While still in the marrow, megakaryocytes give off fragments which are only 2 to 4 microns in diameter. These enter the blood, where they are referred to as platelets. The platelet count varies between 200,000 and 400,000/cu mm of blood.

In their essential task of hemostasis, platelets become swollen spheres that sprout long projections, called pseudopodia. Normally they have a discoid shape and circulate freely. As will be discussed below, in the clotting mechanism the platelets change their shape, cling together, and release substances essential to hemostasis. The life-span of the platelet is thought to be about 4 days.

Plasma

The plasma is the fluid portion of the blood. Under normal conditions, it occupies over half the total blood volume. This percentage varies within a small range, but under pathological conditions it may exceed its usual limits. Like all other values to be studied, the blood-plasma volume is controlled by many homeostatic mechanisms so that, despite variations in the fluid intake, despite the amount of fluid lost by perspiration on hot and humid days, despite even massive hemorrhage, the total blood volume and the ratio between plasma and the formed elements are rapidly restored and kept constant.

Blood plasma is a straw-colored fluid composed of about 91 percent water and 9 percent solids. The solids include a vast variety of substances. Only a few of the more important constituents will be mentioned briefly. Tables 15-1 and 15-2 list normal values.

Plasma Proteins   Of all the blood constituents, perhaps the plasma proteins remain the most constant. In prolonged malnutrition plasma proteins decrease. But the usual dietary variations, mild hemorrhage, and wide changes in bodily activity all fail to produce a significant alteration in the plasma-protein concentration. Beyond question these blood constituents are exquisitely governed by very effective and dynamic homeostatic mechanisms.

Plasma proteins include albumin, globulins, and fibrinogen. Of these, fibrinogen has the greatest molecular weight and albumin the smallest. The concentration of all the plasma proteins together is about 7 g/100 ml of plasma, of which 4.2 g are albumin, 2.5 g globulins, and 0.3 g fibrinogen.

Although albumin is the smallest of the plasma proteins, the molecule is still too large to pass through the capillaries very readily. Thus it, along with the other proteins, contributes to the osmotic pressure of the blood. The osmotic pressure of the blood is an extremely important determinant of the movement of fluid into and out of the blood. As will be explained in some detail later, this osmotic pressure attracts fluid from the tissue spaces into the blood. The hydrostatic pressure of the blood, on the other hand, tends to force fluid out of the capillaries. There is thus a balance. If the concentration of protein in the blood decreases, more fluid will leave than will enter. The result will be accumulation of fluid in the tissue spaces, that is, edema.

There are three major types of globulin, alpha, beta, and gamma. They have many essential functions. Gamma globulins are important antibodies. Globulins also combine with various substances that enter the blood and in this manner provide a carrier to transport them from one point to another.

TABLE 15-1. RANGE OF VALUES FOR THE PRINCIPAL NONPROTEIN ORGANIC CONSTITUENTS OF BLOOD PLASMA

CONSTITUENT	NORMAL RANGE, MG/100 ML	CONSTITUENT	NORMAL RANGE, MG/100 ML
NONPROTEIN NITROGEN	25–40	Creatinine	1–2
Urea	20–30	Uric acid	2–6
Urea nitrogen	10–20	CARBOHYDRATES	
Amino acid nitrogen	4–8	Glucose	65–90
Amino acids	35–65	Fructose	6–8
Alanine	2.5–7.5	Glycogen	5–6
α-Aminobutyric acid	0.1–0.3	Polysaccharides (as hexose)	70–105
Arginine	1.2–3.0	Glucosamine (as poly-	
Asparagine	0.5–1.4	saccharide)	60–105
Aspartic acid	0.01–0.3	Hexuronates (as glucuronic	
Citrulline	0.5	acid)	0.4–1.4
Cystine	0.8–5.0	Pentose, total	2–4
Glutamic acid	0.4–4.4	ORGANIC ACIDS	
Glutamine	4.5–10.0	Citric acid	1.4–3.0
Glycine	0.8–5.4	α-Ketoglutaric acid	0.2–1.0
Histidine	0.8–3.8	Malic acid	0.1–0.9
Isoleucine	0.7–4.2	Succinic acid	0.1–0.6
Leucine	1.0–5.2	Acetoacetic acid	0.8–2.8
Lysine	1.4–5.8	Lactic acid	8–17
Methionine	0.2–1.0	Pyruvic acid	0.4–2.0
1-Methylhistidine	0.1	LIPIDS	
3-Methylhistidine	0.1	Total lipids	385–675
Ornithine	0.6–0.8	Neutral fat	80–240
Phenylalanine	0.7–4.0	Cholesterol, total	130–260
Proline	1.5–5.7	Cholesterol, esters	90–190
Serine	0.3–2.0	Cholesterol, free	40–70
Taurine	0.2–0.8	Phospholipids:	
Threonine	0.9–3.6	Total	150–250
Tryptophan	0.4–3.0	Lecithin	100–200
Tyrosine	0.8–2.5	Phosphatidyl ethanolamine	0–30
Valine	1.9–4.2	Plasmalogens	7–8
Bilirubin	0.2–1.4	Sphingomyelin	10–50
Creatine	0.2–0.9	Total fatty acids	150–500
		Unesterified fatty acids	8–30

TABLE 15-2. RANGE OF VALUES FOR THE PRINCIPAL INORGANIC CONSTITUENTS OF BLOOD PLASMA

ANIONS	CONCENTRATION MEQ/LITER*	CATIONS	CONCENTRATION MEQ/LITER*
Total	142–150	Total	142–158
Bicarbonate	24–30	Calcium	4.5–5.6
Chloride	100–110	Magnesium	1.6–2.2
Phosphate	1.6–2.7	Potassium	3.8–5.4
Sulfate	0.7–1.5	Sodium	132–150
Iodine (total)	8–15†	Iron	50–180†
Iodine (protein-bound)	6–8†	Copper	8–16†

$$* \text{ Milligrams}/100 \text{ ml} = \frac{\text{atomic weight} \times \text{mEq/liter}}{10 \times \text{valence}}$$

† These concentrations are in terms of micrograms/100 ml.

Fibrinogen, as will be explained shortly, plays a primary role in clotting.

The liver is the main site of formation of the plasma proteins. Thus, liver disorders frequently result in inadequate protein formation and a consequent fall in plasma-protein concentration. However, this is not the only cause of decreased plasma protein. Dietary deficiency and kidney diseases, as well as many other disorders, may also be responsible.

Nonprotein Nitrogen   Nitrogen is a major constituent of protein, and thus this element is well represented in the plasma. In addition, there are present other compounds that contain nitrogen. The nitrogen of all these other compounds together constitutes what is termed nonprotein nitrogen, or NPN (Table 15-1). It amounts to about 2 percent of the total nitrogen of the blood. In the plasma, NPN averages 27 mg/100 ml. In the cells it is higher, being on the order of 47 mg/100 ml. Of this average, urea contributes the major component.

The NPN level of the plasma is a good determinant of protein balance. Normally, protein ingestion, protein formation and breakdown, and the excretion of nitrogenous substances all keep pace so that the NPN level of the blood does not vary significantly. However, in various kidney abnormalities the rate of excretion is decreased, and therefore the NPN is elevated. On the other hand, excessive protein catabolism as occurs in cases of infections, fevers, and hyperthyroidism also leads to an elevated NPN level. And, of course, inadequate protein ingestion will produce the same result.

Blood Sugar   Glucose is dissolved in the plasma and delivered to all the cells in the body to supply them with energy. It is one of the most important energy foods. There are excellent homeostatic devices which control carbohydrate metabolism and keep the blood-sugar levels within a relatively narrow range despite wide variations in ingestion and bodily activity. The fasting person has a blood-glucose level of about 80 mg/100 ml of blood. Abnormal variations from this level are associated with widespread disorders, which will be discussed later, or are due to the ingestion of unusual quantities of carbohydrates.

The Electrolytes   The plasma contains varying quantities of the electrolytes such as sodium, potassium, magnesium, chlorides, phosphates, and bicarbonates (Table 15-2). Like the other plasma constituents, these substances are found to exist in very constant concentrations. Even slight variations of potassium, for example, may prove fatal, whereas an increase in the sodium concentration is accompanied by drastic water-balance shifts. The clinically important subjects of acid-base balance and fluid balance will be discussed in Chapter 30.

Blood Functions

Blood functions may be summarized as follows:

1   *Nutrition.* The end products of digestion are absorbed into the bloodstream and carried to the various cells of the body (Chapter 27).
2   *Respiration.* As the blood passes through the lungs, it takes up oxygen and gives off carbon dioxide. The oxygen is then transported to all the cells of the body and carbon dioxide removed (Chapter 22).
3   *Fluid balance.* Fluid is present in the cells, between the cells, and within the blood vessels. Fluid moves in and out of the blood to help maintain this balance (Chapter 30).
4   *Acid-base balance.* The blood contains substances (buffers) which serve to neutralize acids and bases and thus aid in maintaining a normal balance (Chapter 30).
5   *Excretion.* The end products of cellular metabolism are carried by the blood to the kidneys, the lungs, and the sweat glands for excretion (Chapter 29).
6   *Protection.* The blood contains many cells and chemicals which serve a protective function.
7   *Temperature regulation.* The body temperature remains remarkably constant. The blood plays a major role in this regulation (Chapter 19).
8   *Endocrine adjunct.* The endocrine glands secrete hormones directly into the blood, which carries them to various cells of the body (Chapter 31).

HEMOSTASIS

The term hemostasis means, literally, "a standing of blood." Thus it is used in two senses: (1) to imply stagnation or impaired flow of blood and (2) to mean arrest of a flow of blood. The term is used in the second sense most often to refer to all the processes involved in arresting the flow of blood from a break in the circulatory system.

Hemostasis, in this sense, involves four processes: (1) vascular spasm, (2) formation of the platelet plug, (3) blood clotting, and (4) clot retraction.

Vascular Spasm

Severe trauma is often followed by a remarkably small loss of blood. One reason is that the walls of the arteries and arterioles contain smooth muscle capable of contracting. When a vessel is damaged, the area so stimulated responds by contracting, which decreases the lumen of the vessel and thereby impedes the flow of blood. This is called a local myogenic spasm. At the same time a reflex mechanism causes vasoconstriction along a considerable length of the vessel in both directions from the damage. The constriction upstream will, of course, markedly slow the flow of blood; in fact, it may cut it off completely.

As will be discussed presently, the platelets liberate serotonin at the site of injury. This substance, which is 5-hydroxytryptamine, is a vasoconstrictor which adds to the vascular spasm already evoked by local trauma and the reflex mechanism.

Formation of the Platelet Plug

Where the blood vessel is damaged, platelets accumulate, become swollen, sprout long projections, and stick together in an amorphous mass. In so doing, they release so-called releasing agents which initiate similar changes in other nearby platelets. The result is that layer after layer of platelets aggregate to form a plug. This plug fills the opening in the vessel, and if the opening is small, bleeding will cease and a blood clot does not form. The platelet plug simply fills the hole in the vessel but does not occlude the vessel itself; therefore, normal circulation continues unimpeded. This is the mechanism utilized to plug ruptures that occur almost continuously in small vessels.

Platelets contain a high concentration of ATP, which is rapidly converted to ADP. The ADP is extruded, apparently by contraction of the platelet. Contraction is thought to be due to a platelet-contractile protein, thrombosthenin. ADP, in a manner not yet clarified, brings about platelet aggregation. Other substances that do the same are collagen, serotonin, thrombin, and epinephrine.

Clotting of Blood

The clotting of blood involves two basic reactions:

1 Prothrombin is converted to thrombin.
2 Fibrinogen is converted to fibrin.

Fibrin is a gel that becomes deposited in a network which traps the formed elements of the blood, rapidly building up a clot which seals off the opening in the blood vessel and sometimes occludes the blood vessel itself.

Blood clots when there is tissue damage, as when the blood vessel is torn. Blood also clots after it is drawn from the circulation and left to stand. The first sequence is referred to as extrinsic clotting; the second, as intrinsic clotting.

In extrinsic clotting, the initiating substance is called thromboplastin. It is a lipoprotein that seems to be ubiquitous, so that when cells are torn, thromboplastin is liberated. Once liberated, thromboplastin initiates a series of reactions depicted in Figure 15-5.

Figure 15-5. Extrinsic clotting.

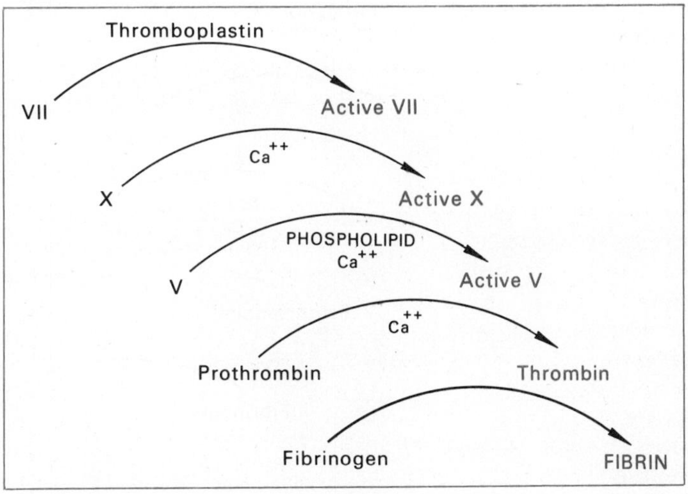

If blood is carefully drawn so as to minimize cell damage, there will be little throm-boplastin to initiate clotting. Yet if the blood remains in contact with a foreign surface, such as glass, it clots. Because all the substances involved are normally found in the blood, this sequence is referred to as the intrinsic system. Something, as yet unidentified in the foreign surface, initiates the sequence shown in Figure 15-6. In both intrinsic and extrinsic clotting, the formation of each active substance catalyzes the next reaction, giving rise to a so-called waterfall, or cascade, sequence.

The reader should be aware of the fact that although a tremendous amount of research has been carried out to elucidate the clotting mechanisms, there is far from universal agreement. In addition, the entire field is confused by diverse terminology. Table 15-3 lists the various names. It is by no means a complete list.

Prothrombin   Prothrombin, the inactive precursor of thrombin, is a protein formed in the liver and transported into the blood in quantities considerably in excess of clotting requirements. Blood clots normally, even when the quantity of prothrombin present is reduced by about 50 percent or more. The normal plasma level is 15 mg/100 ml.

Figure 15-6. Intrinsic clotting.

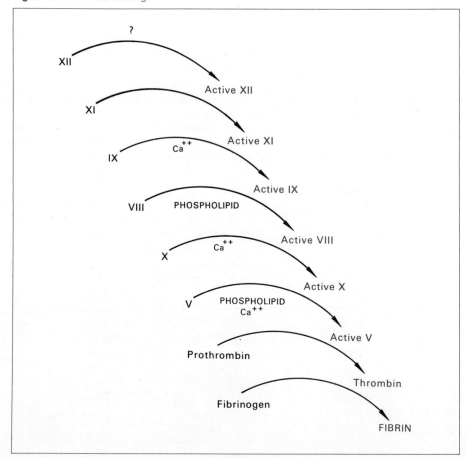

TABLE 15-3. BLOOD FACTORS AND ALTERNATIVE NAMES

FACTOR	ALTERNATIVE NAME
Factor I	Fibrinogen
Factor II	Prothrombin
Factor III	Thromboplastin (tissue), thrombokinase
Factor IV	Calcium
Factor V	Labile factor, proaccelerin, plasma Ac globulin
Factor VI	Serum Ac globulin, accelerin
Factor VII	Stable factor, proconvertin, SPCA (serum prothrombin conversion accelerator), cothromboplastin, autoprothrombin I
Factor VIII	Antihemophilic factor (AHF), antihemophilic globulin, thromboplastinogen, platelet cofactor I, plasma thromboplastic factor A
Factor IX	Plasma thromboplastin component (PTC), Christmas factor, platelet cofactor II, plasma thromboplastic factor B, autoprothrombin II
Factor X	Stuart-Prower factor
Factor XI	Plasma thromboplastin antecedent (PTA)
Factor XII	Hageman factor

The synthesis of prothrombin and several other factors by the liver requires the presence of adequate quantities of vitamin K. These include factors VII, IX, and X. Because vitamin K is fat-soluble, if there is impairment of fat absorption from the intestine, vitamin K deficiency may develop. The consequence is inadequate hemostasis. Hemorrhagic disease is also common in newborn infants due to a diet generally deficient in vitamin K unless it is supplemented.

Fibrinogen The second basic protein essential to clotting, fibrinogen, is also produced by the liver. It is a much larger molecule than prothrombin. Prothrombin has a molecular weight of about 68,000, while fibrinogen has a weight close to 340,000.

Fibrinogen is produced continuously and is catabolized in the blood at the same rate; thus, under normal conditions, the blood concentration remains fairly steady at about 0.3 g/100 ml.

Clot Retraction
The threads of fibrin form a network which traps blood cells and plasma. The clot then undergoes contraction, during which serum (defibrinated plasma) is squeezed out. Contraction is probably the result of shortening of fibrin threads. Fibrin also clings to the edges of the opening in the blood vessel. Therefore, as the threads shorten, the clot becomes smaller and pulls the edges of the blood vessel closer together. The sides of the clot are actually retracted in this process; hence the term clot retraction.

Clot retraction requires the presence of a large number of platelets. Why retraction fails to occur if these platelets are not present is not known.

Clot Lysis
The breakdown, dissolution, or lysis of the clot depends upon the enzymatic digestion of the fibrin; consequently the process is referred to as fibrinolysis. The plasma contains a substance called plasminogen, or profibrinolysin, which is the inactive form of plasmin,

or fibrinolysin, a proteolytic enzyme. Plasmin digests the fibrin in the clot. This system not only removes major clots after the vessel has healed, but it also serves to remove the small clots that apparently occur constantly in peripheral blood vessels as a result of the trauma of everyday life.

The factors that regulate fibrinolytic activity are not known, but it is increased during exercise, which is one reason why exercise is proving to be one of the best procedures for preventing cardiovascular problems.

Prevention of Blood Clotting

Clotting of blood within the vascular system can be fatal. Obviously, there must be effective mechanisms to prevent it. The plasma contains an alpha globulin called antithrombin. Antithrombin does not prevent the formation of thrombin, but the thrombin formed is quickly inactivated before it can cause fibrinogen to be converted to fibrin. The other important intravascular anticoagulant is heparin. Many cells of the body apparently have the ability to secrete heparin. It prevents clotting in several ways. It blocks the formation of thromboplastin, inhibits the formation of thrombin, probably inhibits the fibrinogen-fibrin reaction, and is thought to help antithrombin to inactivate thrombin.

But despite the presence of antithrombin and heparin, blood sometimes does clot in the circulation. One cause is damage to the endothelial wall of the blood vessels. The resulting roughness seems to initiate the clotting mechanisms. Another, and perhaps more important, reason is that the normal electrical charge of the surface is altered. Just how surface charge interacts with the clotting mechanism is still to be clarified.

If blood is removed from the body, it will clot unless steps are taken to prevent coagulation. However, if blood is drawn very carefully into a syringe which has been coated with paraffin, clotting may not occur. The reason is probably that with due caution there is minimal platelet destruction and therefore inadequate thromboplastin formation to initiate the sequence. Another mechanical way to prevent clotting is vigorous stirring with a glass rod. The stirring causes fibrin to be formed, but it adheres to the rod and thus can be removed from the blood. The blood is then said to be defibrinated.

There are many chemicals that will prevent clotting. Oxalates and citrates act as anticoagulants by precipitating calcium. In the absence of calcium, blood does not clot. Other anticoagulants commonly used are heparin, Dicumarol, and hirudin.

Clotting Agents

Often it is necessary to cause the blood to clot. This is true in some forms of surgery, in making skin grafts adhere, and in treating patients who have inadequate blood clotting. Both thrombin and fibrinogen are now commercially available and are used for these purposes.

Blood-clotting Tests

The ability of the blood to clot is evaluated by determining the clotting time, the bleeding time, the prothrombin blood level, and the platelet count.

Clotting Time   The clotting time is determined by drawing blood from a vein into small glass tubes. One millimeter of blood is placed in each, and the tubes are submerged in a water bath at 37°C. The tubes are checked every 30 sec for signs of clotting. The time that it takes for a clot to adhere to the tube wall is taken as the end point. The normal clotting time ranges between 5 and 8 min.

Bleeding Time   In order for the bleeding time to be evaluated, a small wound is made in the fingertip or, preferably, the earlobe. The wound is touched gently with filter paper every 30 sec. When the paper no longer is stained, the time is recorded. The normal bleeding time ranges from 2 to 5 min. The values are so variable, however, that the test is of limited value.

Prothrombin Time   The prothrombin blood level is determined by oxalating the blood as soon as it is drawn. Then an excess of thromboplastin and calcium ion are mixed with the blood. These substances reverse the effect of oxalate and permit the clotting sequence to take place. The time required from the addition of thromboplastin and calcium ion until clotting occurs is called the prothrombin time. Since there is a direct relation between prothrombin time and prothrombin blood concentration, this procedure gives an indication of the latter.

Clotting Abnormalities

Either excessive bleeding or intravascular clotting can prove fatal. Excessive bleeding can result from a deficiency of one or more of the clotting factors. Since prothrombin, fibrinogen, and several of the clotting factors are formed in the liver, liver disease may well cause clotting abnormality. Similarly, since vitamin K is essential for the formation of many of these same factors, vitamin K deficiency generally gives rise to excessive bleeding too.

Hemophilia   No blood disorder could be more inappropriately named. Hemophilia means, literally, "loving blood." In view of the fact that a person suffering with this disorder may readily bleed to death, the very sight of blood is a horrifying experience.

Hemophilia is an inherited disorder. It is sex-linked and seen almost entirely in the male. There is no cure, but it may be controlled by transfusion of normal plasma. The transfusions must be repeated periodically.

Hemophilia is most often caused by a deficiency of the antihemophilic factor, AHF, now known as factor VIII. A deficiency of factor IX and factor XI also causes the disorder.

If two blood samples taken from certain patients thought to have hemophilia are mixed together, the mixture has a normal clotting time. This suggests that the deficiencies in the two bloods differ. In other words, the blood from the true hemophiliac is deficient in AHF; the other blood, in a different factor. The second deficiency is factor IX, also called Christmas factor because Christmas was the name of the patient first shown to have this disorder. So-called Christmas disease is quite rare.

Thrombocytopenia   A condition characterized by less than the normal number of platelets in the blood is termed thrombocytopenia. Instead of the normal 200,000 or 300,000 platelets/cu mm, there will be less than 100,000. In such cases, excessive bleeding occurs.

Thrombi and Emboli   A thrombus is a clot. The term is used clinically to designate an intravascular blood clot which is adherent to the vessel wall at its site of formation. An embolus, on the other hand, means "a plug." It is often a fragment of a thrombus which has broken free and has been carried by the circulating blood elsewhere to become lodged in another vessel. Thrombi and emboli may interfere with circulation, and if the involved vessel is large or supplies a vital organ, death may result.

As mentioned above, intravascular clotting is thought to be due to roughening of the inner lining of the blood vessels. The lining may be roughened by various disorders such as arteriosclerosis or as a result of surgery. In such cases, the anticoagulants are indispensable.

BLOOD TRANSFUSION

There are many circumstances in which it is necessary to transfuse whole blood. Although substitutes such as saline solutions, glucose solutions, and plasma preparations may be used, none takes the place of whole blood, and in many cases whole blood is truly indispensable. Unfortunately, blood cannot be transfused indiscriminately. When some types of blood are mixed together, the formed elements clump together; that is, they agglutinate, which means "to paste." These clumps of cells may fatally block vital vessels.

Antigen, Antibody, Agglutinogen, Agglutinin

Before blood types and the problems of blood transfusion are discussed, it is best to define terms. As already mentioned, an antigen is a foreign substance entering the bloodstream or tissues which stimulates the production of antibodies. In the study of blood groups an antigen is termed an agglutinogen and refers to a substance found in the red cell membrane. An antibody is a protein substance that combines with a particular antigen. In speaking of blood groups, antibody is also given another name, agglutinin. Agglutinin is found in the plasma. The most important classes of red blood cells are those of the ABO system and those of the Rh system.

ABO Blood Group System

Human blood is typed A, B, or O according to what type of agglutinogen is present in the red cell membrane. The A agglutinogen is really a mixture of two different agglutinogens, which are known as A_1 and A_2. Cell membranes that contain no agglutinogen are placed in the O type. There are thus six possible blood types based on what agglutinogen or combination of agglutinogens is present in the membrane: A_1, A_2, B, O, A_1B, and A_2B (Table 15-4).

Agglutinins (that is, antibodies) are present in the plasma of all blood types except those in which the cell membranes contain both A and B agglutinogens. Obviously, in a particular type of blood the plasma normally will not contain the agglutinins that would cause agglutination of the red cells of that blood type. When agglutination does occur, it is because the agglutinogens of the red cell membrane are incompatible with the agglutinin of the blood type; the agglutinogens are "foreign substances" in the plasma of that type of blood. For example, type A blood contains red blood cells with A_1 (or A_2) agglutinogen in their membranes. It does not contain agglutinin anti-A_1 (or anti-A_2) in its plasma but does contain agglutinin anti-B.

Tables 15-4 and 15-5 indicate that:

1 The plasma of type O blood agglutinates the cells of all other blood types. But the cells of this type are not agglutinated by any other plasma.

TABLE 15-4. ABO BLOOD TYPES

AGGLUTINOGEN (RED CELLS)	AGGLUTININ (PLASMA)	FREQUENCY, PERCENT
O	Anti-A, Anti-B	45.0
A_1	Anti-B	31.0
A_2	Anti-B	10.0
B	Anti-A	10.0
A_1B	None	2.9
A_2B	None	1.1

2   The plasma of both A types will agglutinate the cells of the B and AB types. The cells of both A types are agglutinated by plasma of O and B types.
3   The plasma of type B agglutinates the cells of A and AB blood. The cells of type B are agglutinated by the plasma of O and A types.
4   The plasma of both AB types does not agglutinate any cells. The cells, however, are agglutinated by the plasma of all other types.

Principle of Transfusion

The basic principle for transfusion states that if the cells of the blood to be transfused will be agglutinated by the recipient's plasma, the transfusion must not be made. Thus, since the cells of type A blood are agglutinated by B plasma, A blood cannot be given to a B blood patient. On the other hand, since the cells of O blood are not agglutinated by any plasma, O blood can be given to patients with other types of blood. Accordingly, persons with O blood are termed universal donors. The plasma of AB blood does not agglutinate the cells of any blood. Thus persons with this type of blood are termed universal recipients.

Why may type O blood be given safely if its plasma agglutinates the cells of all other groups? The answer is that when blood is administered to a patient, it is greatly diluted by the volume of the patient's blood, so that the injected plasma becomes too dilute to agglutinate the host cells. Thus the basic principle: It is the cellular reaction of the administered blood that is the deciding factor.

Determination of Blood Type

The type into which a patient's blood should be classified can be quickly and easily determined. A small amount of blood is diluted with saline solution, and then a drop of the mixture is added to two test sera, one containing anti-A agglutinin and the other containing anti-B. One then observes whether or not agglutination has taken place. The clumping of the cells can be seen with the naked eye but is usually checked with a microscope (Figure 15-7). Reference to Table 15-5 then indicates the blood type. In the routine procedure A_1 and A_2 are not differentiated.

Rh System

The Rh system consists of many agglutinogens. The most important one in the system is referred to as agglutinogen D. It is also called the Rh factor, after the rhesus monkey in which it was first identified. If the agglutinogen is present in the red blood cell membrane, the person is said to be Rh-positive. About 85 percent of the Caucasian population is Rh-positive. Other races appear to be almost 100 percent Rh-positive.

When Rh-positive blood is transfused into a person with Rh-negative blood (Rh_0), his plasma develops anti-D agglutinin. Generally, there are no ill effects due to this first transfusion, but if another transfusion of Rh-positive blood is given, even years later, serious agglutination may occur due to the anti-D agglutinin in the recipient's plasma.

A similar situation arises when an Rh-negative woman is pregnant with an Rh-positive fetus. If blood from the fetus manages to enter her circulation, her plasma develops anti-D agglutinin. Although this agglutinin may find its way into the blood of the fetus, it is unlikely to be present in quantities capable of affecting the fetus, unless, of course, the mother has been sensitized by an earlier injection or transfusion of Rh-positive blood. In a subsequent pregnancy, however, the anti-D agglutinin, if it enters the blood of an Rh-positive fetus, may cause agglutination. As a result the fetus may die, or a child may be born with a severe disease of the blood, erythroblastosis fetalis. In such cases treatment involves replacing the infant's blood with blood from a normal person.

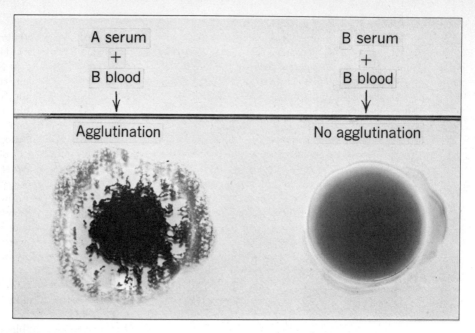

A serum	B serum
+	+
B blood	B blood
↓	↓
Agglutination	No agglutination

Figure 15-7. Blood typing. A blood sample is mixed with sera from known blood types. Whether or not agglutination occurs indicates the blood type of the sample.

Cross Matching

Blood may be typed as previously outlined. In order to ascertain whether or not agglutinogen D is present, serum containing anti-D agglutinin is available. However, in addition to the ABO and the Rh systems, there are many others. The total number of agglutinogens now recognized is very large and continues to grow. Accordingly, cross matching between the donor and recipient blood is carried out before transfusion whenever possible. Because, as emphasized above, it is the reaction of the donor cells that is of vital importance, a suspension of these cells is mixed with the recipient's serum. No agglutination should occur. To be absolutely certain, the reverse procedure is also often done, that is, mixing cells from the recipient with the serum from the donor.

TABLE 15-5. BLOOD TYPING

TYPE	SERUM A (CONTAINS ANTI-B AGGLUTININ)	SERUM B (CONTAINS ANTI-A AGGLUTININ)
O	No agglutination	No agglutination
A	No agglutination	Agglutination
B	Agglutination	No agglutination
AB	Agglutination	Agglutination

QUESTIONS AND PROBLEMS

1. On what basis can blood be considered an atypical specialized type of connective tissue?

2. In the adult male, what are the normal values of the following: (*a*) blood volume, (*b*) hematocrit, (*c*) red blood cell count, (*d*) white blood cell count, (*e*) coagulation time, (*f*) neutrophils, and (*g*) platelets?

3. How does serum differ from plasma?

4. Explain the value of the following to the hemocytologist: (*a*) centrifuge, (*b*) hemocytometer, (*c*) light microscope, and (*d*) anticoagulants.

5. Describe the functions of the circulating red blood cell.

6. Where are the pre- and postnatal hematopoietic centers? What kinds of blood cells do they produce?

7. At what stage of red blood cell development can hemoglobin be identified?

8. What is the normal fate of hemoglobin?

9. Discuss negative feedback as it relates to production of red blood cells.

10. Describe two conditions in which there is an abnormal red blood cell count.

11. Describe and give the major function of each of the five types of white blood cells.

12. Define the following: (*a*) fibrin, (*b*) chemotaxis, (*c*) diapedesis, (*d*) color index, (*e*) bilirubin, (*f*) hypoxia, (*g*) leukemia, (*h*) NPN, (*i*) thrombocytes, (*j*) milliequivalent, and (*k*) hemostasis.

13. Describe the role of the following in blood clotting: (*a*) prothrombin, (*b*) platelets, (*c*) fibrinogen, and (*d*) calcium.

14. What are the eight functions of blood?

15. Explain why edema may develop in patients with serious liver disorders and in patients whose kidneys are damaged to the extent that albumin is excreted in urine.

16. Distinguish between thrombi and emboli, and discuss the significance of each.

17. What mechanisms prevent the formation of clots in intact blood vessels of the human organism?

18. Explain why a child born to a woman with Rh-negative blood may be born with a severe hemolytic disorder.

19. What types of whole blood may be used in transfusion of patients of each of the six blood types in the ABO system?

16 THE HEART

The heart is the pump of the circulatory system. It propels the blood through the body. If the heart stops for even a short time, irreversible changes occur and death quickly ensues. Thus the importance of the heart cannot be overemphasized.

EMBRYOLOGY

The heart, like the other components of the circulatory system—the blood vessels, lymph vessels, and the blood itself—arises from mesenchyme. Initially, it is simply an enlarged blood vessel at the cranial end of the embryo with an unusually thick muscular wall and large lumen. During the fifth week of development this straight tubular organ begins to grow much more rapidly than its investing pericardial cavity (Figure 16-1). The cranial end of the embryo subsequently folds under itself to swing the developing heart toward its ultimate position in the thoracic cavity. The rapid growth of the heart causes the originally straight tube to bulge into first a U shape, then an S shape as both ends of the tube are brought close together and fuse. The initial cranial end of the tube will develop into the pulmonary artery and aorta; the caudal end will develop into the great veins of the heart.

A partition formed by the fusion between the caudal parts of the loop then disappears to create a single ventricle in a three-chambered heart. At about the sixth week, however, an internal longitudinal septum develops into the lumen of this unpaired chamber and divides it into the right and left ventricles. Next a transverse partition develops in the

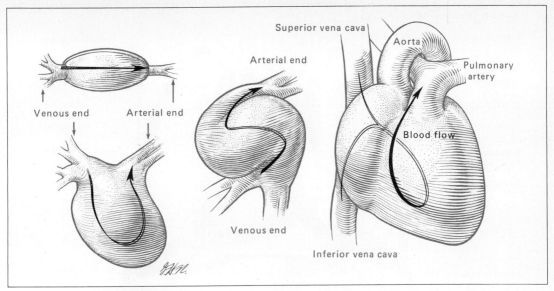

Figure 16-1. Development of the embryonic heart. The arrow indicates the direction of blood flow.

upper portion of the tube and separates the atria from the ventricles. Thus the four-chambered heart is present in the 6- to 7-week embryo.

As this partitioning process progresses, congenital defects may occur if the septal formation is not complete. One such defect is an abnormal communication between the heart chambers, which becomes clinically significant if there is intermingling of blood in the heart chambers. This condition causes the inadequate oxygenation of circulating blood that is seen in a "blue baby."

A septum also forms to divide the arterial end of the heart into its ultimate pulmonary artery and aorta. Maldevelopment of this separation may result in one of these vessels being larger than normal, the other smaller. The smaller vessel may not be capable of transmitting its normal volume of blood.

The pericardium (see ahead) develops and encloses the heart, the visceral layer becoming adherent to the external surface of the developing heart and separated from the parietal layer of membrane by the pericardial cavity.

ANATOMY OF THE HEART

The heart is located a little to the left of the midline in the space between the two pleural cavities (Figure 16-2). This space is called the mediastinum. It is bounded anteriorly by the sternum and posteriorly by the vertebral column. The heart lies centrally within the mediastinum, enveloped in pericardium. The other contents of the mediastinum will be discussed in Chapter 20, on the respiratory system.

The pericardium is a closed, double-walled membranous sac around the heart. The wall closest to the heart is termed the visceral pericardium. It adheres to the heart surface. At the place where the great blood vessels leave the heart, the visceral pericardium turns back, becoming the outer layer of the sac, the parietal pericardium. The parietal pericardium is thickened somewhat into a tough protective membrane. It is attached firmly

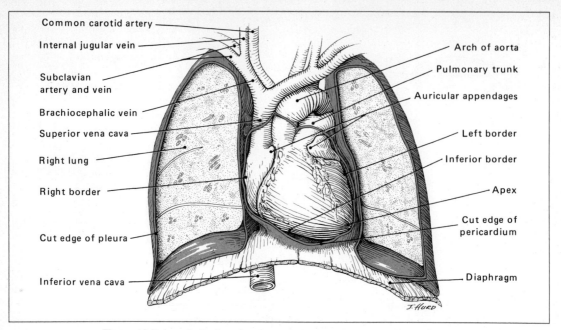

Common carotid artery

Internal jugular vein

Subclavian
artery and vein

Brachiocephalic vein

Superior vena cava

Right lung

Right border

Cut edge of pleura

Inferior vena cava

Arch of aorta

Pulmonary trunk

Auricular appendages

Left border

Inferior border

Apex

Cut edge of
pericardium

Diaphragm

J. HURD

Figure 16-2. Longitudinal section through the thoracic cavity. The mediastinum is the area between the two lungs.

to the central portion of the diaphragm and the posterior aspect of the sternum. Between the two layers is the pericardial cavity. A small quantity of serous pericardial fluid secreted by cells lining the pericardium fills the limited space of the cavity. This fluid film allows for frictionless movement of the heart in the act of beating.

Surface Anatomy

The heart is cone-shaped and approximately the size of a closed fist. It rests obliquely in the thorax and has three borders and three surfaces. The anterior, or sternocostal, surface faces the sternum; the posterior surface (base of the cone) faces the vertebral column; and the inferior, or diaphragmatic, surface rests on the diaphragm.

The heart is divided into right and left halves by a septum. Each half consists of two chambers, an atrium and a ventricle. Thus on the right side there are the right atrium and the right ventricle and on the left side there are the left atrium and the left ventricle. The atria are above and slightly behind the ventricles.

Grooves or sulci on the surface of the heart indicate underlying structures. The grooves contain the major superficial vessels that supply the heart muscle with blood. The atrioventricular sulcus, the deepest of the grooves, partially encircles the heart and lies between the atria and the ventricles. Two interventricular sulci, one on the anterior surface and one on the inferior surface of the heart, lie over the septum dividing the ventricles.

Circulation through the Heart

The heart receives blood by way of the veins and propels it out through the arteries to the body. The right atrium receives deoxygenated blood by way of the superior and inferior venae cavae, while the left atrium receives oxygenated blood from the lungs (Figure 16-3). From the atrium on each side, blood passes through an atrioventricular orifice and into the ventricle on each side. The right ventricle pumps blood to the lungs via the pulmonary arteries, and the left ventricle pumps blood throughout the body by way of the aorta and its branches.

Blood flows through the heart in only one direction because of four valves that guard the openings of the ventricles and prevent backflow.

The valve between the right atrium and ventricle is called tricuspid because of its three leaflets, or cusps. The valve between the left atrium and ventricle is bicuspid (two leaflets) and is also known as the mitral valve.

The cusps of both valves (Figure 16-4) are triangular and attach at their bases to the margins of fibrous connective tissue encircling the atrioventricular orifices. The apices of the cusps project into the ventricles, and several fibrous strands called chordae tendineae

Figure 16-3. Blood flow through the heart. At the top, the direction of flow is illustrated. At the bottom is flow during a single cycle of contraction.

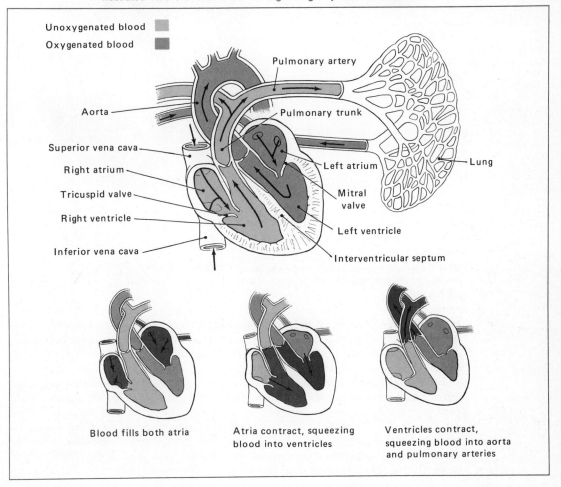

Unoxygenated blood

Oxygenated blood

Pulmonary artery

Aorta

Pulmonary trunk

Superior vena cava

Right atrium

Tricuspid valve

Right ventricle

Inferior vena cava

Left atrium

Lung

Mitral valve

Left ventricle

Interventricular septum

Blood fills both atria

Atria contract, squeezing blood into ventricles

Ventricles contract, squeezing blood into aorta and pulmonary arteries

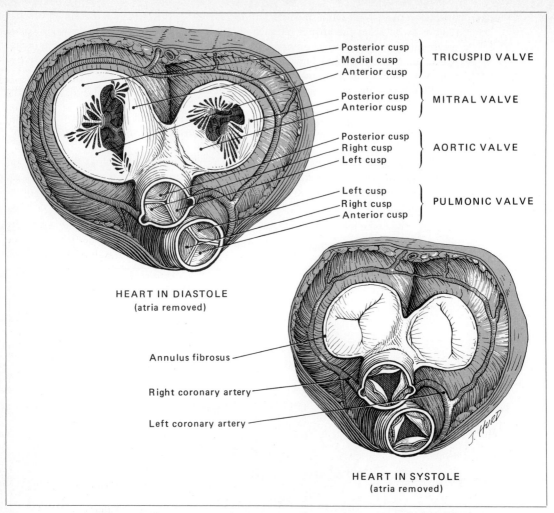

Figure 16-4. Valves of the heart with atria removed. At the left, atrioventricular (A-V) valves are open to permit blood into the ventricles. At the right, semilunar (aortic and pulmonic) valves are open to pass blood into the aorta and pulmonary trunk.

extend from each apex to small, conical muscles (papillary muscles) on the ventricle wall. As the ventricles contract, the blood, surging up against the ventricular surface of the cusps, forces them into apposition and closes the orifice. The chordae tendineae prevent the valves being forced back into the atria.

The aortic and pulmonic, or semilunar, valves guard the openings from the ventricles to the aorta and to the pulmonary artery (Figure 16-4). Each valve has three cusps. Each cusp resembles a teacup that has been cut vertically. The surfaces of the cusps corresponding to the "cut" portion of the teacup are attached to the blood vessel as it originates from the ventricle. When the ventricle contracts, the blood rushing out pushes open the valves. When the ventricle relaxes, blood starts to flow back and fills the "bowl" of the cusps, forcing their free edges into apposition and closing the orifice.

Internal Anatomy of the Heart

The wall of the heart has three layers. The outer layer, or epicardium, is the same as the visceral pericardium; the inner layer, or endocardium, lines the heart as endothelium, while the bulk of the heart wall, between the epicardium and endocardium, consists of cardiac muscle, or myocardium.

Cardiac muscle forms bundles which attach to deeply situated collections of fibrous tissue. The fibrous tissue forms the "skeleton" of the heart and consists of rings surrounding the atrioventricular, pulmonary, and aortic orifices. These rings are collectively termed the annulus fibrosus. Between them there is additional fibrous tissue. The rings provide attachment for the heart valves as well as the heart muscle.

In the right atrium a vertical ridge, the crista terminalis, extends from the superior to the inferior vena cava opening. Slender projections of myocardium, the musculi pectinati, pass at right angles to the crista terminalis (Figure 16-5). On the septum between the right and left atria there is a shallow depression, the fossa ovalis. It marks the site of an opening, the foramen ovale, which during fetal life allowed blood to move directly from the right atrium into the left atrium and thus bypass the nonfunctioning lungs (see Chapter 33). The large coronary sinus (vein) drains blood from the wall of the heart into

Figure 16-5. Internal anatomy of the right side of the heart. The criss-cross ridges in the auricle are musculi pectinati. Those in the ventricle near the papillary muscle are trabeculae carnae. The crista terminalis lies between the superior and inferior venae cavae on the portion of the atrial wall removed for this view of the chambers.

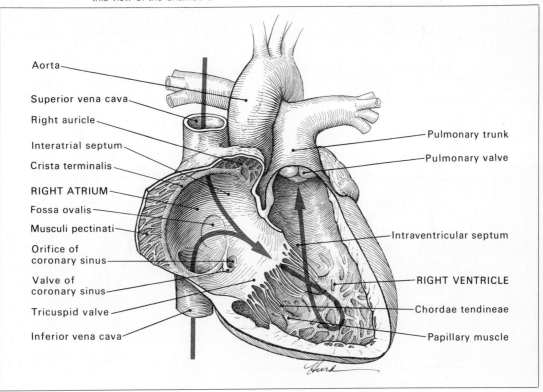

the right atrium. Numerous minute openings for small veins, the venae cordis minimae, lie in all chambers of the heart, but are more prevalent in the right atrium. The interior of the left atrium is smooth with no special characteristics. Two small ear-shaped outpouchings, the auricles, extend from the superior aspects of both atria.

On the interior of the ventricles there are muscular bars or ridges, the trabeculae carnae ("meaty ridges"), which project into the lumen. In addition there are the conical papillary muscles which give attachment to the chordae tendineae. The wall of the left ventricle is about three times as thick as the wall of the right ventricle. The interventricular septum bulges into the right ventricular lumen so that a cross section of the left ventricle is circular in shape and that of the right ventricle somewhat quarter-moon-shaped.

Blood Supply of the Heart

Coronary Arteries Right and left coronary arteries supply the myocardium of the heart (Figure 16-6). They originate from dilatations at the origin of the ascending aorta opposite two of the cusps of the semilunar valve. The coronary arteries or their branches encircle the heart as they course in the atrioventricular groove. The main portion of the left coronary artery is only about 2.5 cm long and divides shortly into circumflex and anterior interventricular branches. The circumflex branch follows the atrioventricular groove to the left. On the posterior aspect of the heart it anastomoses with the right coronary artery, which has followed the atrioventricular groove to the right around the heart.

The anterior interventricular branch of the left coronary artery courses in the interventricular sulcus on the anterior surface of the heart to supply both ventricles. The posterior interventricular branch of the right coronary artery follows a similar course in the interventricular groove on the inferior surface of the heart. The two interventricular branches anastomose with each other around the inferior border of the heart. Additional

Figure 16-6. Coronary blood vessels. Arteries are at the left, veins at the right.

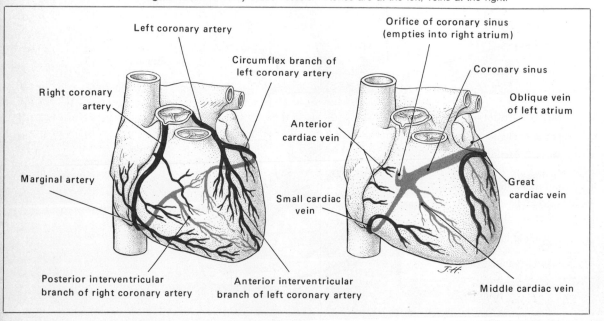

branches of the coronary arteries include a marginal branch of the circumflex, which extends along the left border of the heart to supply the left ventricle, and the marginal branch of the right coronary, which extends along the inferior margin of the heart to supply the right ventricle.

Coronary Veins   Veins accompany the branches of the coronary arteries. The great cardiac vein ascends in the anterior interventricular sulcus. As it reaches the atrioventricular sulcus it becomes the coronary sinus. The middle cardiac vein courses in the posterior interventricular sulcus to empty into the midportion of the coronary sinus.

Among the smaller veins returning blood to the right atrium by way of the coronary sinus are the small cardiac vein and the oblique vein of the left atrium.

In addition to this major pathway, there are other means for the blood from the coronary system to be returned to the general circulation. The anterior cardiac veins carry blood from the anterior surface of the right atrium directly into the right atrium. The venae cordis minimae (Thebesian) veins extend from the coronary capillaries directly to all heart chambers, but most enter the right atrium. There are also arterioluminal and arteriosinusoidal veins that shunt the blood from the coronary arterioles into the heart chambers. Yet most of the coronary blood flow reenters the general circulation via the coronary sinus.

Innervation of the Heart

The heart is supplied with motor nerve fibers from both the parasympathetic and sympathetic divisions of the autonomic nervous system. The parasympathetic fibers are branches of the vagus nerve. Two of these branches arise in the neck region. Additional branches issue from the vagus as it courses through the thoracic cavity. These fibers are inhibitory in function. Fibers of the sympathetic division leave the sympathetic trunk and course to the heart as the superior, middle, and inferior cardiac nerves. The thoracic portion of the sympathetic trunk also sends fibers to the heart. They are acceleratory in effect. All the visceral motor fibers together form the cardiac plexus. Impulses pass from the cardiac plexus to the conduction (Purkinje) system of the heart.

THE HEARTBEAT

The physiology of skeletal and cardiac muscle was considered in Chapter 7. The properties that characterize cardiac muscle are (1) a prolonged repolarization phase, (2) contraction as though it were a syncytium, and (3) inherent rhythmicity. The explanation of the prolonged repolarization is not yet known. A difference between skeletal and cardiac muscle in the conductance of potassium has been reported, but this poses more questions than it answers. The intercalated disks probably account for cardiac muscle contracting as a functional syncytium. The instability of the membrane of the sinoatrial nodal cells (see ahead) gives rise to prepotentials and therefore spontaneous and rhythmic contractions. The rhythm, as will be detailed later, can be modified by the activity of the autonomic nervous system.

Origin and Transmission of the Heartbeat

Histologic and physiologic investigations have proved that the beat of the heart originates in the right atrium.

The **Sinoatrial Node**   Microscopically, a specialized tissue mass may be discerned in the wall of the right atrium at the upper end of the crista terminalis just at the point of entry of the superior vena cava. This mass is composed of interwoven thin strips of modified cardiac muscle. It is termed the sinoatrial node (Figure 16-7) and popularly referred to as the S-A node. The S-A node is supplied with fibers from both divisions of the autonomic nervous system. Impulses which arrive over these pathways modify activity of the node and thus influence the heart rate.

Recordings of action potentials generated by the cells of the S-A node show prepotentials (see Figure 16-23). This finding is consistent with other evidence pointing to the S-A node as the part of the heart that originates the beat. Spread of the action potential from the S-A node activates the rest of the heart muscle. In the embryo, cells that are destined to develop into the S-A node display inherent rhythmicity.

The Atrioventricular Node   The impulse spreads throughout the atria but is not transmitted across the atrioventricular septum, which separates the atria from the ventricles. Instead it activates another specialized mass of tissue lying between the atria and ventricles, the atrioventricular node, or A-V node. This node is made of thin strips of interwoven modified cardiac muscle and is similar, in this respect, to the S-A node. The A-V node is the only normal pathway by which the impulse can be propagated from the atria to the ventricles. If this node is destroyed, as it may be by disease, the atria and ventricles then beat completely independently of one another and at different rates. Such a condition is called complete heart block. On the other hand, if the disease process is not extensive, so that impulses do pass through the node but only after a considerable delay, or if not all the atrial impulses find the A-V node, the disorder is spoken of as partial heart block. In such cases the atria may beat two or three times to each beat of the ventricles, but

Figure 16-7. Conduction system of the heart.

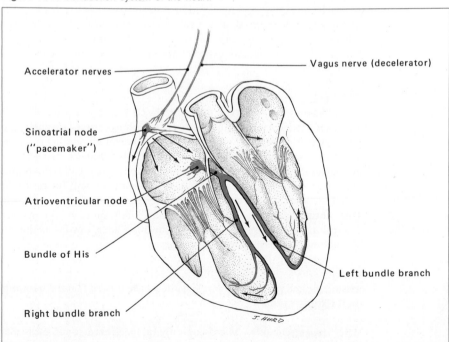

there is a definite ratio between the atrial and ventricular contractions, which indicates that some of the impulses are being transmitted.

Atrioventricular Bundle and Branches The impulse does not fan out concentrically through the ventricles as it does in the atria, because there is a conducting pathway in the ventricles. Figure 16-7 shows that there is a strand of elongated tissue continuous with the A-V node. This is called the atrioventricular bundle, or bundle of His in honor of the man who first described it. The bundle of His splits in two, each part forming a long conducting pathway that passes to the individual ventricles. The fibers that make up the bundles are called Purkinje fibers, again in honor of the man who first described them. The bundles of Purkinje fibers course downward from the A-V node on either side of the septum on the inner surface of the heart. When they reach the apex of the heart, they then turn back toward the base. All along the course of the bundles, the fibers give off small branches. The ends of the Purkinje fibers enter the muscle of the ventricle from the endocardial surface.

By virtue of the Purkinje system, the impulse is quickly propagated to almost all of the muscle that makes up the ventricles. As a result, the ventricles contract practically simultaneously, the result of this contraction being a sharp, rapid rise in intraventricular pressure. Were it not for this specialized conducting system, the impulse would spread slowly from the base to the apex of the heart and thereby produce a rather ineffectual contraction insofar as the pumping of blood is concerned.

Intrinsic Cardiac Rhythm A completely denervated human heart continues to beat at a rate of 70 to 80 beats per min. This is the rate of discharge of the S-A node. The A-V discharges more slowly if it is isolated from external stimuli. Its intrinsic rate is about 50 beats per min. The Purkinje fibers also have an intrinsic rhythm, which is even slower than that of the A-V node. There is thus a gradient, the fastest rate being in the S-A node. For this reason, the S-A node normally sets the pace of the heart, and it is often referred to as the pacemaker. Each time the S-A node fires, the impulse spreads to the A-V node which, in turn, is activated. Before the A-V node can fire again in response to its own slow intrinsic rhythm, another impulse generated by the S-A node again activates it.

Under abnormal conditions, the A-V node may initiate the beat; the impulse then spreads downward through the ventricles and also upward through the atria. Other parts of the heart may also initiate the beat. In such cases, the initiating area is called an ectopic pacemaker.

Even in a normally beating heart there is an occasional impulse initiated by an area of the heart other than the S-A node. This impulse, if it occurs during a time when the cardiac muscle is not in the refractory period, will cause contraction. Such contractions are referred to as ectopic beats, or ectopic systoles. If they occur in addition to the regular beats, they are called extrasystoles. If they simply cause early contraction, then the proper term is premature systole.

ELECTROCARDIOGRAPHY

There is electrical activity associated with the contraction of cardiac muscle which may be recorded by suitable equipment. Such records have proved invaluable both in the physiology laboratory and in the diagnosis and treatment of disease. A complete discussion of electrocardiography is clearly beyond the scope of this book. The basic principles are summarized here because such a study emphasizes many fundamental physiological factors of cardiology. In addition, many paramedical personnel, such as intensive-care unit nurses, are now expected to understand electrocardiography.

Instrumentation

In order to monitor the electrical activity of the heart, a broad variety of instruments have been developed to serve specific purposes. All the modern instruments consist of a system which amplifies the weak potentials emanating from the cardiac muscle. Once sufficiently amplified, the potential may be displayed on a screen or used to activate a pen or a stylus for direct writing.

The instrument used to amplify and record the electrical activity of the heart is called the electrocardiograph. The record so produced is referred to as an electrocardiogram and is abbreviated EKG or ECG.

String Galvanometer One of the first electrocardiographs consisted of a very sensitive string galvanometer. The disadvantage of this instrument is that only one lead can be recorded at a time, and then the record must be developed before it can be studied. For these reasons, the polygraph is most often used at the present time.

Polygraph The polygraph, as the term indicates, is a multiwriting instrument. Ink-writing pens, or styluses with special paper that does not require ink, are generally used. The electrical impulse from the surface of the body is amplified sufficiently to activate the pen, or stylus. The paper moves at a suitable speed under the pen, and therefore the oscillations of the pen describe a continuous record on the moving paper. Usually there are three or more pens available. Each pen along with its amplifying circuitry is referred to as a channel. Because there are several channels, action potentials from several different leads may be recorded simultaneously.

Oscilloscope The cathode-ray oscilloscope is widely used in the operating room, in intensive-care units, and elsewhere to monitor cardiac function. The beam of electrons that is caused to sweep across the face of the screen oscillates in accord with the potential difference between the leads wired to the instrument. Thus, instead of a moving pen that writes on paper, there is a moving beam of electrons displayed on the screen. The beam sweeps from one side to the other describing the electrical activity of the heart as it traverses the screen. Such a system does not produce a permanent record but it may be left on continuously and checked periodically by the physician, nurse, or investigator.

Electrocardiographic Leads

The electrodes may be placed directly on the heart, as is sometimes done in experimental animals, or they may be attached to most any part of the body. The relative position of the two electrodes will, of course, influence the direction and amplitude of the record; therefore a conventional system has been developed. The particular arrangement of the two electrodes is termed a lead.

Electrodes Usually electrodes are made of metal and are slightly concave so as to make good contact with the skin in the regions of the wrists and ankles. A jelly consisting of an electrolyte and an abrasive is first rubbed over the surface of the skin. The abrasive removes dead cells and other accumulations that would interfere with the conduction of the impulse. The electrolyte forms a low-resistance surface between the skin and the metal electrode and thus facilitates the conduction of the impulse. In place of the jelly, thin paper pads presaturated with an electrolyte solution are often preferred because they are inexpensive, less messy than the jelly, and convenient to use. The electrodes are securely held by means of wide rubber straps.

The position of the electrodes on the body and the manner in which they are wired to the recorder will, as already mentioned, influence the record. Three so-called standard leads are therefore described. Today, most electrocardiograms include records taken from 12 leads. They are the three standard leads, the six chest leads, and the three unipolar, or augmented, leads (Figure 16-8).

Standard Leads Lead I requires that the negative terminal of the electrocardiograph be connected to the electrode on the right arm and the positive terminal to the left arm. Thus when the right arm is negative to the left arm, there will be a positive or upward movement of the recording pen. For lead II the negative terminal is connected to the right arm, the positive terminal to the left leg. For lead III, the negative terminal is connected to the left arm, and the positive one to the left leg.

The cardiac impulse is initiated in the S-A node, an area which lies closer to the right arm than it does to the left arm or left leg. The impulse may be thought of as a wave of negativity; therefore when the impulse is initiated, the right arm becomes negative in reference to the left arm or left leg. Because of the conventional wiring of leads I and II, the first deflection of the record is normally positive, that is, upward, in those leads (Figure 16-8). In lead III, the first deflection is also upward, because the S-A node is closer to the left arm than it is to the left leg. Therefore the left arm becomes negative in reference to the left leg.

Chest Leads An indifferent electrode is formed by uniting all the standard leads, that is, from the two arms and the left leg, through resistances of 5,000 ohms each to a central terminal, and this is connected to the negative terminal of the recorder. The potential at the central terminal remains at zero. The positive terminal is connected to an electrode which is placed on various areas of the chest (Figure 16-9). This, then, is the sole determinant of the deflections. In diagnostic work six positions on the chest are recognized. They are labeled V_1, V_2, and so forth (Figure 16-8).

Figure 16-8. Twelve-lead electrocardiogram. Shown are the three standard leads, the three unipolar leads (augmented limb leads), and six chest leads.

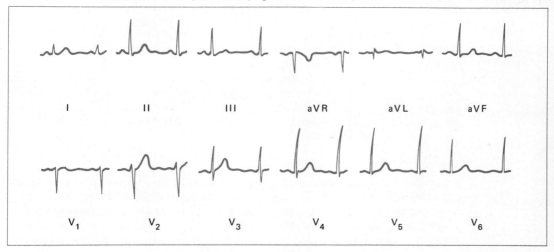

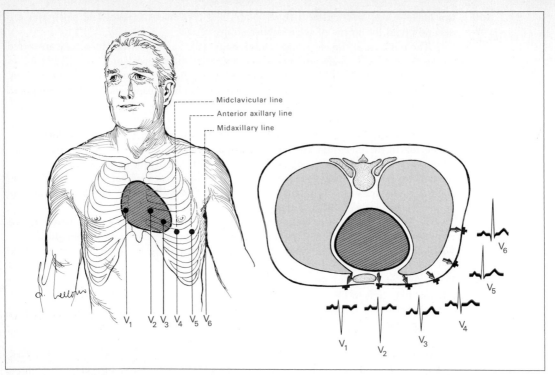

Figure 16-9. Positions for placement of chest leads.

Unipolar Limb Leads For these leads, two of the standard leads are connected through suitable resistance (5,000 ohms) to a common terminal which is connected to the negative terminal of the recorder. The remaining lead goes to the positive terminal. In order to identify the arrangement, the following symbols are used: aVF when the positive terminal is connected to the left leg, aVR when it is to the right arm, and aVL when it is to the left arm. These leads are also referred to as augmented limb leads because the magnitude of the deflections is increased by virtue of this type of wiring.

The augmented leads have the advantage over the standard limb leads that they are controlled almost exclusively by potentials from one particular limb. Thus the resulting record "views" the heart from the root of that limb.

In the aVR lead, the deflections are all inverted (Figure 16-8). This is to be expected because the right arm is connected to the positive terminal. In the standard limb lead, the right arm is connected to the negative terminal. In the aVL lead, the deflections may be upright or inverted. It depends upon the electrical axis (see ahead) of the heart. The same is true of the aVF lead.

The Normal Electrocardiogram

A normal electrocardiogram is depicted in Figure 16-10. The record is seen to be superimposed upon a graph. The lines of the graph have been standardized. The vertical lines indicate time in seconds, the horizontal lines amplitude in millivolts (mv). The interval between two vertical lines measures 0.04 sec. The distance between each horizontal space is 1 mm and is equivalent to 0.1 mv. This means that the instrument

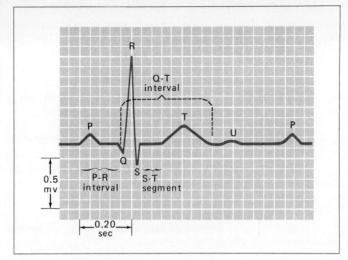

Figure 16-10. The normal electrocardiogram.

must be adjusted so that an input of 1 mv produces a pen deflection of 1 cm. The normal paper speed is 25 mm/sec.

P Wave   The impulse is initiated by the S-A node and spreads over the atria. This is the first event in the cardiac cycle. The P wave represents the electrical activity associated with the spread of the impulse over the atria, that is, the wave of depolarization. The P wave, because of the conventional wiring of the three leads, is normally upward in all three. Note, however, that the amplitude of each wave varies from lead to lead (Figure 16-8). The average duration of the P wave is about 0.08 sec and not more than 0.12 sec (see Table 16-1).

QRS Complex   The wave of depolarization spreads from the S-A node concentrically over both atria. Thus the P wave is a simple, upward, smooth curve. In the ventricles,

TABLE 16-1. NORMAL EKG VALUES

	DURATION, SEC		AMPLITUDE, MV
	AVERAGE	RANGE	
P Wave	0.08	0.06–0.12	Not over 0.3
QRS Complex	0.08	0.06–0.10	Not over 2.5 in any lead. Not under 0.5, average of 3 standard leads
P-R Interval	0.16	0.12–0.20	
T Wave	0.16		Not over 0.5 in any standard lead
Q-T Interval	0.36*	0.31–0.38*	

*At a heart rate of 71 beats/min

however, the impulse spreads out over the Purkinje system to activate both chambers almost simultaneously. But since there are really two waves of depolarization proceeding in different directions, the electrical manifestation is complex. The QRS complex represents this activity. It is the resultant of all the electrical activity occurring in the ventricles. Thus its shape, amplitude, and direction depend upon the position of the heart, the relative mass of each of the ventricles, and the time relationships between their activity. In many records, the Q wave is absent. The duration varies with the heart rate, usually from 0.06 to 0.10 sec.

P-R Interval The P-R interval is measured from the beginning of the P wave to the beginning of the QRS complex. But since the Q wave is so often absent, the interval is termed P-R rather than P-Q. This interval indicates the time which elapses between the activation of the S-A node and the activation of the A-V node. The P-R interval normally measures about 0.16 and is not more than 0.20 sec.

T Wave The QRS complex represents depolarization of the ventricles. Repolarization is represented by the T wave. The amplitude of the T wave is generally about 0.3 mv in the standard leads, but this is highly variable. The duration is about 0.16 sec.

Q-T Interval The Q-T interval is measured from the beginning of the QRS complex to the end of the T wave. It represents electrical systole, that is, the period during which all the electrical activity associated with the contraction of the ventricles occurs. The length of this interval varies with the heart rate. At a normal heart rate of 70 beats/min, the Q-T interval is about 0.36 sec. The faster the beat, the shorter the Q-T interval. For example, at a rate of 120 beats/min this interval may be only 0.28 sec.

S-T Segment The S-T segment is that part of the record which connects the S and T waves. Normally it is level to the base line and is isoelectric, that is, at most only slightly above or below the base line. The significant aspects to look for are its position and shape.

Rhythm The rhythm of the heart should, of course, be fairly regular. The rate of the heart does vary slightly with the respiratory cycle, and this is considered normal. A glance at the electrocardiogram will disclose whether or not there exists an exaggerated variation. Irregular rhythm is termed arrhythmia.

Heart Rate The heart rate may be easily calculated from the EKG. The R wave is usually used because it is so sharp. Since the usual EKG has heavy vertical lines which indicate 0.20 sec, one need only count the number of large spaces between two R waves and divide that number into 300. The figure 300 is used because 0.20 second equals one-fifth of a second and there are 300 fifths of a second in 1 min. Thus, if four large spaces separate two R waves, a rate of 75 is indicated. This simple principle has been incorporated into a rate estimator which is placed on the record with the first line on an R wave. The rate is read off the rule three R waves to the right (Figure 16-11).

Electrical Axis If the three standard leads are connected by hypothetical lines, a figure known as Einthoven's triangle is formed. The heart lies within this triangle. It is therefore possible to calculate, from any two of the leads, the direction of the spread of the electrical impulse during the cardiac cycle. This resultant of the electromotive forces occurring in the heart is termed the electrical axis. Its direction is indicated by an arrow drawn on Einthoven's triangle.

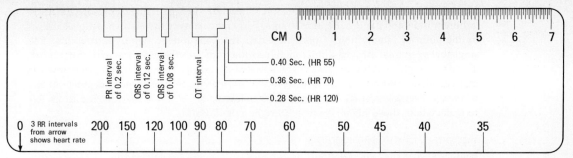

Figure 16-11. Heart-rate estimator. To determine the heart rate from the EKG, the arrow is placed on an R wave. The rate is the point on the scale three R waves to the right.

There are several ways in which to calculate the electrical axis. The record from any two leads may be used. The results may be checked by also plotting the values taken from the remaining lead. Figure 16-12 shows three hypothetical records. An equilateral triangle is also shown wired in accord with standard convention. To plot the data from the records, the upward and downward deflections of each QRS complex are expressed in arbitrary units, usually horizontal lines, or millimeters. For example, in lead I it is seen that the upward deflection is +3, downward −1. The resultant deflection is +2.

Figure 16-12. Calculation of the electrical axis. The net deflections of a QRS complex have been plotted on the equilateral triangle. Here all three leads have been plotted. Only two need be used.

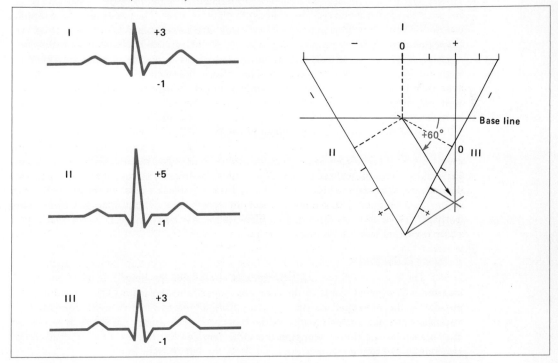

413

Lead III has the same resultant deflection. These are then plotted on the triangle using arbitrary units and paying attention to sign. Perpendiculars are then drawn. A line is drawn to the point of intersection of perpendiculars from the intersection of zero points. The angle between this line and a horizontal base line indicates the electrical axis of the heart. In this example, a value of +60 degrees was obtained. By convention, angles below the horizontal are positive, above negative. The normal range is 0 to +90 degrees. In the adult the average is +58 degrees. Values below 0, or negative, are referred to as left axis deviation; those higher than +90, right axis deviation.

Abnormal Electrocardiograms

Only a few of the more obvious abnormalities that may be detected by electrocardiography will be mentioned.

Abnormal Rates and Rhythms   The rate of the heartbeat and its rhythm are easy to ascertain from the electrocardiogram. The normal rate is about 70 beats/min. A very slow rate, termed bradycardia, unassociated with other cardiovascular abnormalities, is not considered pathological. Increased rate, termed tachycardia, is more often associated with disease. As indicated above, slight variation in the rhythm is normal. Pronounced arrhythmia suggests a cardiovascular disorder. One such example is paroxysmal tachycardia, in which the heart will speed up for a few seconds and then slow down.

The rate may become so great that the heart resembles the rapidly fluttering wings of a bird. For this reason, rates over 200 beats/min are referred to as flutter. The atria are more prone to flutter than the ventricles. Rates up to 350 beats/min have been recorded. Despite the rapid rate the contractions are coordinated, that is to say, the sequence of events is the same as at normal rates, the only difference being one of rapidity. However, contraction may become discoordinated, a state in which small areas of myocardium contract and relax independently of other areas. The rate is very rapid. Rapid discoordinated contraction is termed fibrillation because individual fibers, rather than the myocardium as a whole, contract. Fibrillation of the atria is compatible with life, but ventricular fibrillation is not. A fibrillating ventricle is incapable of elevating pressure high enough to open the aortic valve. No blood is pumped. Atrial fibrillation, on the other hand, prevents the atria from pumping blood into the ventricles, but such atrial pumping probably accounts for only about 25 percent of ventricular filling. The greater percentage is due to venous pressure. Thus, though atrial fibrillation decreases the efficiency of the heart, it is not fatal as is ventricular fibrillation.

Heart Block   First-degree heart block can only be detected by use of the EKG. A P-R interval in excess of 0.20 sec is indicative of this condition. The record will also disclose second- and third-degree heart block. If second-degree block exists, the record will show more P waves than QRS complexes, but there will be a definite ratio between them, perhaps 2:1 or 3:1. In third-degree block there are many more P waves than QRS complexes, and there is no discernible ratio.

Bundle Branch Block   Figure 16-13 shows a record representing bundle branch block (BBB). The duration of the QRS complex is abnormally prolonged (0.12 sec or more), and the S-T segment slants in the direction opposite to the main QRS deflection. The reason for the prolonged complex is the failure of the two ventricles to beat simultaneously. The delay of the impulse in the blocked branch causes the ventricle served by that branch to lag, thus prolonging the QRS complex. By the use of chest leads, it is possible to determine whether the block is in the left or right branch.

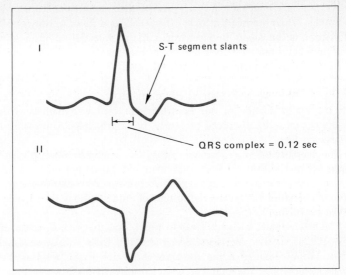

Figure 16-13. Bundle branch block. Note the long QRS complex and the abnormal S-T segment.

Myocardial Infarction An infarct is a region of dead tissue resulting from the interruption of the blood flow. Such an area in the heart cannot undergo depolarization and repolarization; therefore infarction is often represented by an abnormality in the EKG. The two most important findings are (1) elevation of the S-T segment, and (2) inversion of the T wave. In addition, if a record is available of the heart before the infarction occurred, comparison generally shows the Q waves to be more prominent following an infarct (Figure 16-14).

THE CARDIAC CYCLE

The pressure of the blood when it enters the right atrium is very low. The activity in the right ventricle elevates blood pressure to about 30 mmHg, while at the same time

Figure 16-14. Myocardial infarction. The characteristic findings are indicated.

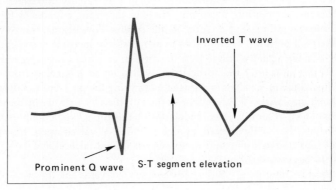

the left ventricle increases the pressure of the blood ejected into the aorta to about 120 mmHg. When the heart muscle contracts, blood flows in only one direction because of the action of the valves. In other words, in the short period of one cardiac cycle, which at a normal heart rate of 72 beats/min occupies less than a second, the valves must open and close rapidly, synchronously, and efficiently.

Physical Principles of Pressure Relationships

Since the major portion of our knowledge of cardiac dynamics is deduced from the pressure curves to be discussed presently, basic principles of pressure relationships will be reviewed first.

The pressure within a chamber represents a ratio between the size of the chamber and the volume of fluid forced into it. If the fluid does not fill the available space, there is no pressure. In order to exert a positive pressure, the chamber must be overfilled, so to speak. Therefore, the greater the volume of fluid in ratio to the capacity of the container, the higher will be the pressure.

There is another factor to bear in mind, and that is the wall of the compartment. If the wall is inelastic, then a small increase in volume will cause a great and sharp rise in the internal pressure. If the container is ideally elastic, then as the volume of fluid increases, the size of the chamber will also grow. If these increments are proportional, the ratio between size and volume will remain constant, and so will the internal pressure. Actually, of course, this degree of elasticity is seldom encountered. In the heart, for example, especially in the thin-walled and highly elastic atrium, a great increase in volume of blood produces only a small elevation in pressure.

The pressures attained in the left side of the heart far exceed those in the right, but aside from this quantitative difference, the events which occur during the cardiac cycle are identical on both sides. Accordingly, the ensuing discussion refers only to the left side of the heart, but a simultaneous and very similar sequence of events takes place in the right chambers.

The Atrial-Pressure Curve

Through much of diastole the pressure within the atrium steadily increases (Figure 16-15). Diastole is the phase of the cardiac cycle during which the muscle is not contracting. Blood flows continuously into the atrium throughout diastole, and this increases the volume of blood in relation to the size of the atrium. Next note the sudden increase in pressure (wave a). This wave is caused by contraction of the atrium, so-called atrial systole. The term systole describes that phase of the heart during which the muscle contracts.

At line 1, the ventricles begin to contract. As a result of this sharp, vigorous contraction, the pressure within the ventricle rises rapidly and quickly becomes greater than the pressure within the atrium. There is, therefore, a tendency for the blood to flow back into the atrium, but as soon as this reverse flow begins, the A-V valves are swept to the closed position. The intraventricular pressure increase is so sudden and sharp that the A-V valve actually bulges into the atrium, and this causes wave b.

Finally (at line 2) the aortic valve opens, blood is ejected from the ventricle, and the force of the A-V valves is released, permitting the pressure to fall in the atrium. From this point until the A-V valves reopen (line 6), there is a steady increase in atrial pressure because of the accumulation of blood during this time. When the pressure in the ventricle falls below that in the atrium, at line 6, the A-V valves open, blood from the atrium rushes into the ventricle, and therefore the intraatrial pressure falls. The cycle is now ready to begin again.

Because there are no valves between the great veins and the atria, blood flows without

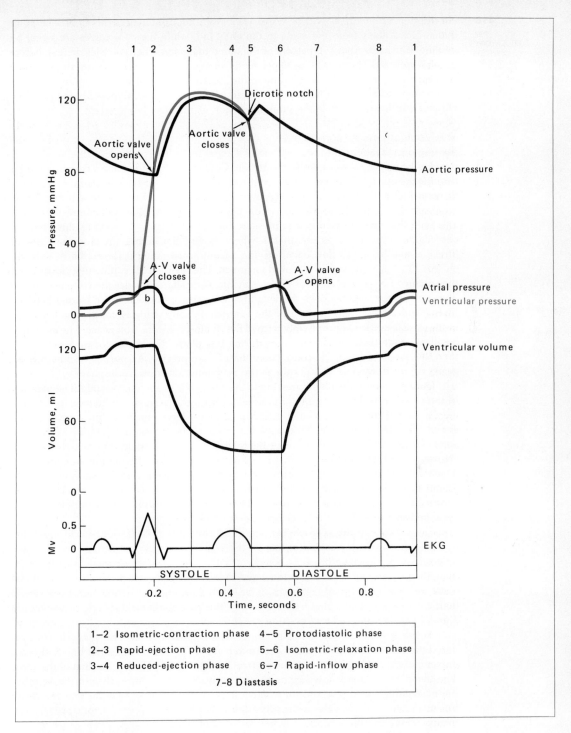

Figure 16-15. The cardiac cycle with the EKG for comparison.

interruption, although, as will be noted later, respiration influences this flow. The point is that blood flows from the veins to the atria and, while the A-V valves are open, on into the ventricles. About 70 percent of the blood that enters the ventricles does so before the atria contract. The remaining 30 percent is pumped into the ventricles by atrial systole. At rest, this contribution is not important, because the ventricles by more vigorous contraction and by a faster heart rate can markedly increase the volume of blood pumped. However, in heavy exercise the pumping action of the atria is important.

The Ventricular-Pressure Curve

Starting once again at the extreme left of Figure 16-15, the pressure in the ventricle is noted to be slightly lower than in the atrium. Consequently, blood flows from the atrium through the open A-V valve into the ventricle. Therefore, as these two chambers steadily fill with blood, there is a progressive increase in pressure. The rise at *a* is caused by atrial systole. At line 1, ventricular systole begins. The ventricular wall contracts forcibly as a unit; thus the intraventricular pressure increases sharply. As soon as this pressure exceeds the intraatrial pressure, that is, where the two lines cross, the blood reverses its direction and begins to flow back into the atrium. In so doing it forces the A-V valves closed. The ventricle now is a closed chamber. The wall is still contracting down upon the volume of blood within. This elevates the pressure sharply. Since the volume of blood does not change, neither does the length of the cardiac fiber. In other words, between lines 1 and 2 the cardiac muscle of the ventricle is contracting isometrically. This part of the cardiac cycle is appropriately termed the isometric-contraction phase. The argument has been made that the cardiac fibers during this phase do shorten to some extent and therefore the term isometric-contraction phase is not precise. Accordingly, some investigators prefer to refer to this period as the isovolumetric-contraction phase.

During diastole, the blood pressure in the aorta falls to about 80 mmHg. The pressure in the ventricle at the beginning of systole is very low, perhaps 5 or 10 mmHg. In other words, during the isometric-contraction phase, the intraventricular pressure must be sharply elevated. Ultimately it exceeds that in the aorta. Consequently, blood forces the aortic valves open and rushes out of the ventricles. The ventricle continues to contract vigorously and, despite the fact that blood is now being ejected, the pressure continues to mount. Because of the great pressure which has been built up between lines 2 and 3, blood leaves the ventricle very rapidly. This can be confirmed by noting the ventricular-volume curve at the bottom of Figure 16-15. Accordingly, the part of the cardiac cycle between lines 2 and 3 is termed the rapid-ejection phase.

The pressure in the ventricle beyond line 3 is seen to fall off. But bear in mind that up until line 4 systole continues; that is, the ventricle is still contracting. The pressure falls off simply because blood is being ejected faster than the ventricle is contracting, thus altering the ratio between volume and capacity. A glance at the ventricular-volume curve indicates that blood continues to be ejected during this period but more slowly than during the previous phase. Consequently, this part of the cardiac cycle between lines 3 and 4 is called the reduced-ejection phase.

At line 4 systole ends. The ventricular wall begins to relax. The ventricle contains very little blood. For this reason a precipitous fall in pressure begins. At line 5, the two curves cross, indicating that the intraventricular pressure has fallen below that of the aorta. The blood in the aorta now begins to reverse its flow and run back into the ventricle but in so doing, slams shut the aortic valve. The short period between the end of systole, line 4, and the closure of the aortic valve, line 5, is termed the protodiastolic phase. Proto means "first." In other words, it is the first part of diastole.

Once again, the ventricle is a closed chamber. The cardiac muscle continues to relax, but since no blood is entering the chamber, the length of the cardiac fiber remains virtually unchanged. Hence, the part of the cardiac cycle between lines 5 and 6 is called the isometric-relaxation phase.

The fall in intraventricular pressure continues until it is lower than that of the atrium. When it is, the two curves cross, at line 6. Now the blood begins to flow into the ventricle from the atrium and, in so doing, it opens the A-V valve. The volume of blood which, between lines 2 and 6, has built up in the atrium now rushes into the ventricle. Once again this fact may be confirmed if the ventricular-volume curve is noted. Appropriately, the part of the cardiac cycle between 6 and 7 is termed the rapid-inflow phase. Lastly, if the heart rate is slow, there will be a short period during which there is relative quiescence throughout. This phase is called the diastasis. The cycle is complete and ready to begin again. Table 16-2 outlines the time relationships of the various cardiac phases.

The Aortic-Pressure Curve

The aorta is very distensible. Therefore, changes in its capacity reflect this distensibility. The pressure in the aorta, at any time, just as in the chambers of the heart, simply expresses this relationship between the capacity of the vessel and the volume of blood contained.

Once again, beginning at the extreme left of Figure 16-15, the pressure in the aorta is seen to be falling progressively. At this time the aortic valve is closed, and the volume of blood ejected into the aorta is dissipated into the vascular tree. The aorta, by its inherent elasticity, closes down upon the volume of blood, but not fast enough to keep the pressure constant. The role played by the distensibility and elastic recoil of the great blood vessels must be appreciated. If they were simply rigid tubes like lead pipes, as soon as the heart ceased to eject blood, the pressure in the aorta would fall precipitously, perhaps even to zero, before the next ejection phase. In other words, it is the elastic recoil of the aorta and other large blood vessels which helps to maintain the blood pressure during diastole.

At line 2, the aortic valve opens and a large volume of blood is ejected rapidly into the aorta, thereby sharply elevating the pressure. The wall of the aorta is stretched by the blood volume. Once again, if the vessels were completely rigid, the capacity would

TABLE 16-2. APPROXIMATE DURATION OF PHASES OF THE CARDIAC CYCLE*

PHASE	DURATION, SEC
DIASTOLE:	
Protodiastolic	0.04
Isometric relaxation	0.08
Rapid inflow	0.09
Diastasis	0.18
Atrial systole	0.09
Total	0.48
SYSTOLE:	
Isometric contraction	0.08
Rapid ejection	0.12
Reduced ejection	0.15
Total	0.35

* Heart rate 72 beats/min

remain constant while the volume rapidly increased, and thus the aortic pressure would soar. The distensibility and elastic recoil of the aorta and the other large blood vessels serves two functions: (1) to maintain the blood pressure during diastole and (2) to buffer the systolic pressure.

When diastole begins, the pressure in the ventricle falls rapidly and the two curves cross. At this point, line 5, there is a tendency for the blood to reverse its flow. However, the aortic valve closes, and so there is a rebound of blood which causes a sudden but transient increase in the aortic pressure. The dip, or notch, observed at this point is called the dicrotic notch.

After the flow of blood in the aorta has once again become smooth and unidirectional, the aortic pressure steadily decreases until the aortic valve opens once again.

The Cardiac Valves

The function of the cardiac valves is to prevent blood from flowing back into the atria from the ventricles, and into the ventricles from the aorta and pulmonary artery. The valves are passive structures that respond to pressure differentials. For example, when the blood pressure in the atria exceeds that in the ventricles, the A-V valves open; when the reverse relationship obtains, the valves close.

The Papillary Muscles   The leaflets of the A-V valves are attached to the chordae tendineae, which are held by the papillary muscles. This system prevents the valves from bulging into the atria during systole when the ventricular pressure increases sharply. Actually the papillary muscles contract along with the ventricular myocardium to prevent this bulging. In cases of coronary occlusion, the resulting ischemia sometimes paralyzes the papillary muscles. In such cases the A-V valves, especially on the left side, bulge into the atria to such an extent that blood regurgitates from the ventricles into the atria. In severe cases this regurgitation is of such a magnitude that the cardiac output falls below levels compatible with life.

THE HEART SOUNDS

Because a large quantity of blood flows through the heart rapidly, not in a steady stream but rather in a stop-and-go pattern, turbulence results causing definite sounds to be produced during the cardiac cycle. The physician uses a stethoscope (*stethos*, "chest"; *skopeo*, "I examine") to listen to these sounds. One can listen to heart sounds without such an instrument by placing the ear close to the chest. This was the accepted procedure before the invention of the stethoscope.

For a record of heart sounds, the phonocardiograph is used. This instrument has a microphone capable of detecting low-frequency sound waves. The microphone converts the sound waves to electrical impulses which are amplified for display on an oscilloscope or to drive a pen for recording. The record is termed a phonocardiogram.

Normal Sounds

Two distinct normal heart sounds can easily be detected. In some cases, a third sound may also be heard. The third sound is actually present in almost all individuals and is considered to be a normal component. However, in many people it is too faint to be detected without special equipment.

First Sound   In Figure 16-16 there is a record of the heart sounds. This tracing serves to place the sounds in their exact relationship to the cardiac cycle. Note that the first heart sound occurs at the beginning of systole. A glance at the pressure curves indicates that at this instant there is a great deal of commotion in the heart which could readily account for the sound.

As soon as the ventricle begins to contract, the reverse flow of blood slams the A-V valve closed. The more vigorously the ventricle contracts, the louder the first sound. A fraction of a second after the A-V valves close, the aortic and pulmonic valves open. Actually, the great pressure in the ventricles snaps the valves open and sets them to vibrating. Such vibration gives rise to sound waves which are heard as a part of the first heart sound.

Second Sound Figure 16-16 shows that the second heart sound begins at the onset of diastole. The major contributor is the closing of the aortic and pulmonic valves. A fraction

Figure 16-16. The heart sounds with the pressure curves for comparison.

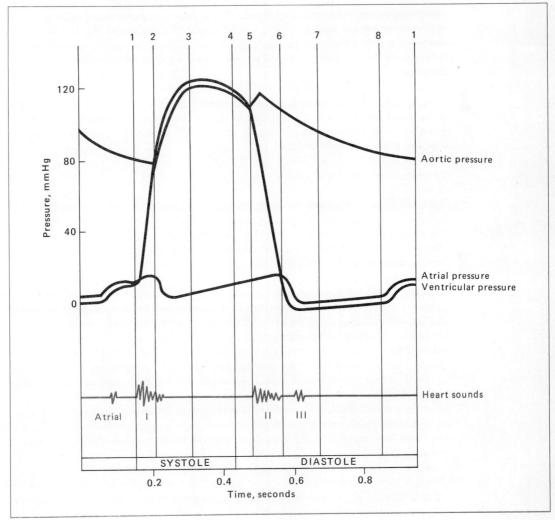

of a second later the A-V valves open, and they also participate in the production of the second sound.

The two major heart sounds serve to identify the particular phase of the cardiac cycle. If one listens to a normal heart, he will hear lub-dub, lub-dub, lub-dub. Table 16-2 lists the length of systole and diastole. Note that, at the normal heart rate of 72 beats per min, diastole is considerably longer than systole. Thus, one hears lub-dub, pause, lub-dub, pause. The lub is the first heart sound and marks the beginning of systole. The dub is the second sound and indicates the beginning of diastole. There is then a brief pause before systole begins again.

The two major normal heart sounds are of great importance to the physician. Their intensity and quality indicate the vigor of cardiac muscle contraction. In the defective heart, they mark the phase of the cardiac cycle and thereby serve to place abnormal sounds correctly. As will be pointed out, a pathologic sound heard in systole indicates a very different condition than an unusual sound detected during diastole.

Third Sound As already indicated, the third heart sound is usually very faint and difficult to hear in the average person without special amplifying equipment. Figure 16-16 shows that this sound is present during the rapid-filling phase. It is thought to be due to the vibrations of the walls of the ventricles occasioned by the rapid inrush of blood from the atria.

Atrial Sound The atrial sound, sometimes referred to as the fourth heart sound, occurs during atrial systole (Figure 16-16). It has such a low frequency—about 20 cycles per second (cps)—that the human ear cannot detect it. The only way it can be demonstrated is by use of the phonocardiograph.

Abnormal Sounds

Gallop Rhythm If the heart is beating very rapidly and if the third sound is audible, three distinct sounds will be heard in rapid sequence. The effect will be similar to that of a galloping horse—thus the name. But since the third sound is not ordinarily heard, a fast heart rate alone does not usually produce a gallop rhythm. In most cases, a gallop rhythm indicates serious heart damage.

Murmurs A murmur is a soft sound which resembles the sound produced by forcible expiration with the mouth held partially open. Cardiac murmurs result from turbulence which causes vibrations and therefore produces sound waves. Such turbulence may be due to a very rapid blood flow, especially when the blood is forced through narrowed and roughened valves. Turbulence also occurs when the valves permit blood to flow back into the ventricles or atria, where it collides with the incoming stream.

Murmurs are, upon occasion, heard in perfectly normal hearts. During severe exercise, the flow of blood through the narrow valves may become so fast as to produce turbulence. But in most cases, murmurs indicate cardiac disorders. Often one or more of the valves are unduly narrowed, a condition referred to as stenosis. In addition, their walls are usually roughened. The rapid flow of blood through these small, roughened orifices results in a murmur.

If the cardiac cycle is kept in mind, the afflicted valve can be determined. For example, if the A-V valves are narrowed, then the murmur will be heard during diastole, notably during the rapid-inflow phase (Figure 16-16). If the aortic or pulmonic valves are stenosed, the murmur will be heard during systole, most conspicuously during the rapid-ejection phase. With experience one can readily differentiate between the sounds of the two A-V valves or between the pulmonic and aortic valves by moving the head of the stethoscope about until the murmur is heard at maximum intensity. This location will indicate the defective valve.

In some cases, the valves may not be narrowed but may close incompletely or insufficiently and thus fail to prevent the backflow of blood. This type of disorder is termed insufficiency, or regurgitation. Again the examiner must listen most carefully in order to place the murmur in relation to the cardiac cycle. For example, if he hears lub-hiss-dub, lub-hiss-dub, he knows that the murmur is occurring during systole. Having established this fact he must, by moving the stethoscope about, ascertain whether the A-V valves are producing the sound or whether the disturbance is in the aortic or pulmonic valves. If the latter, there is probably an aortic or pulmonic stenosis; if the A-V valves are responsible, they are probably insufficient and are allowing blood to regurgitate into the atria during systole.

CARDIAC OUTPUT

The pacemaker has an inherent rhythm of about 72 beats/min. However, this intrinsic cadence can be altered by many diverse forces. Not only can the heart rate change, but also the force of contraction can be varied.

The function of the heart is to propel an adequate quantity of blood to satisfy body needs. During exercise, the oxygen requirements of contracting skeletal muscle increase sharply. This demand can be met only by a greater supply of oxygenated blood. The respiratory system oxygenates the blood; the circulatory mechanism delivers it to the tissues. Here the factors which regulate the rate and force of cardiac contraction will be considered.

Throughout this discussion, the term cardiac output will be employed. By definition, cardiac output is the quantity of blood ejected by one side of the heart in one minute. Under normal conditions, the volume of blood put out by both sides of the heart is the same. Therefore, to calculate the quantity of blood pumped by the heart as a whole, the cardiac output is multiplied by 2. However, the two ventricles can pump different amounts of blood for short periods of time. If the output of the left ventricle is greater than that of the right, there will be a translocation of blood from the pulmonary circulation to the systemic circulation. If the right output is the greater, then blood will be shifted to the pulmonary vessels.

Stroke Volume

The quantity of blood ejected by one side of the heart per beat is the stroke volume. If the stroke volume is known, this value may simply be multiplied by the heart rate to ascertain the cardiac output. In man, the average stroke volume is about 80 ml. Since the heart rate is normally close to 72 beats per min, the cardiac output at rest is approximately 5 to 6 liters per min.

Determination of Cardiac Output

In experimental animals, flowmeters can be placed in the aorta to determine, directly, the cardiac output. In man, except during surgery, indirect methods must be used. The more important procedures for determining cardiac output clinically will be briefly described.

Fick Principle Using the Fick principle, cardiac output may be calculated from the amount of oxygen consumed per minute and the amount of oxygen taken up by each milliliter of blood as it passes through the lungs. These relationships are expressed by the following equation:

$$\text{Cardiac output (ml/min)} = \frac{O_2 \text{ consumed (ml/min)}}{\underset{\text{(ml/ml)}}{\text{arterial } O_2 \text{ content}} - \underset{\text{(ml/ml)}}{\text{venous } O_2 \text{ content}}}$$

Blood oxygen content is sometimes expressed as milliliters percent, that is, the number of milliliters of oxygen per 100 ml of blood. If so expressed, then in order to calculate the cardiac output, the numerator must be multiplied by 100.

To determine the volume of oxygen consumed per minute, the individual breathes from a known volume of oxygen for a period of time, generally about 6 min, and then, by ascertaining the oxygen remaining, the amount consumed can be calculated. A sample of arterial blood is drawn and its oxygen content determined. But because the oxygen content is not the same in the blood of all veins, the so-called mixed venous blood must be used. This is the venous blood just before it enters the right side of the heart. Such a sample is obtained by use of a cardiac catheter, a long small-bore tube that is introduced into the median cubital vein and moved through the vein until the tip enters the right atrium. The procedure is referred to as cardiac catheterization. Once the tip of the catheter is in the atrium, a sample of blood is withdrawn for analysis.

At rest, about 250 ml of oxygen are consumed per minute. The arterial O_2 content is about 0.195 ml/ml blood and the venous O_2 content is about 0.150 ml/ml blood. Accordingly, by use of the Fick equation, the cardiac output is calculated to be 5,556 ml/min.

Dye Method Cardiac catheterization invariably introduces certain difficulties; therefore the dye method is often the method of choice. In this procedure a known quantity of dye, usually Evans blue, known as T 1824, is injected into the median cubital vein. Immediately, small volumes of arterial blood are drawn in series and the dye concentration determined in each sample. Or the arterial blood can be made to pass through a transducer that converts dye concentration to a proportional electrical impulse. The impulse is then amplified to activate a recorder. In either case a curve, such as depicted in Figure 16-17, is obtained.

The secondary rise is due to circulation of the dye through the heart a second time. Thus, the downward limb is extrapolated to the base line. The time embraced by the curve represents the time required for the heart to eject all the dye. From the area under the curve, the average concentration can be determined. If the average concentration per milliliter of blood is known, then the cardiac output can be ascertained by the following equation:

$$\text{Cardiac output (ml/min)} = \frac{\text{dye injected (mg)}}{\text{average concentration (mg/ml)} \times \text{time (min)}}$$

For example, assume that 5 mg of dye is injected. Average concentration under the curve is found to be 0.002 mg/ml. Time equals 0.5 min. Then the cardiac output is 5,000 ml/min $(5/0.002 \times 0.5)$.

Cardiac Reserve

The normal heart, as has been stated, ejects 5 to 6 liters of blood/min per ventricle when the individual is at rest. However, during exercise this output can be increased to 30 or even 35 liters/min. The difference between the rest value and the maximum value is termed the cardiac reserve, that is, up to about 25 liters/min.

Cardiac Index

Cardiac output varies with the body surface area. That is to say, the larger the person, the greater the output. Thus, to compare outputs from person to person, the cardiac index, which is obtained by dividing the cardiac output expressed in liters by the surface area expressed in square meters, is used. The surface area of the average adult is about 1.7 m². Accordingly, the average cardiac index is approximately 3.3 liters/min/m² (5.6/1.7).

Variations in Cardiac Output

A few of the more common conditions which are associated with altered cardiac output will be briefly considered. But first the relation of heart rate to cardiac output will be discussed.

If the curve representing ventricular blood volume is kept in mind (Figure 16-15), the relation between cardiac output and heart rate will be easier to understand. As the heart rate increases, the length of diastole decreases (Table 16-3). The greatest volume of blood enters the ventricle during the rapid-inflow phase. Thereafter the curve representing ventricular volume begins to plateau. Thus, as the heart rate increases from a very low rate to about 90 beats/min, there is very little change in the end-diastolic volume, that is, the volume in the ventricle at the end of diastole (Figure 16-18). Since cardiac output is equal to heart rate times stroke volume, the cardiac output increases as the rate increases because the stroke volume does not decrease proportionately. In the range 90 to 140 there is not much change in cardiac output simply because as the rate increases, the filling time decreases so that in this range, at least, the stroke volume falls proportionately with the increase in rate, if the venous pressure remains constant. Finally, however, at very fast rates the filling time is so short that the stroke volume falls faster than the rate increases. Thus the cardiac output progressively decreases. In brief, changes in heart rate do not always indicate a comparable change in cardiac output (Figure 16-19).

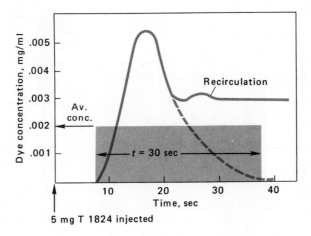

Figure 16-17. Dye curve for determining cardiac output.

TABLE 16-3. LENGTH OF SYSTOLE AND DIASTOLE AT VARIOUS HEART RATES

HEART RATE, BEATS/MIN	SYSTOLE, SEC	DIASTOLE, SEC
50	0.41	0.79
60	0.39	0.61
70	0.36	0.49
80	0.34	0.41
110	0.30	0.25
170	0.23	0.12

Venous Return As Arthur Guyton and others have long insisted, cardiac output is determined by the peripheral circulatory system, not by the heart itself. The heart should not be likened to a mechanical pump which can be regulated to pump more or less fluid into a system of rigid tubes. The heart is, in effect, a demand pump that is, within a certain range, regulated by the demand placed upon it by the amount of blood entering it from the veins, the so-called venous return. This fact is convincingly demonstrated

Figure 16-18. Relation between heart rate and end-diastolic volume. The curve is hypothetical and assumes constant venous pressure and ventricular residue of 50 ml. Time refers to the length of diastole. The numbers on the vertical lines are cardiac output in milliliters.

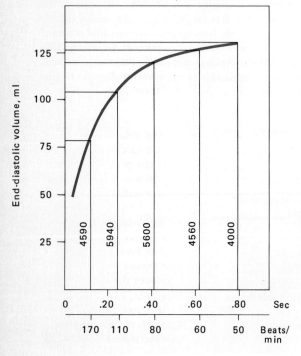

Figure 16-19. Relation between cardiac output and heart rate. Constant venous pressure and ventricular residue are assumed.

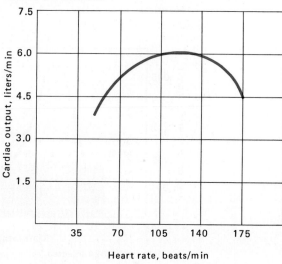

in a simple experiment in which a connection is made between a large artery and a large vein. When this connection is opened, thereby permitting a large volume of blood to flow directly into the venous system, the cardiac output increases almost immediately. And when the connection is closed, the cardiac output immediately decreases to its normal level. The major response of the heart to increased venous return is due to increased stretch of the myocardium during diastole. The increased length of the cardiac fibers causes more vigorous contraction, which increases stroke volume. At the same time, stretch of the right atrium influences the S-A node directly to increase heart rate 10 to 15 percent. The increased heart rate and greater stroke volume combine to augment cardiac output.

The normal human heart can increase its output from the resting level of about 5 liters per min to about 15 liters per min purely as a result of increased venous return. If the venous return increases above this level, then the heart can keep pace only if the sympathetic nervous system increases heart rate and force of contraction. This occurs in exercise permitting a cardiac output of over 30 liters per min in the trained athlete.

The venous return is regulated by many factors that will be considered in Chapter 18. For example, an increase in blood volume, or generalized vasodilatation, or vasoconstriction will increase venous return.

Myocardial infarction, valvular disease, and other cardiac abnormalities reduce the ability of the heart to respond to the demands of venous return. The damaged heart may be able to pump only 5 to 6 liters per min. If the venous return increases over this level, the heart cannot respond. It is said to fail.

As mentioned previously, the normal heart has a cardiac reserve of up to about 2.5 liters per min.

Exercise   The greatest increase in cardiac output occurs during exercise. The reasons for this increase are (1) greater venous pressure, (2) increased heart rate, (3) more vigorous myocardial contraction, and (4) decreased peripheral resistance in the vascular system. The mechanisms of these changes will be discussed later.

Sleep   During sleep, cardiac output decreases below that which prevails when the individual is awake but not exercising. The main cause is the decreased cerebral drive. The heart rate decreases, the force of contraction is less, the venous pressure diminishes; all these decrease cardiac output. Dreaming, however, can reverse these mechanisms and greatly increase the output.

Pathologic Conditions   There are many disorders that alter cardiac output. Only a few will be listed for illustrative purposes. If the heart is defective, the cardiac output may be severely altered. However, even a grossly pathological heart can put out a normal quantity of blood for a time, but there is a progressive deterioration until a critical point is reached at which the output of the heart can no longer keep pace with venous return. The heart is now in failure.

If there is aortic valve insufficiency, for example, part of the blood ejected during systole regurgitates into the ventricle during diastole. This is obviously an inefficient process, for a large quantity of blood must be pumped only to flow back into the heart. If 25 percent of the ejected blood backwashes, in order to supply the normal body needs the heart must pump approximately 25 percent more blood than does the normal heart. The young, vigorous heart can easily handle this load but at a price, and the price is progressive enlargement of the heart (dilatation), thickening of the muscle (hypertrophy), increasingly greater loads, and ultimate death.

If the aortic valve is narrowed (stenosis), in order to pump the normal quantity of blood through the narrow aperture in the usual period of time, the ventricle must contract more vigorously. This, too, constitutes an added load and leads to dire consequences.

There are pathological conditions outside the heart which may eventuate in abnormal cardiac output. For example, as has already been noted, anemia is characterized by a decreased oxygen-carrying capacity of the blood, and, most importantly for this discussion, the viscosity of the blood decreases. As a result of this lowered viscosity, venous return increases, thereby augmenting cardiac output.

In hyperthyroidism, all cells of the body utilize oxygen more rapidly, producing local hypoxia which results in dilatation of the blood vessels. This increases blood flow into the venous system and on into the heart, resulting in a higher cardiac output.

In summary, increased cardiac output follows from increased venous return, which can be caused by any condition which permits more blood to flow into the venous system. Decreased cardiac output results from a decreased flow of blood into the venous system. Most often it is due to abnormalities of the heart itself such as myocardial infarction or valvular disease.

STARLING'S LAW OF THE HEART

As has already been pointed out, the greater the initial length of the cardiac fiber, the more forceful the contraction. The initial length of the cardiac fiber is determined by the quantity of blood which flows into the ventricle during diastole. The force of the contraction then determines the amount of blood ejected by the ventricles during systole. In other words, the stroke output is directly related to the end-diastolic volume up to a critical point (Figure 16-20). This relationship is termed Starling's law of the heart. It is an inherent, self-regulating mechanism that permits the heart to adjust to changing end-diastolic volumes. In other words, the heart will eject the amount of blood that enters over a broad range of volumes. A more detailed investigation of this relationship demonstrated a family of curves (Figure 16-21). The stroke work at various end-diastolic pressures was determined and the experiment repeated while stimulating the stellate ganglion (supplying sympathetic fibers to the heart) or using increasing doses of norepinephrine, a drug which increases the force of myocardial contraction. The important point illustrated

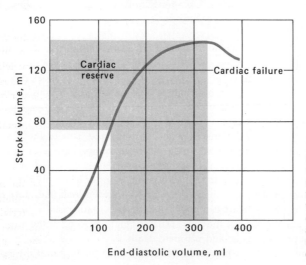

Figure 16-20. Starling's law of the heart. As the end-diastolic volume increases, the ventricles contract more vigorously, thereby increasing the stroke volume. Once beyond the range of normal functioning (shaded area), this relationship is reversed and stroke volume decreases.

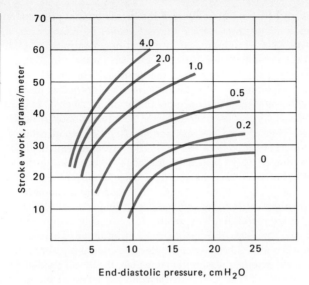

Figure 16-21. Relation between stroke work and end-diastolic volume upon stimulation of the ganglion supplying sympathetic fibers to the heart. Numbers on the lines refer to frequency of stimulation. (*Courtesy of S. J. Sarnoff.*)

by these experiments is that the chemical composition of the blood, nervous activity, or administered drugs may alter the force of contraction, but so long as conditions are kept constant, the stroke work increases as the end-diastolic pressure rises.

Venous Pressure

There are two factors which determine ventricular filling: (1) the length of diastole, which is a function of the heart rate, and (2) venous pressure. If the heart rate is kept constant but the venous pressure is elevated, more blood will be forced into the ventricle than at normal pressures. Clearly, if the venous pressure is high, then even during a very brief diastole a large quantity of blood will be forced into the ventricle resulting, according to Starling's law, in a greater stroke volume. If both stroke volume and heart rate increase, the cardiac output will be markedly raised. During exercise, the venous pressure rises and thus maintains the stroke volume even though the heart rate may increase above 150 beats/min. This causes a greatly increased cardiac output.

Venous pressure is a factor which controls the heart directly by virtue of Starling's law. This relatively simple mechanism is nonetheless highly effective and self-regulatory. As the venous pressure increases, more blood is forced into the ventricle during diastole. Therefore, the chamber contracts more vigorously during the next systole and ejects a greater quantity of blood. Since the heart is now putting out a larger quantity of blood, the veins will be drained more rapidly, and this, in turn, reduces the venous pressure. As the venous pressure falls, so does the cardiac output, until the normal venous pressure is once again established.

Arterial Pressure

The arterial pressure, just like that on the venous side, plays an important role in determining diastolic filling. Let us assume that during diastole circulatory changes have taken place which result in a sudden and sharp elevation of arterial blood pressure. Suppose that the pressure within the aorta just before the aortic valve opens has been about 80 mmHg and that the aortic diastolic pressure is now raised to 100 mmHg. The ventricle will contract with the same force as it did previously, because the initial length of the fiber is unchanged. But because of the higher arterial pressure not as much blood will be pumped out of the ventricle during systole. In other words, the same force of contraction will eject less blood against a head of pressure equal to 100 mmHg than it formerly did against only 80 mmHg. Consequently, at the end of this systole, there will be a considerable residue of blood in the ventricle. This ventricular residue added to the incoming complement of blood will stretch the ventricular fibers, a greater force of contraction will ensue, and therefore more blood will be ejected. To illustrate with figures, assume that normally the ventricle contains 80 ml of blood, of which 70 ml is ejected during systole. Then, during diastole, another 70 ml of blood enters the ventricle. This cycle is repeated until there is a rise in arterial pressure, which, as just explained, may result in a residue of 30 ml instead of the usual 10 ml. During the very next diastole, another 70 ml of blood enters the ventricle. This 70 ml plus the residue of 30 ml gives a total ventricular filling of 100 ml of blood. This quantity of blood will stretch the cardiac fibers more than the usual 80 ml did, and, according to Starling's law, the ventricle will now contract more forcefully. The greater force of contraction will eject, even against the added head of pressure, the customary 70 ml of blood. Therefore, the cardiac output which fell for just one beat because of the elevated arterial pressure is now back to normal. And so long as the arterial pressure remains high, the 30-ml residue will persist. As soon as the arterial pressure falls, the ventricle will eject 90 ml of blood and leave the usual 10 ml as a residue.

These are purely arbitrary figures; as a matter of fact, it takes more than one beat for the necessary readjustments to occur. But the point should be clear that the heart adjusts to the arterial pressure. True, the heart must work harder in order to eject the normal volume of blood against a higher head of pressure, but insofar as the body is concerned, the astonishing fact is that the cardiac output tends to remain normal despite the elevated blood pressure.

Chronic high blood pressure usually leads to cardiac failure. The cardiac output, even in marked hypertension, is within the normal range, but in order to accomplish this feat the heart muscle is overworked. Ultimately it fails.

Cardiac Valve Defects

The point has been made that despite cardiac valve defects, the cardiac output may be within the normal range. This is achieved at the expense of greater myocardial work. The mechanism again depends upon Starling's law. For example, in aortic insufficiency blood regurgitates into the ventricle during diastole. This blood is added to the normal complement of blood during the subsequent diastoles. The end-diastolic volume is increased, and the ventricle contracts more vigorously. Thus the volume ejected is increased, but since some of that volume reenters the ventricle, the difference between the quantity ejected and the amount that regurgitates represents the stroke volume. By virtue of the more forceful contraction, it remains within the normal range.

In aortic stenosis the end result would be the same. The narrow opening exerts resistance to the ejection of blood. Without increased force of contraction not all the volume could be ejected. There would be a residue, greater end-diastolic volume, and thus more forceful contraction which would drive out a normal ejection despite the stenosis.

EFFECT OF BLOOD CONSTITUENTS ON THE HEART

There are many constituents of the blood which can exert a powerful influence on the heart. Fortunately, there are exquisite homeostatic mechanisms which keep these constituents within a very narrow range of variation. When these mechanisms fail, the cardiac results are often fatal.

Plasma Ions

Of all the plasma ions, potassium, sodium, and calcium are the most critical insofar as cardiac function is concerned. This is not surprising in view of the effect that these ions have on the action potential and on contraction. If the concentration of potassium is significantly increased, the heart rate is slowed and the force of contraction weakened. The reason for this action is probably the decreased gradient of potassium between the inside and outside of the myocardial cells. The action potential has less magnitude, and so does the resultant contraction. If the potassium concentration of the plasma increases to a level about three times normal, cardiac arrest generally ensues.

Elevation of calcium ions excites the heart. If the calcium concentration rises high enough, the heart goes into a vigorous, prolonged contraction. Recall that calcium is essential to the initiation of the contractile process. This probably accounts for the effect of excessive calcium ion in plasma.

Sodium ions, like potassium ions, depress the heart. This is unexpected from the standpoint of the mechanism of the action potential. The explanation is thought to be in the competition of sodium and calcium ions. Elevation of sodium seems to prevent calcium from carrying out its contractile role. Conversely, decreased sodium permits calcium to function uninhibited.

The influence of the concentration of the ions of the plasma on cardiac function was discovered many years ago and gave rise to the so-called Ringer's solution. Ringer found that in order to maintain normal function of a perfused heart, specific concentrations of the ions had to be maintained in the perfusing fluid.

Catecholamines

The catecholamines, that is, epinephrine and norepinephrine, are secreted by the adrenal medulla. They are also available commercially. These substances have profound effects on the cardiovascular system. They act directly on the myocardium to increase the rate and vigor of contraction (Figure 16-22). Injection of catecholamines usually first increases the heart rate and then decreases it. The decrease is due to a reflex response to the elevated blood pressure which is also evoked by these drugs. The direct cardiac effect is stimulation.

NERVOUS CONTROL OF THE HEART

Although the beat of the heart is an inherent myocardial property, the rate and the force of contraction can be modified by the activity of the nerves that innervate the heart.

Innervation of the Heart

The autonomic nervous system innervates smooth muscle, cardiac muscle, and glands. Accordingly, all the fibers which supply the heart muscle belong to this part of the nervous system. These nerves may be divided into two groups: (1) the vagus nerve, which carries parasympathetic fibers, and (2) the accelerator nerves, which are sympathetic.

The Vagus   The vagus nerve contains motor fibers which arise from cells in the medulla and pass down to innervate the heart. Some of these axons course to the S-A node, while others supply the A-V node. However, few if any vagal fibers innervate the ventricular musculature. Thus, all the control which the vagus exerts must be directed through the specialized nodal tissue.

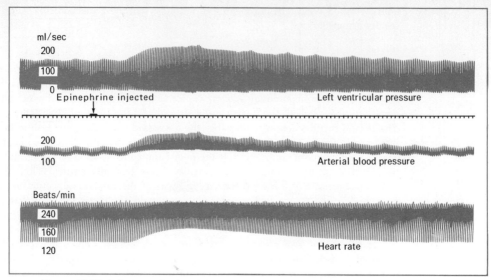

Figure 16-22. Influence of epinephrine on left ventricular pressure, arterial blood pressure, and heart rate.

Stimulation of the vagi causes the release of acetylcholine (ACh), which slows the heart and decreases the force of contraction. If the excitation is intense and prolonged, partial or complete heart block may be produced. In other cases, however, intensive stimulation simply produces cardiac arrest so that the entire heart ceases to beat. Whether heart block or arrest is produced, after a few seconds, contraction once again takes place even though the stimulation is maintained. This phenomenon is known as vagal escape. In other words, the heart escapes from the influence of the vagus.

Membrane-potential studies recorded from the area of the pacemaker go a long way toward explaining how vagal stimulation inhibits the heart. Vagal stimulation is seen to cause hyperpolarization; that is, the resting potential is more negative during stimulation of the vagi (Figure 16-23). Also note in Figure 16-23 that during stimulation the characteristic prepotential does not develop, nor is there an action potential. Contraction, in this case, is completely inhibited.

In normal man, there is a steady propagation of impulses down the vagal fibers to the heart. There is thus a "braking" action upon the heart. This constant vagal activity is termed vagal tone. If both vagi are cut, the heart rate immediately increases. Destruction of the vagal fibers removes the persistent inhibition on the heart, thus permitting the heart to speed up. One should visualize the heart under the influence of a steady stream of inhibitory impulses. If these impulses are decreased, the heart rate will increase; if the impulses are augmented, the heart rate will be diminished. In other words, vagal activity alone is capable of altering the heart rate in either direction.

The Accelerator Nerves Sympathetic neurons innervate both nodes as well as the myocardium of the atria and ventricles. Because stimulation of these neurons releases norepinephrine, which accelerates the heart and increases the force of contraction, they are called accelerator nerves. The function of the accelerator nerves is just the opposite

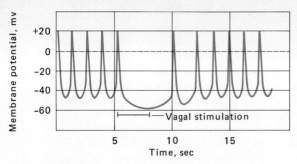

Figure 16-23. Effect of vagal stimulation on the pacemaker potentials of the sinus venosus of a frog heart. Because stimulation causes hyperpolarization of the pacemaker membrane, an action potential does not occur. (*Courtesy of O. F. Hutter and W. Trautwein,* Journal of General Physiology, *39:715, 1956.*)

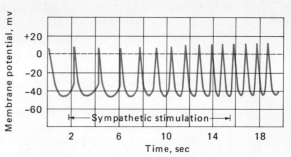

Figure 16-24. Effect of sympathetic stimulation on the pacemaker potentials of the sinus venosus of a frog heart. The slope of the prepotential increases. The membrane thus reaches its threshold sooner, increasing the heart rate. The overshoot also progressively increases. (*Courtesy of O. F. Hutter and W. Trautwein,* Journal of General Physiology, *39:715, 1956.*)

of that of the vagal fibers. There is a steady propagation of impulses along the accelerator nerves. Thus one speaks of accelerator tone as well as vagal tone.

Once again, study of action potentials proves revealing (Figure 16-24). When the accelerator fibers are stimulated, the slope of the prepotentials is seen to steepen. Accordingly, threshold voltage is more quickly attained, and the action potential is generated sooner than it would have been in the absence of stimulation. Also the overshoot progressively increases. The resulting greater magnitude of the action potential probably is the cause of the increased force of contraction.

Action-potential studies during vagal and sympathetic stimulation suggest that the vagus inhibits cardiac muscle by increasing the permeability of its membrane to potassium, thereby increasing the resting potential. Sympathetic stimulation is thought to increase the permeability to sodium, thus permitting a more rapid buildup of the membrane to threshold voltage.

Coordination of Innervation   The heart is under dual control; the vagi slow the heart, and the accelerator fibers speed it. Both exert their effects constantly. It is analogous to a tug-of-war. In this way, heart rate may be varied in four ways, or in any combination: (1) increased activity of the vagi, (2) decreased excitement of the vagi, (3) augmented function of the accelerator nerves, or (4) diminished activity of the accelerator fibers. These influences do not function independently.

The braking and accelerating effects on the heart are integrated in the medulla. Here are located the so-called cardioinhibitor and cardioaccelerator centers. These centers may be stimulated or inhibited. The efferent component of the reflex arc may be considered to originate in the nest of cells which make up these centers. Afferent impulses impinge on these centers reciprocally, bringing about activation of one and simultaneous inhibition of the other, thereby effectively controlling heart rate.

Receptors and Reflex Responses

There are strategically placed receptors sensitive to changes in (1) arterial blood pressure, (2) venous blood pressure, and (3) the chemical composition of the blood.

Pressoreceptors These receptors, as the term indicates, are sensitive to changes in blood pressure. Where the common carotid artery bifurcates into the internal and external carotid branches, there is a noticeable swelling, termed the carotid sinus (see Figure 16-26). The walls of this sinus are lined with pressoreceptors innervated by fibers which join the glossopharyngeal (IX) nerve. These neurons end in synaptic union within the cardioaccelerator and cardioinhibitor centers. There are similar pressoreceptors lining the arch of the aorta in an area called the aortic sinus. Afferent fibers from these receptors join the vagus nerve (X) and pass to the medulla.

A rise in arterial pressure activates the pressoreceptors by stretching them. The impulses are propagated by afferent fibers in the ninth and tenth cranial nerves to the medulla, and then efferent impulses are propagated by the parasympathetic pathways to slow the heart and to diminish the force of ventricular contraction. Diminished cardiac activity tends to reduce the volume of blood ejected per minute, and thus the arterial blood pressure falls. Conversely, in case the arterial pressure is reduced below normal, these reflex arcs speed the heart, increase the force of contraction, and thereby elevate the blood pressure. This inverse relationship between blood pressure and heart rate is known as Marey's law of the heart.

Figure 16-25 shows that the rate of firing of the pressoreceptors varies with the blood pressure, but the important point to observe is that the sensitive part of this curve lies between a pressure of 80 and 120 mmHg. Since the main arterial blood pressure is about 100 mmHg, the pressoreceptors provide a very sensitive regulatory mechanism.

There are also pressoreceptors in the large veins where they enter the right atrium (Figure 16-26). When venous pressure increases, the receptors are fired, and impulses

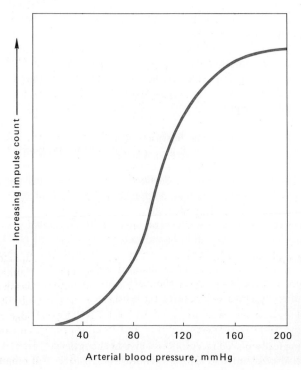

Figure 16-25. Relation between blood pressure and pressoreceptor activity. The greatest change in rate of firing occurs in the normal blood pressure range, that is, 80 to 120 mmHg.

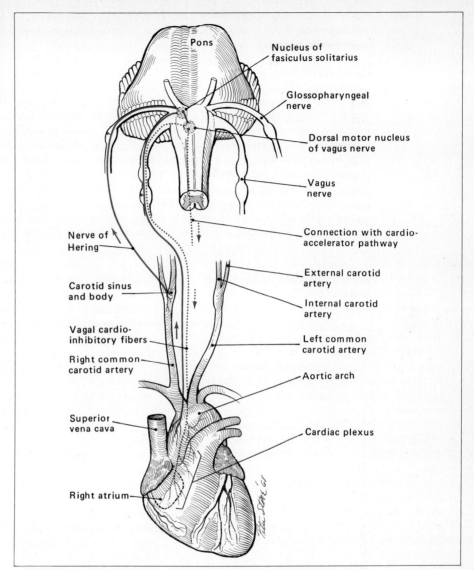

Figure 16-26. Reflex control of the heart. The carotid sinus and aortic sinus (within the aortic arch) contain pressoreceptors and chemoreceptors that initiate reflexes so as to change the heart rate and force of contraction.

pass to the medulla via afferent neurons in the vagus. Here they impinge on the cardiac-control centers to effect an increase in heart rate. This mechanism is termed the Bainbridge reflex. The effectiveness of this arc should be made clear. An increase in venous pressure forces more blood into the ventricle during diastole and, as a result, the stroke volume is augmented. The increased venous pressure through the Bainbridge reflex speeds the heart. Since cardiac output is equal to the product of stroke volume and heart rate, an increased venous pressure markedly elevates cardiac output.

Certain authorities question the existence of this reflex, and others point out that because stretch of the right atrium itself increases heart rate, the Bainbridge reflex is not very important. What is important is that the heart responds to increased venous return by increasing heart rate and stroke volume.

Chemoreceptors   In association with the carotid and aortic sinuses there are several small bodies, 1 to 2 mm in size, which are termed the carotid and aortic bodies. They contain cells sensitive to oxygen and carbon dioxide concentration of the blood which is brought to them through small nutrient arteries. These cells are chemoreceptors. They are innervated by the nerve of Hering. Because these cells respond to hypoxemia and because hypoxemia results in augmented cardiac function, the chemoreceptors of the carotid and aortic bodies were thought to be responsible for the observed changes in heart activity. The fact is, however, that if the hypoxemia is limited to these peripheral chemoreceptors, cardiac function either is unchanged or diminishes. Accordingly, other factors must be responsible. Hypoxemia stimulates cardiac function by a direct action on the medullary cardiac centers and also brings about increased venous return. The chemoreceptors are more important to respiration and are discussed in detail in Chapter 23.

QUESTIONS AND PROBLEMS

1   Describe the developmental defect in the heart of a "blue baby." What symptoms in addition to bluish appearance of its skin will this baby present?

2   Define and give the function of the following: (a) chordae tendineae, (b) mitral valve, (c) pericardial fluid, (d) semilunar valves, (e) foramen ovale, (f) sinoatrial node, (g) standard leads, (h) heart murmur, and (i) pressoreceptors.

3   Explain the relationship between the thickness of the right and left ventricular myocardium and the respective functions of these ventricles.

4   Describe the arteries and veins supplying the myocardium.

5   What characteristic properties of cardiac muscle differentiate it from skeletal muscle?

6   Describe the origin and transmission of the heartbeat.

7   Describe the following components of the normal electrocardiogram, and explain what each represents in terms of electrical activity occurring in the heart: (a) P wave, (b) QRS complex, (c) P-R interval, (d) T wave, and (e) Q-T interval.

8   When a nurse in the coronary-care or intensive-care unit monitors the heart action of a patient, what abnormality of the EKG will tell her that the patient is (a) fibrillating, (b) experiencing first-degree heart block, or (c) experiencing bundle branch block?

9   Describe the cardiac cycle taking into account the following: (a) heart rate, (b) pressure changes, and (c) myocardial activity.

10   In what way does the aorta contribute to the maintenance of systemic blood pressure?

11   What events of the cardiac cycle are represented by the first, second, and third heart sounds?

12   Define cardiac output and describe two methods used to determine this value for man.

13   Discuss the relation of heart rate to cardiac output.

14   What factors contribute to an increased cardiac output during exercise?

15   Discuss Starling's law of the heart as it relates to (a) physical exercise and (b) cardiac valve defects.

16   Describe the effect of the following ions on heart action: (a) potassium, (b) calcium, and (c) sodium.

17   What EKG abnormality might cause the physician to prescribe intravenous administration of a potassium salt for his patient?

18   How do the vagus nerves modify the inherent rhythmicity of the heart?

19   What is the functional relation between the medulla oblongata and the heart?

20   Discuss the response of the heart to hypoxemia.

17 ANATOMY OF THE BLOOD VESSELS AND LYMPH ORGANS

The circulatory system is designed to deliver blood to and from the capillaries, where the exchange of the vital respiratory gases and other metabolic substances occurs. Not only is there a transfer between the capillaries and the cells, but some of the fluid portion of the blood actually leaves the vessels to enter the tissue spaces. This fluid, after nourishing the tissues, drains into the lymphatic vessels, which carry it back to the large veins. Thus in considering the circulatory system it is necessary to include the heart, the arterial vessels, the capillaries, and the venous channels, as well as the lymphatics. The spleen and thymus are largely lymphoid organs and therefore will be described with the lymphatic organs. The tonsils, which are aggregates of lymphoid tissue in the nasal and oral pharynx, will be discussed with the respiratory and digestive system in Chapters 20 and 24.

The structure and function of the heart have already been considered (Chapter 16). Now we shall look at the system of conducting tubes connecting the heart with the rest of the body.

EMBRYONIC DEVELOPMENT

Blood vessels and lymph vessels are derived from mesenchyme. Early formative mesenchymal cells, the angioblasts, are found throughout the embryo in collections called blood islands. The peripheral cells of these collections become endothelial tubes; the centrally lying cells, blood corpuscles. The adult pattern of distributing vessels is essentially laid down at this early stage as adjacent endothelial tubes fuse to form larger vessels. The derivation of the heart has been traced from its origin as a primitive blood vessel in Chapter 16. The blood pathways are predominately the same in all embryos, although variations occasionally occur.

While this primitive network is forming, mesenchyme adjacent to the endothelial tubes lays down additional cells which become the muscular and fibrous components of the vessel walls. Thus arteries come to be distinguished from veins by their heavier muscular coat. Certain veins are further modified by invagination of the inner surface into cuplike folds, which become valves.

ANATOMY OF THE BLOOD VESSELS
Arteries

The arterial vessels carry blood from the ventricles of the heart to the capillaries, in progressively smaller tubes. Arteries vary in cross section from about 25 mm to less than 0.5 mm. The smaller tubes are known as arterioles. Arteries are composed of three layers (Figure 17-1). (1) The tunica intima consists of endothelium resting on an elastic membrane; (2) the tunica media, or middle coat, is a thick layer composed of smooth muscle fibers encircling the tube, and in the larger arteries often including numerous elastic fibers to enhance distensibility; and (3) the tunica adventitia, or outer coat, is made up mostly of loose collagenous fibers.

The walls of large arteries are supplied with blood by minute vessels termed vasa vasorum. The smooth muscle in the wall is innervated by autonomic nervous system fibers of two types, namely, vasoconstrictors, responsible for constriction, that is, narrowing the lumen, and vasodilators, responsible for relaxing and hence enlarging the

Figure 17-1. Cross section of an artery and vein (× 35).

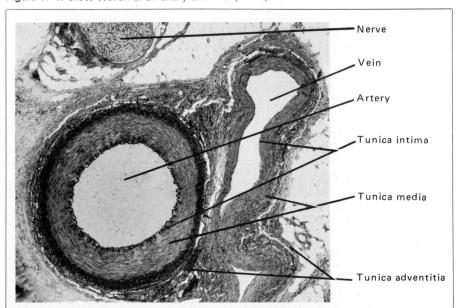

439
ANATOMY
OF
THE
BLOOD
VESSELS
AND
LYMPH
ORGANS

lumen. The action of these vasomotor fibers, as they are collectively known, is more fully described in Chapter 18.

The structure of the arterial wall suggests some things about the functioning of arteries. For instance, because of its elastic fibers, an artery can be stretched lengthwise and distended to accommodate the change in blood volume that accompanies each heartbeat. Second, because of its thick wall, an artery does not readily collapse during times of low blood pressure.

In arterioles the elastic tissue found in arteries is replaced by smooth muscle.

Capillaries

The smallest vessels in the circulatory system are the capillaries. The capillaries have very thin walls, usually only one cell layer thick. It is through this thin wall (endothelium) that the exchange of oxygen, carbon dioxide, and other substances takes place. The endothelium of the capillaries is squamous epithelium held together by a cementing substance. The capillary network is quite complex. There are several channels arranged roughly in parallel. There are also, in places, direct communications between the arterioles and venules. These bypass circuits are particularly in evidence in the muscles and along the gastrointestinal tract. As a consequence, during digestion, for instance, blood is permitted to circulate throughout the capillary network, whereas between meals or during exercise, by selective constriction of muscle within the walls of certain vessels, many of these pathways are closed, and a greater amount of blood is routed directly from arterioles to venules. These alternative pathways are further discussed in Chapter 18.

Veins

After passing through capillary networks, or beds, the blood is collected in the venules, which are the smallest veins. These vessels consist of an endothelial lining reinforced by a few smooth muscle cells encircling the tube and covered by a fairly thick layer of collagenous fibers.

The veins, like the arteries, are composed of three layers. However, their walls are thinner and less elastic than those of arteries of comparable size. In the middle-sized veins the tunica media is encircled by several layers of smooth muscle, and the tunica adventitia is a well-developed fibrous coat that may contain a few smooth muscle fibers. In the large veins, such as those entering the heart, the circular muscular coat is greatly reduced in thickness and the adventitia thickened by longitudinal bands of muscle.

Certain veins, particularly those returning blood to the heart against gravity, as in the extremities, contain valves. Valves permit the blood to flow only toward the heart and are highly important in the maintenance of the circulation. Valves are absent in veins in the cranial, abdominal, and thoracic cavities.

Sinuses

In certain areas, for example, the spleen and lymph nodes, the vascular channels linking the arterial and venous circulation are relatively wide, irregular vessels called sinuses or sinusoids. Sinuses are characterized by (1) wide lumens, (2) sluggish blood flow, (3) channels that join freely, and (4) thin walls that contain reticular fibers but no muscle. They collapse readily when empty. Sinuses are often lined in part by macrophages, especially in the liver, bone marrow, adrenal glands, and pituitary gland.

Anastomosis

Frequently the distal ends of vessels supplying a given structure of the body will unite. Such a junction is called an anastomosis. Anastomoses permit free communication between the vessels involved. They may occur between the terminations of arteries, between the origins of veins, or, in some instances as will be described in Chapter 18, between arterioles and venules.

Anastomoses may occur between small or large vessels. As will be described later, in the brain a circular anastomosis, the circle of Willis, is created by links between branches of the internal carotid artery and the basilar artery.

Because of frequent anastomoses in the circulatory system, the occlusion of a single artery or vein by pathological processes or in surgery does not necessarily eliminate circulation of blood to the part. Thus if an artery supplying a specific portion of the body is tied off during surgery, a new channel may develop to supply blood to the affected area. This alternative routing, which utilizes anastomoses between adjacent arteries, is called collateral circulation.

DIVISIONS OF THE CIRCULATORY SYSTEM

Arteries and veins transmit blood throughout the body. There are two separate circular routes: (1) blood passing from the heart to the lungs and returning to the heart constitutes the pulmonary circulation, and (2) blood passing from the heart to the entire body and back to the heart constitutes the systemic circulation.

Pulmonary Circulation

The large artery arising directly from the right ventricle is the pulmonary trunk. It lies in front of the aorta and ascends about 8 cm to terminate in a T-shaped division, the origin of the right and left pulmonary arteries (Figure 17-2). The pulmonary arteries course to their respective lungs. There the right pulmonary artery divides into three lobar branches, one to each lobe of the right lung, and the left pulmonary artery divides into two branches to supply the two lobes of the left lung. Each of these branches subsequently subdivides to correspond with and supply the bronchopulmonary segments of the lungs. Through repeated division these segmental branches terminate in small arterioles, which ultimately end in capillary plexuses in the walls of the alveoli. Here the essential gaseous exchange between the blood and air occurs.

Pulmonary blood is returned to the left atrium of the heart by the pulmonary veins. Usually each lung sends two pulmonary veins into the left atrium.

Blood carried in the pulmonary arteries differs from that in the systemic arteries in that it is low in oxygen and high in carbon dioxide content. Blood returning to the heart via the pulmonary veins is relatively rich in oxygen and poor in carbon dioxide. In the systemic circulation the opposite is true; that is, the arteries carry blood rich in oxygen to the tissues of the body. Blood returning to the heart in the veins contains a relatively high concentration of carbon dioxide.

Systemic Arterial Supply

Aorta The aorta, larger in caliber than the pulmonary trunk, arises from the left ventricle. Portions of the aorta are given names that correspond to its shape or location on its route to the sacral region. As the ascending aorta it rises to the level of the sternal angle, where it arches posteriorly and to the left to lie adjacent to the vertebral column. As the aorta descends through the thorax, gradually moving toward the midline, it becomes the descending, or thoracic, aorta. At the level of the twelfth thoracic vertebra it passes through the diaphragm, immediately anterior to the vertebral column, and becomes the abdominal aorta. Opposite the sacral promontory it divides into the right and left common iliac arteries.

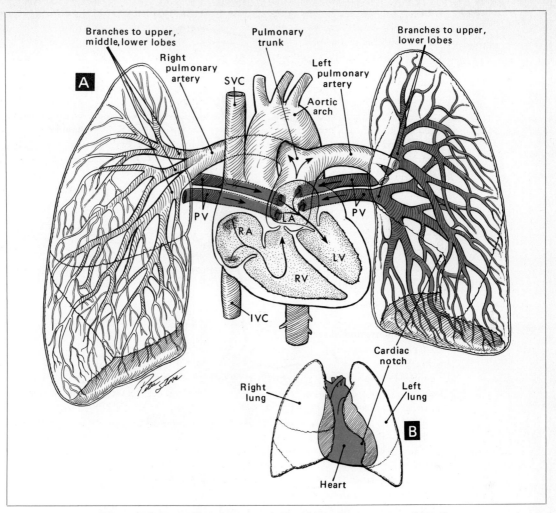

Figure 17-2. Pulmonary circulation. **A.** The right pulmonary artery divides into three major vessels to serve the three lobes of the right lung. Veins are shown in color. **B.** The lungs in relation to the heart. The cardiac notch is a depression in the left lung to accommodate the heart.

Arteries to the Head and Neck Three arteries—the brachiocephalic (innominate), left common carotid, and left subclavian—arise from the arch of the aorta (Figure 17-3). The brachiocephalic artery divides into the right common carotid and the right subclavian. The carotid arteries supply the head and neck. The subclavian arteries send branches to the neck before each becomes an axillary artery, carrying blood to an upper extremity.

   The common carotid arteries ascend on each side of the neck as high as the upper part of the larynx, where they each divide into two branches, the internal and external carotids.

   As the common carotid divides, the lumen of the internal carotid enlarges slightly to form the carotid sinus. Here epitheloid components in the wall form the carotid body.

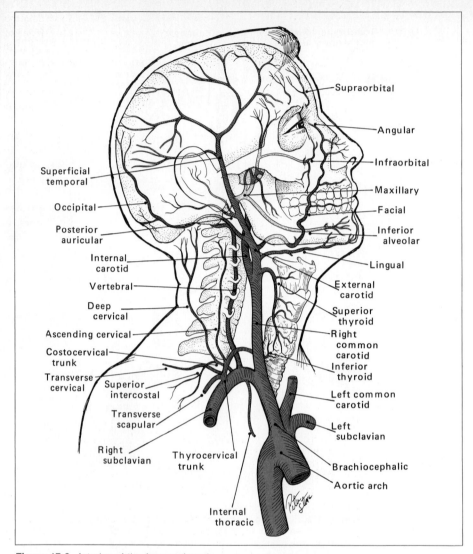

Figure 17-3. Arteries of the face and neck.

Pressoreceptors associated with the carotid sinus and chemoreceptors of the carotid body relay impulses by way of branches of the ninth and tenth cranial nerves to the special centers in the medulla. These impulses act as the afferent limb of important reflex pathways concerned with the control of blood pressure and heartbeat (see Chapter 16).

The external carotid artery on both sides sends branches to the neck, face, and head (Figure 17-3). Its first branch, the superior thyroid, descends alongside the larynx to provide branches to the thyroid gland and larynx; the lingual branch passes forward to supply the tongue; and the facial branch ascends to cross the external surface of the mandible, to supply the anterior portion of the face below the level of the eyes. Two other branches of the external carotid artery, the posterior auricular and the occipital, course posteriorly to supply the area behind the ear and the back of the neck and scalp.

443

ANATOMY
OF
THE
BLOOD
VESSELS
AND
LYMPH
ORGANS

The external carotid terminates by dividing into the superficial temporal and maxillary branches. The superficial temporal artery supplies the area in front of the ear and the side of the head. The maxillary artery courses deep to the mandible to enter the infratemporal fossa. Here it gives off a branch to the lower teeth; one that supplies the bulk of the blood to the meninges; and branches to the muscles of mastication. From the infratemporal fossa the maxillary artery passes into the pterygopalatine fossa, where it sends branches to the upper teeth and the nasal cavity, and, becoming the infraorbital artery, supplies the face immediately below the eye.

The other main branch of the common carotid, the internal carotid, gives off no branches in the neck. It enters the skull through the carotid canal within the temporal bone, where it gives rise to the ophthalmic artery. The ophthalmic supplies all the structures in the bony orbit and sends twigs to the front of the scalp, the area around the eye, and the nasal cavity.

After giving off the ophthalmic branch, the internal carotid divides into the anterior and middle cerebral arteries (Figure 17-4). The anterior cerebral artery supplies the medial surface of the cerebrum and anastomoses with the artery of the opposite side via the

Figure 17-4. Arteries of the brain. The circle of Willis, at the center, joins branches of the basilar and internal carotid arteries.

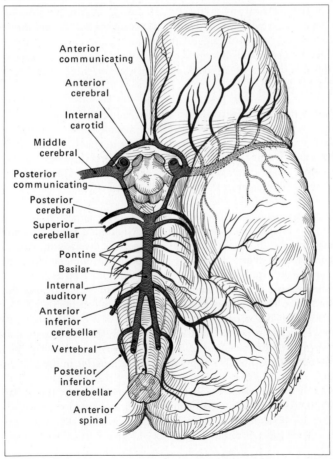

anterior communicating artery. The middle cerebral artery passes between the parietal and temporal lobes in the lateral fissure to supply the lateral surfaces of the cerebral cortex. This branch joins, via the posterior communicating artery, the posterior cerebral branch of the basilar artery (see later). The cerebral arteries and their communicating branches to the basilar artery form an arterial circle on the base of the brain called the circle of Willis.

The vertebral arteries arise from the subclavians as they course over the apices of the lungs in the root of the neck. The vertebral arteries pass deeply and ascend through the foramina in the transverse processes of the cervical vertebrae (see Figure 17-3). They supply the deep neck structures and then enter the cranial cavity through the foramen magnum. Here they provide anterior and posterior spinal arteries to the spinal cord, which course inferiorly along the length of the spinal cord to supply the cord and its meninges. After giving off spinal branches the vertebral arteries continue superiorly and join to form the unpaired basilar artery.

The basilar artery ascends on the ventral aspect of the brainstem, where it gives rise to several branches supplying the cerebellum, the brainstem, and the internal ear. It terminates by bifurcating into the two posterior cerebral arteries. The latter supply the inferior surfaces of the temporal lobes, and the occipital lobes. They communicate with the middle cerebral via the posterior communicating branch to complete the circle of Willis.

The costocervical and thyrocervical trunks both arise from the subclavian (see Figure 17-3). The costocervical trunk divides into branches that supply the upper two intercostal spaces and the deep structures of the neck.

The thyrocervical trunk gives origin to four branches that supply the thyroid gland, superficial structures of the neck, and muscles attaching to the vertebral border of the scapula.

Arteries to the Upper Extremities   The left subclavian arises from the arch of the aorta. The right subclavian is a branch of the brachiocephalic. On each side of the body a subclavian artery courses over the apex of the lung and passes between the clavicle and the first rib to become the axillary artery as it enters the axilla (Figure 17-5). The axillary artery supplies structures on the lateral chest wall and around the shoulder joint. It has six branches. The superior thoracic supplies the first intercostal space, the lateral thoracic supplies the side of the chest, and the thoracoacromial trunk sends small branches to the shoulder region and pectoral muscles. The largest branch of the axillary, the subscapular artery, gives off the scapular circumflex artery, to supply muscles on the scapula, and the thoracodorsal artery, to supply the latissimus dorsi muscle. The last branches of the axillary, the anterior and posterior humeral circumflex arteries, encircle the humerus to supply the deltoid muscle from its deep aspect.

The axillary artery becomes the brachial artery as it leaves the axilla. In the arm the brachial artery provides excellent pressure points to control bleeding of the upper extremity, as it lies adjacent to the humerus. As this artery traverses the arm, it gives rise to the deep (profunda) brachial artery, which curves posteriorly around the humerus, in company with the radial nerve, along the radiospiral groove of the humerus, to supply the triceps muscle. Branches supply the flexor muscles in the arm and contribute to the anastomosis around the elbow joint.

One of the terminal branches of the brachial artery, the radial, sends a recurrent branch to the elbow joint, then courses along the radial side of the forearm, supplying branches to the superficial extensor muscles on the back of the forearm. At the wrist it lies adjacent to the radius (Figure 17-6). Here it is utilized in taking the pulse. The

445
ANATOMY
OF
THE
BLOOD
VESSELS
AND
LYMPH
ORGANS

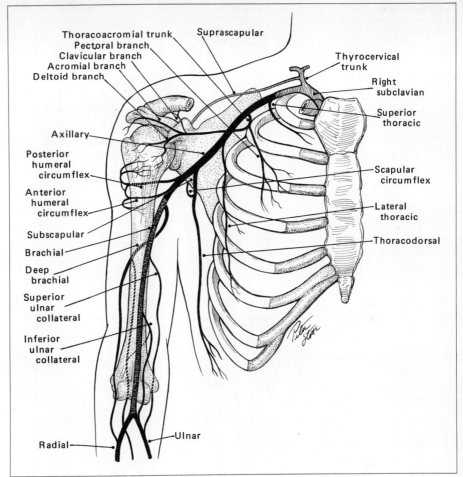

Figure 17-5. Arteries of the shoulder and axilla. The subclavian, axillary, and brachial arteries are segments of a single vessel.

radial artery supplies the main vessel of the thumb and sends a second large branch to the radial side of the index finger. The radial terminates in a superficial and a deep branch. The latter provides the major contribution to the deep palmar arch of the hand.

The second terminal branch of the brachial, the ulnar artery, supplies branches to the elbow anastomosis and passes along the ulnar side of the forearm to the wrist. Its short common interosseous artery divides into anterior and posterior interosseous branches. The former supplies the deep flexor, the latter the deep extensor muscles of the forearm. At the wrist the ulnar artery ends by joining branches of the radial to form the superficial and deep palmar arches. The superficial palmar arch supplies most of the blood to the hand. The three or four common palmar digital arteries that arise from the superficial palmar arch course through the palm. Opposite the webbing of the fingers they give rise to proper digital arteries supplying either side of the fingers. The deep palmar arch sends communicating branches to the superficial palmar arch and perforating branches to the dorsum of the hand.

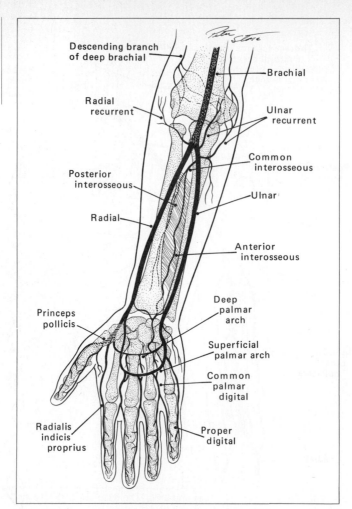

Figure 17-6. Arteries of the forearm and hand.

Arteries to the Thorax The arterial supply to the thorax is from direct branches of the descending aorta and from branches of the internal thoracic artery and costocervical trunk of the subclavian artery (Figure 17-7). Nine or ten pairs of segmental arteries, the posterior intercostal arteries, arise from the aorta and follow the inferior borders of the ribs, to supply intercostal muscles. Branches of the posterior intercostal arteries below the sixth intercostal space continue in an oblique direction to supply the muscles and skin of the lateral abdominal wall. Posterior intercostal arteries to the upper two intercostal spaces are derived from the costocervical trunk.

Anterior intercostal arteries arise from the internal thoracic branch of the subclavian artery. The internal thoracic artery courses along the lateral border of the sternum and gives off branches to the upper six intercostal spaces.

The internal thoracic then terminates as the superior epigastric and musculophrenic branches. The musculophrenic artery supplies anterior intercostal arteries to the lower

447

ANATOMY
OF
THE
BLOOD
VESSELS
AND
LYMPH
ORGANS

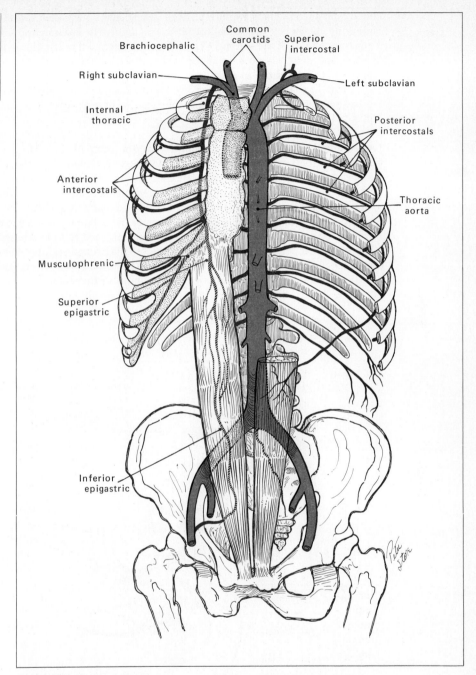

Figure 17-7. Arteries of the thorax. One of the lower intercostal arteries is followed along its oblique course to termination in the abdominal wall. The superior and inferior epigastric arteries anastomose within the rectus sheath.

intercostal spaces. The superior epigastric artery enters the rectus sheath and continues inferiorly to supply the rectus abdominis and then to anastomose with the inferior epigastric branch of the external iliac artery. The internal thoracic also supplies a slender branch, the pericardiophrenic artery, which descends through the thorax between the pleura and pericardium to reach the diaphragm.

Small branches from the anterior surface of the aorta include the bronchial arteries to the lungs, esophageal arteries, and the superior phrenic arteries to the upper surface of the diaphragm.

Arteries to the Abdomen In the abdomen three or four paired lumbar arteries arise from the dorsal aspect of the aorta in line with the posterior intercostal arteries in the thorax. These segmental branches supply muscles of the posterior abdominal wall. Arteries leaving the abdominal aorta to supply viscera include the unpaired celiac trunk, the superior and inferior mesenterics, and the paired renal and gonadal arteries.

As the aorta pierces the diaphragm to enter the abdomen, the celiac trunk branches from it and in turn gives off the left gastric, splenic, and hepatic arteries (Figure 17-8). The left gastric artery sends branches to the esophagus and supplies the stomach along its lesser curvature. The splenic artery, as it courses towards the spleen, gives off branches to the pancreas, short gastric arteries to the fundus of the stomach, a left gastroepiploic artery to the greater curvature of the stomach, and finally terminates in branches to the

Figure 17-8. The celiac trunk and its distributing branches. The edge of the liver has been lifted to expose the gallbladder.

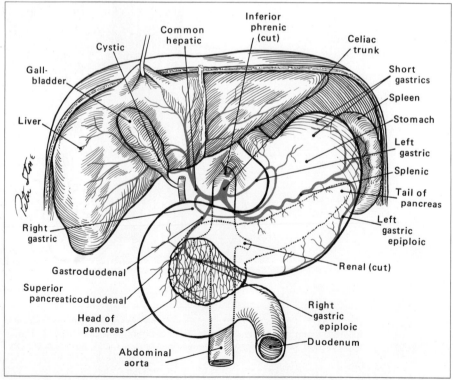

449

ANATOMY
OF
THE
BLOOD
VESSELS
AND
LYMPH
ORGANS

spleen. The hepatic artery sends a right gastric artery to the lesser curvature of the stomach; sends the gastroduodenal artery to the head of the pancreas, the duodenum, and the stomach; and terminates as the right and left hepatic arteries, which supply their respective lobes of the liver.

The superior mesenteric artery arises from the aorta about 2.5 cm below the origin of the celiac trunk (Figure 17-9). It sends 12 to 15 intestinal arteries to the small intestine; the ileocolic artery to the distal portion of the ileum, cecum, and appendix; and the right and middle colic arteries to the ascending and transverse colon, respectively.

The short but large renal arteries supply the kidneys. Passing from each side of the

Figure 17-9. Arteries of the large intestine. The intestinal branches which supply the small intestine are shown cut where they leave the superior mesenteric. Note the contribution of tne internal iliac to the arterial supply of the rectum and anal canal.

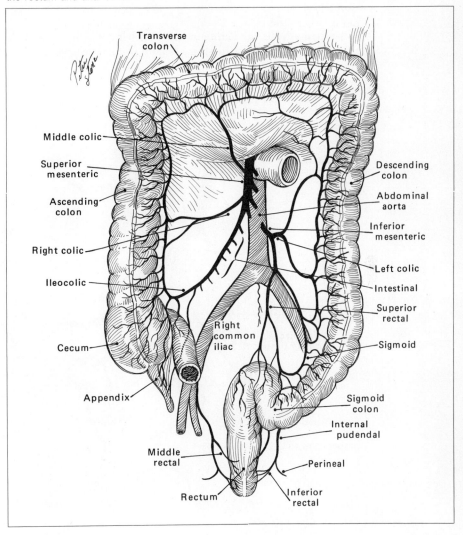

aorta to the hilus of the kidneys, each branches into three or four lobar branches. A detailed description of the distribution to the kidney is given in Chapter 28.

The gonadal arteries (internal spermatic in the male, ovarian in the female) are a pair of small vessels arising from the anterior aspect of the aorta between the renal and the inferior mesenteric arteries. They extend inferiorly into the pelvic cavity, where, in the female, they supply the ovaries. In the male they accompany the ductus deferens as a component of the spermatic cord which traverses the inguinal canal into the scrotum, to supply the testes.

The inferior mesenteric artery (Figure 17-9) gives rise to the left colic artery, which supplies the descending colon; gives off several sigmoid arteries to the sigmoid colon; and terminates as the superior rectal artery, which supplies the upper portion of the rectum.

Opposite the sacral promontory the abdominal aorta terminates as the right and left common iliac arteries. The common iliacs diverge laterally, and each terminates in an external and internal iliac branch.

Arteries to the Pelvis and Perineum   In the pelvic cavity the internal iliac artery bifurcates into an anterior and a posterior trunk (Figure 17-10). The latter gives rise to the iliolumbar

Figure 17-10. Arteries of the pelvis.

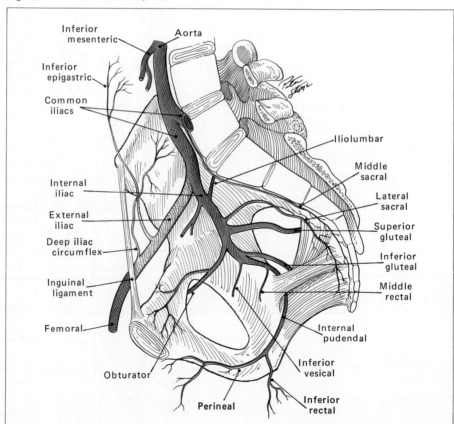

451

ANATOMY
OF
THE
BLOOD
VESSELS
AND
LYMPH
ORGANS

artery, which supplies muscles in the lumbar region and the iliac fossa; two small lateral sacral arteries, which pass to the anterior surface of the sacrum; and the large superior gluteal artery, which leaves the pelvic cavity through the greater sciatic foramen to supply the muscles of the buttocks.

The anterior division of the internal iliac gives off several branches to the pelvic viscera, namely, the uterine to the uterus, the vaginal to the vagina, the middle rectal to the rectum, the superior, middle, and inferior vesicals to the bladder, and the artery to the ductus deferens. The obturator artery from the anterior division passes through the obturator canal to supply the adductor muscles on the medial aspect of the thigh. Terminal branches of the anterior division are the inferior gluteal and internal pudendal arteries. The inferior gluteal artery supplies the gluteal muscles. The internal pudendal artery leaves the pelvic cavity to course through the ischiorectal fossa, where its inferior rectal artery supplies the distal end of the rectum. It then continues into the perineal region as the major blood supply to the external genitalia.

Arteries to the Inferior Extremity   The external iliac artery, which with the internal iliac is a terminal branch of the common iliac artery, courses through the iliac fossa and passes under the midpoint of the inguinal ligament, where it enters the thigh to become the femoral artery. Two branches arise from the external iliac artery (Figure 17-10). The deep iliac circumflex follows the inguinal ligament laterally to supply structures in the iliac fossa. The inferior epigastric passes superiorly to supply the skin and muscles of the anterior abdominal wall.

The femoral artery is the continuation of the external iliac as it passes deep to the inguinal ligament to course through the thigh (Figure 17-11). Small branches—the superficial epigastric, the superficial iliac circumflex, and the superficial external pudendal arteries—arise just below the inguinal ligament to supply the skin in the area of the groin. Two large branches, the lateral and medial femoral circumflex arteries, encircle the upper end of the femur to supply muscles in this area and anastomose with the gluteal branches of the internal iliac. The deep (profunda) femoral artery passes posteriorly to give three or four perforating branches which supply the muscles in the flexor compartment of the thigh. Continuing through the thigh in the adductor canal the femoral artery sends branches to the extensor muscles, then passes through the adductor hiatus to become the popliteal artery.

The popliteal artery, at the back of the knee, gives branches to the anastomosis around the knee and then divides into the anterior and posterior tibial arteries (Figure 17-12).

The anterior tibial artery courses through the anterior compartment of the leg, supplying the extensor muscles of this compartment; at the ankle it becomes the dorsalis pedis. The dorsalis pedis artery supplies the ankle and the dorsum of the foot and terminates in a branch that penetrates muscles in the first interosseous space, to contribute to the formation of the plantar arch of the foot.

The posterior tibial artery courses through the posterior compartment of the leg, supplying the flexor muscles. A large branch, the peroneal artery, parallel to the fibula, sends perforating branches into the lateral compartment to supply the peroneal muscles. The posterior tibial divides at the ankle into lateral and medial plantar arteries, which follow the respective sides of the plantar surfaces of the foot. Each of these vessels supplies branches to the ankle and muscles of the foot. The larger of the two, the lateral plantar, joins with the perforating branch of the dorsalis pedis to form the plantar arch. Similar to the superficial palmar arch of the hand, the plantar arch gives common digital arteries which divide into proper digital arteries to supply the contiguous sides of the toes.

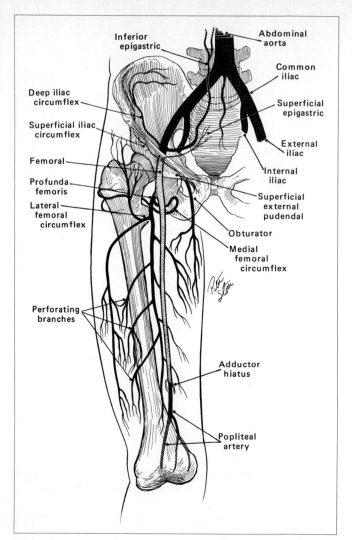

Figure 17-11. Arteries of the thigh. The external iliac, femoral (stippled), and popliteal arteries are a continuous vessel. The internal iliac supplies the medial aspect of the thigh by means of the obturator.

Systemic Venous Drainage

Most of the blood returning to the heart courses through companion veins of the arteries, the venae comitantes. There are usually two veins for each named artery. Such veins have the same name as the artery they accompany and generally drain the area supplied by the artery. They will therefore not be described individually.

The vessels that are not designated as venae comitantes include the large superior and inferior venae cavae, the jugular veins, the azygos veins, the hepatic portal system, and certain superficial veins. The major superficial veins that do not accompany similarly named arteries include those of the face, neck, and upper and lower extremities.

453
ANATOMY
OF
THE
BLOOD
VESSELS
AND
LYMPH
ORGANS

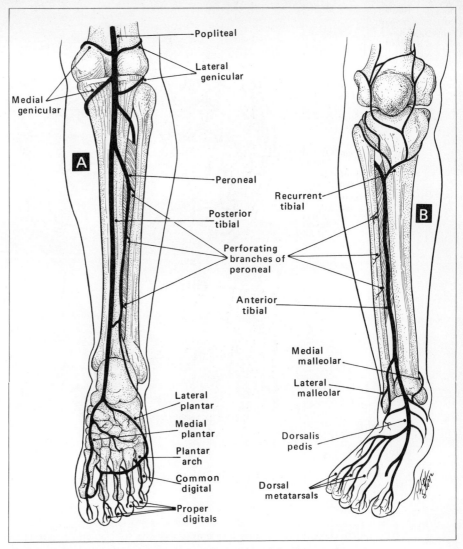

Figure 17-12. Arteries of the leg and foot. **A.** Arteries of the flexor region (posterior) of the leg and sole of the foot. **B.** Arteries of the anterior leg and dorsum of the foot.

Veins of the Head and Neck Venous tributaries in the anterior portion of the face combine in the anterior facial vein; those in the area in front of the ear compose the posterior facial vein; and those in the region behind the ear flow into the posterior auricular vein (Figure 17-13). The posterior facial vein divides into anterior and posterior divisions as it crosses the mandible. The posterior division joins the posterior auricular to form the external jugular vein; the anterior division joins the anterior facial to form the common facial vein.

In the neck the external jugular vein follows an oblique course across the

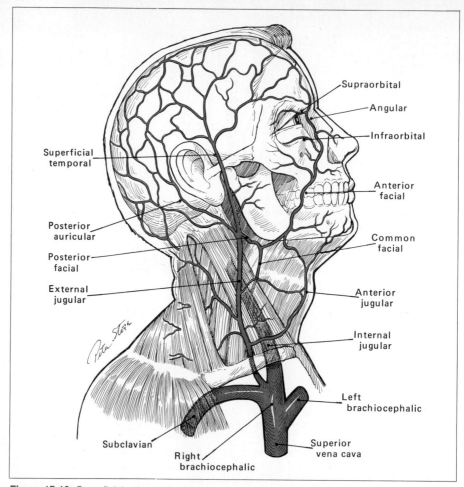

Figure 17-13. Superficial veins of the head and neck.

sternocleidomastoid muscle to empty into the subclavian vein. In the midportion of the anterior part of the neck the common facial passes deeply to drain into the internal jugular vein. A communicating branch from the common facial passes forward to join the anterior jugular vein, which descends adjacent to the midline. The latter vessel passes inferiorly and laterally to empty into the subclavian vein. The internal jugular also receives deep veins draining the face and neck.

Venous Sinuses   The dura mater, which it will be recalled is one of the meninges covering the brain, consists of a double-layered structure. In certain areas between the two layers there are sinuses (Figure 17-14). Most of the venous blood from the brain drains through these dural sinuses, although it may also join the superficial drainage of the scalp (passing through veins that penetrate the skull) or leave through meningeal veins that correspond to meningeal arteries. Among the important venous sinuses are the superior and inferior sagittal sinuses. The superior sagittal sinus is present in the superior border of the falx

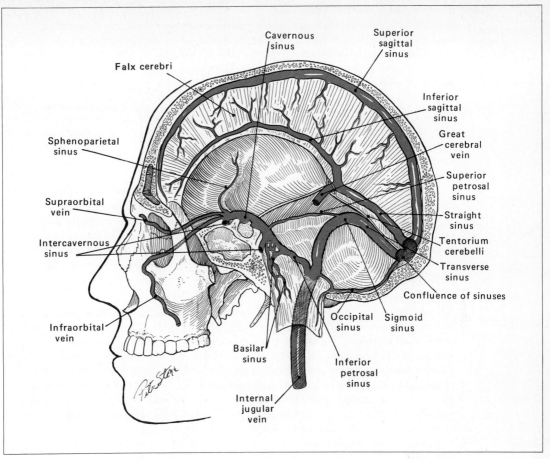

Figure 17-14. Dural sinuses. Superficial veins of the face empty into the cavernous sinus.

cerebri. Associated with this sinus are arachnoid villi or granulations, which project into its lateral aspect. The villi function in the transfer of cerebrospinal fluid from the subarachnoid space into the superior sagittal sinus.

The inferior sagittal sinus is within the inferior free border of the falx cerebri. Blood in this sinus drains into the straight sinus located in the posterior attachment of the falx cerebri to the tentorium cerebelli. The straight sinus also receives blood from the interior of the brain via the great cerebral vein. Both the superior sagittal sinus and the straight sinus meet and empty into the confluence of sinuses, adjacent to the internal occipital protuberance. The transverse sinus extends laterally from the confluence to the inner aspect of the occipital bone. The sigmoid sinus continues from the transverse sinus in an S-shaped course to the jugular foramen, where it leaves the skull to form the internal jugular vein.

Additional smaller dural sinuses include the superior petrosal sinus, which courses along the border of the great wing of the sphenoid bone to empty into the cavernous sinus, which flanks the sella turcica; the inferior petrosal sinus, passing posteriorly along the floor of the middle cranial fossa either to empty into the sigmoid sinus or to pass through the jugular foramen and empty independently into the internal jugular vein; and the basilar sinus, which also drains the cavernous sinus and passes along the floor of the middle cranial fossa to empty into the vertebral plexus of veins. The occipital sinus, which is located in the attached portion of the falx cerebelli, also empties into the vertebral plexus of veins.

Venae Cavae   With the exception of blood draining directly into the heart chambers from the heart musculature, the superior and inferior venae cavae carry all the blood that enters into the right atrium of the heart. The superior vena cava is formed by the junction of the right and left brachiocephalic veins and empties into the superior aspect of the right atrium.

The inferior vena cava, formed by the junction of the common iliac veins opposite the sacral promontory, courses through the abdomen as the companion vein of the abdominal aorta, where it receives blood from the segmental veins draining the abdominal wall. It also receives blood draining the kidneys and testes. Furthermore, just before passing through the center of the diaphragm at the level of the eighth thoracic vertebra, it receives blood from the hepatic veins, which carry blood away from the liver. The inferior vena cava empties into the inferior aspect of the right atrium.

Azygos Veins   The azygos system of veins drains the muscular wall of the thoracic cavity and most of the wall of the abdominal cavity (Figure 17-15). Originating by the junction of the small ascending lumbar veins and the subcostal veins, the azygos vein on the right side and the hemiazygos vein on the left ascend into the thoracic cavity. The azygos vein receives intercostal veins as it ascends along the vertebral column to arch over the root of the right lung. It drains into the superior vena cava. On the left side the hemiazygos vein receives lower intercostal veins and then crosses the midline to empty into the azygos vein. The accessory hemiazygos vein drains the intercostal veins in the middle portion of the left side of the thoracic cage, then drains into the azygos vein before the latter empties into the superior vena cava. The superior intercostal veins on both sides drain the upper two or three intercostal spaces into the brachiocephalic veins.

If the inferior vena cava becomes blocked, collateral circulation for venous blood may be established through the azygos system, since the ascending lumbar veins drain the same area as lumbar branches of the inferior vena cava.

The Hepatic Portal System   Veins ordinarily convey blood from capillaries directly to the heart. If a vein instead carries blood to a second capillary network, the venous pathway is termed a portal system. The hepatic portal system carries blood from the capillaries of the stomach, intestine, spleen, and pancreas to the capillary-like liver sinusoids (Figure 17-16). The portal vein is formed by the junction of the superior mesenteric and splenic veins. The latter in turn receives blood from the inferior mesenteric and pancreatic veins. Coronary, pyloric, and gastroepiploic veins which drain the stomach also empty into the portal vein.

Entering the liver through the porta hepatis, the portal vein breaks up into a series of interlobar vessels to divide subsequently into interlobular and finally intralobular veins. Intralobular veins course through the lobule and form a network of liver sinusoids adjacent to the parenchymal liver cells. These sinusoids drain into the central vein of each lobule. The central veins coalesce into sublobular veins, and these larger veins merge into three or four hepatic veins which drain into the inferior vena cava just before it passes through

457
ANATOMY
OF
THE
BLOOD
VESSELS
AND
LYMPH
ORGANS

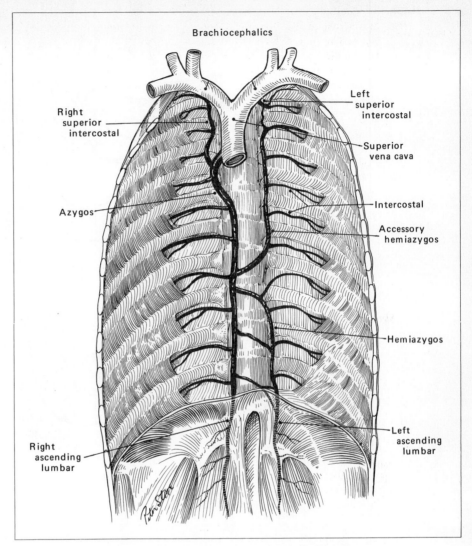

Brachiocephalics

Right superior intercostal

Left superior intercostal

Superior vena cava

Azygos

Intercostal

Accessory hemiazygos

Hemiazygos

Left ascending lumbar

Right ascending lumbar

Figure 17-15. Azygos system. The azygos drains the right half of the thorax and ends in the superior vena cava. The hemiazygos and accessory hemiazygos drain the left half of the thorax and empty blood into the azygos.

the diaphragm to empty into the right atrium. Circulation through the liver is described in greater detail in Chapter 19.

Superficial Veins of the Extremities   Beginning at the venous arch on the dorsum of the hand, the cephalic and basilic veins course proximally on the radial and ulnar sides, respectively, of the forearm (Figure 17-17). The median cubital vein forms a communicating channel between these major veins as it passes obliquely across the antecubital fossa at the elbow. The median cubital vein is the vessel usually used for taking blood samples (Figure 17-18).

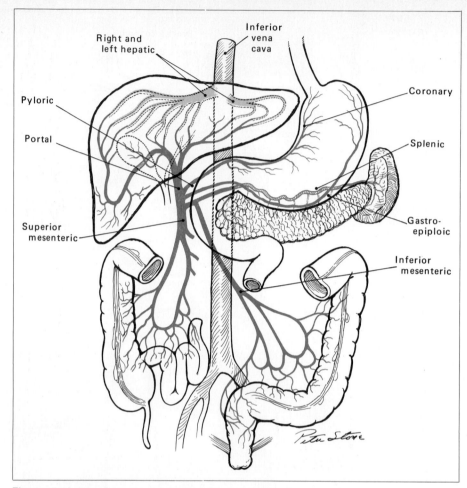

Figure 17-16. Hepatic portal system. Blood is carried from the stomach, intestines, spleen, and pancreas into the liver sinusoids. Hepatic veins convey it to the inferior vena cava.

Just proximal to the elbow the basilic vein passes deeply to empty into the brachial vein. The cephalic vein continues superficially to the front of the shoulder. Here it courses between the deltoid and pectoralis major muscles in the deltopectoral triangle to pass deeply and empty into the subclavian vein.

The long and short saphenous veins originate from the venous arch on the dorsum of the foot (see Figure 17-17). The short saphenous vein passes posterior to the lateral malleolus, ascends on the back of the leg, and at the popliteal fossa, penetrates the deep fascia to empty into the venae comitantes of the popliteal artery.

The long saphenous vein passes in front of the medial malleolus to ascend on the medial aspect of the leg. It swings behind the knee before continuing along the medial side of the thigh. In the inguinal region it penetrates the deep fascia to empty into the femoral vein just before the latter passes under the inguinal ligament to become the external iliac vein.

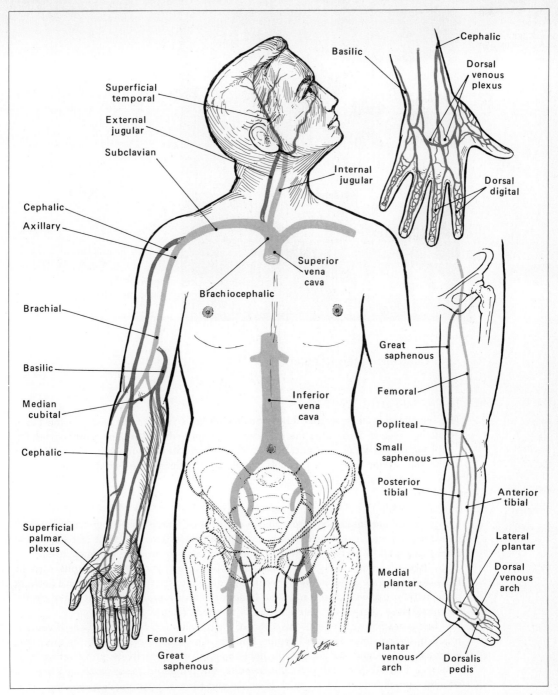

Figure 17-17. Principal veins. Superficial veins are in dark color, deep veins in light color.

459

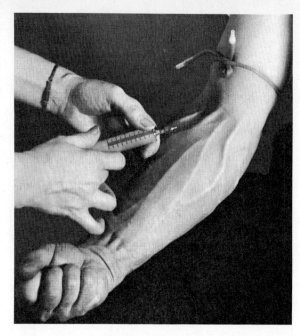

Figure 17-18. The median cubital vein is commonly used for intravenous injection and drawing samples of blood.

Fetal Circulation (see Chapter 33)

THE LYMPHATIC SYSTEM

The lymphatic system returns fluid from the tissue spaces of the body to the blood circulatory system. It consists therefore of (1) lymph, which contains lymphocytes, (2) lymphatic vessels which transport the lymph, (3) aggregates of lymphatic cells or nodules in the wall of the intestines, and (4) lymphoid structures—the lymph nodes, spleen, thymus, and tonsils.

Lymph Nodes

Lymph nodes are small bodies present in the path of lymph-collecting channels which filter the lymph as it returns to the blood circulatory system, ridding it of bacteria and foreign material. A second function of lymph nodes is to produce lymphocytes.

A typical lymph node is a bean-shaped body surrounded by a fibrous capsule. Extensions from the capsule, the trabeculae, go deep into the node. An extensive network of reticular fibers entrap and hold the huge masses of lymphocytes that make up the essential tissue, that is, the parenchyma, of the node (Figure 17-19).

Just beneath the capsule, in the cortex of the lymph node, lymphoid tissue is concentrated in discrete masses called lymph follicles. The centers of the follicles are called germinal centers and are filled with light-staining, young, actively proliferating lymphocytes. Beneath the cortex, in the medulla, the lymphoid tissue is concentrated in strands, the medullary cords, which interlace and form a loose, spongy network.

Each lymph node has several afferent lymph vessels that enter through the capsule into a narrow, relatively cell-free zone, the cortical sinus. The lymph filters slowly through the sinuses and leaves by way of efferent vessels at the hilus.

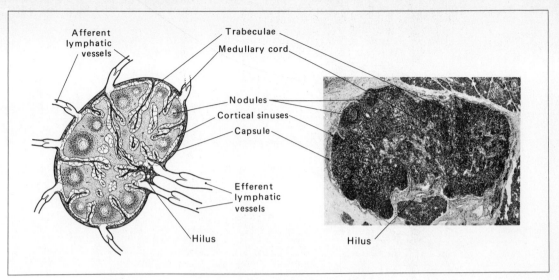

Figure 17-19. Diagrammatic sketch and photomicrograph of a lymph node.

Lymphatic Drainage

Lymphatic vessels begin as minute, blind-ending capillaries which collect lymph from the tissue spaces. These vessels coalesce in larger channels which culminate in the right lymphatic and thoracic ducts, which drain into the brachiocephalic veins. Lymph nodes are interposed along this chain of vessels and filter the lymph as it returns to the circulatory system.

Lymphatic vessels follow the veins of the body, and collections of lymph nodes are grouped at specific sites adjacent to the major veins. These vessels are usually classified as superficial and deep, coinciding with the location of the veins. While lymph nodes and vessels are dispersed throughout the body, only the larger of these structures may be grossly recognized. However, in certain pathological conditions the nodes and vessels become larger and painful, and are easily demonstrable by palpation.

The distribution and the direction of the flow of lymph is of great clinical importance in that, in addition to their role in infection, cancer cells may be transported throughout the body by lymphatic channels and set up secondary sites of cancer growth. The following description locates the major lymphatic vessels and nodes in the body (Figure 17-20).

Extremities In the upper extremity there are a couple of small nodes in the antecubital fossa that drain the hand and forearm. The efferent vessels of these antecubital nodes are directed toward the axilla. From axillary nodes the subclavian lymphatic vessel joins the right lymphatic duct on the right side and the thoracic duct on the left side. In "blood poisoning," lymph vessels, becoming inflamed, are outlined as red streaks, which can be seen to progress slowly up the forearm and arm to the axilla.

Axillary nodes are of special interest to the physician. In addition to filtering lymph from the arm, forearm, and hand, axillary nodes receive most of the lymphatic drainage of the breast and therefore are of prime importance in aiding the spread of breast cancer. If the axillary nodes become invaded by malignancy, a radical mastectomy may have

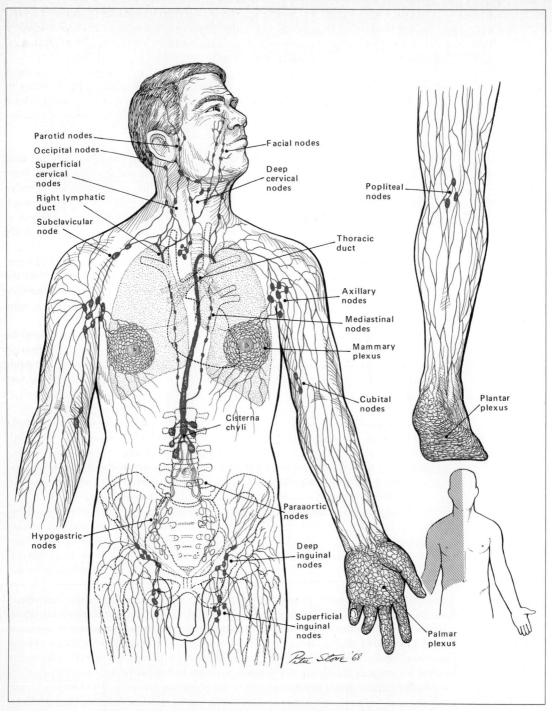

Figure 17-20. Principal lymph vessels and nodes. In the figure at the lower right, cross hatching shows the area drained by the right lymphatic duct.

462

463

ANATOMY
OF
THE
BLOOD
VESSELS
AND
LYMPH
ORGANS

to be performed. This involves not only removal of the breast but all its blood and lymph supply, the underlying pectoral muscles, the axillary vein and its tributaries, and adjacent lymph nodes. This is done to ensure the excision of even the smallest of the lymphatics, which, if left behind, could "seed" the cancer.

In the inferior extremity three or four popliteal nodes in the popliteal fossa receive drainage from the leg and foot. Their efferent vessels extend to the inguinal nodes in the groin, which are scattered along the femoral vein and the inguinal ligament. These nodes filter lymph not only from the inferior extremity but from the adjacent perineal and abdominal regions as well.

Pelvic and Abdominal Lymphatics   Nodes along the internal iliac vein (internal iliac nodes) collect lymph from the pelvic organs, and external iliac nodes receive lymph from the inguinal nodes. Both sets of iliac nodes in turn direct their efferent vessels to nodes flanking the abdominal aorta. This latter group also drains the bulk of the digestive tract, and their efferent vessels end in a saclike enlargement, the cisterna chyli, adjacent to the celiac artery. Essentially all the lymphatic drainage of the lower half of the body passes to this reservoir.

From the cisterna chyli the large thoracic duct ascends to pierce the diaphragm, traverse the thorax, and enter the neck. It receives tributaries from the left side of the thoracic cavity and, as it passes into the neck, joins the jugular lymphatic trunk draining the left side of the head and neck before emptying into the left brachiocephalic vein.

Lymphatics of the Head and Neck   A superficial series of cervical nodes lies along the inferior border of the mandible, the posterior border of the sternocleidomastoid muscle, and the clavicle. Nodes adjacent to the mandible frequently become inflamed and palpable in throat infections. Additional lymphatics accompany the deep veins draining the face and head.

The short right lymphatic duct drains the right upper quadrant of the body and is formed by the jugular trunk on the right side, the right subclavian trunk draining the right extremity and breast, and the right mediastinal trunk. The last corresponds to a similar trunk on the left side which drains into the thoracic duct.

The mediastinal trunks receive lymph principally from (1) pulmonary nodes located on the surface of the lung, (2) hilar nodes present at the hilus of the lung, (3) tracheobronchial nodes located at the bifurcation of the trachea into the primary bronchi, and (4) tracheal nodes lying to either side of the trachea.

The physiological aspects of lymph circulation are discussed in Chapter 18.

Spleen

The spleen is the largest lymphoid organ in the body (Figure 17-21). It is situated under the diaphragm in the upper left portion of the abdominal cavity.

The spleen has a dense, tough, musculofibrous capsule with many trabeculae —extensions of the capsule that pass deeply into the gland. The parenchyma is the splenic pulp, which is of two types, red and white. The more abundant red pulp, found throughout the spleen, is arranged in indistinct plates or cords of cells which loosely blend with the white pulp. The white pulp is concentrated in round masses of lymphoid cells termed the splenic nodules, or Malpighian corpuscles. These cells possess germinal centers which surround a central arteriole.

The terms red and white pulp refer to the appearance of each region in a fresh state. The white pulp appears pale because of the heavy concentration of white blood cells,

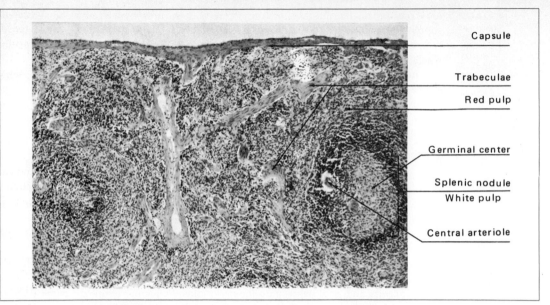

	Capsule
	Trabeculae
	Red pulp
	Germinal center
	Splenic nodule
	White pulp
	Central arteriole

Figure 17-21. Portion of the spleen (× 60). White pulp is confined to splenic nodules.

while the red pulp is filled with red blood cells. In addition to lymphocytes the blood sinuses of the spleen harbor many monocytes and macrophages. In fact, this is the largest concentration of macrophages in the body. The spleen is thus part of the reticulo-endothelial system (see Chapter 3). The chief function of its macrophages is to ingest and destroy senile red blood cells. Hence the spleen is sometimes referred to as the "graveyard" of the red blood cell. Circulation of blood through the spleen and the spleen's importance as a reservoir of red blood cells are discussed in Chapter 19.

Thymus

The thymus is located behind the sternum in the upper portion of the mediastinum, the area between the two lungs. The thymus usually consists of two lobes, each of which is divided into many tiny lobules. Every lobule consists of a cortex and, in the center, a medulla. The cortex is filled with small lymphocytes (often called thymocytes). The medulla is lighter-staining, and usually found within it are one or more prominent eosinophilic corpuscles, called thymic corpuscles, or Hassall's corpuscles. These corpuscles are of hyaline material at the center, surrounded by concentric rings of connective tissue interspersed with flattened epithelial cells.

The thymus is relatively large during prenatal life and reaches its greatest absolute size at about age 6, when it begins a gradual decline in size, becoming atrophied at maturity. The thymus is a source of lymphocytes and furthermore is believed to release a substance that starts up production of lymphocytes in the spleen and lymph nodes. It has been suggested that the thymus regulates immune reactions in early childhood.

Tonsils

The tonsils are aggregations of lymphoid tissue found under the epithelium that lines the pharynx and oral cavity. They release lymphocytes into the blood and are therefore

465

ANATOMY
OF
THE
BLOOD
VESSELS
AND
LYMPH
ORGANS

believed to have a protective function against bacteria. The best-known tonsils are the accumulations on each side of the throat associated with the palate, the palatine tonsils. The tonsils on the posterior wall of the nasopharynx—the pharyngeal tonsils—are called adenoids when they are enlarged, as sometimes occurs in childhood.

QUESTIONS AND PROBLEMS

1 Why is the circulatory system considered to be a closed system?

2 Describe the anatomical elements that make possible dilation and constriction of the vessels of this system.

3 Where is endothelium found in the body? How does it differ from epithelium?

4 What are the distinctive anatomical features of arteries and veins?

5 How do blood vessels receive their blood supply? Nerve supply?

6 In histological structure, how is the aorta different from the inferior vena cava?

7 What vessel is commonly selected to receive intravenous medication?

8 What is a convenient artery for taking the pulse?

9 Where would pressure be applied to control hemorrhage of the lower extremity?

10 Where does the azygos system begin and end, and what area does it drain? In obstruction of the inferior vena cava, what role can the azygos system play?

11 What is a portal system? What vessels make up the hepatic portal system?

12 What are the organs of the lymphatic system? Where are they located?

13 What is the relationship of the lymphatic system to the circulatory and digestive systems?

14 Why are the axillary lymph nodes removed in a radical mastectomy?

15 Why is swelling likely to occur in the upper extremity of a patient after a radical mastectomy?

18 CIRCULATION

Although men have slaughtered animals and themselves for millions of years and thus have had ample opportunity to observe the flow of blood, it was not until the seventeenth century that an English physician, William Harvey, first accurately described the circulation of blood. The explanation of this time lag will be left to the medical historian. Our task is to explain general physiological principles and mechanisms that govern blood circulation. In the next chapter, aspects of circulation peculiar to special regions of the body will be considered.

BASIC PRINCIPLES
To understand the physiology of circulation, the student must be conversant with the anatomy of the circulatory system and also with certain basic principles of physics.

Components of the Circulatory System
The anatomy of the circulatory system was described in Chapter 17. Here it is important to reemphasize that although the arteries, veins, and capillaries that make up the circulatory system vary in diameter, length, capacity, and composition of their walls, and although some vessels are in series and others in parallel, the fact remains that there is a continuous flow of blood from the heart into the arterial system and from the venous system back to the heart. In between there are major differences in pressure, in rates of flow, and in the volume of blood contained in the various components.

The systemic circulation is a high-pressure system which contains about 75 percent of the total blood volume. The volume of blood and the pressure of the blood in the pulmonary system are much less. As already mentioned, blood may be shifted from one system to the other by variations in the output of the two ventricles. The veins in both systems are very distensible and can, therefore, accommodate a wide range of volumes.

Blood does not circulate through a simple, one-channel system connecting one side of the heart with the other. As depicted in Figure 18-1, there are many parallel routes through which the blood may flow from the major arterial vessels to the veins. Because of these parallel routes, pressure in one part of the circulation may be drastically altered without necessarily altering the pressure elsewhere.

In this complex system the heart functions as the primary pump, the large arteries as conduits, the arterioles as resistance vessels, the capillaries as exchange mechanisms, and the veins as capacitance vessels.

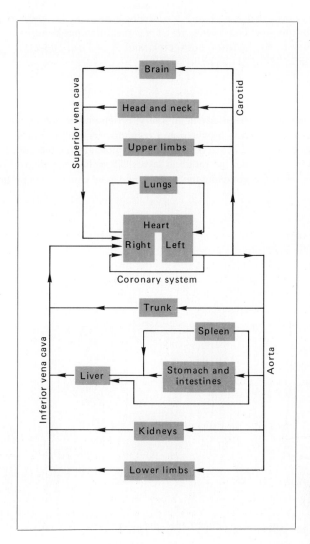

Figure 18-1. Diagrammatic representation of the circulation of the blood. Note the many parallel routes.

Physical Relationships

The circulation of blood involves the interrelation between (1) cardiac output, (2) caliber of the vessels (cross-sectional area), (3) velocity of blood, (4) resistance to flow, (5) distensibility and elasticity of the vessels, (6) blood viscosity, and (7) blood volume.

At any point within a tube filled with a circulating fluid, force is exerted at right angles to the direction of flow. This force is termed lateral pressure. This is the pressure we measure and call blood pressure. Resistance means opposition to force. In the circulatory system, there is resistance to the flow of blood and therefore opposition to the force driving the blood. The relationship between flow, pressure, and resistance may be expressed as follows:

$$\text{Flow} = \frac{\text{pressure}}{\text{resistance}} \tag{1}$$

This equation says that the greater the driving pressure, the lower the resistance, the greater will be the flow. The driving pressure is the difference between the lateral pressure at any two points, for instance, between P_1 and P_2 of Figure 18-2.

By rearranging Eq. (1) to be

$$\text{Pressure} = \text{flow} \times \text{resistance} \tag{2}$$

one can see that the pressure, say in an artery, is determined by the flow (cardiac output) and the resistance which is provided and regulated by the arterioles.

To understand resistance, it is necessary to appreciate that fluid does not flow through a tube as an intact cylinder but rather in layers. These layers slide over one another with little drag, but the layer touching the wall of the vessel experiences far greater drag. This

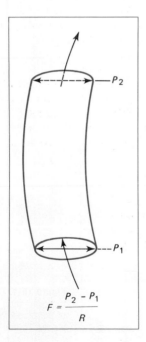

Figure 18-2. Lateral pressure.

may be observed from the bank of a river. The water close to the bank hardly moves at all, while the water out in the middle of the river flows much more rapidly. What this means is that the smaller the cross-sectional area of the vessel, the greater will be the drag, or resistance. And, obviously, the longer the vessel and the greater the viscosity of the fluid, the greater will be the resistance.

These relationships may be expressed as follows:

$$\text{Resistance} = \frac{\text{viscosity} \times \text{length}}{\text{radius}^4} \tag{3}$$

If this equation is substituted for resistance in Eq. (1), then:

$$\text{Flow} = \frac{\text{pressure} \times \text{radius}^4}{\text{viscosity} \times \text{length}} \tag{4}$$

And when Eq. (4) is rearranged:

$$\text{Pressure} = \frac{\text{flow} \times \text{viscosity} \times \text{length}}{\text{radius}^4} \tag{5}$$

Now one can readily see how important the radius of the blood vessel is in regulating blood pressure, since change in radius is raised to the fourth power in its effect on blood pressure, as will be discussed below.

Finally, the velocity of flow in a closed system is inversely related to the radius (Table 18-1), and the pressure is directly proportional to the radius at any point in the system.

Note that blood flow in the capillaries is 1,000 times slower than in the aorta, a situation which occurs simply because the total cross-sectional area of all the capillaries is 1,000 times greater than that of the aorta. Only across the capillary wall does fluid and solute exchange occur; thus the slowing of the flow in this part of the circulatory system contributes to the thoroughness of that transfer.

Blood Pressures

In the veins, blood flows relatively smoothly at a pressure which does not vary much during the cardiac cycle. In the arteries, the situation is quite different. During systole the arterial pressure rises, and during diastole it falls (Figure 18-3). The peak pressure is termed the systolic pressure. The lowest level reached just before ventricular ejection starts is termed diastolic pressure. The difference between systolic and diastolic pressures

**TABLE 18-1. VELOCITY OF BLOOD FLOW VS.
TOTAL CROSS-SECTIONAL AREA OF VESSELS**

PART	AREA, CM^2	VELOCITY, CM/SEC
Aorta	4.5	40.00
Arteries	20.0	9.00
Arterioles	400.0	0.45
Capillaries	4,500.0	0.04
Veins	40.0	4.5
Vena cava	18.0	10.00

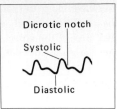

Figure 18-3. Recording of
arterial blood pressure.

is termed the pulse pressure. The average pressure during one cardiac cycle is called the mean pressure. Normal values for the young adult are systolic, 120 mmHg; diastolic, 75 to 80 mmHg; and mean pressure, about 95 mmHg.

Influence of Elastic Walls

Blood vessels are not rigid; they are elastic. That is to say, they will distend and they will recoil. The most elastic vessel is the aorta. If the aorta were distensible enough to enlarge proportionately with the volume of blood ejected during systole, without an increase in tension, there would be no increase in pressure.

During diastole, the elastic aorta closes down upon the volume of blood as it flows into the larger arteries. Therefore, the capacity of the vessel decreases with the decrease in volume. Were the aorta rigid, the fall in pressure during diastole would be precipitous. The elasticity of the aorta, then, may be said to maintain the diastolic pressure. To put it another way, the magnitude of the pulse pressure varies inversely with the distensibility and elasticity of the aorta and large vessels.

The distensibility and elasticity of the aorta and large vessels, in combination with the resistance offered by the arterioles, account for the fact that although the flow is intermittent (pulsatile) in the large vessels, by the time it reaches the capillaries it is no longer pulsatile; it is steady.

In the normal young adult the aorta and larger arteries are highly distensible. As one grows older, the distensibility diminishes; in some cases, these vessels become truly rigid. As a result, the systolic and diastolic arterial pressures which normally differ by only 40 mmHg may vary by more than 100 mmHg in the individual with arteriosclerosis (hardening of the vessels).

Methods of Measurement

The methods for measuring blood flow and blood pressure have improved in accuracy and convenience to the point that these parameters are routinely determined clinically and in research, although blood pressure measurements are still far more common than are measurements of blood flow.

Blood Pressure The most direct method of measuring blood pressure involves placing a cannula (a large-bore needle) directly into the vessel. The cannula then either is connected to a mercury manometer which can be arranged with a kymograph for continuous recording, or can be connected to a transducer for electronic recording. The latter is preferable since the great inertia of the mercury does not permit accurate pulse pressure tracings. The cannulation method is the one of choice in experimental work using anesthetized animals. It is also used clinically, usually during surgery or during diagnostic cardiac catheterization procedures.

The method most often used for routine examination is the auscultatory method. The term auscultation means the detection and study of sounds arising from various organs. To determine arterial blood pressure one utilizes the sounds emanating from an artery which has been compressed by a cuff. The instrument used to determine pressure by this method is called the sphygmomanometer. It consists of a cuff which is wrapped snugly about the arm just above the elbow. The cuff is connected to a mercury manometer, then inflated until the pressure exceeds the systolic blood pressure. Next, the pressure is slowly reduced while the examiner listens with a stethoscope to the sounds coming from the artery, usually the brachial artery. When the pressure in the cuff falls just below systolic pressure, some blood will flow, there will be turbulence, and the sound will be heard. When there is no flow, there is no sound. When the first definite sound is heard, the pressure is noted. This is the systolic pressure because only at the peak pressure of systole is the pressure great enough to force blood under the inflated cuff.

The pressure in the cuff continues to fall. Next a murmur develops which finally changes in intensity. The pressure at this point of change is taken as the diastolic pressure. As the cuff is further deflated, the murmur disappears completely. Because the difference between the change in intensity is sometimes difficult to detect, some clinicians accept the disappearance of sound as the diastolic pressure. The sound disappears because after the cuff pressure is below diastolic pressure, blood flows throughout the cardiac cycle, there is no turbulence, and thus no sound. The sounds detected in this method are referred to as the sounds of Korotkov.

The venous pressure, just like that in the arteries, may be measured either directly or indirectly. The direct procedure is to insert a hypodermic needle into the vessel and then connect it to a manometer, but a water manometer is used rather than a mercury manometer. The explanation lies in the fact that arterial pressures are so much higher than venous levels. The venous pressure is normally not more than 15 to 25 cm of water (cmH_2O). Water, which is about 13.5 times lighter than mercury, is used for accuracy.

The indirect method for determining venous pressure is not nearly so accurate as the indirect arterial blood pressure technique but nonetheless does afford a worthwhile approximation. A quick way of obtaining the approximate venous pressure is to have the patient lie flat on his back. The examiner then slowly elevates the subject's hand until the veins on the back of the hand collapse. The height of the hand above the subject's heart is measured (Figure 18-4). Since blood has a specific gravity close to that of water, the distance gives an indication of the venous pressure expressed in centimeters of water. In other words, if the veins on the back of the hand collapse when the hand is 25 cm above the heart, the venous pressure is said to be approximately 25 cmH_2O.

Blood Flow The first crude determinations of blood flow were made following decapitation of an experimental animal, or after severing a limb. Later the rotameter was developed. This instrument is inserted in the circulatory system so that blood flows through it. In so doing, a float is caused to rise. The greater the flow, the higher the rise of the float. Such an instrument, with appropriate circuitry, can be used for continuous electronic recording. The disadvantages are that the vessel must be cut to insert the instrument, an anticoagulant must be used to prevent clotting, and the instrument itself alters the blood flow to some extent.

More recently, two instruments have been developed which do not have these

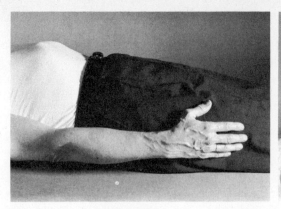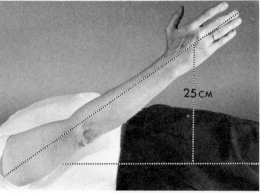

Figure 18-4. Determination of venous pressure. The venous pressure is approximately equivalent to the distance above heart level at which the veins of the hand collapse. (*Photographs by R. Thompson.*)

disadvantages. One is called the electromagnetic flowmeter. It is not inserted into a vessel, but rather placed around the vessel. Blood in the vessel flows between poles of a strong magnet in the flowmeter. Electrodes in the instrument are positioned on two sides of the vessel perpendicular to the magnetic lines of force. Flow of blood in the vessel generates a voltage difference between the two electrodes; the greater the flow, the greater the difference. Thus a continuous and accurate record may be obtained.

The second new instrument is called the ultrasonic flowmeter. This instrument operates on the principle that sound waves travel downstream faster than they do upstream. The instrument sends a sound downstream, then rapidly changes to send one upstream. The difference between the sound waves in the two directions is electronically determined; the greater this difference, the greater the flow rate.

REGULATION OF ARTERIAL BLOOD PRESSURE

The four vital determinants of arterial pressure are (1) cardiac output, (2) peripheral resistance, (3) distensibility of the large vessels, and (4) blood volume. In Chapter 16 cardiac output was considered in detail and should be reviewed at this time.

Autoregulation of Blood Flow

In the previous chapter, emphasis was placed upon the fact that the heart has an inherent rhythmicity and that, by virtue of the Starling relationship, it has an inherent mechanism for regulating its output in accord with the venous return. These inherent mechanisms are fundamental. They permit the heart to vary cardiac output over a broad range. The extrinsic factors, such as the autonomic nervous system and humoral substances in the blood, are important, too, but the student should understand that they function simply to extend the range of the heart. The same is true of the circulatory system. It too is influenced by the autonomic nervous system and by humoral substances, but it is quite capable of adjustments in the complete absence of these extrinsic influences which, once again, simply expand the range.

Rapid Autoregulation Figure 18-5 depicts a curve characteristic of rapid autoregulation. Blood flow is seen to increase rather proportionately to the increase in arterial pressure

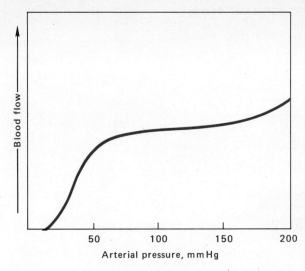

Figure 18-5. Autoregulation. Over a broad range of pressures blood flow remains relatively constant. This means that resistance to flow increases in step with driving pressure.

until a pressure of 50 or 60 mmHg is reached. Then over a broad range of pressures there is little or no increase in flow. At very high pressures the flow again increases. The existence of the plateau, that is, the relative independence of flow to changing pressure, indicates autoregulation. This type of autoregulation occurs very rapidly so that if the pressure is suddenly elevated, there will be an increase in flow for a minute or two, but then flow returns to the plateau level. As will be discussed, there is another mechanism of autoregulation that takes much longer to develop.

There is still little unanimity as to the mechanism of rapid autoregulation. However, most investigators believe that oxygen plays the key role. Decreased oxygen supply results in rapid dilation of the local arterioles, metarterioles, and precapillary sphincters. Thus resistance diminishes and blood flow increases. Similarly, if the blood oxygen concentration remains constant but the blood flow decreases, there will be dilation, thereby returning the local blood flow to normal.

Other investigators contend that the substance responsible for vasodilation is not oxygen but something else. They are willing to concede that oxygen lack triggers the sequence of events but think that some other substance, a specific vasodilator, is liberated as a result of the oxygen lack. The most likely candidate for this role is adenosine, which can be detected in increased concentration in the venous blood of an organ after blood flow has been decreased. Other substances to which a similar role has been attributed are histamine, carbon dioxide, and hydrogen ions.

A model can be constructed with tubes, semipermeable membranes, and pinchcocks (Figure 18-6) that will autoregulate; that is, the blood flow will first increase when the inflow pressure is rapidly increased, but then the pressure will return to the plateau level (Figure 18-6). Since there are no dilator substances nor any tissues capable of responding to them, such autoregulation depends upon mechanical means.

The model, to autoregulate, must have a nondistensible membrane around a permeable "capillary bed." Then, as the pressure within the capillary increases, there is greater outpouring of fluid, but because of the nondistensible surrounding membrane, the pressure between the capillary and the membrane increases sharply. This pressure squeezes the capillary, thereby increasing resistance to flow.

Beyond question the model is simple, and it works without assuming any nervous

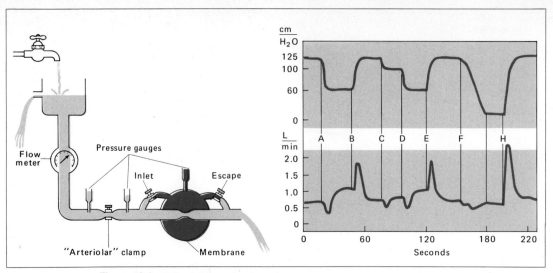

Figure 18-6. At the left is a drawing of an apparatus constructed to test the mechanical hypothesis of autoregulation. At the right is a record obtained. A rise in driving pressure (at B, E, and H) brings an immediate rise in flow, but flow soon returns to the base level, demonstrating autoregulation. (*Courtesy of Simon Rodbard.*)

or humoral control. Unfortunately, such a demonstration does not rule out the possibility that nervous and humoral factors play primary roles in the living organism.

Slow Autoregulation The number of vessels and their size can change and do change in response to metabolic needs. Such changes probably contribute importantly to so-called conditioning. For example, if exercise is vigorous, prolonged, and repeated over a period of days or weeks, the number and size of the vessels increase, thus resulting in improved blood supply. To put it another way, as a result of these vascular changes, there is greater blood flow with the same head of pressure. Once again, the stimulus for increased vascularity appears to be the balance between oxygen demand and supply.

The same mechanism is responsible for the excellent autoregulation that results from a very slow but progressive increase in pressure. In such experiments, after the pressure reaches about 50 or 60 mmHg, blood flow becomes constant and changes hardly at all until the pressure exceeds 200 mmHg. Apparently, as the pressure is raised and the blood flow increases, the plethora of oxygen causes a decrease in the size and number of blood vessels, thus increasing resistance and thereby maintaining flow constant despite higher pressure.

Reactive Hyperemia An example of autoregulation is observed when the blood supply to an area is stopped for a minute or two and then allowed to flow. As shown in Figure 18-7, flow following the period of occlusion is much greater than that before the period of occlusion. The greater flow is referred to as reactive hyperemia. Slowly flow returns to the preocclusion level. If the extra volume of blood is calculated during enhanced flow, it will be found to be just about equal to the amount of blood lost to the tissues during the period of occlusion. During occlusion, of course, there is no oxygen supplied to the tissues; thus after the small amount in the stagnant blood is used, oxygen utilization falls

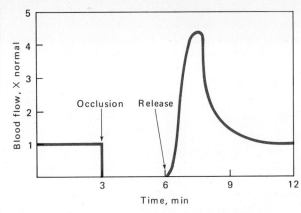

Figure 18-7. Reactive hyperemia. After a period of occlusion, blood flow markedly increases, making up for the loss of flow during occlusion.

to zero. After flow is reestablished, oxygen utilization is much increased, and once again the extra usage just makes up for the period of deprivation.

Blood Volume

Normally the blood volume remains remarkably constant at about 5 liters. But it can change and does so in many circumstances, as will be discussed subsequently. If the blood volume increases, venous pressure and venous return increase; therefore cardiac output increases, which elevates arterial blood pressure. At the normal mean arterial blood pressure of about 100 mmHg the urine output is about 1 ml per min. As the pressure rises, so does urinary output. An increase of pressure to 150 mmHg, for example, may increase the urine output some four times. Conversely, as arterial pressure falls, so does urinary output, and as will be explained in Chapter 29, urine output ceases when arterial blood pressure falls below a mean value of about 60 mmHg. The point is that changes in blood volume alter cardiac output and arterial pressure, and that blood volume itself is closely regulated by renal function.

In summary, the overwhelming current evidence strongly suggests that the role of the circulatory system is to provide the cells of the body with an adequate supply of oxygen and other nutrients. It also removes the end products of metabolism. If tissue needs are not met, local vasodilatation occurs. Vasodilatation increases venous return, which results in augmented cardiac output. At the same time, sensitive mechanisms maintain an adequate blood volume to meet circulatory needs. In short, as with most feedback systems, the end result of the system controls the system itself. In this case, the end result is the supply of oxygen to the tissues.

Vasoconstriction

The walls of the arterioles include a highly developed smooth muscle layer which is capable of varying the diameter of the arteriolar opening. The arteries are also capable of some constriction, but the major variation in peripheral resistance occurs in the arterioles. The importance of peripheral resistance, and therefore the arterioles in regulating blood pressure, is underscored by the fact that the pressure, as explained above, varies with the fourth power of the radius of the arterioles. That is to say, a small change in diameter causes a large change in pressure.

The arterioles are innervated by neurons of the sympathetic nervous system. Activation of these neurons, in most instances but not all, evokes contraction of the smooth muscle and therefore a decrease in the diameter of the arteriole. This contraction and decrease in diameter of an arterial vessel is termed vasoconstriction as opposed to contraction of a vein, which is referred to as venoconstriction. Because vasoconstriction increases resistance, the arterioles are called resistance vessels.

The ends of the postganglionic sympathetic fibers liberate norepinephrine, which is directly responsible for constriction of the vessels. Other chemical agents that have the same effect are epinephrine, angiotensin, and serotonin.

Vasodilatation

Usually, if a vasoconstrictor neuron is severed, the arteriole innervated by that neuron dilates. This indicates that there is normally a continual propagation of impulses which maintain the arteriole in a state of partial constriction, referred to as vasomotor tone. Vasodilatation, then, can be evoked simply by decreasing the activity in the vasoconstrictor neurons. The arteriolar smooth muscle relaxes, and the blood pressure enlarges the vessel. However, vasodilatation in some areas also results from the action of vasodilator neurons. These include (1) sympathetic neurons that innervate arterioles of skeletal muscle, (2) sympathetic neurons that innervate the coronary arterioles, (3) parasympathetic neurons that travel in cranial nerves VII, IX, and X to innervate arterioles of the tongue and salivary gland, and (4) the parasympathetic neurons that travel in the pelvic nerves to innervate arterioles of the genital organs.

Sympathetic neurons that cause vasodilatation of skeletal muscle arterioles are called cholinergic. Acetylcholine is a potent vasodilator. The sympathetic neurons that innervate the coronary arterioles are adrenergic, but there is considerable question as to how norepinephrine, in this instance, evokes vasodilatation if indeed it does. More likely, stimulation of these sympathetic fibers causes vasodilatation secondarily to increased cardiac activity.

Vasodilatation evoked by activity of the cranial parasympathetic neurons may be due to the formation of bradykinin, which is produced during secretion by salivary and sweat glands. Bradykinin, a peptide, is a potent vasodilator.

Vasodilatation is also caused by histamine, elevated carbon dioxide content, lowered oxygen content, a rise in body temperature, and a decrease in pH.

Influence of Vasomotor Activity on Flow

Earlier in this chapter the relations between flow, pressure, and resistance were discussed. The greater the pressure, the greater the flow, and as resistance increases, flow decreases. Vasoconstriction, then, would be expected to decrease flow if the pressure is held constant, and vasodilatation would increase it. Figure 18-8 illustrates these relationships. First note the great change in flow at any arterial pressure. For example, at 100 mmHg the flow through a small constricted vessel may be only 1 ml per min, whereas the flow is seven times that amount when the vessel is dilated. Next observe that flow may cease completely. This cessation occurs when the vessel closes. The pressure at which a vessel closes is referred to as the critical closing pressure. As seen in Figure 18-8, the critical closing pressure varies with the degree of vasomotor tone.

Law of Laplace The sudden closure of a vessel at a certain critical pressure is explained by the law of Laplace. This law also explains, for example, why an aneurysm often ruptures. An aneurysm is a pathological dilatation of a vessel.

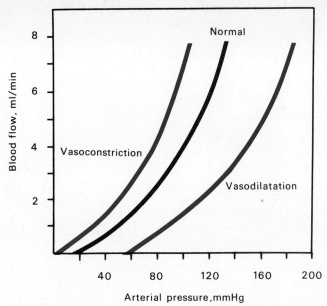

Figure 18-8. Relation of flow to pressure. Vasoconstriction decreases flow; vasodilatation permits it to increase. The pressure at which a vessel closes (zero flow) is the critical closing pressure.

Figure 18-9 illustrates the Laplace relationships. The law states that

$$T = Pr$$

where T = tension in the vessel wall
P = transmural pressure
r = radius of the vessel

Wall tension is defined as the force per unit length tangential to the vessel wall which tends to pull apart a theoretical longitudinal slit in the vessel. Transmural pressure is the difference between the pressures on the two sides of the vessel. In effect, the blood pressure is equal to the transmural pressure since the pressure on the outside of the vessel is negligible.

In simple language, the law of Laplace says that the smaller the radius of the vessel, the less the wall tension needed to support any particular pressure. This explains why the very fragile capillaries can support high blood pressures. It also explains why, as the vessel gets smaller and smaller as a result of vasoconstriction, a point is reached at which the tension in the wall, which tends to close the vessel, exceeds the blood pressure, which tends to hold the vessel open. At that point the vessel suddenly closes. Only when vasomotor tone decreases or the blood pressure increases does the vessel reopen.

In summary, vasoconstriction elevates arterial blood pressure, greatly decreases flow in the constricted vessel, and can actually stop flow altogether.

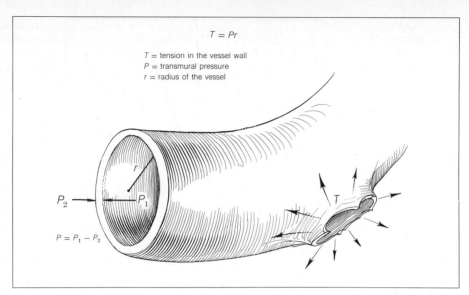

Figure 18-9. Law of Laplace: $T = Pr$, where $T =$ wall tension, $P =$ transmural pressure, $r =$ radius of the vessel.

Vasomotor Center

The activity of vasoconstrictor neurons as well as the vasodilators is coordinated at various levels in the nervous system. Even when the spinal cord is transected, a degree of coordination still persists, indicating that the cord itself may be considered to be a vasomotor center. The most highly developed and most important center, however, lies in the medulla in close association with the centers for cardiac regulation. This medullary vasomotor center can be divided into clusters of cells that evoke vasoconstriction and others that bring about vasodilatation. Thus one speaks of vasoconstrictor and vasodilator centers. But in recent years, the concept of a vasodilator center has lost favor. The theory now is that the medullary control of vasomotor tone is affected solely by variations in the vasoconstrictor discharge and not by alterations in vasodilator nerve activity. The custom is to speak of medullary pressor and depressor regions in order to make clear that they affect the vasoconstrictor tone and hence the arterial pressure by change in activity of the spinal vasoconstrictor neurons.

Reflex Regulation

Impulses from many parts of the body are propagated to the vasomotor center in the medulla to influence arterial blood pressure. Thus, blood pressure may vary in response to a variety of external stimuli.

Carotid and Aortic Sinuses By far the most important reflex regulation of arterial blood pressure involves the carotid and aortic sinuses. In these sinuses, as explained previously, are pressoreceptors which respond to changes in arterial blood pressure. Impulses from the pressoreceptors are propagated to the vasomotor center in the medulla which in turn regulates vasoconstriction and thus peripheral resistance. For example, if the blood pressure increases, the flow of impulses to the vasomotor center increases (see Figure 16-25). These inhibit the center, vasomotor tone decreases, vasodilatation results, and the arterial blood pressure falls. There is thus a self-regulatory mechanism which serves to maintain the constancy of the blood pressure. It is an essential and effective homeostatic mechanism.

In summary, an elevation in arterial blood pressure evokes two important alterations: (1) slowing and weakening of the heartbeat and (2) vasodilatation. The first alteration diminishes cardiac output; the second decreases peripheral resistance. The end result of both, then, is to lower the arterial blood pressure. Conversely, a fall in arterial blood pressure reflexly stimulates faster heartbeat and vasoconstriction, thereby increasing cardiac output and peripheral resistance (Figure 18-10). The end result is the raised arterial blood pressure.

Carotid and Aortic Bodies   The carotid and aortic bodies are sensitive to oxygen, carbon dioxide, and hydrogen ion content of the blood. Low oxygen, high carbon dioxide, and low pH all activate the chemoreceptors. As a result, there is a flow of impulses to the vasomotor center, which stimulates it, causing a sympathetic discharge that increases peripheral resistance and arterial blood pressure. The chemoreceptors activated in this way do not raise cardiac output but only effect the vascular changes. However, there are other chemoreceptors, probably in the medullary centers, which do influence cardiac function.

Higher-Center Control

Reflexes that regulate arterial blood pressure are complete at the medullary level. That is to say, the brainstem may be transected and yet the organism survive and blood pressure be maintained and respond in the usual fashion. However, higher centers in the brain are capable of altering the arterial blood pressure. The two most important higher centers which control blood pressure are in the hypothalamus and cerebral cortex. Direct stimulation of the hypothalamus produces cardiac and vasomotor changes which depend upon the area stimulated. The posterior area evokes an increase in heart rate and peripheral resistance, the anterior area just the opposite.

Figure 18-10. Influence of pressure at the carotid sinus on arterial blood pressure and heart rate.

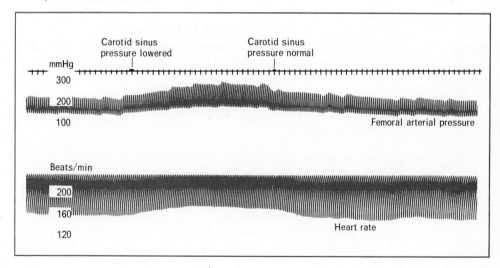

Stimulation of the motor area of the cerebral cortex produces not only movement but also vasodilatation in the contracting muscles with vasoconstriction elsewhere. Also, many emotional experiences, which undoubtedly involve the cerebral cortex and hypothalamus, are accompanied by significant alterations in blood pressure.

Humoral Agents

The blood contains many substances, including ions, the catecholamines, aldosterone, and angiotensin, which influence either cardiac output or peripheral resistance and thereby alter circulatory dynamics.

Ions The influence of the blood ions on the heart was discussed in Chapter 16. Table 18-2 summarizes their effect on peripheral resistance.

Catecholamines Both norepinephrine and epinephrine are secreted by the adrenal medulla. Norepinephrine causes vasoconstriction, as does epinephrine. But whereas the former causes constriction in virtually all arterioles, epinephrine evokes vasodilatation in the arterioles that supply skeletal muscle.

Aldosterone Because aldosterone, a hormone secreted by the adrenal cortex, regulates the quantity of salt and water in the blood, it can alter the blood volume, which is a major determinant of arterial blood pressure, as already explained. The secretion of aldosterone varies in response to blood sodium concentration, blood volume, and cardiac output. A decrease in one or more of these values evokes increased aldosterone secretion. Aldosterone acts on the kidneys, causing retention of salt and water, which increases blood volume and consequently the cardiac output. There is thus a superb homeostatic feedback mechanism which plays a major role in the maintenance of arterial blood pressure.

Kinins The kinins are polypeptides that are split off from plasma globulin. The kinins bring about vigorous vasodilatation. They probably play a role in regulating blood flow in the skin, but little is known about the control of kinin formation; therefore their function in regulating circulation remains uncertain.

Angiotensin When hypertension is discussed later in this chapter, angiotensin will be considered in greater detail. Suffice it to say here that it is formed in the blood by the action of renin, which comes from the kidneys under certain conditions. Angiotensin causes vigorous vasoconstriction and thus elevates blood pressure.

TABLE 18-2. EFFECT OF IONS ON THE ARTERIOLES

ION INCREASE	EFFECT
Potassium	Dilatation
Sodium	Dilatation
Calcium	Constriction
Magnesium	Dilatation
Hydrogen	Dilatation

Histamine Histamine has an interesting dual action. It dilates the arterioles and constricts veins. The result is that venous return increases, thereby increasing cardiac output. But at the same time the pressure in the capillaries increases, forcing fluid out of the capillaries into the tissue spaces. This action accounts for the typical swelling associated with allergies. Histamine exists ubiquitously, so that whenever tissue is damaged, it is released.

CAPILLARY FUNCTION

The major purpose of the circulation is to deliver oxygen and nutrients to the tissues and to remove carbon dioxide and other products resulting from the metabolic processes. The movement of these substances from the vascular system to the tissues occurs only through the capillaries. The arteries and veins are but conducting systems to and from the capillaries. There are approximately 3.6 billion capillaries in the human circulatory system. Thus, although each capillary has but a minute cross-sectional area, the total such area of capillaries is very large, of the order of 4,500 cm^2. This may be compared with that of the aorta and other parts of the circulatory system (Table 18-1). Because the velocity of blood flow varies inversely with the cross-sectional area, flow in the capillaries is very slow.

The wall of a capillary is extremely thin, composed of but a single layer of cells. The capillary wall is semipermeable. Capillaries in the liver are very permeable; those in the brain are much less so. The degree of permeability can apparently be altered. For example, a deficiency of vitamin C, and also of adrenocortical steroids, results in increased capillary permeability.

Arrangement of the Capillaries

The arrangement of the capillary bed is quite complex, as shown in Figure 18-11. There are many side branches, ramifications, and shunts. The blood may flow from the arteriole, through an arteriovenous shunt, directly to a venule, thus bypassing the capillary completely. There appear to be at least two different types of arteriovenous shunts: (1) a short channel called an arteriovenous anastomosis and (2) a longer channel, a metarteriole, from which capillaries branch. Thus, blood from an arteriole may take several routes: (1) from

Figure 18-11. Diagrammatic representation of a capillary network. All the vessels except the true capillary are enclosed in smooth muscle capable of contraction.

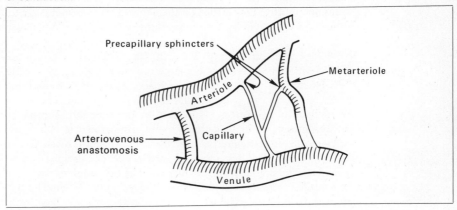

an arteriole to a venule through an arteriovenous anastomosis, (2) from an arteriole into a metarteriole and on through this channel into a venule, (3) from an arteriole into a metarteriole, into a capillary, then to a venule, and (4) from an arteriole through a capillary into a venule. Whether these long metarteriole thoroughfare channels exist everywhere in the body or are limited to certain areas is still not clear.

No matter whether the capillary springs directly from an arteriole or from a metarteriole, in its transition its first part is surrounded by a layer of smooth muscle capable of closing the vessel. When this precapillary sphincter relaxes, the blood pressure causes the capillary to dilate. The arteriovenous anastomosis, the metarteriole, and the precapillary sphincter are all capable of constriction. By virtue of these mechanisms the blood can be shunted in various ways. The capillary, other than the precapillary sphincter area, does not have smooth muscle and cannot change its diameter, except passively, in response to pressure gradients.

Capillary Circulation

The determination of the blood pressure within the capillary bed is a difficult technical exercise. There are reasons to believe that the pressure in the capillaries close to the arterioles is about 25 mmHg and about 10 mmHg close to the venules. The average, then, or what might be called the normal functional mean capillary pressure, is about 17 mmHg.

All evidence indicates that there is not a smooth, steady flow of blood through the capillaries but an intermittent one. The intermittent flow, termed vasomotion, is caused by the rhythmic opening and closing of metarterioles and precapillary sphincters. The cycle occurs eight or ten times per minute. When the vessels are closed, the pressure approximates that of the venules; when open, that of the arterioles.

The flow of blood through a capillary bed is apparently controlled by the oxygen demands of the cells served by that capillary bed. The greater the imbalance between oxygen demand and oxygen supply, the greater the blood flow. The precapillary sphincter seems to be sensitive to this imbalance, which may be brought about by a fall in the blood oxygen concentration or by increased cellular oxygen utilization. There is thus autoregulation. A demand for oxygen increases capillary flow, which brings oxygen to the cells. The oxygen demand is satisfied, the sphincters close, flow stops, and demand builds up again. The result is rhythmic vasomotion. Why there is not a steady flow that just satisfies oxygen needs is not understood.

When blood flows through the arteriovenous anastomosis or through the thoroughfare channels, there is no exchange of nutrients with the tissue spaces as there is in the so-called true capillaries. For this reason, they are referred to as nonnutrient vessels. The purpose of this nonnutritive flow is thought to be to warm the skin. It is controlled by the nervous system and integrated by the temperature centers in the hypothalamus.

Movement of Fluid through the Capillary Wall

Plasma, with the exception of the large protein molecules, is capable of moving through the capillary wall to enter the tissue spaces. Similarly, the fluid that normally occupies the tissue spaces can pass into the capillary circulation. These transfers depend upon the resultant of two forces: (1) hydrostatic pressure and (2) osmotic pressure.

The movement of fluid through any semipermeable membrane depends upon the gradient of hydrostatic pressure between the two sides. As stated earlier, the hydrostatic pressure (blood pressure) within the capillary at the arteriolar end is thought to be about 25 mmHg. The pressure on the outer side of the capillary wall, that is, the pressure of the so-called interstitial fluid, is now believed to be −7 mmHg, that is, 7 mmHg below atmospheric pressure. The effective hydrostatic pressure, then, which is the sum of the two, is 32 mmHg.

The osmotic pressure of the plasma, the so-called plasma colloid osmotic pressure, is of the order of 28 mmHg. The interstitial fluid contains much less protein; thus its colloid osmotic pressure is only about 4.5 mmHg. The effective osmotic pressure, then, would be 23.5 mmHg.

The effective hydrostatic pressure of 32 mmHg forces fluid out of the capillary at the arteriolar end. The effective osmotic pressure attracts fluid into the capillary at that point with a pressure of 23.5 mmHg. There is, therefore, a resultant force of 8.5 mmHg (32 − 23.5) in the outward direction. As a result, fluid flows from the capillary through the membrane into the tissue spaces.

The pressure relationships at the venule end of the capillary are quite different. Now the blood pressure is only about 10 mmHg. Assuming all other values are the same, that is, the effective osmotic pressure and the interstitial fluid pressure, then we have an effective hydrostatic pressure of 17 mmHg (10 + 7), and an effective osmotic pressure of 23.5 mmHg. This means that there is movement of fluid into the capillary under a resultant force of 6.5 mmHg (23.5 − 17.0).

If the forces just outlined exist, there will be movement of fluid out of the capillary at the arteriolar end and movement back into the capillary at the end close to the venule. E. H. Starling was the first to predict such circulation; hence the concept is referred to as the Starling hypothesis.

Edema

Of great importance is the consequence of variations in the normal pressures that control fluid movement. If the capillary blood pressure becomes abnormally high or if the colloidal osmotic pressure of the plasma decreases, fluid will pour out of the capillary along its entire length. Normally, slightly more fluid leaves than returns. The excess is returned to the circulation by the lymphatics. But this drainage system can be overwhelmed so that fluid accumulates in the tissue spaces, a condition termed edema. Everyone is familiar with the edema that follows a sharp blow delivered to a part of the body. Trauma either tears the fragile capillaries or increases their permeability. In either case, because the pressure of blood in the capillaries is greater than the fluid pressure surrounding the capillaries, fluid moves out causing edema, in this case traumatic edema.

Truly great edema is caused by obstruction of the lymphatic vessels as occurs in elephantiasis, a condition resulting from the invasion of the lymphatics by the organism filaria. These organisms clog the lymphatics, preventing tissue space drainage; thus fluid continues to accumulate. The legs become grossly swollen and resemble those of an elephant; hence the name.

Another cause of edema is protein deficiency. Large protein molecules do not pass through the capillary wall. They are thus responsible for the osmotic force that opposes the outpouring of fluid from the vascular tree. When the protein concentration of the blood decreases, fluid enters the tissue spaces in such quantity that it overtaxes the lymphatic drainage and nutritional edema results.

Cardiac edema causes the swollen feet and ankles common in people who have heart disease. In the terminal stages of heart failure there is even more generalized edema, and fluid may accumulate in the abdominal cavity, causing marked swelling. This condition is termed ascites. Cardiac edema is produced by abnormally high venous pressure. When the heart fails, blood accumulates behind it, venous and capillary pressure increase, and edema results.

Movement of Solutes through the Capillary Wall

When plasma leaves the capillary to enter the tissue spaces, it carries with it all the dissolved solutes. These are now in an excellent position for transfer into the cell. But there is also solute transfer independent of plasma movement. Such transfer occurs through the capillary wall and is dependent almost entirely on concentration gradients. The concentration of various nutrients, for example, glucose, is higher in the circulating blood than it is in the tissue fluid. It therefore diffuses through the capillary wall into the tissue spaces. On the other hand, the end products of metabolism are in higher concentration in the tissue fluid than in the plasma and thus diffuse into the capillary blood.

THE LYMPHATIC SYSTEM

Fluid diffusing out of the capillaries contains extremely low concentrations of protein. The amount of protein in this fluid may theoretically be zero, but under many conditions there is some. The permeability of the capillaries can be increased by slight trauma, anoxia, and other factors so as to permit some protein to leak out. And, even though fluid diffuses back into the capillary, little protein is carried with it. In short, under a variety of conditions protein accumulates in the tissue spaces. If there were no method for the removal of this protein, because of its osmotic effect fluid would accumulate in the tissue spaces, and progressive swelling would result. The main function of the lymphatic system is to remove this protein.

The lymphatics draining the lower extremities, lower trunk, and the left side of the chest and head pass to the thoracic duct, which empties into the left brachiocephalic vein. Lymphatics from the right side of the head, right arm, and right chest form the right lymphatic duct, which empties into the right brachiocephalic vein (see Figure 17-20).

The walls of lymphatic capillaries are very thin and highly permeable, so that protein and large foreign particles enter easily. The larger lymphatic vessels do have a layer of smooth muscle and are apparently contractile, but the control and function of lymphatic contractility remain to be clarified.

Lymph Formation

By definition, only when fluid enters the lymphatic system does it become lymph. Accordingly, three fluids are involved: (1) blood plasma, (2) interstitial fluid, and (3) lymph. The forces responsible for the formation of interstitial fluid from blood plasma have been discussed. The question to be considered here is how interstitial fluid moves into the lymph capillaries.

For a time it was thought that the lymphatics transferred fluid into the vessels by an active process. This view has now been abandoned. The passage of fluid from the tissue spaces into the lymphatic capillaries is a purely passive act dependent upon pressure differences.

A positive tissue pressure, one would think, would collapse the lymphatic capillary rather than drive fluid into it. Connective tissue attachments keep the vessels patent. Because they cannot collapse, the external force causes fluid to enter the lymphatic capillary.

Lymph Flow

The lymph must move through a series of vessels and nodes to flow, ultimately, into the venous system. The pressure in the veins at the point of entry is generally very low, but interstitial pressure, which is responsible for the formation of lymph, is usually no higher, and on the average lower. Obviously, therefore, there must be other mechanisms responsible for the flow of lymph.

In man at rest no more than about 100 ml of lymph per hour flows through the thoracic duct; during excerise this quantity increases sharply. Muscle massage is responsible. The contracting muscles squeeze the surrounding tissue and thus elevate interstitial pressure, and they also squeeze the lymphatic vessels. The elevated interstitial pressure forces more lymph into the vessels, and the squeezing of the vessels, like stepping on a hose, forces the fluid in both directions. The valves, however, prevent much backflow; therefore the main result is forward flow. Muscle massage, then, as a lymphatic pump, is very effective.

Another pump for the lymphatic vessels is provided by the pulsation of the arteries that lie in association with them. Each arterial pulsation squeezes the lymphatic vessel and moves the lymph just as muscle massage does. Arterial pulsation probably is the essential pump for lymph flow when a person is at rest and also for the lymphatic vessels that lie encased in bone. During exercise, the pulse rate is greater and so is the pulse pressure; both would increase lymph flow.

The pressure within the thorax is generally negative, that is, below atmospheric pressure. During inspiration, the greatest negativity is reached, as will be explained later. There is thus a pressure differential between the thorax and other parts of the body. The lymphatic vessels join the venous system in the thorax. This pressure differential, then, aids in moving the lymph, and the greater the differential, the greater the flow. Again, during exercise inspiration is deeper, the rate of breathing faster, and thus the lymph flow greater.

In summary, the rate of lymph flow is regulated by (1) interstitial pressure, (2) muscle massage, (3) arterial pulsation, and (4) intrathoracic pressure.

Any substance that increases the rate of lymph flow is termed a lymphatogogue. Most of them work by increasing interstitial fluid pressure either by increasing blood volume, causing arteriolar dilatation, or causing venular constriction. All increase the hydrostatic pressure in the capillaries and thus the outward flow of fluid, which elevates interstitial fluid pressure.

Lymph Nodes

Figure 17-19 depicts a lymph node. Its function has been described as that of removing foreign matter before the lymph joins the blood. In this way the spread of infective agents is prevented. The lymph nodes also form antibodies in response to many of these foreign particles. The antibodies are secreted by the lymph nodes, enter the circulation, and thus combat the infectious process.

In summary, the lymphatic vessels and nodes provide a system for draining the tissue spaces, for removing leaked protein and foreign matter, and for protection.

VENOUS CIRCULATION

The blood flows from the capillary beds, or from the arteriovenous anastomoses, into the venous system for transport back to the heart. But any idea that the veins are simply passive conduits should be immediately put aside. They are vessels which can contribute to the flow of blood, and they are vessels which can greatly change their capacity; thus they strikingly influence circulatory dynamics. In discussing cardiac output, the point was emphasized that venous pressure is a major factor.

Venous Pressure

Figure 18-12 illustrates the blood pressure in different parts of the circulatory system. There is, of course, a progressive decrease in pressure; the highest is in the aorta and the lowest in the veins. Were it otherwise, the forward flow of blood would cease. But the decrease is not described by a straight line. Note that in the arteries there is very

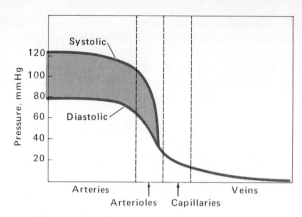

Figure 18-12. Blood pressure in vessels of the circulatory system.

little decrease, whereas in the arterioles the drop in pressure is precipitous. The determinants are (1) total cross-sectional area and (2) individual vessel diameter. In a closed system the lateral pressure of a circulating fluid is directly proportional to the cross-sectional area. The total cross-sectional area of the arterioles is not very much greater than that of the arteries; but there are many more arterioles than arteries. The smaller the vessel diameter, the greater the resistance. Thus the high resistance in the arterioles, not counterbalanced by a comparable increase in total cross-sectional area, results in a sharp pressure drop, about 70 mmHg (95 − 25).

In the capillaries, however, there is a huge increase in total cross-sectional area, and so despite the great number of tiny vessels the pressure drop is nominal, only about 15 mmHg (25 − 10). Blood thus enters the veins at a pressure of about 10 mmHg and flows into the heart close to 0 mmHg. Venous pressures and venous flows can vary, however.

Regulation of Venous Circulation
The same physical principles as those discussed at the beginning of this chapter are applicable, of course, to venous circulation. However, in analyzing venous circulation, the low pressures, the thin vessel walls, their ability to contract and relax, and the great capacity of the system must all be considered.

Output of the Left Ventricle   The flow of blood through the veins must ultimately reflect the output of the left ventricle of the heart. If the caliber of the arterioles and capillaries remains the same and the output of the left ventricle increases, there will be an elevation of pressure all along the line, including in the veins.

Vasomotor Activity   Alterations in the diameter of the arteries and arterioles influence venous pressure, but this influence is usually only temporary. For example, if there is generalized vasoconstriction, quite obviously the greater resistance will cause a greater-than-normal pressure drop as the blood passes through the constricted area, but the entering pressure, that is, the pressure of blood entering the capillaries and veins, is now higher; thus the venous pressure returns to approximately the normal range. Another reason for the temporary nature of the decrease is that when the hydrostatic pressure in the capillaries is decreased by the vasoconstriction, fluid will flow from the tissue spaces into the capillaries rather than out of them; thus venous flow is improved.

For the two reasons just mentioned, vasomotor activity, which markedly alters arterial pressure, leaves the venous pressure unaltered. In addition, as will be explained next, the veins are also capable of venomotor activity which can compensate for changes in input volume.

If such relative independence of the arterial and venous system did not exist, an interesting but fatal cycle would be established by vasoconstriction. If vasoconstriction resulted in a decrease in venous pressure, there would be less filling of the heart, lower cardiac output, and decreased arterial blood pressure. This would decrease venous pressure further, and so forth.

Venomotor Changes The veins are capable of venomotor activity in response to nervous stimulation as well as in response to the direct action of various substances. The changes are consistent with those in the arterioles: Constriction is primarily due to sympathetic activity, and dilatation results from inhibition of the sympathetics. Catecholamines constrict the veins, whereas acetylcholine dilates them. Reflexes initiated by the presso-receptors in the carotid and aortic sinuses control the veins as they do the arterioles. Thus venoconstriction usually accompanies vasoconstriction.

Vasoconstriction involves primarily the arterioles, which are short vessels with little capacity; thus change in their diameter alters resistance and therefore the pressure upstream. But venoconstriction involves entire lengths of long veins. The change then is fundamentally one of capacity. To be sure, the resistance to the flow of blood through the veins is increased, but the influence in a low-pressure system is minor in contrast to the effect of greatly altered capacity. For example, assume that hemorrhage causes loss of a liter of blood. The veins, in response to pressoreceptor reflexes, constrict and reduce their capacity by a liter. This takes up the slack, so to speak, and the normal volume of blood enters the heart, thereby maintaining the cardiac output. Because of this function, the veins are called capacitance vessels.

The importance of venous return to the heart in regulating cardiac output was explained in Chapter 16. Venomotor activity, by changing the capacity of the veins, which hold over 50 percent of the total blood volume, plays a major role in cardiovascular dynamics.

Intrathoracic Pressure The role of intrathoracic pressure on lymph flow has already been considered. It has the same effect on venous flow. During quiet inspiration the intrathoracic pressure is about −6 mmHg. During quiet expiration, it is about −4 mmHg. With maximal respiratory efforts these values may reach −40 mmHg during inspiration and +40 mmHg during expiration.

The great veins that pass into the thoracic cavity have very thin walls, and therefore a change in intrathoracic pressure influences the pressure within the veins. Thus, during inspiration the gradient of pressure in the inferior vena cava between a point in the abdominal cavity and in the thoracic cavity (Figure 18-13) is greater (5 mmHg) than during expiration (2 mmHg). During exercise, the intrathoracic pressure may be −20 or −40 mmHg, resulting in an even greater gradient. The greater the gradient, the greater the flow of blood, as explained at the beginning of this chapter.

To be sure, during forced expiration, intrathoracic pressure increases and reverses the gradient. Were it not for the one-way venous valves, flow in the veins would also reverse. The valves prevent this, but venous flow does stop. However, during the subsequent inspiration, rapid forward flow again occurs. In other words, with wide changes in intrathoracic pressure, venous flow to the heart becomes pulsatile; blood spurts into

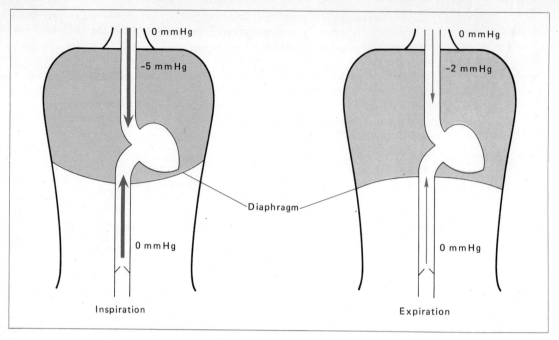

0 mmHg		0 mmHg
-5 mmHg		-2 mmHg
Diaphragm		
0 mmHg		0 mmHg
Inspiration		Expiration

Figure 18-13. Influence of intrathoracic pressure on venous return. The gradient between pressure within the thorax and outside it is larger during inspiration than during expiration, and venous return varies correspondingly.

the heart during inspiration and stops during expiration. The overall flow, however, increases with increased respiratory activity, as occurs in exercise.

Visceral Reservoirs The viscera, such as the liver, the spleen, and the intestine, are capable of holding a considerable volume of blood. By alterations in the caliber of the inflow and outflow vessels, the size of these visceral reservoirs can be significantly changed. Together they can add to, or subtract from, the general circulation a liter or more of blood. This represents about 20 percent of the total blood volume, and therefore the visceral reservoirs play an important role in venous return and cardiac output.

Muscle Support and Massage Veins course in association with skeletal muscle or between bone and skeletal muscle. These structures serve to support the thin-walled vessels, keeping them from bulging; the bulging of the veins permits blood to pool instead of returning to the heart. The great importance of skeletal muscles in supporting the veins is demonstrated in cases of paralysis or in normal subjects held upright in such a way that the leg and abdominal muscles are relaxed. Under such circumstances, the veins have no support; they bulge, blood pools, venous return diminishes, cardiac output falls, arterial blood pressure decreases, the blood supply to the brain becomes inadequate, and unconsciousness results. In other words, normal skeletal muscle tone serves to support the veins.

The muscles, by contracting, also improve venous flow. The bulging of the contracting muscle squeezes the associated vein, just as it does lymphatic vessels, and because of the one-way valves, blood is forced toward the heart. The greater the physical effort, the greater the contraction and thus the greater the volume of blood pumped through the veins. Here, then, is still another reason why during exercise the venous return to the heart is augmented.

In summary, vasodilatation, venoconstriction, visceral reservoirs, and muscle massage all augment venous return, which results in greater cardiac output.

Gravity   Gravity opposes the venous flow in the veins below the level of the heart when one is in the upright position. The role of gravity is emphasized by the following values: in a person who is lying down, venous pressure in the ankle veins is about 12 mmHg, but when he is in the upright position, it is about 90 mmHg simply because of the force of gravity acting on the long column of blood between the ankle and the heart. If this high pressure were permitted to persist, there would be a great outpouring of plasma through the thin-walled capillaries. As a matter of fact, this is exactly what occurs if the person is supported upright and his muscles are relaxed. But normally there is movement, contraction of muscles, which pumps the blood toward the heart and reduces the venous pressure in the leg veins; during walking, the pressure may fall lower than 25 mmHg.

The gravitational effects are exaggerated during high-speed maneuvers. In pulling out of a dive in flight or in making a sharp turn at high speed, gravitational forces may be increased many times. If the direction of the force is away from the heart, the flow of blood toward the heart will be opposed. Most young, well-trained individuals can withstand an opposing force up to five times the normal force of gravity (5 g). Forces greater than this so lower the venous return to the heart that unconsciousness, termed blacking out, occurs. To prevent this result, pressure suits which build up sufficient pressure on the veins to oppose the gravitational forces are worn.

Right Side of the Heart   Resistance to venous flow is determined by the length and caliber of the veins and, most importantly, by the right side of the heart. Normally, the right side of the heart pumps a volume of blood equal to the venous input. But if the right side of the heart is failing or if resistance in the pulmonary tree is enough to decrease the right ventricular output, then blood will accumulate in the right ventricle, and the pressure in the right atrium and veins will increase. Also, if the right A-V valve is stenosed, there will be increased resistance to the flow of blood into the right ventricle, and venous pressure will be elevated.

Normally the pressure in the right atrium is about 0 mmHg, that is, atmospheric pressure. If the right side of the heart is in failure, if there is stenosis of the right A-V valve, or if the venous return is greatly increased, right atrial pressure may rise as high as 20 mmHg. On the other hand, if the right ventricular output is sharply increased or if hemorrhage significantly reduces venous return, right atrial pressure approaches intrathoracic pressure, that is, about −4 mmHg.

VARIATIONS IN BLOOD PRESSURE

Blood pressure in various parts of the circulation vary from the so-called normal resting values under a wide variety of circumstances in health and disease. A few such conditions will be considered in order to underscore and review the physiology of circulation. First, however, now that the key mechanisms that regulate the circulation have been discussed, an attempt will be made to integrate them.

Fundamental to an understanding of circulation is the realization that the cardiovascular system is inherently capable of regulating itself over a broad range in accord with tissue needs. When the cells utilize more oxygen than is being delivered, a transient state of hypoxia exists. This causes vasodilatation, the venous return increases, and the cardiac output becomes greater, thereby pumping more blood to the tissues. Superimposed upon this inherent mechanism are many systems that extend the range, so to speak, of the circulation. Rapid, minute-to-minute adjustments result from the intervention of the nervous system. The next line of defense is movement of fluid into or out of the circulation at the capillary level. The third line of defense is provided by the kidneys, which regulate blood volume and which secrete renin. Renin brings about the formation of angiotensin, a powerful vasoconstrictor.

Exercise

During exercise there is an increased heart rate, an elevation of the arterial and venous pressure, and an increased cardiac output. The rise in arterial pressure is due entirely to greater cardiac output. The overall peripheral resistance decreases because of great vasodilatation in the muscles. This is shown by the fact that the cardiac output may increase to as much as five to seven times the resting level while the arterial blood pressure is only slightly elevated, rarely as much as two times the resting value. Since pressure is equal to flow times resistance, if the pressure increases only by a factor of 2 while the flow, that is, the cardiac output, increases by a factor of say 6, then the resistance must be but one-third normal.

In the athlete both the heart rate and stroke volume increase during vigorous exercise, but in the nonathlete almost all the augmentation of cardiac output results from an increase in heart rate (Figure 18-14). In either case, there must be an elevated venous pressure either to increase the stroke volume or to prevent it from decreasing in the face of decreased filling time occasioned by the fast heart rate. The increased heart rate with augmented force of contraction is probably due to activation of the accelerator fibers. In addition, catecholamines secreted by the adrenal medulla act directly on the myocardium to improve contraction. In all probability impulses from higher centers and from the contracting muscles are propagated to the medullary centers and are primarily responsible for the accelerator stimulation. The resulting vigorous contraction reduces the residue normally remaining in the ventricle at the end of systole and increases the stroke volume.

As emphasized several times, in order to maintain or increase stroke volume while the heart is beating rapidly, there must be faster diastolic filling of the ventricle. This can result only if the pressure differential between the atrium and ventricle increases. The greater venous return increases the atrial pressure, and a more vigorous atrial contraction increases it further. A decreased ventricular residue lowers the ventricular pressure at the end of systole. In addition, there is evidence that during vigorous cardiac activity the pressure in the ventricle at the beginning of the rapid-inflow phase may become negative. All these alterations increase the A-V pressure gradient which drives the blood into the ventricle. At very fast heart rates, atrial contraction and the period of negative ventricular pressure would occur simultaneously, thus creating the maximal A-V pressure differential.

Hypertension

The term hypertension has reference to high blood pressure in the arterial system. Until the heart begins to fail as a result of long-standing hypertension, the venous pressure is usually within the normal range. Chronic systolic pressure in excess of about 160 mmHg is considered abnormal. Despite the chronic high arterial blood pressure, the pressoreceptors continue to function. If external pressure is applied to the carotid sinus of a hypertensive patient, his blood pressure will fail, just as it does in a normal individual.

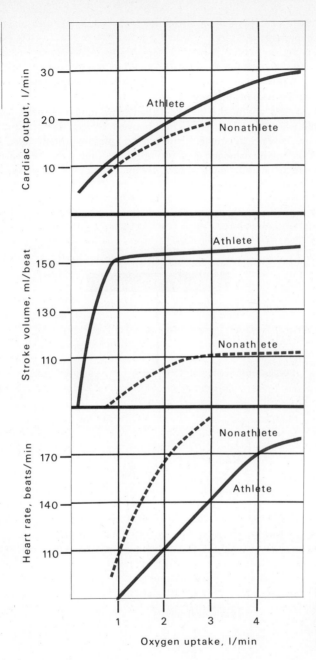

Figure 18-14. Changes in heart rate, stroke volume, and cardiac output in exercise. Note the differences in responses for the athlete and the nonathlete.

Action potential studies disclose that immediately following the onset of hypertension there is vigorous firing in the vagus, but this progressively diminishes. In other words, at first the pressoreceptors respond as they always do when the pressure is elevated. But progressively they cease to do so, thus ultimately becoming acclimated to the high pressure. It is as though the regulator has been reset.

If the cause of hypertension is unknown, the patient is said to have essential, or idiopathic, hypertension.

Kidney Disease In many types of renal disorders the blood flow through the kidneys is decreased. Hypertension usually results. A tremendous amount of research has fairly well disclosed the responsible sequence of events.

There are cells within the kidney, called juxtaglomerular cells, which secrete the enzyme renin when there is an imbalance between oxygen demand and supply, that is, when the blood flow is inadequate. Renin acts on a plasma protein called angiotensinogen (Figure 18-15). Angiotensinogen is a glycoprotein. Under the influence of renin, it is converted to angiotensin I, which in turn becomes angiotensin II. A so-called converting

Figure 18-15. Homeostatic regulation of blood flow through the kidneys. Anything that diminishes flow sets off the sequence shown. Hypertension results.

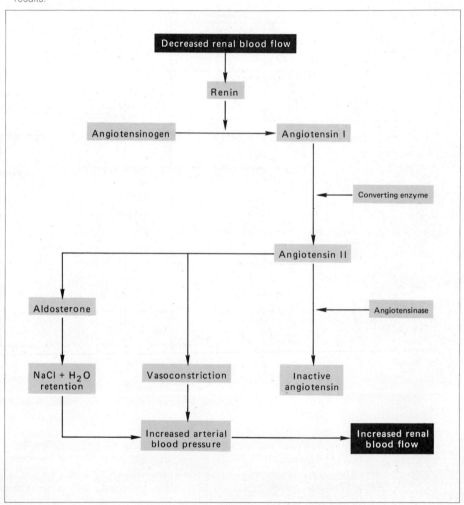

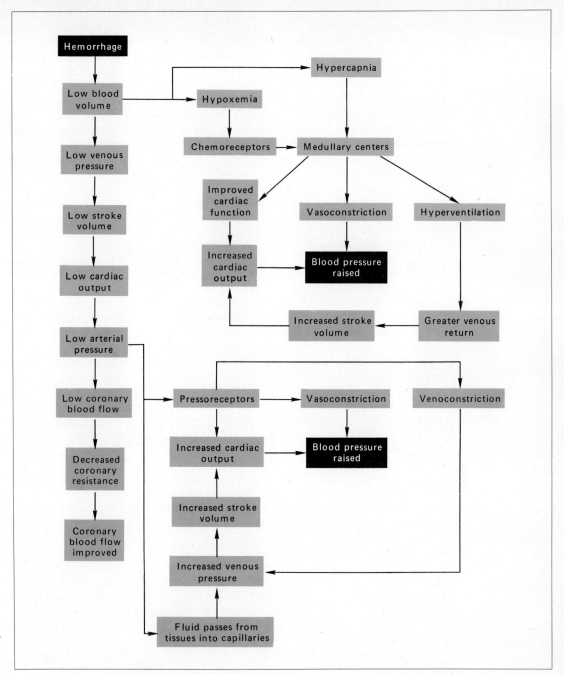

Figure 18-16. Homeostatic control of blood loss due to hemorrhage. In the coronary circulation arteriolar resistance is lowered to correct diminished blood volume. In the systemic circulation pressoreceptors and the medullary centers are triggered to cause vasoconstriction, venoconstriction, and an increase in heart rate.

enzyme is responsible for the latter conversion. Angiotensin II, as shown in Figure 18-15, has at least three actions: (1) it is a pressor agent that acts on the arterioles, evoking constriction, (2) it increases the level of sympathetic activity, and (3) it stimulates the adrenal cortices to secrete aldosterone. Angiotensin II influences the sympathetic nervous system at several levels. It acts on the central nervous system, on the adrenal medullae, on sympathetic ganglia, and at postganglionic nerve terminals. All these actions increase sympathetic activity. Thus there is a direct action of angiotensin II on the arterioles to increase peripheral resistance and elevate blood pressure, and an indirect action via the sympathetic nervous system which improves cardiac output and further increases peripheral resistance, thereby elevating arterial blood pressure even more.

Aldosterone, as explained above, brings about increased blood volume, which augments venous return and therefore cardiac output. All these mechanisms cause hypertension, which drives more blood through the kidney. In this way, the kidney needs are satisfied, but the price is chronically elevated blood pressure.

Hemorrhage

Several hundred milliliters of blood may be lost without serious alterations in blood pressure. However, if more than about 25 percent of the blood volume is lost, pressure falls precipitously, and the removal of approximately 40 percent usually proves fatal. Figure 18-16 shows the sequence of compensatory reactions which occur as a result of hemorrhage and which serve to maintain pressures within a range compatible with life. If the bleeding is halted before more than about 35 to 40 percent of the blood volume is lost, these mechanisms normally suffice to return the pressures and volumes to within their normal ranges.

Severe hemorrhage causes an immediate drop in arterial blood pressure. The pressoreceptors improve cardiac function and cause generalized vaso- and venoconstriction. These alterations are rapid and effective. In addition, the adrenal medulla secretes catecholamines to reinforce the cardiovascular adjustments. The low blood volume results in hypercapnia and hypoxemia. Hypoxemia stimulates the carotid and aortic chemoreceptors; hypercapnia, the respiratory center. The effect of increased respiratory activity is to aid venous return.

If the cardiovascular adjustments are adequate to maintain life, the blood volume gradually returns to a normal value. Because of the low pressure in the capillaries, fluid moves into the circulation from the interstitial space. Also erythropoietin secretion increases, more red cells are produced, and the blood cell count slowly is restored to normal.

Circulatory Shock

Any condition that results in an inadequate blood supply to the tissues is termed circulatory shock. In most cases there is a decrease in the arterial blood pressure. The venous pressure, in some instances, is also very low. Circulatory shock may result from cardiac failure or from hemorrhage. It may also be caused by (1) massive vasodilatation due to substances such as histamine acting directly on the arterioles or (2) great loss of plasma into the tissue spaces due to trauma.

QUESTIONS AND PROBLEMS

1 What are seven factors which contribute to the circulation of blood?

2 Discuss the effect of the following on blood pressure: (*a*) increased cardiac output, (*b*) hemorrhage, (*c*) cardiac failure, (*d*) arteriosclerosis, and (*e*) polycythemia.

3 Why is arterial blood flow pulsatile while capillary flow is steady?

4 Describe the auscultatory method of obtaining blood pressure, and discuss the clinical importance of this measurement.

5 What chemical substances are capable of increasing arteriolar resistance, and what is the ultimate effect upon blood pressure?

6 Discuss the role of the autonomic nervous system in controlling arteriolar resistance.

7 How does the medulla oblongata function in controlling vasomotor activity?

8 Describe the various routes through capillary beds that may be taken by arteriolar blood.

9 What evidence is there that autoregulation of blood flow operates within the capillary bed?

10 What factors control the exchange of fluid between capillaries and tissue spaces? Under what conditions will you see an imbalance in this transfer?

11 What controls the exchange of solutes and gases at the capillary level?

12 If the lymphatics in the right arm become nonfunctional, what effects would this produce in the affected arm? Explain.

13 Discuss the factors which regulate the rate of lymph flow.

14 Explain why hypertension is frequently a complication of some renal disorders.

19 CIRCULATION THROUGH SPECIAL REGIONS

There are regions of the body in which the circulation differs from that observed generally or has special properties which warrant closer attention. For example, the coronary circulation not only is of vital importance to the well-being of the whole organism but is controlled by factors specific to its needs. Likewise, the delicate brain tissue is assured a steady supply of blood by elaborate self-regulating mechanisms.

CORONARY CIRCULATION

Blood that flows through the heart chambers is not utilized by heart tissue. The heart, like all other organs, has its own highly specialized circulatory apparatus, in this case, the coronary system. The term coronary means literally "a crown," and this implies encirclement. The coronary vessels do, in fact, encircle the heart (see Chapter 16).

The importance of the coronary system cannot be overemphasized. If the blood supply to one kidney is shut off, that kidney will die, but the body can survive with the other one alone. If the circulation to a limb is interrupted, the extremity may require amputation, but the person continues to live. However, if the blood supply to the heart is blocked, the cardiac cells die, contraction ceases, and therefore the entire circulation of vital blood comes to a standstill. A steady and adequate cardiac circulation must be maintained.

The two coronary arteries spring from the aorta just beyond the aortic valve, that is, from the ascending aorta. Thus, as soon as the blood leaves the left ventricle, a part of it enters the coronary arteries for distribution to the cardiac muscle. Unfortunately, the coronary arteries are functionally end arteries. This means that the two vessels are more or less independent of each other. In most organs of the body, the various blood vessels meet with one another so that if one is blocked, another vessel may still nourish the tissues. But in the heart, if one of the coronary arteries is suddenly occluded, no blood is delivered to the heart muscle ordinarily supplied by that artery. That part of the heart dies. Of course, if a small branch is plugged, only a small part of the heart muscle will become nonfunctional, and life may still be sustained. However, if a large area of the heart is destroyed, the organ may no longer be capable of doing its work.

The two coronary arteries and their branches divide into numerous arterioles and then into capillary beds, where the actual exchange of vital substances takes place. These capillaries then reunite into larger and larger vessels. At the cardiac veins, they finally join into a single trunk, the coronary sinus, which empties the blood into the right atrium (see Figure 16-6). Blood that supplies the left ventricle, for the most part, flows through the coronary sinus. Most of the blood from the right ventricle flows through the anterior cardiac veins to enter the right atrium. The remainder of the blood that does not return to the right atrium via either the coronary sinus or the anterior cardiac veins is carried by the venae cordis minimae (thebesian veins), which have connections to all the heart chambers.

Determination of Coronary Flow

In man, coronary blood flow must be determined by indirect methods. The so-called Kety method is used. It is based on the Fick principle which was discussed in Chapter 16. To determine coronary flow, the subject breathes nitrous oxide (N_2O) for a 10-min period. Blood samples are taken during this period from an artery and by means of a catheter from the coronary sinus. The concentration of N_2O is determined in each sample and plotted (Figure 19-1). From this plot the average arteriovenous difference can be easily

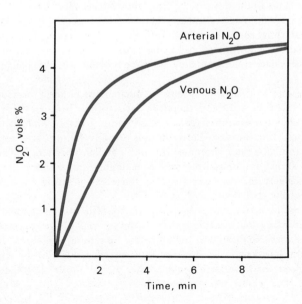

Figure 19-1. Kety method for determining coronary blood flow. (See text for explanation.)

determined by measuring the arteriovenous difference each minute for the 10-min period. Next, for the Fick equation, one needs to know the amount of N_2O taken up by the cardiac tissue during the inhalation period. This is achieved by assuming that at the end of the 10-min period the N_2O in the blood is in equilibrium with that in the tissue. Previously run experiments using samples of cardiac muscle provide the so-called proportionality factor, which is the amount of N_2O cardiac tissue absorbs at a particular blood concentration. The proportionality factor is then multiplied by the arterial blood concentration at the end of the 10-min period. This gives the amount of N_2O taken up by the heart expressed in terms of milliliters of N_2O per 100 g of heart. Assuming the normal heart in the adult male weighs about 350 g, the total coronary flow can be calculated.

Using this method, a resting flow of about 250 ml/min is found in man. This is about 5 percent of the cardiac output. The distribution of blood through the right and left coronary vessels varies. Available figures, which probably are not too reliable, suggest that more blood flows through the right artery than the left in about 50 percent of the population, in 30 percent the flow is equal, and in the remaining 20 percent the flow through the left artery is greater.

Coronary flow must increase in accord with the work of the heart. Cardiac output can increase to four or five times the resting level, that is, up to 20 to 30 liters/min. But this increased output generally occurs during exercise, when the arterial blood pressure is higher than normal. Therefore the work of the heart, because it is pumping blood against a higher head of pressure, increases some eight to ten times the resting level. The coronary flow does not increase this much. To be sure, the amount of oxygen removed by the heart tissue per unit volume of blood increases a bit, but this does not make up the difference, which means that this level of work can be sustained only if the heart functions with greater efficiency, i.e., more work per unit of oxygen.

Coronary Flow during the Cardiac Cycle

Blood flow through the coronary vessels, as it is through all vessels, is determined by the resultant of two factors: (1) the force driving the blood, that is, the aortic pressure, and (2) the resistance to the flow. In the heart, the vessels vary resistance by constriction and dilation, but in addition, resistance is also strikingly altered by intramural pressure. Intramural pressure is, as the term indicates, the pressure within the wall of the heart. This pressure increases during systole and decreases during diastole. Because the coronary branches lie embedded in the wall, they are directly influenced by changes in intramural pressure.

During the cardiac cycle there are sharp fluctuations in aortic pressure and marked changes in intramural pressure and, therefore, resistance to flow. For these reasons, blood flow is not steady. As shown in Figure 19-2, it varies greatly. Note that at the extreme left of this figure, the coronary blood flow is steadily falling. The aortic pressure is diminishing at the same rate. These two lines are parallel. This is to be expected because the ventricle is at rest, intramural pressure is negligible, and therefore the coronary flow is the function solely of aortic pressure. As the aortic pressure decreases throughout diastole, so does the coronary flow.

At line 1 the coronary flow falls precipitously. The fall begins just before line 1. At this time, the atria are contracting. At line 1 the ventricles begin to react. The pressure within the heart rises sharply, and therefore the intramural pressure increases proportionately. This intramural pressure opposes the flow of blood in the coronary vessels, and as can be seen, the flow decreases markedly. It may diminish to the point at which

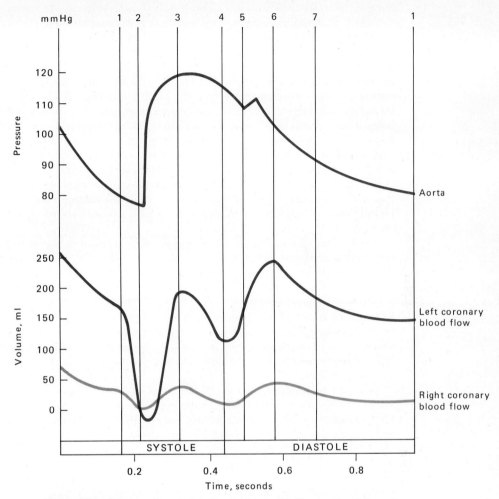

Figure 19-2. Coronary flow during the cardiac cycle. The great variations in aortic pressure and in left ventricular intramural pressure account for the fact that left coronary blood flow varies more than does right coronary blood flow. (Values extrapolated from data obtained from dogs.)

there is no forward flow at all, and as shown, at the peak of intramural pressure the flow may actually be reversed, the blood being squeezed back into the aorta. But at line 2 the aortic valve opens. Now the pressure in the aorta rises and opposes the great intramural pressure and overcomes it, and once again blood flows through the coronary vessels. The curve, therefore, rises until line 3 is reached.

At line 3 aortic pressure begins to fall off, but the ventricle is still contracting. This is the reduced-ejection phase. In short, the intramural pressure remains high while the aortic pressure falls; therefore, the coronary flow diminishes. At line 4, systole ends, the ventricle relaxes, and intramural pressure decreases. The more the intramural pressure decreases, the greater the coronary blood flow becomes. At line 6, the ventricle is completely relaxed, intramural pressure is once again at a minimum, and consequently the coronary blood flow is determined solely by the aortic pressure. The cycle is ready to begin again.

But in pathological conditions, the situation is quite different. To illustrate this point, let us assume that there is a stenosis of the aortic valve. The opening between the left ventricle and the aorta is markedly constricted. In order to drive the normal quantity of blood through this small aperture, the ventricular muscle must contract and therefore squeeze the coronary branches abnormally. This, as just explained, will cut down the coronary blood flow. Now assume, further, that the aortic valves do not close completely. Blood, instead of being forced through the coronary vessels, will regurgitate into the ventricle. Such a valvular disorder will markedly decrease the coronary blood flow. Lastly, in order to cope with this condition, the heart does more work and demands more oxygen. This situation is present despite the fact that the coronary supply is reduced. The heart, under these conditions, is not being adequately nourished and cannot survive long.

The coronary flow for the right ventricle is also included in Figure 19-2. Because the pressures developed within the right myocardium are so much less than within the left, the fluctuations of flow are proportionately less.

Regulation of Coronary Flow

The arterioles of the coronary system are innervated by both branches of the autonomic nervous system, and they also respond to the composition of the blood, but the major regulators of coronary flow are the mechanisms discussed above for autoregulation, as is shown by the fact that the coronary flow in a completely denervated heart increases with oxygen demand.

When the heart rate is at the resting level, that is, about 72 beats/min, approximately 65 percent of the oxygen carried to the heart in the arterial blood is removed. For reasons to be discussed later, this is about all that can be effectively extracted. In other words, the only way increased oxygen needs can be satisfied is by increasing blood flow and cardiac efficiency. In fact, that is exactly what happens. As the work of the heart increases, so does the blood flow. Once again, there is a lively argument as to the mechanism. Everyone agrees that increased flow results from decreased coronary resistance, but whether the vasodilatation results directly from oxygen lack or from the liberation of a dilator substance in response to the oxygen lack is the question. Adenosine, a dilator substance, has been shown to be liberated when the work of the heart is increased or when the blood flow is diminished, but whether it is liberated in sufficient quantities to account for the dilatation is doubted. How efficiency is increased is not known.

Influence of Coronary Innervation The action of the autonomic nerves on the coronary vessels is still being debated. Most authors state that sympathetic stimulation causes coronary vasodilatation, and the parasympathetics have very little influence one way or the other. But an interesting experiment casts doubt upon the primary action of the sympathetic fibers. In this experiment ventricular fibrillation was first induced in an experimental dog. Then coronary blood flow was measured directly before and during sympathetic stimulation. The reason for the fibrillation was to avoid the influence of ventricular systole on mean coronary blood flow. When the sympathetic fibers are stimulated, coronary flow first decreases, then increases. The decrease is interpreted to be a direct vasoconstrictor effect of sympathetic stimulation. The secondary increase is thought to be due to the expected vasodilatation in response to increased cardiac metabolism. Other than academic interest in the direct effect of sympathetic stimulation, the important point is that sympathetic stimulation increases cardiac activity and coronary blood flow does increase.

The innervation of the coronary vessels would appear to be unimportant since autoregulation of flow provides a rapid, effective mechanism. Whether sympathetic stimulation can be responsible for coronary spasm resulting in angina pectoris is unresolved.

Influence of Drugs The larger coronary arteries, as well as the arterioles, dilate in response to amyl nitrite and nitroglycerin. Thus, amyl nitrite and nitroglycerin have long been used in the treatment of coronary disease to bring about improved coronary circulation.

Agents which consistently cause constriction of the coronary arterioles are posterior pituitary pressor extract, angiotensin, and acetylcholine (ACh).

Abnormalities of Coronary Flow

Inadequate coronary blood flow, whether due to disease of the coronary vessels or to vasospasm, accounts for about 30 percent of all deaths in the United States.

Angina Pectoris The term angina means "to strangle," and pectoris refers to the chest. A person suffering with angina pectoris does indeed think his chest is being strangled. The pain is great and spreads rapidly to the left shoulder and arm. It results whenever there is a significant discrepancy between cardiac oxygen demand and supply. Most healthy persons are unable to exercise vigorously enough to bring about such discrepancy and resultant pain, but when the cardiac patient places a load on his heart by excessive exercise or emotion, angina pectoris develops.

The exact mechanism of the pain is not clear, but there is general agreement that oxygen deprivation causes the release of some substance that acts on the sensory nerve endings in the myocardium.

Coronary Occlusion If there is sudden occlusion of a coronary artery that supplies a large area of heart muscle, the resulting ischemia will not only cause the pain of angina but may well prove fatal. On the other hand, if occlusion takes place slowly, the person may survive because during slow occlusion the collateral circulation of the heart develops sufficiently to permit normal flow.

Coronary occlusion, in most cases, is caused by thrombus formation. Thrombi may form due to a clotting abnormality but usually are caused by atherosclerosis, a disorder which alters the walls of the vessels. The intima is roughened, which favors clotting, and parts of the lining (so-called plaques) may break off, blocking a vessel.

If occlusion involves a relatively small vessel, there will be an infarct, but cardiac function may continue to be adequate. The infarcted area is ultimately replaced by scar tissue, which is quite strong and can support high pressures. At the same time, the healthy cardiac tissue takes over the pumping requirements so that cardiac output returns to normal, and if the scar area is not too extensive, exercise may be indulged in without production of pain.

Coronary occlusion may involve a large vessel so that a sizable area of the myocardium undergoes infarction because of ischemia. Several things may happen. In the first place, cardiac output probably fails to keep up with venous return; that is, the heart goes into failure. The failure of the heart results from two causes. First there is a reduced amount of the normal myocardium available for pumping. Second, not only does the infarcted area not work, but before scar tissue forms it actually bulges during each systole. This means that energy for pumping blood is dissipated in bulging the weakened area.

As the heart fails, pulmonary congestion occurs if the infarct is in the left ventricle and systemic edema if it is in the right ventricle. Pulmonary congestion, if severe, can in itself cause death.

Sometimes the infarcted area ruptures. The patient does not quickly bleed to death, as might be imagined, because the blood is contained by the pericardium. However, blood in the pericardium prevents the heart from expanding during ventricular filling. This is termed cardiac tamponade. As a result, cardiac output falls even more.

Finally, a large infarct results in fibrillation which, if not converted, quickly causes death.

CEREBRAL CIRCULATION

The brain contains numerous centers which cannot function for more than a short period of time in the absence of oxygen. If the blood supply to the brain is halted for longer than about 6 to 10 min, irreversible changes take place. Clearly, the blood supply to the brain must be assured, and there are excellent mechanisms designed to guarantee cerebral circulation.

Figure 17-3 shows the main vessels which carry the blood to and from the brain. As has already been pointed out, there are four arteries that supply the brain, namely, the two internal carotids and the two vertebrals. The latter unite to form the basilar artery. Branches of the basilar artery then unite with branches of the internal carotid to form the circle of Willis (see Figure 17-4). By virtue of this arrangement, an ample blood supply to the brain is assured even after blockage of one of the vertebral or carotid arteries. In experimental animals, both carotids may be tied off without seriously impairing the cerebral circulation. Since man normally assumes the upright position, there is no great problem of venous return as is experienced from areas below the level of the heart. The force of gravity actually assists the venous circulation and enhances cerebral drainage. But, at the same time, blood must be pumped to the brain against gravitational force. Therefore, any decrease in cardiac output or in the arterial blood pressure will markedly affect cerebral circulation.

There is still another factor which must be considered: The brain itself is enveloped by the tough and inelastic dura, and both the brain and its coverings are surrounded by the skull. If the blood pressure within the cerebral vessels rises significantly, serious consequences will ensue. In short, the blood supply to the brain must be steady and the pressure maintained within a relatively narrow range.

Determination of Cerebral Flow

Blood flow through the brain is determined by the same method as described for coronary flow, that is, the Kety procedure. In this case, blood samples are taken from the carotid artery and the jugular vein. Again N_2O is used.

In healthy young adults, in the supine position, the cerebral blood flow is about 55 ml/100 g of brain per min, which means approximately 750 ml/min for a brain of average weight. This is 15 percent of the normal cardiac output. As in any other part of the circulation, the flow represents the resultant between the driving force, that is, the perfusion pressure, and the resistance to the flow. The driving force is the arterial blood pressure at the entrance into the cranial vault. Resistance is supplied by the cerebral vessels, by the intracranial pressure, and by the venous pressure. The venous pressure is generally very low because of the influence of gravity, but the intracranial pressure can be an extremely important factor.

Regulation of Cerebral Flow

As shown in Figure 19-3, autoregulation is very effective. In the cerebral circulation, both oxygen and carbon dioxide play important roles. Figure 19-4 shows quite clearly that cerebral blood flow is very sensitive to the concentration of carbon dioxide in the arterial blood. Over a broad range, cerebral blood flow is directly proportional to arterial carbon dioxide, so that if the pressure of this gas is doubled, blood flow is doubled.

Beyond question, both increased CO_2 and decreased O_2 cause cerebral vasodilatation. The questions to be answered are (1) are these direct effects, and (2) if so, what is the mechanism? Both hypercapnia and hypoxia lead to the accumulation or formation of acids. In the case of hypercapnia, CO_2, H_2CO_3, and H^+ all increase. Hypoxia is accompanied by an intra- and extracellular accumulation of lactic acid. Because of the low buffer capacity of the cerebrospinal fluid (CSF), the increased P_{CO_2} or lactic acid causes a relatively large decrease in pH of the CSF. Accordingly, many authorities have concluded that the brain extracellular pH is the main factor controlling cerebral blood flow. No doubt this is an important factor, but the total available evidence indicates that H^+ concentration is not the only factor controlling cerebral vessel resistance.

Neurogenic Control   Although activation of the sympathetic nervous system produces some vasoconstriction of the cerebral arterioles, the changes are so slight in relation to those evoked by the respiratory gases that the sympathetic cerebral vasomotor innervation

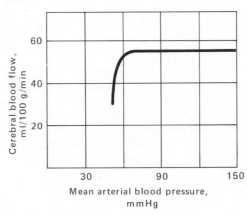

Figure 19-3. Autoregulation of cerebral blood flow. Autoregulation is so successful that an increase in blood pressure from 60 to 150 mmHg brings about almost no change in cerebral blood flow.

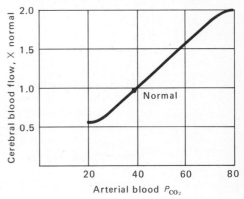

Figure 19-4. Effect of carbon dioxide pressure on cerebral blood flow. Over a broad range cerebral blood flow increases at a constant rate with arterial blood carbon dioxide content.

is probably without physiological importance. Likewise, although vasodilator nerves have been demonstrated, their importance remains in doubt. In short, the cerebral circulation does not seem to participate in the central control of peripheral resistance. Even the catecholamines appear to be without effect.

Prostaglandins   Prostaglandins are found in many tissues, and they are available commercially. Considerable research is being carried out to determine their physiological role or roles, and they are being used clinically for a variety of purposes. They also promise to provide a new and effective birth-control procedure. A preliminary report now suggests that caution should be exercised because these substances apparently cause vigorous constriction of small cerebral blood vessels. Flow may be decreased as much as 40 percent in the dog. Whether prostaglandins play any physiologic role in regulating cerebral flow remains to be seen.

Extrinsic Factors   Pressoreceptors and chemoreceptors are strategically located in the carotid sinuses and bodies to guard the cerebral circulation. These receptors test the pressure and analyze the chemical composition of the blood just before it enters the brain. Any deviation from the normal is, in this way, detected, and rapid and appropriate corrections are made which result in a return to normal levels.

For example, when one assumes the upright position, gravity draws the blood away from the brain and tends to pool it in the lower extremities. This trend if unchecked will deprive the brain of sufficient oxygen, and the person will quickly lose consciousness. However, before this situation occurs, the pressoreceptors in the carotid and aortic sinuses fire into the medulla. Through appropriate nervous pathways, these impulses are propagated to the heart and to all the arterioles throughout the body. Heart rate is increased, which tends to augment cardiac output, and the peripheral resistance become greater. The combination of higher cardiac output and greater peripheral resistance elevates the blood pressure, thus overcoming the force of gravity and maintaining the cerebral blood supply within normal limits.

Likewise, if for any reason the oxygen concentration in the blood should diminish (this may occur as a result of certain pulmonary disorders), the chemoreceptors in the carotid and aortic bodies increase arterial blood pressure. As a result, more blood flows through the brain. And so, even though each unit of blood is deficient in oxygen, more units reach the brain to compensate for the oxygen deficiency.

Influence of Elevated Intracranial Pressure

An increase in intracranial pressure squeezes the cerebral vessels and increases cerebral vascular resistance; therefore, cerebral blood flow decreases unless the arterial blood pressure also rises. And it does. As shown in an experiment in which the intracranial pressure was progressively increased, the arterial blood pressure keeps step (Figure 19-5). This is known as the Cushing reflex. The major mechanism is the influence of carbon dioxide on the vasoconstrictor center. As the intracranial pressure is elevated, less blood flows through the brain. Carbon dioxide from brain metabolism accumulates. This acts on the vasoconstrictor center to evoke generalized vasoconstriction, which elevates the arterial blood pressure. As a result, arterial pressure is once again higher than intracranial pressure, and the cerebral circulation is maintained.

Abnormal Cerebral Blood Flow

By virtue of the mechanisms just described, the cerebral blood flow is maintained within the normal range despite significant alterations in flow elsewhere. Occasionally, however, these mechanisms are overwhelmed.

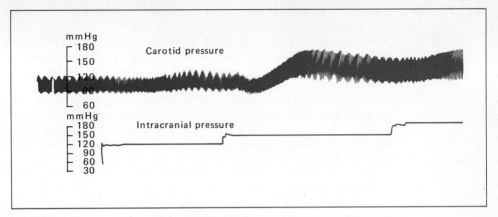

Figure 19-5. Cushing reflex. Elevation of intracranial pressure reflexly increases the carotid blood pressure. This reflex maintains cerebral blood flow in cases of high pressure within the cranium.

Stroke If a cerebral vessel is blocked by a thrombus or if there is rupture of such a vessel, inadequate blood will be delivered to the cerebral capillaries, and brain damage may result from the ensuing inadequate oxygenation. As a result of brain damage, various patterns of neurological disorders occur. Neurological derangement of cerebrovascular etiology is termed a stroke.

Fainting Less severe interference with cerebral blood flow can cause a person to lose consciousness, that is, to faint. A combination of factors may so lower the systemic blood pressure that there is inadequate cerebral flow. For example, profound emotion, probably acting through the hypothalamus, causes such generalized vasodilatation, especially in the muscles, that the arterial blood pressure plunges and so does the cerebral blood flow. Fainting may also occur in a person who is required to stand motionless for prolonged periods. In such cases, the lack of movement allows blood to pool in the veins below the level of the heart and there is inadequate venous return, low cardiac output, low arterial blood pressure, and reduced cerebral blood flow. This sequence is enhanced in very hot weather, when the demands of temperature regulation increase skin blood flow and thus further reduce arterial blood pressure.

Headache Although headaches are common, not a great deal is known concerning the underlying mechanisms. Whether elevated pressure in the cerebral circulation can cause headaches is still not certain, but hypertensive patients often do report headache. On the other hand, there is a current theory that ascribes the pain of so-called migraine headache to pulsation of arterial blood through grossly dilated cerebral vessels. The throbbing pain is thought to be due to the periodic stretch of these vessels. Migraine headache is generally preceded by a series of symptoms that are attributed to cerebral ischemia due to vigorous vasoconstriction of the vessels. Just why the vessels first constrict and then grossly dilate remains to be explained.

Cerebrospinal Fluid Circulation

Cerebrospinal fluid (CSF) acts as a protective cushion for the delicate nervous tissue it surrounds. It buffers the brain against trauma. The CSF fluid volume can be altered. In this way, it compensates for other intracranial volume changes to maintain normal intracranial pressures.

The entire CSF system normally holds only about 125 ml. Close to 750 ml of fluid are formed per day in man. When one is in the recumbent position, the pressure of the fluid is about 10 to 13 mmHg. In the sitting position, the pressure in the lumbar area is about 15 to 20 mmHg.

To determine CSF pressure, a spinal needle is inserted into the spinal canal below the lower end of the cord. Pressure then may be measured using a sensitive pressure transducer, or the fluid may simply be allowed to rise in a vertical tube. In the latter case, the height would be a measure of pressure in terms of mmH_2O. To convert to mmHg, the value is divided by the specific gravity of mercury, which is 13.6.

The CSF pressure is directly proportional to the rate of flow and inversely related to the resistance. In the vascular system, flow is a function of cardiac output, and resistance is determined by the diameter of the arterioles. In the cerebrospinal system, flow is a function of the rate of formation of the fluid; resistance is determined by the rate of absorption of the fluid out of the system. Resistance may be increased by blockage, as occurs in cases of brain tumor. Increased pressure also results when the rate of formation exceeds the rate of absorption.

Formation of CSF Projecting into each of the four brain ventricles is the choroid plexus, richly supplied with capillaries (Figure 19-6). These are sometimes referred to as arterial capillaries, because they spring directly from arteries. The hydrostatic pressure in them is higher than in true capillaries. Because of this high pressure, about 50 mmHg, fluid leaves throughout their length. This imbalance between hydrostatic and osmotic pressure accounts for the formation of most of the CSF. There are reasons to believe, however, that the formation is not wholly a passive process. A comparison of CSF and plasma (Table 19-1) reveals differences which have led to the conclusion that active transport into and out of the CSF occurs. In addition, the high pressure which can develop in cases of occlusion of CSF flow further supports this contention. CSF is also formed by the regular capillary beds that supply the arachnoid.

Not only is there a striking difference in the concentration of various substances in the CSF in comparison with the plasma, but the rate of transfer of these substances from plasma to CSF markedly differs. Because some substances do not penetrate the cerebral capillary walls and others do so only very slowly, one envisions a barrier between the blood and the brain tissue commonly referred to as the blood-brain barrier. The mechanism by which these capillaries differ in this respect from others throughout the body is not known.

TABLE 19-1. COMPARISON OF CSF AND PLASMA

CONSTITUENT	CSF	PLASMA	CSF/PLASMA
Calcium	2.3 mEq/liter	5.0 mEq/liter	0.46
Chloride	123.0 mEq/liter	96.0 mEq/liter	1.28
Glucose	60.0 mg/100 ml	80.0 mg/100 ml	0.75
Phosphorus	2.5 mEq/liter	2.5 mEq/liter	1.00
Potassium	2.5 mEq/liter	5.0 mEq/liter	0.50
Protein	0.03 g/100 ml	7.0 g/100 ml	0.004
Sodium	227.0 mEq/liter	141.0 mEq/liter	1.61

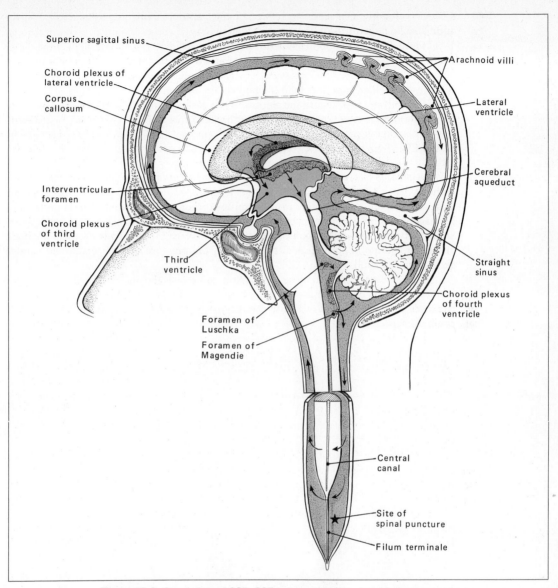

Figure 19-6. Circulation of CSF. CSF is formed in the choroid plexuses and circulates within the ventricles and subarachnoid space (colored area). It is reabsorbed by arachnoid villi into the dural sinuses. Arrows show the direction of circulation.

The blood-brain barrier apparently serves as a highly important protective device for the extremely sensitive neural tissue. On the one hand, it prevents various substances such as the catecholamines, steroids, and penicillin from entering except in minute quantities. On the other hand, it maintains, probably by active secretion, the all-important ionic concentration of the fluid bathing the neural tissue. As might be expected, water and the respiratory gases move through the barrier with ease.

507

CSF Circulation CSF circulates through the ventricles and the subarachnoid space ultimately to be returned to the venous blood by way of villi extending from the arachnoid layer through the dura into the superior sagittal sinus (Figure 19-7). The villi are the major site of fluid transfer from the subarachnoid space into the venous system. Because the sinuses cannot collapse, the pressure is negative within the sinuses of a person who is standing. It is, therefore, less than the hydrostatic pressure of the CSF. Thus fluid moves from the arachnoid space into the sinuses. It is a passive process. Some fluid is also taken up by the regular capillary beds of the arachnoid. Since true lymphatics are not present, the perivascular spaces serve as lymphatics to remove whatever protein may accumulate in the CSF.

Hydrocephalus Excessive fluid usually elevates the pressure sufficiently to damage the nervous tissue. This condition is termed hydrocephalus, which means "water in the cranium." Hydrocephalus may result from overproduction of fluid, obstruction to flow, or inadequate absorption. In most cases, obstruction to flow is the cause. Fluid accumulates in the ventricles and creates considerable pressure which progressively destroys the brain tissue. In children, before cranial sutures fuse, the pressure may become great enough to enlarge the skull.

   If the disorder is caused by an obstruction, sometimes the obstruction can be removed or a bypass created. Also, if the cause is excessive production of fluid, much of the choroid plexuses can be destroyed. Clearly, the diagnosis should be made early, before severe brain damage results.

Figure 19-7. Coronal section through the superior sagittal sinus. The subarachnoid space is filled with CSF. It enters the sinus through arachnoid villi.

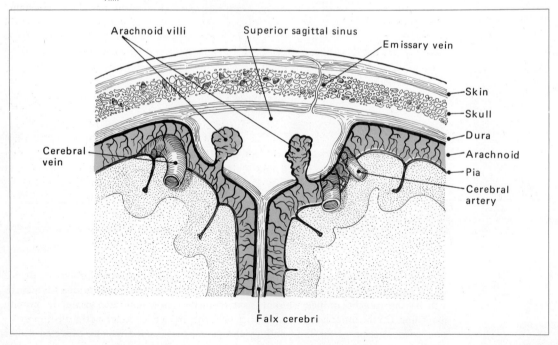

Obstruction is diagnosed by the so-called Queckenstedt test. In this procedure, both jugular veins are firmly pressed. As a result, the venous pressure is elevated, and this prevents the return of CSF into the venous system. The CSF pressure increases sharply. But if there is occlusion, the pressure, measured in the lumbar region, will not increase. If there is no blockage, the pressure in that area increases comparable with the increase in pressure in the ventricles.

PULMONARY CIRCULATION

The blood is pumped by the right ventricle into the pulmonary trunk. This artery immediately bifurcates into two main branches, one of which passes to the left lung and the other to the right lung. These branches in turn give off smaller arteries and arterioles and finally the typical network of capillaries. The pulmonary capillaries then re-form into venules and veins, and finally four large pulmonary veins enter the left atrium of the heart (see Figure 17-2).

Pulmonary Blood Pressure

In the normal, healthy man, the same amount of blood must flow through the pulmonary system per minute as flows through the greater, or systemic, circulation. As previously stated, the left side of the heart pumps 5 liters/min of blood. Thus, the right ventricle must eject a comparable amount. In view of the fact that the cardiac output is one of the major determinants of blood pressure, one might think that the pulmonary arterial blood pressure would be about the same as that in the brachial artery. But this is not the case. The pulmonary blood pressure is very low. In the pulmonary artery, the systolic pressure is about 22 mmHg and the diastolic level usually 5 to 10 mmHg.

The first conclusion to be drawn from these facts is that the peripheral resistance in the pulmonary circulation is considerably less than it is in the systemic circulation. This must be so because the arterial blood pressure is determined by (1) cardiac output and (2) peripheral resistance. Since the cardiac outputs in the two systems are identical, it follows that the peripheral resistance in the lungs is far less than elsewhere in the body.

There is another observation to be made from these same data. Note that although the cardiac output in both sides is the same, the pulse pressure, that is, the difference between the systolic and diastolic pressure, is quite different. The pulse pressure in the general circulation is about 40 mmHg (120/80), but it is only about 14 mmHg (22/8) in the pulmonary system. As already discussed in detail, the pulse pressure reflects three factors: (1) cardiac output, (2) peripheral resistance, and (3) distensibility of the vessels. The cardiac outputs of the two sides of the heart are identical. It has been concluded that the peripheral resistance in the lungs is low. And now the conclusion that the distensibility is markedly greater in the pulmonary system than it is elsewhere becomes inescapable. The distensibility and elastic recoil buffer the systolic and support the diastolic pressure. These results are dramatically exhibited in the pulmonary system.

Regulation of Pulmonary Flow

The pulmonary system is short and highly distensible. Pressures within the lungs and thorax change markedly with respiration, and these pressures are reflected in the pulmonary system. In addition, the blood flows from the pulmonary veins into the left side of the heart, where pressures also vary greatly. All these factors strongly influence pulmonary flow.

Intrathoracic Pressure The alterations in intrathoracic dynamics as a result of respiration markedly influence pulmonary blood flow and pressure. During deep inspiration the intrathoracic pressure is lowered, and so blood is encouraged to flow from the venae cavae into the right side of the heart. Hence, more blood is pumped by the right ventricle. But, at the same time, the lungs are expanded, and consequently the pulmonary vessels are stretched, their capacity thereby being greatly increased. So, although more blood is pumped into the pulmonary system, the greatly increased capacity causes a sharp depression in pulmonary blood pressure, and very little blood enters the left side of the heart during inspiration. That is to say, during inspiration the blood pools in the pulmonary vessels. But during expiration the lungs decrease in size, and the blood in this reservoir is therefore squeezed into the left side of the heart; thus the output of the left ventricle is elevated. Carotid arterial blood pressure consequently varies, and these oscillations are related to breathing. Figure 19-8 shows that during inspiration arterial pressure progressively falls and then during expiration it rises, for reasons just explained.

Vasomotor Activity Although pulmonary vessels are capable of vasomotor changes, these alterations seem to be of little importance to the pulmonary circulation. The great elasticity

Figure 19-8. Influence of respiration on arterial blood pressure. The lower tracing is blood pressure at the carotid sinus. During inspiration carotid pressure falls. During expiration it rises because of the alterations in intrathoracic pressure.

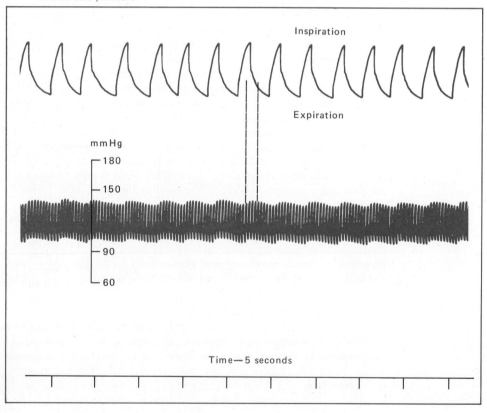

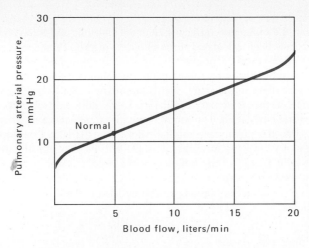

Figure 19-9. Effect of blood flow on pulmonary arterial pressure. Because of the distensibility of pulmonary vessels, an increase in output of the right ventricle raises pulmonary pressure only slightly.

of the vessels and the changes which occur as a result of respiration outweigh any vasomotor modifications which may occur.

Resistance of the Left Side of the Heart   The normal left side of the heart easily accommodates as much blood as is presented to it. However, if the left ventricle is failing, it contracts feebly, leaving a large residue in the ventricle at the end of each systole. Hence, there is a resistance to the flow of blood from the pulmonary veins into the left side of the heart. Blood pools in the lungs. This pooling of blood increases the pulmonary pressure. Consequently, the normal quantity of blood is once again forced into the left side of the heart, but the pool remains in the pulmonary system. In other words, the pulmonary pressure has been elevated. As the left side of the heart grows weaker, the pulmonary pool and pressure progressively increase.

The same process will result from any abnormality in the left side of the heart which exerts resistance to the flow of blood from the pulmonary system into the heart. Such a condition could be, for example, mitral stenosis, that is, a narrowing of the mitral valve which lies between the left atrium and ventricle.

The output of the right ventricle or the resistance of the left side of the heart may increase considerably and thereby increase the pulmonary blood volume with very little change in pulmonary blood pressure because of distensibility of the pulmonary system. But once these vessels have been distended and once all previously dormant vessels have been expanded, further volume changes are accompanied by comparable increases in pulmonary blood pressure (Figure 19-9). As can be seen, when the output of the right side of the heart is increased four times, that is, from 5 to 20 liters/min, the pulmonary arterial pressure increases only about two times.

Because of this buffering power of the pulmonary circulation, that is, the ability to accommodate larger volumes of blood without a comparable rise in pressure, failure of the left side of the heart does not influence the right side until it becomes very severe. Then, however, pulmonary pressure does increase to as much as 40 mmHg. The work that the right ventricle must do to pump blood against this head of pressure causes it to hypertrophy and eventually fail.

511

Pulmonary Capillary Dynamics

The pressure in the pulmonary artery is 22 mmHg systolic and 5 to 10 mmHg diastolic. Mean pressure has been calculated to be about 13 mmHg. Pulmonary venous pressure is only about 5 mmHg; thus the pulmonary capillary pressure must be somewhere between these two values, probably about 7 mmHg. But the colloidal osmotic pressure of the blood is the same everywhere in the body, that is, approximately 28 mmHg. Accordingly, the balance of forces in the pulmonary capillary is vastly different from that in the capillary beds of the greater circulation. The osmotic pressure dominates and results in the movement of fluid into the pulmonary capillary through its length. Consequently, the lung alveoli are kept dry, and any fluid that does get into them is quickly absorbed into the blood. However, if failure of the left side of the heart is severe enough to elevate the pressure in the pulmonary capillaries above about 23 mmHg, edema formation does occur. Fluid not only accumulates in the lung alveoli but also in the thorax, a condition referred to as hydrothorax.

Abnormal Pulmonary Blood Flow

A few conditions that result in abnormal pulmonary blood flow have already been mentioned, for example, a failing left side of the heart, which elevates pulmonary pressures and may cause pulmonary edema and hydrothorax. A few other conditions will be mentioned for illustrative purposes.

Lung Collapse As will be explained later, if air is permitted to enter the pleural cavity, the lung will collapse. Collapse of a lung or a part of a lung is termed atelectasis. When this occurs, the blood that normally would flow through that lung is shunted to the lung which is not collapsed. Normally, the thoracic (intrapleural) pressure is below atmospheric pressure, but when air is admitted, pressure rises to that of the atmosphere. Consequently, the pressure surrounding the pulmonary vessels in the collapsed lung is somewhat higher than that in the normal lung. This provides enough resistance to divert the blood to the lower-pressure area in the normal lung. Probably of even greater importance is the squeezing and kinking of the vessels in the collapsed lung. This markedly increases the resistance to flow.

Pulmonary Embolism Intravascular blood clots may break free as emboli and be carried to the lungs to become trapped in a pulmonary vessel. If both pulmonary arteries are blocked, death ensues; but if only one side is blocked, the other lung can accommodate the normal right ventricular output. Unless proper therapy is instituted, the clot, once lodged, will grow, resistance to flow will increase, and ultimately the right ventricle will fail.

Emphysema The term emphysema means "inflation," that is, overdistention of the alveoli. In this condition not only are they distended, but the walls that separate individual alveoli break down so that a group of individual alveoli become one large cavity as air escapes into the interalveolar spaces and is trapped there. When these walls are destroyed, the small pulmonary vessels in them are also destroyed. As a result, pulmonary resistance increases. Thus, severe emphysema not only interferes with the exchange of respiratory gases but also decreases the blood flow, increases the resistance, and may result in right ventricular failure.

HEPATIC PORTAL CIRCULATION

The circulation in the liver is unusual in that it is composed of both arterial and venous blood. The arterial blood enters the liver via the hepatic artery, and the portal vein brings venous blood from the viscera. Blood from these two sources meets in the liver lobules, to be mixed in the sinuses. The mixed blood then is collected by the hepatic veins, which ultimately drain into the inferior vena cava.

The blood that enters the liver via the portal veins has already drained from the intestines; thus it contains low oxygen tension but has a high concentration of the substances absorbed by the intestinal tract following digestion. On the other hand, the hepatic artery blood is saturated with oxygen essential for the integrity of the liver. This role is clearly shown when the hepatic artery is tied off. In such experiments, necrosis of the liver cells occurs despite the ample portal vein flow.

Pressure Relationships

The pressure in the portal vein is no more than 10 mmHg; in the hepatic veins it is close to 0. These figures mean that in the liver sinuses the pressure is probably about 6 or 7 mmHg. This low pressure has two implications: (1) the resistance in the hepatic arterioles must be higher than in most arterioles, and (2) the effective hydrostatic pressure responsible for driving fluid across the endothelium of the hepatic sinuses is much less than in most capillary beds. From the second implication the conclusion may be reached that there is little or no movement of fluid out of the hepatic portal circulation, as is the case with the pulmonary capillaries. This is not true, however, because the endothelium of the liver sinuses is highly permeable, so much so that proteins can readily diffuse through it. Thus the colloidal osmotic pressure is low, too; therefore, protein and fluid do leak out. In view of the fact that the liver has vital metabolic functions, this facile exchange of nutrients across the vascular membrane is advantageous. That there is leakage of protein out of the liver sinuses is verified by the fact that the lymph coming from the liver has as much protein as does plasma.

Function of the Hepatic Portal System

Normally, the total blood flow through the liver is about 1,500 ml/min, that is, approximately 30 percent of the cardiac output. Of the 1,500 ml, close to 1,100 ml arrives via the portal vein and the remainder by the hepatic artery.

Blood in the portal vein, during digestion, contains substances in high concentration that are absorbed from the intestines. But in passing through the liver, many of these substances are altered and their concentration brought closer to the values found in the general circulation. The hepatic portal system may then be said to provide a protective buffering function. By another mechanism it provides another protective function. The liver sinuses are lined by Kupffer cells, which very rapidly remove bacteria and perhaps other foreign material from the blood. Once in the Kupffer cells the bacteria are digested. This function is highly important because many bacteria pass from the intestines into the portal blood. Hepatic blood is practically bacteria-free, which indicates that the liver is a highly effective filter.

As mentioned earlier, the liver is an important blood reservoir. The capacity of its vascular system is normally equal to 500 or 600 ml. But the liver and the vessels are quite distensible, so that if the pressure relationships change, the volume may increase to a liter or more. The main determinant of liver blood volume is the pressure in the hepatic veins. If the pressure there rises, there will be greater resistance to outflow, and the liver volume will increase. Actually there are three sets of contractile structures in the hepatic portal circulation: (1) afferent arterioles controlling flow into the liver sinuses from the hepatic artery, (2) afferent venules controlling flow from the portal vein into the liver sinuses, and (3) efferent venules controlling the outflow from the liver sinuses by way of the hepatic veins. All three are thought to be under the control of the sympathetic nervous system, but as just stated, the pressure and resistance in the hepatic veins seem to be dominant.

Abnormal Hepatic Portal Circulation

Cirrhosis of the liver, a degenerative disorder, is characterized by the replacement of normal liver tissue by fibrous tissue. The fibers progressively shorten around the blood vessels and thus increase the resistance to flow from the portal vein. As a result, capillary pressure in the intestines rises high enough to cause fluid loss into the intestine. Such fluid loss can be great enough to prove fatal. This is particularly evident in cases in which a blood clot blocks the portal vein.

If the right side of the heart is in failure, there will be elevated pressure in the inferior vena cava and back through the portal system. This causes outpouring of fluid from the liver and the intestine into the abdominal cavity. Fluid in the abdominal cavity is termed ascites.

SPLENIC CIRCULATION

The splenic artery supplies the spleen. In the spleen the blood may flow directly into a capillary bed, then into venous sinuses, and on into the splenic vein which joins the portal vein. Or the blood may ooze out of the capillaries and pass through the pulp of the spleen, which is in essence a filtering system, before it reenters the circulation via the venous sinuses of the spleen.

Contraction of the Spleen

In many animals the capsule of the spleen contains smooth muscle which contracts in response to sympathetic stimulation. This forces a sizable quantity of red cell–rich blood into the circulation. When the smooth muscle relaxes, the spleen is capable of considerable expansion, thereby serving as a blood reservoir. In man, however, there are few muscle fibers in the capsule; therefore investigators have tended to assume that in man the spleen must not have the same function. There is evidence to suggest, however, that sympathetic stimulation in man does result in the expulsion of considerable quantities of blood from the spleen. How this is accomplished is not yet known, but the thinking is that direct constriction or dilation of the splenic vessels can do the same as capsule contraction or relaxation.

Function of the Spleen

The spleen has at least three functions: (1) to serve as a blood reservoir, (2) to filter the blood, and (3) to produce blood cells, at least during fetal life.

The splenic circulation has a capacity, in man, of about 500 ml. Under sympathetic stimulation, this capacity is reduced, forcing some 300 ml into the general circulation. More important, however, than the volume of blood is the erythrocyte count in the expressed blood. It is so high that the total blood hematocrit may be increased 3 to 4 percent. Sympathetic stimulation occurs in response to hemorrhage and may be triggered by the carotid and aortic pressoreceptors. There is also sympathetic stimulation during vigorous exercise. This provides additional blood volume to increase venous return and thus cardiac output. The red cells provide additional oxygen-carrying capacity.

The blood sinuses of the spleen are lined with reticuloendothelial cells which are phagocytic. These cells effectively remove debris left by the breakdown of blood cells, hemoglobin, bacteria, and other foreign particles.

CIRCULATION IN THE SKIN

In the study of circulation through special regions the recurring theme has been self-needs of that region. Thus there is autoregulation, and in some cases the needs of a particular region will influence the entire circulation. For example, the kidneys in ensuring their own circulation can dangerously elevate the systemic blood pressure. Circulation in the skin is also influenced by local needs, but these needs are usually subservient to other demands, particularly body temperature.

The fact that some 2.5 liters/min can flow through the skin has important circulatory consequences. When one exercises vigorously, particularly in a hot, humid, environment, both the muscles and the skin demand large quantities of blood. So long as the cardiac reserve is normal, the demands are met, although, of course, even in the young, well-trained athlete, collapse is not unknown. The problem is critical in the cardiac patient with sharply reduced cardiac reserve. The slightest exertion in hot weather may throw him into acute failure.

Variations in Skin Blood Flow

The total blood flow through all the skin of the body varies tremendously—from a dribble to a remarkable 50 percent of the cardiac output, that is, 2.5 liters/min. Nutritive needs of the skin require very little blood; thus the variation is in response to other factors, and generally the flow far exceeds that required for nutrition.

Regulation of Skin Blood Flow

Figure 19-10 shows in a diagrammatic fashion the unique arrangement of the blood vessels in the skin. In addition to the usual capillary beds, there are arteriovenous anastomoses and there are venous plexuses in the subcutaneous tissue. The arteriovenous anastomoses control the flow of blood from the arteries into the venous plexuses. The walls of these anastomoses are muscular, innervated by sympathetic vasoconstrictor fibers, and capable of vigorous constriction that can reduce the flow into the plexuses to very low volumes.

There is evidence of autoregulation of flow through the skin, but by far the dominant regulation occurs in response to changes in external temperature. When the air temperature increases, skin blood flow increases; in cold weather, skin blood flow becomes minimal. The mechanism for these changes involves both central and local factors.

Under normal conditions, the internal temperature of man and most other mammals is about 37°C (98.6°F). Clothing, heating, and air-conditioning all help to maintain this temperature, but even in the absence of such means, mammals are quite capable of maintaining their internal temperature within a very narrow range despite wide variations in the external environment (Figure 19-11). Under normal conditions, body temperature varies less than 1°C. Body temperature regulation was considered in Chapter 11; here we are concerned only with variations in skin blood flow due to temperature changes.

Skin blood flow is regulated by sympathetic constrictor fibers and sympathetic dilator fibers, that is, adrenergic and cholinergic fibers. As mentioned earlier, the activity of these nerves is integrated by the hypothalamus. The responsible nuclei in the hypothalamus

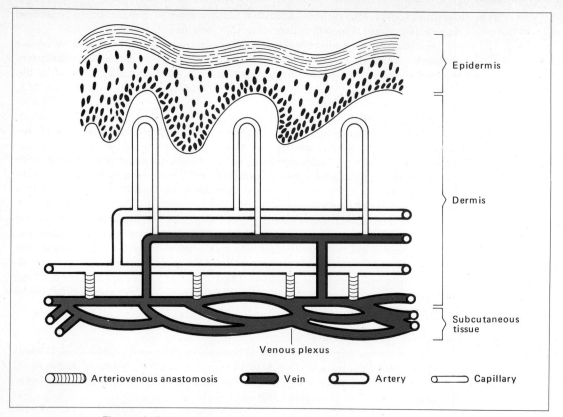

Figure 19-10. Blood vessels of the skin. Schematic cross section. By means of arteriovenous anastomoses and precapillary sphincters blood can be routed into or away from the capillary beds.

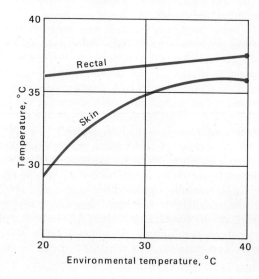

Figure 19-11. Body temperature regulation. A change in external temperature causes a marked change in skin temperature, but internal body temperature remains relatively constant.

have been shown to be sensitive to the temperature of the blood, and they also receive impulses from skin temperature receptors.

Norepinephrine secreted by the adrenergic fibers is thought to act directly on the smooth muscle of the arterioles, arteriovenous anastomoses, and precapillary sphincters of the skin vessels to cause constriction. Just how acetylcholine (ACh), secreted by the cholinergic fibers, functions is a matter of conjecture. One hypothesis holds that the ACh acts upon the sweat glands, evoking not only sweat but also the secretion of kallikrein, an enzyme which splits bradykinin from globulin. The bradykinin then causes vasodilatation. To be sure, bradykinin is a potent vasodilator, but the stumbling block of this hypothesis is that if bradykinin is inhibited, dilatation still occurs.

Low temperatures can cause vasoconstriction of the skin vessels directly, but if the temperature goes to 0°C or lower, the vessels dilate. The mechanism of these opposite changes is not understood completely, but the results are obvious. Until the fall in external temperature becomes severe, the local direct constrictor effect prevents undue lowering of blood temperature. But at very low temperatures, probably because of interference with the metabolic processes responsible for muscle contraction, the vessels dilate and blood flows and warms the skin, thus preventing cell damage. This is why persons exposed to very low temperatures sometimes report suddenly feeling warm and then freeze to death. If the cold is such that skin vessels dilate, there will be a huge volume of blood exposed to the cold, blood temperature will plunge, and body temperature will fall below a level compatible with life.

Skin Color and Temperature

The color, or hue, of the skin is influenced by the rate at which blood flows through the capillaries. If the blood moves slowly, the hemoglobin will surrender much of its oxygen. Reduced hemoglobin is blue. Conversely, fast-moving blood does not have time to give up much of its oxygen and remains red.

The intensity of the color depends partially upon the thickness of the layer of blood flowing through the skin. In other words, the intensity of the color is a function of the diameter of the capillary. Thus, in a Caucasian, four variations are possible: scarlet due to fast flow through dilated vessels, deep blue due to slow flow through dilated vessels, light pink due to fast flow through constricted vessels, and ashen due to slow flow through constricted vessels. The latter is referred to as pallor.

The temperature of the skin also depends upon the rate of blood flow. If the flow is slow, the blood will lose its heat to the external environment, and the skin will feel cold. Conversely, with a fast flow of blood the skin is hot. Obviously, a fast flow through dilated vessels produces the greatest heat.

Abnormal Skin Circulation

In some individuals there is an abnormal sensitivity to cold causing local vasoconstriction. When the external temperature dips below about 40 or even 45°F, vigorous constriction occurs, especially in the fingers and toes. The result is that there is inadequate blood flow to satisfy nutritional needs, and severe pain ensues. In some instances this phase is followed by sudden dilation, so that the extremities become red and the pain becomes even more acute. In other cases, the constriction persists until the hands are warmed, at which time the same painful response is experienced. This sequence is referred to as Raynaud's disease. It may progress to the point that gangrene occurs. Treatment consists either of moving to a warmer climate or in having the sympathetic fibers that serve the area surgically destroyed.

Many elderly people complain of cold extremities; ulceration of the skin may occur, or even gangrene. This is usually caused by progressive clogging of the peripheral arteries by plaques characteristic of arteriosclerosis. Once this has occurred, there is no effective treatment.

SKELETAL MUSCLE CIRCULATION

Because the energy requirements of muscle increase so sharply during exercise, blood flow through muscle likewise greatly increases. This change has a profound effect on cardiac output. For example, at rest, skeletal muscle blood flow averages about 5 ml/100 g of muscle per min. In exercise it can exceed 50 ml/100 g per min. The result is greatly increased venous return, which brings about augmented cardiac output. This is the major reason for increased cardiac output in exercise.

Regulation of Skeletal Muscle Blood Flow

As has been emphasized many times, the primary cause of vasodilatation of the arterioles is local ischemia. When the muscle fibers become active, they sharply increase their demand for oxygen. The available oxygen in the cells and in the surrounding fluids is depleted, and either by a direct influence of hypoxia or due to the liberation of a substance such as adenosine, vasodilatation occurs. Other substances may play a role, but the fact remains that even in muscle deprived of all autonomic innervation, vasodilatation occurs. In addition to dilatation of the arterioles, there is also generalized opening up of capillary beds. At rest, the capillaries close down, periodically reopening to allow minimal blood flow, a process called vasomotion. This is due to smooth muscle sphincters and also to the Laplace effect (see page 476). But when the metabolic needs increase and when the arterioles dilate, all the capillaries are opened, permitting great flow through the muscles and into the venous system.

Superimposed on this inherent system are sympathetic vasoconstrictor and vasodilator fibers. The vasoconstrictors are adrenergic, whereas the vasodilators are cholinergic. The cholinergic fibers cause marked vasodilatation.

The sequence of events is thought to be as follows: the very thought of exercise activates the sympathetic vasodilator fibers. Therefore increased skeletal muscle flow and increased cardiac output occur even before the onset of exercise. Once the muscles begin to contract, the local mechanisms maintain the increased flow. Because of skeletal muscle vasodilatation, because all the capillary beds are open, because there is vasoconstriction in nonactive parts of the body, such as the intestine, and because of elevated cardiac output and consequently increased arterial blood pressure, blood flow to the contracting muscle is high and satisfies metabolic needs.

RENAL CIRCULATION

The kidneys not only function as important excretory organs but also play an essential role in circulatory dynamics. The unique aspects of the renal circulation will be discussed when kidney function is considered in Chapter 29. Suffice it to say here that about 1,300 ml of blood flows through the kidneys per minute. The reason that such a large fraction of the cardiac output flows through the kidneys is the relatively low resistance in the renal circulation. The renal vessels are under the control of the sympathetic nervous system and participate in general circulatory adjustments. On the other hand, as already explained, the kidneys can markedly influence the general circulation by the secretion of renin.

During exercise, or as a result of hemorrhage or diminished cardiac output, there is renal vasoconstriction, which elevates the resistance and thus diminishes the renal blood flow. Pathological conditions such as hemorrhage and heart failure require immediate steps to elevate the blood pressure, or the diminished renal flow may result in renal failure and eventual death.

QUESTIONS AND PROBLEMS

1 Discuss the role of the following in the various hypotheses for rapid autoregulation of blood flow: (*a*) oxygen, (*b*) vasodilator substances, and (*c*) mechanical means.

2 Explain why the skin over the coccygeal region grows red when a person who has been lying supine is turned onto his side.

3 What is the disadvantage for the heart that coronary arteries are functionally end arteries?

4 Explain how intramural pressure of the heart and aortic pressure affect blood flow through coronary vessels.

5 What drugs might be prescribed to relieve the inadequate coronary blood flow that causes angina pectoris?

6 What is myocardial infarction?

7 What are the mechanisms which maintain cerebral circulation?

8 Discuss the formation and circulation of cerebrospinal fluid.

9 Define and discuss the importance of the blood-brain barrier.

10 Discuss the effect of left ventricular failure on pulmonary flow and on the exchange of respiratory gases.

11 Describe the pathology of emphysema and of pulmonary embolism.

12 Discuss the protective functions of the hepatic portal system.

13 Discuss the role of the spleen during hemorrhage.

14 How does the skin function as a temperature-regulating organ?

PART 4
THE RESPIRATORY SYSTEM

20 ANATOMY OF THE RESPIRATORY SYSTEM

Respiration is the process of gaseous exchange within the body and has two phases, external and internal respiration. External respiration is the exchange of gases in the alveoli or air sacs of the lungs. Oxygen in the inspired air is exchanged for carbon dioxide across the wall of the alveolus. Oxygen is needed in the multitudinous metabolic processes of the cells; carbon dioxide is a waste product of those processes. Air passageways from the atmosphere to the lung alveoli include the nose, pharynx, larynx, trachea, and bronchial tree. Air must be forced along these passageways, and the mechanism of this transport is described in Chapter 21.

Internal respiration occurs at the cellular level in the body. Oxygen entering the bloodstream at the alveoli of the lung is transported by the circulatory system to capillaries throughout the body. There it passes from the blood through the capillary walls into the tissue fluid which surrounds the cells. It then crosses the plasma membrane of the cell. Carbon dioxide passes in the opposite direction into the tissue fluid. This exchange of oxygen and carbon dioxide is internal respiration. The carbon dioxide then enters the capillaries and is transported by venous blood to the lungs where it is exhaled.

EMBRYOLOGY

In the 4-week-old embryo the earliest rudiment of the trachea and lungs appears as an outgrowth, a diverticulum, along the ventral side of the primitive foregut (see Figure 4-7).

The proximal portion of this diverticulum hollows out to form the developing trachea, lined by an extension of endoderm from the primitive alimentary canal. The distal end grows caudally as a solid cord, lengthens considerably, and subsequently becomes tubular as it forms the trachea. Condensations of mesenchyme surrounding this tube will develop into the muscles and cartilages of the trachea and bronchi.

Distally the tracheal tube divides into paired buds, the primordia of the bronchi. These extend into the body cavity, pushing the lining of the cavity before them. This lining will ultimately surround the entire lung as the visceral pleura. The endoderm lining the tracheal and bronchial tubes forms the epithelium of the respiratory system.

Bronchial primordia continue to divide, and the multibranched tree continues to expand until over thirty subdivisions have occurred. These branches eventually develop into grapelike clusters of tiny sacs, the alveoli of the definitive lung.

Occasionally the terminal divisions of the developing lung do not fully expand. In this condition, congenital atelectasis, the involved alveoli are not able to function in the exchange of oxygen and carbon dioxide. This condition may involve only a portion of the lung or its entirety. If this lack of inflation is extensive, it is not compatible with life, and the baby dies soon after birth.

An abnormal communication sometimes develops between the trachea and esophagus. If this tracheoesophageal fistula is large, fluid swallowed by the baby is aspirated into the lungs. For the infant to survive, this defect must be surgically corrected at birth.

HISTOLOGY

The respiratory system consists of two specialized portions, a series of conducting passages and the functional respiratory apparatus. The conducting portion serves essentially to transmit air in and out of the lungs and includes the nose, pharynx, larynx, trachea, bronchi, and bronchioles. The respiratory portion provides for the essential function of the system, namely, the exchange of oxygen and carbon dioxide between the blood and air. It consists of respiratory bronchioles, alveolar ducts, and alveoli.

Conducting Passages

Perhaps the most characteristic histological feature of the conducting passages is that they are all lined with ciliated columnar epithelium (Figure 20-1). Because the nuclei in adjacent cells are at different levels, it appears as if the epithelium were stratified, but it is really a single layer. Hence the name pseudostratified ciliated columnar epithelium is used to designate this epithelium. In place of this awkward term we shall refer to it as respiratory epithelium. Many goblet cells (mucous secretory cells shaped like a wine goblet) are interspersed among the ciliated epithelial cells. Their secretion, together with that of numerous other mucous glands, covers the epithelium with a thin coat of mucus which entraps dust and other foreign particles. This material is moved toward the mouth and nose by the constant beating action of the cilia. In inflammatory conditions, secretion of the mucous cells increases and necessitates frequent expectoration as sputum, or phlegm.

Another common feature of the conducting passageways is the presence of hyaline cartilage in the walls of the larger tubes. In the larynx, trachea, and larger bronchi it essentially encloses the tube. As the bronchi decrease in size, the cartilage appears as broken plates or segments, and there is less cartilage and more smooth muscle. Finally, in the bronchioles the cartilage completely disappears, but the circular muscle layer is greatly increased and provides a relatively rigid open tube.

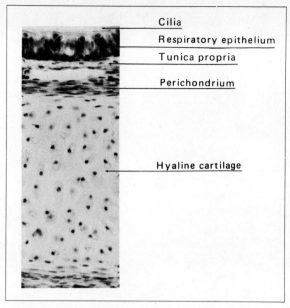

Cilia

Respiratory epithelium

Tunica propria

Perichondrium

Hyaline cartilage

Figure 20-1. Section through a tracheal ring (× 300).

Respiratory Apparatus

In the respiratory portion of this system the cells display further adaptations. In the respiratory bronchioles, muscle in the wall of the tube is reduced to a few fibers, the epithelium becomes cuboidal with only occasional cilia, and goblet cells are completely absent.

Extending from the respiratory bronchiole are alveolar ducts, which are thin-walled fibroelastic tubes with no visible epithelium and few scattered muscle cells (Figure 20-2). The alveolar duct widens considerably into an atrium which terminates in clusters of air sacs or alveoli. The latter are extremely thin-walled, distended spaces and make up the bulk of the lung. The alveoli are the site of external respiration.

There is a common wall between adjacent air spaces. This wall consists of capillaries supported by a few elastic and collagenous fibers, across which gaseous exchange occurs. No epithelium is visible under the light microscope; however, flattened, attenuated alveolar epithelial cells are demonstrable with the electron microscope.

ANATOMY OF THE RESPIRATORY SYSTEM
Nose

The nose includes the external nose and the nasal cavity. The nasal cavity extends from the external nares (nostrils) backward to the nasopharynx.

The skeleton of the nose is formed by bone and cartilage, both of which give attachment to muscles of facial expression. The nasal bones form the bridge of the nose. The lower portion of the nose is formed by a series of cartilages (Figure 20-3). The lateral nasal cartilages form the sides of the nose and abut in the midline with the anterior end of the cartilaginous portion of the nasal septum, which lies just below the skin of the

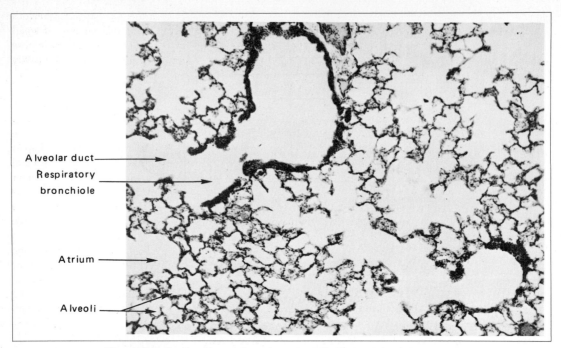

Alveolar duct

Respiratory bronchiole

Atrium

Alveoli

Figure 20-2. Section through lung tissue (× 70). To the right of the respiratory bronchiole is a small bronchiole.

Figure 20-3. Nasal cartilages. At the left, the external nose. At the right, the nasal septum.

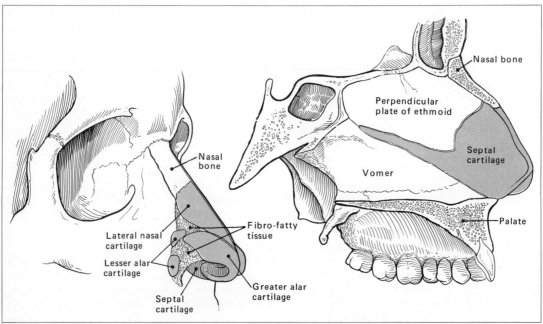

Nasal bone

Lateral nasal cartilage

Lesser alar cartilage

Septal cartilage

Fibro-fatty tissue

Greater alar cartilage

Nasal bone

Perpendicular plate of ethmoid

Septal cartilage

Vomer

Palate

nose. The greater alar cartilages partially encircle and form the supportive framework of the external nares, or nostrils. The lesser alar cartilages, a variable number of small cartilaginous plaques, are located along the nasolabial groove at the junction of the nose with the cheek. In a skull, with these cartilages lacking, the opening into the nasal cavity is a diamond-shaped opening termed the piriform aperture.

Nasal Cavity

The roof of the nasal cavity is the cribriform plate of the ethmoid bone. Above this is the cranial cavity. The floor of the nasal cavity, separating it from the oral cavity, is the hard palate. Osseous components of the nasal cavity have been described in Chapter 5. The cavity itself is divided into two compartments along the midline by the nasal septum. The nasal septum is formed by the septal cartilage anteriorly and the vomer and perpendicular plate of the ethmoid bone posteriorly. Damage from trauma to the nose frequently disrupts the attachment of the cartilage to the bony septum and results in a crooked nose. The septum may also be deflected, or deviated, and block or impede the flow of air through the nasal cavity.

The lateral walls of the nasal cavities are very irregular (Figure 20-4). Plates of bone enveloped by mucous membrane jut into the nasal cavities. These are the conchae, or turbinate bones, and three are present: (1) superior nasal concha, (2) middle nasal concha, and (3) inferior nasal concha. The conchae project over three channels, or meatuses, along the lateral wall, which are named after the overhanging concha. Thus on the lateral wall of each nasal cavity the superior, middle, and inferior nasal meatuses are seen. They extend back to the nasopharynx.

The mucous membrane lining the cavity contains a dense vascular plexus which acts as a radiator to warm inspired air as it passes through the meatuses. The radiant surface is increased by the curved conchae. Mucus secreted by cells of the lining epithelium adds moisture to the inspired air.

The nasal cavity also contains olfactory epithelium, in which are found the receptor endings of the olfactory nerve. Olfactory epithelium is limited to a small area over the uppermost portion of the nasal septum and the adjacent superior concha.

Inflammation of the nasal mucosa, or rhinitis, occurs most frequently with the common cold. It is characterized by a swelling of the nasal mucosa and an increased secretion from the glands opening onto the mucosal lining.

Nasal inflammation may extend through openings into the paranasal sinuses, causing sinusitis, or through the nasolacrimal duct to the conjunctiva of the eye, resulting in conjunctivitis. The latter accounts for the reddened, watery eyes which frequently accompany a head cold.

Skin of the face continues into the nostril and lines the vestibule, the area just inside. Short stiff hairs, the vibrissae, project from this skin into the lumen of the nostril. They function to screen out foreign particles from the inspired air.

The space above the vestibule lying in front of the conchae is the atrium. The sphenoethmoid recess is the small area above the superior meatus.

Paranasal Sinuses

The paranasal sinuses are air cavities within certain bones adjacent to the nasal cavity (Figure 20-5). These are the ethmoid, frontal, sphenoid, and maxillary bones. Mucous membrane lining these sinuses is continuous with the lining of the nasal cavity.

The ethmoid sinuses consist of a series of small air sacs within the upper portion of the ethmoid between the frontal and sphenoid bones. The sacs occupy the area between the orbital and nasal cavities.

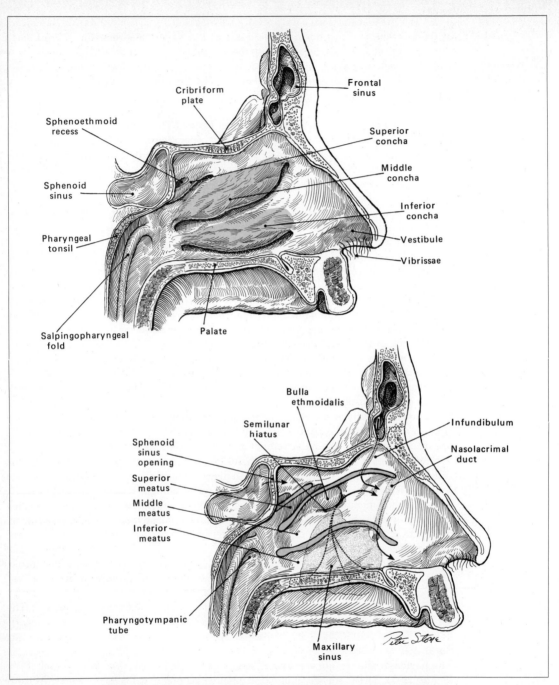

Figure 20-4. Nasal cavity. At the top, structures of the lateral wall. At the bottom, drainage pathways of the paranasal sinuses and the nasolacrimal duct, with conchae removed.

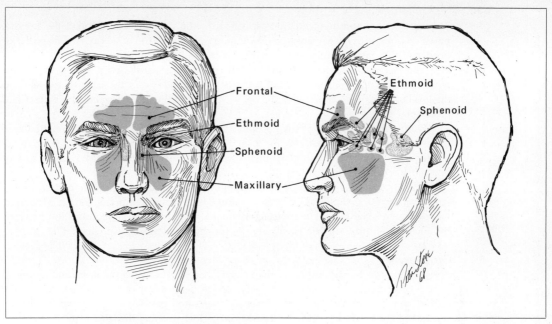

Figure 20-5. Paranasal sinuses.

The frontal sinuses cause the bulging of the frontal bone, or supraciliary ridges, above the eyebrows. They are not present at birth but develop during childhood and become prominent after puberty as secondary sexual characteristics of the male.

The sphenoid sinuses occupy the body of the sphenoid bone and lie inferior to the sella turcica. They are usually bilateral.

The maxillary sinus is by far the largest of the paranasal sinuses. It lies in the maxillary bone between orbital and oral cavities. Sockets holding the upper molar teeth project into this sinus, and thus a tooth abscess may cause infection and inflammation of the sinus.

Fluid secreted by the mucous membrane lining the sinuses drains through small apertures onto the lateral wall of the nasal cavity (see Figure 20-4). The sphenoid sinus drains into the sphenoethmoid recess. The ethmoid sinuses drain into this recess as well as into the superior meatus. The opening of the maxillary sinus is a C-shaped slit, the hiatus semilunaris. This is present at the periphery of an ethmoid air cell, the bulla ethmoidalis, which bulges into the middle meatus. The frontal sinus also drains, by the funnel-shaped infundibulum, into the hiatus semilunaris of the middle meatus. The nasolacrimal duct passes through the maxillary bone, and through this duct lacrimal fluid flows into the inferior meatus (see Chapter 12).

Drainage of the sinuses may be blocked by an allergic or inflammatory swelling of the mucous membrane. As a result, pressure built up within the sinuses may cause severe headaches. If sinusitis is a chronic condition, it may be necessary surgically to open the wall of the sinus. Such a procedure is commonly referred to as "draining the sinuses."

Nasopharynx

The nasal cavity communicates posteriorly with the nasopharynx through the internal nares, or choanae. These openings are bounded by bones of the nasal cavity. The posterior border of the nasal septum separates the right and the left choanae.

The nasopharynx, located immediately behind the nasal cavity, is the uppermost portion of the muscular pharyngeal tube, which extends from the base of the skull to the level of the cricoid cartilage in the neck. Muscles of the pharyngeal wall are described in Chapter 6.

On the posterior wall of the nasopharynx there is an aggregate of lymphatic tissue termed the pharyngeal tonsil (Figure 20-4). If it beomes inflamed and enlarged, the tissue is called the adenoids. The enlargement may actually block the internal nares, so that one must breath through the mouth.

The cartilaginous portion of the pharyngotympanic (eustachian) tube opens into the lateral wall of the nasopharynx. It raises a mound of mucous membrane, the torus tubarius, where it enters the wall. The pharyngotympanic tube forms a communication between the pharynx and the middle ear cavity. Its function is described in Chapter 13.

Oropharynx

The oropharynx, or midportion of the pharynx, extends between the nasal and laryngeal portions. Situated behind the oral cavity, between the level of the soft palate and the aditus (entrance) of the larynx, it serves as a common passage for food and air. At the termination of the oropharynx, food passes posteriorly to enter the laryngopharynx, which is continuous with the esophagus, and air passes through the larynx to reach the trachea. The junction of the oropharynx with the oral cavity is described in Chapter 24.

Larynx

The larynx, or voice box, is the air passageway from the oropharynx to the trachea (Figure 20-6). It is modified for the production of sound. The walls of the larynx, lined by respiratory epithelium, are formed by cartilages, muscles (see Chapter 6), and ligaments.

Three unpaired cartilages (epiglottic, thyroid, and cricoid) and three paired cartilages (arytenoid, corniculate, and cuneiform) make up the skeleton of the larynx. During swallowing, the larynx is elevated so that the arytenoid cartilages, at the posterior aspect of the aditus, are tipped forward against the epiglottis, at the anterior boundary of the aditus. Without this closure of the larynx, food can pass into the larynx, where it will elicit violent coughing.

The broad laminae of the thyroid cartilage form most of the wall of the larynx. The laminae meet in front of the neck as the laryngeal prominence, or "Adam's apple." The thyroid cartilage is so named because the thyroid gland covers its lower portions. The cricoid cartilage is attached by connective tissue and muscle to the highest tracheal ring. Posteriorly the upper surface of the cricoid articulates with the base of the paired arytenoid cartilages. Each of the latter has two important processes. The blunt muscular process at the side of the base gives attachment to most of the intrinsic muscles of the larynx. The anteriorly directed, spinelike vocal process gives attachment to the vocal cord on each side of the larynx.

The aryepiglottic and cricothyroid membranes contribute to the walls of the larynx. The former covers the aryepiglottic muscle, which extends between the arytenoid and epiglottic cartilages to form the side of the aditus. The cricothyroid membrane completes the wall of the larynx between the inferior border of the thyroid cartilage and the cricoid cartilage.

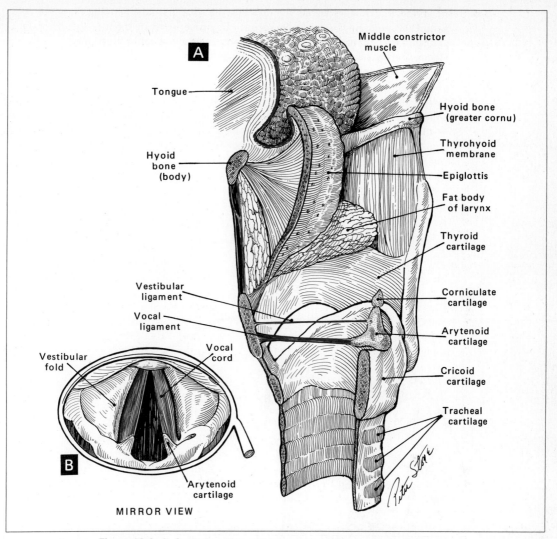

A
Tongue

Hyoid bone (body)

Vestibular ligament

Vocal ligament

Vestibular fold

Vocal cord

B

Arytenoid cartilage

MIRROR VIEW

Middle constrictor muscle

Hyoid bone (greater cornu)

Thyrohyoid membrane

Epiglottis

Fat body of larynx

Thyroid cartilage

Corniculate cartilage

Arytenoid cartilage

Cricoid cartilage

Tracheal cartilage

Figure 20-6. A. Sagittal section of the larynx in relation to the tongue and trachea. **B.** The vocal folds, relaxed as in quiet breathing, as seen by means of a laryngoscope.

The interior of the larynx, or laryngeal cavity, is divided into three parts by two pairs of horizontal folds. The upper pair are the vestibular, or false, vocal folds and the lower pair the true vocal folds or cords. The portion of the larynx superior to the false vocal folds is the vestibule. The area between the false and true vocal folds is the ventricle. The third portion of the larynx is below the true vocal folds and is continuous inferiorly with the trachea.

The prism-shaped vocal cords extend anteroposteriorly from the vocal process of each arytenoid cartilage to the inner aspect of the thyroid cartilage. Any movement of the arytenoid cartilages affects the opening, the rima glottidis, between the vocal cords. In quiet respiration the cords lie so that the space between them is the shape of an elongated triangle with the apex at the front. In heavy or rapid breathing, this aperture widens. During swallowing, the cords are tightly closed to block the passage of anything that may have entered the larynx.

The forward and backward movement of the arytenoid cartilages changes the pitch of the voice. The vocalis muscles enclosed within the vocal cords, by effecting fine adjustment in the tension on the cords, inflect or modulate the voice. (See Chapter 21 for more detail about production of sound.)

Figure 20-7. A. Bronchopulmonary tree. Numbers on the tertiary bronchi refer to segments identified in **B. B.** Bronchopulmonary segments.

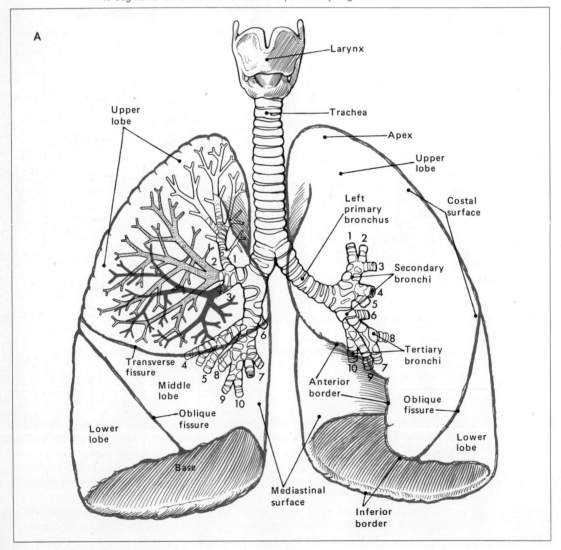

Inflammation, as from excessive use of the vocal cords, limits their normal contraction and causes hoarseness. A massive inflammation may result in laryngitis with temporary loss of voice. Chronic irritation as from heavy smoking may cause a permanent change in voice quality.

Trachea

The trachea is a muscular tube in the wall of which are embedded 16 to 20 C-shaped bands of cartilage, the tracheal rings. The trachea extends from the cricoid cartilage to about the level of the fourth or fifth thoracic vertebra, where it bifurcates into the two primary bronchi. The trachea is held open by cartilages of the tracheal wall so that air can pass through it. The trachealis muscle completes the posterior wall of the trachea, closing the gap between the ends of the tracheal rings.

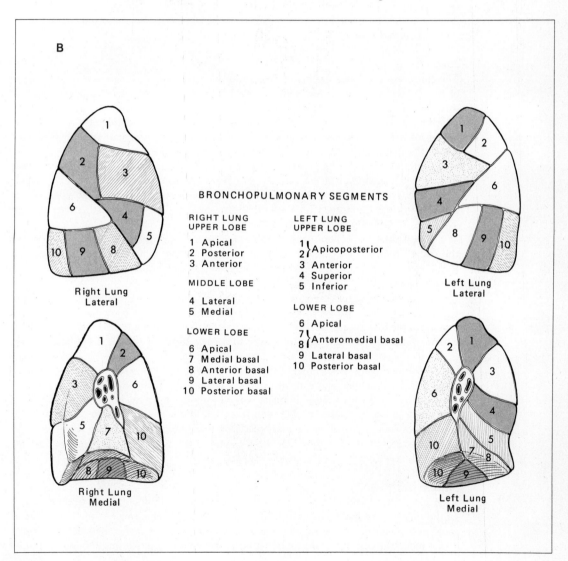

B

BRONCHOPULMONARY SEGMENTS

RIGHT LUNG
UPPER LOBE

1 Apical
2 Posterior
3 Anterior

MIDDLE LOBE

4 Lateral
5 Medial

LOWER LOBE

6 Apical
7 Medial basal
8 Anterior basal
9 Lateral basal
10 Posterior basal

LEFT LUNG
UPPER LOBE

1⎫
2⎬Apicoposterior
3 Anterior
4 Superior
5 Inferior

LOWER LOBE

6 Apical
7⎫
8⎬Anteromedial basal
9 Lateral basal
10 Posterior basal

Right Lung
Lateral

Left Lung
Lateral

Right Lung
Medial

Left Lung
Medial

If the airway above the trachea is blocked, an alternative airway can be created in the trachea by tracheostomy, that is, by cutting a hole between the tracheal rings, after which a tube is usually inserted.

Bronchial Tree

The bronchi are branching tubes that extend from the trachea deep into the lungs. Walls of the larger bronchi have a series of incomplete rings of cartilage embedded in their walls. As the bronchi diminish in size, the cartilages become small plaques; at the bronchiolar level, they disappear completely.

The trachea first divides into two primary bronchi which pass to the right and left lungs (Figure 20-7). The right primary bronchus divides into three secondary bronchi which extend to the three lobes of the right lung. The left primary bronchus terminates into two secondary bronchi for the two lobes of the left lung. Secondary bronchi subsequently divide into tertiary, or terminal, bronchi which pass to the bronchopulmonary segments in each lung.

Lungs

The lungs are the basic organs of respiration. There are two lungs, one on each side of the thoracic cavity. The lungs are shaped like bisected cones with the apices pointing upward and fitting into the space bounded by the upper ribs, sternum, and vertebrae. The lungs are concave at the base and ride on the dome-shaped diaphragm. The inferior border of the base extends like a wedge into the costodiaphragmatic recess, the place where the diaphragm arises from the inferior margin of the rib cage.

Lying between the lungs in a space termed the mediastinum are the heart, great vessels, esophagus, part of the trachea, the bronchi, and some nerves. The lung surface facing this space, that is, facing medially, is the mediastinal surface, and at its center is an area called the hilus, which admits the vascular and respiratory structures serving the lung. Collectively these structures form the root of the lung.

At the anterior border of the lung the mediastinal surface bends sharply to become the costal surface. This surface is convex. It follows the contour of the rib cage back to the vertebral column, where at the blunt posterior border of the lung the mediastinal surface begins.

There are deep grooves, or fissures, which separate each lung into lobes. The oblique fissure of each lung passes downward and toward the midline to demarcate an upper lobe from the lower lobe. On the right lung there is in addition a transverse fissure which extends forward from the oblique fissure to form a third, or middle, lobe between the upper and lower lobes.

The functional subdivisions of the lobes of the lung are termed bronchopulmonary segments. Each of the segments surrounds a tertiary bronchus and its branches and is supplied by a branch of the pulmonary artery. Pulmonary veins, however, do not drain a single segment but rather lie between them and drain adjacent segments. Because of these subdivisions, bronchopulmonary resection—surgical removal of lung segments—has become a preferred procedure for certain conditions, as pathological lung conditions frequently involve only one or two segments.

Pleural Membranes

Each lung is enclosed in a sac consisting of a double-walled serous membrane known as the pleura. The inner (visceral) layer adheres firmly to the lung. At the hilus it folds back to become the parietal layer, which lines the thoracic cavity. Between the layers is the pleural cavity. It should be understood that the lung is within this double-walled sac, not within the pleural cavity. The lung is like a fist pushed deeply into a balloon, except that the pleural layers are in close contact. The space between the layers is potential rather than actual. Movement between the layers is facilitated by a lubricating serous fluid secreted by cells of the membrane.

Pleuritis, or pleurisy, is an inflammation of the pleural membranes. It usually causes an increase of fluid in the pleural cavity and may result in layers adhering to each other (pleural adhesions). Normal respiratory movements will then be painful, and to separate the layers, surgery may be required.

QUESTIONS AND PROBLEMS

1 What anatomical features of the larynx, trachea, and larger bronchi serve to maintain a patent airway?

2 Describe how the medial wall of the nasal cavity is modified for greater surface area. What is the functional significance of this modification?

3 Locate the paranasal sinuses. Where do they drain?

4 Trace the course of an infection in the sore throat to the middle ear and to the mastoid air cells.

5 Explain the prominence of the "Adam's apple" in the male.

6 What surgical measure is possible if the airway is closed in the larynx or above?

7 How is knowledge of the bronchopulmonary segments helpful to someone assisting a patient with postural drainage?

8 Define the following terms:
   a Mediastinum
   b Pleural cavity
   c Alveolus
   d Bronchopulmonary segment
   e Secondary bronchus

9 Name the contents of the mediastinum and pleural cavity.

21 VENTILATION

The purpose of ventilation is to freshen the air in the lungs, that is, to replace lung air with air that has more oxygen and less carbon dioxide.

MOVEMENT OF AIR

During inspiration at rest about 500 ml of air moves with each breath through the passageways to enter the lungs. From this volume of air, about 15.6 ml of oxygen diffuses into the blood and, at the same time, approximately 12.5 ml of carbon dioxide diffuses in the opposite direction. During expiration this altered air leaves the lungs. The cycle is repeated about 16 times per min, which is the normal respiratory rate. In other words, the average-sized individual utilizes about 250 ml of oxygen per min (15.6×16) and expels some 200 ml of carbon dioxide (12.5×16), while moving 8,000 ml per min through his lungs.

Inspiration

Inspiration means "breathing in." The mechanism responsible for the movement of air into and out of the lungs must be understood. During inspiration the capacity of the thoracic cage increases.

The importance of diaphragmatic movements in breathing follows from the fact that the surface area of the diaphragm is about 270 cm^2. Thus the diaphragm needs to descend only 1 cm in order to enlarge the thoracic capacity by 270 cm^3. In quiet respiration the diaphragm moves about 1.5 cm, but with greater respiratory efforts it may move 7 cm. A diaphragmatic descent of 1.5 cm will enlarge the thoracic capacity 405 cm^3 (1.5 × 270). In other words, the diaphragm alone accounts for about 80 percent of the inspired volume (400/500).

The other 20 percent is accounted for by rib movement. During inspiration the ribs swing upward because of their hinge-type articulation with the vertebral column and the arrangement of the muscles. When the ribs and the sternum are raised, the capacity of the thorax is increased (Figure 21-1). As the thorax enlarges, so do the lungs.

Mechanism of Inspiratory Movements   In inspiration both the diaphragm and the external intercostal muscles contract. Contraction of the former lowers the dome. Contraction of the latter elevates the ribs. The diaphragm is innervated by the phrenic nerves. These originate at the third, fourth, and fifth cervical levels, where they synapse with axons from cell bodies in the medulla. The external intercostals also are innervated by fibers that synapse with axons from the medulla. For most inspiratory demands, contraction of the diaphragm and intercostals suffices. However, during great inspiratory efforts, the scalene and the pectoralis minor muscles contribute by raising the rib cage to its ultimate position.

Surface Tension   The two pleural layers are in intimate contact, and they are moist. Thus, all along the surfaces of these layers there is molecular attraction which opposes separation

Figure 21-1. Respiratory mechanism. **A.** Forced expiration, with contraction of the internal intercostal muscles. **B.** Normal inspiration. The external intercostal muscles and the diaphragm contract and together increase the volume of the thoracic cavity.

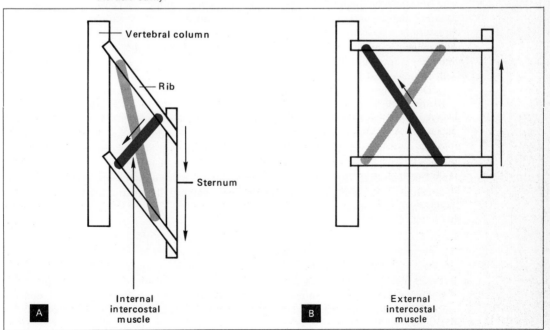

of the layers. The force exerted by this surface tension is considerable. Some idea of its magnitude may be gained by attempting to separate two flat, wet plates of glass. Surface tension causes the lung surface to follow the thoracic wall.

The volume of fluid in the pleural space is extremely difficult to measure. It has been reported to be about 2 ml for each lung. Under normal circumstances there are only a few milliliters of fluid present, but its presence is essential for lubrication. During inspiration and expiration the two pleural surfaces slide against one another, and were it not for the pleural fluid, adhesions would form.

The inner surfaces of the alveoli are also moist. Consequently, there is surface tension between the alveolar walls which tends to collapse the lungs and to oppose their expansion. There is, however, a substance present in the lungs called surfactant which decreases this surface tension. Surfactant is a lipoprotein mixture. The important constituent is lecithin, a phospholipid. Were surfactant absent, surface tension within the lung would become greater as the lung became smaller during expiration. This could then lead to collapse of the lung or a part of it, a condition referred to as atelectasis.

Babies, especially premature ones, are sometimes born with inadequate quantities of surfactant. When this is the case, they experience respiratory difficulties which may prove fatal. This disorder is termed hyaline membrane disease.

Intrapleural Pressure   Even in the absence of surface tension between the thoracic and lung walls, the two would not separate very much during inspiration, because as soon as the thoracic wall begins to pull away from the lung wall, the pressure within the pleural space decreases. This follows from Boyle's law, which states that the pressure of a gas is inversely proportional to its volume as long as the temperature is kept constant. The pleural space is closed. As it enlarges, the volume of the gas therein must also increase, and according to Boyle's law its pressure decreases. The pressure in the pleural space, that is, the intrapleural pressure, is normally below atmospheric pressure. Even at the end of expiration the intrapleural pressure is about -4 mmHg because the thoracic cage is larger than the lungs. Thus, even at that moment the lungs are stretched. The negative pressure is a measure of their elasticity plus surface tension in the resting position. If an opening is made in the chest wall to permit air to enter the intrapleural space, the intrapleural pressure will be equal to the atmospheric pressure, and the lung on that side will contract, that is, collapse.

During normal inspiration the expansion of the thoracic cage causes the intrapleural pressure to decrease progressively from -4 mmHg to about -6 mmHg. The lungs are expanded because of the surface tension and the intrapleural pressure. The pressure within the lungs, that is, the intrapulmonic pressure, decreases due to the enlarged volume.

Because the intrapulmonic pressure falls below atmospheric pressure, air moves into the lungs. The lungs and chest do not expand because of the entry of air; it is just the opposite. The difference between atmospheric and intrapulmonic pressure is the force that drives air into the lungs. The greater this difference, the greater the volume and the more rapidly it moves.

Expiration

During expiration the thoracic cage capacity decreases. This permits the lungs, which are stretched during inspiration, to contract. Both the intrapleural and the intrapulmonic pressure increase. The intrapulmonic pressure now rises above atmospheric pressure. Once again there is a pressure differential between the air in the lungs and the outside air. But now the relationships have been reversed, and therefore air moves out of the lungs. During passive expiration, the contraction of the lungs stays ahead of the thoracic movements; thus the intrapleural pressure remains subatmospheric.

Very different pressure relationships prevail during forced expiration. There is active muscular contraction which rapidly and forcibly decreases the size of the thoracic cage. The thoracic wall drives the lung wall so that the intrapleural pressure increases sharply.

Mechanism of Expiratory Movements The usual, so-called quiet expiration simply involves inhibition of the inspiratory neurons. They cease to fire; therefore the external intercostals and the diaphragm relax. This permits the rib cage to be lowered while the dome of the diaphragm comes up into the thoracic cavity. But during vigorous expiration there is contraction of the internal intercostal muscles, which pulls the ribs downward. In addition, the abdominal muscles contract, increasing intraabdominal pressure. This pushes the diaphragm even higher into the thoracic cavity, increasing intrathoracic pressure that much more.

Respiratory Air Volumes

The average, normal individual has a lung capacity of about 6,000 ml. This total is made up of several components (Figure 21-2). The quantity of air normally inspired with each breath, while at rest, is termed tidal volume, or tidal air. As already mentioned, it averages about 500 ml. Figure 21-2 indicates that there is air in the lungs at the beginning of a normal inspiration and space for additional air at the end of inspiration.

By a forceful effort, at the end of a normal expiration, an additional 1,000 ml can be expelled from the lungs. This additional quantity is termed the expiratory reserve volume. Similarly, after a normal inspiration, one can continue to inspire and move an additional 3,000 ml of air into the lungs. The difference between the lung volume at the end of a normal inspiration and at the end of a maximal inspiration is termed the inspiratory reserve volume.

Because the lungs cannot be completely collapsed by thoracic movements, there is still some air in the lungs even after a maximal expiration. This quantity, termed the residual volume, amounts to about 1,500 ml.

Lung Capacities

Various combinations of the respiratory air volumes make up the so-called lung capacities (Figure 21-2). All combined equal total lung capacity, which, as already mentioned, is about 6,000 ml. The total amount of air that can be moved into or out of the lungs is termed the vital capacity. It is less than the total lung capacity because of the residual volume, which cannot be moved. Vital capacity averages about 4,500 ml and includes the tidal volume, the inspiratory reserve volume, and the expiratory reserve volume.

At the end of normal expiration there is about 2,500 ml of air left in the lungs. This is termed the functional residual capacity and is made up of the expiratory reserve volume and the residual volume.

Finally, from resting position, that is, at the end of expiration, about 3,500 ml of air can be drawn into the lungs. This amount is the inspiratory capacity and consists of the tidal volume and the inspiratory reserve volume. All these values are summarized in Table 21-1.

Dead Space

No exchange of oxygen or carbon dioxide takes place in the respiratory passageways from the nostrils down to the terminal bronchioles. The volume of air contained by these

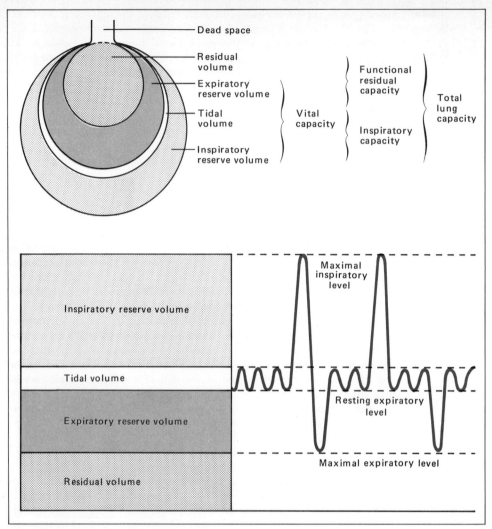

Figure 21-2. Respiratory air volumes and lung capacities. (*Reproduced with permission of Dr. Julius Comroe.*)

passageways is called the anatomical dead space. On the average, the volume of the anatomical dead space amounts to about 150 ml. Thus, if 500 ml of air is inspired, only about 350 ml reaches the alveoli, where the exchange of gases occurs. During expiration the first 150 ml expired will be found, upon analysis, to contain the same percentages of respiratory gases as does room air. That is to say, since this air did not come into contact with the alveoli, there has been no gas exchange, and thus its composition remains unchanged. The anatomical dead space can be varied only moderately, by bronchiolar contraction or dilation.

The physiological dead space, also called virtual or effective dead space, is defined as the space within the lungs and passageways which, just prior to expiration, contains

TABLE 21-1. LUNG VOLUMES AND CAPACITIES

TERM	VOLUME, ML
Total lung capacity	6,000
Residual volume	1,500
Expiratory reserve volume	1,000
Tidal volume	500
Inspiratory reserve volume	3,000
Vital capacity	4,500
Functional residual capacity	2,500
Inspiratory capacity	3,500

air of the same composition as room air. The anatomical and physiological dead spaces are about the same in the normal person. But should some of the alveoli become non-functional or should pulmonary circulation become inadequate, the physiological dead space would be larger. The physiological dead space, when such abnormalities exist, may be a liter or more, compared with the anatomical dead space of 150 ml. Obviously, in such a person, the normal tidal volume of 500 ml would never suffice.

Anatomical dead space is generally measured by use of a nitrogen meter and recorder. The subject is fitted with a mask which is attached to the nitrogen meter. A record is thus obtained of the nitrogen content of the inspired and expired air. Now the subject is caused to inspire pure oxygen. The recording pen falls to the zero line because there is no nitrogen in the inspired air. During the first part of expiration there is still no nitrogen in the expired air, but then the percentage of nitrogen rises sharply to a plateau. The nitrogen, of course, comes from the alveolar air. From the record, the tidal air and anatomical dead space can be easily determined (Figure 21-3).

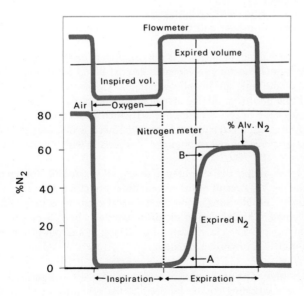

Figure 21-3. Determination of anatomical dead space. A single deep breath of pure oxygen is taken and then expired through a nitrogen meter and a flowmeter. The position of the square wave in the lower curve indicates the volume of anatomical dead space which is read from the upper curve. *(Reproduced with permission of Dr. Julius Comroe.)*

The physiological dead space may be calculated if the concentration of one of the respiratory gases in the expired air and in alveolar air, as well as the volume of the expired air, is known. For example, if the amount of carbon dioxide in the alveolar air is 6 percent, the amount in the expired air is 4 percent, and the volume of expired air is 500 ml:

$$\frac{500}{6} = \frac{x}{4} = 334$$

Thus, 334 ml represents the quantity of alveolar air in the expired air. The physiological dead space then is 166 ml (500 − 334).

The anatomical dead space is increased if the subject breathes through a tube. For example, if one breathes through a tube with a capacity of several hundred milliliters, quite clearly the total dead space will be greater than the normal tidal volume and, unless greater respiratory efforts are made, there will be no renewal of air in the alveoli.

Nonrespiratory Air Movements

Coughing   Usually the movement of the cilia suffices to keep the air passageways clear. If this mechanism is unequal to the task, a reflex cough occurs. In coughing, first about 2 liters of air is inspired, and then the glottis is closed and a forceful respiratory effort is made. This causes a sharp increase to as much as 100 mmHg in intrapulmonary pressure. In fact, the pressure in the entire thoracic cage is increased and, as a result, the membranous back wall of the trachea collapses when the glottis is suddenly opened. Due to the great differential between atmospheric and intrapulmonary pressure, air is driven through the passageways, and because the trachea has been narrowed by the collapse of the membranous back wall, the velocity of the air movement through the constricted trachea is very great. This high-velocity, high-pressure blast of air often serves to remove the obstruction.

Sneezing   A sneeze has been described as an upper respiratory tract cough. A large volume of air is inspired and then expelled with great force through an open glottis. Momentary resistance is provided in the mouth and nasal passageways. The resistance is then suddenly diminished, and the blast of air serves to clear the mouth and nose.

Yawning   The mechanism of yawning is not completely understood. The reflex is thought to be triggered by inadequately oxygenated arterial blood. During normal breathing not all the alveoli of the lung are equally ventilated. Consequently, the blood that passes in contact with the poorly ventilated alveoli is not completely oxygenated. The resulting yawn produces a very deep inspiratory effort which flushes the lung, so to speak, and completely ventilates all the alveoli. Another hypothesis is that the great respiratory effort squeezes the stagnant blood out of pulmonary vessels in which it may have accumulated.

Hiccuping   Occasionally, the diaphragm develops spasmodic contractions. Why this occurs is not known. As a result of the spasmodic contraction, a volume of air is suddenly and rapidly drawn into the lungs, causing the vocal cords to vibrate and to produce the characteristic sound. The inhalation of air containing about 5 percent carbon dioxide usually halts hiccuping. The classic cure of breathing in and out of a paper bag depends upon the accumulation of carbon dioxide.

Production of Sound   The production of sound, or phonation, requires three mechanisms: (1) a force to set a vibrating part in motion, (2) the vibrating part, and (3) one or more resonators which serve to reinforce the vibrations. In the piano, for example, the force is provided by the fingers which strike the keys, the strings are the vibrators, and the sounding board serves as a resonator. In man, the force is the blast of air which is ejected from the lungs and the vibrating bodies are the true vocal cords, while the pharynx, mouth, and nose collectively act as resonators.

The volume and force of air movement through the larynx are controlled by the respiratory muscles, including the diaphragm. First a volume of air is inspired; then the vocal cords are brought together. Contraction of the expiratory muscles builds up the intrapulmonic pressure. The vocal cords then separate, which permits a blast of air to flow by. This rapid flow of air creates a negative pressure, which may seem odd, but one need only recall how a loose shower curtain swings *in* toward the faucet when the water is flowing rapidly to see that such a relationship does indeed create a negative pressure. The negative pressure between the vocal cords causes them to close, the pressure below them once again builds up, and they are forced open. In other words, the flow of expiratory air causes the vocal cords to vibrate and therefore to produce sound.

The stronger the blast of air, the greater the amplitude of the vibrations. The amplitude of vibrations determines the intensity, or loudness, of the sound.

The pitch of a sound is a function of the rate of vibration of the vocal cords. The rate of vibration is regulated by the tension of the cords and the length of the vibrating surface. Both can be varied. Therefore, the human voice has a tone range of several octaves.

The resonating apparatus—pharynx, mouth, and nose—contributes to the quality, or timbre, of the sound. These resonating structures also contribute to articulation, that is, the formation of vowels and consonants. The role of the resonators in regulating the quality of the sound is made apparent when one has nasal congestion, has a defect in the palate, or has lost his teeth.

TESTS OF PULMONARY FUNCTION

Many tests of pulmonary function have been designed. Some of these procedures, however, are so complex or require such elaborate equipment that they are seldom used outside the sophisticated research laboratory. Only a few of the more common tests will be described.

Vital Capacity

Vital capacity can easily be measured with a volume recorder such as a spirometer (see page 640). The subject simply inspires maximally and then expires as much as he can into the spirometer. The average value is 4.5 liters, but this varies considerably from person to person depending upon size and several other factors. In most instances there is no record of the patient's vital capacity through the years or prior to the present illness. Therefore, the obtained value is compared with a predicted value. The prediction is based upon a consideration of height and age. Equations are available for this purpose. Usually a decrease in vital capacity indicates a reduction in functioning lung tissue as occurs with patients with pneumonia, carcinoma, fibrosis, or pulmonary congestion. On the other hand, the decrease may be due simply to an abnormality that restricts thoracic movements.

Maximal Breathing Capacity

The maximal breathing capacity is defined as the greatest volume of gas that can be breathed per minute. This value is obtained by instructing the subject to breathe as deeply and as rapidly as possible for 15 sec into a recording system. As little resistance as possible should exist in the apparatus. Even then each laboratory should conduct its own series of normal values in order to calibrate the equipment. If the subject cooperates so as to give a maximal effort and if the machine is well calibrated, the test is highly significant.

By the use of equations which take into consideration surface area, age, and sex, maximal breathing capacity may be predicted and compared with the measured value.

A young, healthy, male adult can move as much as 170 liters/min of air in this 15-sec test. However, for longer periods of time not much more than 100 liters/min can be sustained. Nevertheless, in comparison with the resting level of 8 liters/min, this is a tremendous feat and illustrates the great reserve of the pulmonary system.

Although a reduction in the maximal breathing capacity is not indicative of any single disorder, it does suggest a reduction in muscular force or increased airway resistance. If the maximal breathing capacity is reduced out of proportion to the decrease in vital capacity, an obstruction of the airways, or emphysema, is indicated. The ratio between vital capacity and maximal breathing capacity is called the air-velocity index. This index is obtained by dividing the percent of predicted maximal breathing capacity by the percent of predicted vital capacity.

Timed Vital Capacity

Usually the same information can be obtained from the so-called timed-vital-capacity test as from the maximal-breathing-capacity evaluation. And since this test is far easier and less exhausting for the patient, it is usually preferred. The subject simply makes a maximal forced expiration into a spirometer. The apparatus must be equipped with a device that records the total expired volume and also that emitted in 1, 2, and 3 sec. The normal person is able to expire at least 83 percent of his vital capacity in 1 sec, 94 percent in 2 sec, and 97 percent in 3 sec. In cases of obstructive pulmonary disease, these values will be reduced.

Arterial Blood Measurements

The final proof whether or not the respiratory system is performing its task lies in the status of the respiratory gases in the arterial blood. To be sure, normal arterial values may be maintained, and usually are, in the face of pulmonary dysfunction but at a price of greater respiratory effort and sharply reduced respiratory reserve. For this reason, pulmonary-function tests as well as arterial blood measurements must be carried out and evaluated in the light of one another.

ARTIFICIAL RESPIRATION

Since artificial respiration may prove lifesaving and because deprivation of oxygen to the brain for more than a few minutes causes irreversible damage, artificial respiration should be effectively utilized without delay. It is indicated in any condition in which breathing has stopped but circulation continues.

Many methods of artificial respiration have been designed. For years the so-called Schafer (prone-pressure) method was taught but was eventually shown to be ineffective. Then the arm-lift–back-pressure method was developed, but while it is certainly more effective than the Schafer procedure, unless there are contraindications, mouth-to-mouth breathing is now the method of choice.

Mouth-to-Mouth Resuscitation

As the term indicates, one simply places his mouth to that of the victim and forcibly breathes air into his lungs. The advantage of the method lies in the ease and quickness with which it can be done and also in the very large volume of air that can be moved through the victim's lungs. The minute volume achieved by any method is the decisive factor to take into consideration. In mouth-to-mouth breathing, one quickly inhales a large volume of air and then breathes forcibly into the victim's mouth. Usually the normal recoil of his thoracic cage and lungs will suffice to expire the air. Thus, in most cases, it is necessary to breathe into the victim's mouth only at a rate comparable to the normal rate, that is, about 16 times/min, or perhaps slightly faster.

Another advantage to this method is that the air forced into the victim's lungs contains some carbon dioxide. This prevents loss of carbon dioxide from his blood due to hyperventilation and aids in speeding the return to spontaneous breathing.

The disadvantage of the method is that the victim is generally on his back and therefore there is danger that his tongue may fall back into the back of his throat and obstruct the airway. Also in some cases, the condition of the patient may make mouth-to-mouth contact impossible or undesirable.

Arm-Lift–Back-Pressure Method

In this procedure the victim is placed face down in the prone position. The elbows are bent and the hands placed one upon the other. The face is turned to one side with the cheek upon the hands. The arms are drawn upward and forward for inspiration (Figure 21-4). This serves to expand the chest. For expiration, pressure is applied to the back. The arm-lift–back-pressure method, if carried out properly, can move about a liter of air and thus is quite effective.

Mechanical Aids

The best type of apparatus for artificial respiration is one that forces a controlled volume of air into the lungs and then withdraws it by negative pressure. Thus, there is no danger of overfilling the lungs and perhaps causing rupture. In addition, air is forced into and out of the lungs. This moves a large volume even when partial obstruction exists. One danger, however, is increasing intrapulmonary pressure enough to impede venous return to the heart. If such positive pressures are not long maintained, there is no problem, and the subsequent negative pressure will enhance the venous return. But prolonged positive intrapulmonary pressure can prove fatal.

Figure 21-4. Artificial respiration by the arm-lift–back-pressure method. At left is the inspiratory phase; at right, the expiratory phase. (*Photographs by B. Salb*)

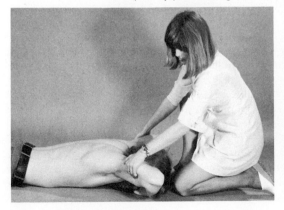 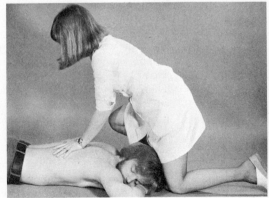

The usual resuscitator used for artificial respiration is the push-pull type. Another type of apparatus is indicated when prolonged mechanical aid to breathing is necessary, for example, with patients whose respiratory muscles have been paralyzed. In such cases, a tank respirator is preferred because it leaves the head free and room air is breathed. The patient's body lies inside the tank with his head protruding. An airtight diaphragm is placed around his neck. At the other end of the tank or underneath the tank there is a diaphragm that is moved back and forth by mechanical means. The movement of this diaphragm, just like a human diaphragm, changes the pressure in the tank. When the pressure in the tank becomes positive, the thorax is compressed and air is forced out of the lungs. When the tank pressure is negative, the thorax expands and inspiration occurs. The pressure and rate can be varied according to the needs of the patient.

ABNORMALITIES OF VENTILATION

A wide variety of abnormalities reduce the effectiveness of the breathing mechanism. Only a few will be mentioned.

Paralysis

Paralysis of the respiratory muscles interferes with ventilation. The degree of interference depends upon which muscles are involved. The movement of air is normally the result of coordinated activity of several muscles. The paralysis of some of them does not necessarily make it impossible to move adequate quantities of air. For example, the diaphragm alone can suffice, or the abdominal muscles can be utilized to move a flaccid diaphragm.

Paralysis of the respiratory muscles may be caused by damage to the respiratory center in the medulla, spinal cord injury, or poliomyelitis, which is an acute viral disease. Generally the virus afflicts the motor neurons of the spinal cord, which weakens or paralyzes the respiratory muscles. If adequate ventilation becomes impossible, the tank respirator must be used.

Pneumothorax

If air enters a pleural cavity—a condition termed pneumothorax—the lung collapses. The collapsed lung takes no further part in breathing movements. A lung is sometimes collapsed purposely for the treatment of tuberculosis. The collapsed lung rests, which apparently is helpful in combating the infection. Once the opening into the pleural cavity is closed, the air therein contained is slowly absorbed and the lung expands and becomes active once again.

Bronchial Obstruction

The bronchus may be blocked by a foreign body, an inflammatory structure, an acute swelling, a tumor, or merely a mucous plug. It may also be occluded by external pressure caused by some abnormal process such as malignant disease of the thyroid gland. The alveoli beyond the site of blockage collapse because the air in them is absorbed and a new supply cannot enter.

Pulmonary Fibrosis

In the condition known as pulmonary fibrosis the elastic tissue of the lungs is replaced by fibrous tissue as a result of certain disease processes. The lungs are thus less distensible and less elastic. Respiration becomes difficult, and the maximal breathing capacity is sharply reduced. There is a large residual air volume with a small tidal volume. Consequently, the blood is inadequately oxygenated, and blood carbon dioxide increases. Both changes, as will be explained later, stimulate deeper and faster breathing movements. In other words, breathing is labored, a condition termed dyspnea.

QUESTIONS AND PROBLEMS

1 What is the respiratory rate for an average-sized person at rest? How much oxygen is breathed per minute?

2 What muscles are involved in maximal inspiration?

3 Which nerve activates the diaphragm? Where does the respiratory impulse originate?

4 Describe the pathology of hyaline membrane disease.

5 What is the role of intrapleural pressure in maintaining proper lung functioning?

6 Why does intrapulmonic pressure fall during inspiration? What happens as a result?

7 Why is the term "passive" used to describe normal expiration?

8 What are the components of total lung capacity?

9 Of what diagnostic value is a test of vital capacity? Of arterial blood measurement?

10 Explain why there is a large residual air volume and small tidal volume in pulmonary fibrosis.

11 Explain why physiological dead space is increased both when some of the alveoli become nonfunctional and when pulmonary circulation is inadequate.

12 What is the difference in mechanism between a sneeze and a cough?

13 Describe the production of sound, identifying the force, the vibrating element, and the resonators.

14 What tissue in the body is most sensitive to oxygen deprivation?

22 DIFFUSION AND TRANSPORT OF RESPIRATORY GASES

The first step in respiration is ventilation, the constant replenishment of the air in the lungs. The next step is the movement of oxygen from the alveolar air to the blood and the movement of carbon dioxide in the reverse direction. The blood leaves the lungs low in carbon dioxide and saturated with oxygen. Throughout the body, in the capillary beds there is again movement of the respiratory gases: oxygen out of the blood into the cells and carbon dioxide from the cells into the blood. These vital processes must now be explained.

DIFFUSION OF RESPIRATORY GASES

At one time, physiologists thought that the epithelial membranes separating the alveolar air and the pulmonary blood possessed a secretory ability which served to move the gases. This concept is no longer tenable. The factors which determine the diffusion of the respiratory gases through these membranes are (1) partial pressure of the gas, (2) permeability of the membranes, (3) size of the alveolar surface area, (4) rate of pulmonary circulation, and (5) the chemical reactions in the blood.

Physical Principles

Blood pressure is expressed in millimeters of mercury. The pressure of gases is designated in the same units. Atmospheric pressure simply reflects the weight of a column of air

549

DIFFUSION
AND
TRANSPORT
OF
RESPIRATORY
GASES

above the earth. Air possesses mass, and thus is attracted by gravity. It presses down on the earth's surface, and at sea level it will support a column of mercury 760 mm high. Therefore, it is said that the normal atmospheric pressure at sea level is 760 mmHg. But what is the pressure of oxygen in the air? And what influence does temperature have on volume and pressure of the gas? In order to answer these questions, the pertinent gas laws will be considered briefly.

Dalton's Law of Partial Pressure Let us assume that a volume of oxygen is introduced into a bag and the pressure measured and found to be 100 mmHg. Now if an equal volume of nitrogen is added to the same bag, it too will exert a pressure equal to 100 mmHg. It is true that both gases together create a pressure of 200 mmHg. But each gas has the same pressure whether it occupies the chamber exclusively or in company with other gases. Phrased more succinctly, Dalton's law states that, in a mixture of gases, each one exerts the same pressure as though it alone occupied the total volume. Consequently, the pressure of each gas in a mixture is referred to as partial pressure. The term tension may be used as a synonym for partial pressure.

Boyle's Law If a million molecules of gas are contained in a small space, they will clearly exert a greater pressure than if they are allowed to occupy a much larger chamber. Boyle's law states that the pressure of a gas is inversely proportional to its volume so long as the temperature is kept constant.

Charles' Law The pressure that a gas exerts in a closed chamber is a function of molecular movement. Molecules move rapidly and at random. The faster they move, the greater the pressure exerted by the gas. The rate of molecular movement is a function of temperature. At absolute zero ($-273°C$), all molecular activity ceases. Therefore, at that temperature a volume of gas would exert no pressure. The higher the temperature, the greater the pressure. Charles' law states that the pressure of any gas is directly proportional to its absolute temperature so long as the volume remains constant. This can be verified by a simple illustration. One has only to check the pressure in a tire the first thing in the morning and then again after the car has been driven fast, especially on hot roads, to prove that the pressure varies with the temperature of the air within the tire.

Boyle's and Charles' laws may be combined to express the relation of volume, pressure, and temperature as follows:

$$\text{Volume} = \frac{\text{temperature}}{\text{pressure}}$$

Partial Pressure of the Respiratory Gases

A gas moves from a high-pressure area to a region of lower pressure, just as water flows downhill. The differential between the pressure of oxygen in the lungs and in the blood suffices to drive the gas through the membranes and into the blood. Carbon dioxide diffuses from the blood through the membranes into the alveoli because the partial pressure of this gas is higher in the blood than it is in the alveolar air. Using Dalton's law, the magnitude of these pressures can be calculated, but first the percentage composition of the gas mixtures must be known.

Figure 22-1 shows the simple instrument used to collect a sample of alveolar air. One simply breathes through the tube until the tidal volume has almost completely been expired. Then the valve is turned, and the very last part of the expired volume is trapped in a bag. This is the alveolar-air sample. It can be studied with the aid of the Haldane

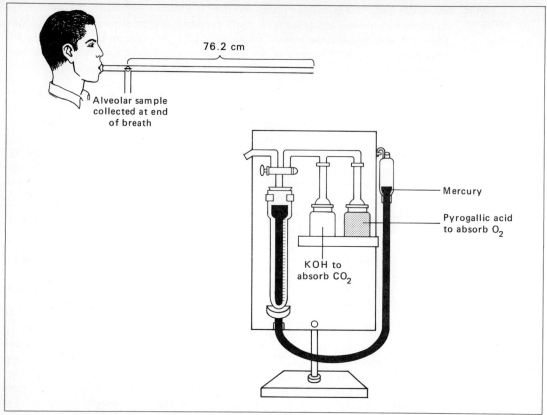

76.2 cm

Alveolar sample
collected at end
of breath

Mercury

Pyrogallic acid
to absorb O_2

KOH to
absorb CO_2

Figure 22-1. Analysis of alveolar air. A subject expires most of a breath through
a tube. The valve is then turned to divert the remainder into a collecting
container. The sample is directed into a Haldane gas analyzer, where its volume
is measured as well as its content of carbon dioxide and oxygen.

gas analyzer (Figure 22-1). This instrument permits the volume of the sample of air to
be measured. Then the air is passed over a chemical which removes the carbon dioxide
(CO_2). The volume of the sample is measured again. The difference between the first
and second volumes is, of course, an index of the amount of CO_2 originally present.
Next, the gas is run through a chemical which absorbs the oxygen (O_2). For the third
time the volume is measured. The difference between the second and third measurements
equals the volume of O_2 contained in the sample.

Table 22-1 shows the composition of atmospheric and alveolar air. If the percentage
of each gas that makes up air is known, the partial pressure exerted by that gas can be
calculated. For example, air contains 20.94 percent oxygen, therefore the partial pressure
of oxygen (P_{O_2}) in air is about 159 mmHg when the total atmospheric pressure is
760 mmHg (760 × 0.2094). In alveolar air there is considerable water vapor. The temperature of the body remains constant, and thus the vapor emitted from the moist surfaces
of the lungs remains constant. It exerts a pressure of about 47 mmHg. The composition
of alveolar air is expressed in terms of dry air; therefore, before multiplying atmospheric
pressure by the oxygen percentage, the partial pressure of water vapor must be subtracted.

TABLE 22-1. PARTIAL PRESSURES OF THE RESPIRATORY GASES

GAS	ATMOSPHERIC AIR		ALVEOLAR AIR		ARTERIAL BLOOD	VENOUS BLOOD
	PERCENT	mmHg	PERCENT	mmHg	mmHg	mmHg
O_2	20.94	159.1	14.2	101	100	40
CO_2	0.04	0.3	5.5	39	40	46
N_2	79.02	600.6	80.3	573	573	573
Total	100.00	760.0	100.0	713	713	659

For example, if the percentage of oxygen in dry alveolar air is 14.2 percent, its partial pressure is equal to $760 - 47 = 713$, which is the pressure of dry air, times 0.142, that is, about 101 mmHg. To be sure, there is also some water vapor in atmospheric air, but its partial pressure is generally so small that it is ignored.

Table 22-1 also gives the partial pressure of each gas in the arterial and venous blood. Note that the venous blood has a partial pressure of oxygen (P_{O_2}) of 40 mmHg, whereas the P_{O_2} in the alveolar air is about 101 mmHg. The venous blood, of course, enters the right side of the heart and is then pumped through the pulmonary system. As the blood enters the pulmonary capillaries, it has a P_{O_2} of 40 mmHg; therefore oxygen rapidly diffuses from the alveolar air into the blood, quickly reaching a P_{O_2} of 100 mmHg (see Figure 22-3). Again, as the blood enters the pulmonary capillaries, it has a partial pressure of CO_2 (P_{CO_2}) of 46 mmHg, whereas the P_{CO_2} in the alveolar air is only 39 mmHg. CO_2, therefore, rapidly diffuses from the blood into the alveolar air.

If one voluntarily hyperventilates, the percentage of oxygen in the alveolar air increases, and the percentage of carbon dioxide decreases. Accordingly, the partial pressure gradients for each increase. Carbon dioxide will, as a result, leave the blood more rapidly, and the amount left in the blood will be below normal, a condition termed hypocapnia. But because the arterial blood is normally almost completely saturated with oxygen, there is not much change in the blood oxygen content as a result of hyperventilation.

Membrane Permeability

Between alveolar air and blood there are two membranes, that of the lung epithelium and that of the capillary wall. But in addition to these membranes, the gases must diffuse through the layer of surfactant that lines the alveolar surface and through connective tissue and fluid that are in the space between the lung epithelium and the capillary wall (Figure 22-2). The two membranes plus the various layers of fluid and connective tissue as a unit are referred to as the respiratory membrane.

The respiratory membrane exerts resistance to the movement of the respiratory gases, resistance that is expressed by the diffusion coefficient, also referred to as the diffusion capacity. The diffusion coefficient is defined as the amount of gas which passes across the membranes per minute per millimeter difference in tension (partial pressure) of the gas on the two sides of the membrane. In simpler language, the greater the rate of diffusion at any particular pressure, the greater the diffusion coefficient.

The diffusion coefficient for oxygen is about 20 ml/min per mmHg. The CO_2 coefficient is 20 times greater, or 400 ml/min per mmHg. At rest, the values for both are great enough to permit equilibration to occur soon after the blood enters the pulmonary

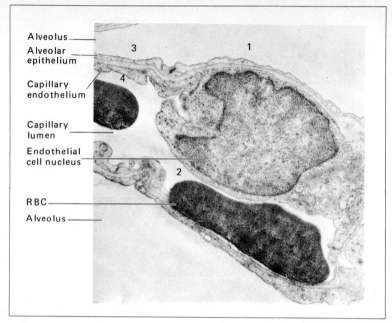

Alveolus

Alveolar
epithelium

Capillary
endothelium

Capillary
lumen

Endothelial
cell nucleus

RBC

Alveolus

Figure 22-2. Respiratory membrane. Gases travel variable distances in diffusing across the respiratory membrane. For instance, compare the pathway between 1 and 2 with that between 3 and 4. (*Courtesy of Michel Campiche.*)

capillaries and long before the blood leaves the capillaries. For example, the pulmonary capillaries are estimated to hold about 60 ml of blood. If the cardiac output is 6,000 ml/min, the blood must be renewed in the pulmonary capillaries 100 times per min. In other words, the blood takes about 0.6 sec to pass through the pulmonary capillaries. At rest, the oxygen uptake is 250 ml/min. In 0.6 sec, 2.5 ml of oxygen must diffuse into the blood. At the arteriolar end of the pulmonary capillary, the P_{O_2} difference between alveolar air and the blood is 60 mmHg (100 − 40). The diffusion coefficient of O_2 is 20 ml/min per mmHg. According to the diffusion coefficient, 0.2 ml O_2 will diffuse each 0.6 sec per mmHg; with a driving force of 60 mmHg, 12 ml would diffuse. This is far greater than the 2.5 ml that actually does diffuse, but as oxygen enters the blood, the P_{O_2} in the blood increases, and therefore the driving force decreases. As a result, the uptake is greatest at the beginning of the capillary and progressively lessens (Figure 22-3). Within the first third of the capillary, almost the entire oxygen load diffuses into the blood, which provides a large reserve factor. Carbon dioxide equilibrates even faster.

The diffusion coefficient is a function of membrane permeability and also reflects alveolar surface area and the pulmonary blood flow. During exercise the diffusion capacity increases three- to fourfold because (1) more alveoli become functional, thereby increasing the surface area; (2) each alveolus dilates, which also increases the surface area; and (3) there is an increased number of active capillaries. Greater pulmonary blood flow also contributes by maintaining high partial pressure gradients. If blood flow is sluggish, there is more time for equilibrium to be reached; thus the partial pressures of each gas on either side of the respiratory membrane approach one another and thereby diminish the driving force of diffusion.

553

DIFFUSION
AND
TRANSPORT
OF
RESPIRATORY
GASES

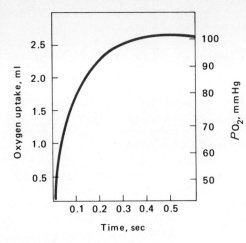

Figure 22-3. Uptake of oxygen in the pulmonary capillaries. The high diffusion coefficient of oxygen is reflected in the rapid rise of the line to its peak.

Function of the Residual Air

In the previous chapter, mention was made of the fact that at the end of a normal expiration, about 2,500 ml of air remains in the lungs. Of this quantity, about 1,000 is the expiratory reserve volume and 1,500 the actual residual volume. A normal breath has a volume of about 500 ml, of which only 350 ml reach the alveoli. This 350 ml mixes with the 2,500 ml normally present. One can immediately see that the large volume of residual air buffers the incoming new air, which is high in oxygen and low in carbon dioxide, and thereby prevents large oscillations in the partial pressure of these gases.

TRANSPORT OF OXYGEN

The quantity of oxygen carried in physical solution in the blood is extremely small. If oxygen were carried in physical solution alone, man would require over 200 liters of blood in order to survive. The normal blood volume, it will be recalled, is about 5 liters. By far the greater percentage of the oxygen is carried in chemical combination.

Hemoglobin

As has already been pointed out, the characteristic color of the blood is a function of the iron protein pigment hemoglobin (Hb). Oxygen enters into a loose combination with hemoglobin by virtue of the iron contained within the molecule. Hemoglobin combines with oxygen as follows:

$$Hb + O_2 \rightleftharpoons HbO_2$$

One molecule of O_2 unites with each atom of iron. A molecule of hemoglobin contains four atoms of ferrous iron. If the ferrous iron is oxidized to the ferric state, no oxygen is carried.

Oxygen Capacity The maximum quantity of oxygen which the blood can carry is termed the oxygen capacity and is determined by the amount of hemoglobin present. One gram of hemoglobin can combine with 1.34 ml of oxygen. The blood of man contains approximately 15 g of hemoglobin per 100 ml of blood. Therefore, the oxygen capacity of 100 ml of blood is 20 ml (15 × 1.34). This is commonly expressed in volumes percent (vol percent). Thus, it is said that the oxygen capacity of the blood is 20 volumes percent (20 ml O_2/ 100 ml blood).

Oxygen Dissociation Curve Oxygen unites with hemoglobin in a very unstable manner. The quantity of oxygen which combines is determined by its partial pressure. If the partial pressure is increased, more oxygen unites with hemoglobin; when the partial pressure falls, the hemoglobin surrenders a portion of the oxygen.

The dissociation curves of human blood are shown in Figure 22-4. Such curves are derived by the use of a series of tonometers. A tonometer is simply a tube in which a small quantity of blood is introduced along with a mixture of gas. Let us assume that in a series of tonometers there is only blood and pure oxygen but that in each tonometer the oxygen pressure varies. It may be 10 mmHg in the first, 30 in the next, 50 in the third, and so forth. The tonometers are then slowly rotated so as to spread the blood in a thin layer along the inner surface and thus encourage equilibrium between the pressure of oxygen in the tonometer and in the blood. The tonometers are rotated in a water bath heated to 37°C in order to simulate body temperature. After a period of time sufficient to allow equilibration, the blood is analyzed and the oxygen content determined. If the blood is found to contain 5 vol percent, for example, it is said to be 25 percent saturated, since 20 vol percent equals 100 percent saturation. In this way, the curve representing zero CO_2 in Figure 22-4 is constructed. The procedure is repeated with varying quantities of oxygen in the tonometer. Furthermore, varying amounts of carbon dioxide are introduced so that the marked influence of this gas on oxygen saturation can also be plotted.

An understanding of the oxygen dissociation curves is essential. When a person breathes air at normal atmospheric pressure, the partial pressure of oxygen in the alveolar air is in excess of 100 mmHg. The pressure of this gas in the arterial blood will then

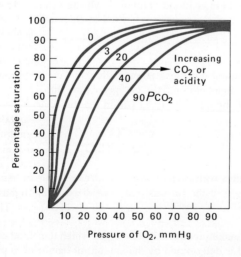

Figure 22-4. Oxygen dissociation curves. Four of the curves were constructed with varying amounts of carbon dioxide present. An increase in blood acidity shifts all curve values to the right.

555
DIFFUSION
AND
TRANSPORT
OF
RESPIRATORY
GASES

be below but close to this figure (Table 22-1). The oxygen dissociation curves show that at 100 mmHg the blood is almost completely saturated, regardless of the carbon dioxide pressure. Actually, the normal carbon dioxide pressure in the venous blood is about 40 mmHg. The appropriate curve for this value (Figure 22-4) indicates that the blood must leave the lungs about 97 percent saturated. To put it another way, each 100 ml of blood, as it leaves the lungs, contains 19.4 ml of oxygen (0.97 × 20).

These curves show that increasing the P_{O_2} in the lungs beyond 100 mmHg is of no particular advantage. The saturation may be raised to 100 percent, but since blood is normally 97 percent saturated, the gain is clearly not great. Conversely, the P_{O_2} can be reduced considerably below normal without depriving the blood of much of its load. Even when the pressure is cut in half (reduced to 50 mmHg), the saturation is still over 80 percent. In short, the oxygen saturation curves mean that the P_{O_2} in the lungs can vary over a relatively wide range and still the blood will be adequately saturated.

When the blood reaches the capillaries, some of its oxygen load passes through the capillary wall and enters the tissue spaces. It does so because the pressure of oxygen in the tissues is less than it is in the blood. Due to the pressure gradient, oxygen moves from the blood into the tissues. The actual pressure of this gas in the tissues varies widely, depending on the activity of the particular area. But let us assume for the moment that it is about 35 mmHg. Consulting the dissociation curve marked "40" (Figure 22-4), note that the blood will now be only 70 percent saturated. It will contain about 14 vol percent of oxygen. That is to say, each 100 ml of blood has surrendered to the tissues about 6 ml of oxygen. If the tissue being supplied is muscle which now begins to contract vigorously, it will utilize oxygen at an accelerated rate, the P_{O_2} pressure will drop, and more oxygen will leave the blood. Thus the venous blood returning from actively contracting muscle may contain only 8 vol percent of oxygen or even less.

Utilization Coefficient   The percentage of the total oxygen load given up to the tissues, that is, utilized by the tissues, has been termed the utilization coefficient. Blood leaves the lungs containing 20 vol percent and returns with 15 vol percent; thus 5 vol percent is utilized, and the overall utilization coefficient is 25 percent (5/20 × 100). During exercise, venous blood from muscles, as just mentioned, may contain only 8 vol percent. The utilization coefficient then is 60 percent. In exhaustive exercise, which can be maintained only for short periods of time, the venous blood from active muscles may be almost depleted of oxygen; that is, the utilization coefficient approaches 100 percent. But if the oxygen content of blood returning to the right side of the heart is analyzed, the oxygen utilization coefficient is found rarely to exceed 75 percent. This simply means that while the utilization rate of oxygen by the muscles is very high, in other tissues it is much lower. Blood in the right side of the heart is, of course, a mixture from all parts of the body.

Influence of Carbon Dioxide   The quantity of carbon dioxide contained in the blood markedly influences the oxygen saturation at any given partial pressure (Figure 22-4). For example, if the partial pressure of oxygen is 50 mmHg, note that at a carbon dioxide pressure of 40 mmHg the blood will be about 80 percent saturated. But if the carbon dioxide is only 20 mmHg, then the saturation is close to 90 percent. As will be discussed presently, the carbon dioxide content of the blood varies, being highest in the blood as it leaves the systemic capillaries and lowest in the lung capillaries. This change favors the uptake of oxygen in the lungs and encourages the divorce of oxygen from hemoglobin in the systemic capillaries.

Role of Acid Metabolites The third factor determining the saturation of hemoglobin with oxygen is the pH of the blood. An increase in the acidity shifts the curves to the right. That is to say, for any P_{O_2} the percent saturation will be lower. For example, at a P_{O_2} of 40 mmHg and a P_{CO_2} of 40 mmHg, the blood is about 75 percent saturated with oxygen (Figure 22-4). If the blood acidity increases and the curves shift to the right, then at the same partial pressures for oxygen and carbon dioxide the hemoglobin may now only be 65 percent saturated with oxygen. Thus, just as increased carbon dioxide causes greater dissociation of hemoglobin, so do the acid metabolites, which are end products of cellular metabolism. In summary, when tissues become more active, additional oxygen can be quickly made available without alterations in circulatory or respiratory dynamics for three reasons: (1) lowered P_{O_2} in the tissues, (2) increased P_{CO_2}, (3) increased acidity.

Function of Diphosphoglycerate Although 2,3-diphosphoglycerate (DPG) was identified in erythrocytes some fifty years ago, only recently has its role in oxygen release by hemoglobin been clarified. In general, DPG strikingly lowers the affinity of hemoglobin for oxygen. ATP has been shown to do the same. What role do these compounds play in oxygen transport?

One possible role has been identified in terms of the difference between the way hemoglobin functions in the newborn and in the adult. The oxygen affinity of human fetal and newborn blood is higher than that of adult blood. This means that for any given P_{O_2}, newborn blood has a higher oxygen saturation than does adult blood. A difference in reaction of newborn hemoglobin with DPG from the way it reacts with adult hemoglobin is thought to be the explanation.

Another role is in response to hypoxia. Under these conditions the level of DPG increases, thereby decreasing the affinity of hemoglobin for oxygen, thus making more O_2 available to the tissues. This change occurs within a few hours and probably plays a role when one resides at high altitude or when there is inadequate oxygen of the blood due to some lung abnormality. Also, when the level of hemoglobin decreases, DPG concentration increases. The result, once again, is to compensate for the decreased oxygen-carrying capacity by shifting the oxygen dissociation curve to the right, thereby making more oxygen available to the tissues.

Finally, there is great clinical impact in the observation that blood stored longer than 1 week has an abnormally low DPG level and the level does not return to normal until at least 24 hr after transfusion.

Oxygen in Physical Solution

Thus far, only the transport of oxygen by hemoglobin has been considered. At the partial pressures normally existing, by far the greater amount of oxygen in the blood is combined with hemoglobin, but even at these pressures there is a small amount simply dissolved in the plasma. At a P_{O_2} of 100 mmHg, about 0.3 ml of oxygen is dissolved in 100 ml of plasma. This figure is to be compared with the 20 ml of oxygen combined with hemoglobin. In other words, under normal conditions the amount of oxygen carried in physical solution is too small to be considered. But as the P_{O_2} is increased, a very different situation develops.

After hemoglobin is fully saturated, an increase in P_{O_2} cannot increase the amount of oxygen carried by hemoglobin, but the amount dissolved in plasma increases proportionately with the increase in atmospheric pressure. By increasing P_{O_2} two or three times normal, sufficient oxygen enters physical solution to be effective in satisfying tissue needs. Such pressures can be obtained in a so-called hyperbaric chamber, a chamber large enough to contain a person and in which the pressure of air, or of pure oxygen, can be increased.

557

DIFFUSION
AND
TRANSPORT
OF
RESPIRATORY
GASES

For example, a patient with severe anemia, so severe that inadequate oxygen is transported by the hemoglobin present, would be markedly benefited by increasing the amount of oxygen in his plasma. In recent years, treatment in hyperbaric chambers has been extensively investigated and found to be of remarkable value in many disorders in which there is inadequate oxygenation of tissues.

Excessive oxygenation (hyperoxia) can lead to oxygen poisoning. There is still not complete agreement as to the mechanism of oxygen poisoning. In some way hyperoxia inactivates oxidative enzymes, and therefore cellular metabolism becomes deranged. How hyperoxia inactivates oxidative enzymes is the question. One theory is that hyperoxia decreases blood flow by virtue of the autoregulation mechanism. Despite the decreased flow, oxygen needs are satisfied because of the high concentration of oxygen in the blood, but other substances required for cellular metabolism are not supplied adequately by the diminished blood flow.

Whatever the mechanism, the dangers are very real. The results of oxygen poisoning range from pulmonary edema, to retrolental fibroplasia, which is an ingrowth of fibrous tissue into the vitreous humor of the eye to cause blindness, to death. Oxygen poisoning is always a danger when because of respiratory difficulties premature infants are placed in a pure oxygen environment.

TRANSPORT OF CARBON DIOXIDE

As the blood flows through the capillaries, each 100 ml picks up about 4 ml of carbon dioxide while giving up 5 ml of oxygen. When tissues become more active, these values increase. At rest, the total amount of carbon dioxide in the venous blood is about 53 vol percent. It is carried in three ways: (1) combined with hemoglobin, (2) as bicarbonate, and (3) in physical solution (Table 22-2).

Combination with Hemoglobin

Oxygen combines with the iron in hemoglobin, whereas carbon dioxide combines with an NH_2 group. There is thus no competition for a common binding site and so both gases can be carried simultaneously, although for other reasons, the partial pressure of each influences the amount of the other carried by hemoglobin.

The combination of carbon dioxide and hemoglobin is referred to as carbamino-hemoglobin, or more simply as carbhemoglobin. The reaction is:

$$CO_2 + HbO_2^- \rightleftharpoons HbCO_2 + O_2$$

TABLE 22-2. TRANSPORT OF CARBON DIOXIDE

FORM	ARTERIAL BLOOD		VENOUS BLOOD		A-V DIFFERENCE	
	VOL PERCENT	PERCENT OF TOTAL	VOL PERCENT	PERCENT OF TOTAL	VOL PERCENT	PERCENT OF TOTAL
Bicarbonate	43.5	88.8	46.3	87.3	2.8	70.0
Hemoglobin	3.0	6.1	3.8	7.2	0.8	20.0
Solution	2.5	5.1	2.9	5.5	0.4	10.0
Total	49.0	100.0	53.0	100.0	4.0	100.0

Approximately 20 percent of the carbon dioxide that enters the blood in the tissue capillaries is carried combined in this way with hemoglobin (Table 22-2).

Transport as Bicarbonate
Before carbon dioxide can enter the erythrocyte, it must pass through the plasma where it reacts with water:

$$CO_2 + H_2O \rightleftharpoons H_2CO_3 \rightleftharpoons HCO_3^- + H^+$$

In the plasma this reaction proceeds very slowly; therefore, much of the carbon dioxide continues on into the erythrocyte where the same reaction takes place, but at a very much greater rate. The reason for the difference in the speed of these reactions is the presence of a catalyst, carbonic anhydrase, in the erythrocytes.

Because far more bicarbonate is formed in the erythrocytes than in the plasma, bicarbonate ions diffuse through the erythrocyte membrane into the plasma. Chloride ions diffuse in the opposite direction to maintain ionic equilibrium. This movement of chloride ions into the erythrocytes is termed the chloride shift.

Hemoglobin facilitates the formation of bicarbonate in the erythrocyte. In the above equation note that H$^+$ is an end product. If this ion can be removed as it is formed, the reaction will continue to proceed to the right, resulting in the continuing formation of bicarbonate. In the erythrocyte the following reaction occurs:

$$H^+ + HbO_2^- \rightleftharpoons HHb + O_2$$

In this way the H$^+$ is removed and more bicarbonate is formed. The liberated oxygen diffuses out of the blood. Of the 4 ml of carbon dioxide which diffuses into the blood from the tissues, about 70 percent is carried as bicarbonate.

Transport in Solution
Carbon dioxide is highly soluble. The amount that goes into physical solution is a function of its partial pressure. The P_{CO_2} changes from 40 to 46 mmHg in the tissue capillaries. The amount of carbon dioxide in solution changes from about 2.5 to 2.9 vol percent. Thus, of the 4 ml of carbon dioxide about 10 percent is transported in solution.

Carbon Dioxide Dissociation Curve
In considering the transport of oxygen, the term oxygen saturation was used. Because oxygen is carried almost exclusively by hemoglobin, the degree of saturation of the hemoglobin with oxygen is an accurate measure of the quantity of oxygen carried. However, the carbon dioxide combined with hemoglobin represents only a part of the total transported; therefore, the degree of hemoglobin saturation would not necessarily present an accurate estimate of the quantity of carbon dioxide carried. For this reason, the term carbon dioxide content, which is the amount of the gas present in the bloodstream, expressed in volumes percent, is used. Figure 22-5 shows that the carbon dioxide content varies with the partial pressure of the gas.

At rest, the P_{CO_2} in the arterial blood is about 40 mmHg. The dissociation curves reveal that at this P_{CO_2} the arterial blood contains about 49 vol percent of carbon dioxide. In the venous blood the P_{CO_2} is normally 46 mmHg, and the content of carbon dioxide is close to 53 vol percent. Note that the carbon dioxide dissociation curve does not plateau as does the oxygen dissociation curve at high pressures. This means that the greater the

559

DIFFUSION
AND
TRANSPORT
OF
RESPIRATORY
GASES

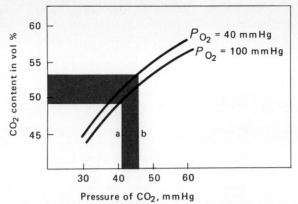

Figure 22-5. Carbon dioxide dissociation in relation to partial pressure of oxygen. The normal pressure of carbon dioxide lies between a and b, a representing the pressure in arterial blood and b, in venous blood. The carbon dioxide content of the blood at any partial pressure is higher when the oxygen partial pressure is lower.

production of carbon dioxide, the more the venous blood is capable of carrying, and thus it constitutes an excellent transport mechanism.

Role of Oxygen

Just as carbon dioxide influences the transport of oxygen, so does oxygen affect the transport of carbon dioxide. The carbon dioxide dissociation curves show that at any P_{CO_2}, the amount carried by the blood is greater at low P_{O_2} levels than at high levels. Thus, when tissues become active and use more oxygen, the resulting lower P_{O_2} permits the blood to carry a greater load of carbon dioxide. In short, at least two local mechanisms aid in the removal of carbon dioxide from active tissues without alteration in cardio-vascular or respiratory dynamics: (1) increased P_{CO_2} and (2) decreased P_{O_2}.

The reciprocal influence of oxygen and carbon dioxide on the transport of these gases occurs because carbon dioxide makes the blood more acid, thereby increasing the dissociation of oxygen and hemoglobin, and because the liberation of oxygen frees hemoglobin which combines with H^+, forming more bicarbonate.

Respiratory Quotient

The blood, in passing through the lungs, takes up 5 vol percent of oxygen and gives off 4 of carbon dioxide. The ratio between CO_2 output and O_2 uptake is referred to as the respiratory quotient, or the respiratory exchange ratio, and is expressed as:

$$\text{Respiratory quotient (RQ)} = \frac{CO_2 \text{ output}}{O_2 \text{ intake}}$$

Using the normal values, the RQ would be 0.8 (4/5). Actually the RQ may vary between 0.7 and 1.0, and in some cases be higher than 1.0. More will be said about this when metabolism is considered, but suffice it here to say that the RQ depends upon the type of substrate being metabolized. For example, if only carbohydrate is involved, the

RQ will be 1.0, because in the metabolism of carbohydrate one molecule of CO_2 is produced for each molecule of O_2 utilized. In the metabolism of fat this one-to-one relationship does not hold, and the RQ approaches 0.7. Thus, as will be seen, by determining the RQ an indication of the type of foodstuff being metabolized is obtained.

QUESTIONS AND PROBLEMS

1 List the five factors which determine the diffusion of gases across the epithelial tissue of the alveoli and the endothelial tissue of the pulmonary capillaries.

2 When one hyperventilates, why is there no increase in passage of O_2 from the alveoli to the blood?

3 Define (a) Dalton's law of partial pressure, (b) Boyle's law, and (c) Charles' law.

4 What factors are necessary to increase the diffusion coefficient of O_2 in the lungs?

5 Explain why hypocapnia develops if a person hyperventilates.

6 What is a respiratory membrane?

7 Why is oxygen uptake greater at the beginning than at the terminal third of a pulmonary capillary?

8 What volume percent of oxygen would be found in an individual with 12 g of hemoglobin per 100 ml of blood?

9 Why is there an increase in the diffusion capacity of the respiratory membrane during exercise?

10 How can you explain that the oxygen partial pressure may vary over a wide range and have little effect on hemoglobin saturation?

11 What is the function of hemoglobin? What symptoms would appear in someone with a low hemoglobin?

12 What is the oxygen capacity of human blood?

13 What is a hyperbaric chamber? What advantage is gained by placing an anemic individual in one?

14 Discuss the relation of bicarbonate ions to the chloride shift.

15 In what way does hyperoxia threaten the premature infant?

16 In what chemical forms is carbon dioxide transported from the tissues to the lungs?

17 Describe the reciprocal influence of oxygen and carbon dioxide on the transport of these gases.

18 What is the meaning of respiratory quotient?

23 REGULATION OF VENTILATION

At rest about 8 liters of air moves through the lungs per minute. These 8 liters contain approximately 1.6 liters of oxygen of which some 250 ml enters the blood. Five to six liters of blood flows through the pulmonary and tissue capillaries per minute. The arterial blood entering the tissue capillaries contains close to 20 volumes percent (vol percent) of oxygen. About 5 vol percent is given up to the tissues. The basal metabolic processes produce 4 vol percent of carbon dioxide, which is then transported to the lungs and expired.

When one becomes more active, more oxygen is required. It can be supplied by several mechanisms. At the tissue level there may simply be a greater utilization of available oxygen. Thus, instead of the tissues extracting only 25 percent of the oxygen load, that is 5 vol percent, almost all the 20 vol percent carried by the blood may be extracted. If this still fails to satisfy the demands, local vasodilatation results in greater blood flow and thus the deliverance of more oxygen per unit time. Greater cardiac output increases the supply even more. The cardiac output can increase from 5 liters/min up to about 35 liters/min. But unless the blood continues to leave the lungs with a full load of 20 vol percent of oxygen and, at the same time, with only about 49 vol percent of carbon dioxide, the circulatory alterations will be of little value. Fortunately, ventilation is beautifully attuned to cellular requirements for oxygen and the elimination of carbon dioxide. During vigorous exercise, ventilation can increase from the basal 8 liters/min to over 100 liters/min. As a result of greater gas exchange, faster blood flow, and increased ventilation, the concentration of oxygen and carbon dioxide in the arterial blood remains remarkably constant over a wide range of physical activity. The important question, as always, is how?

THE RESPIRATORY CENTER

Breathing, like contraction of the heart, is rhythmic. But whereas heart rhythm depends upon the contraction of a single muscle, breathing involves many. In addition, these are skeletal muscles, devoid of inherent rhythmicity. If their innervation is destroyed, they are paralyzed. Breathing requires the coordination of many muscles. There is a cluster of cells, a so-called center, responsible for this coordination and rhythm.

Location of the Respiratory Center

The brainstem may be transected above the pons without arresting breathing. However, if transections are made progressively lower, a point is finally reached at which breathing stops. As a result of these experiments and others using discrete stimulation of various areas of the brainstem, the conclusion has been reached that the respiratory center is located in the upper two-thirds of the medulla and extends up into the pons.

The concept of the respiratory center as a discrete, circumscribed cluster of specialized cells has been seriously questioned. Stimulation experiments indicate that one part of the medullary reticular formation is responsible for inspiration and another part for expiration. Thus, it is the custom to speak of an inspiratory and an expiratory center in the medulla. However, the available evidence indicates that the cells responsible for inspiration and expiration are intermingled. Anatomically, then, one should not visualize separate centers, but from a functional viewpoint this is permissible.

Axons of the cells which compose the respiratory center end in synaptic union with cranial and spinal nerves. By virtue of this arrangement, the muscles of the face, throat, chest, and diaphragm are coordinated for respiratory purposes.

Rhythmicity of Breathing

That there is a rhythm to breathing is obvious. At the resting rate of about 16/min, each cycle of inspiration and expiration takes a bit less than 3.8 sec. At this rate, inspiration lasts about 1.6 sec and expiration 2.2 sec. The rhythm of breathing, as stimulation of the reticular formation has shown, can be altered by changing the rate or depth of breathing, or both.

Whether the basic rhythmicity of breathing is due to an inherent property of the respiratory center or to factors which influence these cells is the question. For some time it was held that if the center were completely divorced from extrinsic forces, apneusis, which is a sustained inspiration, would result. More sophisticated experiments cast considerable doubt on this conclusion, and now most authorities believe that the respiratory center possesses an inherent rhythmicity, a rhythmicity that can be markedly altered by extraneous factors. There is, however, no evidence for the presence of pacemaker neurons which possess an intrinsic rhythm. The characteristic rhythm is thought to be due to self-reexcitation between the inspiratory and expiratory neurons.

CHEMICAL REGULATION OF VENTILATION

Many nerves impinge upon the respiratory center and markedly alter its activity. In addition, the chemical composition of the blood and CSF is of major importance.

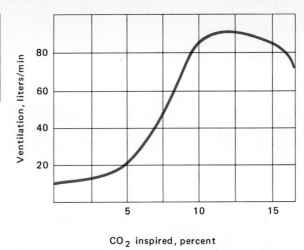

Figure 23-1. Influence of carbon dioxide on ventilation. At very high partial pressures carbon dioxide depresses the respiratory center.

Carbon Dioxide

Carbon dioxide powerfully influences the respiratory center (Figure 23-1). Normally there is only about 0.04 percent carbon dioxide in the inspired air. No measurable alteration in ventilation occurs until the inspired air contains at least 1 percent. When it contains 4 percent, ventilation is about double. At higher concentrations, ventilation increases very sharply until a maximum of about 80 or 90 liters/min is reached. Most individuals can tolerate about 10 percent of carbon dioxide in the inspired air, but higher concentrations produce great discomfort and then, at about 20 percent, depression and unconsciousness. Forty percent causes death.

In terms of P_{CO_2} in the blood, an increase in P_{CO_2} of as little as 2.5 mmHg doubles ventilation. Normal P_{CO_2} in the arterial blood is 40 mmHg. If this increases to 60 mmHg, ventilation increases over tenfold.

The mechanism by which carbon dioxide stimulates certain neurons but inhibits others has only recently been studied. The results suggest that the carbon dioxide action is due to a resulting fall in pH, not to molecular carbon dioxide per se. The fall in pH increases the conductance of chloride ion. If the chloride equilibrium potential (calculated from the Nernst equation) is less negative than the membrane potential, the membrane is depolarized, that is, excited. If the chloride equilibrium potential is more negative than the membrane potential, the cell is hyperpolarized, that is, inhibited.

Hydrogen-Ion Concentration

As the hydrogen-ion concentration of the blood increases (lowering pH), ventilation increases. Figure 23-2 shows this relationship. The relative increase in ventilation is not as great as when carbon dioxide is the stimulus. In addition, the pH of the blood is so effectively controlled that great variations occur only in disease. Thus it may be thought the hydrogen-ion concentration plays, at best, a minor role in regulating ventilation. But the facts indicate otherwise. An increase in the hydrogen-ion concentration of the CSF increases ventilation markedly. Certain areas of the medulla have been shown to be highly sensitive to changes in hydrogen-ion concentration.

The so-called blood-brain barrier (see Chapter 19) is impermeable to hydrogen ions, but carbon dioxide passes through easily. The CSF is not buffered; thus carbon dioxide

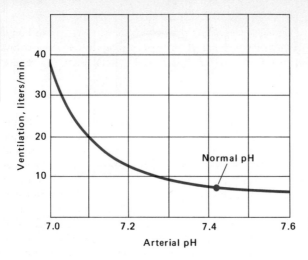

Figure 23-2. Influence of pH on ventilation. Acidity stimulates respiration.

present forms carbonic acid. Theoretically, doubling P_{CO_2} doubles hydrogen ions. In the CSF the alteration in hydrogen ions reaches at least 95 percent of this maximum.

In summary, the blood pH does not give an accurate picture of the influence of hydrogen-ion concentration in the regulation of ventilation. Undoubtedly a significant part of the effect of carbon dioxide on ventilation operates by lowering the hydrogen-ion concentration in the cerebrospinal fluid.

Oxygen

Hypoxemia does not stimulate the respiratory center and may actually depress it. However, as shown in Figure 23-3, ventilation increases when the percent of oxygen in the inspired air is decreased. But this increase is due completely to activation of the chemoreceptors in the carotid and aortic bodies (see ahead). When these receptors are denervated, hypoxemia results in decreased ventilation.

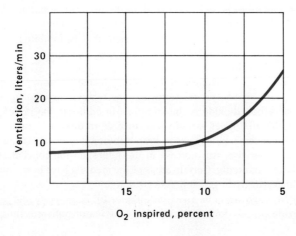

Figure 23-3. Influence of oxygen on ventilation.

REFLEX REGULATION OF VENTILATION

The respiratory center receives impulses from many areas of the body, including the cerebral cortex. Those from the cerebral cortex are essential to the voluntary control of breathing and also probably play an important role in the increased ventilation associated with exercise.

Hering-Breuer Reflex

The cells that make up the inspiratory center fire impulses to the inspiratory muscles. The thorax expands and the lungs follow. In the lungs there are receptors sensitive to stretch. Consequently, as the lungs expand, these receptors fire more and more rapidly. The impulses so initiated are propagated by afferent neurons in the vagus nerve of each lung to the inspiratory center, which finally is inhibited. Impulses may also go directly to the expiratory center to fire it. Thus, inspiration ends and expiration occurs. This reflex was described by Hering and Breuer in 1868 and still bears their names.

Stimulation of the inspiratory center causes increased rate and depth of breathing, and the inspiratory center becomes more difficult to inhibit. Thus, the lungs must stretch more before there is a sufficient barrage of impulses to stop inspiratory activity.

The Hering-Breuer reflex may be demonstrated by stimulating the central end of the cut vagus nerves in an experimental animal (Figure 23-4). Stimulation activates the

Figure 23-4. Hering-Breuer reflex. Stimulation of afferent vagal fibers causes apnea.

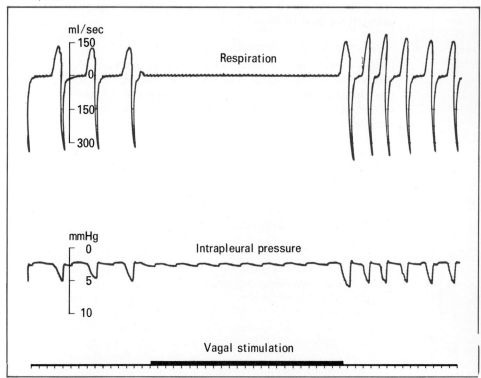

afferent fibers which normally propagate the impulses from the stretch receptors in the lungs. As a result inspiration is inhibited, and apnea, which is absence of breathing, continues as long as the stimulation is maintained.

The major function of the Hering-Breuer reflex would seem to be to contribute to respiratory rhythm. When the reflex is destroyed, rhythmic breathing continues, but the pattern is different. In addition, in the absence of the reflex, the inspiratory center is harder to inhibit than when the reflex is functional. This function is probably even more important when the inspiratory center is under neural drive from other areas of the body.

Pneumotaxic Reflex

In the pons there is a cluster of cells which has been labeled the pneumotaxic center. Whether this is really a discrete center or simply a part of the total respiratory center is open to question. Its function has also been questioned. Nonetheless, many authorities believe that there is a discrete pneumotaxic center and that, under some circumstances, it does function importantly. For example, after the vagi are cut, breathing first becomes slower and deeper, as stated before. But after a period of time, breathing returns to normal. This has been interpreted to mean that, under these circumstances, the pneumotaxic center takes the place of the Hering-Breuer reflex.

The theory is that when the inspiratory center fires, impulses go not only to the motor neurons serving inspiratory muscles but also to the pneumotaxic center (Figure 23-5). This center then fires impulses back to the respiratory center, resulting in inhibition of inspiration and perhaps activation of expiration. Experiments in which action potentials are recorded directly from the pneumotaxic center strongly suggest that the pneumotaxic reflex does contribute to the breathing mechanism, especially under conditions which demand maximal ventilation.

Carotid and Aortic Sinus Reflexes

Stimulation of the pressoreceptors in the aortic and carotid sinuses by excessive blood pressure not only slows the heart and decreases peripheral resistance but also depresses breathing. Conversely, a fall in blood pressure causes hyperventilation. At least a part of the hyperventilation which characterizes severe hemorrhage and circulatory shock is thought to be due to this pressoreceptor response.

Carotid and Aortic Body Reflexes

Stimulation of the carotid and aortic bodies produces effects opposite to those of activation of the pressoreceptors; that is, there is an elevation in blood pressure and increased ventilation. From a quantitative standpoint the carotid and aortic sinus mechanisms are far more important in the control of circulation than in the regulation of ventilation. On the other hand, the carotid and aortic body reflexes are of much greater significance to respiration than to circulation.

Oxygen Mention was made above that hypoxemia causes increased ventilation. In view of the fact that hypoxemia depresses the respiratory center, whatever influence decreased oxygen content has on breathing must be exerted through the chemoreceptors in the aortic and carotid bodies.

Ventilation is not significantly increased until the P_{O_2} in the arterial blood decreases below about 60 mmHg. Under normal conditions, even during severe exercise, the P_{O_2} does not fall that low. This means that other mechanisms are so much more sensitive and effective that even when oxygen is being utilized rapidly, the increase in ventilation suffices to keep the arterial P_{O_2} from decreasing significantly.

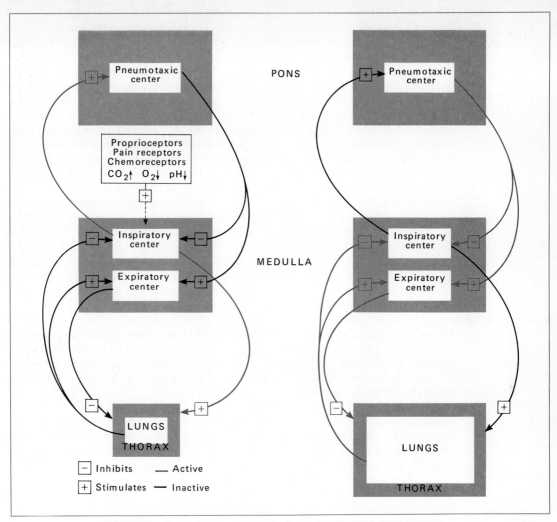

Figure 23-5. Control of respiration. At the beginning of inspiration (left) the inspiratory center activates inspiratory muscles and sends impulses to the pneumotaxic center as well. Expansion of the lungs activates stretch receptors, which eventually inhibit the inspiratory center and stimulate the expiratory center (right). The pneumotaxic center sends impulses having the same effects. Superimposed on this basic mechanism are impulses from joint proprioceptors, pain receptors, and chemoreceptors of the carotid and aortic bodies.

Oxygen regulation of ventilation via the chemoreceptors must be looked upon as a secondary line of defense. Under specific circumstances, such as residence at high altitude or other conditions that produce hypoxemia, the chemoreceptors do function to augment ventilation. Moreover, the chemoreceptors constitute the only mechanism the body has to respond to hypoxemia and in such circumstances are of primary importance. If one studies action potentials recorded from the nerves which innervate the chemoreceptors, a much greater sensitivity is indicated than is suggested by measurement of ventilation. As stated above, significant alteration in ventilation only occurs after the P_{O_2}

falls below about 60 mmHg. Increased firing of the nerve to the carotid sinus is detected at a much higher P_{O_2}. These findings reinforce the conclusion that the chemoreceptors serve as a second line of defense; they are a defense which is normally overshadowed by other mechanisms but which can, if need be, regulate breathing.

These findings are not too hard to explain. As soon as chemoreceptor drive augments ventilation, carbon dioxide is "blown off," the blood carbon dioxide level decreases, and respiration is depressed. The loss of carbon dioxide also decreases the hydrogen-ion concentration, which also inhibits breathing. Contrast this sequence with an increase in arterial carbon dioxide. This directly stimulates the cells of the respiratory center and raises the hydrogen-ion concentration of the CSF to stimulate breathing even more. Hyperventilation raises the P_{O_2} in the alveolar air, but since arterial blood is normally about 97 percent saturated, hyperventilation can have little effect in increasing the P_{O_2} of arterial blood. Thus, oxygen cannot act as a "brake" to breathing when carbon dioxide increases, as carbon dioxide and hydrogen ions do when P_{O_2} falls. In short, there must be a really major drop in P_{O_2} to drive breathing in the face of this braking action.

Recent studies of the carotid body disclose that the sensitive cells are the so-called glomus cells. Interestingly, they have a dual innervation, one afferent and the other efferent. When the glomus cells are stimulated, they liberate ACh, which fires the afferent neurons. However, the afferent neurons can, apparently, be fired directly by appropriate stimuli. The role of the efferent neurons remains uncertain. They probably modulate the response of the glomus cells to various stimuli.

Carbon Dioxide and Hydrogen Ions Under carefully controlled conditions, hypercapnia and increased hydrogen-ion concentration can be demonstrated to activate the chemoreceptors in the carotid and aortic bodies. However, the central effect of these substances is so much greater that this response of the chemoreceptors is normally of little importance in the regulation of ventilation.

Temperature There is some evidence that the chemoreceptors respond to the temperature changes in the blood. Hyperthermia causes hyperventilation. Again, hyperthermia, acting centrally on the hypothalamus, causes hyperventilation, especially in animals that pant; thus the significance of the chemoreceptor reflex is unknown.

In summary, the chemoreceptors constitute the only mechanism for response to hypoxemia. In addition, they constitute a second line of defense to furnish respiratory drive when the primary mechanisms fail. However, when the chemoreceptors in the aortic and carotid bodies are denervated, ventilation decreases about 30 percent. This seems to indicate that the chemoreceptors normally exert a constant facilitating influence on the respiratory center.

Joint Reflexes

There are proprioceptors in the muscles and tendons which are activated by the movement of the joint. The influence of the impulses from these receptors on ventilation is so great that the so-called joint reflex must play a major role in regulating ventilation in exercise.

The suggestion has been made that the proprioceptors are sensitive to changes in the metabolism of the muscle. However, if the leg is left attached to the body only by the nerve, the infusion of dinitrophenol, which increases cellular metabolism, has no influence on ventilation. On the other hand, if the joint of the limb attached only by the nerve is bent rhythmically, ventilation does increase. This means that the proprioceptors respond to movement, not to metabolism.

Pain Reflexes

In the conscious subject, painful stimulation usually produces hyperventilation. Even in the anesthetized animal, stimulation of almost any afferent trunk produces hyperventilation (Figure 23-6). Apparently afferent pathways send collaterals to the respiratory center.

Advantage is taken of the pain reflexes in several circumstances. For example, the newborn baby is slapped to initiate breathing. Likewise, when breathing has been arrested, it may sometimes be reinstated by evoking pain.

Protective Reflexes

A number of protective reflexes are able to block breathing temporarily. One well-known reflex of this type originates in the mucous membrane of the nose. Irritating substances, such as ether, chloroform, ammonia, and hydrogen sulfide, slow or stop breathing. Another protective reflex occurs during swallowing. The nasopharynx is closed by the soft palate, the glottis is closed, and the larynx raised. At the same time breathing is arrested.

VOLUNTARY VARIATION OF BREATHING

Breathing is unique in that not only is it controlled by exquisitely attuned automatic mechanisms but it may be varied voluntarily, in contradistinction to circulation, for example. The reason that breathing is responsive to the will is that skeletal muscle is involved. This voluntary element, however, is not absolute. That is to say, the breath cannot be held indefinitely, nor may hyperventilation be voluntarily continued for indefinite periods of time.

Higher-Center Influence

The fact that breathing can be modified at will indicates that the respiratory center is "wired" to the cerebral cortex, so to speak. Experiments in which the cerebral cortex is directly stimulated confirm these connections. Similarly, alterations in ventilation have been evoked by stimulation of the thalamus, the posterior commissure, the hypothalamus, and the cerebellum.

Breath Holding

Obviously, breathing can voluntarily be stopped. During the period of apnea, carbon dioxide accumulates in the blood, hydrogen ions accumulate in the CSF, and hypoxemia

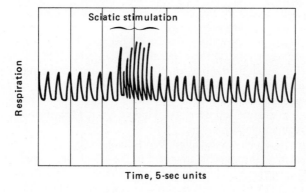

Figure 23-6. Influence of pain on respiration. Stimulation of afferent nerves containing pain fibers increases respiration.

develops. Together they act as a great stimulant to initiate breathing. Ultimately, this drive is sufficient to overcome the volitional inhibition, and breathing once again begins. If one hyperventilates before holding his breath, he can then hold it longer, because the hyperventilation lowers the carbon dioxide content of the blood. Most people have difficulty maintaining apnea for much longer than 1 min.

Hyperventilation

Voluntary hyperventilation causes a progressive lowering of the carbon dioxide content of the blood and the hydrogen-ion concentration of the CSF, but, of course, it has only a minor effect on the oxygen saturation of the arterial blood. If hyperventilation is continued for several minutes, the individual becomes dizzy, perhaps even giddy, and then fatigue makes it impossible to continue the effort.

Dizziness and giddiness are caused by inadequate blood supply to the brain. Carbon dioxide acts on the vasoconstrictor center to maintain generalized vasomotor tone. During hyperventilation, the lowered carbon dioxide results in loss of constrictor tone, decrease in resistance and blood pressure, and a decrease in blood flow to the brain. The decrease in blood carbon dioxide also shifts the oxygen dissociation curve to the left. As a result, the quantity of oxygen made available to the tissues for any P_{O_2} is reduced. The combination of inadequate flow and decreased oxygen cause dizziness.

In some cases, prolonged hyperventilation results in muscle spasm and pain, a condition termed tetany. Tetany is caused by inadequate ionized calcium in the blood. When the carbon dioxide content of the blood is reduced, so is the hydrogen-ion concentration. The amount of ionized calcium in solution is a function of the pH. The higher the pH, the lower the ionized calcium. If hyperventilation has raised the blood pH sufficiently, tetany may be seen.

REGULATION OF VENTILATION DURING EXERCISE

One of the most fascinating, and yet unsolved and thus perplexing, fields of study is the response of the breathing mechanism to exercise. As has been repeatedly emphasized, ventilation is so remarkably attuned to tissue oxygen demands that the concentrations of the respiratory gases in the arterial blood rarely change during prolonged, vigorous exercise.

Magnitude of the Ventilatory Changes

Up to a point, increased tissue needs can be met with little or no change in ventilation. At rest, the venous blood contains only about 5 vol percent less O_2 than the arterial blood. Because arterial blood normally contains about 20 vol percent, 15 vol percent remain in the venous blood. During exercise, this reserve can be drawn upon. In vigorous exercise the muscles may use almost the entire 20 vol percent. This increased extraction results from (1) lower tissue P_{O_2}, (2) increased tissue P_{CO_2}, and (3) higher temperature which, like increased P_{CO_2}, shifts the oxygen dissociation curve to the right. Not only is more oxygen extracted from each milliliter of blood, but more blood per unit time flows through the capillaries of contracting muscle because of (1) local vasodilation evoked by locally produced carbon dioxide and decreased blood oxygen, (2) increased cardiac output, and (3) elevated arterial blood pressure.

At the resting level of ventilation, that is about 8,000 ml/min, about 5,600 ml of air containing 1,120 ml of oxygen is presented to the alveoli per minute. At rest, only about 250 ml of oxygen enter the blood each minute. So theoretically about a fourfold increase in oxygen extraction by the tissues could be supplied with no increase in ventilation.

Ventilation, however, does increase. And the increase in ventilation does not occur only after the local mechanisms have first been utilized. To the contrary, at the very beginning of exercise ventilation increases. In fact, the very thought of exercise usually suffices to initiate changes.

In very vigorous exercise, the oxygen utilization increases from 250 ml/min up to 4 to 5 liters/min. The carbon dioxide expired rises from the resting level of about 200 ml/min to approximately 6 liters/min. This means that the respiratory quotient, which is usually about 0.82, rises over unity to about 1.5, or even 2.0. In other words, more carbon dioxide is blown off than is produced by the cells. Finally, ventilation increases from 8 to over 100 liters/min.

Neural Regulation

All current evidence indicates that impulses from the cerebral cortex and from the proprioceptors in the contracting muscles and joints are responsible for the major altera- tions in exercise hyperventilation. Some authorities concede that the respiratory gases, acting centrally or reflexly, bring about fine tuning, so to speak.

Figure 23-7 displays possible mechanisms responsible for exercise hyperventilation. The very thought of exercise apparently suffices to send impulses to the respiratory center. The major stimulus comes when exercise begins. Greater impulse flow from the motor cortex increases the effort of exercise and, at the same, increases the effort of breathing. But in addition, there is a flow of impulses back from the proprioceptors. These two flows are thought to be responsible for driving ventilation from 8 liters/min up over 100 liters. The other mechanisms, hypercapnia, hypoxemia, lactic acid, elevated body temper- ature, and high venous pressure, may enter the picture but, at best, as a final regulator to bring about perfect attunement between oxygen demand and supply.

ABNORMALITIES OF RESPIRATION

A few conditions that cause abnormal breathing were mentioned in Chapter 21. Those, such as paralysis, had to do with the mechanical interference with the movement of air. Here, more complex abnormalities will be discussed.

Cheyne-Stokes Breathing

Cheyne-Stokes breathing, also referred to as periodic breathing, is characterized by a periodic increase and decrease in the depth of breathing. The rate may also wax and wane, but the outstanding change is the periodic variation in amplitude (Figure 23-8). This type of breathing sometimes occurs when there is increased intracranial pressure, or when a person is under the influence of a narcotic. Apparently the sensitivity of the respiratory center is dulled so that it requires a greater stimulus to activate it. The result is a period of apnea followed by a period of hyperpnea. The stimulus, which is probably carbon dioxide in the fluids of the respiratory center, also waxes and wanes, as does the P_{CO_2} in the alveolar air (Figure 23-8). The correlation is such that during apnea carbon dioxide builds up, finally drives ventilation, then the hyperventilation lowers the carbon dioxide abnormally, and apnea again occurs.

The above sequence is typical of an oscillating, or inadequately dampened, feedback mechanism. The wonder is that more homeostatic feedback systems produce a steady end result, rather than an oscillating one.

Cheyne-Stokes breathing is commonly seen in patients with failing hearts. In these cases, the mechanism is different, although again interference with the feedback circuit is the problem. In cardiac patients, there is slowed circulation, increased pulmonary blood

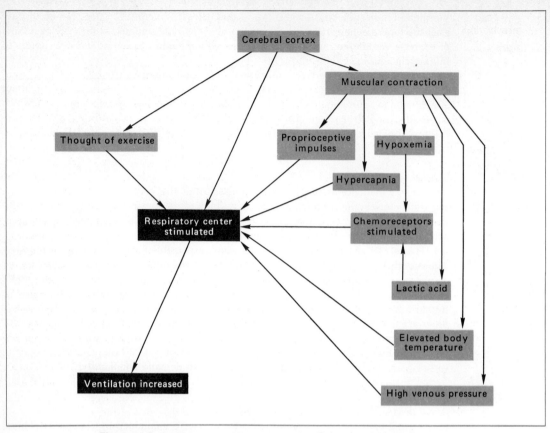

Figure 23-7. Respiratory response to exercise. The cortex initiates the response, and proprioceptors in muscles and joints augment it. Only in very vigorous exercise do the respiratory gases in the arterial blood change enough to play a role.

volume, and pulmonary edema. All three delay the feedback of information from lung to respiratory center, information that is carried in the form of carbon dioxide. In other words, the respiratory center alters ventilation and, as a result, the P_{CO_2} of the alveolar air is altered, but the pulmonary edema slows diffusion, and the large pulmonary blood volume and the sluggish circulation cause a delay in getting the altered blood to the respiratory center. Thus, hyperventilation develops and continues too long, abnormally lowering the P_{CO_2} of the alveolar air. And the cycle begins again.

Dyspnea and Orthopnea

Labored, apprehensive breathing is termed dyspnea. If there is labored breathing in one body position but not, or considerably less, in another, the condition is termed orthopnea.

The dyspneic patient feels as though he is not satisfying his respiratory needs. He has what has been described as "air hunger." The cardinal finding in this disorder is apprehension. Most often dyspnea occurs when the patient must make a greater breathing effort to maintain normal blood gas values. Dyspnea usually results when the tidal-

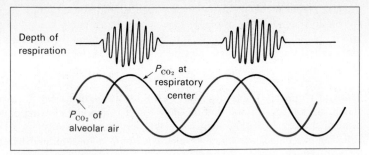

Figure 23-8. Cheyne-Stokes breathing. The feedback mechanism is unable to maintain a constant depth of respiration.

volume–vital-capacity ratio increases. Normally, in young adults, the tidal volume is about 500 ml and the vital capacity about 4,500 ml. The normal ratio, then, is 1:9. In abnormal conditions, such as congestive heart failure, there is accumulation of fluid in the lungs which diminishes the vital capacity. The tidal volume usually increases. Therefore, the combination of elevated tidal volume and decreased vital capacity changes the ratio so that it may be 1:3. The greater this fraction, the more severe the dyspnea. To put it another way, as tidal volume approaches vital capacity, thereby decreasing ventilatory reserve, dyspnea worsens. Even though the respiratory gases in the arterial blood may be kept in the normal range, there is still apprehension. When the respiratory gases are not kept normal, the apprehension becomes maximal.

Orthopnea is common in cardiac patients. They usually find that their dyspnea is less if they sit up than if they lie flat. Physicians and nurses are familiar with the fact that such patients require two or three pillows for comfortable sleep. The mechanism is pulmonary congestion, which interferes with the diffusion of the gases. Thus deeper and faster breathing is required to make the gradients steeper in order to keep the blood gas concentrations normal. Pulmonary congestion is maximal when lying flat, when the force of gravity does not act on the venous return. But when sitting up, gravity does impede venous return and thus helps to diminish pulmonary congestion somewhat.

Cyanosis

Cyanosis, that is, blueness of the skin, occurs when there is too much reduced hemoglobin in the blood flowing through the capillaries. In view of the fact that the primary function of respiration and circulation is to deliver an adequate supply of oxygen to the tissues, cyanosis is a sign of failure of either respiration or circulation, or both. And since it is such a readily observable sign, it has great diagnostic value.

Inadequate oxygenation is termed hypoxia. It may refer to inadequate oxygenation of the air, the blood, or the cells. Hypoxemia, on the other hand, specifically signifies a decreased oxygen saturation of the blood. A complete absence of oxygen is termed anoxia; absence of oxygen in the blood, anoxemia. This extreme condition is rarely, if ever, encountered.

Hypoxemic Hypoxia This seemingly pleonastic term implies that the hypoxemia is due to an inadequate uptake of oxygen by the blood. It may be the result of pneumonia, in which the accumulation of fluids in the lungs impairs the transfer of oxygen from the alveolar air into the blood, or may be the result of respiratory paralysis, drowning, or exposure to low oxygen pressures.

Anemic Hypoxia   A reduction in the hemoglobin concentration of the blood causes anemic hypoxemia. The available hemoglobin is generally completely saturated in the arterial blood, but because there is so little hemoglobin, there is little oxygen available to the cells. In such cases there is insufficient reduced hemoglobin to give skin the blueness of cyanosis.

Stagnant Hypoxia   Many disorders of the cardiovascular system, such as heart failure, shock, arterial spasm, and emboli, may result in marked slowing of the flow of blood through the capillaries. In such cases the tissues extract virtually all the oxygen contained in the blood, and there results a plethora of unsaturated hemoglobin and thus intense cyanosis.

Polycythemia   Cyanosis, at least a mild cyanosis, is often seen in patients with polycythemia, that is, an abnormal increase in red blood cells. This emphasizes the point that has already been made, namely, that cyanosis is caused by excessive amounts of reduced hemoglobin in the blood. In polycythemia there is so much hemoglobin that the amount not completely oxygenated is sufficient to give rise to a mild cyanosis. In addition, the great number of cells increases the viscosity of the blood and slows the flow.

Pneumonia

A disease characterized by inflammation of the lungs is termed pneumonia. More specifically, it is a disorder in which the alveoli contain excessive fluid. Pulmonary edema may be caused in several ways. Most often it is due to an invasion of pneumococci, to the aspiration of foreign substances, or to a failing heart which elevates the pulmonary capillary pressure sufficiently to cause edema. Because of the fluid there is serious interference with the diffusion of the respiratory gases. The magnitude of the interference is clearly shown in cases in which but one lung is involved. In such cases, blood coming from the normal lung is about 97 percent saturated, whereas the blood coming from the diseased lung may be less than 60 percent saturated with oxygen.

Emphysema

This disorder has already been mentioned. It is characterized by the breakdown in alveolar septa so that in the place of many small alveoli, there are fewer larger ones. Not only are the septa lost, but so are the blood vessels normally present in them. In addition, the breakdown of alveoli allows air to escape into interalveolar spaces, where it is trapped. As a consequence, there is (1) greatly reduced surface area for the diffusion of respiratory gases, (2) a decrease in the number of pulmonary capillaries, and (3) a decrease in vital capacity. With reduced surface area and number of pulmonary capillaries, diffusion of the gases decreases. In addition, destruction of the pulmonary vessels increases pulmonary resistance to blood flow and contributes to pulmonary hypertension.

The cause of emphysema is not really known, but smoking is certainly implicated.

Asthma

The term asthma is derived from a Greek term meaning "to pant." Asthma, or more accurately, bronchial asthma, is caused by an allergy to airborne substances. In addition, there is thought to be a psychological component. Whatever the cause, there is edema in the bronchioles as well as bronchial spasm. The result is a decrease in the bronchial lumen, and therefore increased resistance to air flow, and dyspnea.

QUESTIONS AND PROBLEMS

1   What mechanisms bring an increased supply of oxygen to an actively contracting muscle?

2   What is the respiratory center, and where is it located?

3   How do you explain the cause and mechanism of rhythmicity of breathing?

4   What is the Hering-Breuer reflex? What is its major function? Trace the nerve pathways involved.

5   What is the relation between cardiac output and respiratory rate?

6   How would you expect the pressoreceptors in the aortic and carotid sinuses to react to a condition of shock due to severe loss of blood? How would you expect the chemoreceptors in the aortic and carotid bodies to react?

7   What is the relation between muscle and tendon proprioceptors and ventilation?

8   In hypoxia, does lack of oxygen stimulate hyperventilation? Why?

9   What prevents a person from voluntarily sustaining apnea?

10   Why is prolonged hyperventilation accompanied by dizziness? By tetany?

11   What type of hypoxia accompanies a severe case of pneumonia?

12   In what conditions is Cheyne-Stokes breathing commonly seen? Why?

13   Describe the pathology of the respiratory tract in (a) pneumonia, (b) emphysema, and (c) asthma.

PART 5
THE DIGESTIVE SYSTEM

24 ANATOMY OF THE DIGESTIVE SYSTEM

The digestive system consists of the alimentary canal, through which food passes to be digested, and the glands that secrete digestive juices into the canal. The digestive organs are the mouth, stomach, small intestine, and large intestine. The accessory organs are the salivary glands, liver, gallbladder, and pancreas.

In man the digestive system involves elaborate organs, and each region of the system is highly specialized to perform specific functions in digestion. For example, the mouth and its structures are adapted for mastication, salivation, and swallowing of food. The secretory organs along the canal introduce enzymes that convert the broken-up food into a form that can be absorbed through the mucous membrane of the canal into the blood-stream. Finally, the alimentary canal eliminates the residue of ingested food as fecal material.

EMBRYONIC DEVELOPMENT

As described on page 77, the developing flat embryonic disk of the early embryo is rapidly converted into a hollow cylinder lined with endoderm internally and covered by ectoderm externally. This endothelium-lined cavity is the primitive digestive tract, or gut. The central portion is termed the midgut. Anteriorly a tube develops which is known as the foregut. The tube that develops posteriorly is called the hindgut. In the midgut there is an opening into the attached yolk sac, or blastocoele. As the embryo grows, the yolk sac becomes smaller. Finally, at about the fifth week, the yolk sac constricts at its point of attachment and seals off that opening into the midgut.

The cephalic blind end of the foregut expands into the primitive pharynx with five paired vertical ridges, the branchial arches, on its lateral walls. (See Figures 4-6 and 4-7.) On the surface of the embryo the branchial grooves separate the arches and extend inwardly into the pharyngeal cavity to enlarge as the branchial pouches. The pouches give rise to several structures, namely, the thymus, parathyroid, palatine tonsil, and auditory cavity.

Caudal to the pharynx a second expansion of the foregut develops into the stomach with the narrow, tubular esophagus connecting the two structures. Caudal to the stomach the tubular gut forms the coiled intestines.

As both fore- and hindgut elongate, they come into contact with the surface ectoderm. In the head region this ectoderm forms a pitlike depression, the oral fossa, which is separated from the pharynx by a thin membrane. At about the fourth week this pharyngeal membrane breaks through to form the oral cavity. Likewise the blind, caudal end of the hindgut meets a shallow depression of ectoderm, the anal fossa. Upon rupture of the intervening cloacal membrane, the anal opening of the digestive tube is established.

Many of the glands of the body, such as the thyroid, parathyroid, and salivary, as well as the liver and pancreas, arise out of the primitive foregut. These structures appear initially as hollow buds of endodermal tissue which grow into the mesenchyme surrounding the gut. The liver, pancreas, and salivary glands develop an elaborate excretory duct system which retains its connection with the gut lumen, while the thyroid and parathyroid lose their connection and become ductless (endocrine) glands.

Liver

As early as the fourth week, the liver can be identified as a hollow outgrowth of endoderm from the midgut (Figure 24-1). Initially the liver diverticulum is a thick-walled vesicle, or pouch, from which two solid masses, or cords, of cells appear as the primordia of the right and left lobes. As these cords expand, they incorporate the vitelline veins from the yolk sac into their substance to form a network of sinusoids among the developing liver cells. The vitelline veins contribute to the formation of the portal and hepatic veins of the adult. The original liver vesicle also develops into the gallbladder and its associated ducts.

As the liver rudiment continues to grow, it invades a sheet of mesoderm, the septum transversum, lying across the primitive body cavity. This septum divides the thoracic from the abdominal cavity and contributes the major portion of the diaphragm. The liver capsule and its fibrous extensions are derived from this structure. The liver continues to grow rapidly until at 9 weeks it occupies most of the abdominal cavity and represents about 10 percent of the total weight of the embryo. At birth the liver contributes 5 to 6 percent of the total body weight and in the adult only 2 to 3 percent.

Gallbladder

During embryonic development, the gallbladder is intimately associated with the liver. The caudal portion of the original hepatic diverticulum from the foregut gives origin to the gallbladder. Initially the stalk of attachment of the diverticulum is solid, but later it becomes canalized as the cystic duct.

Pancreas

Although the pancreas is a single gland in the adult, it arises during the fourth week from two separate endodermal outgrowths called the dorsal and ventral pancreas. The

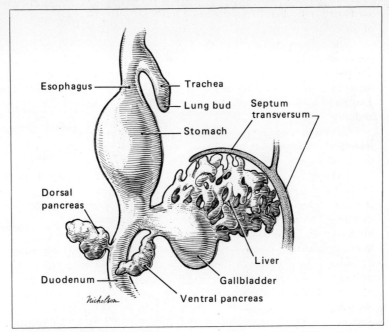

Figure 24-1. Digestive organs of a 4-week-old embryo. Both liver and pancreas develop from evaginations of the foregut.

dorsal pancreas is an outgrowth of the midgut near the hepatic diverticulum, while the ventral pancreas arises directly from the liver diverticulum. The two primordia come in contact and fuse. The larger dorsal portion gives rise to all the definitive pancreas except a small part of the head and neck which come from the ventral part. The main excretory pancreatic duct is, however, derived from the ventral pancreas, while the smaller accessory duct develops from the dorsal pancreas.

Early in development there is no differentiation of pancreatic cells. At about the third month, some secretory cells appear arranged in alveoli as exocrine cells, while other clusters of cells become the endocrine, islet cells. It is interesting to note that both the exocrine and endocrine cells are derived from the same embryonic cell type.

DIGESTIVE ORGANS
Oral Cavity

The oral cavity (Figure 24-2) extends from the lips to the oropharynx. It is lined with stratified squamous epithelium. It contains the teeth and tongue, and the openings of the parotid, submandibular, and sublingual salivary glands. The oral cavity may be subdivided into the mouth cavity proper and the vestibule. The vestibule is the area between the teeth and cheeks or lips. The mouth cavity lies internal to the teeth and communicates posteriorly with the oropharynx.

The roof of the oral cavity is formed by the palate, which is hard toward the front and soft toward the back of the cavity. The bones of the hard palate consist of the palatine processes of the maxillae anteriorly and the horizontal portion of the palatine bones posteriorly.

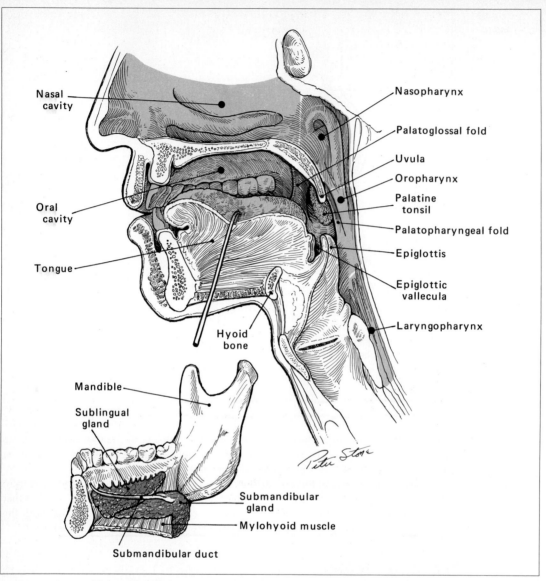

Nasal cavity

Oral cavity

Tongue

Hyoid bone

Mandible

Sublingual gland

Submandibular gland

Mylohyoid muscle

Submandibular duct

Nasopharynx

Palatoglossal fold

Uvula

Oropharynx

Palatine tonsil

Palatopharyngeal fold

Epiglottis

Epiglottic vallecula

Laryngopharynx

Figure 24-2. Sagittal view of oral, nasal, and pharyngeal cavities. The lower drawing shows the relation of the sublingual and submandibular glands to the floor of the mouth.

The soft palate is formed primarily by muscles. The uvula is a small midline conical projection which hangs down from the posterior border of the soft palate.

The lips develop from both sides of the face toward the midline. If the palatine processes of the maxillary bones fail to fuse, a cleft to one side of the midline may persist in the upper lip (harelip) or in both the upper lip and the palate (cleft palate). The latter condition creates an opening between the oral and nasal cavities.

Tongue The tongue is a fibromuscular organ located partially in the oral cavity and partially in the pharynx. It is shaped like a short boot turned upside down. The sole, corresponding to the dorsum of the tongue, is the portion visible when a person's mouth is widely opened. The root of the tongue is anchored to the hyoid bone. In addition to subserving taste functions, the tongue assists in speech, mastication, and swallowing. The bulk of the tongue is formed by muscles, which are described in Chapter 6.

In mastication the tongue functions to push the food between the teeth, and once the food has been broken up by the teeth, the tongue forms a bolus, or ball, of food which it propels into the oropharynx.

The dorsum of the tongue is covered by numerous minute projections, the papillae, which vary in shape and distribution. Small, slender, threadlike filiform papillae, by far the most numerous, are dispersed over the entire dorsal surface; larger, mushroom-shaped fungiform papillae are fewer in number and are scattered throughout the dorsum of the anterior two-thirds of the tongue, while at the back 10 to 12 large, circular, and dome-shaped vallate papillae converge into a wide "V."

Taste buds (see Figure 13-13) may be present on any of the papillae but are most numerous at the base of the vallate. Rounded swellings are present at the root of the tongue. These are aggregates of lymph nodules called the lingual tonsil.

Malformation of the tongue may result in its incomplete separation from the floor of the mouth. In this condition, ankyloglossia (tongue-tie), the tongue is not freely movable, and speech may be severely affected.

Teeth Teeth arise as outgrowths of the primitive mouth (buccal) epithelium and the surrounding connective tissue. They develop in the alveolar processes of the maxillae and mandible and are named from their position in each quadrant of the jaw. Teeth are unique among body structures in that the child's first set of 20, the deciduous teeth, is shed and replaced by 32 permanent teeth.

The deciduous teeth (Figure 24-3) develop before birth and erupt through the gingivae (gums) at rather regular intervals during early childhood. They appear in each quadrant beginning at the midline in the following sequence: central and lateral incisors (four in each jaw), cuspids (two in each jaw), and first and second molars (four in each jaw). They begin to erupt at about 6 months and are all present by about $2\frac{1}{2}$ years. The teeth of the lower jaw usually appear earlier than those of the upper. The deciduous teeth are shed in their order of appearance between the sixth and eleventh years of age, to be replaced by permanent teeth.

The first permanent tooth is the first molar. It appears at about the age of 6. Other permanent teeth appear between ages 11 and 21 (Figure 24-4). In addition to the incisors, cuspids, first molars, and the first and second premolars (which replace the first and second deciduous molars), the permanent set is increased by the second and third molars. The third molars, or "wisdom" teeth, expected between ages 17 and 25, frequently become impacted and cannot erupt.

The fibroelastic periodontal membrane holds the teeth solidly in the sockets of the jawbones. A tooth consists of an exposed crown and one or more buried roots (Figure 24-5). The junction of the crown and root, the neck, is surrounded by the gingival margins. Enamel, the hardest substance in the body, covers the crown, while cementum, a hardened type of bone, covers the roots. Beneath the enamel is a thick layer of dentine, another hard, calcified substance much like enamel but exquisitely sensitive. Beneath the cementum is a thinner layer of dentine. The dentine surrounds a central cavity of the tooth which is filled with pulp. Pulp is connective tissue laced with nerves and blood vessels which supply the tooth. Dentine is the main substance of the tooth.

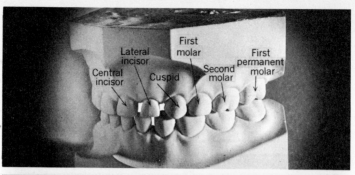

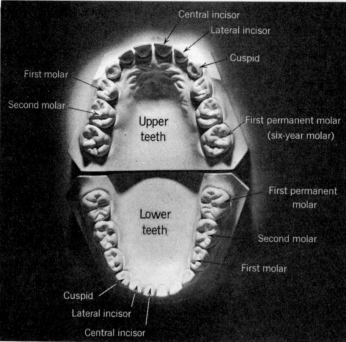

Figure 24-3. Primary teeth. All these are deciduous teeth except the first permanent molar, which appears at about age 6.

Caries is localized decay in the dentine, enamel, or cementum. Exposure of the dentine or, more rarely, the pulp brings the pain of toothache.

Salivary Glands

The salivary glands are not true digestive organs but are accessory organs, as are the liver and pancreas. However, for the reader's better visualization they will be described here rather than with the other accessory organs.

The salivary glands are the sublingual, submandibular (also called submaxillary), and parotid glands. They are large, paired glands (see Figure 24-2). The ducts of the salivary glands empty into the oral cavity. The salivary glands are of the branched tubuloalveolar type.

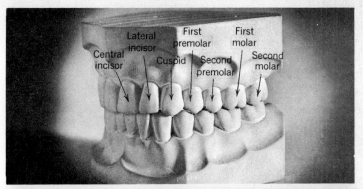

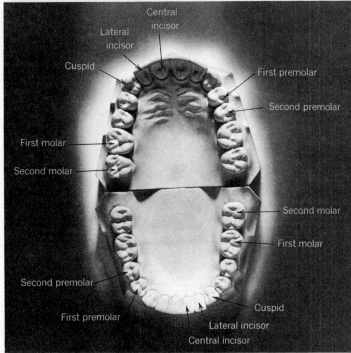

Figure 24-4. Permanent teeth. The first molar is the 6-year molar.

The sublingual gland is in the floor of the mouth. Its location is demarcated by a transverse ridge, the sublingual fold, passing between the inner surfaces of the mandible. A series of 12 to 15 small ducts open along the length of this ridge. This gland is largely composed of mucous-type cells with some serous cells interspersed throughout.

The submandibular gland, on the internal surface of the mandible, is predominantly serous in type with only occasional mucous cells (Figure 24-6). A single duct crosses the floor of the mouth, deep to the sublingual fold, to open at the medial end of the fold.

The parotid is the largest of the salivary glands. Its major portion lies superficial to the ramus of the mandible, but a deep or retromandibular segment passes to the internal surface of the mandible. The duct of the parotid extends anteriorly to cross the masseter

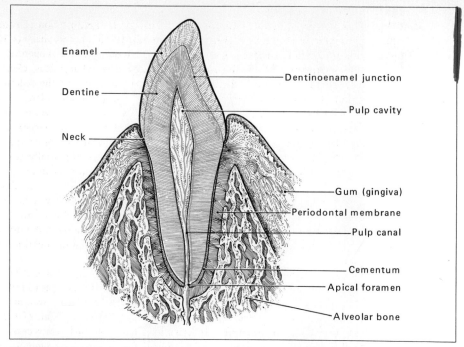

Figure 24-5. An incisor and related structures. Longitudinal section.

muscle. It pierces the buccinator muscle and fat pad and opens into the oral cavity opposite the second upper molar. The opening of the duct of the parotid may be felt with the tongue as a small bump on the inside of the cheek. The parotid is a pure-serous-secreting gland.

Saliva secreted by these glands aids swallowing by moistening the food. Ptyalin, an enzyme component of saliva, initiates the digestion of carbohydrates (see Chapter 25).

Figure 24-6. Section through a submandibular gland (× 210).

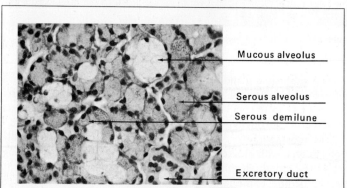

Pharynx

The pharynx (Figure 24-2) does not contribute to the digestion of food but merely acts as a passageway to the esophagus. The middle portion of the pharynx, or the oropharynx, opens anteriorly into the oral cavity. In the lateral wall of this opening (fauces) two folds of mucous membrane, the palatoglossus and palatopharyngeus, overlie muscles of the same name. The folds demarcate the bed, or location, of the palatine, or true, tonsil. Clinically these folds are called the anterior and posterior pillars, respectively.

At the level of the hyoid bone, at the base of the tongue, the oropharynx divides into two tubes, the anterior larynx and the posterior laryngopharynx. During swallowing, the larynx is elevated, and the posterior segment of the superior aperture, the aditus of the larynx, comes into apposition with the epiglottis to partially close the larynx. This keeps food from entering the air passageways. If this occlusion does not occur and food is aspirated into the larynx, a nervous reflex initiates violent coughing.

The anterior wall of the lower portion of the oropharynx is formed by the root of the tongue as it attaches to the hyoid bone. Three folds of mucous membrane, the median and two lateral glossoepiglottic folds, extend between the dorsum of the tongue and the epiglottis. The small fossa between the medial and lateral folds is known as the vallecula.

The laryngeal portion of the pharynx has no special characteristics. It extends from the termination of the oropharynx at the level of the laryngeal aditus to the origin of the esophagus at the level of the cricoid cartilage. Its anterior wall and the posterior wall of the larynx is a common structure. Between the lateral wall of the pharynx and the upper portion of the larynx is the small, pocketlike piriform recess. If in swallowing a pill one has the feeling it is "caught in the throat," it may be temporarily lodged in this recess.

Esophagus

The alimentary canal continues from the pharynx as the esophagus, which traverses the posterior mediastinum of the thorax. The esophagus passes through the esophageal hiatus in the central tendon of the diaphragm to join the stomach at the level of the tenth thoracic vertebra.

The esophagus functions simply as a passage for food from the pharynx to the stomach. It is characterized by a heavy layer of stratified squamous epithelium and an unusually well-developed muscularis mucosae (see Figure 3-6). Esophageal mucous glands are irregularly spaced in the submucosa. Since it is not covered by peritoneum, its outer layer is called the tunica adventitia (see ahead).

As the esophagus traverses the thorax, it is surrounded by the esophageal plexus, formed by fibers of the vagus nerve. The musculature of the esophagus is skeletal in its upper third and smooth in its lower two-thirds. Swallowing, or deglutition, thus may be voluntarily initiated, but once begun, smooth muscle takes over, and this action becomes involuntary (see Chapter 25). Food is propelled through the esophagus by peristaltic activity.

An engorgement of the esophageal veins at the junction of the esophagus and stomach occurs if venous return through the liver is impeded. A surgical solution is to establish a portacaval anastomosis (anastomosis between the portal vein and inferior vena cava). If engorgement persists over a long period of time, esophageal varices (tortuous, dilated veins) will develop. This is a serious problem in that they may be easily ruptured by a severe coughing episode or by any vigorous movement of the diaphragm.

Histology of the Alimentary Canal

The wall of the alimentary canal from the esophagus to the anal canal has the same basic structure. Typically there are four encircling layers, or tunics. In certain organs these layers are increased or reduced in size. Furthermore, where special functions are involved, there are minor modifications of the basic structure. These variations will be touched on in the anatomical description of the organ. The following is the basic structure of the wall of the alimentary canal from the lumen outward:

1 Tunica mucosa (mucous membrane).
   a Epithelium, either stratified squamous or simple columnar.
   b Lamina propria, a delicate areolar connective tissue layer usually infiltrated with lymphocytes and lymph nodules.
   c Muscularis mucosa, a thin layer of smooth muscle which forms the boundary between mucous membrane and submucosa.
2 Tunica submucosa, a loose areolar connective tissue layer containing larger blood and lymphatic vessels, nerves, and glands in certain regions.
3 Tunica muscularis, a double layer of smooth muscle, the inner sheet lying circularly, the outer sheet running longitudinally, with nerve and vascular plexuses between the layers. The tunica muscularis controls the diameter of the intestine and propels its contents toward the anus.
4 Tunica adventitia or serosa, the fibrous outer coat. This tunic carries large vessels and nerves. If the organ is covered by peritoneum (see later), this layer is called the serosa; otherwise, tunica adventitia.

Peritoneal Cavity

The innermost layer of the abdominal wall consists of a serous-secretory membrane. At several sites within the abdominal cavity this membrane is reflected (drawn away) from the abdominal wall to cover organs. The portion of the membrane extending to the organ is a mesentery, or visceral ligament. The portion covering the organ is termed visceral peritoneum; that adhering to the abdominal wall is parietal peritoneum. The peritoneal cavity is the space between the visceral and parietal layers. Because of this arrangement the inner surface of the abdominal cavity is smooth and slick, as is the surface of the covered organs within it. These two well-lubricated layers of peritoneum facilitate movement of organs in the abdomen.

    The mesenteries provide some support for their respective organs. Furthermore, vessels and nerves serving the organs pass between the two layers of the mesenteries. The vessels and nerves branch from major trunks lying along the posterior wall of the abdomen behind the peritoneum.

    Visceral ligaments are also present between certain organs of the abdominal cavity. For example, the lesser omentum connects the liver and the stomach. The lesser omentum arose during embryonic development as the liver evaginated from the primitive foregut, pushing visceral peritoneum before it.

    The peritoneal cavity is not limited to the abdominal cavity but extends into the pelvic cavity.

Stomach

The stomach extends between the esophagus and the duodenum. It lies under the diaphragm in the upper left portion of the abdominal cavity. The anterior abdominal wall is in front of the stomach, and the quadratus lumborum muscle, the inferior vena cava, the aorta, and the pancreas are among the structures behind it.

Two curvatures, the lesser and the greater, are recognized. A double-layered visceral ligament, the lesser omentum, originating at the liver, is attached along the length of the lesser curvature and a portion of the duodenum. It connects the stomach with the liver and, besides assisting to support the stomach, provides a route for the branches of the celiac artery and the left gastric and hepatic arteries to reach their destination. The two layers of the lesser omentum separate along the lesser curvature of the stomach and pass across the gastric surfaces. The ligament becomes the serosa on these surfaces. At the greater curvature the layers rejoin and continue inferiorly as the greater omentum, which drapes the intestines, passes upward over the transverse colon, and continues posteriorly as a component of the transverse mesocolon. The portion lying between the stomach and the colon is termed the gastrocolic ligament. Arteries passing between the layers of the greater omentum include the short gastric branches and the left and right gastro-epiploic branches.

The stomach is divided into the body, cardiac, fundic, and pyloric regions (Figure 24-7). The cardiac region is the small area that surrounds the esophageal opening. In

Figure 24-7. Stomach, duodenum, and pancreas. Coronal section.

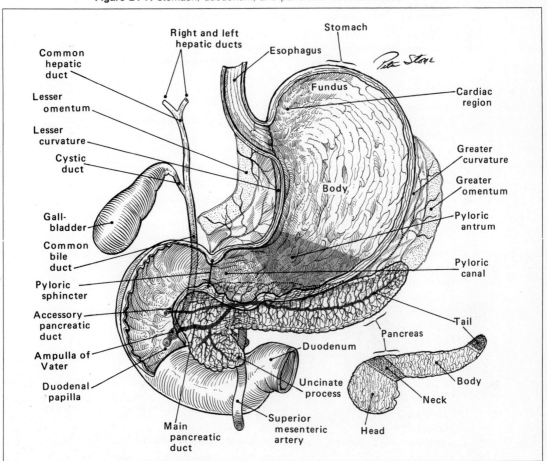

the fundic region the stomach dilates so markedly that it rises above the level of the esophageal opening. Below is the body, the main portion of the stomach, which extends to the pyloric region. As the stomach approaches the duodenum, it narrows considerably, becoming the pyloric canal, which extends to the pylorus, the junction of the stomach and duodenum. At this point the muscular wall thickens into the pyloric sphincter, which guards the entrance to the duodenum.

The stomach, although conforming to the general pattern of the alimentary canal, has several special features. The mucosa and submucosa are thrown into heavy longitudinal folds, the rugae, which are flattened when the stomach is full.

The mucosal surface is pitted by many small openings. These gastric pits are the site of the gastric glands, which extend the entire thickness of the mucosa. Gastric glands are simple or coiled tubular and are named cardiac, fundic, or pyloric glands depending on their location in the stomach.

The fundic glands are simple straight glands located in the fundus and body of the stomach (Figure 24-8). These are the most numerous of the gastric glands. They are lined with three types of cells. Mucous (neck) cells surround the constricted upper portion of the gland where it empties into the gastric pit. The remainder of the gland is lined by

Figure 24-8. Fundic glands of the stomach. Section through a portion of the tunica mucosa.

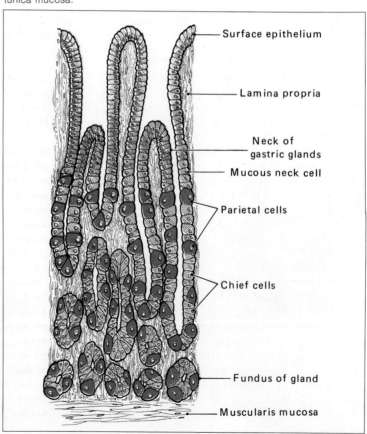

Surface epithelium

Lamina propria

Neck of gastric glands

Mucous neck cell

Parietal cells

Chief cells

Fundus of gland

Muscularis mucosa

cuboidal chief cells, which secrete pepsinogen, a precursor of the enzyme pepsin, and the larger, wedge-shaped parietal cells, which produce hydrochloric acid. Cardiac glands occupy a narrow zone around the esophageal opening and are lined only with mucous cells. Pyloric glands, also containing mucous cells, are found in the pyloric region. Both pyloric and cardiac glands are coiled tubular glands.

Another adaptation of the stomach occurs in its muscular tunic. It will be recalled that in the alimentary canal the tunic usually consists of an outer layer with longitudinal orientation and a second layer arranged circularly. In the stomach there is a third layer of muscle, oriented obliquely and lying internal to the circular layer. It serves to reinforce the curvatures of the stomach.

The pylorus may become partially occluded by a tumor or by hypertrophy of the pyloric sphincter. Such pyloric stenosis results in retention of food in the stomach. Surgery may be needed to correct the condition.

Small Intestine

The small intestine extends approximately 4.5 m and is therefore the longest part of the alimentary canal. It is relatively small in caliber. The first 30 cm of this tube are designated the duodenum. The next 1.5 or 1.75 m are the jejunum, and the final 2.5 or 2.7 m are the ileum. The small intestine joins the large intestine at the ileocecal junction (see Figure 24-10).

The duodenum, lying behind the peritoneum, has a C-shaped course as it passes posteriorly and around the head of the pancreas (see Figure 24-7). This first part of the duodenum is particularly vulnerable to ulcers, as it receives the full impact of the highly acid chyme driven from the stomach by peristaltic waves upon relaxation of the pyloric sphincter.

The jejunum and ileum are suspended in the abdominal cavity by a mesentery. This extension of the peritoneum passes from its attachment on the posterior wall at the duodenal flexure, or bend, to the right iliac fossa at the junction between the ileum and the large intestine (see Figure 24-11). At the flexure where the duodenum joins the jejunum the mesentery is thickened by connective tissue and muscle fibers. This is the ligament of Trietz. Thus while the duodenum is fixed in position, the jejunum and ileum, due to the mesentery, possess mobility. Twelve to fifteen intestinal branches of the superior mesenteric artery course between the layers of the mesentery to supply the small intestine.

Throughout the length of the small intestine there are many villi, that is, fingerlike processes rising out of the mucosa and projecting into the intestinal lumen (Figure 24-9). At the center of each villus is a lymphatic channel called a lacteal, which is important in the absorption of fats. A network of blood capillaries surrounds the lacteal.

The mucosa contains many straight tubular intestinal glands, or crypts of Lieberkühn, which extend the depth of the mucosa. Both villi and glands are covered by a layer of simple columnar epithelium.

In the columnar epithelium are certain cells specialized for mucus secretion (goblet cells), absorption (lining cells), and enzyme secretion (Paneth cells), as well as argentaffin cells, which release serotonin, a substance with widespread physiological effects.

Although several digestive enzymes such as lipase, amylase, and proteinase are produced in the small intestine, only the Paneth cell appears to be involved in their manufacture. Even here the evidence is indirect. For example, these rather large round cells, located in the depths of the intestinal crypts, contain reddish-orange granules. After a period of fasting there are more of the granules, while following a meal there are fewer, suggesting that these cells have poured digestive enzymes into the digestive tract by the release of their granules.

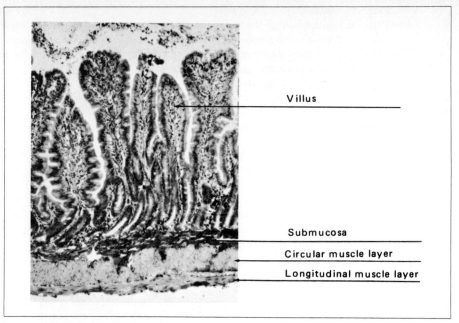

Figure 24-9. Villi lining the small intestine (× 100).

The most numerous cells of the mucosa are the so-called intestinal lining cells, adapted for rapid absorption of soluble nutrients from the digestive tract into the bloodstream. Their free ends facing the lumen are modified with a somewhat thickened border known as the striated border, so called because of its faintly striped appearance when seen under high magnification. With the electron microscope these delicate striations can be seen as closely packed microvilli intimately involved in nutrient absorption.

In the submucosa of the duodenum there are large clusters of mucus-secretory cells termed Brunner's glands. In the submucosa of the ileum there are aggregates of lymph nodules called Peyer's patches.

In much of the small intestine, especially the jejunum, the mucosa is thrown into folds, termed circular folds, or folds of Kerckring. These folds encircle the lumen and thus lie at right angles to the direction of the tube. The folds serve to increase the surface area of the lumen.

Large Intestine

The first part of the large intestine is the cecum. The ileum empties into the cecum by way of the ileocecal orifice, guarded by the ileocecal valve (Figure 24-10). The cecum begins in a blind pouch slightly below the ileocecal orifice. The vermiform appendix, a tubelike extension from the cecum, hangs 7 to 10 cm below. If the appendix becomes infected and acutely inflamed (appendicitis), it must be removed to prevent rupture and spread of the infection to the peritoneum, leading to peritonitis.

From the cecum, the ascending colon passes upward to the level of the liver and makes a right-angle turn (the hepatic flexure) to pass obliquely across the abdomen as the transverse colon (Figure 24-11). As the transverse colon reaches the spleen it makes a second, inferiorly directed turn (the splenic flexure) to become the descending colon.

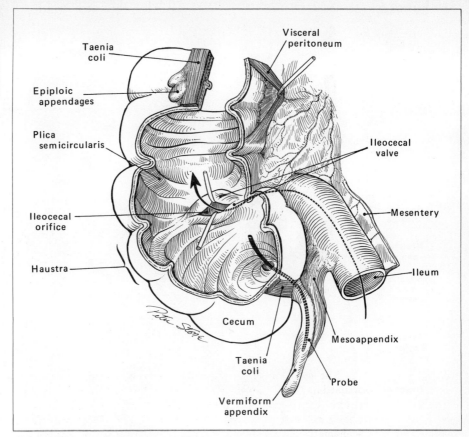

Figure 24-10. Ileocecal junction. Openings into the cecum are shown by arrow and probe. The pressure of chyme is sufficient to open the ileocecal valve.

When this portion of the colon reaches the left iliac fossa, it describes an S, during which it is termed the sigmoid colon. It crosses the brim of the pelvis and enters the pelvic cavity, where it becomes the rectum.

Three anatomical characteristics of the colon differentiate it from the small intestine: the haustra, taeniae coli, and epiploic appendages. The taeniae coli are three heavy, thick longitudinal bands of muscle which represent the external longitudinal muscular coat along this part of the alimentary canal. They extend the length of the colon. At the appendix and again at the rectum, the three bands fan out into a uniform layer of muscle. Because the bands are shorter than the colon itself, they pucker the wall of the colon to form small pouches along its entire extent. This gives the colon a scalloped appearance. The pouches are called haustra. The mucosa between adjacent haustra lies in folds that half encircle the lumen of the tube. The epiploic appendages are small tabs of fat on the external surface of the colon.

Rectum

The rectum is the tube extending 10 to 15 cm beyond the pelvic brim. The rectum lies adjacent to the sacrum and follows its curvature. The anterior surface of the rectum

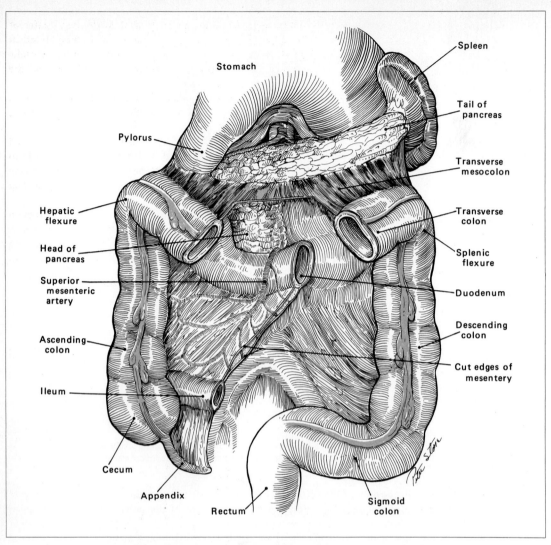

Figure 24-11. Posterior abdominal wall. Most of the transverse colon has been removed to expose the duodenum.

is covered with peritoneum folded back from the uterus or urinary bladder. The enclosure that results is termed the rectovesicular pouch in the male and the rectouterine pouch in the female. At the sides, the peritoneum covering the rectum is attached firmly to the sacrum, and this connection aids in holding the rectum in position.

Histologically the rectum is similar to the colon except that in the rectum the taeniae coli fan out and enclose the tube in an outer longitudinal layer of uniform thickness.

The rectum acts as a reservoir for the semisolid feces. Internally the lumen is modified by several transverse mucosal folds known as rectal "valves." These help to support the fecal mass as the rectum fills.

Anal Canal

The last 5 or 7.5 cm of the large intestine form the anal canal, which terminates at the surface of the body as the anus (Figure 24-12). The anal canal is maintained at approximately a 90-degree angle with the rectum by the normal tonus of the puborectalis muscle. In defecation (evacuation of the bowel) this muscular "sling" relaxes, and the rectum straightens to permit easy passage of the stool.

The mucosa of the anal canal forms a series of longitudinal folds, the anal columns. At their inferior extent, adjacent anal columns are joined together by a small transverse fold to form the anal valves.

An internal and an external muscular sphincter guard the anal opening. The internal anal sphincter consists simply of a thickening of the circular layer of smooth muscle in the wall of the digestive tube near its termination. The external anal sphincter is a ring of skeletal muscle in the anal triangle. Its superficial portion attaches to the skin. Its normal tonus wrinkles the skin around the anus.

The veins at the junction of the rectum and anal canal or at the anocutaneous junction may become varicosed. If this occurs, those at the former are called internal hemorrhoids; at the latter, external hemorrhoids. These varicosities may be traumatized from stretching of the anal canal with each bowel movement. In this condition they frequently burst and bleed during defecation.

Figure 24-12. Anal canal. When the rectum is empty, the rectal valves overlap and help support the fecal mass, which accumulates in the sigmoid colon prior to defecation.

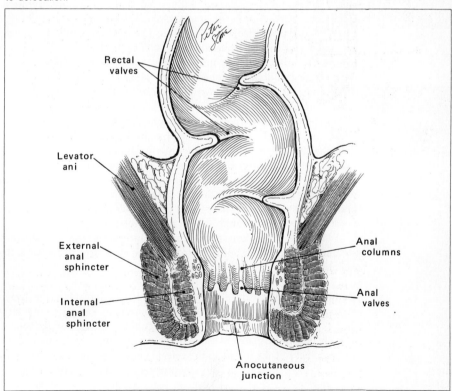

If for any reason the rectum is removed surgically, the distal end of the colon is joined to a surface opening made in the lateral abdominal wall in a procedure called a colostomy. Subsequent evacuation of the bowel then occurs through this artificial opening and into a plastic bag affixed to the margins of the opening.

Innervation of the Alimentary Canal

All the organs of the alimentary canal are served by both sympathetic and parasympathetic fibers. Sympathetic fibers are transmitted by splanchnic nerves from the sympathetic trunk. Parasympathetic fibers issue from the sacral nerves (pelvic splanchnic branches) and the vagus. The two kinds of fibers lie in plexuses along the major vessels of the abdominal aorta. The fibers following the celiac and mesenteric arteries pass to a plexus located in the submucosa of the gastrointestinal tract (the submucosal, or Meissner's, plexus) and to a plexus found between the circular and longitudinal layers of the tract's muscular walls (the myenteric, or Auerbach's, plexus).

ACCESSORY DIGESTIVE ORGANS
Liver

The liver is the largest organ in the body. It is roughly wedge-shaped, with the base of the wedge on the right. The superior surface is convex and underlies the entire right dome of the diaphragm and approximately one-third of the left dome. The posterior and inferior surfaces are not clearly demarcated and together are often called the visceral surface (Figure 24-13). On the right side this surface is in contact with the colon at the right colic flexure, and also the kidney, adrenal glands, and the duodenum. On the left side it is in contact with the stomach and spleen.

The liver is divided into four lobes. The right and left lobes are separated by fissures on the inferior surface of the liver. Parallel to these fissures there are two shallow fossae for the gallbladder and the inferior vena cava. The right lobe is further subdivided into the caudate and quadrate lobes, but the separation is apparent only on the inferior side of the liver. Between these two minor lobes is the porta hepatis, the place where veins, arteries, nerves, and lymphatic vessels enter or leave the liver.

Most of the liver is covered by visceral peritoneum. An exception is the crown-shaped "bare area," which touches the diaphragm. Ligaments of peritoneum connect the liver with the diaphragm and with the anterior abdominal wall. The coronary ligament, for example, extends from the margins of the bare area and attaches to the undersurface of the diaphragm and so helps to hold the liver in position. From the anterior surface of the liver a sheet termed the falciform ligament extends forward and attaches to the anterior abdominal wall. Its inferior free border encloses the ligamentum teres, a fibrous cord which was the umbilical vein during fetal life.

The lesser omentum leaves from the inferior surface of the liver and extends to the stomach. Its right free border surrounds the structures penetrating the liver at the porta hepatis.

Histology   A thin capsule of connective tissue surrounds the liver and sends fibrous strands into the liver mass to partition it into many irregular units termed lobules (Figure 24-14). Each lobule consists of plates of cuboidal cells arranged radially around a central vein. Between the plates of cuboidal cells are sinusoids through which blood flows from the periphery of the lobule to the central vein. The sinusoids are lined with endothelium, in which are scattered stellate macrophages termed Kupffer cells.

Two vessels and a duct course together in the connective tissue between lobules.

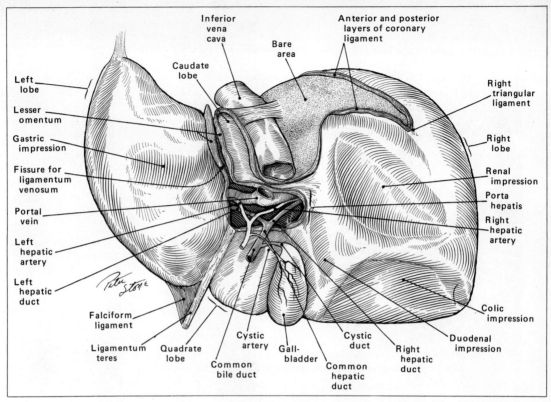

Figure 24-13. The liver. Posteroinferior aspect.

They are together termed the hepatic triad and include branches of the hepatic artery and portal vein and a small bile duct.

The typical hepatic cell is polyhedral in shape and enclosed in a delicate plasma membrane. Its nucleus is large and round. In the cytoplasm are fat droplets, glycogen, and protein granules. A cross section through a lobule suggests that the hepatic cells are arranged in cords extending outward from the central vein like spokes of a wheel. Actually this is an illusion, and a three-dimensional view shows that the cells of a lobule in fact resemble a honeycomb.

Circulation   The blood which bathes the hepatic cells is derived from two sources. Approximately three-fourths comes from the portal vein, which drains the digestive system and therefore is rich in nutrients. The remainder enters by way of the hepatic artery and is rich in oxygen. Terminal branches of these two vessels empty into the sinusoids of the liver. Thus the liver cells are bathed in a mixture of venous and arterial blood. The blood finds its way past the cells and into the central vein. From the central veins of each lobule the blood flows into larger, sublobular veins which empty into the hepatic veins. The hepatic veins in turn empty into the inferior vena cava close to the place where it pierces the diaphragm. Note that while there are two sources of blood into the liver, there is only one pathway out.

One of the important functions of the liver cell is to produce bile, a thick yellow-green

597

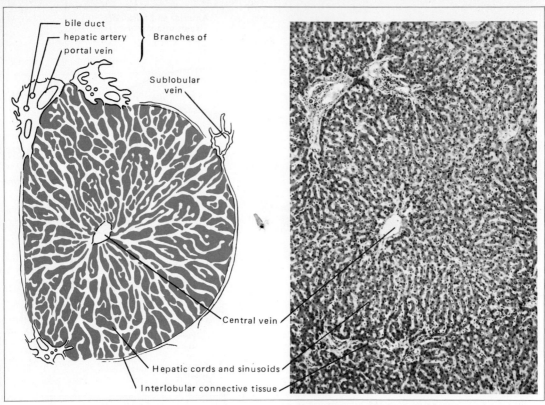

Figure 24-14. A liver lobule. The three structures at the upper left of the drawing form a hepatic triad. (Photomicrograph × 57.)

Image labels:
- bile duct
- hepatic artery
- portal vein
- Branches of
- Sublobular vein
- Central vein
- Hepatic cords and sinusoids
- Interlobular connective tissue

fluid which plays an essential part in digestion of fats (see Chapter 25). The bile is transported from the hepatic cells into bile capillaries located between the cells. Bile follows these capillaries toward the periphery of the lobule (therefore, in a direction opposite that of blood) and empties into the interlobular bile duct of the hepatic triad. The interlobular ducts coalesce and ultimately form the right hepatic duct, draining the right lobe of the liver, and the left hepatic duct, draining the left lobe. The right and left ducts combine in the common hepatic duct, which leaves through the porta hepatis. The common hepatic duct joins the cystic duct, through which bile enters and leaves the gallbladder.

Gallbladder

The gallbladder (Figure 24-7) is a pear-shaped, hollow saclike organ about 7 to 10 cm long. It lies in a shallow fossa on the inferior surface of the liver, to which it is attached by loose connective tissue. Its duct, the cystic duct, joins the common hepatic duct from the liver to form the common bile duct which drains into the duodenum.

The principal function of the gallbladder is the storage of the bile constantly being secreted by hepatic cells. When food containing fat enters the small intestine, the gallbladder contracts, forcing bile through the cystic and common bile ducts and into the duodenum.

The wall of the gallbladder has an outer serous coat, a middle muscular coat, and an inner mucosal coat. The mucosal lining is thrown into folds, or rugae, especially when the gallbladder is empty. Simple columnar epithelium cells devoid of goblet cells line the lumen.

Pancreas

The pancreas is an oblong, flattened gland, about 15 cm in length, which lies in an oblique plane between the duodenum and the spleen (see Figure 24-7). It is on the posterior body wall, behind the parietal peritoneum. A head, neck, body, and tail can be identified. The head lies in the concavity formed by the duodenum and is firmly attached to the duodenum. The neck, extending to the left, crosses in front of the aorta at the level of the superior mesenteric artery. The body, the major part of this gland, supplies a bed for the stomach. The tail of the pancreas touches the spleen.

The pancreas is divided by connective tissue into lobules. The blood and nerve supply of the lobules lies in this connective tissue. The pancreas is composed of both exocrine and endocrine cells. The exocrine portion, which manufactures digestive enzymes, consists of tubuloalveolar glands lined with serous-secretory cells. In this respect the pancreas closely resembles the salivary glands. The cells that secrete enzymes are termed acinar

Figure 24-15. Section through the pancreas ($\times$ 185). The acinar cells are exocrine cells. The enzymes they produce are carried off through excretory ducts. The islet cells are endocrine cells and release secretions into the bloodstream by way of capillaries surrounding the islets.

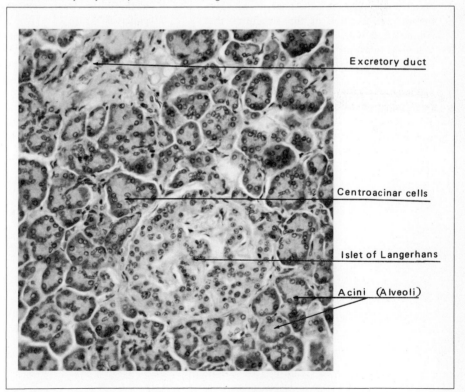

Excretory duct

Centroacinar cells

Islet of Langerhans

Acini (Alveoli)

cells and are found at the terminal ends of the glands. They empty their secretions into a duct system that eventually communicates with a large excretory duct, termed the main pancreatic duct. This extends the length of the pancreas and merges with the common bile duct just before entering the duodenum. In some instances there is an accessory pancreatic duct lying parallel to the main duct and with its own outlet into the duodenum.

The endocrine portion of the pancreas consists of cells lying in small clusters known as the islets of Langerhans (see Figure 24-15). The cells of the islets are arranged in cords surrounded by wide blood capillaries. With ordinary staining methods the cytoplasm of islet cells appears clear and nongranular. With special stains the cytoplasm of some cells appears to contain brownish pink granules. Other cells show larger red granules. The former cells are termed beta cells and are known to produce insulin, a hormone necessary for the proper metabolism of glucose. Insufficient production of insulin results in diabetes mellitus (see Chapter 31). In this condition the glucose content of the blood rises markedly. The excess is not reabsorbed by the kidney tubules but rather is excreted in the urine. The cells with red granules in the cytoplasm are known as alpha cells. They manufacture glucagon, which raises blood sugar. The hormones made by the islet cells are discharged into the bloodstream via the surrounding blood capillaries.

QUESTIONS AND PROBLEMS

1. Name the secretory products of the salivary glands, liver, pancreas, and glands of the stomach, small intestine, and large intestine.
2. Name the openings into the pharynx, and give the relative position of the structures that communicate with it.
3. What is a sphincter muscle? Name the sphincters of the alimentary canal, and give their function.
4. What structures traverse the diaphragm?
5. Differentiate between visceral ligaments and skeletal ligaments macroscopically and microscopically.
6. Discuss the differences in structure between the small and large intestine. How do these lead to the differences in function?
7. Describe the anatomical relation of the liver to the biliary tract, stomach, duodenum, and pancreas.
8. What are the distinguishing structural features of the liver capillary bed? What is unusual about blood circulation through the liver?
9. What kind of gland is the pancreas? Describe its histology in relation to its functions.
10. Describe the structure of a tooth. Name the deciduous teeth, and give the approximate age of their appearance.

25 DIGESTION

The alimentary tract, which was described in the previous chapter, extends from the mouth to the anus. Food is placed in the mouth and then, by a series of alimentary activities, it is moved through the entire length of the tract. During this passage, the food is digested, and the residue is prepared for elimination.

The term digestion embraces all the processes by which food is converted into an assimilable form, into end products that may be absorbed from the lumen of the gut to enter the blood. Digestion begins in the mouth and continues in the stomach and small intestine. Very little digestion occurs in the large intestine.

INTAKE OF FOOD

The regulation of food intake is so precise that despite widely changing activities, and therefore energy expenditure, weight normally remains constant. Apparently there are several mechanisms including those located in the nose, mouth, pharynx, stomach, and intestine. But the major regulator appears to be in the hypothalamus.

In speaking of food intake, three terms should be understood. Hunger implies a desire for food in general. Appetite means a craving for a particular food. One need not be hungry to have an appetite. For example, after ingesting a large helping of meat and potatoes one is hardly hungry, but he still may have a strong appetite for a particular dessert. Satiety means a lack of desire for food. It results from the adequate intake of food.

Gastrointestinal Regulation of Food Intake

The smell of food, taste, chewing, salivation, and swallowing all play minor roles in regulating food intake. This is shown by experiments in which an opening is made in the esophagus so that the food, after being swallowed, is prevented from entering the stomach. Although hunger is certainly not satiated, eating ceases, at least momentarily, after a quantity of food has passed through the mouth.

Hunger Contractions After the stomach has emptied and the food has been digested, the stomach undergoes rhythmic contractions which give rise to a sensation of hunger. These contractions are often referred to as hunger pangs. The longer the period of time following a meal, the more frequent the contractions and the greater their amplitude. To a limited extent, at least, they are controlled by the level of glucose in the blood as shown by the fact that once established they can be inhibited by the intravenous administration of glucose. Interestingly, habit plays a role. For example, if a person is in the habit of eating lunch, hunger contractions will occur at lunchtime even though he has had a large breakfast. If lunch is skipped for several days without changing food intake at the other meals, the hunger contractions at lunchtime become less vigorous and may not occur at all.

Gastric and Intestinal Distention Distention of the stomach and the duodenum reduces hunger. This is apparently the first line of defense, so to speak. Long before the food can be digested and the products absorbed, eating ceases. Even if an opening is made just below the stomach so as to prevent the food from entering the small intestine, eating ceases. Likewise, mechanical distention of the stomach has the same effect. Removal of the stomach, however, does not eliminate this reaction, and experiments have shown that distention of the duodenum functions in the same way to cause cessation of eating.

Nervous System Control of Food Intake

The hypothalamus has the major role in controlling food intake. The parts of the hypothalamus which function in this respect are the ventrolateral nuclei and the ventromedial nuclei. Stimulation of the lateral nuclei causes voracious, continuous eating. Stimulation of the medial ones causes even a starving animal to refuse to eat. Thus, the lateral nuclei are referred to as the feeding center and the medial nuclei as the satiety center. If a lesion is placed in the medial nuclei, the animal never experiences satiety and therefore continues to eat vigorously and continuously. The result, of course, is incredible obesity.

Just how the feeding center gives rise to hunger is not known. Hunger contractions can be only a small part of the total picture, because hunger still persists even after removal of the stomach. Obviously, the drive for food is intense and must exist at the conscious level. Once food has been taken in and digested, activity of the satiety center inhibits the feeding center. This is shown by experiments in which a cut is made between the medial and lateral nuclei. Such animals do not experience satiety and continue to eat and to become obese.

Most attention has been directed to the role of glucose in activating the satiety center. There is a complete feedback system. After eating, the blood glucose level rises due to ingestion, and then it progressively decreases as the glucose is either utilized or removed from the blood to be converted into other substances. As the blood glucose falls, firing of the satiety center decreases, thereby progressively reducing inhibition of the feeding center. Finally the feeding center brings about food intake, the blood glucose level is elevated, and the satiety center fires and inhibits the feeding center. Other food substances probably act in an analogous way. These would be fats and amino acids.

Higher-Center Control The limbic system is certainly involved and probably accounts for weight divergencies noted in emotionally disturbed individuals. It is thought that the feeding center gives rise to adrenergic neurons and it is the norephinephrine which stimulates the higher centers. These centers, in some manner, give rise to the sensation of hunger. This is indicated by experiments in which the lateral nuclei are destroyed, which results in anorexia. Now when norepinephrine is injected into the lateral ventricles, the animal immediately begins to eat and will overeat if the injections are continued.

"Set" of the Hypothalamus In the prior discussion of body temperature control the point was made that the hypothalamus is "set" to maintain a certain temperature. Changes in this "set" result in fever or hypothermia. The same appears to be true for food intake, and the mechanism is the same, namely, the concentration of calcium and sodium ions in the extracellular fluid in the hypothalamus. A hyperphagic response occurs when calcium concentration is increased in the ventromedial region. Increasing sodium ion concentration in the lateral hypothalamus has the same result.

MASTICATION

The function of the mouth, at least insofar as alimentation is concerned, is the ingestion of food, the reduction in its size by the action of the teeth, the mixing of particles with saliva, and then the passage of the food into the pharynx in order to be swallowed. To be sure, saliva contains an enzyme which initiates the digestion of at least one of the foodstuffs, namely, carbohydrate, but, in the final analysis, the mouth is more concerned with ingestion than digestion.

Function of the Teeth

The term mastication simply means "chewing," a process by which substances taken into the mouth are reduced in size. For this purpose the teeth are virtually indispensable. But the teeth serve other purposes, and many animals employ them for protection as well. For example, in some of the dangerous snakes, there is a canal in the tooth connected to a gland which elaborates a poison. Thus, the snake can bite and allow the poison to pass down through the tooth and into the victim. Interestingly enough, it is this arrangement which precipitated the invention and design of the hypodermic needle.

Prehension and Division The word prehension means "grasping" or "taking hold of." Division means simply "separating" a part from the whole. For these functions the sharp-edged anterior teeth are used. For example, when an individual eats a sandwich, he grasps it with his anterior teeth and succeeds in dividing a part from the whole. Obviously, if the anterior teeth are lacking or ineffective because of malposition or decay, prehension and division are not possible. The individual then must cut or tear the food into bits before placing it in his mouth.

Masticatory Efficiency By masticatory efficiency is meant the ability to reduce large particles to smaller size with a standard number of chews. The absence of but one tooth, even a wisdom tooth, reduces masticatory efficiency. But although this is undoubtedly true and though it is clear that people with few or defective teeth cannot reduce the size of food particles efficiently, this deficit does not seem to be a serious handicap. Apparently, insofar as digestion is concerned, if the bulk of the food is reduced to a size compatible with swallowing, that is sufficient.

Abnormalities of the Teeth

The term caries means literally "decay." Caries develop in the permanent dentition in the average person at a very early age. The permanent teeth do not usually appear in the mouth until the sixth year of life; yet the average child of seven has at least one cavity. The startling and lamentable fact is that in the United States the average 30-year-old person has at least one-half of his teeth missing, filled, or defective. Other dental conditions are included in this statistic, but caries is the major contributor.

The present status of our knowledge of the genesis of dental caries can be summarized by saying that at least three factors play major roles: (1) the presence of fermentable carbohydrates, (2) bacteria in the mouth, and (3) the physiochemical characteristics of enamel. Fermentable carbohydrates are acted on by oral bacterial enzyme systems, so that organic acids are produced. These, in turn, destroy the teeth. Everyone has bacteria in his mouth, and there is no practical way of reducing this flora. The numerous available mouthwashes have scarcely any influence on the oral bacterial population. Since we consume a large amount of carbohydrate, it will usually be present in the mouth, particularly right after meals. However, the time during which fermentable carbohydrates are in a position to react with the bacteria to produce acids which attack the teeth may be limited. If one effectively rinses the mouth soon after eating, the concentration of carbohydrate about the teeth is strikingly reduced, and the incidence of dental caries is diminished.

But the third factor mentioned above, the physiochemical characteristics of enamel, offers the greatest possibilities of reducing caries. Clearly, if the enamel can be strengthened to the extent that it successfully resists the acid formed from the combination of fermentable carbohydrates and bacteria, dental caries will not develop. There is excellent evidence that small amounts of fluorides render the teeth remarkably resistant to formation of dental caries. Painting the teeth of a child periodically with a weak solution of a fluoride salt gives some resistance. But by far the better procedure is to incorporate minute quantities of this chemical, about one part per million, in the public drinking water. In communities where this has been done, the incidence of dental caries has been reduced as much as 60 percent.

The term malocclusion refers to the misalignment of the biting surfaces of the teeth. Great strides have been made in the treatment of such cases. Today, the science of orthodontics, which is chiefly concerned with the study of the arrangement of the teeth in the jaws and their effect on dentofacial growth, has developed complicated wires and bands and bars which can be fixed to the teeth and which cause the teeth to shift about so that a correct alignment is encouraged.

SALIVATION

Saliva is essential for swallowing, it cleans the teeth, it lubricates the oral mucous membranes, and it serves several other purposes.

Production and Composition of Saliva

The three pairs of major salivary glands, parotid, submandibular, and sublingual, were described in the previous chapter. These glands are composed of serous and mucous cells in varying percentages. Thus, the composition of the saliva depends on the relative contribution from each of the glands.

On the average, about a liter of saliva is produced per day. The quantity varies greatly, of course, with climatic conditions, fluid and food intake, and oral activity. If one chews gum all day, for instance, the saliva production will be greater than the average figure just cited. On the other hand, during sleep virtually no saliva is produced.

The composition of saliva varies under many conditions. In general, saliva may be said to contain the following: (1) water (about 98 percent); (2) most of the salts found in the plasma; (3) the enzymes ptyalin and (in some cases) maltase; (4) an insoluble substance which precipitates with phosphates to form the tartar of the teeth; (5) mucin; and (6) epithelial cells and bacteria. The reaction of saliva is usually slightly on the acid side.

Mechanism of Salivary Secretion

Saliva is generally hypotonic to serum. In some cases it may be isotonic and in rare cases it is even hypertonic. In hypotonic saliva the concentrations of sodium and chloride ions are lower than in the serum, while concentrations of potassium and bicarbonate are frequently higher than serum levels. The current concept for the secretion of saliva envisions that the acinar cells produce a primary fluid that is essentially isotonic to serum. Then as this fluid flows through the excretory ducts of the salivary glands, it is modified. The excretory ducts reabsorb sodium and chloride in excess of water so as to produce a hypotonic fluid into which potassium and bicarbonate are secreted. This is a convenient concept which, if accurate, closely simulates the way the kidneys form urine. However, there is a considerable body of evidence which suggests that saliva may not be formed in this way, at least not under all circumstances. The subject is still under investigation.

Function of Saliva

When food is taken into the mouth, it is cut and ground by the teeth and thoroughly mixed with saliva. The saliva functions in three ways: (1) as a solvent, thus making taste possible, (2) to initiate digestion, and (3) to lubricate the food so that it can be swallowed. Lack of saliva causes dry mouth, termed xerostomia.

As has already been pointed out, the taste buds can be stimulated only by substances which are in solution. If dry salt, for example, is placed on the dried tongue, there is no sensation of taste until saliva is produced and dissolves some of the salt.

Digestion Saliva contains the enzyme ptyalin. Ptyalin, also termed amylase, is an enzyme which catalyzes the reduction of the higher, or more complex, carbohydrates to simpler forms. There are many intermediate steps between a substance such as starch and the end product, glucose. Ptyalin does not ordinarily see this breakdown to completion, but merely initiates it.

Lubrication To confirm that saliva is important in swallowing, the student should swallow several times so as to reduce the amount of saliva in the mouth. If he now places a piece of dry bread in the back of the mouth, he will experience great difficulty in downing it. As a matter of fact, even without the bread, if he simply attempts to swallow continually, he will find that after a while it becomes extremely difficult, if not impossible.

Saliva flows even when there is no food in the mouth. It serves to lubricate the mouth and lips. In very dry climates and during a fever, little saliva is produced, and its rapid evaporation causes excessive drying of the lips and the oral mucosa. Therefore, in these cases the saliva may fail to protect the mucous membranes. If other measures are not instituted, the lips may dry and fissure. Under normal conditions, however, the saliva suffices to keep these structures soft and pliable. This is referred to as a demulcent action.

Cleansing Action   The length of time that fermentable carbohydrates reside in the mouth is an important factor in caries formation. There is, of course, no better way of decreasing this time than by vigorous rinsing, but the steady flow of saliva assists greatly. The value of saliva is clearly shown in cases where salivation is deficient on one side. In such instances, food accumulates about the teeth on that side and the incidence of caries is much higher than on the normal side.

Protection   Saliva also serves an important protective function. There are many substances in the secretion which give to it a fairly potent buffering capacity. Thus, if acids or bases (lemon or baking soda) are introduced into the mouth, the oral structures are protected by the saliva which quickly neutralizes these substances.

Saliva has also been shown to exert a bactericidal effect on many organisms. This protective function is of value, since bacteria are constantly being introduced into the mouth. Finally, saliva aids in ridding the mouth of foreign objects which may appear there.

Secretion of saliva occurs in response to stimulation of the autonomic innervation to the glands or in response to drugs that mimic the actions of the autonomic innervation. Salivary flow results mainly from stimulation of the parasympathetic neurons, but sympathetic stimulation also causes salivation. The composition of the saliva resulting from sympathetic stimulation differs from that evoked by parasympathetic action. In general, sympathetic stimulation produces a saliva higher in organic content and also higher in potassium, calcium, and HCO_3.

Regulation of Salivation

When a person thinks of food, sees food, or smells delectable aromas, impulses from higher brain centers flow down into the brainstem to activate the autonomic fibers. Consequently, saliva is secreted.

Food chemically stimulates the taste buds. Impulses are then initiated by these taste buds and conducted to the brainstem to set off reflex secretion of saliva. The composition of the food eaten determines, to a large extent, the quantity and quality of the saliva. If foods containing starch, such as bread, are chewed, there is a copious secretion of a serous saliva. On the other hand, protein, such as meat, brings forth a smaller amount of thick mucous saliva. The precise mechanisms which account for these differences are not entirely clear, but whatever the explanation, the results prove admirable. This is so because in the case of starch a large quantity of ptyalin serves to begin the digestion of that type of carbohydrate. But there is no enzyme in the saliva which breaks down protein. Therefore, the saliva serves only a lubricative function.

If one chews tasteless paraffin, which therefore does not evoke the salivary secretions by chemical means, saliva nonetheless pours forth. In this case it results in response to mechanical stimulation which also sets up a reflex mechanism. When one first begins to chew gum, the flavoring agents produce a chemical result. However, after prolonged chewing, little flavor remains. The copious flow of saliva is now due to mechanical reflexes. For the same reason, when the dentist is filling the teeth or cleaning them, there is usually a profuse salivary flow.

DEGLUTITION

After the food has been grasped and divided by the anterior teeth, it is ground into fine particles by the posterior dentition. At the same time it is tossed about by the tongue and thoroughly mixed with saliva. Saliva makes it possible to swallow the mass of food, now termed the bolus, which means a "lump." The act of swallowing is termed deglutition.

The Mechanism of Swallowing

Deglutition is a complex process that involves voluntary and reflex mechanisms. It is generally considered to occur in three stages, sometimes referred to as the voluntary, pharyngeal, and esophageal stages.

First Stage The food is ground and rolled into a bolus which has been thoroughly soaked with saliva. The tongue then directs the bolus to the back of the mouth and forces it to enter the pharynx. All these acts are voluntary. Once it is forced into the pharynx, the voluntary phase of the swallowing has ended. From now on all the mechanisms are purely reflex. The reflex control of the second and third stages is initiated by the voluntary act of the first phase.

Second Stage During this phase of deglutition, the bolus passes through the pharynx to enter the esophagus. This term is most descriptive, since it means "I carry food," which is precisely what it does. The passage of food through the pharynx is, in reality, rather precarious. There are many orifices into and out of the throat. There is the opening of the trachea, through which air, but not food, must pass. There are the entrances of the pharyngotympanic tubes, which connect the middle ear and the pharynx. Finally, there are the openings into the mouth and nasal passageways. But the bolus of food must be directed into the esophageal opening and into none of the others. In order to accomplish this feat, a series of reflex muscular responses seal off all the other openings while leaving the esophagus free to receive the bolus.

The bolus is prevented from reentering the mouth by the tongue and neighboring structures. The soft palate is raised to bar the bolus from the nasal passages. The orifices to the pharyngotympanic tubes are guarded by muscles which prevent entrance. The trachea is protected by elevation of the larynx and the movements of the arytenoid cartilages. All these actions occur as a result of a series of reflexes integrated in the so-called deglutition, or swallowing, center in the medulla and lower pons. The sensory impulses travel over afferent neurons in the fifth and ninth cranial nerves. The motor fibers are in the fifth, ninth, eleventh, and twelfth cranial nerves.

Third Stage In this phase, the food traverses the esophagus to enter the stomach. Man usually eats in the upright position; therefore, the force of gravity assists in transporting the bolus through the esophagus. However, one need not rely on gravity. As a matter of fact, it is possible to swallow and pass the bolus, or even liquids, through the esophagus while being held upside down.

The esophagus is composed chiefly of muscles which contract in wavelike fashion along the length of the tube. These are called peristaltic waves or movements. The term peristalsis means "clasping and compressing" and describes the contraction of one part of the tube, then contraction below it and relaxation of the originally constricted segment, and so forth (Figure 25-1).

Peristalsis of the esophagus is under reflex control. Both the afferent and the efferent neurons are in the vagus.

Mechanism of Peristalsis

Many tubes composed of smooth muscle exhibit peristalsis. Besides this inherent property, however, there is a neural factor responsible for the peristalsis seen in the alimentary

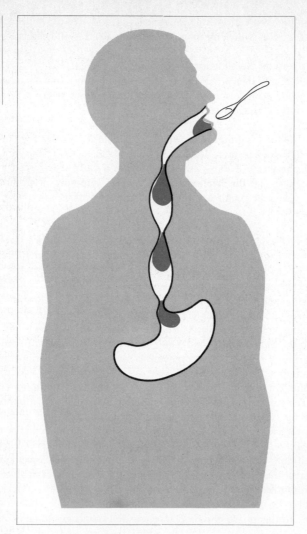

Figure 25-1. Peristalsis in the esophagus. Sequential contractions of the esophagus are responsible for moving the bolus from the pharynx to the stomach.

tract. The neural influence is exerted through the intramural nerve plexuses, that is, the submucosal and myenteric plexuses in the wall of the tract. These plexuses are composed of autonomic fibers. When transmission through the nerve plexuses is blocked by drugs, such as atropine, peristalsis still occurs, but the force of the individual contractions and the velocity of the peristaltic waves are markedly diminished. For example, when the plexuses are blocked, peristaltic waves move at a velocity of less than 1 cm/sec. When the plexuses are functional, the velocity of peristalsis may reach 3 or 4 cm/sec. Further, stimulation of the parasympathetic fibers serving the plexuses increases the force and velocity of peristalsis.

If a piece of intestine is removed and stimulated near the center, peristalsis will occur in waves that spread in both directions from the center. In the intact organism, however, the wave of peristalsis generally moves forward, that is, in a direction away from the mouth and toward the anus. This so-called analward direction of peristalsis in the alimentary tract is thought to be due to the organization of the nerve plexuses.

Abnormalities of Deglutition

Any impairment in swallowing is termed dysphagia (*phagia*, "to eat"; *dys-*, "difficult"). Thus the term simply implies that there is difficulty in eating because of some abnormality in the swallowing mechanism. The impairment may involve any or all stages. One speaks of dysphagia of the first stage, of the second stage, and so forth. If the tongue is paralyzed, for example, the bolus cannot be directed into the pharynx. There would then be dysphagia of the first stage. If, through a nervous system disorder, reflex mechanisms are interfered with or if muscular coordination is lost, then clearly the second stage of swallowing may be seriously impaired. Finally, should the esophagus be damaged, as often happens when one inadvertently or otherwise swallows lye, it may not be possible to pass food through the tube. This illustrates dysphagia of the third stage.

The Cardia

The cardia, or cardiac sphincter, as it is also known, keeps the passageway between the esophagus and the stomach closed. The cardia is generally in tonic contraction. But this contraction is never vigorous and is easily overwhelmed by very low pressures. Yet the cardia suffices, under normal conditions, to prevent regurgitation from the stomach into the esophagus. The contents of the stomach are very acid and contain proteolytic enzymes which would destroy the lining of the esophagus were significant regurgitation to occur. Belching, the eructation of gas from the stomach, requires sufficient pressure to open the sphincter. Cardiospasm, which is forceful enough to prevent the passage of food from the esophagus into the stomach, sometimes occurs. This is referred to as achalasia, which means the inability to relax.

Regurgitation of the stomach contents into the esophagus proves very irritating, a condition called heartburn because the burning sensation seems to emanate from the heart. The cause is failure of the cardia to prevent regurgitation, and there is interesting evidence that this results, in many cases, from smoking. Smoking brings about a decrease in cardia sphincter pressure, and direct measurements show that when this pressure falls, there is regurgitation with the expected increase in acidity in the esophagus.

THE STOMACH

When the bolus of food enters the stomach, it forces any other food that might be present there toward the sides. For this reason, the newly entered food continues under the action of ptyalin for a considerable time before the gastric juices can penetrate to it and stop the ptyalin action. The more food eaten, the longer this activity will continue.

The stomach acts as a reservoir, a place where a great mass of food may be received to be moved slowly into the intestine. But, in addition, the stomach also possesses important digestive functions.

Accommodation of the Stomach

The stomach, like all smooth-muscle bags, accommodates itself to an increase in contents without an increase in pressure, at least until a critical point is reached, after which the pressure mounts sharply.

Accommodation occurs for two reasons. In the first place, according to a basic physical principle described by the law of Laplace, the greater the diameter of a tube, the less pressure required to distend it. Second, smooth muscle progressively relaxes as it is

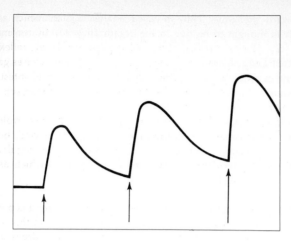

Figure 25-2. Accommodation of the stomach. Food entering the stomach (arrows) brings an immediate rise in intragastric pressure. Within moments the smooth muscle of the stomach wall relaxes sufficiently to return the pressure almost to its base level.

stretched. Thus, if the intragastric pressure is suddenly increased by injecting an amount of fluid or solid, within a few seconds the muscle relaxes sufficiently to return the pressure close to its former value (Figure 25-2).

The quantity that the stomach can hold varies with eating habits. If one regularly ingests large quantities of food at one sitting, the stomach will become progressively stretched until it can contain truly remarkable quantities. Under more normal circumstances, however, the upper limit is about 1 liter. The ingestion of greater amounts causes a sharp rise in intragastric pressure, which results in considerable pain.

Stomach Movements

Gastric movements serve two functions: (1) to mix the contents of the stomach and (2) to move the food toward the small intestine. Mixing movements differ from peristalsis in that the contractions are not synchronized in a way that causes an orderly wave which would move the food analward. Rather, small segments of the stomach contract, and thus the contents are tossed and turned and mixed.

Emptying Peristaltic waves in the stomach propel the mass toward the opening into the small intestine. Fluids pass through the stomach very rapidly, but the more solid substances remain longer. The rate of emptying depends upon (1) the relative pressures in the stomach and duodenum, (2) the state of contraction or relaxation of the pyloric sphincter, and (3) the fluidity of the gastric contents.

Although the stomach accommodates to greater volumes, there is some increase in intragastric pressure as food accumulates. This growing pressure tends to force the food from the stomach into the small intestine. Conversely, a filled duodenum will oppose the movement. Superimposed on these pressure relationships are the peristaltic waves which can elevate the gastric pressure sufficiently to pump the gastric contents into the intestine.

As has been observed, peristalsis, in the stomach as elsewhere, is fundamentally regulated by impulses traveling over sympathetic and parasympathetic fibers in the basic nerve plexuses. Stimulation of sympathetic fibers inhibits stomach movements; parasympathetic activity increases them. When food enters the stomach, not only is the inherent contractile mechanism activated to start peristalsis, but impulses from the stretched muscle initiate a reflex contraction of the stomach to reinforce the peristalsis. The greater the stretch, the more forceful the contractions.

Emptying of the stomach can be inhibited by mechanisms which reflect activity in the small intestine. As the small intestine becomes engorged by the material which is pumped into it from the stomach, a reflex, called the enterogastric reflex, is initiated by the stretched small intestinal wall. The efferent limb of the reflex causes gastric inhibition. In addition, certain foods with a high fat content cause the small intestine to secrete a hormone, enterogastrone, which is carried by the blood to the stomach where it causes inhibition of contractions.

By virtue of these mechanisms, the small intestine in effect controls the amount of material pumped in. Thus the small intestine is not only protected, but the length of time substances are permitted to remain in the intestine is also regulated. Fats, which are only slowly digested, thereby are allowed a longer stay, which assures complete digestion.

Vomiting   Vomiting is a complex act coordinated by the vomiting center located in the medulla. Impulses from many regions of the body impinge upon this area. Stimulation of the semicircular canals may have the same result. Various drugs also activate the vomiting center. When the center is stimulated, a great inspiratory effort is made. Suddenly, there is a forceful contraction of the abdominal muscles which elevates the pressure surrounding the stomach. The cardia then relaxes so that the high intragastric pressure expels the gastric contents through the esophagus, the pharynx, and the mouth.

Composition of Gastric Juice

There are about 35 million gastric glands in the stomach. The juice from these glands pours directly into the stomach cavity. Gastric juice contains water, mucin, pepsin, hydrochloric acid (HCl), rennin, and a gastric lipase. The neck cells produce mucin, the chief cells elaborate pepsinogen, the precursor of pepsin, and the parietal cells secrete HCl.

The most unusual component of the gastric secretion is the HCl. The pH of the juice is close to 1. When secretions are collected directly from the parietal glands, the pH is found to be less than 1. In the plasma the hydrogen-ion (H^+) concentration is 5×10^{-5} mEq/liter, but in the gastric juice it is about 150 mEq/liter. Quite obviously, the secretion of H^+ in the gastric juice must be an active process.

There is still not agreement as to how the gastric glands produce and secrete HCl. Figure 25-3 offers one version in a highly simplified form. Note that in this scheme water and CO_2 diffuse into the cell. There, under the influence of carbonic anhydrase, H_2CO_3 is formed. This compound then ionizes to HCO_3^- and H^+. The bicarbonate ions diffuse out of the cell into the blood while the hydrogen ions are actively transported into the gastric juice. At the same time chloride ions are actively transported by the cells from the blood to the juice. In the juice, H^+ and Cl^- combine to from HCl.

There are several consequences of this secretion of HCl. In the first place, the parietal cells take up CO_2 rather than excrete it. This means that the arterial blood has a higher CO_2 in this case than does the venous blood. Second, while CO_2 molecules are being removed from the blood, they are being replaced by HCO_3^-. CO_2 molecules are potentially acid; HCO_3^- is alkaline, and thus the pH of the venous blood is markedly higher than it is elsewhere in the venous system. As a result, the pH of the urine rises whenever the gastric glands are sufficiently stimulated. This normally occurs as a sequel to eating and is referred to as the postprandial alkaline tide.

Pepsin is formed when the secreted pepsinogen comes into contact with the existing pepsin in the acid medium of the stomach. Pepsin is a potent proteolytic enzyme that functions optimally at the pH of the stomach, that is, 1.6 to 2.4. By its action proteins are converted to proteoses and peptones.

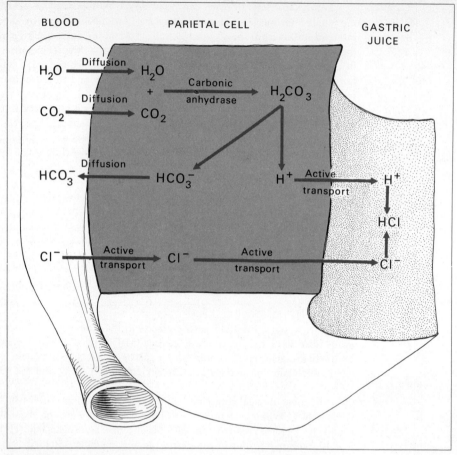

Figure 25-3. Formation of gastric juice hydrochloric acid (HCl). Note that arterial blood gives up potentially acid carbon dioxide (CO_2) and venous blood acquires alkaline HCO_3^-. Thus venous blood pH is markedly higher here than elsewhere in the body.

Mucin serves to buffer the strong acid, or at least to form a protective barrier between it and the stomach mucous membrane.

Rennin is an enzyme which acts specifically upon the casein in milk to cause the formation of curds. A curd is a coagulated semiliquid glob of milk. Such formation causes the milk to be retained in the stomach long enough for digestion to occur. Very little rennin is secreted by the adult stomach. In addition, this enzyme does not function very well in a highly acid medium. In the child's stomach the acidity is not nearly as great as it is in the adult.

Lipase appears in the gastric juice only in small quantities and may initiate the digestion of fats. But since the concentration of lipase is so much greater in the small intestine, this gastric function is normally of little relative importance.

The gastric juice probably also serves a protective function. The high acidity is capable of killing many types of bacteria.

Regulation of Gastric Secretion

The quantity of gastric juice formed varies with the type and amount of food ingested. On the average diet about 2 liters of juice is secreted daily. The gastric glands are under both nervous and hormonal control. The regulation of gastric secretion may be subdivided into cephalic, gastric, and intestinal phases.

Cephalic Phase The thought, sight, smell, or taste of food evokes gastric secretion. Afferent impulses are propagated to the cerebral cortex, from where efferent impulses descend to the dorsal motor nuclei of the vagi and then via the vagi to the stomach. As a result, the gastric glands are stimulated to secrete. Thus, even before the food reaches the stomach, secretion has begun.

Gastric Phase When food actually reaches the stomach, the flow of juice becomes very copious. Several mechanisms are involved. First the direct stimulation of the stomach mucosa evokes a small amount of secretion, mostly mucus. Second, the food initiates impulses in the mucosa which are propagated by afferents in the vagus to the medulla and then back in the vagal efferents to the gastric glands. And third, a hormone termed gastrin is produced by the stomach wall. Gastrin is then carried by the bloodstream to the gastric glands, causing secretion.

Gastrin is most copiously liberated in response to meat extracts; thus food with a high protein content causes gastrin to be secreted. As a result, a highly acid gastric juice, which is essential for the digestion of protein, is elaborated. Substances that evoke secretion are termed secretagogues.

Intestinal Phase The presence of the food mass in the small intestine, as mentioned earlier, causes the secretion by the intestinal wall of the hormone enterogastrone. Enterogastrone not only inhibits gastric movements, but it also inhibits gastric secretion. However, since gastric secretion continues for several hours after gastric emptying, some investigators have postulated that there must be intestinal hormones which stimulate gastric secretion, but none has yet been identified.

Gastric Abnormalities

Ulcer An ulcer is a cavity, an area that is unduly worn or which has disintegrated. Ulcers occur in various parts of the gastrointestinal tract, particularly in the stomach and the duodenum. If the ulcer is due to the digestive action of pepsin, it is referred to as a peptic ulcer. The major question that has fascinated physiologists for years is why the gastric juice does not digest the stomach itself. No doubt the layer of mucus is very important, but it does not seem to be the whole story.

Figure 24-8 shows that the mucosa has a surface layer of tall, columnar epithelial cells. These cells are fused together at their apices, forming tight junctions. Very few hydrogen ions normally manage to get through these junctions. But various things can happen to permit them to enter. Among these are (1) too little mucus, (2) excessive gastric acid secretion between meals, and (3) the presence of detergents. The naturally occurring detergents are bile salts and lecithin, both secreted into the small intestine. If they are regurgitated back into the stomach in adequate quantities, the mucosal barrier is broken and hydrogen ions enter freely.

Once hydrogen ions enter the mucosa, the smooth muscle is irritated and caused to contract vigorously, which gives rise to pain. Also, histamine is secreted. This dilates local blood vessels, causes edema, and in combination with the acid can produce hemorrhage.

Fortunately the cells on the surface of the mucosa reproduce themselves at an astonishing rate. In the human stomach about half a million cells per minute are shed and replaced. This means that the surface lining of the stomach is completely renewed every 3 days!

Interestingly, aspirin can breach the mucosal barrier and produce bleeding, and this effect is greatly enhanced by the presence of alcohol in the stomach. Clearly the two should not be taken together.

Treatment of gastric ulcer generally consists of eliminating the cause. Psychoanalysis, section of the vagi, and diet have all been in vogue. Very recently a new treatment has been suggested. Certain analogs of the naturally occurring prostaglandins have been found to inhibit gastric secretion in man without causing any side effects. The main inhibition appears to be of acid secretion. Since the prostaglandins can be given orally, they may prove very effective in preventing ulcers or permitting an ulcer to heal.

Gastritis Gastritis, simple inflammation of the gastric mucosa, may result from a number of causes. Among the more common is vitamin deficiency, especially of the vitamin B complex. Gastritis may also be caused by the excessive ingestion of alcohol, or simply by overeating.

Failure to Empty Spasm of the pyloric sphincter or absence of gastric contractions results in a failure to empty. This occurs in some vitamin-deficiency states, and in cases of intestinal irritation.

THE SMALL INTESTINE

The mass of partially digested food, now called the chyme, passes from the stomach into the small intestine. As will be discussed presently, the intestine mixes and moves the food and secretes a juice that is essential to digestion. In addition, the pancreas and the liver contribute secretions that aid in digestion. The highly acid chyme is progressively neutralized by pancreatic, biliary, and intestinal secretions, all of which are alkaline. By the time the mass reaches the ileum, the pH is close to 7.

Movements of the Small Intestine

The intestine, like the stomach, is capable of contractions which serve to mix the food (segmentation) and also of peristaltic contractions which move the food analward. In the small intestine the peristaltic wave moves about 20 cm/min.

Composition of Intestinal Juice

The mucosa of the small intestine secretes a fluid termed succus entericus, or simply, intestinal juice. It is secreted by intestinal glands.

Pure intestinal juice, that is, fluid which has not been contaminated by intestinal debris and secretions from other parts of the alimentary tract, contains two enzymes: (1) enterokinase, which activates trypsin (see page 615), and (2) a weak amylase. But interestingly, the epithelial cells of the intestinal mucosa do contain high concentrations of enzymes for the digestion of polypeptides, disaccharides, and fats. Intestinal juice, which is secreted at a rate of about 3,000 ml/day, may function primarily to carry substances to be digested to the epithelial cells, where, while these substances are being absorbed, they undergo final digestion by the enzymes in those cells.

In addition to the intestinal juice, Brunner's glands secrete large quantities of mucus. The mucus serves to protect the intestinal wall from the highly acid and proteolytic gastric juice.

Regulation of Intestinal Juice Secretion

Distention of the small intestine by the entrance of the chyme is the effective stimulus for the outpouring of intestinal juice. The greater the quantity of the chyme in the intestine, the greater the secretion.

Secretion may also be enhanced by activity of the parasympathetic nervous system. Sympathetic stimulation inhibits secretion, especially of mucus; thus the mucosa is not so well protected, and duodenal ulcers may result.

THE PANCREAS

The pancreas secretes a juice which contains very potent enzymes essential for digestion.

Composition of Pancreatic Juice

The pancreatic juice contains at least three enzymes: (1) trypsin, (2) amylase, and (3) lipase. In addition, the pancreatic secretion contains a considerable quantity of bicarbonate ion, which makes it alkaline and which functions to neutralize the highly acid gastric juice. The pancreatic enzymes function best at a pH much higher than is optimal for pepsin. The pH of the intestinal contents progressively increases until it is almost neutral in the ileum.

Sodium, potassium, calcium, and magnesium concentrations in pancreatic juice reflect the concentrations of these cations in the plasma. This is in striking contrast to saliva, in which the electrolyte concentrations in the secretions are quite different. Pancreatic juice is isotonic with plasma; saliva is hypotonic.

Just as the mechanism responsible for the remarkable secretion of H^+ in gastric juice continues to concern investigators, so does the mechanism responsible for the secretion of high concentrations of HCO_3^- in pancreatic juice. As a matter of fact, the two mechanisms seem to be the same but operate in reverse directions. The gastric parietal cells secrete HCl in the juice and sodium bicarbonate in the venous plasma. The pancreatic cells secrete sodium bicarbonate in the juice; therefore, they must put an equivalent amount of acid into the venous plasma. The pancreatic cells secrete a juice containing the various enzymes and an aqueous juice containing the electrolytes. These two secretions are then mixed to make up the final product.

The pancreas secretes trypsinogen. When trypsinogen reaches the intestinal tract, it is converted into trypsin by trypsin already present and by enterokinase. Enterokinase is produced by the intestinal mucosa. It is a proteolytic enzyme, that is, it acts upon protein and its derivatives. Trypsin too is a proteolytic enzyme. Some protein reaches the small intestine, having escaped digestion by pepsin in the stomach. But for the most part, the chyme that enters the small intestine contains peptones and proteoses. Trypsin acts upon these protein derivatives and succeeds in reducing them to the end products of protein digestion, the amino acids.

In view of the fact that trypsin can activate trypsinogen to form more trypsin, there must be some mechanism present in the pancreas to prevent the activation of trypsinogen; otherwise the proteolytic enzyme would destroy the pancreas and the pancreatic ducts. In fact, there is a so-called trypsin inhibitor formed in the very same cells that produce trypsinogen. Within the cell it prevents trypsinogen activation, and when trypsinogen is secreted, so is the trypsin inhibitor. Thus, the ducts of the pancreas are protected. Only when trypsinogen reaches the intestine is the inhibitor overwhelmed, and active trypsin results.

Amylase is an enzyme that speeds the conversion of carbohydrates to the simple sugars. Ptyalin is the salivary amylase that initiates this process. The pancreatic amylase continues the work so that the intermediary sugars—maltose, sucrose, and lactose—are formed. The pancreatic juice also contains amylases which act specifically on these intermediary sugars to convert them to the simple sugars glucose, fructose, and galactose.

The only carbohydrate in milk is the disaccharide lactose. In order to convert it to the monosaccharides glucose and galactose, the enzyme lactase is necessary. Lactase is present in most infants, but it is absent or exists in insufficient quantities in a surprising number of adults. Recent studies suggest that in populations that drink milk, many adults do have lactase. How this occurs is not known. However, a significant number of adults cannot digest milk, and when it is ingested, they suffer from flatulence and diarrhea, a fact that most hospital dietitians are amazingly reluctant to recognize.

Pancreatic lipase catalyzes the conversion of the fat molecule into its component fatty acids and glycerol. The bile salts make it possible for the lipase to act on fat and to bring about this conversion (see page 617).

Regulation of Pancreatic Secretion

About a liter of pancreatic juice is formed per day. The pancreas is under both nervous and hormonal control. Stimulation of the vagi and the splanchnics causes secretion of pancreatic juice, but hormonal control of secretion seems to be more important.

One of the most potent pancreatic stimulants is secretin, a hormone produced by the intestinal wall. The acid chyme is the effective stimulus for secretin production. The intermediary protein products, that is, the peptones and proteoses, act on the mucosal wall to evoke the secretion of still another hormone, namely pancreozymin. The secretin-induced pancreatic juice has a much lower enzyme concentration than does the juice which results from nervous stimulation or from the influence of pancreozymin. Accordingly, there is the concept that secretin is primarily responsible for the volume of the juice, while the other stimulants control the enzyme concentration.

The copious flow of a pancreatic juice with a high bicarbonate concentration is of great value. The chyme that enters the duodenum is very acid. The duodenal wall is not as well protected as is the gastric mucosa. In addition, the pancreatic enzymes function at a pH close to neutrality. For these two reasons, the acid chyme must be neutralized. The acidity, as already mentioned, evokes secretion of secretin, which acts on the pancreas to produce the high-bicarbonate solution. Bicarbonate reacts with hydrochloric acid as follows:

$$HCl + NaHCO_3 \longrightarrow NaCl + H_2CO_3$$

The H_2CO_3 dissociates, producing CO_2 and H_2O. In short, HCl is effectively buffered.

Abnormalities of Pancreatic Secretion

If the pancreatic duct becomes blocked, pancreatic juice is still secreted, but it lingers in the pancreas and in the duct, both of which it destroys. This gives rise to acute pancreatitis that can prove fatal.

The failure of pancreatic juice to reach the intestine has profound results on digestion. The lack of trypsin, amylase, and lipase means that proteins, carbohydrates, and fats will not be digested and malnutrition will ensue.

THE BILIARY SYSTEM

The bile is formed in the liver. It is then transported by the hepatic duct from the liver to a point where the hepatic and cystic ducts join to form the common bile duct (see Figure 24-7). The common bile duct unites with the pancreatic duct before entering the small intestine. The opening of the common bile duct is guarded by the sphincter of Oddi. When this sphincter is closed, the bile formed by the liver flows down the hepatic duct and then through the cystic duct to reach the gallbladder.

Composition of Bile

Bile is a thick yellow-green fluid of slightly alkaline reaction. It is formed from cholesterol, and some unconverted cholesterol remains in it. Bile contains, in addition to water, many inorganic salts which probably exert a buffering function to neutralize the acid chyme. There is also present a nucleoprotein, which accounts for the high viscosity. But the most important constituents are the bile salts and the bile pigments.

Lipase in the pancreatic juice cannot effectively exert its influence upon fat unless the bile salts are present to lower the surface tension. This enables the water-soluble lipase to come into more intimate contact with the fat molecule. Bile itself contains no enzymes. The fatty acids and glycerol which result from fat digestion can be adequately absorbed only by the intestinal wall in the presence of the bile salts. If the common bile duct is ligated so as to prevent the bile from entering the intestinal lumen, about one-half of the ingested fat will appear in the feces, a condition termed steatorrhea. This fat fails to be absorbed because of the absence of the bile salts. Many of the vitamins are fat-soluble, notably A, D, E, and K. If bile is not present, these vitamins are poorly absorbed.

The bile pigments, bilirubin and biliverdin, are derived from the red blood cells. They represent end products of red-cell disintegration and are excreted from the body in the bile. If there is malfunction of the liver so that the bile pigments are not eliminated as rapidly as they are formed or if the bile system is blocked, the pigments accumulate in the blood to discolor the skin. This gives rise to a yellow appearance, a condition termed jaundice.

Bile, like the pancreatic juice, has a high bicarbonate concentration which becomes even higher as the flow rate increases. The electrolyte composition of bile is also very similar to that of pancreatic juice.

Regulation of Bile Secretion

Between meals, bile is formed at a relatively slow rate. It passes down the hepatic duct and then backs up the cystic duct into the gallbladder. The gallbladder has a capacity of about 50 ml. Much of the water and electrolytes contained in the bile are absorbed by the gallbladder so that the bile salt, bile pigment, and cholesterol are concentrated five to ten times during storage there. On the average, about 700 ml of bile is secreted by the liver per day, but this can vary from 250 to over 1,000 ml.

The secretion of bile is an active process. The best evidence for active secretion is the secretion pressure, which, in the case of bile, can reach 20 mmHg. The pressure within the liver blood capillaries is less than 5 mmHg; thus bile can hardly represent simply a filtration product.

The most potent stimulus for the production of bile is bile salt. About 80 percent of the bile salts which reach the intestine are reabsorbed and pass into the bloodstream to be returned to the liver, where they excite further production of bile (Figure 25-4).

In addition to the bile salts, vagal stimulation and secretin increase bile secretion.

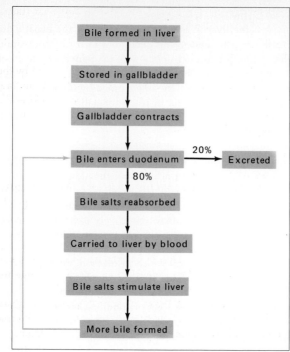

Figure 25-4. Regulation of bile secretion. Contraction of the gallbladder starts a self-sustaining mechanism for bile secretion. After digestion is complete, the sphincter of Oddi closes, breaking the cycle.

Radioisotope studies have disclosed that the total amount of bile salts in man is about 3.6 g. Following an average meal, about twice this amount of bile salts is secreted, which means that the total bile salt pool must circulate at least twice following each meal. About 20 percent of the bile salts is synthesized each day. Since the total pool remains relatively constant, this amount must be lost each day.

The contraction of the gallbladder and the relaxation of the sphincter of Oddi are controlled by the hormone cholecystokinin. Cholecystokinin, which is very similar and probably identical to pancreozymin, is secreted into the bloodstream by the small intestine in response to the presence of fat. The hormone causes contraction of the gallbladder and relaxation of the sphincter.

There is thus a self-regulating mechanism. The chyme in the intestine promotes the secretion of cholecystokinin. As a result, bile enters the intestine. The bile salts are reabsorbed and stimulate the copious production of more bile, which flows directly to the intestine. After digestion is complete and the intestine emptied, the sphincter closes, bile is no longer absorbed, bile production slows, and the gallbladder fills.

Abnormalities of the Biliary System

Cholesterol crystals may settle out of the bile to form particles, or stones, of varying sizes. Such gallstones result when the bile becomes unduly concentrated with cholesterol or when there is an infectious process in the biliary system. Other gallstones are composed of calcium carbonate, and some have been found which are about 25 percent cholesterol and the remainder, bile pigment.

Cirrhosis, in which the areas of the liver are replaced by fibrous tissue, in most cases causes reduced production of bile. In addition, the limited liver function may prevent adequate excretion of bilirubin so that jaundice results.

Ileocecal Valve

The outlet of the small intestine into the large intestine is guarded by the ileocecal valve. The contraction of this sphincter is under the control of a reflex initiated in the cecum. Stretch of the wall of the cecum initiates impulses which cause ileocecal contraction. This arrangement, then, permits the cecum to regulate the inflow of chyme in accord with the volume already present.

THE LARGE INTESTINE

Very little digestion, if any, occurs in the large intestine. Substances that are not absorbed in the small intestine are generally waste products. These accumulate in the large intestine and are prepared for expulsion.

The Intestinal Flora

Many microorganisms are ingested with the food. Most succumb to the strong acid and enzymatic action in the stomach, but some survive, and in the large intestine they find the environment favorable for fermentation and putrefaction.

The action of a living organism in the conversion of a complex substance into simpler components is termed fermentation. Usually, as a result of bacterial fermentation, various gases result. In abnormal cases these gases cause painful distention.

The term putrefaction refers specifically to the conversion of protein substances into various compounds including indole, skatole, phenol, hydrogen sulfide, and ammonia. All are characterized by a pungent odor.

Movements of the Large Intestine

Movements of the large intestine are similar to those in the small intestine. There are contractions which do not move the contents analward but instead serve to break it up and mix it. In addition, there are the so-called mass movements, vigorous peristalsis, which do drive the contents to lower segments. Mass movements take place two or three times a day.

The residue of a meal is generally turned into feces and is ready for expulsion within 12 hr after ingestion. This figure varies greatly, however. Fluids move through the stomach rapidly, carbohydrates more slowly, proteins still more slowly, and fatty foods may linger much longer. On the average, it takes about 6 hr for the chyme to traverse the small intestine and another 3 or 4 hr in the large intestine.

Not all the residue in any one meal is excreted at the same time. Part of the residue remains to be mixed with that of subsequent meals, so that a part of the feces may represent a meal taken several days before.

Feces

The term feces means "dregs," that is, worthless residue. The quantity of feces formed varies with the quality and quantity of food ingested. On the average diet it amounts to about 200 g/day. Even during complete starvation, some fecal material is formed, which consists mostly of water, bacteria, and epithelial cells.

Feces contain about 75 percent water. The water, to a great extent, is independent of the amount of fluid imbibed. The maintenance of this constancy is a function of the large intestine. Apparently, the colon can either add water or absorb it from the feces and normally keeps the water content fairly constant.

The color of feces is determined by the bilirubin concentration. Bilirubin is converted by bacteria in the large intestine to urobilinogen, which is then oxidized to urobilin. These are the pigments which color the feces. In disorders of the biliary system the feces appear white, clay-colored, or gray.

Defecation

As fecal material accumulates, the rectum becomes distended. As a result, peristaltic waves occur in the colon which drive the colonic contents toward the anus. When the intrarectal pressure reaches about 40 mmHg, there is adequate propagation of impulses to the sacral level of the spinal cord to fire the parasympathetic neurons which travel in the nervi erigentes. The motor impulses cause vigorous peristaltic waves and, at the same time, relaxation of the internal anal sphincter, a circular mass of smooth muscle. Expulsion of the feces, termed defecation, will occur unless the external anal sphincter is kept contracted. This sphincter is composed of skeletal muscle, which is under voluntary control.

The defecation reflex may be initiated and assisted through voluntary efforts. This is accomplished by contractions of the abdominal muscles, which result in an increase in intraabdominal pressure. At the same time, the epiglottis is closed and a forceful expiratory effort is made. The intrapleural pressure, which is thus greatly increased, forces the diaphragm to descend. This, plus simultaneous contraction of the abdominal muscles, greatly increases intraabdominal pressure. The pressure, which may reach 100 mmHg or more, squeezes the rectum and assists defecation.

Abnormalities of Defecation

Constipation is a condition in which the evacuations from the rectum are inadequate so that fecal material accumulates in the large intestine. A cathartic is a drug used to relieve this abnormality. Cathartics work in different ways. Castor oil, for example, increases fecal evacuation by irritating the intestinal tract. This causes greater peristaltic activity. There is too little time for fluid absorption, and so the feces remain semifluid. Epsom salts are not readily absorbed by the intestine; thus by osmosis they inhibit the absorption of water so that the feces remain fluid. Liquid petroleum also cuts down fluid absorption and, in addition, lubricates the passageways.

Excessive elimination of a semifluid feces is termed diarrhea. It may result from inflammation caused by specific bacteria, or it may be due to excessive use of cathartics, the ingestion of irritating or spoiled foods, emotional upsets, or fever.

Secretions of the Large Intestine

The large intestine secretes a juice that has a pH of about 8. This alkalinity serves to neutralize the acid end products of fermentation. Actually the fecal mass is not completely neutralized. The surface has a pH close to 7, but at the center the pH is closer to 5.

Under normal conditions, the secretions in the large intestine are mostly mucus. Like the small intestine, the large intestine has a mucosa lined with intestinal glands. But unlike the small intestine there are no enzymes in the epithelial cells. There are only goblet cells which produce mucus in response to parasympathetic stimulation. The mucus serves not only to protect the wall of the large intestine but also to bind the fecal material together in a mass that can be easily moved along and expelled.

SUMMARY OF DIGESTIVE PROCESSES

The discussion in this chapter, for the sake of simplicity and clarity, has been organized according to the various subdivisions of the alimentary tract. Now the digestion of each of the types of food will be summarized. Tables 25-1 and 25-2 list the major digestive enzymes and hormones.

Digestion of Carbohydrate

The human diet contains carbohydrate polysaccharides, disaccharides, and monosaccharides. Carbohydrate is usually absorbed in the small intestine as monosaccharides.

TABLE 25-1. THE MAJOR DIGESTIVE ENZYMES

ENZYME	SOURCE	FUNCTION
Ptyalin	Salivary glands	Begins carbohydrate digestion
Pepsin	Gastric glands	Proteins to polypeptides
Rennin	Gastric glands	Clots milk
Gastric lipase	Gastric glands	Fats to glycerides
Enterokinase	Duodenal mucosa	Trypsinogen to trypsin
Trypsin	Pancreas	Proteins to polypeptides
Pancreatic lipase	Pancreas	Fats to fatty acids
Pancreatic amylase	Pancreas	Starch to disaccharides
Peptidases	Intestinal glands	Peptides to amino acids
Maltase	Intestinal glands	Maltose to glucose
Lactase	Intestinal glands	Lactose to glucose and galactose
Sucrase	Intestinal glands	Sucrose to glucose and fructose

TABLE 25-2. THE MAJOR GASTROINTESTINAL HORMONES

HORMONE	SECRETION STIMULATED BY	FUNCTION
Gastrin	Vagus, protein, stomach distention	Stimulates gastric juice secretion
Enterogastrone	Acid and fat in intestine	Inhibits stomach contractions and gastric juice secretion
Secretin	Acid and peptides in duodenum	Stimulates bicarbonate-rich pancreatic juice secretion
Pancreozymin*	Peptides in duodenum	Stimulates enzyme-rich pancreatic juice secretion
Cholecystokinin*	Fat in duodenum	Stimulates gallbladder contraction
Enterocrinin	Chyme in intestine	Stimulates secretion of intestinal juice

*Probably the same hormone.

Thus, the digestive processes must break the polysaccharides and disaccharides down to the simpler form by the enzymes in the various secretions. Starches are the polysaccharides. Cellulose is also present in large quantities, but there are no enzymes in the alimentary tract capable of digesting this substance; thus it is not utilized, but rather excreted in the feces. The major disaccharides are sucrose and lactose. There are many monosaccharides, that is, carbohydrates which cannot be hydrolyzed to a simpler form. They may contain from 3 to 10 carbon atoms. Examples are glucose and the simpler alcohols.

When food is contemplated, seen, smelled, or tasted, the secretion of saliva is evoked. During chewing the food is well mixed with saliva and its enzyme, ptyalin, begins to digest starch to maltose. The amount of starch converted to maltose depends upon how long the food is kept in the mouth. Usually only about 5 percent of the starch is converted.

The bolus is then moved into the stomach where the action of ptyalin continues until the strong acid in the stomach reaches it and inactivates it. By that time perhaps 40 percent of the starch is converted to maltose. The other carbohydrates remain unaltered in the mouth and stomach. Thus, when the chyme enters the small intestine it contains some starch, maltose, lactose, sucrose, and other disaccharides and monosaccharides ingested in the food.

In the small intestine, pancreatic and intestinal juices complete carbohydrate digestion. The epithelial cells of the intestine contain many enzymes, including lactase, sucrase, and maltase, which convert lactose, sucrose, and maltose to monosaccharides. The average diet produces monosaccharides consisting of 80 percent glucose, 10 percent fructose, and 10 percent galactose.

Digestion of Protein

Protein in the diet of human beings is obtained from meat and vegetables. Proteins, which are very large molecules composed of many amino acid units, must be digested to release the amino acids for absorption. This process begins in the stomach, not in the mouth.

Pepsin is the essential enzyme for protein digestion. It functions best at a pH of about 2 or 3. The secretion of hydrochloric acid in the gastric juice provides this favorable environment. Pepsin is also capable of digesting collagen, which is the major constituent of the connective tissue of meat. Until this connective tissue is destroyed, the proteolytic enzymes cannot reach the meat protein.

Pepsin converts the large protein molecules into proteoses, peptones, and polypeptides. In the small intestine these intermediary products are converted to the basic amino acids primarily by the pancreatic enzyme trypsin. In addition, the intestinal epithelial cells contain enzymes capable of completing the conversion to amino acids. By these processes over 90 percent of the ingested protein is converted to amino acids, but some molecules of protein escape, and others are hydrolyzed only to proteoses, peptones, and polypeptides.

Digestion of Fat

Most diets contain the so-called neutral fats, which are triglycerides. In addition, there are varying quantities of phospholipids. As with protein, there are no enzymes in the mouth capable of digesting these products; thus the process does not begin until the food reaches the stomach. And even here, the gastric lipase functions so poorly in the acid medium that, for all practical purposes, the digestion of fat can be considered not to begin until the chyme reaches the small intestine.

As was mentioned earlier, in order for the enzymes in the watery secretion to reach the fat molecules, the bile salts are necessary to reduce the surface tension. In the process,

the globules of fat are broken into much smaller globules, a process termed emulsification.
Lipase in the pancreatic secretion and another lipase in the intestinal epithelial cells succeed in converting fat to glycerol and fatty acids, the forms necessary for intestinal absorption. The stool contains virtually no fat under normal conditions; but if there is a digestive abnormality, such as the presence of too little pancreatic juice or bile, the quantity of fat in the stool increases sharply.

QUESTIONS AND PROBLEMS

1 Describe the homeostatic mechanism involving glucoreceptors which is responsible for regulating appetite.
2 Define (a) caries, (b) mastication, (c) prehension, (d) malocclusion, (e) bolus, (f) peristalsis, (g) dysphagia, and (h) chyme.
3 What can be done to protect the mucous membranes of the mouth at times when there is insufficient saliva?
4 Which stage or stages of deglutition are voluntary and which reflex?
5 What ordinarily prevents a bolus of food from being regurgitated in the esophagus?
6 In the "fight or flight" phenomenon, what happens to the peristaltic activity of the stomach and intestine? Explain why this occurs.
7 What factors determine the stomach's rate of emptying?
8 What is the pressure relation between the stomach and the duodenum?
9 List the substances that make up the gastric juice, intestinal juice, and pancreatic juice, and describe the function of each.
10 Briefly discuss the cause of gastric ulcers and of gastritis.
11 Describe the relation of gastric parietal cells and pancreatic cells to the pH of the blood.
12 How is bile transported to the duodenum and the gallbladder?
13 What are the functions of bile? What is its relation to red blood cells?
14 How does bile fresh from the liver differ from bile which has been stored in the gallbladder?
15 Describe the digestive processes that reduce complex carbohydrate to absorbable monosaccharides, and proteins to amino acids.

26 ABSORPTION AND UTILIZATION

The chain of complex processes by which the basic foodstuffs are transformed into their elemental compounds has been described. The end products of digestion must now be absorbed from the lumen of the gastrointestinal tract into the blood, which carries them to various parts of the body for storage, transformation into other compounds, or oxidation for the production of energy. Following absorption, there is utilization of these substances in various parts of the body. Utilization embraces a study of intermediary metabolism and calorimetry.

ABSORPTION

There is no doubt that some substances are absorbed from the gastrointestinal tract in accordance with physicochemical laws. This means that the passage of a particular compound from the lumen into the bloodstream or into the lymph system is a function of concentration and pressure gradients. Likewise, high pressure within the intestinal tract speeds the absorptive processes.

Although the concentration and pressure gradients undoubtedly play a role, all absorption cannot be explained on this basis alone. There are active absorptive mechanisms capable of conveying substances from the intestine into the blood or lymph regardless of concentration or pressure. In other words, the intestinal wall is capable of doing work. This is clearly shown by the fact that when absorption is going on, the intestine uses more oxygen than when it is dormant. In addition, substances can be absorbed even from hypotonic solutions, that is, when the concentration is lower in the intestine than in the blood. Finally, the rate of absorption of related substances such as glucose and fructose varies significantly.

Site of Absorption

The stomach is not designed for absorption. However, lipid-soluble substances are absorbed in limited quantities. As mentioned in the previous chapter, alcohol is absorbed by the stomach, as are certain lipid-soluble drugs.

Anatomic and physiologic evidence points irrefutably to the small intestine as the principal site of absorption. The small intestine is a very long tube containing millions of villi, which create a huge surface area. Absorption is so efficient in the small intestine that very little absorbable substance reaches the large intestine.

The large intestine, like the stomach, is completely devoid of villi. But if the concentration and pressure gradients are favorable, absorption purely by diffusion may take place from the lumen of the large intestine. Under normal circumstances most of the end products of digestion are absorbed by the small intestine. In addition, virtually all the digestible material has been converted to absorbable end products before the food reaches the large intestine. Thus very little, if any, absorption takes place in the large intestine. It does, however, regulate the concentration of water in the feces. Water, for this purpose, may move in either direction. Usually, about 500 ml of chyme per day enter the colon. The average fecal load is approximately 200 g. If 75 percent of the feces is water, the amount of water is 150 ml. Thus, the colon must absorb about 350 ml of water per day. The colon actively absorbs sodium, and its concentration in the feces is about 25 percent of the plasma level or less. Potassium, however, is actively transported by the colon into the feces, where the concentration may reach over 100 mEq/liter. The significance of this potassium excretion is not known.

Villi

The total surface area of the small intestine, because of the villi, is huge. This favors the transfer of substances by simple diffusion. In addition, there are closely packed microvilli. Under the light microscope they appear as a striated border. The villi increase the surface area of the intestine, and the microvilli greatly increase the surface area of each villus.

The end products of carbohydrate and protein digestion, that is, monosaccharides and amino acids, are absorbed by the microvilli of these epithelial cells. These substances then diffuse through the epithelial cells and the basement membrane and pass into the blood capillaries.

The end products of fat digestion, that is, fatty acids and glycerol, being lipid-soluble, easily enter the epithelial cells of the villi.

The villi have another role besides increasing the surface area. They are capable of vigorous movements, and such movements create pressures which aid absorption. This mechanism has been termed the villous pump. The villi also serve to stir the food lying next to the intestinal mucosa. In this way new surfaces are constantly presented to the mucosa to facilitate absorption. A hormone, villikinin, which is purported to be released by the presence of chyme, has been described. Villikinin is said to augment the movement of villi.

Absorption of Carbohydrates

The normal end products of carbohydrate digestion are the monosaccharides. If a large quantity of the disaccharide sucrose is ingested, some of it may not be converted to monosaccharides and will be absorbed as unchanged sucrose. Interestingly, such absorbed sucrose undergoes no metabolic alteration after absorption and is simply excreted unchanged in the urine.

The evidence is conclusive that the absorption of the hexoses involves active transport. This is made clear by the observation that sugars which have the same physical diffusion rate are absorbed by the intestine at very different rates. Further, pentoses, which are smaller molecules and which diffuse more rapidly than hexoses, are actually absorbed by the intestine more slowly. If glucose is given an arbitrary absorption rate of 100, then the rate for galactose is 110 and for fructose about 43.

In some manner, monosaccharide absorption is tied to sodium absorption, as shown by the fact that when sodium uptake is blocked, no carbohydrate is absorbed.

Absorption of Amino Acids

The end products of protein digestion are the various amino acids. These are absorbed in the small intestine. Some protein escapes digestion or is digested only to the dipeptide state. These large molecules can be absorbed only in very small quantities, probably by the process of pinocytosis.

Amino acids are absorbed by active processes, as shown by the fact that during amino acid absorption, oxygen utilization of the intestine increases. Further, if the mucosa is deprived of oxygen, the rate of absorption is reduced. Finally, the rate of absorption of different amino acids varies.

The mechanism of the active transport of amino acids is thought to be similar to that responsible for the absorption of monosaccharides. That is, they seem to be actively transported across the luminal membrane along with sodium.

Absorption of Fats

Fats are digested to glycerol and fatty acids and absorbed by the epithelial cells of the villi. These end products are reconverted within the cells to the triglyceride fat molecule. The triglyceride molecules are carried by the tubules of the endoplasmic reticulum to the base of the cell. There they are discharged as tiny fat droplets termed chylomicrons. Fats that contain fatty acids with long chains (over 12 carbon atoms) enter the lymph vessels of the villi. The lymph vessels then assume a milky appearance and for this reason are called lacteals. The lymph carries the chylomicrons through the lymphatics to enter the venous blood via the thoracic duct.

Fats that contain fatty acids with chains less than 12 carbon atoms long enter the portal bloodstream. The liver quickly removes much of this fat so that a diet high in fats containing short-chain fatty acids does not result in a milky plasma.

Electron microscopic studies indicate that large fat molecules can be absorbed by pinocytosis without first undergoing digestion and resynthesis.

Fatty digestion and absorption are normally so complete that none of the ingested fat reaches the feces directly. There is, however, about 8 percent of the ingested fat in the stool, but this comes from bile fats and intestinal secretions. That is to say, the fat is first absorbed completely; part of it is then converted into bile fats and intestinal secretions which end up in the stool.

Impairment of digestion or absorption results in a high fat content in the feces (steatorrhea). This may be due to an inadequate secretion of lipases or of bile. Bile, by lowering surface tension, facilitates not only digestion but also absorption. There is a simple clinical test for determining whether steatorrhea is due to impairment of digestion or absorption. The fat triolein, labeled with ^{131}I, is given orally. The blood and stool are then checked for radioactivity. In the normal patient the blood activity rises rapidly, reaching a peak in about 3 or 4 hr. The peak level reaches at least 8 percent of the dose

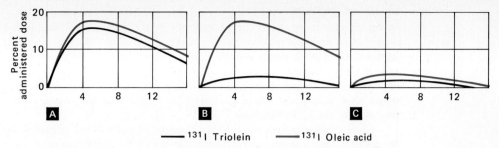

Figure 26-1. Test for digestion and absorption of fat. **A.** Normal curves for triolein and oleic acid. **B.** Oleic acid is normal but triolein is low. This means that absorption of fatty acid is normal but digestion of fat is inadequate. **C.** Failure of both digestion and absorption.

(Figure 26-1). In the stool not over 2 percent is found at 24 to 48 hr. But if too much activity is found in the stool and too little in the blood, obviously there is either inadequate digestion or inadequate absorption. The fatty acid oleic acid, labeled with ^{131}I, is then given in the same way, and the blood and stool are checked once again. If fatty acid values are normal while the whole-fat values are abnormal, the defect is in digestion. If both are abnormal, there is malabsorption.

Absorption of Vitamins

Vitamins are organic compounds present in various foodstuffs. As will be discussed in the next chapter, they are essential for normal growth and health and are very effective in remarkably small quantities. Some of the vitamins are water-soluble, others are fat-soluble. The water-soluble vitamins are quickly absorbed. But if there is fat malabsorption, there will be decreased absorption of the fat-soluble vitamins, that is, A, D, E, and K.

The absorption of vitamin B_{12} is unique. It is a water-soluble vitamin, but it has a large molecule, probably too large for simple diffusion. For absorption to take place, a mucoprotein, called the intrinsic factor, is required. The intrinsic factor is secreted by the stomach, but the responsible cells have not been identified nor has the structure of the factor. Though the intrinsic factor is secreted in the stomach, no absorption of vitamin B_{12} takes place there. The site of absorption is the lower half of the small intestine. There appears to be binding of intrinsic factor and the vitamin. Then, if adequate calcium is present, the bound vitamin is absorbed. Once absorbed, the vitamin is released from the intrinsic factor and enters the portal bloodstream.

Absorption of Electrolytes

The monovalent electrolytes are absorbed far more easily than are the polyvalent ions. In all probability, the monovalent ions sodium and potassium are actively transported by the intestinal mucosa. Chloride probably is too, but most of the absorbed chloride moves across the intestinal epithelium by virtue of the electrical gradient set up by the movement of sodium and potassium.

Of the polyvalent ions, calcium is the most important, and it too is actively absorbed. The rate of absorption by the intestinal mucosa is regulated by the parathyroid hormone secreted by the parathyroid glands (see Chapter 31). Vitamin D also plays a role and appears to be essential for normal absorption of calcium.

Absorption of Water

As already mentioned, water can be absorbed by the stomach, but the major absorption takes place in the small intestine. In the small intestine about 10 liters of water is absorbed per day, and this volume can be markedly increased. The small intestine can easily absorb water at a rate of at least 1 liter/hr. In fact, water can be absorbed so fast that if sufficient amounts are presented to the intestine, there will be such dilution of the blood that hemolysis will occur. The large intestine is also capable of water absorption, but not to the extent of the small intestine.

Actually, water moves in both directions in all parts of the intestine and stomach, but under normal conditions the movement out of the lumen exceeds movement into it. Water moves in accord with the osmotic gradient. There is no evidence that water is transported actively. As solutes are actively absorbed, an osmotic gradient favorable for the movement of water in the same direction is created. When the fluid content of the food is very low or when solute absorption is impaired, there will be net movement of water into the intestine because the osmotic gradient is in that direction.

CARBOHYDRATE METABOLISM

The intricacies of the various chemical reactions essential for the utilization, storage, or conversion of the various end products of digestion lie within the province of biochemistry, and no attempt will be made here to consider the subject in full detail. Insofar as carbohydrates are concerned, the point has already been made that although small amounts of disaccharides are absorbed, they are not metabolized but rather are rapidly and completely excreted in the urine. The monosaccharides glucose, galactose, and fructose undergo a series of essential interconversions.

Blood Sugar

Within an hour after the absorption of the monosaccharides, only glucose circulates in the blood in any significant amount. The others, galactose and fructose, either will have been converted to glucose by the liver cells and then returned to the blood or will have been utilized by the cells. Thus, the so-called normal blood sugar is almost exclusively glucose. If no food has been ingested for about 5 hr, the blood sugar is normally 80 to 90 mg percent. The usual mixed diet will increase the blood sugar to over 100 mg percent. But even on a high-carbohydrate diet, the blood sugar does not exceed about 140 mg percent in the healthy individual.

The blood-sugar level is obviously a function of the interrelations depicted in Figure 26-2. These interrelations are, to a great extent, controlled by insulin, glucagon, thyroxine, the adrenocorticoids, and the catecholamines. They will be considered in later chapters.

The monosaccharides undergo various transformations in the cells. This requires that they move from the blood into the cells, a process that is not wholly passive diffusion. To be sure, the rate of movement into the cells is a function of the concentration gradient which suggests diffusion, but this diffusion is greatly facilitated by insulin. One of the reasons blood sugar increases in diabetes mellitus is the slow transport of glucose into the cells due to the lack of insulin.

Liver Glycogen

The formation of glycogen is termed glycogenesis. The breakdown of glycogen to simpler products is called glycogenolysis. Figure 26-3 shows the interconversions of the monosaccharides to glycogen and of glycogen to glucose in a liver cell. Note that after the monosaccharides enter the cell, an enzyme facilitates the reaction with ATP, a reaction

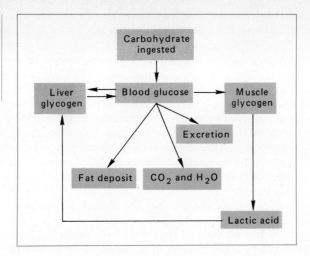

Figure 26-2. Regulation of blood glucose. The level of blood glucose represents a balance between the factors shown.

Figure 26-3. Carbohydrate interconversions in a liver cell. Monosaccharides are converted to glycogen. Glycogen can be converted to glucose which is released into the bloodstream.

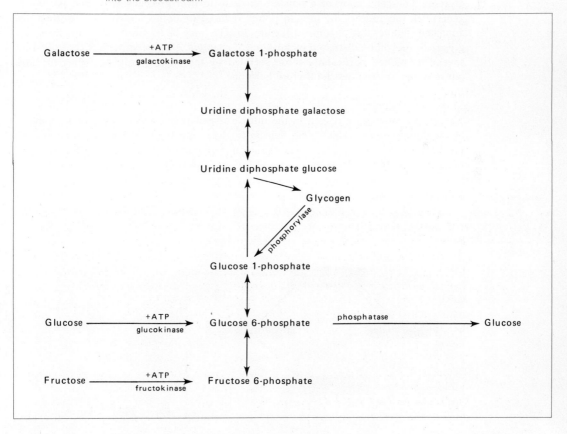

that is termed phosphorylation because the monosaccharide becomes joined to phosphate. For each of the monosaccharides there is a specific enzyme, named for the monosaccharide phosphorylated. Thus, there is a galactokinase, a glucokinase, and a fructokinase.

Glycogenolysis is greatly enhanced by the catecholamines. Epinephrine activates phosphorylase, and glucose is rapidly produced and poured into the blood. This is an important facet of blood-sugar regulation. When the blood-sugar level becomes very low, a pancreatic hormone, glucagon, is secreted. This too speeds glycogenolysis.

The quantity of glycogen in the liver is a function of the diet. In the fasting state there is almost complete depletion of liver glycogen. Following a meal, 4 to 5 percent of the weight of the liver is glycogen. Between meals, then, glycogen in the liver represents a reservoir which can be drawn upon to maintain the blood sugar. As depicted in Figure 26-3, under the influence of the enzyme phosphorylase, glycogen is converted to glucose 1-phosphate, then to glucose 6-phosphate. This compound, under the influence of phosphatase, is split into phosphate and glucose. The glucose diffuses out of the liver cells to enter the blood.

Liver glycogen is formed not only from the simple sugars but also from lactic acid. Lactic acid is produced by active muscles. It is carried by the venous blood to the liver, where most of it is converted into glycogen. Since muscles are capable of converting glucose to glycogen, there is a cycle, first described by C. F. Cori and G. T. Cori, and therefore called the Cori cycle (Figure 26-4).

Liver glycogen is also formed from noncarbohydrate sources, namely, protein and glycerol. The conversion of a noncarbohydrate to liver glycogen is termed gluconeogenesis. In this process the protein is first reduced to amino acids which are then deaminated. The residue of some of the amino acids may be transformed by the liver into glycogen. As will be discussed in subsequent chapters, gluconeogenesis is controlled by several hormones, particularly those secreted by the adrenal cortices, which increase gluconeogenesis, as does the thyroid hormone.

Muscle Glycogen

The concentration of glycogen in the muscle is generally between 0.5 and 1.0 percent of the muscle weight. This is less than in the liver, but the total mass of muscle is far greater than that of the liver. Thus, the greater quantity of glycogen is in the muscles. Liver glycogen truly represents a reservoir for regulation of blood sugar. Muscle glycogen, in contradistinction, represents a source of energy for the processes involved in muscle

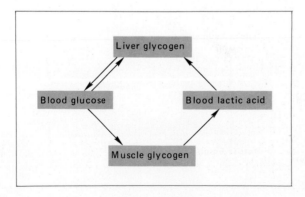

Figure 26-4. Cori cycle.

contraction. During fasting, for example, liver glycogen is rapidly and almost completely depleted, while the muscle glycogen is only slightly reduced.

Glycogenesis, the breakdown of glycogen, in both liver and muscle is believed to involve the same sequence of transformations. Glycogenolysis requires a special phosphatase that is not present in muscle but is present in liver cells. Therefore, liver cells can liberate glucose from glycogen, but muscle cells cannot. Muscle cells, however, utilize glycogen in the process of contraction (see Chapter 7). The end product is lactic acid, which diffuses out of the muscle and is carried by the blood to the liver. There it is converted into glycogen and then into glucose, which is carried back to the muscle. This completes the Cori cycle.

Carbohydrate Utilization

Carbohydrate is utilized for energy. The complete oxidation of 1 mole of glucose releases 686,000 cal. In order for this to happen the glucose molecule must be split and oxidized. The splitting of glucose into two molecules of pyruvic acid is termed glycolysis. Glycolysis involves 10 steps, or chemical reactions, each catalyzed by a specific enzyme. The overall reaction which embraces these 10 steps is

$$\text{Glucose} + 2ADP + 2PO_4^{\equiv} \longrightarrow 2 \text{ pyruvic acid} + 2ATP + 4H$$

Next, the two pyruvic acid molecules are converted to two molecules of acetyl coenzyme A (acetyl CoA) which enter the so-called Krebs cycle, also referred to as the tricarboxylic acid cycle or the citric acid cycle (Figure 26-5). The cycle depicted shows that acetyl CoA combines with oxaloacetic acid to form citric acid. In the process CoA is liberated and thus is again available to react with pyruvic acid to form more acetyl CoA. As the cycle continues, oxaloacetic acid is ultimately produced which, once again, reacts with acetyl CoA to complete the cycle. The overall reaction of the two revolutions of the Krebs cycle (one revolution per molecule of acetyl CoA) is:

$$2 \text{ Acetyl CoA} + 6H_2O + 2ADP \longrightarrow 4CO_2 + 16H + 2CoA + 2ATP$$

Recall that the respiratory chain, depicted on page 39, requires hydrogen ions from the substrate to proceed. The substrate here, of course, is glucose. In the process of glycolysis four hydrogen atoms are released. When pyruvic acid is converted to acetyl CoA, another four are liberated. And in the Krebs cycle, 16 hydrogen atoms are freed. Thus in the degradation of a molecule of glucose, 24 hydrogen atoms are made available for oxidation in the respiratory chain. The end product is water. In the process, ATP is synthesized.

In summary, the utilization of glucose produces carbon dioxide, water, and ATP. The released energy is represented by the molecules of ATP, which are the immediate sources of energy for all cellular processes. For each molecule of glucose utilized, 38 ATP molecules are produced. One mole of glucose, as already stated, releases 686,000 cal of energy during complete oxidation. One mole of ATP represents 8000 cal; thus there are 304,000 cal in the ATP formed as a result of the oxidation of one mole of glucose. This is an efficiency of 44 percent. The remainder of the energy liberated during glucose oxidation is dissipated as heat and is not available for cellular processes requiring energy.

The reactions just described, that is, the Krebs cycle and the oxidation of hydrogen, all take place in the mitochondria. As described earlier, the function of the mitochondria is to produce ATP, which is then made available to the rest of the cell for a variety of purposes.

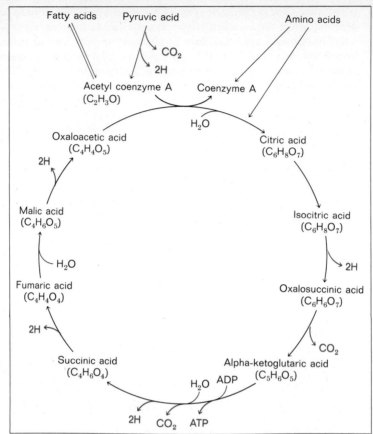

Figure 26-5. Krebs cycle. There are two turns of the cycle for the breakdown products of each molecule of glucose, since one molecule yields two molecules of pyruvic acid.

In order for the Krebs cycle to continue to operate, the hydrogen atoms produced must be oxidized. If they are not, the buildup of hydrogen will ultimately bring the Krebs cycle to a halt. This is merely to say that although oxygen enters the picture late, it is vital. Yet, some energy can be made available to the cell for a short period of time in the absence of oxygen by a process termed anaerobic glycolysis.

Recall that in the initial splitting of glucose to pyruvic acid, two ATP molecules result. Hydrogen and pyruvic acid, under anaerobic conditions, react to form lactic acid which diffuses out of the cell, thereby permitting glucose to be split. Thus, so long as there is glucose available, some energy can be liberated even in the absence of oxygen. However, under these conditions a molecule of glucose produces only 2 molecules of ATP instead of 38, and the process is therefore very inefficient.

Conversion to Fat

That carbohydrate can be readily converted to fat is common knowledge. In this conversion, the first step is again the formation of glucose 6-phosphate which is then transformed to pyruvic acid and finally to fatty acids which combine with glycerol to form fat.

The storage of carbohydrate as fat occurs only when very large quantities of carbohydrate are ingested. Normally, the blood sugar may be maintained within normal limits by conversions to glycogen. But when these stores become full, the excess carbohydrate is converted to fat and stored as such.

Carbohydrate Excretion

Any disaccharide that is absorbed is rapidly excreted in the urine. Glucose, however, is not normally excreted. But if the blood-sugar level rises quite high, in excess of about 160 mg percent, glucose does appear in the urine. The excretion of glucose may be looked upon as still another mechanism to control blood glucose level.

AMINO ACID METABOLISM

Although some protein and peptides are absorbed from the intestine, the end products of protein metabolism, the amino acids, are far more readily absorbed from the intestine. The average concentration of amino acids in the blood, expressed as amino nitrogen, is about 6 mg percent. This level is maintained as the result of amino acid absorption and formation on the one hand and amino acid utilization on the other. The amino acids may be (1) oxidized for energy, (2) used to synthesize proteins, (3) utilized for nonprotein nitrogenous compounds, (4) converted to carbohydrate, (5) transformed into fat, (6) excreted, or (7) stored.

Amino Acid Oxidation

In the liver, the amino acids undergo deamination, which means that the ammonia is split off. Deamination can occur as a result of oxidation catalyzed by amino acid oxidases or by transamination, which is more common. In transamination, the amino group is simply transferred to another substance, forming another amino acid in the process. Ultimately the amino group may be released in the form of ammonia. The liver converts ammonia into urea:

$$2NH_3 + CO_2 \longrightarrow H_2N-\overset{\displaystyle O}{\overset{\|}{C}}-NH_2 + H_2O$$

Urea diffuses from the liver cells into the blood, is carried to the kidneys, and is excreted. Malfunctioning of the liver leads to accumulation of ammonia, which can cause coma and death.

The deaminated product is termed a keto acid. Most of the keto acids can enter the Krebs cycle in the form of alpha-ketoglutaric acid, with the liberation of energy as discussed above. A gram of protein produces about as much ATP as does a gram of glucose.

Protein Formation

In Chapter 2 the mechanism by which cells utilize amino acids to synthesize proteins was discussed. Only 11 of the 21 amino acids usually present in animal proteins can be synthesized by the cells. The others cannot be synthesized or are produced in such small quantities as to prove inadequate. Since these amino acids must be provided by the diet, they are termed essential amino acids, essential in the sense that the diet must contain them to maintain health. The others are termed nonessential, again in the sense that because they are readily synthesized, they need not be in the food.

Amino acids are produced by the addition of the amino radical to the appropriate keto acid. Once present in the cell, they are then molded into the protein molecule according to the RNA template. The number of different proteins which may be synthesized in this way is astronomical.

Nonprotein Nitrogenous Compounds

The amino acids are used to synthesize many nitrogen-containing substances that are not protein. These include creatine, creatinine, urea, ammonia, pyrimidines, the purines, and porphyrins. These are only some of the more important products.

Fat Formation

Protein is a potent source of energy. But if the diet is made up exclusively of protein and provides more calories than are utilized, there will be deposition of fat. This indicates quite clearly that protein can be converted to fat. Partial oxidation of the deaminated residue of many of the amino acids produces fatty acids. The fatty acids then combine with glycerol to form fat.

Amino Acid Excretion

Between meals there is almost no amino acid in the urine, but soon after eating, amino acids are excreted. In other words, there is a threshold for amino acids so low it can be easily exceeded. This serves to prevent undue increase of amino acids in the plasma.

Amino Acid Storage

The amino acids are absorbed into the portal blood, but the liver removes amino acid so rapidly that the blood leaving the liver shows only a moderate increase in amino acids during digestion and absorption. Not only does the liver convert amino acids, as already discussed, but the liver is also capable of storing amino acids as such. The muscles have a similar capacity.

Nitrogen Balance

Many of the conversions of protein involve deamination. The nitrogen made available in this process is used to synthesize a variety of compounds, but ultimately most of them, in one form or another, appear in the urine. Thus, the rate of nitrogen excretion in the urine is an indication of the rate of protein catabolism. The nitrogen content of the ingested food and the nitrogen in the urine may be determined with relative ease. Usually these two values are the same, that is, the organism is in what is termed nitrogen balance.

There are many circumstances in which nitrogen intake and excretion are not in equilibrium. If an individual is on a complete protein fast, he still eliminates nitrogen. Since the output, under these conditions, is greater than the intake, a situation of negative nitrogen balance exists. On a protein fast, about 3 g of nitrogen is excreted. It takes 6.25 g of protein to produce 1 g of nitrogen. Therefore, on a protein-free diet, 18.75 g of body protein must be catabolized. However, if exactly this amount of protein is ingested each day, there will still be a negative nitrogen balance. Considerably more than 18.75 g of protein, actually about 45 g, must be ingested in order to establish equilibrium. The reason for this large amount is that, as already mentioned, the body requires specific amino acids, and not all these are supplied by each gram of protein. Even though controlled feeding experiments have shown that 45 g of protein will maintain nitrogen equilibrium, about twice this quantity is generally recommended.

If carbohydrate is added to the diet, the amount of protein required for equilibrium is less than when no carbohydrate is present. This does not mean that the carbohydrate supplies nitrogen, but rather that carbohydrate is used for energy and thus spares protein from this necessity. For this reason, carbohydrates and also fats are often called protein sparers.

The main causes of negative nitrogen balance are starvation and fever. Positive balance is achieved during growth, pregnancy, and recovery from a wasting disease.

Hormonal Regulation of Protein Metabolism
In brief, hormones secreted by the thyroid gland, gonads, adrenal cortices, hypophysis, and pancreas all influence protein metabolism. Thyroid hormone probably functions indirectly by virtue of its role in increasing cellular metabolism. The sex hormones increase protein deposition. The adrenal cortical hormones decrease protein deposition and increase the blood level of amino acids. Growth hormone seems to have the opposite effect, while insulin inhibits protein synthesis. All these reactions will be discussed in greater detail later.

LIPID METABOLISM
The end products of lipid digestion, glycerol and fatty acids, are absorbed and then may be either converted to various lipid compounds or oxidized for energy. In addition, glycerol may be converted to carbohydrate.

Body Lipids
Blood plasma normally contains about 500 mg percent of total lipid. This includes triglyceride, 165 mg percent; cholesterol, 170 mg percent; and phosphatides, 165 mg percent. In the body as a whole, approximately 10 percent of the total body weight is lipid. This, however, varies greatly. As will be discussed presently, each gram of lipid provides more than twice the number of calories that carbohydrate or protein provides. It is, therefore, an excellent source of energy, and the fat depots represent reservoirs of energy.

Lipid is distributed in the subcutaneous layers, which play an important role in body temperature regulation. These layers effectively insulate the internal environment from external temperature variations. They also protect the organism, to some extent, against mechanical trauma.

Lipid Transport
Recall that lipids are absorbed by the intestinal tract in two ways: (1) the fatty acids and glycerol are resynthesized upon absorption into new triglycerides, which then enter the lymph as chylomicrons, and (2) to a lesser extent, the short-chain fatty acids are absorbed directly into the portal blood. The chylomicrons enter the blood circulation via the lymph, and now, once again, the triglycerides are hydrolyzed to glycerol and fatty acids. This is accomplished by virtue of an enzyme in the blood, lipoprotein lipase. The fatty acids combine with albumin in the blood and in this form are transported. They are referred to as free fatty acids (FFA). The concentration of FFA is normally about 15 mg/100 ml of blood. Fatty acids diffuse very rapidly through cell membranes.

Adipose Tissue
Lipids are stored and metabolized principally in adipose tissue and secondarily in the liver. But adipose tissue cells do far more than serve as passive reservoirs for fat. When the fatty acids enter these cells, they are synthesized, once again, into triglycerides. Fat cells can also synthesize triglycerides from carbohydrate. The triglycerides are stored in

cells which can hold fat up to about 90 percent of their volume. Finally, adipose tissue cells can hydrolyze triglycerides to make glycerol and fatty acids available for subsequent degradation. The reader should realize that fat is a superb source of energy, that there is a steady traffic of lipid products to and from the fat cells, and even while total body weight remains constant, there is a constant turnover of fat in adipose tissue cells.

Role of Hormones in Lipid Metabolism

Later, in Chapter 31, hormones will be considered in detail. Here only brief mention will be made of those which influence fat metabolism. Insulin is of primary importance; in fact, insulin is now said to be the primary regulator of FFA mobilization. As shown in Figure 26-6, blood glucose and FFA are inversely related. When food is ingested, the blood glucose level rises and the FFA level falls. Then as glucose returns to its fasting level, the FFA level rises once again. The sequence of events is that the secretion of insulin is regulated by blood glucose level; when glucose level is high, insulin is secreted. The insulin facilitates transport of fatty acids into the cells, thereby lowering the blood concentration. Conversely, when insulin secretion is low, triglycerides are hydrolyzed and fatty acids pour into the blood.

The catecholamines increase lipolysis and therefore increase FFA release. Adipose tissue is innervated by sympathetic fibers, and thus this part of the nervous system is an important regulator of lipolysis.

Other hormones which play roles are certain adrenal cortical secretions, growth hormone, and the thyroid hormones, all of which directly or indirectly increase fat mobilization.

Synthesis of Fat from Carbohydrate and Protein

There is ample evidence that fat can be synthesized from carbohydrate and protein. Contrary to popular belief, one can store fat on a high-protein diet and, of course, on a high-carbohydrate diet. Both carbohydrate and protein are first converted to acetyl CoA. Then through a series of complex reactions, acetyl CoA and malonyl CoA combine to form fatty acid.

Figure 26-6. The relation between plasma fatty acids and blood glucose after feeding in a normal rhesus monkey. The animal was deprived of food 16 hr prior to zero time. (Reproduced with permission of *Medical World News*.)

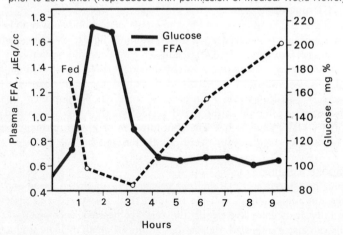

Lipid Oxidation

In the process of digestion, the triglycerides are split to free the fatty acids and glycerol. During absorption, these units are reunited in the cells of the intestinal mucosa and enter the blood or the lymph as triglycerides. If they are to be used for energy, they must be dismantled once again.

Glycerol is converted to glyceraldehyde and enters into a sequence of chemical transformations that eventuates in glucose. Then by means already discussed, the glucose is utilized for energy.

The fatty acids, on the other hand, undergo a process termed beta oxidation (Figure 26-7). As a result of this oxidation, a molecule of acetic acid is split off. Continued oxidation of the fatty acid shortens the molecule by two carbon atoms each time until the entire fatty acid molecule is oxidized.

An end product of beta oxidation of fatty acids is acetyl CoA. In the liver, two molecules of acetyl CoA combine into one molecule of acetoacetic acid, which can be further converted into beta-hydroxybutyric acid and acetone. These three substances are called ketone bodies. An excessive concentration of ketone bodies in the blood is termed ketonemia. Normally, the ketone bodies are catabolized as rapidly as they are formed and thus do not accumulate in the blood much above 1 mg percent. However, when the metabolism of carbohydrate is abnormally low, as in starvation and diabetes mellitus (see Chapter 31), fatty acids are used in abnormal quantities to satisfy energy needs, and ketonemia results. Ketonemia upsets the acid-base balance of the blood in the direction of acidosis. Ketones accumulate in tissues and body fluids, a condition termed ketosis. Acetone, a volatile substance, can often be detected in the breath of a person in ketosis. The kidney's effort to correct the ketonemia leads to the presence of ketone bodies in the urine, a condition known as ketonuria.

LIVER FUNCTION

The liver, as was mentioned in Chapter 24, is a large, important organ without which life cannot be sustained. In many sections of this book the roles that the liver plays in various functions have been mentioned. Here those activities will be summarized so as to provide an overall, integrated view of liver function.

The liver enters into circulatory, digestive, and metabolic mechanisms.

Circulatory Function

In view of the fact that approximately 1,500 ml of blood (about 30 percent of the cardiac output) flows through the liver each minute, alterations in hepatic flow will clearly influence general circulation.

Changes in the pressure relationships between the blood entering and leaving the liver result in marked changes in the volume of blood within the bloodstream at any

Figure 26-7. Beta oxidation of fatty acids.

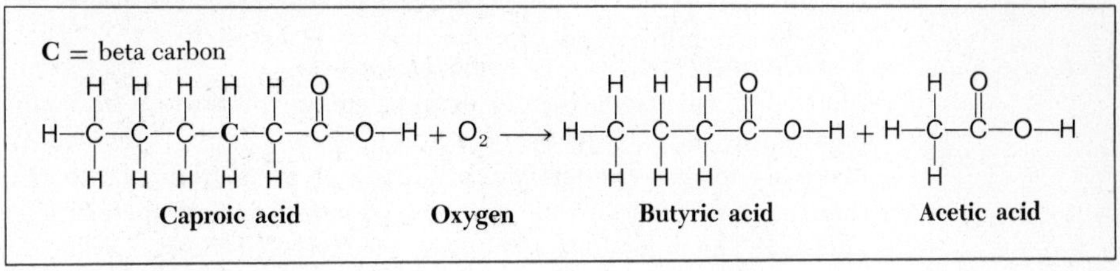

particular time. By this mechanism the liver can add to or remove from the general circulation approximately 500 ml of blood. For this reason the liver is an important blood reservoir. An alteration in circulating blood volume can, quite obviously, strikingly influence cardiac output and arterial blood pressure.

Biliary Function

The role of bile in digestion was discussed in Chapter 25. Bile is formed by liver cells and is secreted into the tiny bile canaliculi. These canaliculi empty into bile ducts which form into progressively larger ducts and finally terminate in the hepatic duct. As already explained, there is a feedback mechanism which regulates the production of bile by the liver so that during digestion the amount of bile produced increases rapidly. The amount produced then decreases once digestion is finished.

Metabolic Function

In addition to the previously discussed role that the liver plays in carbohydrate, fat, and protein metabolism, the liver also functions as a storehouse for various vitamins, particularly vitamin A. Other vitamins stored by the liver are D, E, and B_{12}. This reservoir enables one to withstand diets low in these vitamins for remarkably long periods of time. For example, experimental animals on diets devoid of vitamin A survived for over a year without exhibiting signs of vitamin A deficiency.

The liver also is important for normal iron metabolism. Most of the iron in the body is in the liver in the form of ferritin, which is an iron-protein complex. The hepatic cells contain the protein apoferritin. This reacts with iron brought to the liver by the blood to form ferritin. By virtue of a negative-feedback system regulated by the level of iron in the blood, iron is either stored in the liver as ferritin or released from ferritin to the blood. In this way the blood level of iron is maintained constant.

Finally, the liver houses metabolic functions essential to coagulation. Many of the factors, such as fibrinogen, prothrombin, accelerator globulin, and factor VII, are produced by the liver in quantities regulated by coagulation needs.

Filtration Function

While the blood flows through the liver, it is filtered. By this process virtually all the bacteria picked up from the intestinal circulation are removed. Filtration is effected by the Kupffer cells which line the venous sinusoids. Kupffer cells are very phagocytic. They not only remove bacteria but other particulate matter as well, such as the remains of dead erythrocytes or other cells.

CALORIMETRY

When food is placed in an atmosphere of oxygen within a metal chamber and then ignited, approximately the same quantity of heat is produced as when the same amount of food is burned, that is, oxidized, within the body. Not only is the heat production comparable, but the end products of this combustive process are the same.

The Calorie

The most convenient way to express energy production and utilization in the living organism is in terms of the calorie. The small calorie is defined as the amount of heat necessary to raise the temperature of 1 g of water from 14.5 to 15.5°C. This is the unit that was used earlier in the chapter in connection with the energy produced by glucose and ATP. The numbers are large and inconvenient for discussion of calorimetry. For that reason the large calorie, abbreviated Cal, which is 1,000 times greater than the small calorie (cal), is most commonly used. It is defined as the amount of heat necessary to raise the temperature of 1 kg of water from 14.5 to 15.5°C. The discussion of calories in this and the next chapter refers to these large calories.

The values obtained when 1 g of each of the three food types is placed in a carefully regulated bomb calorimeter (Figure 26-8) and ignited are interesting to compare with those obtained from metabolism in the body:

FOOD, 1 g	CALORIMETER VALUES, CAL	BODY VALUES, CAL
Protein	5.3	4.1
Carbohydrate	4.3	4.1
Fat	9.5	9.3

Note that the values for fat and carbohydrate are practically the same whether burned in the calorimeter or in the body. The slight difference is probably due to failure of absorption plus a small experimental error. The large discrepancy in the protein values stems from the fact that protein is not completely oxidized in man.

Direct Calorimetry

To determine the heat production directly, the subject must be placed in a closed chamber through which water circulates. If the temperature of the water entering and leaving is determined and if the total quantity of water which passes through the chamber is known, the amount of heat taken up by that volume of water can be calculated. These values along with other calculations disclose the number of calories produced by the subject in a period of time.

The direct method is a useful procedure for making highly accurate studies. However, the equipment is so large and so expensive that this method is usually not feasible. In the clinic, the overwhelming majority of determinations are made by means of the indirect method.

Figure 26-8. Bomb calorimeter. To determine the caloric value of food, the material is burned in the calorimeter, and the heat produced is measured.

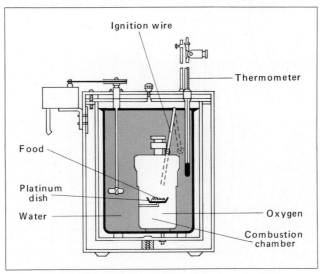

Indirect Calorimetry

Since oxygen ultimately enters into all processes involving heat production, a measurement of oxygen consumption can be used to ascertain heat production. The apparatus usually consists of a spirometer filled with oxygen. Within the spirometer is a canister of soda lime (Figure 26-9). When the subject inspires, he takes in pure oxygen. When he expires, the carbon dioxide is quickly and completely removed by the soda lime. A pen attached to the spirometer writes on a slowly moving drum of paper. During inspiration there is an upward stroke; during expiration, a downward stroke. However, the crest of each stroke is higher than that of the previous one so that there is a fairly even slope to the record. The slope indicates the rate of oxygen utilization.

In practice, the subject breathes into the apparatus for about 7 min. A 6-min period is then taken for the calculations. The paper is ruled and calibrated so that the actual quantity of oxygen used by the subject per minute can be easily determined from the slope. Since there is a direct relation between oxygen utilized and heat produced, there

Figure 26-9. Indirect calorimetry with a recording spirometer. Energy utilization is estimated from the volume of oxygen used per unit time. The subject breathes into the apparatus while lying down. As oxygen is used, the volume in the chamber decreases and the pen traces a rising line. Expired air passes through soda lime, where carbon dioxide is extracted.

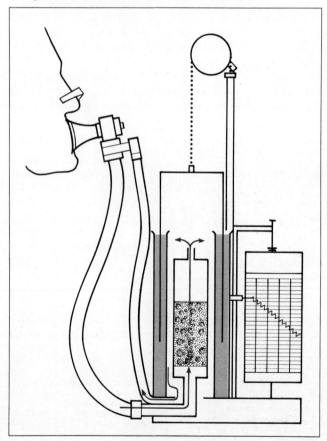

is also a calibration of the paper in calories per hour. Consequently, the technician can easily determine the heat production directly from the record. However, the student should understand the basis for this relationship. For this purpose, a simple calculation using the oxygen consumption will be presented.

The indicated volume of oxygen utilized must be corrected to standard conditions of 760 mmHg pressure and 0°C. This can be done by using equations combining Boyle's and Charles' laws, or prepared tables may be consulted. The corrected oxygen volume is then multiplied by the factor 4.825, which represents the caloric equivalent of 1 liter of oxygen. Consider the following reaction:

$$C_6H_{12}O_6 + 6O_2 \longrightarrow 6CO_2 + 6H_2O$$

One gram-molecular weight of glucose is 180 g. According to the above equation, this quantity must react with 192 g of O_2. Avogadro's principle states that 1 gram-molecule of any gas has a volume of 22.4 liters. Therefore, 6 gram-molecules will occupy 134.4 liters (6 × 22.4). In other words, 134.4 liters of O_2 react with 180 g of glucose. Therefore, 0.75 liter of O_2 must react with each gram of glucose (134.4/180). Direct calorimetry establishes that each gram of glucose oxidized in the organism liberates 4.1 Cal. Since 1 g of glucose being oxidized utilizes 0.75 liter of O_2, 1 liter of O_2, in reacting with carbohydrate, must produce 5.47 Cal (4.2/.75).

In a similar way the data in Table 26-1 have been obtained. Note that 2.03 liters of O_2 react with each gram of fat, which produces 9.3 Cal. One liter of O_2 in burning fat would produce only about 4.60 Cal (9.3/2.03). For protein, this figure is about 4.23 (4.1/.97). And, as shown in the table, for carbohydrate it is 5.47. These figures represent caloric equivalents of 1 liter of O_2 for each of the foodstuffs. On a mixed diet of all three, the caloric equivalent, as mentioned before, is 4.825. This is the factor that is used to calibrate metabolic charts in terms of calories.

Respiratory Quotient

By definition, the respiratory quotient (RQ) is equal to the volume of CO_2 output divided by the volume of O_2 intake. On a mixed diet it is equal to 0.82. But the equation presented above representing the oxidation of carbohydrate shows that six molecules of O_2 are used and six molecules of CO_2 are produced. In other words, the RQ is 1.0. The RQ for fat is about 0.71 and for protein about 0.8. Obviously, the actual RQ depends upon the makeup of the diet.

The RQ may rise above 1.0. This occurs especially under experimental conditions in which large quantities of carbohydrate are force-fed. The conversion of carbohydrate to fat releases O_2 which is used metabolically, and therefore less O_2 is taken into the

TABLE 26-1. ENERGY RELATIONSHIPS

	CARBOHYDRATE	FAT	PROTEIN
Liters of O_2 utilized per gram	0.75	2.03	0.97
Liters of CO_2 produced per gram	0.75	1.43	0.78
RQ	1.00	0.71	0.80
Calories produced per gram	4.10	9.30	4.10
Caloric equivalent of 1 liter O_2	5.47	4.60	4.23

body during this conversion period and the RQ rises above unity. Conversely, a very low RQ, approaching 0.7, is found in cases in which there is abnormally low carbohydrate utilization.

The RQ varies strikingly during and following exercise. During exercise there is high production of CO_2 and hyperventilation. Thus, the output of CO_2 is high. But at the same time an oxygen debt is being accumulated so that the O_2 consumption does not keep step with the activity. With a high CO_2 output and a relatively low O_2 intake, the RQ rises to 2.0 or even higher. Following exercise the production of CO_2 quickly returns to normal, but the excessive utilization of O_2 continues. Thus the RQ may drop to about 0.5.

BASAL METABOLIC RATE (BMR)

Since heat is produced by the metabolic processes and the quantity of heat liberated varies directly with the rate at which metabolism proceeds, the determination of heat production accurately reflects the metabolic rate. One therefore speaks of the metabolic rate rather than of heat production.

The more active an individual, the faster is his metabolic rate. Conversely, this rate falls to its lowest ebb during sleep. This is the true basal level, but since it is difficult to induce sleep whenever one desires to measure the metabolic rate, the basal level is calculated while the subject is awake but at rest and in the postabsorptive state. In actual practice, the subject is instructed to eat a light meal the night before the determination, to retire early so as to get about 8 hr of sleep, to refrain from excessive exercise for at least 24 hr, and to omit breakfast before the test is conducted. Most metabolic tests are performed early in the morning after the subject has reclined in a quiet, semidark room for about 30 min. The equipment is then attached to the subject, and he is instructed to breathe normally.

Using the standard conditions just mentioned for the determination of the calories produced, the results must now be interpreted. The best agreement among individuals is had when the heat production is expressed in terms of surface area. Very complicated formulas are required to calculate the surface area, but tables and other aids, such as the nomogram shown in Figure 26-10, simplify the task. If the individual's weight and height are known, the surface area may be easily ascertained. The average surface areas for Americans is about 1.6 m^2 for women and 1.8 m^2 for men.

The total heat production per hour is then divided by the surface area. The final figure for comparison is expressed in calories per hour per square meter. Assume that our subject produces 72.5 Cal/hr, that he is a man 1.8 m (6 ft) tall, and that he weighs 79.5 kg (175 lb). Using the nomogram, he is found to have a surface area of 2 m^2. Dividing this into 72.5, the subject is seen to produce 36.25 Cal/hr/m^2.

Normal BMR Values

The BMR is commonly expressed as a percentage of normal. A person is said to have, for example, a BMR of +20 or perhaps −15. These figures mean that his metabolic rate is 20 percent above normal or 15 percent below normal. But what is normal?

The figures are obtained by determining the heat production of a great number of people. After this is done, the values are averaged and then grouped according to age and sex. Table 26-2 gives some representative values. Let us assume that our subject is 33 years old. Using the table, we find that the average man of this age produces 39.5 Cal/hr/m^2. Our subject generated only 36.25 Cal, which is approximately 8 percent below the average. He has a BMR of −8. Is this abnormal? Actually, a variation of only 8 percent

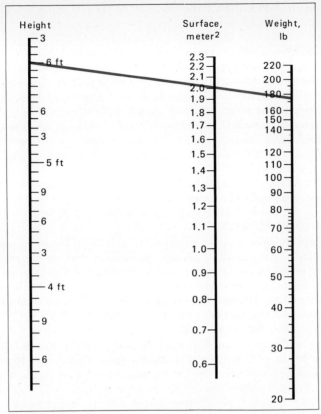

Figure 26-10. Nomogram for determining body surface area from weight and height.

TABLE 26-2. HEAT PRODUCTION, CALORIES PER HOUR PER SQUARE METER

AGE	MALES	FEMALES
6	52.7	50.7
8	51.2	48.1
10	49.5	45.8
12	47.8	43.4
14	46.2	41.0
16	44.7	38.5
18	42.9	37.3
20–24	41.0	36.9
25–29	40.3	36.6
30–40	39.5	36.5
40–50	38.0	35.3
50–60	36.9	34.4
60–70	35.8	33.6
70–80	34.5	32.6

is not regarded as pathologic. Generally, a metabolic rate which is within 15 percent of normal signifies average metabolism.

Abnormal BMR Values

As will be discussed in Chapter 31, the thyroid gland controls the metabolic rate. Therefore, malfunction of this organ usually results in marked alterations in the BMR. Hypothyroidism is associated with rates as low as 30 to 40 percent below normal. Hyperthyroidism, in contradistinction, produces a comparable increase.

The caloric production determined while an individual has a fever will be above the true basal level. Roughly, for each degree of fever, the metabolic rate is increased about 7 percent. For this reason, the temperature of the subject is routinely taken before the BMR is determined.

The BMR falls some 10 percent below normal while the subject is asleep. The BMR is also lower than normal in conditions of malnutrition.

Variations in the Metabolic Rate

Increased activity elevates the metabolic rate (Table 26-3). The heat production in sedentary occupations is about 2000 Cal/day. Strenuous mental activity requires truly small amounts of energy. An experiment in which difficult mathematical problems were undertaken by a group of students elevated the metabolic rate only 3 to 4 percent, and most of this was attributed to muscle movements during the period of the test!

Specific Dynamic Action of Food

One of the conditions required for determining the BMR is that the subject should not have eaten for 10 or 12 hr. This stipulation is included because the ingestion of food elevates the heat production. The rise in metabolism above the basal level due to the ingestion of food is termed the specific dynamic action of food.

Protein has a greater specific dynamic action, by far, than either fat or carbohydrate. If a man who produces, say, 75 Cal/hr is fed protein with a caloric value of 75, his heat production will be found to be not 75 but about 96 or 97 Cal. In other words, there is an increase of about 30 percent. Thus, the specific dynamic action of protein is usually stated to be about 30 percent, for carbohydrate it is close to 6 percent, and for fat, 4 percent.

If amino acids are injected into the bloodstream, the specific dynamic action is almost as great as when protein is ingested. This indicates that the digestive and absorptive

**TABLE 26-3. INFLUENCE OF ACTIVITY
ON ENERGY UTILIZATION FOR
A 72.5-KG MAN**

ACTIVITY	CALORIES/HOUR
Asleep	60
Basal	70
Standing	100
Walking	200
Swimming	500
Running	600

mechanics are not primarily responsible. The extra heat production is apparently associated with processes involved in amino deamination, and it differs for different amino acids. When glucose is metabolized, the extra heat production seems to be associated with glycogen formation.

QUESTIONS AND PROBLEMS

1   What is meant by active transport of digestion products across cell walls?

2   Under what condition can absorption of the end products of digestion from the intestinal lumen be said to be a function of concentration and pressure gradients?

3   What experimental data support the statement that absorption of amino acids is an active rather than a passive process?

4   Define (a) chylomicrons, (b) microvilli, (c) lacteals, (d) steatorrhea, (e) intrinsic factor, (f) glycogenolysis, (g) deamination, (h) the calorie, and (i) ketosis.

5   List the structures involved in the absorption and transport of the end products of fat digestion from the intestinal lumen to the left innominate vein.

6   Briefly explain the relation of bile to fat-soluble vitamins, and of calcium to vitamin B_{12}.

7   In what area of the body is glycogen stored?

8   What normally prevents the blood-sugar value from exceeding 140 mg percent?

9   What is the ultimate use of the ATP formed in the Krebs cycle?

10   Why is anaerobic glycolysis less efficient than aerobic glycolysis?

11   What is the fate of excess carbohydrates in the body?

12   For what purposes are proteins included in the normal diet?

13   Where in the body is urea produced, and what is its ultimate destination?

14   Explain the difference between essential and nonessential amino acids.

15   What is nitrogen balance? What are positive and negative nitrogen balance?

16   Why are fat and carbohydrates considered to be "protein sparers"?

17   In what body organs can amino acids be stored?

18   Gram for gram, which of the three food substances (carbohydrate, fat, protein) yields the most energy?

19   List the respective body functions of fat, protein, and carbohydrate.

20   Define basal metabolic rate. Under what conditions will it be found to be abnormal?

27 NUTRITION

Any of the foodstuffs can provide one's caloric needs. But, as suggested in the previous chapter, there are good reasons to have a mixture of carbohydrate, protein, and fat in the diet, and caloric needs are only part of the picture. The diet must contain various vitamins and minerals and essential amino acids. In this chapter, vitamins and minerals will be briefly discussed, and then a word will be said concerning diets.

VITAMINS

A vitamin is an organic compound essential for normal metabolic activity. In most cases, the vitamin cannot be synthesized by the cells or is produced in inadequate quantities. There are far more vitamins than are discussed below; but those presented are the major ones, and the ones about which the most is presently known.

When vitamins were first discovered, there was no inkling of their structure. For that reason, the various substances were designated simply by letters, such as A, B, etc., more or less in the order in which they were discovered. Some of the vitamins are water-soluble and others fat-soluble. Accordingly, they are often classified in that way. But now that there is a considerable knowledge concerning the chemical nature of the individual vitamins, they should be referred to according to their chemical structure. On the other hand, usage of the letter designation is so common in some instances that to change it is probably not possible.

In all probability the known vitamins represent only a fraction of those that actually exist. Vitamins characteristically are required only in very small amounts, and the results of their lack are so subtle in many cases as to be missed. In other instances, deficiency probably rarely if ever occurs and thus is not detected.

The need for vitamins has long been the subject of vigorous and often acrimonious debate. Food faddists make wild claims, vitamins are freely available without prescription, and the imagination being what it is, stories of remarkable powers of large doses of vitamins abound. All this is flamed by the fact that the actual daily requirements of the known vitamins are difficult to determine. In addition, there is a difference between the minimal requirement and the optimal intake. But more and more evidence is accumulating that vitamin overdose can be deleterious.

Vitamin requirements vary. Growing children have different needs than do adults. The same is true of pregnant women. Also the larger the individual, obviously the greater the requirement. And a person who regularly and vigorously exercises has greater requirements than does the sedentary person. In short, the statements made in this chapter represent the best available information, but the reader should be aware of the lack of knowledge and precision in this field. Table 27-1 lists the various known vitamins and summarizes the available information.

Vitamin A

The precursor of vitamin A is beta-carotene. Vitamin A itself is retinol. This is the form in which it is deposited in the liver.

Source Most of the vitamin A in the diet is supplied by fruits and vegetables. Excellent sources are spinach, lettuce, and carrots. The vitamin is also abundant in liver and milk.

Requirement Vitamin A is expressed, for nutritional purposes, in international units (IU). The requirement of an adult is about 5,000 IU per day, or 3 mg/day.

Toxicity There is some evidence that the ingestion of two or three times the daily requirement of 5,000 IU may be beneficial in terms of improved vigor. However, toxicity may result from excessive intake.

Deficiency If vitamin A is deficient in the diet during the developmental period, growth is retarded. One of the earliest signs of deficiency is night blindness. There are also serious changes in the epithelial tissues. Such changes in the eyes lead to xerophthalmia ("dry eye"). There is increased susceptibility to respiratory infection. In severe deficiency states there is sterility in the male and abnormal fetus formation in the female.

Fate Vitamin A is fat-soluble and thus probably depends, at least in part, on the presence of bile salts for adequate intestinal absorption. The precursor of vitamin A, beta-carotene, is converted to the active form by the intestine and is stored, for the most part, in the liver. The livers of people killed in accidents have been assayed for their vitamin A content and found to contain about 330 units/g. The human liver weighs approximately 1,500 g; therefore approximately 500,000 units of vitamin A are stored in the liver. This amount should suffice for 100 days if the diet contains no vitamin A.

Thiamine

Since vitamin B was first identified, so many related compounds have been found that one refers to the vitamin B complex. This complex contains vitamins that are sometimes

TABLE 27-1. VITAMINS

VITAMIN	FOOD SOURCE	FUNCTION	DAILY REQUIREMENT	RESULT OF DEFICIENCY
A₁, retinol	Fruits, vegetables, liver, milk	Essential for synthesis of visual pigments	5,000 IU 3 mg	Night blindness, epithelial changes, xerophthalmia
B₁, thiamine chloride	Yeast, liver, cereal, peanuts	Coenzyme in cellular respiration	0.5 mg/1000 calories	Anorexia, neuritis, beriberi
B₂, riboflavin	Milk, liver, spinach, eggs	Part of flavoproteins essential to oxidative chains	1.8 mg	Glossitis, cheilosis, dermatitis
Niacin, nicotinic acid	Liver, meat, chicken, yeast	Part of DPN and TPN essential to oxidative chains	20 mg	Dermatitis, diarrhea, dementia, pellagra
Folic acid	Vegetables, eggs, liver, cereals	Coenzyme in blood cell synthesis	0.5 mg	Anemia, leukopenia, sprue, growth retardation
B₆, pyridoxine	Liver, rice, milk, cereals	Participates in active transport of amino acids	2.0 mg	Growth retardation, epileptiform seizures, dermatitis, anemia
Pantothenic acid	Yeast, liver, eggs	Part of coenzyme A	?	Impaired reproduction, graying of hair, adrenocortical insufficiency
B₁₂, cyano-cobalamin	Liver, meat, eggs, milk	Erythropoiesis	2.8 mg	Pernicious anemia
Biotin	Liver, peanuts, eggs, yeast	Part of coenzymes that combine CO_2 with other compounds	?	Dermatitis, loss of hair, incoordination
Choline	Liver, fruits	Source of methyl radicals, lecithin	?	Fatty liver, hemorrhagic kidneys
Inositol	Fruits, nuts, vegetables	Formation of cephalins	?	Loss of hair, fatty liver
C₁, ascorbic acid	Citrus fruits, tomatoes, butter, potatoes	Maintains intracellular substance, collagen synthesis	75 mg	Scurvy, cessation of bone growth, poor wound healing
D₁, calciferol	Fish, fortified milk	Intestinal absorption of calcium	200–400 IU or 10–20 μg	Rickets
E₁, alpha-tocopherol	Wheat germ oil, meat, milk, butter	Cellular antioxidant	?	Infertility in animals, muscular dystrophy, renal tubular changes
K₁, naphtho-quinone	Spinach, other green vegetables	Prothrombin and factor VII synthesis	?	Failure of blood to clot

designated as B_1 or B_2 etc., but the chemical names are preferable. Thiamine is also known as vitamin B_1.

Source Thiamine is distributed widely throughout the animal and vegetable tissues, but the most concentrated source is yeast and cereals. Next in importance are liver, peas, beans, oatmeal, whole wheat, lean pork, and peanuts.

Requirement One mg of thiamine hydrochloride equals 333 IU. The minimum daily requirement for this vitamin is determined in relation to the size of the person and the number of calories utilized per day. The normal adult's minimal requirement is about 0.5 mg/1000 Cal. Recent surveys indicate that a surprising number of Americans consume less than this amount. For this reason, many foods such as bread are enriched with thiamine.

Deficiency Vitamin B_1 deficiency first causes anorexia and indigestion. Next, the nervous system becomes involved. There is degeneration of the neurons, causing pain, then loss of sensation and motor function. Long-continued deficiency of thiamine produces the clinical syndrome termed beriberi. The patient usually first complains of easy fatigue. This is followed by pain in the legs, loss of appetite, headache, enlarged heart, and shortness of breath.

Thiamine plays an essential role as a coenzyme in cellular respiration, specifically in the oxidation of carbohydrates. Since both nervous and cardiac tissue depend almost exclusively on carbohydrate metabolism, these systems are impaired. The impairment in nervous function not only gives rise to polyneuritis but also results in vasodilatation. The venous return is increased and the right ventricle dilated, after which it undergoes hypertrophy and may eventually fail. The gastrointestinal complaints are also believed to be due to inadequate nervous control which results in indigestion, constipation, and loss of appetite.

Fate Thiamine is readily and rapidly absorbed by the small intestine since it is water-soluble. It is stored in the liver but in such small quantities that it can last only a few days when the intake falls below needs. The vitamin is rapidly excreted in the urine. This prevents accumulation of sizable stores.

Riboflavin

Another member of the vitamin B complex is riboflavin, also known as vitamin B_2. Like all components of this complex, it is water-soluble.

Source The more common foods which offer a rich source of riboflavin are milk, broccoli, spinach, eggs, and liver.

Requirement The minimum daily requirement for riboflavin has not been accurately established, but authoritative sources recommend that an adult doing an average amount of work receive about 1.8 mg of riboflavin per day. This vitamin can be synthesized by intestinal bacteria.

Deficiency A deficiency of this vitamin alone rarely occurs. Generally there are also inadequate amounts of other components of the vitamin B complex present as well. A chronic deficiency of riboflavin results in blurred vision, burning and soreness of the eyes and tongue, dermatitis, and cracking and fissuring at the angles of the mouth (cheilosis). In animal experiments there is impaired growth, ataxia, weakness, bradycardia, and respiratory failure.

The rather widespread alterations that occur in deficiency states probably reflect the role riboflavin plays in biological oxidations. Because of these diffuse findings, the diagnosis of riboflavin deficiency is sometimes missed. A good indication is the blood concentration of the vitamin.

Fate Riboflavin is readily absorbed by the intestine. But it is also excreted in the urine in comparable amounts; thus there is very little storage. Only when the intake falls very low is urinary excretion of riboflavin suspended.

Niacin
Niacin is the official designation of the vitamin nicotinic acid.

Source Meat, particularly liver, is the most important source. It is also quite abundant in peanuts, chicken, and yeast.

Requirement The need for niacin varies with the total composition of the diet. In general, however, there is agreement that a daily intake of about 20 mg should suffice. Niacin is also synthesized by bacteria in the colon.

Deficiency The main finding in niacin deficiency is the syndrome termed pellagra. Usually, there is also a deficiency of other B-complex members that contributes to the disorder. Pellagra is characterized by skin disorders which appear on all exposed parts. The skin may crack and then become infected.

Niacin participates in biological oxidations, and yet no oxidation impairment has been demonstrated in tissues from niacin-deficient animals. Accordingly, the mechanism by which niacin leads to pellagra and other findings is not clear.

Folic Acid
Folic acid is actually pteroylglutamic acid, but this vitamin is usually referred to by its simpler name. Some refer to it as vitamin M.

Source Folic acid is concentrated in fresh green vegetables, eggs, and liver. In lesser amounts it also appears in beef, veal, and breakfast cereals.

Requirement The exact body needs of this vitamin are not certain. They are thought to be about 0.5 mg/day. Because folic acid occurs in so many foods and because it is produced by intestinal bacteria, deficiency is unlikely to occur.

Deficiency In animal experiments, folic acid deficiency produces growth retardation, anemia, and leukopenia. In human beings, sprue and megaloblastic anemia seem to be associated with folic acid deficiency. This vitamin is essential for the normal production of both red and white blood cells.

Pyridoxine
Still another member of the vitamin B complex is pyridoxine, which is also known as vitamin B_6.

Source Pyridoxine is found most abundantly in liver, yeast, rice, brain, cereals, and milk.

Requirement The human requirement for pyridoxine is not known. It is thought to be about 2.0 mg/day. The urinary excretion of the vitamin may exceed the intake. This may indicate that there is synthesis by bacteria in the intestine or perhaps elsewhere.

Deficiency Pyridoxine deficiency results in retardation of growth and dermatitis. In severe cases there is anemia and involvement of the nervous system with epileptiform seizures. In human beings, especially in children, pyridoxine deficiency sometimes occurs due to inadequate feeding. There is generally dermatitis, convulsions, and anemia.

Pyridoxine seems to be essential for amino acid metabolism. The primary role of this vitamin seems to be to aid, probably as a carrier, in the active transport of amino acids across cell membranes. The failure of growth could be explained on this basis, but the other findings are more difficult to understand.

Pantothenic Acid
Another component of the vitamin B complex is pantothenic acid.

Source Pantothenic acid is richest in yeast, liver, and eggs. It is also found in meat and milk in lesser concentrations, but because meat and milk are consumed in such large quantities, these represent important sources.

Requirement The human requirement for pantothenic acid is not known.

Deficiency Deficiency of this vitamin does not seem to occur in human beings. Pantothenic acid deficiency may be produced in animals, however. The findings are retardation of growth, impaired reproduction, and graying of hair. All that can be said concerning the metabolic role of pantothenic acid is that it is incorporated into CoA.

Vitamin B_{12}
The usual active form of vitamin B_{12} is cyanocobalamin, which is not unlike hemoglobin. In vitamin B_{12}, cobalt occupies similar positions to those occupied by iron in hemoglobin.

Source Liver, meat, eggs, and milk are the best source of the vitamin. Soil microorganisms appear to be the prime source of B_{12}.

Requirement The daily requirement of about 2.8 μg is for the most part supplied by the diet.

Deficiency This term usually refers to inadequate *absorption* of vitamin B_{12} owing to insufficient secretion of intrinsic factor by the stomach lining. The chief finding is pernicious anemia, characterized by deformed red cells and neurological symptoms such as easy fatigue, soreness of the tongue, and numbness in the extremities.

Other B-complex Vitamins
Other members of the B complex are biotin, choline, and inositol. Biotin is concentrated in liver, yeast, peanuts, chocolate, and eggs. Ten micrograms is considered to be the daily requirement. Deficiency leads to dermatitis, loss of hair, and muscular incoordination.

Inositol occurs in seeds and nuts. Glandular tissues are excellent sources of choline. It also appears in egg yolk. Choline deficiency results in a fatty liver and hemorrhagic kidneys.

Ascorbic Acid

Vitamin C, which is ascorbic acid, is water-soluble.

Source Fresh fruits and vegetables are the best sources of ascorbic acid. Especially good are oranges, lemons, grapefruit, tomatoes, strawberries, cabbage, butter, and potatoes.

Requirement The minimal requirement for an adult is 75 mg/day; however, beneficial results may be derived from ingesting considerably more than this, for example, prevention of the common cold.

Deficiency Deficiency of this vitamin results in lowered plasma levels and the following findings: sore gums, loosening of the teeth, spontaneous subcutaneous hemorrhages, anorexia, and anemia. This nutritional disorder has been termed scurvy. Because of the large reserves of ascorbic acid, it takes 2 to 4 months on a diet deficient in vitamin C to produce scurvy in man.

The metabolic function of ascorbic acid is not clear. It is thought to enter into biological oxidation systems, but little specific information is available. In some manner, ascorbic acid seems to play a role in regulating the colloidal condition of intercellular substances. In ascorbic acid deficiency, even superficial wounds fail to heal. In addition, the walls of the blood vessels become fragile and are easily ruptured. The loosening of the teeth may also be explained on this basis.

Vitamin D

Like vitamin A, vitamin D is fat-soluble. Vitamin D exists in several forms. The irradiation of ergosterol produces calciferol, which is the form of commercial vitamin D. But vitamin D is not used directly by the body; it is first transformed into other substances. The active form is now believed to be 1,25-dihydroxycholecalciferol (1,25-DHCC). 1,25-DHCC is produced by the kidney from vitamin D. For this reason, damaged kidneys often result in inadequate calcium absorption from the intestine with consequent demineralization of the bones.

Source Vitamin D is found principally in fish, especially oily fish such as salmon, sardines, and herring. Although milk has some vitamin D, it is often fortified by irradiation with ultraviolet light.

Requirement One international unit (IU) of vitamin D is equal to 0.05 μg of calciferol. The daily requirements of man are difficult to estimate, because he manufactures the vitamin when exposed to sunlight. The need for the vitamin is greatest during the developmental period, when at least 400 IU/day, or 2 μg, should be provided. During pregnancy, the requirement may be even higher.

Toxicity Excessive intake of calciferol has much the same influence as hyperparathyroidism, that is, demineralization of bone, elevation of plasma calcium, formation of renal calculi, and deposition of calcium in soft tissues.

Deficiency The main function of calciferol appears to be in facilitating the intestinal absorption of calcium. It may also influence calcium metabolism by controlling bone formation or resorption. Whatever the mechanism, vitamin D deficiency leads to defective bone formation. As a consequence there are skeletal deformities characteristic of a disorder termed rickets, which means "to twist." Since children are born with very little reserve vitamin D, this substance must be supplied in adequate amounts in order to assure proper bone formation.

Vitamin E

Vitamin E is a fat-soluble vitamin that has been identified as alpha-tocopherol. It probably functions primarily by preventing oxidation of unsaturated fats. For this reason vitamin E is said to be an antioxidant. In the absence of the vitamin, the quantity of unsaturated fats in the cells falls so low as to produce abnormal cellular function such as rupture of the lysosomes and interference with mitochondrial activity. Vitamin E is also reported to inhibit some actions of the prostaglandins.

Source Vitamin E is most concentrated in wheat-germ oil, but it is also found in meat, milk, margarine, butter, and leafy vegetables. Interestingly, the fish-liver oils which are rich in A and D have little E.

Requirement There is evidence that vitamin E is essential in the human diet, but the daily requirement is unknown.

Deficiency In animals, vitamin E deficiency leads to infertility and muscular dystrophy. The only known finding in man is defective fat absorption.

Vitamin K

Various compounds have been found to have fat-soluble vitamin K activity. All these appear to be forms of naphthoquinone.

Source Vitamin K is present in most foods. The leafy vegetables are especially rich in it. Spinach is outstanding.

Requirement Vitamin K is synthesized by intestinal bacteria; thus one cannot state how much must be ingested. At the present time even the total amount required by the body is not known.

Deficiency Because of bacterial synthesis, vitamin K deficiency does not readily occur. In experimental animals, vitamin K deficiency may be produced by a combination of dietary lack and blockage of bacterial action by sulfonamide. In man, a deficiency sometimes does occur in cases of abnormalities of the biliary system. Since the vitamin is fat-soluble, the absence of bile salts decreases intestinal absorption. A vitamin K deficiency is characterized by a decreased plasma level of prothrombin. As a result, clotting does not readily occur. This disorder most often appears in newborn babies. So effective is vitamin K treatment that in many hospitals it is now routine to give the expectant mother a large dose of the vitamin in order to control hemorrhage that sometimes occurs during difficult delivery. Exactly how vitamin K functions in the hepatic synthesis of prothrombin is not known.

ELEMENTS

Only a few of the elements, the so-called minerals, will be discussed in any detail. The activities of other elements are summarized in Tables 27-2 and 27-3. The metals that enter into enzymatic action are listed in Table 27-4. For more extensive treatment, biochemistry and nutrition textbooks should be consulted.

Potassium

The normal serum potassium concentration ranges from 3.8 to 5.1 mEq/liter. By far the greater fraction of potassium in the body is in the cells.

Source and Requirement Potassium is found in most foods; therefore it is extremely difficult to have a dietary deficiency. The normal intake is about 2 g/day. This would appear to be more than sufficient since the kidneys can decrease potassium excretion to a minimal level of only 1 g/day.

Function The primary role of potassium seems to be in nerve and muscle activity. It also influences body water balance. Elevation of the serum potassium level causes first tachycardia, then bradycardia, ventricular dissociation, and finally cardiac arrest. At the same time there are neurological findings, such as numbness or tingling, weakness, and finally confusion. It is interesting that low serum levels also produce cardiac and neurological abnormalities.

Sodium

In contrast to potassium, the major site of sodium is in the extracellular fluid. Its concentration in the serum is about 136 to 144 mEq/liter. There is only a trace in the cells.

Source and Requirement The average diet contains ample sodium, but it has become the practice to add considerable quantities of salt to most foods. The dietary requirement is very low because the kidney is capable of excreting a urine virtually devoid of sodium. On the other hand, there is sodium in sweat, and when perspiration is copious, the intake of sodium must be increased proportionately. Adults should ingest at least 2.5 g of sodium per day. The average diet contains at least 4 g.

Function Sodium is the major extracellular base and therefore plays an essential role in acid-base balance. It is also an important determinant of blood osmotic pressure and thus enters into body water balance. Decreased body sodium occurs following excessive sweating and in cases of adrenocortical hormone deficiency. This gives rise to weakness, cramps, nausea, and diarrhea. Abnormally high sodium levels are associated with excessive adrenocortical steroids. As a result, there is water retention.

Calcium

Calcium is mainly in bone and teeth. Its concentration in blood serum is 5 mEq/liter. There is only a trace in the cells.

Source and Requirement Milk, cheese, egg yolk, and some vegetables are the foods richest in calcium. The minimal requirement for adults is 1 g/day. This should be doubled for growing children and pregnant women.

TABLE 27-2. MAJOR MINERALS

MINERALS	FOOD SOURCE	FUNCTION	DAILY REQUIREMENTS	RESULT OF DEFICIENCY
Potassium	All foods	Essential for nerve and muscle activity	1–2 g	Muscular and neurological abnormalities
Sodium	Most foods, table NaCl	Acid-base balance, body fluid balance	2.5 g	Weakness, cramps, nausea, diarrhea, dehydration
Calcium	Milk, cheese, egg yolk	Formation of bone and teeth, essential to clotting, contraction, nerve conduction	1 g	Tetany, rickets, bone demineralization
Phosphorus	Milk, cheese, egg yolk, beef	Formation of bone and teeth, phosphate bonds essential for all processes requiring energy	1.5 g	Bone demineralization, deranged metabolism
Iron	Meat, eggs, spinach, prunes	O_2 carrier, hemoglobin	5–15 mg	Anemia, digestive disorders, dry skin
Iodine	Seafood, fortified salt	Formation of thyroid hormones	0.25 mg	Hypothyroidism
Magnesium	Green vegetables	Neuromuscular transmission	300 mg	Vasodilatation, arrhythmia, hyperirritability, spasticity
Manganese	Liver, kidneys	Activates certain enzymes	?	Infertility, menstrual irregularities
Copper	Most foods	Synthesis of hemoglobin, part of several enzymes	2 mg	Anemia
Cobalt	Most foods, tap water	Part of vitamin B_{12}; essential in erythropoiesis	1 mg	Anemia
Zinc	Most foods	Part of carbonic anhydrase, catalyses $CO_2 + H_2O$ reaction, part of other enzymes	?	Retarded growth and sexual maturation, alopecia, skin lesions, anorexia
Fluorine	Water supply	Combines with trace minerals, thus blocking bacterial enzymes	?	Caries, weak bones

TABLE 27-3. OTHER ESSENTIAL ELEMENTS*

ELEMENT	FUNCTION
Hydrogen	Component of H_2O and organic compounds
Carbon	Organic compounds
Nitrogen	Many organic compounds
Oxygen	Component of H_2O and organic compounds
Sulfur	Required for proteins
Chlorine	Main intra- and extracellular anion
Chromium	Glucose metabolism
Vanadium	Growth
Selenium	Liver function
Molybdenum	Enzyme activity

*In addition, boron, silicon, aluminum, nickel, and germanium may have essential functions.

Function The major function of calcium is in the formation of bone and teeth. It is also essential to the clotting mechanism and to normal muscle and nerve activity. There is a constant turnover of calcium salts in bone; thus if the diet is deficient, bone will be demineralized. This often occurs during pregnancy. Low serum levels may lead to tetany and, in children, to rickets. This occurs in dietary deficiencies, poor intestinal

TABLE 27-4. PRINCIPAL ENZYMATIC FUNCTION OF THE ELEMENTS

ELEMENT	ENZYME	FUNCTION
Iron	Succinate dehydrogenase	Aerobic oxidation of carbohydrates
	Aldehyde oxidase	Aldehyde oxidation
	Cytochromes	Electron transfer
	Catalase	Opposes hydrogen peroxide
Copper	Ceruloplasmin	Iron utilization
	Cytochrome oxidase	Terminal oxidase
	Lysine oxidase	Elasticity of aortic walls
	Tryosinase	Skin pigmentation
Zinc	Carbonic anhydrase	$CO_2 + H_2O$ reactions
	Carboxypeptidase	Protein digestion
	Alcohol dehydrogenase	Alcohol metabolism
Manganese	Arginase	Urea formation
	Pyruvate carboxylase	Pyruvate formation
Cobalt	Ribonucleotide reductase	DNA synthesis
	Glutanate mutase	Amino acid metabolism
Molybdenum	Xanthine oxidase	Purine metabolism
	Nitrate reductase	Nitrate utilization
Calcium	Lipases	Lipid digestion
Magnesium	Hexokinase	Phosphate transfer

absorption, nephritis, and hypoparathyroidism. Excessive serum calcium levels result in calcium salt deposition in the soft tissues and cardiac arrest in systole.

Phosphorus
Most of the body phosphorus is in the phosphate form combined with calcium in bone and teeth. Serum contains about 3 to 4 mg percent.

Source and Requirement Phosphorus appears in the same foods as does calcium, particularly in milk. It is also found in beef, beans, and almonds. The diet of growing children and pregnant women should be equal in calcium and phosphorus. For other adults there should be approximately 1.5 times as much phosphorus as calcium.

Function Phosphorus, in the form of calcium phosphate, is deposited in teeth and bone and is essential to these structures. It also plays an important role in muscle contraction. Phosphate bonds are considered to be excellent sources of energy.

Iron
There is only slightly over 4 g of iron in the entire adult body. About half this amount is in the hemoglobin of the blood. The plasma contains 0.08 to 0.19 mg percent.

Source and Requirement Iron is abundant in meat, eggs, beans, peas, whole wheat, spinach, and prunes. The actual content of the various foods may prove misleading, because iron is not taken from all foods in the same percentage. The minimal requirement for adults is about 5 mg/day, but some authorities recommend at least 12 mg daily and even more in menstruating women.

Function Iron not only functions as an oxygen carrier, but also seems to be intimately associated with the vitamin B complex. Many of the disorders characteristic of vitamin B-complex deficiency are also seen when the intake of iron is inadequate. Not only does the iron-poor diet lead to anemia, but there are also digestive disturbances and changes in the epithelium, such as cracking of the corners of the mouth, dryness of the skin, and dullness of the fingernails. Exactly how iron operates in these instances is not clear.

Iodine
There is very little iodine in the body, most of it being concentrated in the thyroid gland. The total iodine concentration in the plasma is about 10 μg percent.

Source and Requirement Iodine becomes of great nutritional importance in areas where the food and water do not contain adequate amounts. The farther one lives from the ocean, the poorer the water and food iodine supply becomes. For this reason, various foods, such as table salt, are fortified with minute but adequate quantities of iodine. The diet should contain 2 to 5 μg of iodine per day per kilogram of body weight. For the average adult this is about 0.25 mg/day.

Function The primary, if not only, function of iodine is in conjunction with the thyroid activity. Iodine is an essential component of the thyroid hormone.

Magnesium
Magnesium is widely distributed throughout the body. The serum magnesium ranges from 1.5 to 3 mg percent. In the red blood cells it is about 4 mg percent. Tissue cells contain about 20 mg/100 g, but most of the body magnesium is in the bone.

Source and Requirement   Magnesium is widely distributed in various foods, being particularly abundant in green vegetables. Deficiency has been produced in animals with diets containing less than 2 parts/million of magnesium. The normal adult diet should contain about 300 mg/day.

Function   In animal experiments, deficiency of magnesium causes vasodilatation, cardiac arrhythmia, hyperirritability, spasticity, and ultimate death in convulsion. Overdosage of magnesium produces central nervous system depression as well as inhibition of neuromuscular transmission. Calcium has an antagonistic effect. The odd finding is that while calcium antagonizes magnesium overdosage, both low magnesium and low calcium levels cause tetany.

Manganese
Manganese is stored mostly in the kidney and liver. It is very low in the blood, being on the order of about 10 μg percent.

Source and Requirement   No definite manganese requirements have yet been set. Since few cases of manganese deficiency are reported the usual diet must contain ample amounts.

Function   Manganese seems to be essential for normal reproductive function. In experimental animal deficiency, males are sterile and females are unable to suckle their young. Excessive manganese is toxic. Cases of hepatic cirrhosis have been reported.

Copper
The blood contains about 100 μg percent of copper. In the body as a whole there is only about 100 mg, most of it in the liver, spleen, and kidneys. Yet its presence is essential.

Source and Requirement   At least 2 mg/day of copper is required. The average diet easily supplies this quantity.

Function   Copper is necessary for the synthesis of hemoglobin. It also forms an important component of some of the enzymes. When animals are placed on a copper-deficient diet, they lose weight and die. Deficiency also occurs in infants who are on an exclusive milk diet. As a result, anemia develops. In chronic animal experiments there are bone and neurological changes.

Other Elements
There is only fragmentary knowledge concerning the metabolic role of other elements such as cobalt, zinc, and fluorine. Cobalt is a component of vitamin B_{12}, and therefore a deficiency of this element may interfere with synthesis of the vitamin. Zinc deficiency in animals results in retarded growth, loss of hair, dermatitis, anorexia, and vomiting. Zinc is a component of several enzymes and thus probably plays an essential role in digestion. The inclusion of fluorine in the drinking water reduces dental caries markedly.

THE NORMAL DIET
The normal diet must fulfill the requirements for (1) calories, (2) protein, (3) vitamins, and (4) minerals. The mineral and vitamin requirements have been outlined.

TABLE 27-5. CALORIC REQUIREMENTS PER DAY

ACTIVITY	CALORIES
Resting	2,000
Sedentary (student)	2,500
Light work (clerk)	3,000
Moderate work (truck driver)	3,500
Heavy work (athlete)	4,000

Caloric Requirements

Table 27-5 gives the approximate calories required by an adult in various occupations. These are very rough figures. The true measure is whether or not the person is gaining or losing weight. Whenever one ingests fewer calories than he utilizes, weight will be lost. The average person, without giving much thought to the problem, is able to maintain his weight without significant change. To obtain some idea of the caloric content of various foods, a few representatives are included in Table 27-6. These are average values. For exact values, textbooks on nutrition should be consulted.

Protein Requirements

Table 27-7 indicates the protein requirements of persons in various types of occupations who therefore have different caloric demands. The carbohydrate and fat composition of the diet are also included. Because protein can be converted into the other two foodstuffs, a pure protein diet suffices, but this may lead to renal or other disorders. Protein is needed for functions the other foodstuffs cannot satisfy. Therefore, when the others are present, protein is spared for its essential tasks. Some fat is always included because it is a high caloric source and also because it gives a sensation of satiety.

Obesity

There is perhaps no subject that concerns more people and about which there is more misinformation than obesity. And yet the facts are so simple as to be, apparently, unbelieved by the majority of people, especially by obese ones. The facts are that if one persists in ingesting more calories than are expended, weight will be gained, and if fewer are ingested than expended, weight will be lost. Quite clearly, then, a person's weight represents the result of the relation of energy input and output.

Role of the Hypothalamus   In Chapter 25, the control of the intake of food was discussed. The point was made that hunger and satiety are regulated by the hypothalamus. This knowledge has given rise to speculation that obese individuals may have a hypothalamic abnormality, but only very rarely has hypothalamic damage been found in such cases. Still, the hypothalamic centers concerned with food intake may conceivably be set at different levels in different people. More likely there are other factors that cause obese people to eat excessively for their level of activity.

Role of the Genes   Undeniably, obesity runs in families. But, of course, the reason may be eating habits rather than genetics. Still the possibility exists that the genes do regulate food intake and food metabolism so that some people must consciously suppress their desire for food in order to maintain a normal body weight.

TABLE 27-6. REPRESENTATIVE CALORIC CONTENTS

FOOD	PORTION	CALORIES
Apple	large	100
Banana	medium	100
Bacon	1 strip	50
Bread	slice	70
Broccoli	cup	15
Broth, chicken	bowl	100
Butter	thin pat	50
Cantaloupe	one-half	60
Cauliflower	1 helping	25
Chicken, fried	1 piece	200
Consomme	1 bowl	25
Corn-on-cob	1	100
Doughnut	1	200
Egg, fried	1	100
Frankfurter and roll	1	300
Grapes	small bunch	40
Ham	slice	150
Lettuce	2 large leaves	5
Orange	medium	70
Peas	1 helping	100
Potato, baked, butter	medium	150
Shrimp	6 medium	90
Spaghetti	1 helping	250
Spinach	half cup	45
Steak	12 oz	800
Tomato	medium	25
Veal cutlet, broiled	9 oz	550

BEVERAGE	PORTION	CALORIES
Coffee or tea, no sugar or cream	cup	0
Coffee or tea, sugar	cup	50
Coffee or tea, sugar and cream	cup	100
Grapefruit juice	4 oz	45
Manhattan	average	150
Martini	average	150
Milk	glass	150
Orange juice	4 oz	50
Scotch, bourbon, rye, gin	per ounce	85
Tomato juice	4 oz	30

Treatment of Obesity The treatment of obesity is well understood. In some manner the obese person must be made to alter the ratio between his energy intake and output. And therein lies the difficulty. The reasons for his excessive eating generally must be discovered and corrected before his obesity can be successfully and permanently treated. For the person to change his eating habits is no simple undertaking as is shown by the tremendous sums of money expended in the United States and elsewhere by obese persons seeking a quick, easy, miraculous reduction of weight. A simple calculation will disclose

TABLE 27-7. PROTEIN, FAT, AND CARBOHYDRATE REQUIREMENTS PER DAY

ACTIVITY	PROTEIN, GRAMS	CARBOHYDRATE, GRAMS	FAT, GRAMS
Resting	75	250	80
Sedentary	95	325	90
Light work	110	350	100
Moderate work	130	400	120
Heavy work	150	500	140

that weight cannot be lost rapidly. Recall that the utilization of 1 g of fat makes 9.3 Cal available. One pound is equal to about 450 g; therefore the utilization of 1 lb of body fat makes available approximately 4185 Cal. The average sedentary adult will utilize in a 24-hr period not much more than 2,000 Cal. In other words, if no food at all is eaten, fat loss will not exceed 0.5 lb/day.

Some diets, of course, so restrict fluid intake that there is a pleasing loss of weight, at least during the first few days on the diet. However, most of that weight loss represents a decrease in body fluid, not in body fat.

One can lose up to 2 lb/week on a low-calorie diet, but such a diet must be balanced; it must contain adequate protein, minerals, and vitamins. And it should include adequate fluid. Food normally contains considerable fluid, and in the metabolism of food, water, as already explained, is an important end product. Thus, when food intake is significantly restricted, greater quantities of fluid, as fluid, must be consumed if normal kidney function and normal body fluid level are to be maintained.

Carefully controlled experiments have shown that once an animal has been obese, it will gain weight faster on the same fat-producing diet than will an animal that has never been obese.

Physiology of Starvation

When no food is ingested, an interesting sequence of events occurs that serves well as a review of various facets of metabolism. Despite popular belief, one can go without any food at all for remarkably long periods of time. Some obese individuals have existed without food for as long as 8 months with no apparent deleterious results and certainly with notable loss of weight.

Carbohydrate cannot be stored in sufficient amounts to meet glucose needs during starvation. The main storage of glucose is in the form of glycogen, but this store is not adequate to supply the needs of the brain for glucose even overnight. Within a few hours after one leaves the dinner table amino acids begin to accumulate in the blood; this indicates increased protein catabolism. In addition the level of FFA rises, too. The fatty acids come from the catabolism of fat stores. The liver and the kidney cortex use amino acids and fatty acids for gluconeogenesis.

The main amino acid used for the synthesis of glucose by the liver is alanine, and yet alanine makes up only 7 percent of the total content of amino acids in the cell proteins. This means that most of the alanine used for gluconeogenesis does not come directly from protein catabolism but rather from the conversion of other amino acids. Figure 27-1 shows the alanine cycle. Note that NH_2 from various amino acids combines with pyruvate in muscle to produce alanine. The circulation carries the alanine to the liver, where the alanine is broken down to pyruvate and NH_2. The NH_2 goes into urea formation and is excreted in this form by the kidneys. The pyruvate gives rise to glucose.

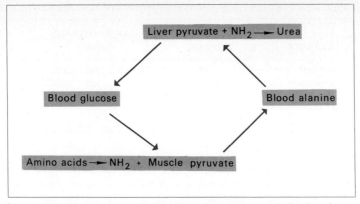

Figure 27-1. The alanine cycle. The formation of glucose in the liver from protein depends mainly upon alanine. But there is only 7 percent of alanine in body protein. Alanine used in this process of gluconeogenesis is synthesized de novo from NH_2 contributed by other amino acids in combination with muscle pyruvate.

The brain normally uses glucose exclusively. But since it requires the energy equivalent of about 100 g of glucose per day, the body's store of protein would quickly be depleted in supplying amino acids for conversion to this quantity of glucose. To be sure, triglycerides can be converted to glucose, but only at the rate of about 16 g per day. The rest would have to come from about 155 g of muscle protein per day. Unless something changed, starvation would result in death in about 3 weeks. What changes is the metabolism of the brain. In some way, still not understood, the brain now uses acetoacetic acid, acetone, and beta-hydroxybutyric acid in place of glucose.

Significantly, children cannot withstand starvation as can adults. During starvation growth stops, and if starvation continues, the child never does attain normal growth even after adequate nutrition is restored. In addition, there are irreversible changes in the nervous system including brain deficit. Along these lines, the importance of breast feeding must be emphasized. There is adequate evidence to point to the fact that there is no substitute for breast feeding, and in its absence the child suffers nutritional disorders.

From a purely biochemical standpoint, the most efficient way to lose weight is by complete fasting, that is, starvation. This is perfectly permissible for a few days, but should never be undertaken for longer periods except under completely competent medical supervision.

In view of the fact that fat loss on a starvation diet is about ½ lb/day, one can immediately see that the claims of popular diets for a greater rate of loss of fat simply cannot be substantiated.

Role of Diet in Disease

Evidence is accumulating that the makeup of the diet may be an important contributory factor in some diseases. For example, a tremendous amount of research has been carried out to discover the role of diet in coronary disease and in atherosclerosis. The results strongly suggest that not only obesity but diets high in fat may cause atherosclerosis and heart disease. Cholesterol seems to be the main villain, although a diet high in saturated fats is also involved. Unfortunately, the average diet in the United States not only is excessive in terms of calories but also contains an abnormal quantity of saturated fats and cholesterol. Even if these facts should be substantiated beyond question, there would be as much difficulty in convincing people to alter the content of their diets as there is in convincing them to alter the quantity of their diets.

QUESTIONS AND PROBLEMS

1 Construct and complete a table with the following column headings: (*a*) vitamin, (*b*) source, (*c*) daily requirement, (*d*) function, (*e*) deficiency, and (*f*) fate.

2 Construct and complete a table with the following column headings: (*a*) mineral, (*b*) source, (*c*) daily requirement, (*d*) function, and (*e*) deficiency.

3 What is the basic cause of obesity?

4 Why is it more difficult to lose than to gain weight?

5 Why is iron so important to the body?

6 What is the relation of diet to heart disease?

7 Are vitamins synthesized by the body?

PART 6
THE EXCRETORY SYSTEM

28 ANATOMY OF THE URINARY SYSTEM

The organs of the urinary system extract nitrogenous waste products from the bloodstream and excrete them from the body. Furthermore, as will be discussed in Chapter 29, urinary organs are essential to the maintenance of a constant internal environment. The two kidneys are the main organs of the urinary system. A ureter transmits urine from each kidney to a common bladder, which acts as a storage receptacle and is drained by a single tube, the urethra.

EMBRYONIC DEVELOPMENT

In the 3-week-old embryo a mass of cells called the nephrotome, or intermediate cell mass, develops just lateral to the somites. From this tissue, which gives rise to the bulk of the urinary system, three pairs of kidneys appear in succession, namely, the pronephros, the mesonephros, and the metanephros. Only the latter is retained as the permanent functional kidney.

The earliest kidney, the pronephros, is never functional in man; in fact, it develops and degenerates during the fourth week of development. Each pronephric segment possesses about a dozen tubules that connect to a collecting tube, the pronephric duct. The more cephalic tubules develop first and degenerate before the most caudal tubules appear. This primitive kidney disappears by the fifth week, but its duct persists and is incorporated into the mesonephros.

The mesonephros develops from the intermediate cell mass segments, caudal to the pronephros, before the latter completely disappears. The mesonephros is a much larger embryonic structure and extends as two prominent longitudinal ridges along the posterior wall of the thorax and abdomen. By the sixth week, as this organ reaches its maximum growth, degeneration of it has already begun. Finally by the seventh or eighth week, all mesonephric tubules are lost except those at its caudal end, which will persist to contribute to the formation of the male reproductive ducts. As mesonephric tubules develop, they join the pronephric duct, which now becomes the mesonephric duct. Into this duct each mesonephric tubule will invaginate, to be surrounded by a network of capillaries. Such an arrangement is a renal glomerulus, which foreshadows the numerous glomeruli that will appear in the permanent kidney. None of these structures are really functional, and the mesonephros soon gives way to the metanephros, or permanent kidney.

The metanephros is derived from two separate sources, the mesonephric duct and tissue derived from the nephrotome. The first indication of the permanent kidney is the proliferation of a tubular outgrowth, the metanephric bud, from the caudal part of the mesonephric duct. Its distal end expands to develop into the pelvis, calyces, and collecting ducts of the functional kidney, while its constricted proximal segment becomes the ureter. A mass of kidney-forming tissue derived from the nephrotome, the nephrogenic cap, comes to rest over the growing distal end of the metanephric bud. From this cap of tissue all the secretory portion of the kidney develops. Into the free end of each secretory tubule a tuft of capillaries invaginates and expands as a functional glomerulus. Thus the unit of the kidney, the nephron, is formed with its two parts, the renal glomerulus and the renal secretory tubule.

KIDNEYS

The kidneys are bean-shaped organs about 1.5 cm long, 7.5 cm broad, and 5 cm thick. They are located in the posterior aspect of the abdominal cavity under the thoracic diaphragm and anterior to the quadratus lumborum muscle (Figure 28-1). The upper borders of the kidneys are at the level of the eleventh and twelfth thoracic vertebrae. The right kidney is lower than the left, being displaced by the liver lying over it.

Anterior to the right kidney are the ascending colon, pancreas, duodenum, liver, and transverse colon. Anterior to the left kidney are the descending colon, spleen, and jejunum. The kidneys are behind the peritoneum ("retroperitoneal") and thus may be approached surgically without entering the peritoneal cavity. They are embedded in a mass of fat and enclosed by the renal fascia. The fat pads not only protect the kidney from mechanical trauma but also hold it in place, as there is no structure specially for that purpose.

The lateral border of the kidney is convex. The medial border is concave and indented in a depression termed the renal hilus. All structures entering or leaving the kidney do so at the renal hilus. The largest structures penetrating the kidney are the renal artery, the renal vein, and the renal pelvis, which is simply the enlarged distal portion of the ureter.

Internal Structure

In Figure 28-2 the kidney has been cut longitudinally for examination of its internal structure. There are two regions. The outermost is the cortex, occupying about one-half of the kidney and itself enclosed by a capsule of connective tissue. Internal to the cortex is the second region, the medulla, divided into 10 or 15 triangular tissue masses, the renal pyramids. The apices of the pyramids face toward the center; the bases, toward the cortex. The apex of a pyramid is known as a papilla. Between the pyramids are extensions of

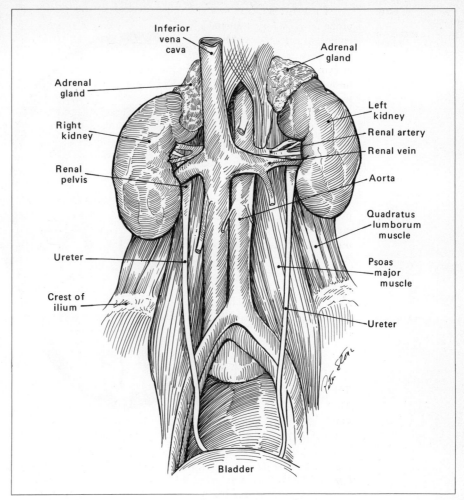

Figure 28-1. The kidneys in relation to structures of the posterior abdominal wall.

cortical tissue termed renal columns. The blood supply also passes through this tissue.

Once past the renal hilus the renal pelvis divides into three to five branches termed major calyces. Each major calyx divides into two or three tubelike minor calyces, which attach to the margins of the papillae. Behind the papillae are the renal pyramids. Each pyramid is a mass of collecting tubules for the urine formed in the functional units of the kidney. Thus urine droplets move along the collecting tubules of a pyramid, pass through the dozen or more tiny holes in its papilla, and enter a minor calyx, a major calyx, and, by way of the renal pelvis, the ureter. From there urine flows to the urinary bladder.

At the renal hilus the renal artery divides into three to five lobar branches, which are directed to different portions of the kidney. As these branches enter their respective areas, they give rise to interlobar arteries, which pass through the renal columns. At the

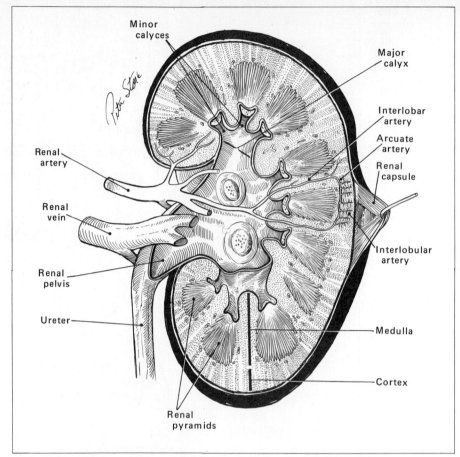

Figure 28-2. A kidney in longitudinal section. At the center are papillae of renal pyramids lying perpendicular to the plane of the section.

junction of the medulla and cortex, a T-shaped division into arcuate arteries occurs. At intervals the arcuate arteries give rise to interlobular arteries, which enter the cortex and divide into a series of afferent arterioles. An afferent arteriole terminates as a tuft of capillaries, or glomerulus (see page 671). The capillaries re-form not into a venule, as one would expect, but into another arteriole, the efferent arteriole. This vessel leaves the glomerulus and breaks into a plexus of capillaries (peritubular capillaries) around the tubule system of the kidney's functional unit. Hairpin loops within this capillary plexus are called vasa recta. The capillaries unite in an interlobular vein, which empties into an arcuate vein. Arcuate veins form interlobar veins, and these converge into the renal vein, which leaves the kidney at the renal hilus and empties into the inferior vena cava.

The functional unit of the kidney is the nephron. It lies for the most part in the cortex (Figure 28-3). The nephron consists of a glomerulus and a renal tubule. The glomerulus is composed of a knot of capillaries as described above. The glomerulus is enclosed by Bowman's capsule, or the glomerular capsule, which is simply the expanded proximal portion of a renal tubule.

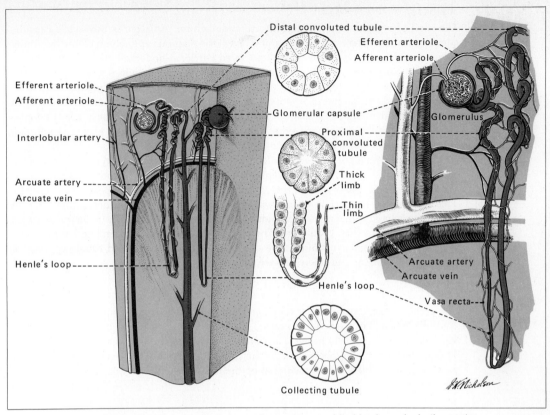

Labels in figure:
- Distal convoluted tubule
- Efferent arteriole
- Afferent arteriole
- Efferent arteriole
- Afferent arteriole
- Interlobular artery
- Glomerular capsule
- Glomerulus
- Proximal convoluted tubule
- Arcuate artery
- Arcuate vein
- Thick limb
- Thin limb
- Henle's loop
- Arcuate artery
- Arcuate vein
- Henle's loop
- Vasa recta
- Collecting tubule

Figure 28-3. Schematic diagram of a nephron and its blood supply. In the center are sections of the tubules.

The renal tubule follows a winding course. The segment arising from Bowman's capsule is termed the proximal convoluted tubule because of its many turns. The tubule then becomes straight and dips into the medulla. Eventually it makes a hairpin turn and returns to the vicinity of the glomerulus. The hairpin turn is called the loop of Henle. The tubule on the way to the loop is the thin descending limb. The tubule on the way back is the thick ascending limb. On the return from the loop of Henle the tubule is again thrown into convolutions, and this segment is named the distal convoluted tubule. It drains into a collecting tubule, which passes through the medulla as a component of a renal pyramid. The course beyond that point has been described.

The glomeruli, approximately a million in each kidney, give the cortex its granular appearance. Delicate streaks are also visible, extending from the medullary border into the cortex. These are called medullary rays and consist of straight portions of the renal tubules.

A distinction is made between nephrons that lie near the medulla and those that lie well within the cortex. The former are referred to as juxtamedullary nephrons. The renal tubules associated with juxtamedullary nephrons go deep into the medulla, while tubules of the other nephrons enter the medulla only slightly or not at all. This distinction is of importance in renal physiology and will be discussed in Chapter 29.

Histology of the Nephron

The capillary walls of the glomerulus are endothelium. Adjacent is Bowman's capsule, composed of basement membrane and a layer of squamous epithelium. Thus the initial filtrate from the bloodstream must cross three barriers—endothelium, basement membrane, and squamous epithelium—on its way into the renal tubule. The glomerulus and Bowman's capsule are together known as the renal corpuscle, or Malpighian corpuscle.

Just outside the glomerulus, smooth muscle cells of the afferent arteriole are enlarged and polyhedral in shape, and their cytoplasm contains granules. These are the juxta-glomerular cells, which produce renin. Renin, as discussed on page 492, plays an important role in some cases of hypertension.

In the proximal convoluted tubules there are thick cuboidal or pyramidal epithelial cells. Their free surfaces contain prominent hairlike processes. This border is termed the brush border and under the electron microscope is seen to be composed of microvilli. This type of cell is specialized for reabsorption of water. The thin descending limb of the tubule is lined with flattened cuboidal or simple squamous epithelium. In this segment microvilli are much less prominent. The epithelium of the thick ascending limb changes abruptly into a regular cuboidal type of cell. The distal convoluted tubule consists of cuboidal epithelium, and microvilli are present.

URETER

The ureter is a narrow, muscular tube that transports urine from the kidney to the urinary bladder. When the ureter is contracted, its lumen is star-shaped in cross section. Its muscular coat is relatively heavy. The outer layer is circular; the inner layer, longitudinal. This arrangement is opposite that of the alimentary tract. The ureter, like all urinary passageways, is lined with transitional epithelium. In the ureter this is four or five cell layers thick.

URINARY BLADDER

The kidney produces urine continuously. The urinary bladder stores the urine (Figure 28-4). It is situated predominately in the pelvic cavity although, as it fills with urine, the bladder expands into the abdomen. Peritoneum of the abdominal wall reflects onto the superior aspect of the bladder and continues onto its posterior surface where, in the female, it reflects onto the uterus, forming the uterovesical pouch. In the male the peritoneum reflects onto the rectum, forming the rectovesical pouch.

The anterior surface of the urinary bladder lies just behind the pubic symphysis. The region between the two is the retropubic space, through which a surgical approach may be made to pelvic organs without disturbing the peritoneal cavity. The urachus, the remnant of the embryonic allantois, extends from the superior surface of the bladder to the umbilicus. Inferiorly the bladder narrows to a neck at the internal urethral orifice, where the urethra begins.

The bladder wall contains a smooth muscle layer, the detrusor muscle, which is important in initiating urination (see Chapter 29).

The interior of the bladder is thrown into a series of folds except at a small, smooth triangular region, the urinary trigone. The openings to the two ureters and the urethra are at the corners of the trigone. The ureters travel on an oblique course through the bladder wall, marked by the ureteric fold on the external surface. Because of this oblique entrance, urine cannot flow back into the ureters and renal pelvis if the bladder becomes overdistended. But the angle does not prevent infections from reaching the kidney from the urinary bladder.

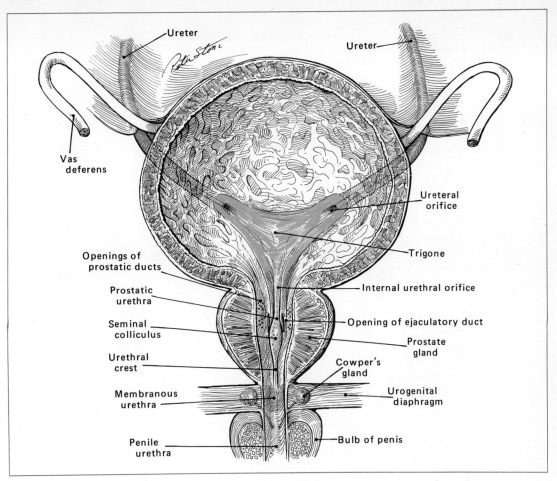

Figure 28-4. Urinary bladder, prostate, urethra, and related structures. Coronal section. The internal sphincter is smooth muscle in the wall of the neck of the bladder. The external sphincter is striated muscle around the urethra where it traverses the urogenital diaphragm.

URETHRA

The urethra is a tube that conducts urine to an external orifice. In the female it is only about 2.5 cm long. It extends from the neck of the bladder through the urogenital diaphragm to an opening about midway between the clitoris and the vagina.

In the male the urethra ia much longer, and prostatic, membranous, and penile portions are present (Figure 28-5). The prostatic portion lies at the middle of the prostate, a muscular and glandular organ that surrounds the male urethra as it leaves the urinary bladder (see Figure 28-4). The superior surface of the prostate blends with the tissue of the neck of the bladder. On the posterior wall of the prostatic portion of the urethra is the urethral crest, a longitudinal ridge. Flanking this crest are two elongated slits where the ejaculatory ducts transmit semen into the urethra (see Chapter 32).

Beyond the prostate the membranous portion of the urethra passes through the

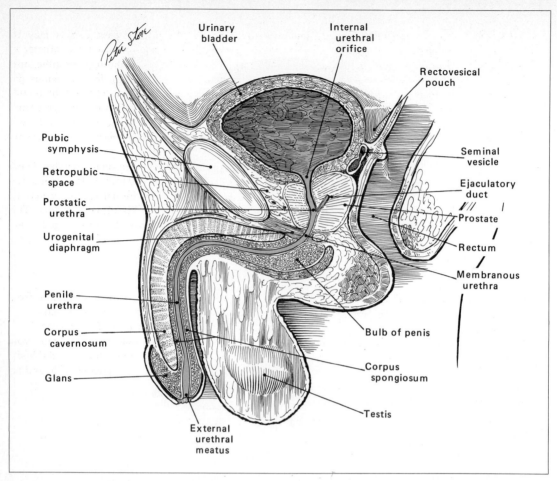

Figure 28-5. The male pelvis. Sagittal section.

urogenital diaphragm to enter the bulb of the penis. The membranous portion is surrounded by muscles of the urogenital diaphragm. The urethra next courses through the corpus cavernosum urethra to reach the glans penis, where it becomes dilated at the external urethral orifice. The entire penile portion is surrounded by erectile tissue.

In both sexes two sphincters are associated with the urethra. The internal sphincter is a thickening of the smooth muscle making up the wall of the neck of the bladder. The external sphincter, composed of voluntary (striated) muscle, surrounds the urethra as it passes through the urogenital diaphragm.

In the female the urethra is lined with transitional epithelium, which becomes irregular, and at the external meatus the epithelium is continuous with the stratified squamous epithelium of the vulva. In the male the prostatic urethra is lined with transitional epithelium. In the membranous portion the epithelium is a mixture of transitional and stratified columnar. In the penile portion the epithelium is stratified columnar except at the external meatus, where it becomes stratified squamous and is continuous with the skin.

CLINICAL APPLICATIONS

Tumors, strictures, and dilatations associated with the urinary passageways may be investigated by pyelography. Retrograde pyelography consists of placing a catheter in the urethra and ureter, passing a radiographic contrast medium through the tube, and, with the fluoroscope, studying the appearance of the passageways. In intravenous pyelography a dye that is radiopaque (that is, opaque to x-rays) is injected into the bloodstream. Pictures taken by means of x-ray at intervals after injection of the dye permit a view of the urinary passageways and an evaluation of renal function. If a kidney is not functioning properly, it will not extract dye from the bloodstream, and its absence will be apparent from the x-ray picture.

Occasionally the soluble minerals in the urine precipitate out in small granules. These granules may coalesce into a small stone, or calculus. Calculi may form in the renal pelvis ("kidney stones") or in the bladder ("urinary stones"). In the former case they may be forced ("passed") by peristaltic action through the ureter into the urinary bladder. This movement is usually accompanied by excruciating pain. If stones become lodged in the ureter, they must be removed by surgery.

QUESTIONS AND PROBLEMS

1 How are the kidneys supported in situ?

2 Discuss the significance of the retroperitoneal location of the kidneys.

3 Compare the structures found at the hilus of the kidney with those at the hilus of the lung both anatomically and functionally.

4 Describe the vascular pattern of the kidney, and compare it with that of the liver.

5 What is a glomerulus? How does it differ from other capillaries in structure and function?

6 What are the components of a nephron?

7 Trace the course of the urinary filtrate from Bowman's capsule to the exterior of the body.

8 Describe the differences between male and female urethras.

29 PHYSIOLOGY OF THE URINARY SYSTEM

The primary function of the kidneys is to regulate the volume and composition of the extracellular fluid. The kidneys are essential homeostatic organs. Changes in the environment, in the food and fluids ingested, and in metabolism threaten the constancy of the internal environment. The kidneys, by their ability to alter the volume and composition of the urine, are able, in most instances, to maintain this constancy.

In this chapter the physiology of renal function and the elimination of urine will be considered. In the next chapter the body fluids will be examined.

RENAL CIRCULATION

The circulation through the kidneys is unique. As described in the previous chapter, there are two capillary beds in the renal circulation. One capillary bed is associated with the glomerulus, and the second with the proximal tubule, the loop of Henle, and the distal tubule. In addition, some of the so-called peritubular capillaries give rise to long hairpin loops, the vasa recta (see Figure 28-3). A clear understanding of these anatomical relationships is essential for comprehension of renal function.

Renal Blood Flow

In man, approximately 1,200 ml of blood flows through the kidneys per minute, or about 600 ml per kidney. But this is a highly variable value. It varies not only from person to person but also in the same person under a wide variety of conditions. During exercise, for example, the renal blood flow is sharply decreased.

The fact that on the average some 1,200 ml of blood flows through the kidneys per minute indicates that under normal conditions the resistance in the kidneys must be relatively low. The left ventricle of the heart ejects about 5,000 ml/min. But not all this output is available to the kidneys, because they lie downstream, so to speak, from the coronary vessels, the carotids, the splanchnics, and the celiac artery. Thus in man, probably only 2,500 ml/min flows in the aorta at the site of the renal arteries. Consequently about half the available blood enters the renal arteries, which indicates that the renal resistance must be about the same as resistance farther down the aorta. Vasoconstriction of the renal afferent and efferent arterioles therefore markedly reduces renal flow.

Renal Blood Pressures

Figure 29-1 portrays the pressures of the renal blood in various segments of the renal circulation. The mean pressure in the renal arteries is about 100 mmHg. The first major pressure drop occurs in the afferent arterioles, so that the blood enters the glomerular capillaries at a pressure of 65 or perhaps 70 mmHg and leaves at a pressure of about 60 mmHg. There is then a very great drop of pressure in the efferent arteriole. Just how great depends upon the state of constriction of that vessel. At any rate, under normal conditions, the pressure probably drops to 18 or 20 mmHg before the blood enters the peritubular capillary. The pressure then decreases about 10 mmHg in that capillary. Thus, the blood returns to the inferior vena cava via the renal vein under a pressure of no more than 10 mmHg. Finally, since the vasa recta take off from the peritubular capillaries, or the efferent arterioles, and return to the venules, the pressure differential at the two ends of the vasa recta can be only a few millimeters of mercury. For this reason, blood flow in these relatively long and very minute tubes is very slow.

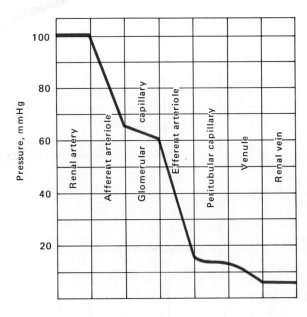

Figure 29-1. Pressures in the renal blood circulation. The areas of major pressure change are the afferent arterioles and efferent arterioles. (*Redrawn from Pitts, Physiology of the Kidney and Body Fluids, Year Book Medical Publishers, Inc., Chicago, 1963.*)

Innervation of the Renal Blood Vessels

Although parasympathetic fibers from the vagus innervate the kidney, they apparently have no significant influence on renal blood flow. Insofar as is known, renal blood flow is regulated exclusively by the thoracolumbar sympathetic supply. Activation of these fibers causes vasoconstriction of the afferent and efferent arterioles. For reasons not yet completely understood, sympathetic activation sometimes causes both arterioles to constrict equally, and at other times, one arteriole to constrict more than the other. The influence of these changes on the formation of urine will be discussed.

Autoregulation of Renal Blood Flow

Figure 29-2 demonstrates that renal blood flow, over a wide pressure range, remains quite constant. In other words, the kidneys exhibit autoregulation. The mechanism by which the kidneys increase resistance to flow as the driving pressure is increased is not known, although there are several theories. In Chapter 17, the various hypotheses purporting to explain autoregulation were discussed. In addition to these, there is in the kidney, some investigators insist, still another mechanism responsible for autoregulation.

The nephron makes a turn in the loop of Henle. As a result, the distal tubule ascends to come into close proximity with the glomerulus, where it passes between the afferent and efferent arterioles. Lining the afferent arterioles at this point are juxtaglomerular cells. These cells contain renin, the substance that was discussed in connection with hypertension (page 492). When arterial blood pressure decreases, flow through renal vessels tends to decrease. As soon as this occurs, renin is secreted. Renin, in the manner previously described, initiates a sequence of events that elevates arterial blood pressure, thereby restoring flow to normal in the kidney. But when arterial blood pressure is changed, as shown in Figure 29-2, renal blood flow remains relatively constant, which means that renal resistance must change. Again the juxtaglomerular cells are thought to play a key role. Exactly how is not known. It is thought that if, for example, the arterial blood pressure is increased, glomerular filtration (see page 679) will increase and there will be a greater flow of fluid through the renal tubules. The greater flow alters the sodium

Figure 29-2. Autoregulation of renal blood flow.

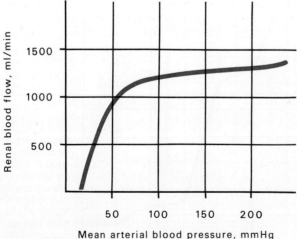

concentration and also the osmolality of the fluid. These changes are thought to influence the juxtaglomerular cells to cause constriction of the afferent arteriole, which increases resistance and maintains renal blood flow constant in the face of an increased arterial pressure.

THE FORMATION OF URINE

Urine is formed by the filtration of plasma through the glomerular capillary. This filtrate then moves through the nephron, during which passage water and solutes are transported out of the tubules into the peritubular capillaries (see Figure 2-4). A few substances are secreted into the tubules from the peritubular capillary. And at least one solute, namely, potassium, passes in both directions. Thus, as the filtrate flows through the nephron, it is considerably modified to form the final product, urine.

Glomerular Filtration

Fluid is filtered by the glomerulus; it is not secreted. That is to say, the process is a passive, not an active, one. In order for the plasma to pass from the glomerular capillary into the tubule, it must traverse three membranes: (1) the capillary endothelial layer, (2) a basement membrane, and (3) the epithelial cells that cover the glomerulus. Whether or not there are pores that pass completely through the three layers is not certain, but the evidence indicates that molecules with a molecular weight of less than about 68,000 can readily pass through. All the plasma proteins are larger; thus little or no protein normally gets into the filtrate.

Glomerular Filtrate Composition   Micropipets may be inserted in the tubules of the kidneys of various animals. In this way samples of the filtrate in the glomerulus have been obtained and analyzed. The results show that the glomerular filtrate is an ultrafiltrate of the plasma, that is, it contains everything plasma does and in almost the same concentration, with the exception of protein. The plasma contains about 7 g percent protein. The glomerular filtrate has less than 0.03 g percent.

Regulation of Glomerular Filtration   Since glomerular filtration is a passive process, it is controlled completely by the resultant of various pressures. As stated above, the blood pressure in the glomerular capillary is about 65 mmHg. This is the pressure that forces plasma through the membrane. On the other side of the membrane, that is, in the tubule, the pressure exerted by the filtrate at that point is thought to be about 12 mmHg. This opposes the filtration process. Also opposing it is the colloid osmotic pressure of the plasma. This osmotic pressure, as the blood enters the capillary, is about 26 mmHg. As the blood flows through the capillary, considerable fluid filters, thereby progressively concentrating the proteins and thus increasing the osmotic pressure. The average osmotic pressure in the entire glomerular capillary, because of this progressive concentration, is thought to be about 30 mmHg. The so-called effective glomerular filtration pressure is 23 mmHg [65 − (30 + 12)].

Quite clearly, changes in the caliber of the afferent and efferent arterioles will alter the pressure relationships in the glomerulus and thereby vary the rate of glomerular filtration. For example, if the afferent arteriole constricts, the blood pressure downstream to that constriction, that is, in the glomerular capillary, will be decreased, and so will the effective glomerular filtration pressure. Thus, the rate of filtration is decreased.

Constriction of the efferent arteriole will have the opposite influence: it will increase the blood pressure in the glomerulus. Therefore, the effective glomerular filtration pressure and the rate of filtration will increase. On the other hand, if constriction of the efferent arteriole is so great as to reduce renal blood flow markedly, then, despite the high blood pressure in the glomerulus, considerable filtrate will be formed over a period of time as the blood moves slowly through the capillary. As a result, the osmotic pressure of the highly concentrated plasma proteins will oppose filtration, and the filtration rate, expressed in milliliters per minute, may decrease. In short, moderate efferent arteriolar constriction increases the filtration rate, but excessive constriction diminishes it.

Finally, constriction of both the afferent and efferent vessels will diminish renal blood flow, but maintenance of high blood pressure in the glomerulus maintains filtration. However, excessive reduction of renal blood flow will decrease filtration.

Because of autoregulation discussed previously the rate of filtration is not much influenced by changes in arterial blood pressure. However, at sustained high pressures there may be a small increase in filtration rate which, as will be explained, can markedly increase urine production (Figure 29-3). On the other hand, should arterial blood pressure fall very low, the effective glomerular filtration pressure will approach zero, and all filtration will stop. This is one of the consequences of circulatory shock.

A significant change in the protein concentration of the plasma will, of course, alter glomerular filtration. If one drinks very large quantities of fluid, the plasma proteins will be diluted, and therefore the plasma colloid osmotic pressure will be reduced and the effective glomerular filtration pressure increased. As a result, the filtration rate will increase. Conversely, dehydration concentrates the plasma proteins and thus reduces the filtration rate.

Glomerular Filtration Rate   As a result of the forces just described, the normal glomerular filtration rate is approximately 125 ml/min. That is to say, all the filtrate formed by the numerous nephrons in the two kidneys working together produces glomerular filtrate at the rate of 125 ml/min. The normal plasma flow through both kidneys is close to 650 ml/min. Of this 650 ml, 125 ml leave the capillary, to be filtered into the glomerulus. The ratio between the plasma flow and the filtration rate is termed the filtration fraction and averages about one-fifth (125/650).

Figure 29-3. Influence of arterial blood pressure on the rate of urine output.

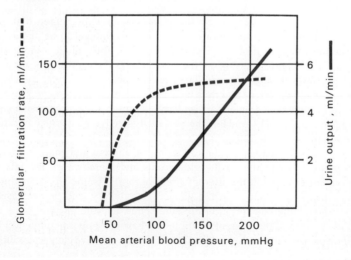

Tubular Reabsorption

As the fluid passes through the lumen of the tubules, there is marked modification of the filtrate. In the first place, since the glomerular filtration rate is 125 ml/min and the rate of urine formation normally is about 1 ml/min, obviously some 124 ml of fluid must be reabsorbed in the nephron and collecting duct. Of the 125 ml, about 100 ml are removed from the proximal tubules so that only 25 ml of filtrate enters the loops of Henle. Here another 7 ml is removed and an additional 12 by the distal tubules. This leaves only 6 ml entering the collecting ducts per minute, where an additional 5 ml is reabsorbed. Thus, the flow of fluid in the various segments of the two kidneys is glomeruli, 125 ml/min; loops of Henle, 25 ml/min; distal tubules, 18 ml/min; collecting ducts, 6 ml/min; renal pelvis, 1 ml/min.

There is no evidence that the reabsorption of water is anything but a passive process. The factors controlling the movement of water will be discussed later.

In the proximal tubule virtually all the glucose, the amino acids, the vitamins, and any protein that leaked through the glomerular membrane are absorbed. These are all transported by active processes. Protein is removed by pinocytosis (see Chapter 2). Although the glomerular filtrate contains less than 0.03 g percent protein, because of the large volume of filtrate, about 30 g of protein leaks into the filtrate per day. One could not long survive were this amount of protein not returned to the circulation.

Other substances, such as urea, sulfates, phosphates, nitrates, sodium, potassium, chloride, and bicarbonate, are also transported, to a limited extent, back to the blood from various parts of the nephron. But potassium can also be transported in the opposite direction by the distal tubules. The reabsorption of potassium, however, is generally much greater than secretion; thus very little potassium remains in the urine.

Tubular Secretion

As already mentioned, the renal tubules are also capable of transporting substances from the blood into the filtrate. This too is an active process. Most of the substances that the tubules handle in this manner, with the exception of potassium, creatinine, NH_4^+ and H^+, are foreign to the body.

Countercurrent Mechanism

According to the classic explanation of the formation of urine, from the time the filtrate enters the proximal tubule until it reaches the collecting duct there is a progressive concentration, resulting in the final, hypertonic urine. In order for this concept to be satisfied, water would have to be actively transported. Second, the tubular cells would have to create and maintain a gradient of about 900 mOsm/liter. There is no evidence that water is actively transported anywhere in the body, nor is there any evidence that any cell can maintain such a great gradient. Finally, micropuncture studies reveal that the filtrate as it enters the proximal tubule is isotonic. In the thin loop of Henle it is highly hypertonic, but then it progressively becomes isotonic once again in the ascending tubule (Figure 29-4). These observations demand modification of the classic concept.

The countercurrent hypothesis has been proposed. It does not require active transport of water, nowhere is there a gradient over about 200 mOsm/liter, and there is good evidence to support it.

Anatomically, the descending, ascending, and collecting tubules are found to be close to one another. In these tubes, the filtrate circulates in such a way that the current in the ascending tubule is counter to that in the descending tubule, as well as to that in the collecting duct. Thus, the countercurrent hypothesis states that in the proximal tubule many substances, including sodium, are actively transported out of the filtrate and, because

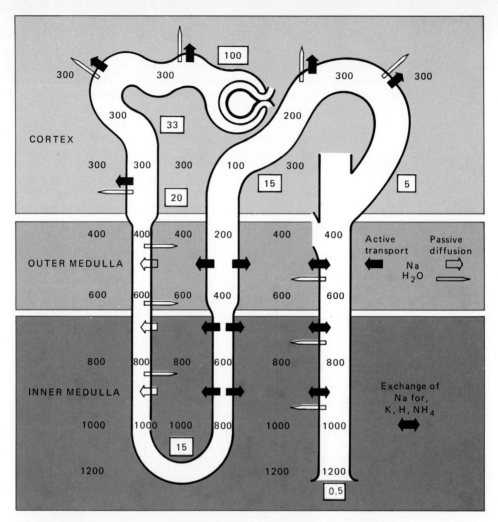

Figure 29-4. Mechanism of urine formation. Boxed numbers indicate the estimated percent of glomerular filtrate remaining in the tubule at that point. Other numbers refer to the concentration of tubular urine and peritubular fluid, in milliosmols per liter. (*Redrawn from Pitts, Physiology of the Kidney and Body Fluids, Year Book Medical Publishers, Inc., Chicago, 1963.*)

of the osmotic gradient so established, water also moves out. This active transport of solute is so great that about 80 percent of the total filtrate leaves the tubule by the time it reaches the thin loop of Henle. To this point, then, the filtrate is still isotonic although markedly decreased in volume (Figure 29-4).

In the descending thin part of the loop of Henle, sodium passively diffuses into the filtrate, and water passively diffuses out. As a result, the filtrate becomes greatly concentrated, reaching a maximum at the bottom of the loop. Another 5 percent of the filtrate is removed, making a total of 85 percent to this point.

Throughout the entire ascending limb, which includes the distal tubule as well as the loop of Henle, there is active transport of sodium out of the filtrate but no movement of water for most of that length. Consequently, the filtrate becomes less concentrated, and in the distal convoluted tubule it actually becomes hypotonic until the subsequent movement of water makes it isotonic. Most of the ascending tubule is impermeable to water. In the last part of the distal convoluted tubule, the membrane is permeable, and another 10 percent of the filtrate is removed.

When the filtrate enters the collecting duct, it is isotonic. Here, as Figure 29-4 indicates, there is additional active transport of sodium, but because the interstitial fluid surrounding the duct is hypertonic, water diffuses out of the filtrate even faster, thus concentrating it. By the time the filtrate reaches the end of the collecting duct, over 99 percent of the fluid has been reabsorbed.

The interstitial fluid concentration in the medullary tissue surrounding the tubules is the key to the countercurrent concept. One should note that this fluid becomes more and more concentrated, reaching a maximum in the innermost portion. The tubules of the juxtamedullary nephrons (those lying at the edge of the medulla) dip down deep into this medullary tissue. It is the active transport of sodium from the ascending tubules that accounts for the progressive increase in concentration of the medullary interstitial fluid. Because the ascending tubules are impermeable to water, water cannot move with the sodium. But the collecting duct is permeable to water. Now, because of the highly concentrated interstitial fluid surrounding that duct, water leaves the duct in accordance with this osmotic gradient, resulting in a concentrated urine.

Not only does this countercurrent concept explain the movement of water without postulating active water transport, but a final examination of Figure 29-4 will disclose that nowhere along the system is there a concentration gradient greater than 200 mOsm/liter, and yet the concentration of the filtrate increases from about 300 mOsm/liter to 1,200 mOsm/liter by the time it passes through the collecting duct.

One who truly understands the countercurrent mechanism recognizes that it cannot operate unless the concentration of the medullary interstitial fluid is maintained. It is maintained by the slow blood flow in the vasa recta. The vasa recta act as countercurrent systems too (Figure 29-5). In this case it is the blood flowing through the long hairpin tubes that is counter. The blood flows down into the medullary tissue and then back up toward the cortex. As it flows down into the medulla, water diffuses out and sodium diffuses in. But as the blood flows up toward the cortex, sodium diffuses out and water moves in. The net result is that the blood carries away water but leaves most of the sodium to recirculate through the medullary interstitial fluid.

The vasa recta maintain the high concentration of sodium and other solutes in the medullary interstitial fluid not only because of their configuration as a countercurrent system but also because the blood flow is very slow. The faster the flow, the more inefficient it becomes. Measurements show that the blood transit time in the cortex is 2 or 4 sec. In the medullary tissue it is about 30 sec. In the dog, the medulla composes about 10 percent of the total kidney tissue, but it receives only approximately 1 percent of the total blood flow.

In the interstitial fluid of the cortex there is no concentration gradient. Thus, the cortical nephrons probably do not form a hypertonic urine. Consequently, the juxtamedullary nephrons must concentrate the filtrate even more than the figures given above would suggest. Since only about one-seventh of all the nephrons are the juxtamedullary type, these units must bear the burden of concentrating the filtrate to such an extent that when it is mixed with the isotonic fluid from the cortical nephrons, the characteristic

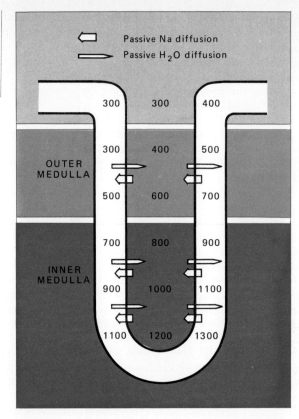

Passive Na diffusion
Passive H$_2$O diffusion

300 300 400

OUTER
MEDULLA

300 400 500

500 600 700

INNER
MEDULLA

700 800 900

900 1000 1100

1100 1200 1300

Figure 29-5. Concentration gradients in the vasa recta and the interstitial fluid of the medulla. The net result is that blood carries away water but leaves most of the sodium in the interstitial fluid.

hypertonic urine results. Interestingly, animals native to the desert have kidneys with a much higher percentage of juxtamedullary nephrons and can concentrate their urine far more than can man.

Composition of Urine

The composition of urine can vary greatly. As emphasized at the beginning of this chapter, by virtue of this variation the composition and volume of the body fluids are maintained. The degree of variation of the composition of urine simply reflects altered body activity, and food and fluid intake. Under normal conditions, intake is fairly constant. Thus fairly meaningful average figures for the composition of urine can be given. Table 29-1 lists representative average values for some of the more common constituents of urine.

REGULATION OF RENAL FUNCTION

Glomerular filtration, as already explained, depends upon the resultant between hydrostatic and osmotic pressures. Changes in the caliber of the afferent and efferent arterioles are the main determinants of the hydrostatic pressure. Hydration and dehydration can alter the osmotic pressure. Thus, the glomerular filtration rate can, and does, vary markedly, strikingly influencing the filtered load of each substance and the rate of urine

TABLE 29-1. PARTIAL COMPOSITION OF URINE COMPARED WITH PLASMA

SUBSTANCE	PLASMA, MEQ/L	URINE, MEQ/L	24-HR URINE, G
Sodium	140	120	4.0
Potassium	4.6	50	3.0
Calcium	4.5	6	0.2
Magnesium	1.9	9	0.1
Ammonia	–	40	1.1
Chloride	103	180	9.5
Bicarbonate	27	17	1.5
Sulfate	1.1	35	1.1
Phosphate	1.8	40	1.2
	MG%	**MG%**	**G**
Glucose	80	0	0
Urea	2.5	1,600	24
Uric acid	4	53	0.8
Creatinine	1.5	100	1.5
pH	7.41	6.0	

formation. But, under more normal circumstances, the volume and composition of the urine reflect tubular activity more than glomerular dynamics.

Water

The excretion of water is under the control of the so-called antidiuretic hormone (ADH) from the posterior lobe of the hypophysis. Under the influence of this hormone, the reabsorption of water may be so great as to reduce the urine output to about 0.35 ml/min. In the complete absence of the hormone, from 15 to 18 ml of urine per min may be excreted. At this rapid flow, the specific gravity of the urine is practically the same as that of blood. That is to say, it is nearly isotonic, rather than hypertonic.

ADH is thought to control water movement by increasing the permeability of the collecting ducts. ADH has been shown to increase the permeability of many types of membranes. According to the countercurrent concept, water moves out of the collecting duct because of an osmotic gradient. The rate at which it moves, for any gradient, is a function of the duct membrane permeability. Under the influence of ADH, permeability increases; thus water moves more rapidly out of the filtrate, thereby decreasing the urine output and increasing the urine concentration.

A small change in the rate of reabsorption of water can, of course, make a big difference in the urine flow rate. Approximately 5 percent of the glomerular filtrate enters the collecting duct. Of the 6 ml of filtrate that enter the duct, about 5 ml is reabsorbed. If only 4 ml were reabsorbed, the urine flow would double!

Adrenocortical and gonadal steroid hormones (see Chapter 31) also influence the rate of urine formation. However, they regulate water reabsorption by virtue of their effect on active sodium transport. The steroids increase the reabsorption of sodium, thereby creating a greater osmotic gradient; thus water is moved too.

Sodium and Potassium

The level of these ions in the extracellular fluid is markedly influenced by the adreno-cortical steroids. Part of this control is due to the action of these hormones on renal function. The adrenocortical steroids increase the reabsorption of sodium and decrease the reabsorption of potassium. Probably the tubular secretion of potassium is also enhanced. The net result is that the more steroid, the greater the excretion of potassium and the lower the sodium excretion. The extracellular fluid concentration of these ions reflects the altered excretion rates so that sodium increases and potassium decreases.

Calcium and Phosphate

These ions, like sodium and potassium, are also actively transported by the tubules, a transport mechanism which is influenced by the hormone of the parathyroid glands. Under the influence of this hormone, calcium reabsorption is increased and phosphate reabsorption is decreased.

Abnormal Urinary Constituents

Substances which do not normally appear in the urine or which are present only in small quantities may be excreted either because of deranged metabolic functions or because of abnormal renal activity. In the first category glucose and the ketone bodies may be mentioned. Normally there is no glucose in the urine, because the rate of reabsorption by the tubules is great enough to remove all the glucose that normally filters through the glomerulus. The filtration rate is about 125 ml/min, and the concentration of glucose in the plasma is about 1 mg/ml. Thus, 125 mg/min enter the tubules and are reabsorbed so that none reaches the urine. But in diabetes mellitus the plasma level may increase to 2 mg/ml or higher. Now more than 250 mg/min of glucose reach the tubules. Consequently, part is not reabsorbed and appears in the urine, a condition termed glycosuria. The same is true of ketone bodies during impairment of carbohydrate metabolism. As was explained in Chapter 26, excessive amounts of fats are mobilized, and ketone bodies accumulate in the blood (ketonemia) and spill over into the urine (ketonuria).

In the second category, that is, as a result of disordered renal function, many abnormal urinary constituents are found. First of all, glycosuria can occur in the face of a normal plasma glucose level if the renal tubules are incapacitated and therefore unable to reabsorb glucose as fast as it is filtered. The more common results of renal disease are the presence in the urine of protein, blood cells, and casts, which are particles of proteinaceous material from the tubules. The diseased kidney usually cannot concentrate urine normally.

The protein that is most often found in the urine is albumin, a condition termed albuminuria. It generally occurs because of increased permeability of the glomerular membrane. If the loss of protein in the urine exceeds the rate at which the liver can produce it, the protein concentration of the blood will decrease. This is a dangerous condition because the decrease in blood osmotic pressure will result in massive edema, low blood volume, and death.

The end products of protein metabolism are normally excreted as rapidly as they form. But the diseased kidney may be so impaired that they accumulate in the blood, a condition termed uremia. Severe uremia produces confusion, muscle contraction abnormalities, convulsions, coma, and death.

The Artificial Kidney

If the kidneys do not adequately regulate the composition of the blood, other means must be taken in order to prevent changes that are incompatible with life. The so-called artificial kidney is now widely used. It consists of a mechanism containing a membrane through which substances from the blood may diffuse. They are then carried away from the other side of the membrane by the constant movement of fluid called the dialyzing fluid. The membrane permits all constituents of the plasma, except protein, to pass through. The dialyzing fluid is adjusted so that there is a concentration gradient for the substances to be removed from the blood. These substances then diffuse out of the blood.

In order for the artificial kidney to be utilized, one catheter is placed in an artery and another in a vein. Arterial blood is then pumped to the kidney; the filtered blood returns via the vein catheter. Heparin is used to prevent coagulation. The time that the patient must be attached to the artificial kidney depends upon the degree of residual function of the patient's own kidneys. As yet a permanent, implantable artificial kidney has not been developed.

Diuretics

A diuretic is a substance that increases urine flow. This may be accomplished either by increasing the glomerular filtration rate or by decreasing the reabsorption of water.

Osmotic Diuretics   Any substance that is freely filtered by the glomerulus but is not readily absorbed by the tubule will cause diuresis if its concentration in the blood, and therefore in the filtrate, is increased. Sucrose is a good example. When this substance appears in the filtrate, it impedes the reabsorption of water because it is not readily reabsorbed. Consequently, the osmotic pressure of the filtrate opposes the reabsorption of water. Glucose will do the same thing if the plasma concentration is raised high enough to exceed the ability of the tubules to reabsorb it.

Xanthines and Alcohol   These substances are thought to increase urine flow primarily by augmenting the rate of glomerular filtration. The mechanism is believed to be dilation of the afferent arteriole, which increases renal blood flow and also elevates glomerular filtration pressure. Alcohol also inhibits the secretion of ADH.

Mercurial Compounds   Compounds containing mercury are commonly used as diuretics. Their action is probably at least twofold. In the first place, since they are not readily absorbed by the tubules, they act osmotically. Second, they impede the reabsorption of sodium in the proximal tubule. This is probably the more important action. Because of the inhibition of sodium reabsorption, water reabsorption is decreased. Mercury decreases sodium reabsorption by inhibiting the responsible enzyme system.

THE CONCEPT OF CLEARANCE

The student may well have wondered, during the preceding discussion, how glomerular filtration rate is determined or even how renal blood flow is monitored in man short of placing a flowmeter on the renal vessels. These and other values can be determined by means of the concept of clearance. Renal clearance is defined as the volume of blood, or plasma, that is completely cleared, or emptied, of any substance in unit time, usually one minute. This is obviously only a theoretical value, but it has broad application. Clearance determinations are most often done in reference to plasma; therefore, the term plasma clearance is proper.

In order to calculate plasma clearance the following values must be ascertained:

1   Concentration of a substance in the urine (U)
2   Concentration of the same substance in the plasma (P)
3   Rate of urine flow (V)

Then

$$\text{Clearance} = \frac{UV}{P}$$

If concentration is expressed in milligrams per milliliter and flow in milliliters per minute, clearance will be expressed as milliliters per minute. This is in line with the definition, namely, the volume (milliliters) of plasma cleared per minute. Actually the units used to express concentration are unimportant so long as the same units are used for both urine and plasma. In the equation they cancel each other out.

Glomerular Filtration Rate (GFR)

If a substance is freely filtered by the glomerulus but is neither reabsorbed nor secreted by the tubule, then the plasma clearance of that substance must be a measure of GFR. In other words, the only way it gets into the urine is through the glomerulus. The amount of plasma cleared of that substance then must be the GFR. The only naturally appearing substance that comes close is creatinine, but a small amount of creatinine is secreted by the tubules into the filtrate. Inulin, a polysaccharide with a molecular weight of 5,000, is neither reabsorbed nor secreted. Its molecular weight is small enough for free glomerular filtration, and it is nontoxic and easy to determine.

To carry out a clearance determination using inulin the substance is infused into the blood at a rate that maintains a constant concentration in the plasma. The bladder is then emptied at zero time. After about 20 min, the bladder is again emptied, the urine volume measured, and the inulin concentration ascertained. During the 20-min period a blood sample is drawn and the inulin concentration of the plasma determined. Assume that the urine flow is found to be 1 ml/min, plasma concentration of inulin 1 mg/ml, and the urine concentration 125 mg/ml. Then

$$C = \frac{125 \times 1}{1}$$

$$= 125 \text{ ml/min}$$

This inulin clearance (C) is 125 ml/min, which means that 125 ml of plasma had to filter through the glomeruli of both kidneys each minute. The GFR is 125 ml/min. This determination is completely independent of the rate of urine flow or the concentration of inulin in the plasma. Since no inulin is reabsorbed, concentration increases as urine flow decreases and vice versa, and therefore UV is constant. Also, as plasma concentration increases, so does urine concentration.

Once the GFR is known, the amount of any plasma constituent that is filtered can be calculated. This is termed the filtered load and is expressed in milligrams per minute. For example, concentration of glucose in the plasma during fasting is about 0.8 mg/ml; therefore the filtered load of glucose is 100 mg/min (0.8 × 125). Yet there is normally no glucose in the urine. This means (1) the clearance of glucose is zero and (2) the tubules reabsorb glucose at the rate of 100 mg/min. Interestingly, if the reabsorption of glucose is blocked by using the drug phlorhizin, the clearance of glucose is the same as inulin, that is, 125 ml/min.

Table 29-2 lists the clearance of various substances. These are average values which can change sharply under varying circumstances. The important point is that if the clearance of a substance is less than that of inulin, it must undergo tubular reabsorption to some extent. If its clearance is zero, reabsorption is complete.

TABLE 29-2. RENAL CLEARANCE VALUES

SUBSTANCE	CLEARANCE, ML/MIN
Glucose	0.0
Bicarbonate	0.5
Sodium	1.0
Calcium	1.2
Chloride	1.3
Magnesium	5.0
Potassium	12.0
Uric acid	15.0
Phosphate	25.0
Sulfate	45.0
Urea	75.0
Inulin	125.0
Creatinine	140.0
Diodrast	570.0
PAH	600.0

Tubular Secretion

If the renal clearance of a substance is greater than that of inulin, at least some of that substance is put into the urine by tubular secretion. Para-aminohippuric acid (PAH) and Diodrast are such substances. The same procedure as described for inulin is used to determine their clearances. That is, PAH or Diodrast must be infused into the blood and the plasma concentration held relatively constant during the collection period. PAH, in man, has a clearance of about 600 ml/min; Diodrast is close to 570 ml/min. Note in Table 29-2 that creatinine has a clearance of 140 ml/min. This is so close to inulin that creatinine clearance is sometimes taken as an approximation of GFR.

Renal Blood Flow

If the concept of clearance has been grasped, the reader will quickly appreciate the fact that if a substance could be found that is completely cleared during one circulation through the kidneys, its clearance would be a measure of renal plasma flow. In other words, the substance would be in the renal arterial blood but not in the renal venous blood. No such substance has been found, but PAH comes close. Only 10 percent of the amount in the arterial blood remains in the venous blood. Knowing this, renal plasma flow can be determined from PAH clearance. If the clearance is 600, then the renal plasma flow is about 667 ml/min (600/0.9). To determine blood flow the hematocrit must be known. If it is 45 percent, then renal blood flow is 1,213 ml/min (667/0.55).

Glomerular Filtration Fraction (GFF)

Inulin and PAH clearances may be determined simultaneously. The ratio of the two equals the glomerular filtration fraction:

$$\text{GFF} = \frac{125}{667}$$

$$= \frac{1}{5}$$

This means that of the total plasma flowing through the kidneys, approximately 20 percent is filtered through the glomeruli.

Tubular Mass

The transport mechanism responsible for the transfer of any substance in either direction through the tubules may be so loaded that it reaches a maximum. To put it another way, the rate of transfer increases with the load presented up to a point. After that point is reached, the transfer rate does not change. The maximal mass of a particular substance that can be transported per unit time is termed the tubular mass (T_m), or better, transport maximum. It is expressed in milligrams per minute.

Earlier in this chapter, the point was made that despite a relatively constant GFR in the face of rising arterial blood pressure, urine output increases sharply (see Figure 29-3). The explanation is that as the GFR increases, even though it is a small increase, the filtered load of each substance increases. For some substances the tubules cannot reabsorb the additional load, and the molecules remain in the tubular fluid, thereby increasing its osmolality, which prevents reabsorption of water. Urine output increases.

Without going into great detail, suffice it to say that in order to determine T_m the filtered load must be increased until the rate of reabsorption or secretion becomes maximal and constant. Knowing the filtered load and the amount excreted, the amount reabsorbed can be easily calculated. For glucose, for example, T_m is about 325 mg/min. Filtered loads greater than that result in glycosuria. As the blood concentration of glucose continues to rise, the clearance of glucose increases. It approximates that of inulin, but, of course, so long as some glucose is reabsorbed, its clearance can never reach that of inulin.

TESTS OF RENAL FUNCTION

There are many ways of evaluating kidney function. The simplest involves measurements of the daily urine volume and the specific gravity of the urine. The daily output of urine is very important. It is normally 1 liter or more per day and should not fall below about 500 ml per day. In acute renal insufficiency it can fall below this level, and there may even be complete lack of urine formation. Obviously, the patient cannot long survive under these conditions unless steps are taken to rid the blood of the end products of metabolism.

The specific gravity of the urine varies greatly depending upon the intake of fluid, but it also depends upon the ability of the kidneys to concentrate urine. Normally the specific gravity of urine is between 1.015 and 1.025. If the specific gravity does not increase as the urine volume decreases, renal insufficiency is indicated.

More specific information concerning renal function is obtained using clearance tests. The final determination of the ability of the kidneys to function comes from an examination of the body fluids. If the composition and volume of the extracellular fluids are normal, the kidneys are doing their job.

MICTURITION

After the urine is formed by the individual nephrons, it flows into the renal pelvis. From the renal pelvis it moves down the ureters to the bladder. The elimination of urine from the bladder is termed micturition, or urination.

Function of the Ureters

The ureters are not merely passive ducts. They are equipped with smooth muscle, which permits peristaltic movements. These movements propel the fluid along. Thus the urine enters the urinary bladder in irregular spurts.

The ureters are capable of inherent movements, as shown in the denervated state. However, the innervation does modify the contractions, and the ureters may undergo spasms which completely block the flow of urine.

Function of the Bladder

The urinary bladder, which is a hollow muscular organ, stores urine and contributes to micturition. In addition, there is some evidence that water and solutes can be passively absorbed from the ureters and bladder, but just how is not yet clear.

The urinary bladder does not behave simply as an elastic bag. Rather, like other smooth muscle viscera, as fluid collects within it, the internal pressure varies very little. In other words, it accommodates to urine as does the stomach to food. Were it otherwise, a back pressure would develop which could impede glomerular filtration.

The bladder wall contains stretch receptors. Impulses which they initiate are propagated via visceral afferent neurons in pelvic nerves to the sacral part of the spinal cord. Parasympathetic neurons are finally activated, and the detrusor muscle, which makes up the wall of the bladder, contracts. This sharply increases intravesicular pressure, the sphincters are forced open, and urine is eliminated (Figure 29-6). The urinary bladder volume necessary to trigger micturition varies greatly. On the average, reflex contraction occurs when the bladder contains about 300 ml of urine. However, this reflex can be inhibited so that much greater volumes can be held. Conversely, the bladder can be emptied voluntarily, by increasing intraabdominal pressure, even when it contains only a few milliliters.

Contraction of the detrusor muscle is but one reflex responsible for urination. Another reflex apparently controls the internal sphincter. This sphincter cannot be voluntarily opened or closed independently of detrusor contraction and relaxation. The external sphincter, on the other hand, can be vigorously closed despite detrusor contraction. Normally, a sequence of reflex actions occurs which causes bladder contraction and relaxation of both sphincter muscles. Following elimination, the sphincters close and the bladder relaxes.

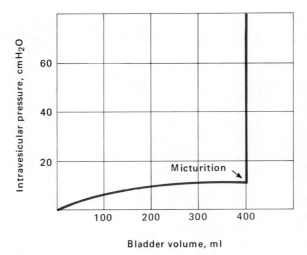

Figure 29-6. Pressure within the urinary bladder. Accumulation of urine increases intravesicular pressure, that is, bladder pressure, only slightly. The micturition reflex, which involves contraction of the detrusor muscle, causes pressure to rise sharply and forces open the sphincters.

Abnormalities of Micturition

Immediately following transection of the spinal cord there is complete areflexia of the bladder. This is part of spinal shock and persists for a variable period of time. It is thought to be due to loss of facilitation of the reflex arc. In some cases, the areflexia is followed by a so-called automatic bladder that fills with a fairly normal volume before it automatically empties due to firing of the reflex. In other cases, spasticity develops so that after but a small volume of urine has accumulated, the bladder empties.

Advanced syphilis may cause destruction of sensory roots of sacral nerves. The motor roots remain intact. There is thus loss of the desire to urinate and a failure of the reflex. Accordingly there is great distention of the bladder, with overflow incontinence.

Great distention of the bladder is to be avoided. Not only may it damage the bladder, but it may cause the bladder to lose its tone so that it cannot contract, forcing a volume of urine to be retained for long periods, a condition favorable to the development of infection. If adequate bladder function is not reestablished or if there is retention or the dribbling of urine (enuresis), it may be necessary to introduce a catheter into the bladder to control the discharge of the urine. This is now a safe procedure, but before the advent of the antibiotics, catheterization of the bladder almost invariably caused severe infection.

QUESTIONS AND PROBLEMS

1   Explain why renal blood flow decreases during exercise.

2   What percent of ventricular output enters a single kidney each minute in someone at rest?

3   Which division of the autonomic nervous system exerts the greatest influence on kidney blood flow?

4   What is the function of the juxtaglomerular cells?

5   Why are proteins considered to be an abnormal constituent of urine?

6   What is the composition of the glomerular filtrate?

7   How is the so-called effective glomerular filtration pressure calculated?

8   Explain how constriction of the afferent renal arterioles would affect the effective glomerular filtration pressure.

9   Explain why the glomerular filtration rate is reduced in dehydrated individuals.

10   What is the difference between tubular secretion and tubular reabsorption?

11   If one consumed a salty meal and drank a great deal of water, what effect would it have upon urine production?

12   Explain the countercurrent hypothesis as it relates to the formation of urine.

13   Is transport of water through tubule cells an active or passive phenomenon?

14   Explain the mechanism by which the excretion of water is controlled by the antidiuretic hormone (ADH) and by adrenocortical and gonadal steroid hormones.

15   Explain why glycosuria and ketonuria can be presenting symptoms in someone with uncontrolled diabetes mellitus.

16   What is a diuretic, and by what mechanisms can it achieve the desired effect?

17   Describe the micturition reflex.

18   What is the capacity of the urinary bladder? What happens when the bladder becomes overdistended?

30 THE BODY FLUIDS

The human body is approximately 59 percent fluid. The exact percentage, as will be explained below, depends upon the percentage of body fat. Thus, a man who weighs about 70 kg and has normal body weight will contain approximately 41 liters of fluid. The largest portion, that is, about two-thirds of this total, is located within the cells and is appropriately termed the intracellular fluid. The remaining one-third is called extracellular fluid. It includes the blood, lymph, cerebrospinal fluid, synovial fluid, specialized fluids (such as that found in the eyes and ears), and the fluid that bathes the cells, termed the interstitial fluid. The volume, composition, and regulation of these fluids are discussed in this chapter.

TOTAL BODY FLUID
The most direct way to determine total body water is simply to weigh the body, then dry it completely and reweigh it. The difference represents total body water. This procedure can be carried out in experimental animals and was also done many years ago on executed human beings. Obviously, a method for measuring total body fluid in the living state is preferable.

The Dilution Principle
The dilution principle is used to determine the volume of the total body fluid as well as that of the various subdivisions. This principle was briefly discussed in Chapter 16.

A substance is administered that distributes itself evenly and completely throughout the fluid compartment being measured and which restricts itself to that compartment alone. A sample of fluid from the compartment is then compared with a sample of fluid containing the same substance in a known volume. In other words, the appropriate substance is administered in a known amount. The same amount is added to a known volume of water and mixed. This is the so-called standard. Samples of body fluid and of the standard are then compared. Thus

$$\text{Body fluid} = \frac{\text{standard concentration} \times \text{standard dilution}}{\text{body fluid concentration}}$$

Quite obviously, the larger the fluid compartment, the greater will be the dilution of the test substance, and thus the lower will be its final concentration in that compartment.

The dilution principle is simple. The difficulty lies in finding a substance that is nontoxic, that distributes itself evenly throughout the compartment being measured, that is restricted to that compartment alone, and that can be conveniently and accurately analyzed. Various compounds have been used to determine the total body fluid, such as so-called heavy water and tritiated water. Heavy water is water that contains hydrogen with an atomic mass of 2, in contrast to regular water that contains hydrogen with an atomic mass of 1. Tritiated water contains hydrogen that has an atomic mass of 3, and it is radioactive. Heavy water can be identified by its specific gravity or by infrared spectrophotometry. Tritiated water emits radiation which can be accurately counted with liquid scintillation counters.

The main question in the use of any of these substances is whether or not they are distributed evenly and completely in all the body fluids. Water with the heavier isotopes of hydrogen may not readily pass through all membranes. However, there is considerable evidence that if sufficient time is permitted for distribution, complete dilution is obtained.

Because of the ease and accuracy of measuring tritium, it is now being used in most clinics and laboratories. After the dose is given intravenously, about 1 hr is allowed for distribution throughout the body fluids. A sample of any body fluid may be used for the analysis. Blood is most conveniently obtained and is, therefore, generally used. A volume of blood and the same volume of the diluted standard are then counted for radioactivity. Thus

$$\text{Total body fluid (ml)} = \frac{\text{standard (counts/min)} \times \text{dilution (ml)}}{\text{blood (counts/min)}}$$

Total Body Fluid Values

The amount of fluid in the body is a function not only of the total body weight but also of the amount of fat present. Fat contains very little water; therefore, as the proportion of fat in a body increases, the proportion of fluid in the total body weight decreases. That is to say, there is an inverse relationship between percent body fat and percent body fluid (Figure 30-1). For this reason, total body fluid is generally expressed in terms of "fat-free tissue," or the "lean body mass."

Body fat is estimated by determining the corporeal specific gravity. Fat has a specific gravity less than 1. Other tissues have a specific gravity greater than 1. Thus the higher the specific gravity of the whole body, the less fat. To determine body fat percentage, the subject is weighed in air, then in water. The difference represents the weight of water displaced. This value is divided into the weight of the subject in air. Since there is a

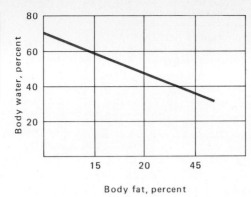

Figure 30-1. Relation between the proportion of water making up total body weight and the proportion of fat in total body weight. The average adult male has about 15 percent body fat and therefore about 59 percent gravity water.

linear relation between body specific gravity and percent body fat, the fat content can be estimated (Figure 30-2).

After the fat content is known, body water may be expressed in terms of the fat-free tissue. Total body fluid averages about 70 percent of the fat-free body weight.

Regulation of Total Body Fluid

The fluid in the body represents a balance between that which enters and that which leaves. Fluid normally enters only via the intestinal tract. Fluid loss takes place in urine, perspiration, feces, and evaporation from the lungs. The amount lost by each of these routes varies with many conditions, particularly with the temperature and activity (Table 30-1).

As was discussed in Chapter 19, the loss of water in the perspiration can vary tremendously, and while exercising in hot, humid weather it can be as much as 1 liter/hr. Under such conditions there will be hyperventilation, and thus water loss in the expired air will also increase. The amount in the feces is generally fairly constant. Thus it falls to the kidneys to vary fluid output and to the thirst mechanism to control fluid intake.

Kidney Function In Chapter 29 the role of ADH in regulating the volume of urine formed was discussed. This hormone can regulate urine volume over a very wide range. If no hormone is secreted, urine output may reach about 18 liters in a 24-hr period. Secretion of the hormone can lower this output to as little as 400 or 500 ml/day.

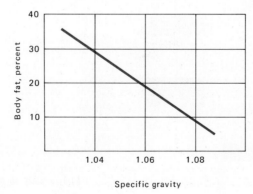

Figure 30-2. Relation of percent body fat to the specific gravity of a body.

TABLE 30-1. LOSS OF BODY WATER, ML/DAY

ROUTE	AT REST	HEAVY EXERCISE
Urine	1,400	500
Lungs	400	600
Sweat	500	5,000
Feces	200	200
Total	2,500	6,300

In the supraoptic nucleus of the hypothalamus, there are specialized cells called osmoreceptors. They are so termed because they respond to the osmotic state of the fluid surrounding them, that is, the extracellular fluid. The osmoreceptors are nerve cells, and they fire impulses through neurons in the pituitary stalk to the neurohypophysis (see Figure 31-3). These impulses stimulate the posterior lobe of the hypophysis to release ADH.

Apparently change in the osmotic state of the extracellular fluid causes fluid to move into or out of the osmoreceptors. If the osmolarity of extracellular fluid is lower than normal, water, by virtue of osmosis, will move into the cells; when the osmolarity of extracellular fluid is high, fluid moves out. The inward movement of fluid is thought to inhibit the osmoreceptors so that they fire less frequently; thus less ADH is released, and therefore the kidneys excrete more water. In summary, there is a negative feedback mechanism that regulates fluid output (Figure 30-3).

Figure 30-4 is illustrative of this mechanism. In this experiment the subject drank 1 liter of water at zero time. Note that urine output increased within a few minutes and remained elevated until an extra liter of urine was excreted, after which the urine flow returned to normal. The order of events in the mechanism is as follows: dilution of the blood by the water, movement of fluid into the osmoreceptors, decreased rate of firing

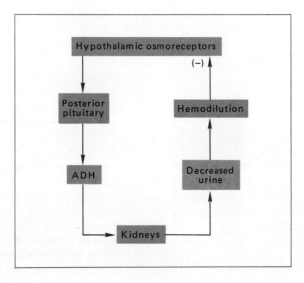

Figure 30-3. Regulation of urine output.

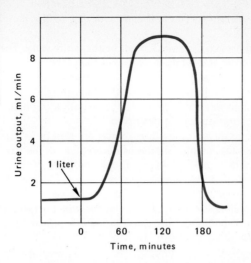

Figure 30-4. Renal response to water intake. A rise in intake is quickly met by an increase in output until the added quantity is excreted.

of the osmoreceptors, reduction in amount of ADH released, decreased permeability of the collecting ducts, and thus greater urine excretion. As a result of the high urine output, the osmolarity of the blood returns to normal.

But the osmolarity of the extracellular fluid is not the only factor involved in urine output. This is shown by the simple experiment in which isotonic fluid is added to the blood. Extracellular fluid volume is increased, but its osmolarity is unchanged. Still urine output increases. Consequently, so-called volume receptors have been postulated, receptors which initiate impulses that are propagated to the supraoptic nucleus to regulate ADH release. More likely, however, no such mechanism exists, but rather the increased blood volume causes circulatory responses discussed earlier. For example, if the blood volume increases, so does the blood pressure. As a result, there is inhibition of sympathetic outflow, and vasodilatation results. Vasodilatation of the afferent and efferent arterioles in the kidney will change glomerular dynamics so that the glomerular filtration rate will increase and thus increase output of urine. Likewise, in severe dehydration arterial blood pressure falls and glomerular filtration decreases.

Thirst A very large percentage of the total amount of fluid ingested is in the food. Virtually every type of food in the diet contains at least 50 percent water. Fruits and vegetables have a considerably higher percentage. But if the water in the food proves inadequate, there is a thirst. The amount of fluid taken as such varies greatly. Some individuals will drink less than 1 liter/day, others drink two or three times as much. When one exercises in a warm, humid climate, the intake may exceed 10 liters/day. For the most part, the volume imbibed is determined by thirst. As a matter of fact, thirst is so attuned to body water that when an animal is dehydrated and then permitted free access to water, almost exactly the amount lost during dehydration is taken in.

The mechanism of thirst has been investigated for many years, yet it is still poorly understood. There is general agreement that at least part of thirst is a local phenomenon, that is, drying of the tissues of the oral cavity. A reduced salivary flow always gives rise to a sensation of thirst. There is ample evidence that changes in the circulation or alterations in the osmotic state of the blood influence the rate of salivary secretion. There is also evidence that ADH diminishes blood flow through the glands and thus decreases salivation. Therefore dehydration would seem to evoke circulatory changes which are reflected in the rate of salivation. The consequent drying of the mouth and pharynx produces thirst.

Oral and pharyngeal drying, however, cannot be the entire story. The hypothalamus is apparently also involved in water intake as it is in food intake. Stimulation of dorsal hypothalamic nuclei in the dog results in drinking. Lesions in the area abolish fluid intake. And these actions appear to be independent of salivation. In other words, activation of the appropriate hypothalamic nuclei gives rise to the sensation of thirst, a sensation that cannot be evoked when those nuclei are destroyed.

The cells that make up the so-called thirst center are said to be osmoreceptors and probably function in the same manner as the osmoreceptors in the supraoptic nucleus. In some manner firing of these cells gives rise to the sensation of thirst.

The interesting aspect of all this is that thirst is satisfied long before the fluid taken in has a chance to be absorbed from the gastrointestinal tract. Part of the relief stems from wetting of the dry oral and pharyngeal membranes. But a major inhibition to thirst is thought to arise from distention of the intestine, which probably initiates impulses that are propagated to the thirst center.

INTRACELLULAR FLUID

The fluid of the body that is encased within the cells is termed the intracellular fluid. As shown in Table 30-2, this is the largest subdivision of the total body fluid.

No substance has yet been found that will distribute itself throughout the intracellular compartment exclusively. Consequently, the volume of the intracellular fluid cannot be determined directly. It is estimated by subtracting the extracellular fluid volume from the total body water volume.

Table 30-3 shows the composition of the intracellular fluid in contrast to the composition of blood plasma and interstitial fluid. The striking difference between the composition of the intracellular and extracellular fluids is immediately apparent. The major difference lies in the concentration of the various ions. These differences, as already explained, can be maintained only by active metabolic processes.

EXTRACELLULAR FLUID

The extracellular fluid consists of all the fluids of the body outside the cells, that is, plasma, interstitial fluid, cerebrospinal fluid, lymph, synovial fluid, and the specialized fluids found in the ears and eyes. The two major divisions that have been most extensively investigated are the blood plasma and the interstitial fluid.

TABLE 30-2. DISTRIBUTION OF BODY FLUID*

COMPARTMENT	PERCENT BODY WEIGHT	VOLUME, LITERS
Intracellular fluid	37	26
Extracellular fluid	22	15
Interstitial fluid	18	12
Plasma	4	3
Total body fluid	59	41

* Assuming 15 percent fat in a 70-kg subject.

TABLE 30-3. COMPARISON OF BODY FLUID COMPOSITION

SUBSTANCE	BLOOD PLASMA*	INTERSTITIAL FLUID*	INTRACELLULAR FLUID*
Sodium	148.0	142.0	9
Potassium	4.3	4.1	145
Calcium	4.0	3.8	0
Magnesium	3.0	2.7	40
Total cations (+)	159.3	152.6	194
Chloride	106.0	114.0	6
Bicarbonate	28.0	30.0	11
Phosphate	2.1	2.0	90
Sulfate	1.1	1.1	20
Protein	17.0	1.0	67
Organic Acids	3.2	3.4	0
Total anions (−)	157.4	151.5	194

*mEq/liter.

The total extracellular fluid volume may be determined with either inulin, thiocyanate, or thiosulfate. On the assumption that there is very little sodium within the cells, sodium has also been used. The results, using these various substances, are not in agreement. Most authorities believe that inulin, a carbohydrate derived from certain plants, provides the most accurate measure. To avoid confusion, the practice now is to refer to the results not as extracellular volume, but rather as "inulin space," "sodium space," etc.

The interstitial fluid volume is estimated by subtracting plasma volume from the total extracellular fluid volume. As explained in Chapter 15, plasma volume may be determined using the dye Evans blue or albumin-bound [131]I. The value obtained in this way for interstitial fluid includes the lymph, cerebrospinal fluid, and other specialized fluids, but no adequate way has yet been found to separate these components.

Table 30-3 shows the composition of interstitial fluid in comparison with the blood plasma and the intracellular fluid. One big difference is the very small protein content of the interstitial fluid. In addition, there are the ionic differences already mentioned.

The differences between the interstitial fluid and the intracellular fluid depend upon active transport mechanisms. But the differences between the interstitial fluid and the blood plasma are simply in accord with the Donnan equilibrium. Donnan showed that if a membrane separates two solutions of electrolytes, one of which contains a nondiffusible ion, the distribution of the diffusible ions in the solutions will be unequal. Without going into the mathematics involved, the higher the concentration of the nondiffusible substance (in this case, a protein), the greater will be the resulting ionic imbalance. In short, there need be no active transport postulated between plasma and interstitial fluid to account for the ionic differences in the two.

ACID-BASE BALANCE

The term acid-base balance has long been used in relation to the constancy of the hydrogen-ion concentration in the body fluids. As has been indicated many times already, even relatively small changes in the hydrogen-ion concentration of the plasma cause widespread alterations or even death.

In Chapter 1 the ways of expressing hydrogen-ion concentration were explained. The point was made that though normality may be used, it rarely is because either very small decimals or negative exponents are necessary. To circumvent this difficulty, hydrogen-ion concentration is generally expressed in terms of pH, which is simply the negative logarithm of the hydrogen-ion concentration. On the pH scale, neutrality is a pH of 7, and the usual range of acid to base is 0 to 14, with 0 highly acid and 14 highly alkaline (see Table 1-2, page 15).

Normal Body Fluid pH

Because the extracellular fluid is more easily available than is the intracellular fluid, its pH is known with greater certainty. Of the extracellular fluids, the plasma is the most easily obtained. The normal pH of arterial blood is generally 7.41; that of venous blood is 7.36. Venous blood is more acid because it carries more CO_2, and since the amount of CO_2 in the venous blood varies with activity, so does its pH. The pH of arterial blood, for this reason, is more constant.

When the pH of arterial blood falls below 7.41, the person is said to have acidosis; if the pH is higher than this value, he has alkalosis. The range compatible with life is from about 7.0 to 7.8.

Although the pH of intracellular fluid cannot be measured directly without altering the cell and thus probably the pH, indirect measurements indicate that intracellular pH is somewhat lower than extracellular fluid pH. In the first place, considerations of the Donnan equilibrium indicate that the bicarbonate concentration of interstitial fluid is lower than in the plasma, and, second, the CO_2, an end product of metabolism, is higher. Both factors make the pH of the intracellular fluid more acid. Under normal conditions of activity and blood flow, the intracellular pH is probably about 7.1. As metabolism increases, the pH falls, and if the blood flow is not adequate to remove the CO_2 as rapidly as it forms, the pH will fall even more, perhaps to as low as 6.0.

There is thus a pH gradient between the intracellular fluid and the plasma. The pH of the interstitial fluid lies between these extremes.

The Buffer Systems

A buffer is something that prevents rapid or great change. An acid-base buffer is one that prevents a rapid or great change in pH. In this connotation, a buffer is a solution consisting of a weak acid or base together with a highly ionized salt of the same acid or base. In the blood there are several buffer systems. The more important are expressed as follows:

$$\frac{B_2HPO_4}{BH_2PO_4} \qquad \frac{BHCO_3}{H_2CO_3} \qquad \frac{Bprotein}{Hprotein}$$

These are the phosphate, bicarbonate, and protein buffer systems.

As an example of how a buffer works, assume that the strong acid, HCl, is added to a solution containing the bicarbonate buffer system:

$$HCl + BHCO_3 \longrightarrow H_2CO_3 + BCl$$

In this case the weak acid H_2CO_3 has been formed. It replaces the strong acid HCl. BCl is a neutral salt. The B stands for any cation, usually sodium or potassium, since these are the major cations present in the blood.

Consider a second example, that of a strong base being added to the same buffer system:

$$NaOH + H_2CO_3 \longrightarrow NaHCO_3 + H_2O$$

Note that a weak base, sodium bicarbonate, has been formed and thus replaces a strong base, NaOH.

These two examples indicate that the addition of an acid or base to a buffer system must alter the ratio between the concentration of each member. When acid is added, the bicarbonate is diminished and the H_2CO_3 is increased. The addition of the base does the opposite. When one member has been completely used up, the system no longer can function as a buffer.

Henderson-Hasselbalch Equation The pH of a solution containing a buffer system depends upon the ratio of the two members of that system. This is expressed by the Henderson-Hasselbalch equation as follows:

$$pH = pK + \log \frac{salt}{acid}$$

This equation says that when the ratio of the salt to the acid is 1, pH is equal to pK (K stands for a constant). The letter p here is used as with H to express a negative logarithm. Thus, pK could also be written as $-\log K$. As used in the Henderson-Hasselbalch equation, pK expresses the so-called acid dissociation constant.

The Henderson-Hasselbalch equation is used to calculate the pH of a solution if the buffer ratio and the pK are known. Likewise, it can be used to calculate the buffer ratio if the pH and pK are known. For example, the pK for the bicarbonate system is 6.1. Using this value, the pH of the system can be calculated at all ratios between the constituents as shown in Figure 30-5. Figure 30-5 also indicates that the bicarbonate system in blood is not a very effective one per se, because the most effective buffering is accomplished when the pH is close to the pK, that is, when the ratio of the buffer members is close to 1. But the pH of blood is in the 7.36 to 7.41 range, which makes the ratio between $BHCO_3$ and H_2CO_3 about 20 to 1. In contradistinction, the pK of the phosphate system is 6.8, which is closer to the pH of blood; thus on this basis it is a more effective buffer.

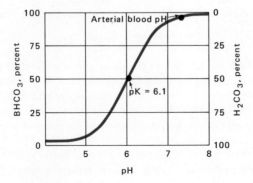

Figure 30-5. The pH of the bicarbonate buffer system at different proportions of the buffer members. This system is most effective when the proportion is 50:50, but the pH then is not that of normal arterial blood.

Protein Buffers Different amino acids have different valences and thus form weak acids of varied dissociation constants. For this reason, no one pK value can be assigned to the protein buffers of the blood. Within the red blood cells there is a high concentration of hemoglobin, which forms an extremely important buffer system. The protein buffers are in large supply in the plasma and in the cells. They are thus admirably situated to act, especially within the cells of the body, where approximately three-fourths of the chemical buffering of the body as a whole occurs. The end products of the various intracellular metabolic processes are, for the most part, acid. The protein buffers in the cells can immediately and directly act to protect the cells from the deleterious effects of a highly acid medium.

Regulation of Acid-Base Balance

As has been stated several times in this and previous chapters, the pH of the blood remains remarkably constant within a very narrow range, despite the wide variety of chemical processes that go on throughout the body. The buffer systems, in concert with respiration and renal function, are responsible. The point was just made that the protein buffers within the cells are vital to the maintenance of the constancy of the intracellular pH and thus may be considered the first line of defense at that point. But when highly acid or alkaline substances do enter the blood, then the buffer systems there are called upon. They are so effective that the infusion of a liter of $1\ N$ HCl could be successfully buffered so that the blood pH would not fall below the lethal level of pH 7.0. Normally, the blood buffers are not so challenged. Their usual problem is with the respiratory gas, CO_2.

CO_2 pours out of all the cells of the body to be concentrated in the venous blood, yet the pH of the venous blood changes only 0.05, that is, from 7.41 to 7.36. This is a very small alteration. As the CO_2 enters the blood, a part of it is quickly buffered by the systems in the plasma. But the greater fraction of CO_2 passes through the plasma to enter the red blood cells. There it reacts as follows:

$$CO_2 + H_2O \longrightarrow H_2CO_3 \longrightarrow HCO_3^- + H^+$$
$$H^+ + HbO_2^- \longrightarrow HHb + O_2$$

Hb stands for hemoglobin. HHb is the acid form, and it is weaker than H_2CO_3; thus by the formation of a weak acid to take the place of a stronger one, pH changes only slightly.

Respiration As explained above, the bicarbonate buffer system is not by itself an effective system because it has a pK of 6.1 and the pH of the blood is considerably higher. However, in combination with respiration it becomes extremely effective.

As acid is added to the bicarbonate buffer, the ratio $BHCO_3 : H_2CO_3$ changes because H_2CO_3 forms. But in the blood, in the presence of the enzyme carbonic anhydrase,

$$H_2CO_3 \xrightarrow{\text{carbonic anhydrase}} H_2O + CO_2$$

The CO_2 then diffuses from the blood into the lungs and is expired. If the rate of CO_2 elimination in this way keeps step with the CO_2 production, the amount of H_2CO_3 in the arterial blood will not change; however, the reaction of the acid with $BHCO_3$ reduces this component of the buffer system. Nonetheless, because of the elimination of CO_2 through respiration, the ratio of the buffer components does not change as much as it otherwise would.

But this is still not the entire story. For as soon as the buffer ratio does change, the pH of the blood decreases, and this acts as a stimulus to breathing so that hyperventilation ensues. Consequently, even more CO_2 is blown off, and the H_2CO_3 decreases further. The end result is that H_2CO_3 is reduced almost as much in this way as $BHCO_3$ is reduced by its reaction with the acid. To be sure, there is now less of each, but the ratio between them is almost normal and, therefore, so is the pH of the blood.

The point must be emphasized that all the buffers in the blood are in balance with one another so that if the ratio of one alters, they all do. Thus, respiration not only makes the bicarbonate system more effective but also increases the effectiveness of all the buffers.

In summary, the CO_2 produced by metabolic processes is buffered in the venous blood, diffuses into the lungs, and is expired. Respiration, as already explained, is beautifully attuned to the metabolic rate; thus the more CO_2 formed, the greater the ventilation. Acid products that get into the blood are also buffered by the various systems and, in addition, react with the bicarbonate system to produce CO_2. If ventilation is not adequate to eliminate the CO_2 as rapidly as it is formed, the concentration of CO_2 in the arterial blood will increase, and this will directly stimulate greater respiratory activity. Thus, the interaction of respiration and the bicarbonate system normally maintains the pH of the arterial blood unchanged.

Finally, as just explained, the presence of excess acid in the blood will alter the bicarbonate concentration. For this reason the blood bicarbonate level gives a good indication of the blood buffering power, and it has been termed the alkaline reserve.

Renal Function The kidneys regulate the composition of the urine. If there is an excess of acid or basic metabolites in the blood, renal excretion of these substances will increase. In addition, the components of the buffer systems are also excreted in variable quantities, thereby serving to maintain favorable ratios.

The effectiveness of the bicarbonate buffer system is further enhanced by renal function. This is so because the kidneys can (1) conserve bicarbonate, (2) excrete excessive bicarbonate, and (3) generate bicarbonate and add it to the blood.

Usually there is very little bicarbonate in the urine. Over 99.9 percent of the filtered bicarbonate is reabsorbed, mostly by the proximal tubules, partly by the collecting ducts. The normal blood level of bicarbonate is about 27 mmoles/liter. If the blood level rises above about 28 mmoles/liter, bicarbonate appears in the urine in ever greater concentration (Figure 30-6). The value of 28 mmoles/liter is spoken of as the renal threshold for bicarbonate. Clearly, this is an effective mechanism for conserving bicarbonate and for excreting large amounts when the bicarbonate blood level rises.

In addition to conserving bicarbonate, there must be a mechanism for producing bicarbonate because the acid end products of metabolism are constantly being formed and buffered by bicarbonate, thereby lowering the bicarbonate store. The kidneys produce bicarbonate as follows:

$$Na_2HPO_4 + 2H_2CO_3 \longrightarrow NaH_2PO_4 + 2NaHCO_3 \qquad (1)$$

reabsorbed ↑ (above $2NaHCO_3$)

excreted ↓ (below NaH_2PO_4)

$$Na_2SO_4 + 2H_2CO_3 + 2NH_3 \longrightarrow (NH_4)_2SO_4 + 2NaHCO_3 \qquad (2)$$

reabsorbed ↑ (above $2NaHCO_3$)

excreted ↓ (below $(NH_4)_2SO_4$)

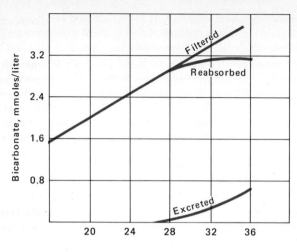

Figure 30-6. Excretion of bicarbonate. Virtually all the bicarbonate is reabsorbed until the plasma level exceeds 27 or 28 mmoles/liter. (*Redrawn from Pitts, Ayer, and Schiess, Journal of Clinical Investigation, 28:35, 1949.*)

Not only are these reactions responsible for the production of bicarbonate, but also they result in the acidification of the urine. Note that in the first reaction there is an exchange of hydrogen and sodium ions. Then the kidneys excrete the resulting compound with the most hydrogen, while reabsorbing two molecules of $NaHCO_3$. In the second reaction, ammonia, which is produced by the kidney, is combined with hydrogen ion to form NH_4^+. This NH_4^+ is then exchanged for sodium and excreted. Again there is elimination of hydrogen ion and conservation of sodium ion.

The transport of both ions must involve active processes. The H^+ comes from the hydration of CO_2 to carbonic acid, which then ionizes to $H^+ + HCO_3^-$. Thus:

$$H_2O + CO_2 \xrightarrow{\text{carbonic anhydrase}} H_2CO_3 \longrightarrow H^+ + HCO_3^-$$

The formation of H_2CO_3 proceeds rapidly in the presence of carbonic anhydrase. This enzyme is present in the kidney tubule cells.

The coordination between respiratory and renal function should now be clear. As acid enters the blood, bicarbonate is used up; but by blowing off carbon dioxide, carbonic acid is also reduced. The kidneys generate bicarbonate and put it into the blood, thereby replenishing the bicarbonate store. But the kidneys do more than this, as mentioned. They eliminate carbonic and organic acids, they exchange basic ions for hydrogen ions, and by the formation and excretion of ammonium salts the kidneys form an acid urine. Thus, by virtue of these actions and their contributions to the buffer systems, there is very effective control of the acid-base balance of the extracellular fluids and, indirectly, of the intracellular fluid.

Disturbances in Acid-Base Balance

Alkalosis and acidosis are terms which are often used inaccurately. The plasma bicarbonate content is a measure of the alkaline reserve. The point was made previously, however, that the bicarbonate level can vary considerably without significantly altering the pH of the blood. In other words, there is confusion as to whether the terms alkalosis and acidosis indicate a change in pH or merely an alteration in the bicarbonate content of the blood. For this reason, the terms compensated and uncompensated alkalosis or acidosis are used.

Compensated Alkalosis   If $NaHCO_3$ is ingested in sufficient quantities, the bicarbonate content of the blood increases. However, respiratory and excretory functions may cause the H_2CO_3 content to rise proportionately so that the ratio of the bicarbonate buffer does not change; therefore, neither will the pH of the blood. In this instance, as measured by the bicarbonate content, there is alkalosis, but since the pH did not change, it is said to be compensated alkalosis. Such a condition not only arises from excessive ingestion of alkaline substances, such as bicarbonate, but also may occur as a result of excessive vomiting with the loss of HCl from gastric juice.

Uncompensated Alkalosis   The ability of the respiratory and excretory mechanisms to compensate has a limit. Consequently, the buffer systems and the respiratory and excretory adjustments may be overwhelmed. When this occurs, there is an increase in blood pH, and uncompensated alkalosis exists.

Compensated Acidosis   In diabetes mellitus, nephritis, and other conditions in which there is an abnormal production or retention of acids, there results a decrease in the bicarbonate content of the blood. The same end result can be obtained by the ingestion of mineral acids or salts such as NH_4Cl. The respiratory and excretory mechanisms function to reduce the H_2CO_3 content of the blood. In this case there is a decrease in bicarbonate, and yet there will be no change in pH; this condition is called compensated acidosis.

Uncompensated Acidosis   If the bicarbonate level in the plasma is so depleted that the pH falls, uncompensated acidosis exists.

QUESTIONS AND PROBLEMS

1   What is the principle of the "dilution technique" in determining total body fluid?
2   If an individual becomes extremely dehydrated as a result of vomiting and diarrhea, what body mechanisms will restore the fluid and electrolyte balance?
3   How does the concentration of body fluids bring about the release or retention of ADH?
4   Would wetting of the oral cavity completely satiate one's desire for water?
5   List the fluids which constitute extracellular fluid.
6   What are the major differences in composition between intracellular and interstitial fluid?
7   Explain why the presence of carbon dioxide in venous blood makes it more acid than arterial blood. What constitutes a lethal pH?
8   Theoretically how would it be possible to increase the pH of blood in a human being?
9   Why can't the pH of blood be determined by using the pK of a protein buffer?
10   What buffer systems exist in the body? Which is considered the most effective? Explain.
11   What is the fate of the carbon dioxide given off into the bloodstream by the cells of the body?
12   Explain how the following might affect the buffering ability of the body: (*a*) severely reduced hemoglobin, (*b*) reduction of respiratory membrane surface, (*c*) impaired renal function.

PART 7
THE ENDOCRINE AND REPRODUCTIVE SYSTEMS

31 THE ENDOCRINE GLANDS

The term endocrine denotes internal secretion. Thus the endocrine glands are clearly differentiated from exocrine glands, such as the sweat glands, which secrete externally, that is, to the surface or outside the body. Glands that secrete the digestive juices are also exocrine glands in that the juices are secreted into the digestive tract which in turn leads outside the body. The endocrine glands, on the other hand, secrete their products into extracellular space around the secretory cells. The secretions are absorbed from the extracellular space into the blood, which then transports them throughout the body. Endocrine glands thus do not possess ducts as do exocrine glands, and are therefore termed ductless glands.

The secretion of an endocrine gland is termed a hormone. This word is derived from a Greek word that means "to arouse or set in motion." Hippocrates used the term to describe certain body secretions that were considered to be vital, to be "rousers" or stimulators. It was first used in the modern sense at the turn of the century in connection with the secretion of secretin. The discovery and study of secretin marks the beginning of the subscience of endocrinology.

At first all "hormones," that is, all internal secretions, were thought to have a stimulatory effect. Later, internal secretions were discovered that inhibit; nonetheless, "hormone" has generally been used for all internal secretions, whatever their function.

But as knowledge grew, terminology became confused. For example, all cells produce carbon dioxide. This enters the extracellular spaces around the cells and then diffuses

710

THE
ENDOCRINE
AND
REPRODUCTIVE
SYSTEMS

into the blood, and it certainly influences various body activities, for example, respiration. Should carbon dioxide and other cellular products be called hormones? Usually they are not so classified on the basis that they are produced generally, and not by a specific endocrine gland.

There are still other complications. What about acetylcholine, which is secreted by many neurons, or even more confusing, norepinephrine, which is secreted by postganglionic sympathetic neurons and also by the medulla of the adrenal gland? Neurons also secrete oxytocin and vasopressin (see page 723). How should all these products be classified? There is general agreement today that the products of neurons should be called neurohumors. In some cases, as with oxytocin and ADH, they function as typical hormones and can be so termed. Norepinephrine when secreted by neurons is a neurohumor; when secreted by the adrenal medulla, it is a hormone.

The endocrine system has a regulatory function. In this respect it is very similar to the nervous system, and there is a growing tendency to consider the two as closely related components of one regulatory mechanism. Not only do they have a similar function, but they influence each other as well. For example, as will be discussed in this chapter, the brain controls the activity of the hypophyseal gland, which, in turn, regulates many of the other endocrine glands. Conversely, various hormones influence the blood levels of the electrolytes such as calcium, potassium, and sodium, and thereby influence neuronal function.

The point was made in earlier chapters that the nervous system functions as a homeostatic mechanism; the same is certainly true of the endocrine system. For example, the blood level of the electrolytes remains remarkably constant despite the intake of each and despite a wide variety of activities. But if the adrenal glands are removed, there are marked changes in sodium and potassium, and if the parathyroid glands are extirpated, the blood level of calcium is markedly altered. The difference in regulatory function between the nervous and endocrine systems is, in most cases, one of rate of response. Nerve action is measured in milliseconds, whereas the endocrines may take seconds and, in some cases, days.

The regulatory action of the various hormones is usually superimposed upon a basal rate, or level. For example, in the absence of growth hormone there is still some growth but not as much as when the hormone is present; in the absence of thyroid hormone metabolic rate falls sharply but then continues at a low, basal level; and in the absence of parathyroid hormone the blood concentration of calcium falls to a new level and is then maintained there.

The amount of most hormones circulating in the blood remains remarkably constant. This means that the rate of production by the endocrine glands keeps in step with the rate of utilization and destruction of the hormone so that the balance remains the same. If a hormone is suddenly used at an accelerated rate, hormone production also increases. This interlocking relationship has been the object of great and fascinating investigation. Not only is there a relationship between the rate of utilization and production of each hormone, but the level of one hormone may well influence the levels of others.

As each endocrine gland is discussed, the chemical structure of the secreted hormones will be mentioned. In general, hormones seem to be steroids, peptides, proteins, or glycoproteins.

Finally, the way in which hormones interact is of considerable interest and importance. They may have a synergistic, permissive, or inhibitory relationship. A synergistic relationship is cooperative or complementary, as when two or more hormones acting together evoke a greater response than if the response evoked by each were added together. A permissive relationship is one in which the presence of one hormone is essential for another to evoke its given response. In practice, it is often difficult to differentiate between synergistic and permissive relationships.

An inhibitory relationship occurs as a result of the interaction of two or more hormones on the target organ, not as a result of a chemical blockage between the hormones. A typical example is seen in the amount of glycogen in the liver. Insulin increases liver glycogen; epinephrine and glucagon both decrease it.

ROLE OF CYCLIC AMP IN ENDOCRINOLOGY

A major advance in endocrinology has resulted from the discovery of cyclic-3',5'-adenosine monophosphate, commonly referred to as cyclic AMP. The first steps leading to the discovery of this substance were taken about twenty-five years ago by Earl Sutherland, Jr. In 1971 he received the Nobel prize in physiology and medicine, not only for his discovery, but also for his meticulous elaboration of the sweeping role that this substance has throughout the body, especially in the sequence of events by which hormones exert their influences.

Sutherland began his investigations by attempting to ascertain how epinephrine increases blood sugar. He was able to conclude that only the initial step, that is, the transformation of glycogen into glucose 1-phosphate, is controlled by epinephrine. This step is catalyzed by the enzyme phosphorylase. Sutherland and his coworkers soon found that there is an active and an inactive form of phosphorylase. The investigators demonstrated that somehow the hormone converts the inactive form to the active form. Further studies showed that the cell membrane is essential for this action, that the cell membrane has an enzyme that produces a heat-stable factor, and that this factor is cyclic AMP (Figure 31-1). The enzyme was found to be adenyl cyclase. It catalyzes the reaction by which cyclic AMP is formed from ATP. Finally, still another enzyme was found, this one called phosphodiesterase, that inactivates cyclic AMP.

Concept of the Second Messenger

The tremendous importance of the Sutherland discoveries lies in the concept of the "second messenger." The first messenger is the hormone, not only epinephrine, but many others as well (Table 31-1). The hormone first binds with a specific receptor site, for example, on a liver cell in the case of epinephrine or on the renal tubule cells in the case of parathyroid hormone. This specific binding on the cell membrane increases the

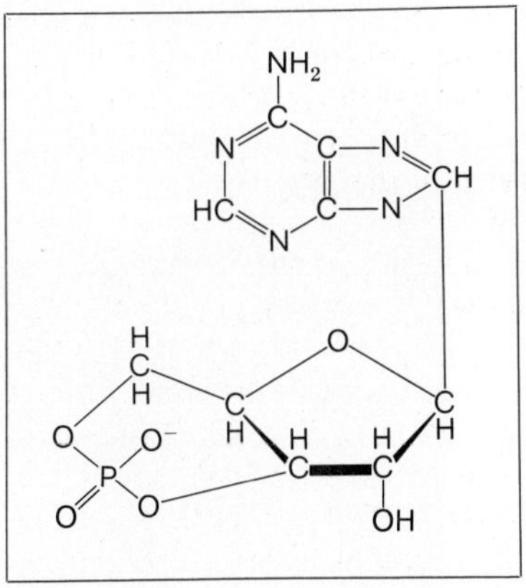

Figure 31-1. Adenosine 3',5'-phosphate (cyclic AMP).

712

THE
ENDOCRINE
AND
REPRODUCTIVE
SYSTEMS

TABLE 31-1. HORMONES THAT USE CYCLIC AMP AS A SECOND MESSENGER

HORMONE	CYCLIC AMP ACTION
Epinephrine	Glycolysis in liver and heart Lipolysis in fat cells Amylase secretion from salivary gland
Norepinephrine	Discharge frequency of Purkinje cells Acetylcholine release in nerve Melatonin release in pineal
Glucagon	Glycolysis in liver Lipolysis in fat cells
Melanocyte-stimulating	Darkening of frog skin
Parathyroid	Phosphaturia in renal cortex Calcium resorption in bone
ACTH	Steroid production in adrenal
Luteinizing hormone	Steroid synthesis in corpus luteum
Vasopressin	Water resorption in adrenal medulla
Thyroxine	Tachycardia

activity of the adenyl cyclase in the cell membrane. How this occurs is not yet known. The enzyme diffuses from the membrane into the cell to come into contact with the cytoplasmic ATP. Cyclic AMP results (Figure 31-2). It then spreads throughout the cell, causing the cell to carry out functions characteristic of that particular cell. In this sequence the hormone carries the first message and cyclic AMP the second; hence it is referred to as the second messenger.

Figure 31-2. The cell membrane contains the enzyme adenyl cyclase essential for the conversion of ATP to cyclic AMP. The hormone initiating the sequence is the first messenger; cyclic AMP the second.

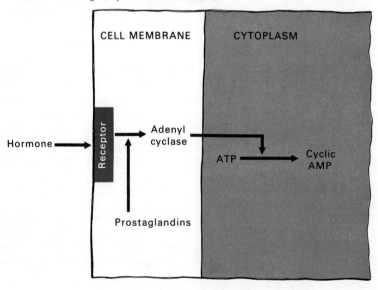

TABLE 31-2. ACTIONS OF CYCLIC AMP

TISSUE	ACTION
Liver	Increased glycogenolysis Increased phosphorylase Decreased glycogen synthetase Increased protein kinase Induction of tyrosine transaminase Induction of PEP carboxykinase Induction of serine dehydratase Increased amino acid uptake Increased ketogenesis
Adipose	Increased lipolysis Increased amino acid uptake Increased clearing-factor lipase
Anterior hypophysis	Increased release of ACTH, TSH, GH, and LH
Epithelial	Increased permeability to water
Pancreas	Increased release of insulin
Thyroid	Increased release of thyroid hormone
Cardiac muscle	Increased contractility
Smooth muscle	Increased tension Hyperpolarizes membrane potential
Adrenal	Increased steroidogenesis
Bone	Increased calcium resorption
Kidney	Increased phosphaturia Increased renin
Nerve	Increased acetylcholine release
Gastric mucosa	Increased HCl secretion
Leukocytes	Increased histamine release
Platelets	Decreased aggregation
Uterus	Increased amino acid uptake
Parotid	Increased amylase release

Table 31-2 lists some of the actions of cyclic AMP. The increased level of cyclic AMP persists for only a short period of time for at least two reasons: (1) it diffuses out of the cell, and (2) it is inactivated by phosphodiesterase. This enzyme degrades it to an inert form of AMP.

Clearly, there are many unanswered questions in this sequence, but the student can readily appreciate its importance in our basic understanding of endocrinology. Cyclic AMP probably has other functions as well. For example, there is good evidence that it plays a role in neuronal transmission.

Prostaglandins

In the early 1930s a new substance was found in human semen. The first observation was that this substance caused uterine contractions. Because it was thought to be produced by the prostate gland, it was termed prostaglandin. Actually, it comes from the seminal vesicles. Now we know that there are several closely related substances which are present in many body fluids and tissues.

714

THE
ENDOCRINE
AND
REPRODUCTIVE
SYSTEMS

Prostaglandins are fatty acids. The prime site for formation is thought to be the cell membrane. They are extremely potent substances which produce effects in remarkably small doses. They have a very wide spectrum of actions including behavioral and central nervous system effects as well as actions which appear to mimic or inhibit many of the known hormones. The suggestion has been made that all these actions may be due to an interaction with adenyl cyclase. The current concept is that prostaglandins serve as regulators of hormonal action by modulating cyclic AMP formation.

RADIOIMMUNOASSAY

Advances in endocrinology demand methodology for measuring plasma levels of the various hormones. Such methods must be accurate and require only small amounts of plasma. This has been particularly difficult for the protein hormones. There is now such a procedure termed radioimmunoassay. The method is relatively simple, highly accurate, and permits measurement of very small amounts of hormone.

In essence the procedure is as follows: (1) the protein hormone to be assayed is first labeled with a radioactive isotope, (2) the labeled hormones are then added to an antibody specific for that hormone, (3) a small amount of plasma is then added to the mixture. The protein hormone in the plasma competes with the radioactive hormone for reactive sites on the antibody; therefore the antibody will contain some radioactivity. The amount of activity depends upon the amount of hormone present in the plasma; the more hormone present the lower the radioactivity.

HYPOPHYSIS

In the sixteenth century Vesalius named this gland the pituitary because he thought it had something to do with the phlegm or mucus of the nose and throat. The term pituitary means "phlegm." When the true function of the gland was determined, the term hypophysis, "to grow under," was substituted; but the gland is still often called the pituitary.

The hypophysis is often referred to as the master gland of the endocrine system because it secretes many hormones that control other endocrine glands.

Anatomy and Histology of the Hypophysis

The hypophysis is a small round structure about 13 mm ($\frac{1}{2}$ in.) in diameter. It is located in the floor of the cranial cavity in a small deep fossa (sella turcica) of the sphenoid bone and, in the adult male, weighs about 0.6 g (Figure 31-3). The infundibular stalk of the hypophysis penetrates the dura mater and attaches to the hypothalamus. Embryologically the hypophysis is derived from two sites: (1) an outpouching from the roof of the oral cavity which develops into the adenohypophysis, that is, the glandular portion of the hypophysis, and (2) an outpouching from the developing forebrain which becomes the neurohypophysis, that is, the neural portion of the hypophysis.

The adenohypophysis is the larger of the two portions. It consists of the (1) pars distalis, which contains most of the functional cells of the gland; (2) pars tuberalis, a thin epithelial plate that surrounds the infundibular stalk; and (3) pars intermedia.

The neurohypophysis consists mostly of the pars nervosa. In addition there is the infundibulum (composed of the infundibular stalk and the median eminence).

An older but still used terminology divides the hypophysis into anterior and posterior lobes, the anterior lobe being equivalent to the pars distalis, and the posterior lobe embracing the pars intermedia and the pars nervosa and infundibular stalk of the neurohypophysis. These relationships can be summarized as follows:

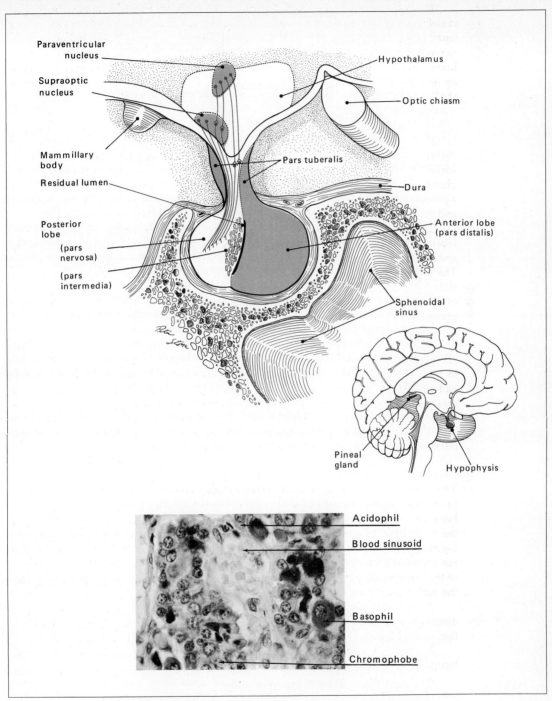

Paraventricular nucleus
Supraoptic nucleus
Mammillary body
Residual lumen
Posterior lobe
(pars nervosa)
(pars intermedia)
Hypothalamus
Optic chiasm
Pars tuberalis
Dura
Anterior lobe (pars distalis)
Sphenoidal sinus
Pineal gland
Hypophysis
Acidophil
Blood sinusoid
Basophil
Chromophobe

Figure 31-3. Hypophysis. Hypothalamic nuclei associated with the hypophysis are shown. The three cell types present in the anterior lobe are visible in the photomicrograph ($\times$ 420).

715

716

THE
ENDOCRINE
AND
REPRODUCTIVE
SYSTEMS

ADENOHYPOPHYSIS

Pars distalis   ⎫

Pars tuberalis   ⎬   **ANTERIOR LOBE**

Pars intermedia ⎭

NEUROHYPOPHYSIS   ⎫

Pars nervosa   ⎬   **POSTERIOR LOBE**

Infundibular stalk   ⎭

Median eminence

Anterior Lobe The anterior lobe is characterized by blocks, or cords, of epithelial cells arranged around many thin-walled sinusoidal blood vessels. By the staining reactions of the cytoplasm the cells can be differentiated into basophils (dark-staining), acidophils (pink-staining), and chromophobes (faintly staining). A nucleolus is present in the nucleus of each cell, and there is a distinct nuclear membrane enclosing fine chromatic granules. In normal adults the percentage of each type varies considerably but is approximately 15 percent basophils, 35 percent acidophils, and 50 percent chromophobes. The number of cells, especially of chromophobes, increases sharply in pregnancy.

Blood supply to the anterior lobe is by means of the superior hypophyseal arteries, which issue from the internal carotid arteries and those of the circle of Willis. The superior hypophyseal arteries enter the infundibular stalk to form a capillary plexus, which gives rise to capillary loops. Blood from these vessels flows in the portal venules to the sinusoids of the anterior lobe (Figure 31-4). There is thus the blood circuit known as the hypothalamicohypophyseal portal system between the hypothalamus and the anterior lobe of the hypophysis. This system plays an extremely important role in the function of the hypophysis.

Posterior Lobe The posterior lobe is characterized by many small cells termed pituicytes, resembling neuroglia (connective tissue cells of the nervous system). They have short branching processes often terminating on or near blood vessels. The stroma of the lobe largely consists of branching nerve fibers but no true nerve cell bodies. Considerable connective tissue from the capsule penetrates the gland and carries with it an abundant blood supply. The posterior lobe's blood supply is different from that of the anterior lobe. The superior hypophyseal artery supplies the anterior lobe, and the inferior hypophyseal artery serves the posterior lobe.

The hypothalamic nuclei give rise to neurons that pass through the stalk to terminate in the posterior lobe. This pathway plays an essential part in the function of the lobe, in contrast to the anterior lobe, which is controlled by neurohumors carried by the blood in the hypothalamicohypophyseal portal system. Both lobes receive sympathetic fibers concerned with blood flow rather than hormonal secretion.

Hypothalamic-Hypophyseal Relationship

Endocrinology, since the turn of the century, has been the fastest-developing subdivision of physiology, and the progress made since the last edition of this book was published in 1969 continues that record. The most dramatic and significant progress has been made in the hypothalamic-hypophyseal relationship.

Observations have been made as long as history has been recorded which have suggested that environmental factors, operating through the nervous system, influence many activities which we now know to be hormonally controlled. For example, the Greeks

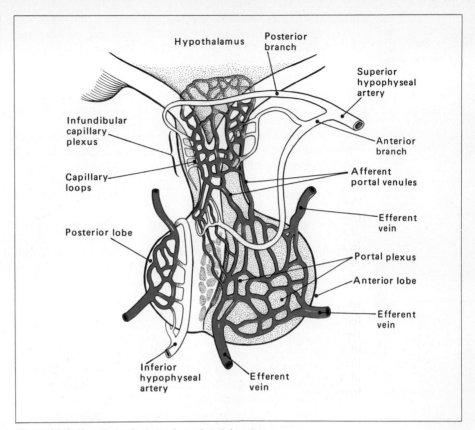

Figure 31-4. Hypothalamicohypophyseal portal system.

noted that when food was plentiful and the weather favorable, sheep would breed twice a year. Prolonged exposure to cold increases thyroid activity. And all sorts of emotional factors influence menstruation and fertility. As knowledge of the physiology of the hypophysis developed, a linkage between the brain and the gland became obvious, but a mistake by the anatomists confused the issue greatly. They knew that the hypothalamus and the hypophysis were in communication via blood vessels called the portal system, but they insisted that blood flowed in this system from the gland to the hypothalamus. In 1937, however, G. W. Harris, in England, theorized that the hypothalamus secretes substances that influence the anterior lobe of the hypophysis, and that these substances are carried by the portal system, which means that the blood flows in an opposite direction from that stated by the anatomists. His subsequent work, and that of many others, proved the correctness of this concept.

By the 1960s enough information had accrued to indicate that the hypothalamus secretes specific substances, commonly called releasing factors, which regulate the release of the various hormones of the anterior lobe of the hypophysis. By 1969, at least one of the releasing factors, thyroid-releasing factor, had been synthesized. This notable feat was achieved almost simultaneously by the Gullemin group at Baylor and the Schally group at Tulane. Schally and his colleagues have also been able to determine the structure of another hypothalamic substance, the so-called luteinizing hormone–releasing factor.

718
THE
ENDOCRINE
AND
REPRODUCTIVE
SYSTEMS

Finally, evidence is now rapidly accumulating which shows that some, and perhaps all, of the hypophyseal hormones are controlled not only by releasing factors but also by inhibiting factors.

The following section summarizes the state of knowledge at the present writing, but to remain current in this fast-developing field, the student must turn to the original literature.

Physiology of the Hypophysis

The hypophysis has three lobes, all of which secrete hormones. The hormones of the anterior lobe are (1) growth hormone, (2) prolactin, (3) thyrotropin, (4) adrenocorticotropin, (5) follicle-stimulating hormone, and (6) luteinizing hormone. The pars intermedia secretes the melanocyte-stimulating hormone. The posterior lobe is concerned with two hormones: (1) oxytocin and (2) the antidiuretic hormone. These will be discussed individually.

Growth Hormone   The growth hormone (GH) is also referred to as somatotropin (STH). In 1970, Choh Hao Li, at the University of California in San Francisco, announced an outstanding feat, the synthesis of pure human GH. It contains no less than 188 amino acid units of 20 different kinds joined in a chain with two cross links. The molecular weight is 21,500, which makes it the largest protein ever synthesized.

Functions of Growth Hormone   GH stimulates growth, increases protein synthesis, decreases carbohydrate utilization, and increases fat catabolism. Exactly how GH stimulates growth is not completely understood, but it is undoubtedly tied in with the influence this hormone has on protein synthesis. GH facilitates the transport of many amino acids through the cell membrane. Once in the cell, they are available for protein synthesis. The augmented protein synthesis could be due to the increased number of amino acids available, to increased formation of RNA, or to increased activity of RNA after it is formed.

As a result of the suppression of carbohydrate utilization under the influence of GH, blood glucose increases. This stimulates the secretion of insulin. If GH is given in large doses over a long period of time or if there is a tumor of the hypophysis which results in excessive secretion of GH, the persistent high blood glucose level eventuates in the exhaustion of the insulin-producing cells, and diabetes mellitus results. For this reason GH is said to have a diabetogenic effect.

Regulation of Growth Hormone Secretion   The development of the radioimmunologic method for assaying GH has greatly advanced knowledge of its secretion and the control of its secretion. Whereas heretofore we thought that GH was secreted at a fairly steady rate, or at least that the blood concentration remained constant, we now know that secretion varies as does the blood level. During sleep secretion markedly increases. It is also increased by protein deficiency. On the other hand, emotional deprivation in the infant can decrease secretion and stunt growth. The level of GH in the plasma is normally about 2 μg/liter during the day, rising to peaks of 10 or 15 μg/liter during sleep and even higher during protein starvation.

Secretion is controlled by two factors from the hypothalamus: (1) the growth hormone–releasing factor (GRF) and (2) the growth hormone–release inhibiting factor (GRIF). But exactly how these factors are regulated remains to be clarified. The nutritional status of the body no doubt plays an important role. During protein deficiency, as already noted, GH secretion increases as it does when the blood-sugar level falls. Exercise increases secretion. In all probability they all exert their influence via the hypothalamus, but this remains to be confirmed and detailed.

Abnormalities of Growth Hormone Secretion If secretion of GH is insufficient during the formative years or if the gland is removed during those years, growth is limited and dwarfism results. Conversely, oversecretion during the growth period results in gigantism—in persons reaching 7 or even 8 ft in height. Even after the period of growth, marked changes occur upon removal or with insufficient secretion of GH. These changes are primarily in protein, carbohydrate, and fat metabolism.

Excessive GH in the adult causes bone growth at sites where cartilage persists and stimulates the growth of soft tissue. This condition, acromegaly, is characterized by a large protruding jaw; large hands, feet, nose, and supraorbital ridges; a large tongue; and enlarged internal organs.

Because GH is a protein and because the composition of protein differs according to species, only human GH can be used in therapy. For obvious reasons there is a limited supply of the hormone available, but preliminary tests indicate that, when used, it is effective in preventing dwarfism.

Normal growth requires other hormones as well as GH, particularly thyroid hormone and insulin. Inadequate secretion of either will stunt growth, even in the presence of adequate GH.

Prolactin The lactogenic factor secreted by the hypophyseal anterior lobe, prolactin, is also referred to as luteotropin, or the luteotropic hormone (LTH), not to be confused with the luteinizing hormone (see page 721). For some time GH and prolactin were thought to be one and the same hormone. Now the evidence is unequivocal that they are discrete and separable. LTH, too, may differ from prolactin, but to date they appear to be the same.

Functions of Prolactin The major function of prolactin is to stimulate secretion of milk. Its other function, if indeed LTH and prolactin are one hormone, is to stimulate the corpus luteum of the ovary to secrete (see page 743).

Prolactin levels begin to rise early in pregnancy and peak just before delivery. The level of prolactin in the normal, nonpregnant female is about 10 μg/liter of blood plasma (in men it is about 6). Just before delivery it reaches 200, yet lactation does not occur during pregnancy despite these high levels. High levels of estrogen and progesterone are thought to block the end-organ effect of prolactin. These fall sharply at delivery, thereby permitting prolactin to stimulate secretion of milk. Active nursing maintains high prolactin blood levels. In fact each act of suckling produces transient increases in the prolactin level.

There is a high level of prolactin in fetal and infant blood, and it is also present in the blood of adult males. What function it has in these cases is not known. However, following hypophysectomy the weight of the prostate gland decreases sharply. Also, there is some evidence that elevated prolactin levels in the male depress sperm count and libido.

Now that prolactin can be measured in the plasma by the radioimmunoassay procedure, many reports are appearing concerning variations under a wide variety of conditions. One such report indicates that the level of prolactin increases sharply soon after the onset of sleep. Even more interesting is the observation that prolactin levels in both men and women increase sharply during sexual excitement and especially at orgasm. The significance of these findings remains to be determined.

Regulation of Prolactin Secretion As with GH, there appear to be two substances secreted by the hypothalamus which regulate prolactin release, one that stimulates release and another that inhibits it. These are referred to as prolactin-releasing factor (PRF) and

720

THE
ENDOCRINE
AND
REPRODUCTIVE
SYSTEMS

prolactin-release inhibiting factor (PRIF). Obviously, suckling influences the secretion of these factors, but other than that little is known concerning their regulation.

Thyrotropin   The basophilic cells of the hypophyseal anterior lobe secrete thyrotropin, also referred to as the thyroid-stimulating hormone (TSH). It is a polypeptide with a molecular weight of about 10,000.

Functions of Thyrotropin   As the term indicates, the major function of TSH is to stimulate the thyroid gland. Consequently, the gland synthesizes and secretes more thyroxine and triiodothyronine, the two active forms of circulating thyroid hormones. TSH is thought to increase the amount of cyclic AMP in the thyroid cell. This second messenger then increases all the known activities of the cells.

Regulation of Thyrotropin Secretion   The secretion of TSH is regulated, at least in part, by thyrotropin-releasing factor (TRF). As mentioned earlier, the purification and synthesis of TRF has been recently achieved and represents a major advance. The isolation and purification of TRF required no fewer than 5 million sheep hypothalami. This yielded only 1 mg of TRF! Analysis showed that it contained only three amino acids, histidine, glutamic acid, and proline in equal amounts, but there was too little TRF available to determine the sequence of these amino acids. The solution came about by synthesizing each of the six possible tripeptides containing the three amino acids. Actually none of these possessed any physiological activity until an acetyl group was added to the so-called *N* terminus. When this was done, the tripeptide in the sequence glu-his-pro showed all the physiological characteristics of the natural TRF. Actually, a few minor alterations still had to be made before the synthetic material was identical to the natural TRF, which is now known to be pglu-his-pro-NH$_2$ (pglu is pyroglutamic acid). Thus, not only was the structure known, but ample quantities were easily available.

There is a feedback control method involving TRF, TSH, and the thyroid hormones. In some manner, when the concentration of circulating thyroid hormones increases, the secretion of TSH is inhibited. There may also be some inhibition of the secretion of TRF. The mechanism of these inhibitions remains to be clarified.

Many external factors influence the secretion of TRF and therefore of TSH and the thyroid hormones. Exposure to cold increases secretion and thus raises the BMR. Various emotional states can result in either an increase or decrease.

Adrenocorticotropin   The basophilic cells of the hypophyseal anterior lobe secrete adrenocorticotropin, often referred to as adrenocorticotropic hormone (ACTH). ACTH is a polypeptide with 39 amino acids.

Functions of Adrenocorticotropin   ACTH acts upon the adrenal cortices. When there is too little ACTH, the cortical layers atrophy. Hyperplasia occurs when excessive quantities of ACTH are given. Under the influence of ACTH all the adrenal cortical hormones are secreted, but the effect is greater on the glucocorticoids than it is on aldosterone and the adrenal androgens. In addition, ACTH has a direct lipolytic action. Fat is mobilized and the fatty acids level in the blood increased.

Regulation of Adrenocorticotropin Secretion   The hypothalamus secretes corticotropin-releasing factor (CRF). This was first reported in 1955. Since then all attempts to purify CRF have failed. It is known to be secreted by the median eminence of the hypothalamus and to regulate the release of ACTH. As usual, there is a feedback mechanism involving

CRF, ACTH, and the adrenal cortical hormones. The inhibition exerted by the adrenal cortical hormones is thought to be both at the hypothalamic and the hypophyseal level (Figure 31-5). Probably the adrenal hormones themselves do not inhibit secretion, but rather some alteration brought about by the adrenal hormones is responsible. This may well be true for all hormonal feedback systems.

The CRF-ACTH system is activated by a wide variety of means. Most forms of stress do so, for example, pain, intense heat or cold, hypoxia, or emotional trauma. Just how the resulting outflow of adrenal hormones assists the organism in the face of stress has never been satisfactorily explained.

Follicle-Stimulating Hormone The basophilic cells of the anterior lobe secrete follicle-stimulating hormone (FSH), a glycoprotein. Its function is to stimulate the growth of the ovarian follicles. In the absence of FSH, follicles develop to an immature state but do not go on to mature and ovulate. In the male, FSH acts on the seminiferous tubules to stimulate the production of sperm.

Luteinizing Hormone In the female the luteinizing hormone (LH), which is a glycoprotein secreted by basophil cells of the anterior lobe, contributes to the function and development of the ovarian follicles after the initial action of FSH. More important, LH brings about ovulation, and it is essential for the formation of the corpus luteum. Still another function of LH is to stimulate the secretion of estrogens. It thus has four distinct roles, and all four require the synergistic action of FSH.

In the male, LH is known as interstitial cell–stimulating hormone (ICSH). It acts on the testicular interstitial cells to produce the androgen testosterone. Testosterone, in turn,

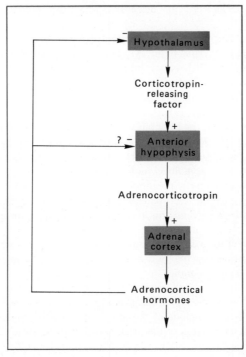

Figure 31-5. Regulation of adrenocortical hormone level in the blood.

722

THE
ENDOCRINE
AND
REPRODUCTIVE
SYSTEMS

inhibits the secretion of ICSH, probably by acting on the hypothalamus to inhibit the output of LRF.

Regulation of the Gonadotropins   Both of the gonadotropins, that is, FSH and LH, are thought by some authorities to be controlled by one hypothalamic substance, LRF. Others contend that there is also a follicle-stimulating–hormone-releasing factor. However, LRF has now been purified and synthesized, and it regulates both gonadotropins, which is the reason for considering only one releasing factor. The structure of LRF is pglu-his-tryp-ser-tyr-gly-leu-arg-pro-gly-NH$_2$.

Just how LRF, the gonadotropins, and the sex hormones estrogen, progesterone, and testosterone are all interrelated is still unclear. In the female there is a waxing and waning of the sex hormones and of the gonadotropins which relate to the menstrual cycle. Yet, LRF appears to be secreted at a relatively constant rate throughout the cycle. In addition to this confusing finding, there are observations that estrogens and progesterone may have a positive-feedback effect as well as a negative one. And as already indicated, prolactin is also involved. One can speculate as to possible interrelations, but that would only confuse the picture further.

Melanocyte-Stimulating Hormone   Melanocyte-stimulating hormone (MSH), often called intermedin, is a polypeptide secreted by basophilic cells of the pars intermedia. In lower vertebrates this hormone causes a dispersion of the pigment granules, resulting in darkening of the skin. If the pigment granules are not dispersed, they cluster around the cell nucleus and the skin appears light. MSH also brings about darkening of the skin in man, especially after exposure to sunlight. In some manner light regulates the secretion of MSH, probably via the hypothalamus.

The secretion of MSH probably accounts for characteristic pigmentation associated with various disorders such as Addison's disease (in which the adrenal cortices atrophy), and in pregnancy.

Very little is known concerning the regulation of MSH, but clearly sunlight plays a role. Light, acting through the visual pathways, could influence the hypothalamus, which is now thought to secrete an MSH-release inhibiting factor (MRIF). MRIF seems to be either tocinamide or tocinoic acid, which are ring structures of oxytocin. Whether or not there is also an MSH-releasing factor is not known.

Oxytocin   The posterior lobe of the hypophysis is known as the neurohypophysis. This has reference to the large number of nerve fibers which originate in the supraoptic and paraventricular nuclei of the hypothalamus and terminate in the lobe. In contradistinction, the anterior lobe has very few nerve fibers and is known as the adenohypophysis. There are two hormones associated with the posterior lobe, but neither is formed there. Rather they are formed primarily in the hypothalamus, and then they travel down the nerve fibers to the neurohypophysis. When the nerve fibers fire, the hormones are released from the lobe.

Oxytocin is a polypeptide with eight amino acids: tyrosine, proline, glutamic acid, aspartic acid, glycine, cystine, leucine, and isoleucine. It is formed mostly in the paraventricular nuclei of the hypothalamus.

Functions of Oxytocin   The term oxytocic refers to rapid childbirth. Oxytocin is so named because it causes vigorous contractions of the uterus, thus expelling the fetus.

Oxytocin also causes contraction of the myoepithelial cells. These cells are arranged around the mammary alveoli and ducts in such a way that contraction forces the milk out. There is also some evidence that oxytocin is carried by the circulation directly from the neurohypophysis to the adenohypophysis, where it stimulates the release of prolactin. Whether oxytocin and the prolactin-releasing factor are one and the same is not clear. At any rate, prolactin, along with other hormones, causes the breasts to secrete milk; oxytocin is responsible for milk ejection.

Regulation of Oxytocin Secretion and Release Oxytocin is thought to accumulate in the posterior lobe bound to some substance which prevents diffusion out of the lobe. Then when the hypothalamohypophyseal fibers fire, a transmitter substance is released at the neuron endings. This transmitter substance, which has still to be identified, releases the hormone from its bond, permitting it to diffuse out of the lobe.

There are apparently many ways in which these fibers can be activated. Suckling of the breast initiates impulses which impinge upon the paraventricular nuclei. This occurs whether the woman is pregnant or not. Thus, during sexual intercourse, stimulation of the breasts and other erogenous areas has the same effect. At orgasm, the high titer of oxytocin in the blood causes, or at least strongly contributes to, uterine contractions. The contractions are thought to act like peristalsis in moving sperm upward to the ovaries. This is not to say that the sperm cannot make their way up the uterus and through the fallopian tubes in the absence of an orgasm and concomitant contractions. The contractions simply facilitate the movement.

Antidiuretic Hormone (ADH) The other hormone of the neurohypophysis, ADH, is also a polypeptide with eight amino acids: tyrosine, proline, glutamic acid, aspartic acid, glycine, cystine, phenylalanine, and arginine. Note that the first six are the same as those in oxytocin. ADH is formed mostly in the supraoptic nuclei of the hypothalamus.

Functions of ADH The primary function of ADH is to regulate the rate of urine production. As already described in Chapter 29, ADH acts on the renal collecting ducts, influencing their permeability. Water moves passively in response to the osmotic gradient. If the gradient remains constant, the rate of movement is a function of permeability. ADH can alter urine production from about 500 ml/day to over 15 liters.

In high concentration, ADH causes contraction of smooth muscle. One of the results is increased peripheral resistance and therefore elevated blood pressure. Because of this action, ADH is also referred to as vasopressin. But the amounts required for the pressor effect are so much greater than for antidiuresis that the role of ADH in circulation dynamics is probably nil.

Regulation of ADH Secretion and Release ADH is handled in the same way as oxytocin insofar as its secretion in the hypothalamus and release from the posterior lobe are concerned. The primary regulator of this system appears to be the supraoptic nuclei which are, in effect, osmoreceptors. Increase in the osmotic pressure of the blood is detected by these receptors, which fire the nerve fibers, causing ADH to be released. The result is conservation of water and progressive decline in the blood osmotic pressure. There is thus a self-regulating feedback system which keeps the blood osmotic pressure constant despite great variations in water gain and loss (Figure 31-6).

Many emotions influence ADH secretion. Alcohol decreases its secretion, which accounts for the diuresis associated with drinking. Actually the amount of fluid put out by this mechanism is greater than the amount taken in, especially when hard drinks rather

724

THE
ENDOCRINE
AND
REPRODUCTIVE
SYSTEMS

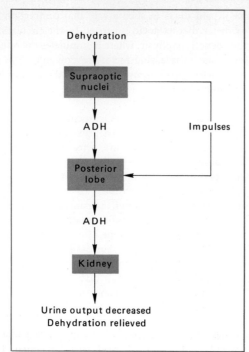

Figure 31-6. Function of the antidiuretic hormone (ADH). ADH is formed in the supraoptic nuclei of the hypothalamus and released by the posterior lobe of the hypophysis.

than beer are consumed. The result is general dehydration, which, at least partially, accounts for the hangover. The prevention or remedy is to consume adequate quantities of nonalcoholic beverages before, during, or after the drinking bout.

Coherin The bovine posterior hypophyseal lobe has been reported to contain still another hormone, named coherin, which is said to induce a coherent, organized pattern of electrical cycles in successive segments of the jejunum. Whether this is truly another hormone and whether it is secreted in human beings remains to be determined.

Abnormal Hypophyseal Function Endocrine abnormalities can generally be classified according to whether the endocrine gland secretes too little or too much hormone. Because the hypophysis secretes so many hormones, abnormalities stemming from its under- or overproduction may be widespread and complex.

   If there is hyposecretion of hormones of the anterior lobe, the condition is termed panhypopituitarism. The result is atrophy of the thyroid, adrenal cortices, and gonads. This leads to widespread alterations. During the developmental period dwarfism results from panhypopituitarism. In the adult there is great wasting and profound weakness in addition to the alterations to be expected from fading activity of the adrenals, thyroid, and gonads. This condition is known as Simmonds' disease.

   Hyperactivity of the anterior lobe produces a variety of abnormalities depending upon the stage of life in which it occurs and the specific cells which become overactive in the lobe. As noted earlier, hypersecretion of GH causes gigantism or acromegaly. If there is a tumor involving primarily the basophils, the main result is hyperactivity of the adrenal cortices due to excessive ACTH production. This disorder is known as Cushing's syndrome.

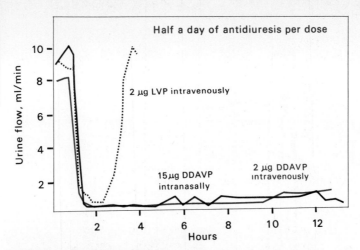

Figure 31-7. Comparison of ADH (L-vasopressin, or LVP) with the new analog, 1-deamino-8-D-arginine vasopressin (DDAVP). Note the prolonged antidiuresis even when given intranasally. (*Reproduced with permission of Medical World News.*)

The major disorder associated with posterior lobe dysfunction is diabetes insipidus, a condition in which abnormally large volumes of urine are excreted. The disorder is termed insipidus to differentiate it from diabetes mellitus. Insipidus means flat, or tasteless. The implication is that the urine does not contain sugar as it does in diabetes mellitus.

A new ADH analog has been reported. It is 1-deamino-8-D-arginine ADH. It is said to be more potent than ADH and to have fewer undesirable side effects. In addition, its antidiuresis lasts longer (Figure 31-7). The substance can be given by the usual intravenous route, or more conveniently as an intranasal spray.

THYROID

Just as the hypophysis was misnamed "pituitary" because an erroneous function was attributed to it, so did early writers assign a fanciful one to the thyroid. Its position in the neck gave rise to the assumption that it must secrete a lubricant for the larynx. The name thyroid, from a Greek word that means "like a shield," is derived from the thyroid cartilage, which has the shape of a long, or oblong, shield.

Anatomy and Histology of the Thyroid

The lobulated, butterfly-shaped thyroid gland, located in the base of the neck, straddles the trachea and has two large lateral lobes connected by an isthmus of glandular tissue of variable size. The isthmus is at about the level of the third tracheal ring, and the lateral lobes, which are molded to the trachea and larynx, rise as high as the thyroid cartilage (Figure 31-8). The thyroid originates as a ventral outgrowth from the floor of the primitive pharyngeal cavity between the first and second branchial pouches and later migrates caudally into the neck.

In man the gland is about 5 cm long and 3 cm wide. It weighs approximately 30 g, but this may vary 5 or 10 g depending upon age, sex, diet, and climate.

Blood reaches the thyroid via the right and left superior thyroid arteries and the

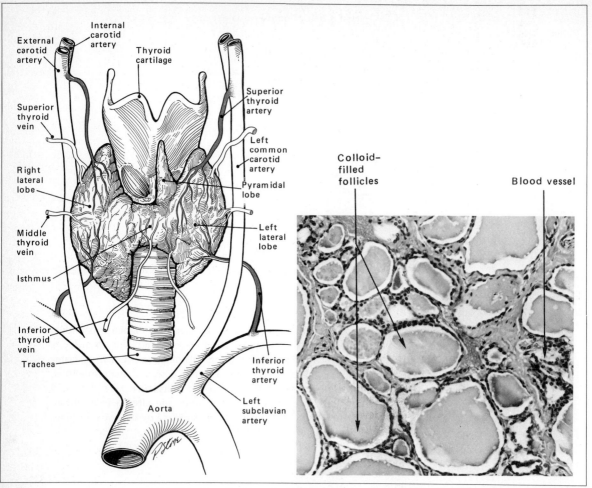

External carotid artery

Internal carotid artery

Thyroid cartilage

Superior thyroid vein

Superior thyroid artery

Left common carotid artery

Right lateral lobe

Pyramidal lobe

Middle thyroid vein

Left lateral lobe

Isthmus

Inferior thyroid vein

Trachea

Inferior thyroid artery

Aorta

Left subclavian artery

Colloid-filled follicles

Blood vessel

Figure 31-8. Thyroid gland. Colloid-filled follicles characteristic of the thyroid are the circular areas in the photomicrograph ($\times$ 135).

right and left inferior thyroid arteries. These vessels then give rise to arterioles that break up into a very extensive capillary network. The gland is drained by the superior and middle thyroid veins into the internal jugular vein and by the inferior thyroid vein into the innominate vein. The blood flow is very high, averaging 3 to 7 ml/min per g of tissue. In contrast, the liver, for example, has a flow of about 1 ml/min per g.

The thyroid receives its innervation from the postganglionic sympathetic fibers which arise in the superior and inferior cervical ganglia, and parasympathetic vagal fibers which arise in the superior and inferior laryngeal nerves. The innervation controls blood flow but not secretion of hormone.

The highly vascular thyroid gland lies in a heavy connective tissue sheath formed in part by the deep cervical fascia. Beneath this covering, the true thyroid capsule, a thin connective tissue investment, sends trabeculae into the gland to divide it into many irregular lobules. Within the lobules are the follicles, the structural unit of the gland.

Each thyroid follicle, or vesicle, is composed of a layer of simple cuboidal epithelium which surrounds a clear hyaline mass of eosinophilic colloid, containing the thyroid hormones.

Physiology of the Thyroid

Thyroid colloid is a glycoprotein. Its main constituent is thyroglobulin which is synthesized by the follicular cells.

Thyroid Hormones Several amino acids are released from thyroglobulin. The most important is thyroxine. It normally appears in the blood and is considered to be the major hormone of the thyroid gland.

Another amino acid, triiodothyronine, is found in the blood in extremely small amounts (less than 5 percent of the total thyroid hormone). Yet it is physiologically active and is now considered to be a true product of the thyroid gland and one of the thyroid hormones. Thyroxine is generally referred to as T_4, and triiodothyronine is called T_3.

Thyroid Iodide The most characteristic function of the thyroid gland is its ability to take up and concentrate iodine. The thyroid can extract iodine from the blood only in the iodide form. Iodide moves into the gland both by diffusion and by active transport. The active transport of iodide is under the control of TSH; this is one important manner in which TSH regulates thyroid function.

Iodide exists in food in many forms. It is readily absorbed from the intestinal lumen into the blood. As the blood circulates through the thyroid gland, the iodide is effectively removed by an active transport mechanism. The iodide is said to be trapped by this iodide pump. The mechanism is so effective that the concentration of iodide in the gland can be as much as 300 times greater than that in the blood. Normally it is some 50 times greater.

In the thyroid gland iodide is oxidized to elemental iodine, which then combines with the amino acid tyrosine to form monoiodotyrosine and diiodotyrosine. These unite in various combinations to form T_3 and T_4 (Figure 31-9).

The thyroid gland also secretes thyroglobulin into the follicles. This globulin has a molecular weight of 680,000. It represents the storage form of the thyroid hormones. The thyroid cells produce proteinases which split T_3 and T_4 from the globulin. These then diffuse into the blood. In the blood the hormones are immediately bound to plasma globulin. As the blood circulates, the hormones are slowly freed for diffusion to the various body cells.

TSH, probably working through cyclic AMP, controls all facets of thyroid activity including iodide uptake, synthesis of the hormones and thyroglobulin, synthesis of the proteinases, and therefore the rate of release of the hormones into the blood.

Function of Thyroid Hormones The major function of the thyroid hormones is to control the metabolic rate. Removal of the thyroid causes the basal metabolic rate to drop 30 to 40 percent below normal. Administration of thyroid hormone returns the rate to normal, whereas excessive doses elevate the rate considerably above normal.

The thyroid hormones are also an important factor in growth. Inadequate secretion during the formative years results in stunted growth. Overproduction of the hormone, however, does not cause gigantism; in fact, that condition may retard growth too.

Nervous system activity is influenced by the thyroid hormones. In cases of hyperthyroidism there is hyperexcitability, nervousness, and rapid responses. Hypothyroidism results in mental retardation and sluggishness. The thyroid hormones, probably by their

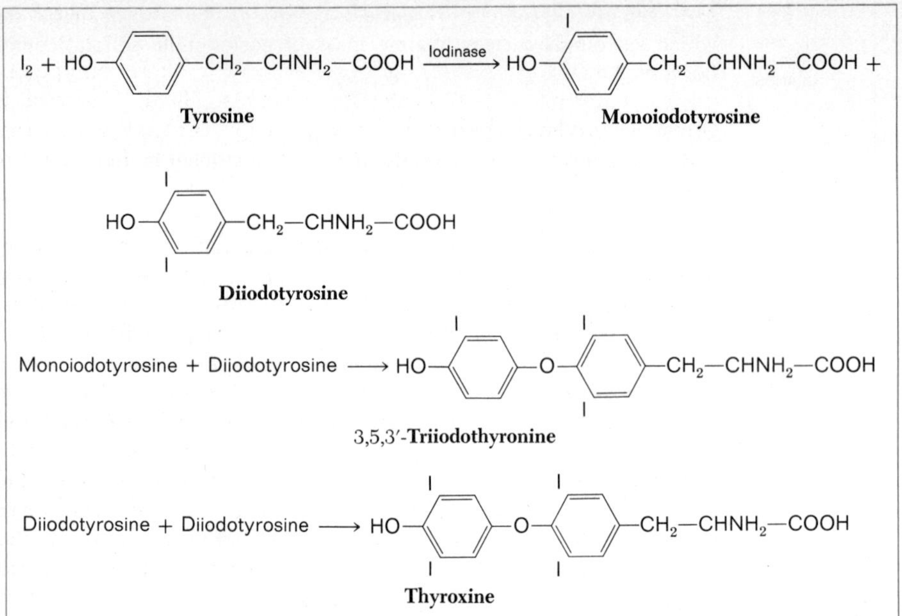

728

THE
ENDOCRINE
AND
REPRODUCTIVE
SYSTEMS

Figure 31-9. Formation of the thyroid hormones.

effect on metabolism, also influence circulation, water and electrolyte balance, and protein, carbohydrate, and fat metabolism.

The effects of the thyroid hormones have been known for years, but the manner in which these iodinated amino acids bring about these changes is still not certain. Some authorities believe the hormones to be regulative oxidative enzymes. The thyroid hormones cause the mitochondria to swell, and this is thought, in some indirect way, to alter the metabolic function of the mitochondria. In short, T_4 undoubtedly is involved in fundamental cellular processes, but the exact nature of this involvement still remains to be elucidated.

Calcitonin In the previous edition of this book we described a substance secreted by the thyroid gland, termed thyrocalcitonin, and a substance secreted by the parathyroid glands, called calcitonin. The evidence is now firm that there is only one such hormone; it is secreted by the thyroid gland, not by the parathyroids, and it is now termed calcitonin.

The discovery of calcitonin proved to be more than a little embarrassing to endocrinologists since the thyroid gland has been under extremely active investigation for at least 80 years and only recently was the observation made that the thyroid secretes a substance different from the iodothyronines. Yet, for many years, histologists have described two types of cells. It is the so-called C cells that produce calcitonin. Needless to say, once the presence of this hormone was established, the entire armamentarium of techniques was brought to bear with spectacular results insofar as the hormone itself and its actions are concerned. Calcitonin is a polypeptide consisting of a single chain of 32 amino acids. The sequence of these amino acids has been established, and the substance has been synthesized.

Function of Calcitonin   Detailed discussion of calcium and phosphate metabolism will be deferred until the parathyroid gland is considered. Suffice it to state here that the primary function of calcitonin is to act on bone to inhibit resorption. This means that less calcium leaves bones to enter the extracellular fluids. The result is lowering of the blood calcium level. As will be noted presently, this action is exactly the opposite of that of the parathyroid gland, and there is thus a dual mechanism for regulating calcium metabolism.

Regulation of Calcitonin   The secretion of calcitonin is controlled by the level of the blood calcium. As the concentration of calcium in the blood increases, the secretion of calcitonin increases. As a result, the outflow of calcium from bone is diminished, and blood calcium decreases (see Figure 31-12).

Regulation of Iodothyronines   The hypothalamus, anterior lobe of the hypophysis, and thyroid gland are components of an effective feedback mechanism. The hypothalamus secretes the neurohumor TRF. TSH regulates the thyroid. Then, in turn, the iodothyronines inhibit the hypothalamus and, probably, the hypophysis. As a result, the level of circulating iodothyronines remains relatively constant. Many factors can alter this level. Environmental conditions that influence the nervous system such as cold and high altitude cause more TRF to be secreted and may result in hyperthyroidism. Likewise there are conditions in which greater quantities of iodothyronine are metabolized. The increased metabolism momentarily lowers the circulating level of the hormone, thus frees the hypothalamus and anterior hypophyseal lobe from inhibition, and, as a result, stimulates the thyroid gland by additional TSH.

Abnormal Thyroid Function   Hypothyroidism during the formative years causes stunted growth, retarded mental development, slow pulse, low body temperature, late eruption of teeth, and general lethargy. The disorder is termed cretinism; the victim, a cretin. If there is normal thyroid function until after growth and development are complete and then hypothyroidism ensues, the BMR falls, the patient fatigues easily, and there is characteristic puffiness (especially around the eyes), dry skin, and poor hair growth. This condition is termed myxedema.

Hyperthyroidism causes an elevated BMR, increased heart rate, emotional instability, sweating, hyperactivity and, in some cases, exophthalmos, which is a bulging of the eyes. Hyperthyroidism is referred to as thyrotoxicosis, or Graves' disease.

Tests of Thyroid Function   Thyroid activity is generally assessed in three different ways. The oldest and simplest procedure is determination of the BMR. A BMR within the range of -15 percent to $+15$ percent is normal. Another procedure is to determine the blood protein-bound iodine. The normal level is 4 to 8 $\mu g/100$ ml of plasma. The third test involves the use of ^{131}I (radioactive iodine). ^{131}I in minute quantities (10 microcuries) is administered orally. The radioactivity of the thyroid gland is then determined 24 hr later. The normal thyroid, at the end of 24 hr, contains between 15 and 45 percent of the administered dose. The hyperactive gland contains more, the hypoactive one less.

Very recently a new procedure was announced which uses the radioisotope technetium $-$ 99 m (^{99m}Tc) in combination with some very sophisticated detection and camera equipment. According to the proponents the patient is positioned in front of the camera, after which the isotope is given intravenously. A record is then obtained during the next 15 sec which is said to suffice for the diagnosis. If this proves to be the case, a major advance will have been made.

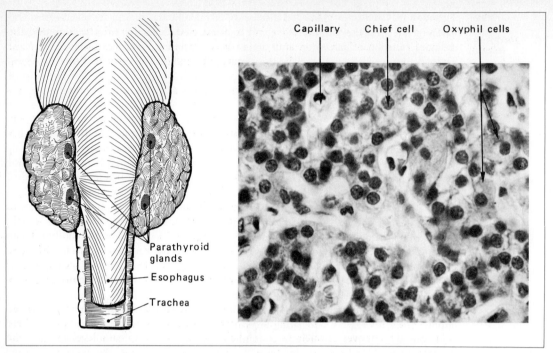

Capillary   Chief cell   Oxyphil cells

Parathyroid
glands
Esophagus
Trachea

Figure 31-10. Parathyroid glands. These glands are embedded in thyroid tissue. Note the abundance of chief cells in the photomicrograph (× 420).

PARATHYROIDS

The fact that the parathyroid glands are situated so close to the thyroid gland obscured their function until it was demonstrated that the tetany that resulted from thyroidectomy was due to removal of the parathyroid tissue rather than to removal of the thyroid. As a result of that discovery, the true endocrine nature of the parathyroids has become clear.

Anatomy and Histology of the Parathyroids

Two minute, yellow-brown bodies, the parathyroid glands, are embedded in each lateral lobe of the thyroid (Figure 31-10). Embryologically, these glands develop from the third and fourth branchial pouches. Each gland is covered by a thin capsule that sends fine septa into the gland, carrying with them an abundant blood and nerve supply. A network of reticular fibers supports clusters or cords of glandular cells composed of two types of epithelial cells: (1) principal, or chief, cells and (2) oxyphil cells. The prominent chief cells far outnumber the larger oxyphil cells. The form of the nuclei of the chief cells further divides these into clear cells, which have a large, light-staining nucleus and nongranular cytoplasm, and dark cells, which have a smaller, dark nucleus and granular cytoplasm. The oxyphil cells are large cuboidal cells appearing singly or in small groups scattered throughout the gland whose cytoplasm contains very fine eosinophilic granules. Occasionally small colloid-filled follicles may be present.

The parathyroids receive their blood supply from branches of the inferior thyroid arteries. The blood leaving the glands drains into the thyroid veins.

Both sympathetic and parasympathetic fibers innervate the parathyroid glands.

Physiology of the Parathyroids

Early, crude experiments indicated that removal of the parathyroid glands resulted in marked alterations in calcium and phosphorus metabolism while extracts of the gland reversed these changes. It thus became obvious that the glands secrete a true hormone.

Parathyroid Hormone   The active substance secreted by the parathyroid glands is a polypeptide containing 84 amino acids. Very recently, the amino acid sequence of the biologically active region of the molecule was determined. Apparently all the activity lies in the first 34 amino acids. Their sequence is shown in Figure 31-11. The hormone is referred to as either the parathyroid hormone (PTH) or parathormone.

Calcium and Phosphate   Because the major role of PTH is to regulate blood level of calcium and phosphate, the physiology of these substances will be briefly reviewed.

Calcium and phosphate are derived from the diet. Dairy foods are the richest in calcium. Phosphate appears in these foods too, but it is also abundant in meats as well. Phosphate is readily absorbed by the intestine, but calcium is not. The amount of calcium absorbed is a function of vitamin D and PTH. Both increase transfer from the intestine to the blood. The regulation is of calcium, but phosphate is influenced secondarily. If calcium is not absorbed, it forms insoluble calcium phosphate and thus prevents phosphate too from being absorbed.

About 80 percent of the absorbed calcium ends up back in the intestine. It is secreted with the digestive juices. The other 20 percent exits via the urine. At the same time there is a constant turnover of calcium into and out of bone. Blood calcium level and also the phosphate level are kept constant by regulation of absorption from the intestine, urinary excretion, and the balance of calcium and phosphate ions in the bones. The normal plasma level for calcium is about 10 mg percent; for phosphate, expressed in terms of phosphorus, it is about 4 mg percent.

Bone is the great reservoir of calcium and phosphate. Bone is not a static tissue; rather there is a constant formation and demineralization. Cells responsible for the deposition of calcium and phosphate in bone are termed osteoblasts. The absorption or demineralization of bone is the responsibility of the osteoclasts. When bone is being formed, the osteoblasts secrete alkaline phosphatase. This enzyme is thought, in some

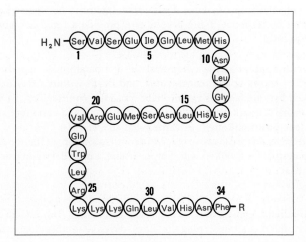

Figure 31-11. Amino acid sequence of the biologically active region of the parathyroid hormone molecule.

732

THE
ENDOCRINE
AND
REPRODUCTIVE
SYSTEMS

manner, to increase the local concentration of phosphate, which then precipitates with calcium into the bone matrix. The blood alkaline phosphatase level provides an indication of the rate of bone formation. It is elevated in growing children, for example.

In a way not yet understood, bones respond to the load placed upon them; the greater the load, the thicker and stronger they become. The load, or stress, activates the osteoblasts to form new bone. An example is seen in the athlete whose bones are heavier than in the sedentary individual.

An interesting relation between heparin and bone has recently been reported. Heparin by itself is capable of causing rapid resorption of bone. The mechanism is not clear. Bone acid phosphatase and collagenase increase after heparin is administered, but the sequence of events remains to be clarified. The observation is important because heparin is now frequently administered to patients to prevent blood clotting.

Functions of the Parathyroid Hormone PTH has three sites of action: (1) the intestine, (2) the bones, and (3) the kidneys. The overall result of PTH administration is elevation of the blood calcium level and usually a decrease in the phosphate level. It is thought to act through cyclic AMP. PTH enhances absorption of calcium and phosphate from the intestine. At the same time it acts on the renal tubular cells to decrease the reabsorption of phosphate and increase the reabsorption of calcium. And in bone, PTH is believed to function by converting osteoblasts to osteoclasts. As a result, demineralization occurs which pours both calcium and phosphate into the blood. The excessive phosphate is excreted by the kidneys. The overall result of these actions is to elevate the calcium level in the blood and lower the phosphate level.

The level of magnesium also appears to play an important role. Low magnesium levels lead to impaired PTH synthesis or release. At any rate, hypocalcemia is a frequent complication of magnesium deficiency in man, and analysis shows that the cause is decreased PTH secretion.

Regulation of Parathyroid Function The parathyroid glands are among the few endocrines free of hypophyseal control. The secretion of PTH is regulated by the level of calcium ion in the serum. When the level falls, more PTH is secreted; when it rises, less PTH is secreted. There is thus a negative-feedback system between serum calcium level and the secretion of hormone (Figure 31-12). In the absence of the parathyroid gland the serum calcium level does not fall below about 7 mg percent. Further, and importantly, this level is kept relatively constant. Apparently, then, the hormones function to raise the calcium level up to about 10 mg percent and provide a rapid homeostatic means of maintaining this level.

Abnormal Parathyroid Function Removal of the parathyroid glands or inadequate secretion of PTH causes tetany, convulsions, and death within a few days. Tetany is a disorder marked by spontaneous intermittent muscular contractions, tremors, and muscular pain. In man, in cases of hypoparathyroidism, the hands and feet are usually held in a characteristic position known as carpopedal spasm.

Hyperparathyroidism causes excessive demineralization of bone. As a result, the bones develop cystlike areas filled with fibrous tissue, a disorder known as osteitis fibrosa cystica generalisata.

ADRENALS

There are two adrenal glands, and each gland is, in reality, two endocrine glands. The adrenal cortex and the adrenal medulla have different embryologic origins, and the hormones they secrete have quite different chemical structures and functions.

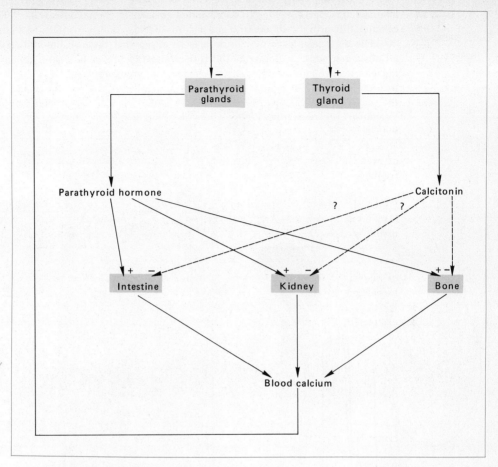

Figure 31-12. Regulation of blood calcium. Parathyroid hormone increases blood calcium by acting on the intestine, kidneys, and bone. The mechanism of calcitonin action is still unclear.

Anatomy and Histology of the Adrenal Glands

The adrenal glands cap the upper pole of each kidney. They are enclosed in the renal fat and have an extensive blood supply. The adrenal gland is subdivided into an outer cortical and an internal medullary portion, each of which secretes distinct hormones and has an entirely separate embryonic origin.

The adrenal cortex is of mesodermal origin. It has its beginning in a thickening of peritoneal epithelium near the primitive kidney. These masses of cells continue to expand but retain their relation to the developing kidney. The adrenal medulla, on the other hand, is derived from neural crest cells that differentiate into the sympathetic ganglia. Cells destined to form the adrenal medulla subsequently leave the ganglia and migrate toward and become incorporated in the center of the developing adrenal cortical tissue. Migration of the medullary cells ceases in late fetal life as they become completely surrounded by cortical tissue.

The cortex makes up most of the adrenal gland and is divisible into three rather poorly defined zones: an outer, narrow zona glomerulosa; a middle, heavier zona fascicu-

734

THE
ENDOCRINE
AND
REPRODUCTIVE
SYSTEMS

lata; and an inner, irregular zona reticularis adjacent to the medulla (Figure 31-13). The eosinophilic cortical cells are arranged in cords that are continuous through all three zones.

The outermost zona glomerulosa is so designated because the cuboidal cells are arranged in groups, or glomeruli, rather than in parallel rows. In this region dark-staining nuclei are surrounded by rather homogeneous foamy cytoplasm.

Figure 31-13. Adrenal glands, with a photomicrograph of a section through adrenal tissue.

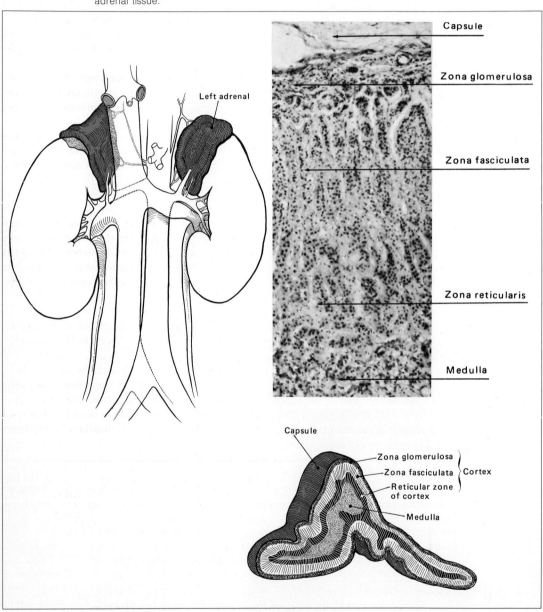

The thick zona fasciculata is characterized by solid parallel cords of cuboidal cells. The cytoplasm of cells in this midzone appears vacuolated (spongy) because of the presence of empty lipid vacuoles. Such cells are called spongiocytes.

In the zona reticularis, cords of cells anastomose, creating a network (reticulum) of cords. Cells in this zone stain unevenly because of the variable amounts of fat and yellow granules in the cytoplasm.

The boundary between the cortex and the medulla is poorly defined. However, cords of cuboidal cells surround the blood vessels of the highly vascular medulla. The medullary cells are basophilic cells, as contrasted to the eosinophilic cortical cells. Their cytoplasm is filled with fine granules which become dark brown when treated with chrome salts or chromic acid. This is the chromaffin reaction which results from the presence of the hormone epinephrine or its precursors.

The adrenal medulla is closely allied with and especially responsive to the autonomic nervous system. Innervation to the medulla is unique in that preganglionic sympathetic nerve cell fibers pass directly to the cells of the medulla. In all other structures that they supply, the fibers synapse with postganglionic neurons.

Arterial blood reaches the adrenal gland both from the aorta and from small branches of the phrenic and renal arteries. The blood flows from the capsular plexus through the cortical layers into the sinusoids of the medulla. From here the blood enters the central adrenal vein and then returns to the vena cava.

Adrenal Cortical Hormones

The three layers of the adrenal cortex secrete several different hormones. All of them are steroids. These substances may be divided into three categories: (1) those that primarily influence sodium and potassium, referred to as mineralocorticoids and secreted principally by the zona glomerulosa; (2) those that primarily influence carbohydrate metabolism, referred to as glucocorticoids and secreted principally by the zone fasciculata; and (3) sex hormones secreted by the zona reticularis.

Functions of the Mineralocorticoids   The principal mineralocorticoid is aldosterone (Figure 31-14), although corticosterone and deoxycorticosterone have similar, but far less potent, functions.

Aldosterone causes many cells to transport sodium. It is thought to do this by increasing the synthesis of messenger RNA, which results in the formation of enzymes responsible for sodium transport. Whatever the precise mechanism, the influence of aldosterone on the cells that make up the renal tubules brings about increased reabsorption of sodium, chloride, and water and decreases the reabsorption of potassium. The primary

Figure 31-14. Chemical structure of three adrenal cortical hormones.

Corticosterone   Hydrocortisone   Aldosterone

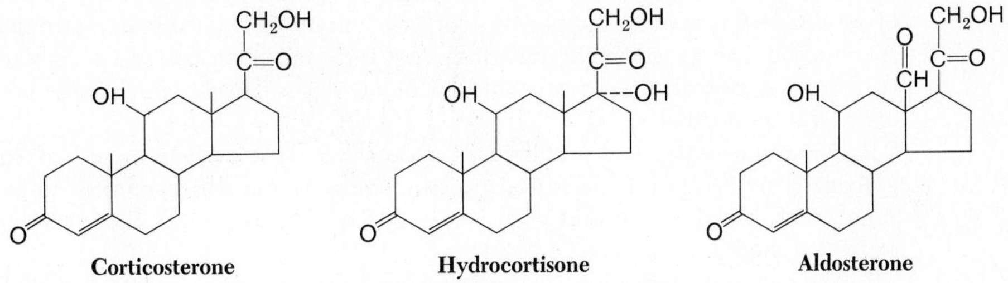

736

THE
ENDOCRINE
AND
REPRODUCTIVE
SYSTEMS

effect is thought to be on sodium transport. As more is transported, an electrical gradient is established which encourages movement of negative chloride in the same direction and the repulsion of the cation potassium. However, this is probably not the entire story. At any rate, under the influence of aldosterone, sodium is retained and, because of the osmotic gradient, water is retained, too. The net result is that blood concentration of sodium increases (hypernatremia) somewhat, and the total volume of extracellular fluid increases greatly, which means that the total quantity of sodium in the body has increased significantly. Interestingly, aldosterone, probably because of sodium retention, also increases thirst, so even more water is taken in, and ADH is released, which decreases urine output.

There are many consequences of the increased sodium and fluid. First, because of the retention of sodium, hydrogen ion is excreted in the urine; alkalosis results. Second, the expansion of the extracellular fluid volume may result in increased cardiac output, elevated arterial blood pressure, and edema. Third, the loss of potassium decreases extracellular potassium (hypokalemia), and since the action potential of nerve and muscle is so dependent upon the potassium gradient, the results can be widespread. In general there will be hyperpolarization resulting in paralysis.

If the adrenal glands are removed, large quantities of sodium are lost, as are chloride and fluid; potassium and hydrogen are retained. Cardiac failure and death occur within a few days.

Regulation of Aldosterone Secretion There are, apparently, several mechanisms which regulate the secretion of aldosterone and thus the levels of sodium, potassium, chloride, hydrogen, and bicarbonate in the blood, as well as the volume of the extracellular fluid (Figure 31-15). First, decreased sodium concentration in the blood perfusing the adrenal cortex sharply increases aldosterone secretion. Increased potassium concentration has a similar effect. Second is the renin-angiotensin system discussed previously (see page 294). In brief, diminished blood flow to the kidneys causes renin to be released. Renin stimulates the formation of angiotensin in the blood. Angiotensin, in turn, evokes aldosterone secretion. Third, ACTH increases aldosterone secretion to some extent. Its major influence is on the glucocorticoids. And, fourth, there is some evidence that there is a substance termed adrenoglomerulotropin secreted by the brain, perhaps by the pineal gland, that stimulates aldosterone.

Functions of the Glucocorticoids The major glucocorticoid is hydrocortisone, also referred to as cortisol. Cortisol initiates a sequence of events resulting in an elevated blood-sugar level. Blood sugar increases because the rate of gluconeogenesis increases and the utilization of glucose by the cells decreases. Gluconeogenesis is augmented because the hormone facilitates the transport of amino acids into the liver cells. In addition it stimulates increased production of liver cell enzymes required for gluconeogenesis. How cortisol diminishes glucose utilization is not certain. It may be by virtue of suppression of an enzyme system or simply by blocking entrance of glucose into peripheral cells. At any rate, just as excessive quantities of growth hormone can cause diabetes, so can cortisol. The mechanism is persistent high blood glucose levels, which lead to exhaustion of insulin production.

Perhaps directly, but probably as a consequence to the influence of cortisol on gluconeogenesis, there is decreased protein synthesis and increased protein catabolism. As a result, blood amino acid level is raised. Even less is known about the mechanism by which the hormone mobilizes fat.

The glucocorticoids also have an important anti-inflammatory effect which has broad

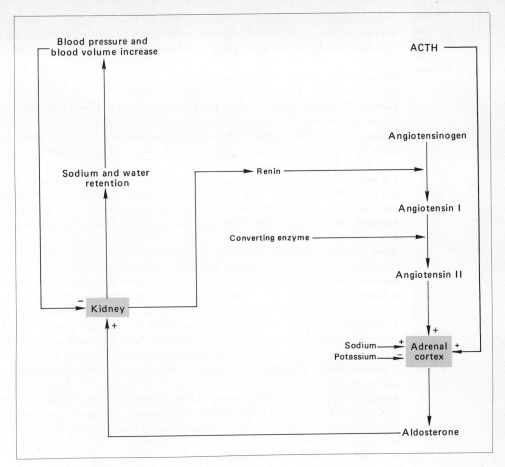

Figure 31-15. Regulation of aldosterone secretion.

significance in clinical medicine. These hormones are used in cases of rheumatoid arthritis and other conditions in which inflammation plays a major role. The corticoids are thought to combat inflammation in two ways: (1) by stabilizing the lysosome membrane and (2) by decreasing capillary permeability. When the lysosome membrane breaks down, enzymes are freed which have great destructive ability. Decreasing the capillary membrane permeability helps to prevent movement of fluid into the inflamed area.

Regulation of Glucorticoid Secretion The major control over the secretion of the glucorticoids is exercised by ACTH. Cyclic AMP is the mediator. As mentioned previously, ACTH is regulated by the hypothalamic substance CRF. In turn, the glucocorticoids inhibit the secretion of CRF and ACTH. There is thus the typical self-regulating feedback mechanism.

Functions of the Adrenal Sex Hormones Normally, the zona reticularis secretes hormones which are similar or identical to male sex hormones. The female hormones are secreted in very small quantities, but under abnormal conditions they may increase.

738

THE
ENDOCRINE
AND
REPRODUCTIVE
SYSTEMS

Just what physiological role these secretions have is difficult to assess. They may supplement the secretions of the gonads, or they may have other functions. An interesting recent report suggests that the adrenal androgens increase the libido in the female.

Abnormal Adrenal Cortical Function

Addison's Disease   The failure of the adrenal cortices to produce an adequate quantity of steroids results in Addison's disease, which is characterized by increased serum potassium, decreased serum sodium, hypoglycemia, low blood pressure, dehydration, acidosis, renal failure, and skin pigmentation resulting from melanin deposition.

Cushing's Syndrome   The excessive production of the adrenal steroids causes the expected alterations in carbohydrate and electrolyte metabolism. In addition, obesity, especially in the back of the neck, gives rise to a characteristic "buffalo hump." The face becomes round ("moon face") and is usually flushed. There is general weakness.

Adrenogenital Syndrome   Cushing's syndrome may or may not be associated with sexual changes, depending on whether or not there is excessive secretion of adrenal androgens or, in some cases, estrogen. If there is, there is usually exaggeration of the male characteristics. A young boy may undergo maturity with enlarged penis, lowering of the voice, development of chest hair, and sexual desire. Male adults show overgrowth of body hair, enlargement of the penis, and increased sexual drive. Adult women usually undergo virilism. The breasts atrophy, the clitoris enlarges, body hair increases, and baldness may develop. Young girls usually become sexually mature with precocious menstrual cycles and development of the breasts. In some cases, adult males show feminization.

Aldosteronism   Excessive secretions of aldosterone usually result from an adrenal cortical tumor. In any event, the excessive hormone causes hypokalemia, elevated blood sodium level, retention of water, increased blood volume, hypertension, and edema.

Adrenal Medullary Hormones

There are two adrenal medullary hormones: (1) epinephrine and (2) norepinephrine. In man, approximately 80 percent of the total catecholamine production by the adrenal medulla is epinephrine. In contradistinction, adrenergic nerves liberate norepinephrine. The biosynthesis of the catecholamines is shown in Figure 31-16. Note that norepinephrine is first produced and then is converted to epinephrine in a reaction catalyzed by phenylethanolamine-N-methyltransferase (PNMT). Julius Axelrod, who received the Nobel prize in 1970, and his coworkers made the very interesting observation that the adrenal cortex is essential for the conversion of norepinephrine to epinephrine. As a matter of fact, they showed that several of the key enzymes depend upon the integrity of the cortex. If an animal is hypophysectomized, PNMT, tyrosine hydroxylase and dopamine-β-oxidase all decrease. As shown in Figure 31-16, these enzymes are essential for catecholamine synthesis. Administration of ACTH returns levels to normal.

Functions of the Adrenal Medullary Hormones   Epinephrine increases cardiac output but decreases peripheral resistance. In spite of this resistance, the cardiac effect is usually great enough to elevate the blood pressure. Norepinephrine increases blood pressure because it is a potent vasoconstrictor.

Both catecholamines induce release of free fatty acid, stimulate the nervous system, and cause increased heat production. They also elevate the level of blood sugar, but in

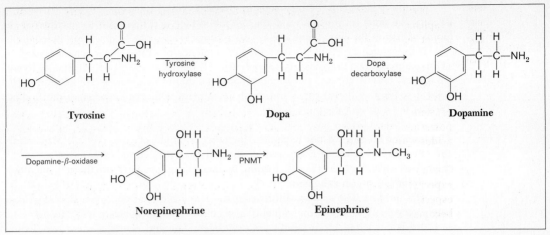

Figure 31-16. The biosynthesis of catecholamines. PNMT is phenylethanolamine-*N*-methyltransferase.

this respect epinephrine is far more powerful than norepinephrine. This hyperglycemic effect results from accelerated conversion of liver glycogen to glucose. Most of these effects are mediated through cyclic AMP.

Activation of the sympathetic fibers that innervate the adrenal medulla increases the secretion of the catecholamines. Since the ratio of the two can be changed, the cells responsible for the secretion of each are probably separately innervated. Increased medullary activity is part of a generalized sympathetic response to many stimuli.

Tumors of the medullary cells cause excessive secretion of the catecholamines. Such a tumor is termed a pheochromocytoma, and the amount of hormone secretion resulting may be 500 times the normal secretion. As a result, there is hypertension, hyperglycemia, glycosuria, and a markedly elevated BMR.

PANCREAS

The pancreas is well known to the gourmet as "sweetbread," which is the original meaning of the word pancreas. The ancient Greeks so named this organ because of its meaty, fleshy character. Because the pancreas functions both as an important exocrine organ of digestion and as an endocrine gland, it has already been discussed in Chapter 24. The islets of Langerhans (see Figure 24-15) are responsible for the secretion of hormones.

Pancreatic Hormones

The islets of Langerhans secrete two hormones: insulin and glucagon. The primary function of each is to regulate carbohydrate metabolism. Insulin causes a decrease in blood sugar; glucagon increases it.

Insulin   Insulin is produced by the beta cells of the islets of Langerhans. It is a simple protein of 51 amino acids, consisting of two peptide chains held together by disulfide bridges. Insulin is initially synthesized as a single polypeptide with 84 amino acids. This is known as proinsulin. It has little biologic activity. Under the influence of a trypsinlike enzyme, the molecule is split at two points. The larger unit is insulin.

740

THE
ENDOCRINE
AND
REPRODUCTIVE
SYSTEMS

Insulin is stored in the beta cells in granules. When the beta cells are activated to secrete, the granules move toward the cell membrane and fuse with it, and then at the point of fusion the membrane opens, permitting the granule to discharge its contents.

If the source of insulin is destroyed, a sequence of events occurs characterized by (1) hyperglycemia, (2) glycosuria, (3) polyuria, (4) dehydration, (5) polydipsia, (6) hyperphagia, (7) fatigue, (8) loss of weight, (9) acidosis due to ketonemia, (10) ketonuria, (11) loss of sodium, and (12) coma and death. Treatment with insulin reverses the findings leading to coma and death.

Functions of Insulin Insulin expedites the transport of glucose into the cells. Apparently the rate at which glucose is utilized by the cell is controlled to a significant extent by the rate at which it enters the cell. Insulin, by facilitating transport into the cell, augments the rate of glucose metabolism.

The hormone also acts on cellular enzymes responsible for the conversion of glucose into glycogen. Thus, under the influence of insulin, glucose not only enters the cell faster but also is converted more rapidly into glycogen. The end result of these two actions is a lowering of the blood-sugar level and an increase in liver and muscle glycogen.

Insulin expedites the transfer of amino acids into the cell and their incorporation into peptides. Likewise this hormone augments fatty acid synthesis.

The above explanation of the way in which insulin works makes clear the sequence of events that occurs after beta cell destruction. Because of the inability of the cells to take up glucose at the normal rate in the absence of insulin, glucose accumulates in the blood, resulting in hyperglycemia. When the blood sugar exceeds about 160 mg percent, glucose is excreted in the urine (glycosuria). Glucose in the renal filtrate osmotically opposes the reabsorption of water, causing polyuria, dehydration, loss of sodium, and ultimately polydipsia (thirst). The loss of glucose and the changes in protein and fat metabolism result in hyperphagia (excessive eating), fatigue, and weight loss. The disturbed fat metabolism is reflected in the excessive conversion of fatty acids to ketone bodies which build up in the blood (ketonemia), appear in the urine (ketonuria), and, if unchecked, cause acidosis, coma, and death.

Glucagon The alpha cells of the islets of Langerhans secrete glucagon, also referred to as the hyperglycemic factor because of its ability to increase blood sugar. Glucagon is a small protein consisting of a chain of 29 amino acids.

Functions of Glucagon Glucagon causes an increase in blood sugar by accelerating the conversion of liver glycogen to glucose. In this respect it acts just like epinephrine, although it is thought that the two hormones produce the same end result by somewhat different means. It should be emphasized that glucagon and insulin do not antagonize or block one another. Rather they have opposite end results so that one elevates blood glucose and the other depresses it.

Glucagon has another effect that may win for it important use in clinical medicine. It has been found to be a potent cardiotonic agent. That is, like digitalis and the catecholamines, glucagon increases heart rate and stroke volume, and therefore cardiac output. To be sure, large doses must be given, but there are apparently few if any side effects, and therefore it is now being evaluated for treatment of patients in acute heart failure.

Regulation of Pancreatic Hormonal Secretion

The level of blood sugar determines the rate of secretion of insulin and glucagon. There is thus a double-headed control mechanism very similar to the one that regulates the level of blood calcium. When the blood-sugar level falls, glucagon is secreted; when it rises, insulin is secreted. The net result of these two negative-feedback mechanisms is that the blood sugar remains within a relatively narrow range despite great alterations in the amount of ingested carbohydrate.

Recent studies show that the mere presence of glucose around, or even in, the beta cells does not lead to insulin secretion. For this to occur, glucose metabolism must be unhindered. This means that the metabolism of glucose is involved in the act of secretion of insulin. Exactly why has not yet been determined. Either a particular metabolite is an effector agent in the discharge of the beta granule, or the secretory process requires, and is linked to, a reaction in one of the pathways of glucose metabolism.

When a load of glucose is given intravenously, secretion of insulin quickly follows. The rate of secretion reaches a peak and then starts to decline, only to rise once again (Figure 31-17). The first peak is thought to be due to the liberation of preformed insulin; the second rise, to new synthesis.

The hypophysis does not secrete a hormone that directly regulates the secretion of the pancreatic hormones. GH, however, does influence carbohydrate metabolism in that it raises the blood glucose level and thereby stimulates the liberation of insulin. ACTH, by stimulating the secretion of the glucocorticoids, also brings about hyperglycemia and the release of insulin.

Abnormal Pancreatic Hormonal Function

As described above, inadequate secretion of insulin results in a sequence of events characterized primarily by hyperglycemia. This syndrome is called diabetes mellitus. The term diabetes implies a large urine output, while mellitus suggests that this urine is sweet, owing to the glucose.

A tumor in the islets of Langerhans may cause overproduction of insulin, and this results in hypoglycemia. If the blood-sugar level falls below about 60 mg percent, the individual appears to be inebriated. When it is lower than about 50 mg percent, convulsions occur. The problem obviously is to maintain adequate blood sugar. This can sometimes be accomplished by diet, but more often the tumor must be removed.

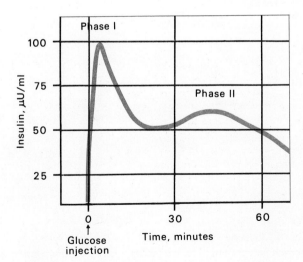

Figure 31-17. Secretion of insulin in response to glucose injection. The first peak is preformed insulin; the second, newly synthesized insulin.

742

THE
ENDOCRINE
AND
REPRODUCTIVE
SYSTEMS

GONADS

The term gonad means "seed" and thus has reference to the organs which produce the ovum and the sperm. In the female the gonads are referred to as the ovaries; in the male, they are the testes. These structures not only produce the reproductive cells but also are important endocrine glands. In the next chapter reproduction is discussed, and the anatomy and histology of the gonads are presented there. The discussion here is limited to the endocrine function of the ovaries and testes.

At birth the ovaries of the newborn girl contain only immature follicles, which are referred to as primary, or primordial, follicles. There are about 400,000 primary follicles in the two ovaries. At puberty, the primary follicles ripen or mature, usually one each month. The mature follicles are sometimes referred to as the graafian follicles. Each contains an ovum. About midway between two menstrual periods, the mature follicle expels the ovum, which then enters one of the fallopian tubes and is carried to the uterus.

Immediately following ovulation, a spot of blood persists in the ruptured follicle. The follicle is at this time termed the corpus hemorrhagicum. This form quickly gives way to a follicle which contains a yellow substance, and it is now called the corpus luteum. If pregnancy does not occur, the corpus luteum undergoes regression, is replaced by fibrous tissue, and becomes a corpus albicans.

Ovarian Hormones

The major female sex hormones produced by the ovaries are the estrogens and progesterone. In addition, small amounts of androgens are secreted and also a substance termed relaxin.

Estrogens   Estrogens are secreted not only by the ovaries but also, in small amounts, by the adrenal cortex. Also, during pregnancy the placenta puts out large quantities of these substances (see Chapter 33). The estrogens are steroids. Figure 31-18 shows the major ones along with progesterone and the male hormone, testosterone, for comparison.

Figure 31-18. Structure of the estrogens, progesterone, and testosterone.

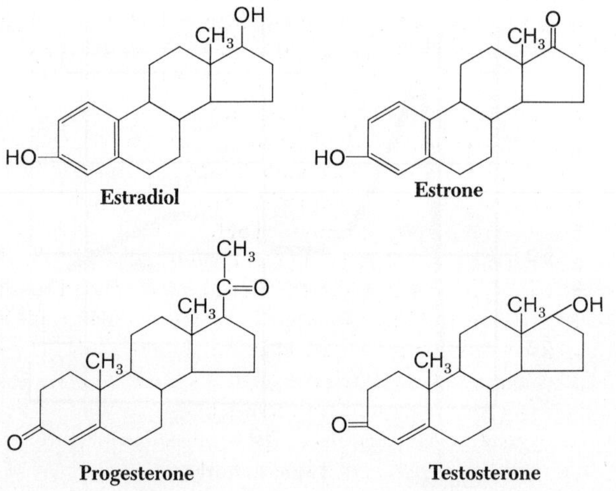

The most potent is beta-estradiol. In the ovary, estrogens are secreted by the theca interna and perhaps the granulosum cells of the developing follicle, and by the corpus luteum.

Functions of the Estrogens Before puberty only very small amounts of estrogens are secreted by the ovaries. At and after puberty, under the stimulus of the gonadotropins, estrogens pour forth and bring about changes in the sexual organs characteristic of sexual maturation. In addition to enlargement of the vagina, uterus, and fallopian tubes, there are deposition of fat in the mons pubis, growth of pubic hair, broadening of the pelvis, and development of the breasts due mostly to deposition of fat.

The estrogens stimulate skeletal growth, but they also bring about cessation of lengthening of the long bones. In the absence of these hormones, growth is slower but continues longer so that the final height is greater.

The role that the estrogens, indeed that all the female sex hormones, have in sexual interest and drive, that is, the so-called libido, is difficult to assess. The contraceptive pill which contains various mixtures of these substances increases the libido in some women but markedly reduces it in others. Interestingly, the androgen secreted by the adrenal cortex appears to have the most stimulating effect.

Progesterone The cells of the corpus luteum are responsible for the secretion of progesterone. As is the case with estrogens, the adrenal cortex secretes minute quantities of progesterone. The placenta, in the last 6 months of pregnancy, puts out large amounts. There are other steroids secreted by the ovaries which are very similar to progesterone, for example, the progestins mentioned by some writers, but the amounts of these substances are negligible.

Functions of Progesterone In general, progesterone finishes what the estrogens begin. Its main effect on the uterus is to cause thickening and increased vascularization of the endometrium. In the breast, estrogens cause deposition of fat and growth of the duct system; progesterone is responsible for the development of the secretory apparatus, that is, the alveoli. Neither stimulates milk production or secretion. That is the function of prolactin. Progesterone plays its most important role during pregnancy (see Chapter 33).

Regulation of the Female Sex Hormones This is an area of physiology with many more theories than facts. The problem lies in the complexity of interrelating the two sex hormones, three gonadotropins, and two or more hypothalamic factors. Confusing the issue even more is the observation that the ovarian hormones apparently exert a typical negative-feedback effect when they are present in the blood in large quantities but a positive-feedback effect in small quantities.

With those caveats, an attempt will be made to integrate the various substances and mechanisms. Most authorities recognize that there are three gonadotropins: (1) follicle-stimulating hormone (FSH), (2) luteinizing hormone (LH), and (3) luteotropic hormone (LTH). FSH starts the development of the primary follicle and the secretion of estrogens. LH furthers the development and secretion, and then there occurs a sharp increase in LH, the so-called LH surge, which is responsible for ovulation. After the expulsion of the ovum, the follicle undergoes luteinization to become the corpus luteum, which secretes considerable quantities of estrogens and progesterone. LTH now enters the picture to maintain secretions from the corpus luteum.

Also, as described previously, the current evidence indicates that there is but one releasing factor responsible for regulating the release of the FSH and LH, and two others that control LTH. Undoubtedly estrogens and progesterone act upon both the hypo-

744

THE
ENDOCRINE
AND
REPRODUCTIVE
SYSTEMS

thalamus and the hypophysis to influence secretion of the releasing factors and the gonadotropins, but much additional work is necessary to fill in the details.

The unique feature is that there is a waxing and waning of the sex hormones, an oscillation. Recent work strongly suggests that the hypothalamus is responsible, but there are undoubtedly a host of factors that impinge upon the hypothalamus that remain to be discovered.

Androgen   The ovaries produce small quantities of androgen, which is a masculinizing hormone. Very little is known concerning the function of androgen in the female.

Relaxin   In various animals, the ovaries secrete one or more polypeptides which relax the symphysis pubis and uterus, and soften the cervix. Because of their relaxing role, the polypeptides, as a group, are referred to as relaxin. Just what role they play in women is not known, but there is some evidence that they potentiate the effect of estrogen and progesterone on the endometrium.

Ovulation

Figure 31-19 shows the surge of LH which occurs the day before ovulation and which is responsible for expulsion of the ovum. The LH is thought to activate adenyl cyclase of the steroid-secreting cell of the theca interna, causing an increase in cyclic AMP. As a result steroid synthesis and secretion are increased. The resulting progesterone stimulates synthesis and secretion of the so-called ovulatory enzyme which increases distensibility, permitting the follicle to swell until rupture occurs.

Ovulation takes place about 14 days before the onset of the next menses. Once the ovum is expelled from the ovary, it enters one of the fallopian tubes to be carried toward the uterus. Fertilization generally occurs during this transit, after which the fertilized egg enters the uterine lumen to become implanted in the wall. Knowledge of the time of ovulation is important for the so-called safe-period method of contraception (see Chapter 33) and also for successful artificial insemination.

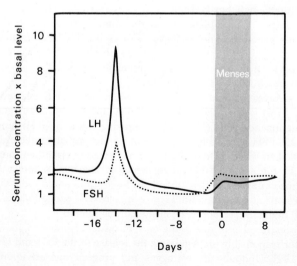

Figure 31-19. Composite curves of basal body temperature and concentrations of LH and FSH in sera during single menstrual cycles from 37 women. (*From A. R. Midgley, Jr., and R. B. Jaffe, Journal of Clinical Endocrinology and Metabolism, 28:1699, 1968.*)

Menstrual Cycle

In women there is a cycle of approximately 28 days during which the involved hormones undergo marked alterations in secretion and, most importantly, the uterine endometrium undergoes cyclic changes. Although the term menstruation refers to the monthly periodicity of the phenomenon, women generally use the term in reference to the part of the cycle during which there is vaginal discharge of blood and the remnants of the endometrium. This phase is more accurately termed the menses. Following this phase, the estrogen level in the blood is low (Figure 31-20). Under the prodding of FSH, however, more estrogen is secreted. This, in turn, causes the endometrium to begin to thicken and become more vascular. As the estrogen level increases, FSH is inhibited, and LH is stimulated. LH brings the ovum to full maturation and evokes ovulation and the formation of the corpus luteum with the consequent estrogen and progesterone secretion. Because of the positive-feedback mechanism, the level of progesterone in the blood rapidly rises.

Figure 31-20. Menstrual cycle. Estrogen and progesterone cause endometrial thickening. A sudden decrease in the level of the hormones causes menstruation.

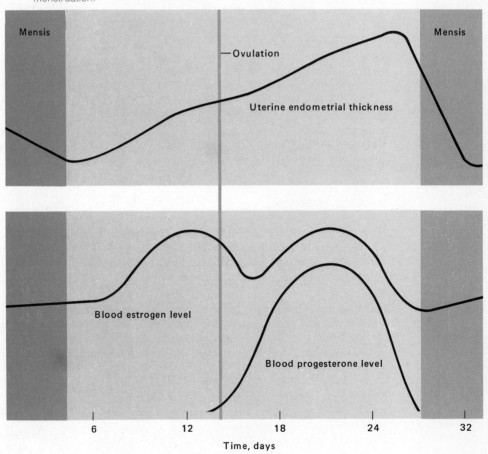

746
THE
ENDOCRINE
AND
REPRODUCTIVE
SYSTEMS

Under the stimulus of estrogen and progesterone, the endometrium reaches its peak of thickening and vascularization. As noted in Figure 31-20, the blood estrogen level, which had fallen with the demise of the follicle, once again increases as the corpus luteum becomes functional. Thus, for several days the blood level of both progesterone and estrogen is high. Ultimately, for reasons not yet explained, the secretion of FSH and LH diminishes, the corpus luteum ceases to secrete, and the levels of estrogen and progesterone fall rapidly. The endometrium then breaks down and is sloughed off along with considerable bleeding. In the absence of inhibition, the anterior lobe once again begins to secrete FSH, and the cycle begins anew.

During the menses about 35 ml of blood is lost. The total amount of fluid is greater, however, because an equal amount of serous fluid is mixed with the blood. Normally the blood does not clot, because of the presence of fibrinolysin.

Abnormal Ovarian Function

Amenorrhea   As the term indicates, this is a condition characterized by the complete absence of menstruation. It may be due to a number of causes. If the menstrual flow has been well established and regular, the most common cause for the sudden cessation is pregnancy. But if the cycle has never been initiated, hormonal imbalance is suspected. As might be anticipated, such women also exhibit other abnormalities. The breasts fail to develop, and a rather boyish figure persists.

Dysmenorrhea   Dysmenorrhea is the term applied to a difficult or painful menstruation. The chief complaints are cramps, low abdominal pain, aching of the thighs, headache, nausea, and perhaps emotional lability. There is no simple explanation of this syndrome, but a hormonal imbalance probably is a significant factor.

Hypergonadism   Excessive production of the female sex hormone usually does not give rise to any alterations that bring a person to the clinic. For these reasons it is usually stated that hypergonadism rarely occurs.

Testicular Hormones

The interstitial cells of Leydig, which are situated between the seminiferous tubules in the testes, secrete the male sex hormones, referred to as androgens. There are at least two such hormones, both steroids. Testosterone (Figure 31-18) is the major product, and because the other hormone is secreted in such minute amounts, testosterone is usually considered to be the only male hormone.

Functions of Testosterone   The male hormone is secreted by the male fetus. Interestingly, if the testes are removed from the fetus early in development, female sexual organs develop rather than those characteristic of the male. Similarly, injections of testosterone during pregnancy cause the female fetus to develop male organs. In short, the embryonic development of male sexual organs depends upon the presence of adequate quantities of testosterone regardless of the chromosomal makeup. In the absence of the hormone, female organs develop in both the male and female fetus.

During the eighth or ninth month of fetal life, the testes descend from the abdominal cavity through the inguinal canal into the scrotum. This descent does not occur in the absence of testosterone. The role of this hormone is clearly shown in cases of male children born with undescended testes. Administration of testosterone usually evokes prompt descent.

During childhood small amounts of testosterone are secreted and are responsible for much of the muscular development that characterizes the developing boy in contradistinction to young girls. At puberty, the secretion increases sharply with widespread results. The penis, testes, and scrotum all enlarge and continue to grow for several years, usually until about age 21 or older. In addition there is characteristic lowering of the voice due to enlargement of the larynx. If a young boy is castrated, the voice remains high. This was the procedure used some time ago to produce so-called castrati, male sopranos, much cherished in many operas.

Like the female sex hormones, testosterone acts on bones, causing them to thicken and to lengthen. However, it also brings about closure of the epiphyses and thus termination of growth. The result is that when testosterone increases at puberty there is a spurt in growth, but this growth soon terminates. Thus the male castrated before puberty, called a eunuch, generally grows to be taller than the average.

After puberty, the influence of testosterone on muscle development becomes pronounced. The hormone is now often used for muscular development in cases where normal development is below par or expectations.

Testosterone has an influence on hair growth and distribution. At puberty pubic hair grows, as does hair on the chest and other areas of the body. Also baldness characteristically develops in many men. Eunuchs do not become bald. This and other lines of evidence indicate that baldness requires adequate testosterone secretion. However, some men, despite such adequate secretion, do not become bald. Therefore there must be another factor, probably genetic.

Testosterone controls, to some extent, sexual interest and drive. It undoubtedly provides the underlying motivation, but, in man at least, many other factors are superimposed upon it and confuse the issue. At any rate, testosterone and various analogs are used to treat a variety of sexual inadequacies, sometimes with success.

The problem of using a simplistic approach to human behavior, especially sexual behavior, is well illustrated by attempts to explain homosexuality on the basis of hormonal titer. In a recent study from the famed Reproductive Biology Research Foundation, directed by William H. Masters, a significant difference in plasma testosterone level was found in homosexuals as compared with control heterosexuals. In fact, if the population was arranged according to the Kinsey classification with 0 for exclusively heterosexual to 6 for exclusively homosexual, a very nice gradient was found with progressively less testosterone moving from 0 to 6. For example, the controls had an average level of 689 mg/100 ml of plasma; group 6, only 264 mg/100 ml. This is impressive, but the denouement comes when group 5 and 6 men are given testosterone without any obvious effect upon their sexual preference and activities. Thus, again, one wonders if the hormonal level is a cause or an effect.

Regulation of Testosterone Secretion   In the male, as in the female, there are three gonadotropins: FSH, thought to be primarily concerned with the development of the spermatogenic mechanism; LH, with testosterone production; and LTH, which is not known to play any role in the male. Recent work shows that FSH also stimulates testosterone secretion and that it acts synergistically with LH, which in the male is referred to as interstitial cell-stimulating hormone (ICSH).

Testosterone inhibits the secretion of the gonadotropins, ICSH more so than FSH. Another hormone, called inhibin, or the "X" hormone, inhibits FSH. Inhibin may be

748

THE
ENDOCRINE
AND
REPRODUCTIVE
SYSTEMS

estrogen, which is known to be secreted to some extent by the testes. Whether the inhibition is exerted by testicular hormones at the hypothalamic level on the releasing factor or at the hypophyseal level on the gonadotropins remains to be clarified.

Most authorities contend that testosterone remains at a relatively constant concentration in the blood in contradistinction to the female sex hormones, which wax and wane. If this is the case, one wonders what accounts for the difference.

Testosterone secretion increases at puberty, reaches a peak at about 21 years of age, and then slowly declines so that at age 50 the level is about half that of age 21. This is said to account for the declining sexual interest in the male with age. However, there are interesting observations which show that the rate of spermatogenesis is a function of frequency of ejaculation; the more frequent the ejaculation, the greater the rate of sperm production. Also, vigorous sexual activity increases testosterone secretion. There is thus a vicious cycle that may well account for the declining level of testosterone with age, a cycle that could possibly be broken in a variety of ways.

Abnormal Testicular Function

Cryptorchism This is a condition in which the testes remain in the abdominal cavity and fail to descend. The testes in this position are capable of producing and secreting androgen, and such persons develop normal sexual characteristics. They do not, however, produce viable sperm. Thus they are potent but not fertile.

Hypogonadism The complete absence of testes prevents the development of characteristics associated with manhood. The external genitalia remain infantile, the voice does not deepen, and there is very little growth of hair on the body or face. Thus the victim retains a juvenile appearance. Growth, however, may be excessive. Such persons are termed eunuchs. If the testes are present but do not secrete adequate androgenic substances, the result may also be a failure of sexual maturation and little or no sexual drive.

Hypergonadism Benign tumors involving the interstitial cells have been reported. In such cases there is excessive production of androgen and, as a result, an overdevelopment of the sexual organs and the secondary sexual characteristics. These end results rarely bring the person affected to the clinic, and thus such cases are said to be rare.

PINEAL GLAND

The pineal gland, also referred to as the epiphysis, is located in the roof of the third ventricle of the brain (Figure 31-21). It is connected by a stalk to the posterior commissure and habenular commissure. The gland itself contains neuroglia similar to those found in the adrenal medulla and the neurohypophysis. It is innervated by preganglionic sympathetic fibers.

Until about 1955, the pineal gland interested philosophers more than it did endocrinologists. At a recent meeting of pinealologists, the chairman suggested that the reason for this may be that endocrinologists do not have much respect for philosophers! However, he went on to point out other possible reasons, including that animals survive pinealectomy and that even the observable effects of pinealectomy disappear after a period of time and the animal appears to be perfectly normal once again. Further, during the period of time that the effects are apparent, the pinealectomized animal cannot be brought to its normal status by administering crude pineal extracts or even the more refined com-

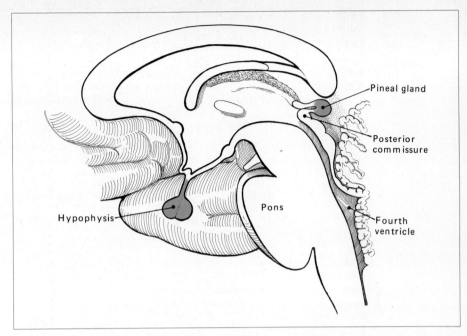

Figure 31-21. Pineal gland. Sagittal section through the diencephalon and brainstem.

pounds which have recently become available. These and other facts have, until the past decade or two, left the pineal gland relatively unnoted. Even today, few textbooks of physiology give it much attention. Fortunately, a growing band of dedicated and distinguished scientists have persevered and established the pineal gland as an endocrine organ and, in all probability, an extremely important one.

Although the gland lies in close association with the brain, it is exclusively innervated by the peripheral autonomic nervous system. These sympathetic postganglionic fibers originate in the superior cervical ganglia. In other words, as already mentioned, the pineal is an end organ of the peripheral sympathetic system. The neurotransmitter, most probably norepinephrine, diffuses through the basement membrane lining the pineal parenchyma to stimulate the so-called pinealocytes. The pinealocytes extrude their products into the blood circulating in the capillaries. The hormones may directly or indirectly get into the cerebrospinal fluid as well.

Pineal Gland Hormones
Whether there is one hormone or more associated with the pineal is still being debated. Several substances, all indoles, can be extracted from the gland. These include melatonin, serotonin, and 5-hydroxyindole. In addition, other biogenic amines, norepinephrine, dopamine, and histamine are found. Melatonin, however, is considered to be the major, if not the only, pineal hormone.

Figure 31-22 illustrates the steps in the biosynthesis of melatonin. Each reaction has, as usual, a specific enzyme. The final step, that is, the conversion of acetylserotonin to melatonin, involves the enzyme hydroxyindole-O-methyltransferase (HIOMT), found

750

THE
ENDOCRINE
AND
REPRODUCTIVE
SYSTEMS

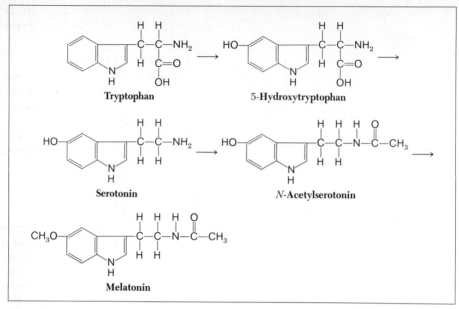

Figure 31-22. Biosynthesis of melatonin.

only in the pineal gland and considered to be an important key to the function of the organ.

Functions of Melatonin

Current thinking is that the site of action of the melatonin is the brain itself. Melatonin becomes concentrated within the midbrain, the hypothalamus, and the telencephalon.

The growing body of evidence points to the conclusion that melatonin acts on the brain by influencing the synthesis or release of one or more neurotransmitters. In other words, it is thought to modulate the secretion of the releasing factors of the hypothalamus. If this is so, then we have still another factor to contend with in the growing chain of hormones that are interrelated.

Effect on the Gonads Starting in the early 1950s reports began to be published which showed that the pineal gland exerts an inhibitory influence on the gonads. The assumption was then made that melatonin inhibits the gonadotropins. This has now been amply confirmed. Recent data show that melatonin specifically inhibits the secretion of LH, while another substance from the pineal gland, methoxytryptophol, reduces the output of FSH. When these substances are introduced into the cerebrospinal fluid, the inhibition is exercised via the hypothalamus rather than directly on the hypophysis. For this reason, as well as many others, pinealologists are most eager to establish whether or not the pineal secretes into the cerebrospinal fluid. Whatever the mechanism, the fact that pineal secretion blocks ovulation has important implications.

Effect on the Adrenal Cortex Adrenal ascorbic acid concentrations are reduced following pinealectomy, and the animals secrete an increased amount of corticosterone. These results suggest that the pineal normally exerts a tonic inhibitory influence on the secretion of

ACTH. This field is even more controversial than is the role of the pineal in gonadal activity, but the present thinking is that principles synthesized in the pineal gland inhibit the secretion of ACTH. Whether they influence any of the other hypophyseal secretions remains to be established.

Effect on Temperature Regulation Pinealectomy abolishes the normal daily oscillations of body temperature. These oscillations are tied to environmental light and darkness. Whether it is melatonin or some other pineal product that influences the hypothalamic temperature centers has not been established.

Other reported effects of the pineal relate to sleep and behavioral patterns. Overall, the pineal gland may be looked upon as a neuroendocrine transducer in that it converts a neural input to an endocrine output.

Regulation of the Pineal Gland

The function of the pineal gland is strikingly controlled by environmental lighting. Light influences the synthesis of melatonin. Animals kept in constant light end up with about one-half the enzyme activity (HIOMT) of those in continuous darkness. The effect of lighting on pineal HIOMT has been noted in immature rats, as well as in oophorectomized and hypophysectomized animals. These observations indicate that the responses of HIOMT to environmental lighting do not require the presence of the gonads or hypophysis.

Information about environmental lighting reaches the pineal via the retina and the inferior accessory optic tract, which then leaves the primary optic projections just behind the optic chiasma to run through the medial forebrain bundle and terminate in the medial terminal nucleus. The light messages reach the preganglionic sympathetic fibers of the spinal cord by a yet unknown pathway to synapse with the superior cervical ganglia. The sympathetic fibers that innervate the pineal liberate norepinephrine, which is thought to have at least two actions: (1) to stimulate adenyl cyclase activity and (2) to enhance tryptophan transport into the cell. Tryptophan is the primary substance for melatonin synthesis (Figure 31-22). Adenyl cyclase catalyzes formation of cyclic AMP that is thought to hasten the conversion of serotonin to acetyl serotonin. The overall effect of light, then, is to increase the synthesis of melatonin. The environmental pattern of day and night, light and darkness, is reflected in the output of melatonin.

Experiments with blinded animals indicate that there must also be an extraretinal pathway for light to influence the pineal. The responsible extraretinal receptor appears to be the Harderian gland, which lies behind and around the vertebrate eye. Its function is unknown, but it contains large amounts of protoporphyrin which fluctuate under different lighting conditions. Removal of this gland in blinded rats completely destroys the pineal response to light.

Relation of Pineal Activity to Biological Rhythms

Many of the circadian rhythms are regulated by environmental light and darkness. The influence of light on the pineal may not be responsible for all observed biological rhythms, but it certainly is involved in some of them.

One important rhythm that may well be modulated by the pineal is the menstrual cycle. At the very least, another dimension has been added to the already complex interrelations between hypothalamic, hypophyseal, and gonadal secretions.

TABLE 31-3. THE MAJOR HORMONES

HORMONE	ORIGIN	CHEMISTRY	CONTROL	FUNCTION
Adrenocorticotropin	Anterior hypophysis	Polypeptide	CRF	Adrenocortical hormone synthesis and secretion
Follicle-stimulating	Anterior hypophysis	Glycoprotein	FSH-RF	Ovum and sperm production
Growth (GH); somatotropin (STH)*	Anterior hypophysis	Protein	Blood sugar, hypothalamus	Stimulates growth, elevates blood sugar, increases protein synthesis
Luteinizing (LH); interstitial cell–stimulating (ICSH)*	Anterior hypophysis	Glycoprotein	LRF	Follicular development, ovulation, sex hormone secretion
Prolactin	Anterior hypophysis	Protein	PIF, PRIF	Milk production
Thyrotropin; thyroid-stimulating (TSH)*	Anterior hypophysis	Glycoprotein	Thyroxine, TRF	Thyroid hormone synthesis and secretion
Melanocyte-stimulating (MSH); intermedin*	Pars intermedia	Polypeptide	?	Skin pigmentation
Antidiuretic (ADH); vasopressin*	Hypothalamic supraoptic nuclei	Polypeptide	CNS	Decreases urine formation
Oxytocin	Hypothalamic paraventricular nuclei	Polypeptide	CNS	Uterine contraction, milk ejection
Thyrocalcitonin, calcitonin*	Thyroid	Polypeptide	Serum Ca^{++}	Lowers serum Ca^{++}
Thyroxine	Thyroid	Amino acid	TSH	Increases metabolic rate, required for normal growth
Parathyroid (PTH)	Parathyroids	Polypeptide	Serum Ca^{++}	Raises serum Ca^{++}, lowers serum PO_4
Aldosterone	Adrenal cortex	Steroid	Angiotensin, Na, K, ACTH	Increases blood Na^+, decreases blood K^+, increases body fluid
Corticosterone	Adrenal cortex	Steroid	ACTH	Elevates blood sugar, antistress, antiinflammatory, antiallergy
Hydrocortisone	Adrenal cortex	Steroid	ACTH	Elevates blood sugar, antistress, antiinflammatory, antiallergy
Epinephrine	Adrenal medulla	Catecholamine	CNS	Increases cardiac output, raises blood sugar, elevates BMR
Norepinephrine	Adrenal medulla, nerve endings	Catecholamine	CNS	Vasoconstriction
Glucagon	Pancreas	Polypeptide	Blood glucose	Decreases liver glycogen, increases blood sugar
Insulin	Pancreas	Polypeptide	Blood glucose	Lowers blood sugar, increases liver and muscle glycogen
Estrogen	Ovary	Steroid	FSH, LH	Development of female reproductive organs and secondary sex characteristics
Progesterone	Ovary	Steroid	LH	Vascularization of uterine endometrium, development of breast alveoli
Testosterone	Testes	Steroid	ICSH	Development of male reproductive organs and secondary sexual characteristics
Melatonin	Pineal	Indole	Light	Inhibits hypophyseal function
Erythropoietin	Kidney	Glycoprotein	Blood oxygen	Stimulates erythropoiesis

*Alternative name.

THYMUS

The thymus gland is located in the anterior mediastinum. Its anatomy was described in Chapter 17. It is relatively large at birth, reaches its greatest absolute size at about 6 years of age, and thereafter diminishes in size. The thymus contains lymphocytic-like cells termed thymocytes, and therefore a prevalent hypothesis is that the gland plays a role in immune reactions during childhood.

The thymus secretes a hormone termed thymosin, which stimulates lymphoid organs to form lymphocytes that function as antibodies. If the thymus is removed in young animals, they do not reject tissue transplants. In a variety of disorders such as rheumatoid arthritis, hemolytic anemia, and myasthenia gravis, the thymus is found to be greatly enlarged. In these cases, it is thought that the thymus secretes excessive thymosin which causes a reaction akin to rejection of normal tissue within the body.

QUESTIONS AND PROBLEMS

1 On what basis is a substance classified as a hormone, neurohumor, or exocrine substance?

2 Differentiate between endocrine and exocrine glands. Give an example of each.

3 Explain why the rate of an endocrine gland's response differs from the rate of response characteristic of the nervous system.

4 Describe the negative-feedback system responsible for maintaining the blood level of adrenocortical hormones.

5 What is the relation of suckling to oxytocin?

6 Why is thyroxine sometimes prescribed for obese patients?

7 How do the glucocorticoids elevate the blood sugar level?

8 Compare diabetes insipidus and diabetes mellitus in terms of pathology and character of urinary output.

9 How are blood pressure and viscosity affected by aldosterone?

10 What is the relation of parathormone to tetany and to serum calcium?

11 Which part of the pancreas secretes hormones, and what are the hormones and their functions?

12 If the islet cells of the pancreas are nonfunctioning, what sequence of events follows if the individual is deprived of insulin?

13 What organs in both male and female are capable of producing estrogens?

14 List the four ovarian hormones and describe the function(s) of each.

15 What are two causes of amenorrhea?

16 What effect does cryptorchism have on sperm production?

17 Explain why the thymus gland in an individual with rheumatoid arthritis is found to be greatly enlarged.

32 ANATOMY OF THE REPRODUCTIVE SYSTEM

The generative or reproductive organs function in procreation. They consist in the male of the penis, testes, ductus deferens, seminal vesicles, and prostate, and in the female of the vagina, uterus, uterine tubes, and ovaries. Additional structures in the female include the labia majora, labia minora, and the clitoris.

The reproductive organs are located in the pelvic cavity and the perineum. The pelvic cavity contains the uterus, uterine tubes, and ovaries of the female, and in the male the seminal vesicles, prostate, and a portion of the ductus deferens. The pelvic cavity lies internal to the innominate bones, and its floor is a curving musculofascial sheet, the pelvic diaphragm. Muscles contributing to the formation of the floor of the pelvic cavity are described in Chapter 6.

The serous peritoneal cavity extends into the pelvis, and therefore the reproductive organs as well as the rectum and urinary bladder have partial peritoneal coverings and visceral ligaments associated with them.

The external genitalia are essentially attached to a musculomembranous partition inferior to the pelvic cavity, known as the urogenital diaphragm, which passes between the ischiopubic rami. The muscular components of the urogenital diaphragm are described on page 158. The vagina in the female and the urethra in both sexes pierce the urogenital diaphragm as they pass from the pelvic cavity to open to the outside of the body.

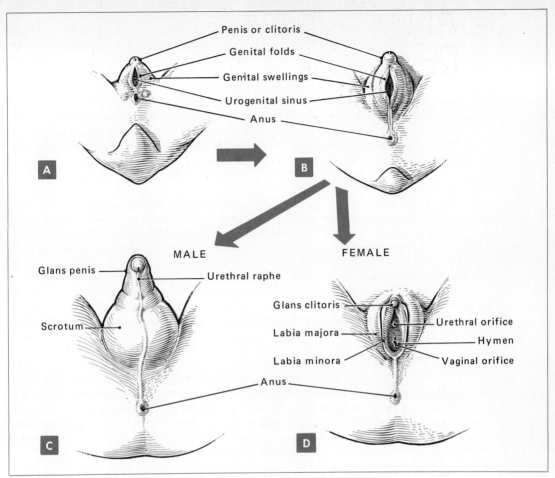

Figure 32-1. Embryonic development of external genitalia. **A, B.** Early undifferentiated stages. **C, D.** Differentiation into male and female genitalia.

Embryonic Development

Although sex is determined at fertilization, the sex of an early embryo can be learned only by chromosomal analysis. Six-week-old embryos of both sexes have a similar prominent external genital tubercle, while internally they both possess mesonephric and paramesonephric ducts (Figure 32-1). Both male and female embryos have genital ridges still undifferentiated into ovaries or testes. By the seventh week the genital tubercle in both sexes has assumed a cylindrical shape and is called the phallus. It develops a bulblike swelling, the glans, on its tip. Two bilateral ridges, the labioscrotal swellings, appear at the root of the phallus, and the rudiments of the external genitalia are now present on embryos of both sexes.

Male Differentiation

During the eighth week of development, cords of cells originating from the germinal epithelium covering the genital ridges penetrate deeply into the primitive gonad and grow

756

THE
ENDOCRINE
AND
REPRODUCTIVE
SYSTEMS

toward the mesonephric duct. These become the testis cords and will develop into seminiferous tubules of the adult testes. Tubules of the mesonephric kidney join the testis cords which lead into the mesonephric duct. From this duct the epididymus, ductus deferens, and the seminal vesicles will develop. Thus the discarded excretory duct of the mesonephros is salvaged to become the functional excretory passageway of the male genital system. The paramesonephric (Müllerian) duct, except for a few unimportant vestigial structures, disappears in the male.

The phallus elongates as the penis and encloses a ventral depression, the urethral groove, to create an enclosed tube, the penile urethra. The two swellings at the base of the phallus fuse in the midline to form the scrotum, the future external receptacle for the testes.

Female Differentiation

The female embryonic sex gland differs from the male in that the proliferating sex cords are less prominent and the germ cells are larger and arranged in clusters. Cells near the center of a cluster continue to enlarge as primordial germ cells, each of which will develop into an oogonium, and the remaining cells form the primary ovarian follicle.

The paired mesonephric ducts degenerate, but the pair of paramesonephric (Müllerian) ducts develop cranially into the uterine tubes and fuse caudally to form the uterus and vagina. The phallus undergoes only slight development to become the clitoris. The urethral groove fails to be enclosed in the phallus but remains patent to open into the vestibule. The labial swellings do not fuse but remain in their original position to become the labia majora.

Descent of the Gonad

The testes and ovaries occupy entirely different positions in the adult from those in the embryo. Initially they appear as elongated bodies extending caudally from the diaphragm. By the end of the tenth week the accelerated growth of the cephalic part of the body over the caudal part results in a relative downward shift of the more slowly growing gonad toward the lower abdominal or pelvic cavity. In the female the ovaries remain approximately in this position, but the male gonads descend further.

During the third week of gestation the testes move caudally from their position high in the abdomen toward saclike depressions of the anterior abdominal wall on each side of the midline. Each depression then evaginates as the processus vaginalis and carries with it the components of the abdominal wall into the developing scrotum. Further growth of the embryo results in a lateral elongation of these evaginations to form the definitive inguinal canals. The testes now move along the same pathway as the processus vaginalis, normally reaching the scrotum during the eighth month of development.

The exact cause of descent is not known, but hypophyseal hormones and testosterone initiate the process, and a ligament of the testis, the gubernaculum, probably plays an important role. The gubernaculum testis is a ligamentous cord that extends from the caudal extremity of the testis through the inguinal canal to attach to the floor of the scrotum. The growth of this structure fails to keep pace with the rapid growth of the rest of the body, and it may in fact shorten, pulling the testis into the scrotum as the gubernaculum atrophies and finally disappears.

MALE REPRODUCTIVE ORGANS
Scrotum and Testis

The scrotum is a pouch of skin that contains the two testes and their spermatic cords. Inside the scrotum each testis is enveloped by a double-walled serous sac derived from the processus vaginalis, an evagination of the abdominal wall during embryonic development. The testes produce sperm and the hormone testosterone.

The testis is a flattened, oval body surrounded by a connective tissue capsule, the tunica albuginea. Inward extensions from the tunica form septa which divide the testis into 250 to 400 lobules. Within each lobule are one to three tightly packed, convoluted tubules, the seminiferous tubules (Figure 32-2). The tubules contain sperm cells in various stages of development, arranged in several layers. The cells adjacent to the basement membrane are spermatogonia and are medium in size and round. The cells of the next layer toward the lumen of the tubule are primary spermatocytes, derived from spermatogonia. These are the largest sperm cells and undergo reduction cell division (meiosis) to secondary spermatocytes with a haploid number of chromosomes (see Chapter 33). Bordering on the lumen of the tubule are spermatids, daughter cells of the secondary spermatocytes. Each spermatid undergoes metamorphosis into a spermatozoon. Sustentacular, or Sertoli, cells are embedded between the various germ cells. They nourish the cytoplasm-poor spermatozoa.

Figure 32-2. A. Epididymis and testis. Testis is shown in sagittal section. **B.** A single lobule of the testis.

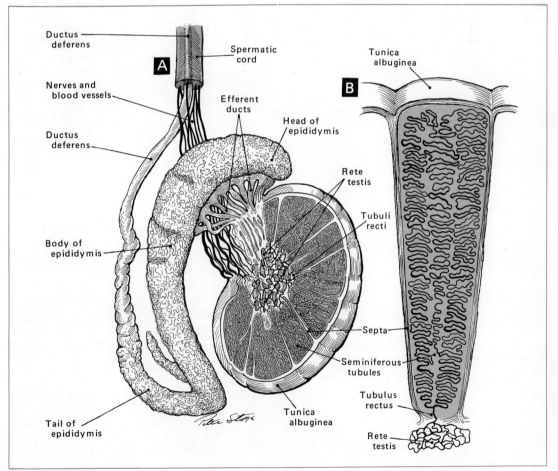

758

THE
ENDOCRINE
AND
REPRODUCTIVE
SYSTEMS

Between the seminiferous tubules are clusters of epithelioid cells, termed interstitial cells, which secrete testosterone (Figure 32-3).

The seminiferous tubules are drained by short, straight vessels, tubuli recti, into a network of anastomosing tubules known as the rete testis located at the posterior upper margin of the testis.

Male Genital Duct System

The male genital duct system originates in the seminiferous tubules and terminates in the urethra. Sperm from each testis move from the seminiferous tubules through the tubuli recti and into the rete testis. The tubuli recti and the rete testis are lined with epithelium. Some epithelial cells of the rete possess a central flagellum, which may help move the sperm along. From the rete testis the sperm move into efferent ducts (ductuli efferentes), which are 10 to 15 long spiral tubes lined with ciliated epithelium. The efferent ducts convey the sperm out of the testis and into the ductus epididymis, a comma-shaped body

Figure 32-3. Cross section through seminiferous tubules. Mature sperm are at the center of a tubule. Interstitial cells can be seen between tubules (× 265).

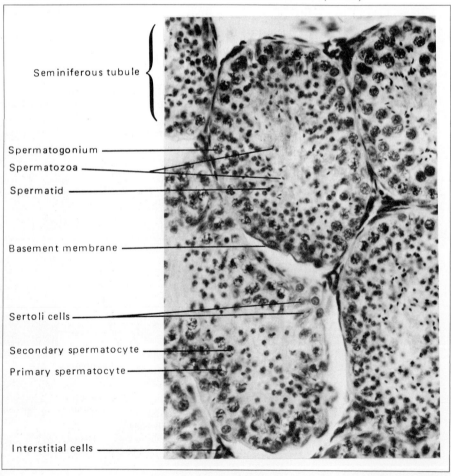

posterior to the testis (Figure 32-2). The epididymis is a tortuous tube which if uncoiled would be 4 to 5 m in length. It can be divided into a head, body, and tail. The epididymis serves as a reservoir for the sperm. It is lined with pseudostratified columnar cells. The walls contain smooth muscle which during ejaculation propels the sperm toward the urethra.

The ductus epididymis becomes less coiled and its wall thickens considerably toward its tail, where it becomes the ductus deferens. The ductus deferens has a heavy circular layer of smooth muscle which moves sperm toward the urethra by peristaltic action. Its lumen is lined with pseudostratified columnar epithelium. The ductus deferens penetrates the prostate gland, joins the duct of the seminal vesicle (see ahead), and the tube, now called the ejaculatory duct, empties into the urethra. The urethra from this point to its termination is a conduit for both semen and urine. Semen consists not only of sperm but of fluid contributed by the accessory glands, namely, the seminal vesicles, prostate gland, and bulbourethral glands. The function of the fluid portion is discussed in Chapter 33.

In summary, the testis, ductus epididymis, ductus deferens, and ejaculatory duct are paired organs. The urethra is a single organ.

Spermatic Cord

As the testis moves from the posterior abdominal wall to the scrotum during embryonic development, it draws the ductus deferens, arteries, veins, and nerves along with it. They traverse the inguinal canal in the lower anterior abdominal wall and form the spermatic cord (Figure 32-4). The cord consists of (1) the internal spermatic artery, which is surrounded by a meshwork of autonomic nerves, (2) veins draining the testis, (3) lymphatics draining the testis, and (4) the ductus deferens. All these are enclosed by the cremaster muscle and by layers of fascia, all of which issue from the abdominal wall.

Accessory Male Glands

Seminal Vesicles   The seminal vesicles are two highly coiled, blind-ending saccular glands which are attached to the posterior aspect of the urinary bladder (Figure 32-5). The ducts of the seminal vesicles penetrate the prostate and merge with the ductus deferens to form the ejaculatory duct which terminates in the urethra.

The seminal vesicles are lined with a highly convoluted mucous membrane of either simple or pseudostratified columnar epithelium which secretes a viscous fluid.

Prostate Gland   The prostate gland, about the size of a walnut, embraces the upper portion of the urethra (Figure 32-5). Surrounding the gland is a musculofibrous capsule, extensions from which divide the gland into lobules. Each lobule is lined with cuboidal or columnar epithelium and possesses a duct to convey the thin, milky secretion of the epithelium into the urethra.

The prostate is incompletely divided into two lateral lobes and a median lobe. The median lobe may hypertrophy, especially late in life, and so obstruct urinary flow that surgical removal of the gland is necessary.

Bulbourethral Glands   Bulbourethral glands, or Cowper's glands, are two pea-sized structures one of which is located on each side of the membranous portion of the urethra inferior to the prostate gland (Figure 32-6). They are tubuloalveolar glands and secrete a mucoid substance. A duct from each gland conveys this substance to the urethra and adds it to the semen.

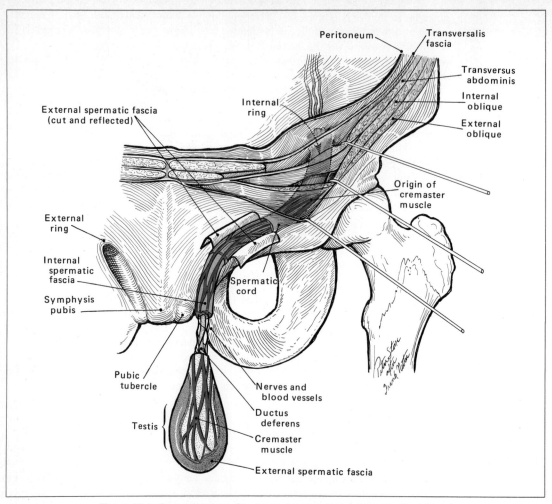

Figure 32-4. Spermatic cord. The internal and external rings are the internal and external apertures, respectively, of the inguinal canal. Spermatic fasciae and the cremaster muscle have their origin in the abdominal wall.

Penis

The penis, the male organ of copulation, consists of a body or shaft which terminates distally in an expanded portion, the glans (Figure 32-6). Skin loosely attached to the body of the penis covers the glans as a free flap known as the prepuce, or foreskin. Circumcision is the surgical removal of the prepuce and is performed shortly after birth.

The shaft of the penis is composed of three cylindrical bodies, each surrounded by a capsule of dense, unyielding connective tissue. Around the three bodies is another sheath of connective tissue and, finally, a covering of skin. The tissue of the three bodies is spongy. With erotic stimulation the spaces of this tissue become filled with blood, causing the penis to become erect (see Chapter 33). The dorsal bodies are called corpora cavernosa penis, and each corpus has an artery coursing through its center. The ventral body is

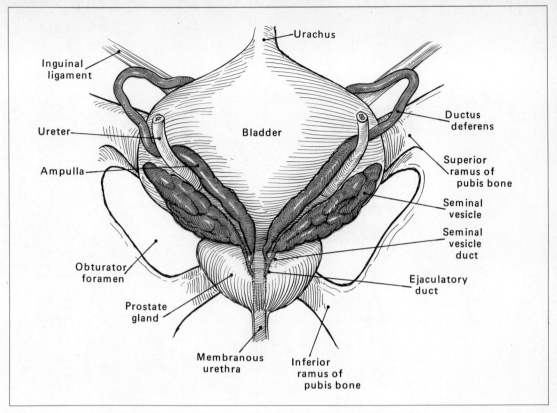

Figure 32-5. Seminal vesicles. Posterior view. The vesicles empty into the prostatic urethra.

the corpus spongiosum. In addition to its cavernous tissue the corpus spongiosum contains the penile portion of the urethra. Distally it extends beyond the corpora cavernosa to become the glans. Within the glans the urethra widens (the fossa navicularis) and terminates in a longitudinal slit at the external urethral meatus.

Occasionally the corpus spongiosum fails to completely surround the urethra. If this occurs, the external opening of the urethra may be present along the ventral aspect of the shaft (hypospadias) or along the dorsum of the penis (epispadias).

Proximally, the penis extends into the perineal region. The corpora cavernosa diverge to become the crura of the penis. The corpus spongiosum becomes the bulb of the penis. All three bodies attach to the inferior layer of the urogenital diaphragm. The crura attach laterally along the ischiopubic rami. Each of these bodies is covered by a thin sheet of muscle (see page 159).

FEMALE REPRODUCTIVE ORGANS

The ovaries are paired, almond-shaped glands lying on the wall of the pelvic cavity, one on each side of the uterus. They produce ova, progesterone, and estrogen.

The uterus, uterine tubes, and ovaries are covered and supported by a number of ligaments (see Figure 32-8). The largest of these is the broad ligament, a sheet of

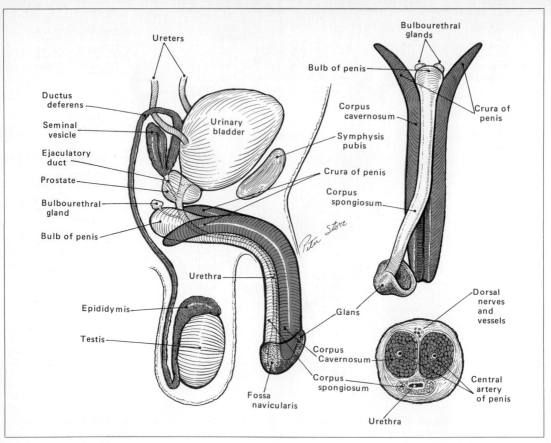

Figure 32-6. Diagrammatic sketch of the male reproductive system. Upper right: a dissected penis. Lower right: cross section of a penis.

peritoneum that drapes these organs anteriorly and posteriorly and attaches to the pelvic wall. The modified portion of this covering which encloses the ovaries continues to the lateral wall of the pelvis as the suspensory ligament and supplies a pathway for vessels and nerves to reach the ovaries. In addition, a fibrous cord, the ovarian ligament, extends from the ovary to the lateral margin of the uterus.

A cross section of the ovary (Figure 32-7) reveals an outer cortex and an inner medulla, with a poorly defined boundary between them. The peritoneal covering of the ovary is modified on the inner surface as a thin germinal epithelium. Beneath this is a connective tissue layer, the tunica albuginea. Throughout the cortex below these two layers are many ovarian follicles. In the medulla of the ovary are nerves, lymph vessels, and large, coiled blood vessels, along with a less compact connective tissue.

Follicular Growth   Follicles are abundant in the ovary. The smallest are just beneath the tunica albuginea. Each is composed of an ovum around which is a layer of flat follicular cells. At menarche, that is, a woman's first menstrual cycle, one of these primary follicles from the pair of ovaries grows to maturity each month. Menarche occurs most often

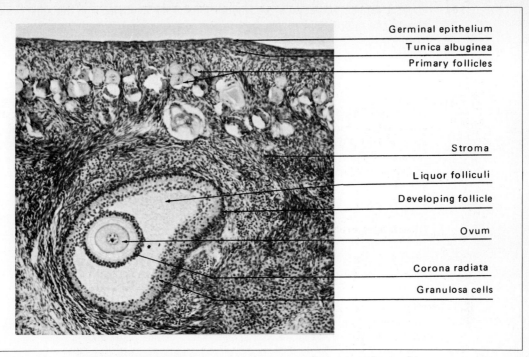

Germinal epithelium
Tunica albuginea
Primary follicles

Stroma

Liquor folliculi

Developing follicle

Ovum

Corona radiata

Granulosa cells

Figure 32-7. Section through the cortex of an ovary showing primary and maturing follicles (× 88).

between the age of 12 and 15. Menstrual cycles usually continue until between the age of 45 and 50, when a woman's reproductive life ends. Before menarche each ovary contains over 200,000 primary follicles. Yet only about 400 follicles in the two together reach maturity during her lifetime. Many thousands of follicles become active but then undergo regression. They are termed atretic follicles. For this reason a cross section of an ovary usually discloses follicles in various stages of development and regression.

When the follicle begins to develop, a clear membrane, the zona pellucida, forms around the ovum. Follicular cells increase in number and size and arrange themselves in layers around the ovum. A fluid collects within the follicle. This liquor folliculi eventually forces the ovum and its follicle cells to the edge of the follicle, where they form a mound of cells termed the cumulus oophorus. The remaining cells, arranged on the periphery of the follicle, are the stratum granulosum. During this period the ovum has increased in diameter from about 35 to 150 μ. The follicle has increased in diameter from 120 to 6,000 μ (6 mm). The ovum (primary oocyte) meanwhile undergoes reduction cell division. The daughter cells are a secondary oocyte and a polar body. Each has 23 chromosomes (see Chapter 33).

After about 10 to 14 days of growth the follicle is mature and ready for ovulation. A mature follicle is termed the graafian follicle. Ovulation consists of the rupture of the graafian follicle and release of the secondary oocyte into the peritoneal cavity. Because of peristaltic action of the uterine tube the ovum is drawn into the tube. There the cilia of cells lining the tube sweep it toward the uterus. The menstrual cycle is, on the average, 28 to 30 days long. Ovulation occurs about midway in this period.

764

THE
ENDOCRINE
AND
REPRODUCTIVE
SYSTEMS

The role of hormones of the anterior lobe of the hypophysis in initiating follicular growth and in promoting ovulation has been discussed in Chapter 31.

Corpus Luteum   After ovulation, blood and serous fluid fill the collapsed follicle and a clot forms. The clot, the corpus hemorrhagicum, is surrounded by the remaining cells of the stratum granulosum, which invade and gradually reabsorb it. The granulosal cells enlarge and arrange themselves into cords of cells which assist in transforming the collapsed follicle into the corpus luteum (yellow body). If pregnancy does not occur, the corpus luteum reaches its maximum development in about 2 weeks and degenerates into a white fibrous scar, the corpus albicans (white body). If pregnancy occurs, the corpus luteum continues to grow and reaches its maximum development in the second month of gestation, after which it recedes until childbirth, when it rapidly becomes a corpus albicans.

The corpus luteum is an endocrine gland within the ovary. One of its hormones, progesterone, maintains the mucosal lining of the uterus during pregnancy. It also stimulates development of the mammary glands.

Uterine (Fallopian) Tubes

The two uterine (fallopian) tubes lie between the layers of the broad ligament. From attachments at the uterus they extend laterally to the wall of the pelvis, then turn down-

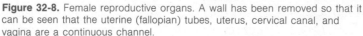

Figure 32-8. Female reproductive organs. A wall has been removed so that it can be seen that the uterine (fallopian) tubes, uterus, cervical canal, and vagina are a continuous channel.

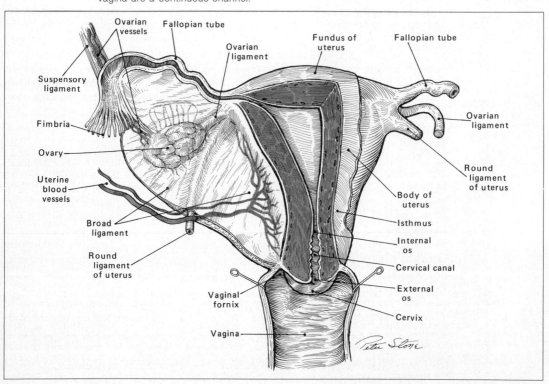

ward and terminate near the ovaries (Figure 32-8). The distal open end of each tube has a fringe of fingerlike processes, the fimbriae, the longest of which may touch the ovary. The uterine tubes serve to conduct the ovum to the uterus.

The mucosa of the tube is arranged in longitudinal folds covered by simple columnar epithelium, some cells of which are ciliated. The muscular tunic becomes progressively heavier toward the uterus and consists of a well-developed inner circular layer and a thinner, outer layer, arranged longitudinally. Contractions by these layers of muscle plus movement of the cilia moves the ova through the tubes.

Unlike the male peritoneal cavity, that of the female opens to the outside by way of the uterine tubes, uterus, and vagina. There is therefore a portal of entry for infectious agents. Peritonitis from this source is a common complication of abortion performed without sterile techniques.

Fertilization of the ovum usually occurs in the upper third of the uterine tube, and implantation in the uterus. If the fertilized egg instead implants in the tube, therapeutic abortion must be performed, as the tube is not able to accommodate the expansion of the developing embryo. On rare occasions the sperm traverses the entire tube and fertilizes an ovum which then drops into the peritoneal cavity. Implantation may then occur in the peritoneum. Because the blood supply there is inadequate, such embryos most often die at an early stage. Implantation outside the uterus constitutes ectopic pregnancy.

Uterus

The uterus (Figure 32-8) is a pear-shaped organ about 7.5 cm long and 5 cm wide at its broadest point. It lies behind the urinary bladder and in front of the rectum in the lowermost portion of the pelvic cavity. Implantation of the fertilized egg, or zygote, and development of the embryo occur in the uterus. The lowermost portion of the uterus, the cervix, projects slightly into the vagina. Above it is a slight constriction, the isthmus. The main portion of the uterus, the body, is above the isthmus. The fundus, the uppermost segment, lies above the level of attachment of the uterine tubes. Normally the axis of the uterus is about 90 degrees from that of the vagina. Deviation from this angle, referred to as a "tipped" uterus, may create tension on the ligaments of the uterus and cause discomfort or even severe pain.

Attachments of the Uterus As the peritoneum extends into the pelvic cavity, it covers the superior and posterior surface of the urinary bladder and, from the floor of the pelvic cavity, reflects onto the anterior surface of the uterus to form the uterovesical pouch. Continuing over the uppermost portion of the uterus, it covers the posterior surface of this organ and reflects onto the rectum to form the rectouterine pouch. Extension of the peritoneal lining of the pelvic cavity onto the vagina is of clinical importance. An excess of fluid in the peritoneal cavity (ascites) may be relatively easily drained at this site. A trochar (large needle) inserted in a superior direction through the posterior fornix of the vagina immediately enters the peritoneal cavity where the fluid collects, and it can be aspirated.

At the lateral margins of the uterus the peritoneal coverings fuse and extend to the floor and lateral walls of the pelvic cavity as the broad ligament of the uterus. Vessels and nerves pass between the peritoneal layers forming the broad ligament to reach the uterus and uterine tubes. As we have seen, the uterine tubes and ovaries are also located between the layers of the broad ligament (Figure 32-8).

Connective tissue condensations between the peritoneum and musculature of the floor of the pelvis attach to the cervix of the uterus and provide the major support of this

766
THE
ENDOCRINE
AND
REPRODUCTIVE
SYSTEMS

organ. During parturition, any loss of this support may result in downward displacement (prolapse) of the uterus, and, if severe, the organ may actually emerge at the vulva.

Additional uterine support is provided by the round ligament, a fibrous cord. From the midportion of the lateral margin of the uterus this structure extends between the layers of the broad ligament to the midportion of the inguinal ligament. Here it follows a course similar to the spermatic cord in the male. It enters the internal inguinal ring, traverses the inguinal canal, and emerges through the external inguinal ring. The round ligament thereupon spreads out to attach diffusely throughout the connective tissue of the mons pubis and labium majus.

Histology The uterus has the typical tunics of a hollow organ, but these layers have special names (Figure 32-9). The mucosal lining of the uterus is the endometrium and consists of simple columnar cells, some of which are ciliated. Tubular, coiled uterine glands penetrate deeply into the underlying loose connective tissue (tunica propria). The heavy muscular tunic is termed myometrium and is arranged in three layers, an inner longitudinal layer, a thick, oblique middle layer, and a thin outer longitudinal layer. Externally a typical fibrous layer, or serosa, covered by peritoneum, surrounds the fundus and body of the uterus.

Endometrial Changes The endometrium undergoes changes associated with the menstrual cycle, which is under the control of hormones of the pituitary and the ovary. Histological changes of the endometrium during a menstrual cycle of about 28 days are separated into three stages:

1   The proliferative phase (day 5 to 15), which follows the cessation of the menstrual flow. The denuded epithelium begins to regenerate and increase in height from

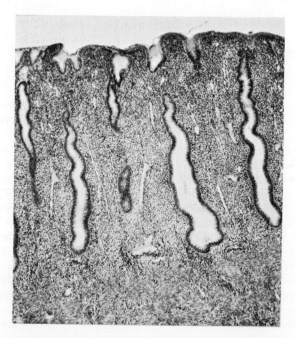

Figure 32-9. Section through the uterus in the proliferative phase. The surface of the uterine cavity (top) is covered with epithelium. Beneath is endometrium with uterine glands (elongated bodies). Below this is myometrium, which appears swirled ($\times$ 42).

cuboidal to columnar cells. The uterine glands are short and straight with wide lumens. Short, small arteries begin to invade the tunica propria. The dominant hormone controlling this phase is estrogen.

2   The secretory phase (day 16 to 28), during which the endometrium thickens markedly as the uterine glands increase greatly in size and become coiled and distended with secretions. The arteries become coiled and increase in length until they almost reach the epithelial layer. The hormone progesterone is the dominant influence during this stage.

3   The menstrual phase (day 1 to 4), which is ushered in by a sudden constriction of the coiled arteries on about day 27 or 28, causing a loss of blood to the endometrium. Ischemia is produced by the arterial constriction and is followed by necrosis, or death, of the endometrial tissue. Menstrual flow is initiated by a sudden relaxing of the coiled arteries, which sends blood rushing into the capillary beds damaged by ischemia. The weakened mucosa ruptures, exposing open glands, arteries, and veins as the endometrium is sloughed off. The resulting discharge carries with it glandular secretions, blood, blood clots, desquamated epithelial cells, and bits of mucosal tissue. The epithelial membrane and parts of the uterine glands are lost, leaving a raw surface. As the initial heavy flow subsides, the epithelium from the ruptured glands regenerates over the raw surfaces to again envelop the area with cells, and seals off the discharge. Coincident with this proliferation, stromal cells of the endometrium multiply to complete the regeneration of the endometrium, and the menstrual cycle is completed.

Vagina

The vagina is a musculomembranous tube that extends internally about 10 cm, where it attaches to the uterus. The lining of the vagina is composed of stratified squamous epithelium. This lining is thrown into numerous transverse folds, or rugae. There are no glands in the vagina. A highly vascular fold, the hymen, surrounds and partially blocks the entrance. Rupture of the hymen, which occurs, for instance, during the first experience of sexual intercourse, causes bleeding. Proximally the cervix of the uterus projects into the vagina. The vagina serves to receive the sperm and to give passage to the fetus in childbirth.

External Genital Organs

The external genital organs consist of the labia majora, labia minora, vestibular glands, and clitoris (Figure 32-10). The structures are collectively termed the vulva.

Over the pubic symphysis is a mound of fatty tissue, the mons pubis, covered by coarse hair. The mons pubis blends posteriorly with two hair-covered folds of skin, the labia majora, which flank the vaginal opening, urethra, and clitoris and join posteriorly. During delivery of the fetus an incision may be made in this area to enlarge the vaginal opening. This posteriorly directed incision, an episiotomy, prevents the tearing of tissue if the vaginal opening is too small for the head of the fetus. A surgical incision is easily closed, whereas a tear may not be.

Internal to the labia majora and to the sides of the vaginal orifice are the two smaller, highly vascular folds termed the labia minora. Between them is a space called the vestibule, into which the vagina and urethra open. Deep to each labium is a mass of erectile tissue. Together the masses comprise the bulb of the vestibule, which is homologous to the bulb of the penis. The bulb of the vestibule attaches to the urogenital diaphragm.

A number of glands open into the vestibule. The most important are the greater vestibular glands (Bartholin's glands), which secrete a mucoid lubricating fluid, assisting

768

THE
ENDOCRINE
AND
REPRODUCTIVE
SYSTEMS

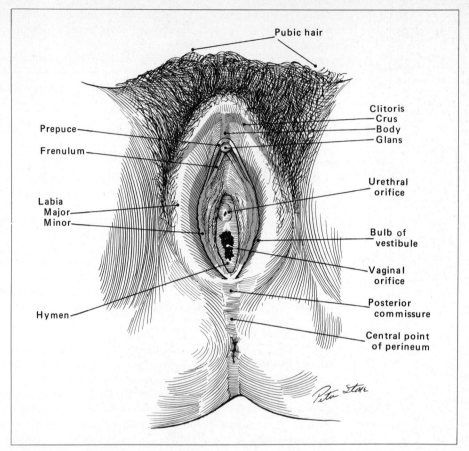

Pubic hair

Clitoris
Crus
Body
Glans

Prepuce

Frenulum

Urethral
orifice

Labia
Major
Minor

Bulb of
vestibule

Vaginal
orifice

Posterior
commissure

Hymen

Central point
of perineum

Figure 32-10. External genitalia of the female. Subcutaneous structures are in color.

the entrance of the penis in sexual intercourse. These glands are similar in structure to the bulbourethral glands of the male.

The clitoris is a small organ at the anterior extent of the vestibule. The clitoris is the homologue of the penis. It is composed of two cylindrical masses, the corpora cavernosa clitoris. Most of it is enclosed by a covering, the prepuce, formed by the joined labia minora. The portion so covered is termed the body; that exposed is the glans. The clitoris is covered by stratified epithelium containing many sensory nerve endings. It is capable of engorgement and erection upon tactile stimulation. The clitoris is important for sexual arousal of the woman. Proximally the corpora become the crura of the clitoris. The crura attach to the inner aspect of the ischiopubic rami. The clitoris does not contain the urethra as the penis does in the male. The urethra has its own orifice posterior to the clitoris.

QUESTIONS AND PROBLEMS

1 Describe the descent of the testes. If they fail to descend, what is the consequence? How can this condition be corrected?

2 Trace the course of sperm from the seminiferous tubules to the exterior of the body. Trace the course of an ovum from the graafian follicle to the site of implantation.

3 Discuss the relation of the prostate gland to the urinary bladder and the urethra. What is the primary symptom of hypertrophy of the prostate?

4 What structures traverse the inguinal canal in the male? In the female? Where does inguinal hernia occur?

5 Describe the normal position and location of the uterus. What is the structural relationship of the broad ligament, uterus, uterine tubes, and ovaries?

6 Describe the histological changes of the endometrium during the menstrual cycle.

7 How does the epithelium of the vagina differ from that of the uterus?

8 Which cells secrete estrogen, progesterone, and testosterone?

9 Differentiate between the following: (*a*) gonad and gamete, (*b*) mitosis and meiosis, (*c*) fertilization and ovulation, (*d*) menarche and menopause, (*e*) corpus luteum and corpus albicans.

33 REPRODUCTION

A very important characteristic of all living matter is the ability to reproduce itself. The primitive ameba carries out this function simply by dividing in two, with the subsequent growth and development of each half. As the phylogenetic scale is ascended, reproduction becomes increasingly more complex. In higher mammals, including man, it includes the processes of gametogenesis, sexual intercourse, fertilization and pregnancy, lactation, and neonatal development.

The capacity to take part in reproduction begins in puberty, which usually occurs between the twelfth and fifteenth years. Puberty is brought about by a marked increase in the secretion of sex hormones—testosterone in the male and estrogens and progesterone in the female.

The penis, scrotum, and testes of the male enlarge. The testes begin to produce mature reproductive cells. The voice deepens. Hair grows over the pubis, the face, and often the chest. Body fat is lost and skeletal muscles develop, creating the characteristic male form, in which hips are narrow in relation to the shoulders.

In women the most dramatic change is, of course, the menarche, the onset of menstruation, which accompanies the production of mature reproductive cells. Other changes include enlargement of the uterine tubes, uterus, and vagina. Fat is deposited in the breast and on the hips, and the breast acquires the alveoli and lobules essential for milk production.

GAMETOGENESIS

A gamete is a male or female reproductive cell, that is, either the spermatozoon or the ovum. The production of the gametes is termed gametogenesis. Involved is the important process called meiosis, briefly mentioned in Chapter 3. This process will now be considered in greater detail.

By meiosis, or reduction division, the number of chromosomes in each sex cell is reduced by one-half. If the reduction of chromosomes did not occur, the zygote would possess double the normal number of chromosomes. Since the fertilized cell divides by mitosis, all subsequent cell generations would result in an organism with a double set of chromosomes. This excess genetic material would be doubled in the next generation, and the next, etc. Such a catastrophe is avoided because of this unique type of cell division. Meiosis simply reduces the number of chromosomes by one-half, to the haploid number. Thus the union of male and female gametes restores the normal complement of chromosomes, and normal development can follow.

Meiosis occurs in two stages during spermatogenesis. By proliferation of the spermatogonia lining the seminiferous tubules of the testis, a reserve supply of large, round, germ cells called primary spermatocytes is produced (Figure 33-1). Each has the normal (diploid) number (46) of chromosomes, 23 of which are derived from the male parent and 23 from the mother. In the first cell division that follows, these 23 pairs of homologous (similar) chromosomes separate so that a chromosome of each pair migrates to opposite poles (ends) of the cell. A nuclear membrane is then formed which encloses each set of 23 chromosomes. The cytoplasm of the primary spermatocyte constricts, and two daughter cells (secondary spermatocytes) are formed, each with half the usual number of chromosomes. The second stage involves cell division by mitosis, that is, each daughter cell has the same number of chromosomes as the parent cell. The cells now undergo changes (metamorphosis) and become spermatozoa. During metamorphosis, the volume of cytoplasm is greatly decreased, and the nucleus is crowded into one end of the cell, the head region; a small intermediate neck portion appears, and the remainder of the cell is transformed into a very thin tail, or flagellum, capable of propelling the sperm through the genital passageway by its whiplike beating action.

In the development of the mature ovum (oogenesis) the sequence of events is essentially the same (Figure 33-1). Oogonia multiply by mitosis to produce primary oocytes. The latter undergo reduction division to give rise to secondary oocytes with the haploid number (23) of chromosomes. When the secondary oocyte divides, the division of the cytoplasm is unequal. Only the cell receiving the bulk of the cytoplasm becomes the mature functional ovum. The cell receiving the smaller amount of cytoplasm is the relatively tiny, nonfunctional polar body. The polar body plays no future role in reproduction. From each oogonium, therefore, only one functional ovum is formed, in contrast to four mature sperm from each spermatogonium.

Thus reduction division produces haploid gametes, each of which expresses its own individuality by having a distinctive set of chromosomes. The genetic makeup of any gamete will depend on its complement of paternal and maternal chromosomes, as well as any mixing or interchanging of chromosomes or their segments that might occur. Such combinations of genes are almost innumerable, and hence the variety of the offspring is virtually inexhaustible.

SEXUAL INTERCOURSE

Reproduction requires the union of the sperm and the ovum. In order for this conjugation to occur, viable sperm must be delivered into the vagina. Sexual intercourse accomplishes

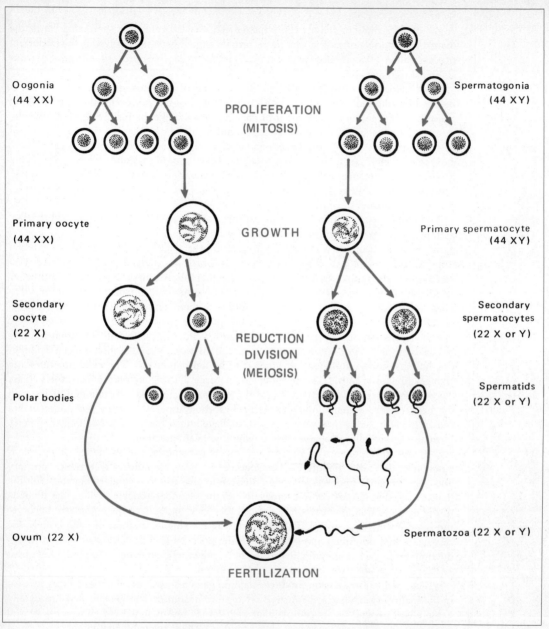

Figure 33-1. Gametogenesis in man. Oogonia and spermatogonia start with a full complement of chromosomes, that is, 44 autosomes and two sex chromosomes (oogonia are XX, spermatogonia are XY). Meiosis produces the haploid number, 22 autosomes and 1 sex chromosome, in the secondary oocyte and spermatocytes. In the fertilized ovum two haploid sets are united and the normal complement (diploid number) restored.

this purpose. However, a woman may become pregnant in the absence of sexual inter-course simply by the introduction of the male ejaculate deep into the vagina. This procedure is termed artificial insemination. This technique has advanced to the point where most attempts are successful.

The Male Role in Sexual Intercourse

Normally, to effect fertilization the penis is inserted into the vagina. Then, as a result of an autonomic reflex mechanism, the sperm are expelled from the male into the vagina.

Erection   The penis, which normally is flaccid and small, becomes firm and enlarged in the process termed an erection. The size of the erected organ varies considerably from man to man. It averages about 15 cm in length and 3 to 4 cm in diameter.

Stimuli that can reflexly elicit an erection are classified as (1) psychogenic and (2) reflexogenic. Auditory, visual, olfactory, gustatory, tactile, and imaginative stimuli are examples of those that act psychogenically. Stimulation of various parts of the body, including the genitals, causes reflexogenic erection.

Efferent neural impulses for erection arise from parasympathetic fibers in sacral cord roots S_2, S_3, and S_4. They make up the nervi erigentes. Activation of these neurons causes opening of anastomoses between arterioles and the vascular spaces in the two corpora cavernosa and the corpus spongiosum (see Figure 32-6). As a result, blood rushes into these spaces, causing them to become greatly distended. When the penis is flaccid, these spaces contain very little blood.

There are valvelike structures called polsters which contain smooth muscle located at the anastomoses between the arterioles and the vascular spaces. The polsters cause blood to be shunted away from the corpora directly into the veins when the penis is flaccid. When these valves open, the rate of arterial inflow is temporarily greater than venous outflow and the penis swells. At full erection, the rates of inflow and outflow are equal, causing the organ to remain erected and rigid.

Polsters are also present in the erectile tissue drainage veins, and they may decrease venous outflow at the same time as arterial inflow is increasing. However, if, as some writers state, venous outflow were occluded, the penis would become cool, cyanotic, and edematous. This does not normally occur. The erected penis is warm, red, and not edematous. This means that a good circulation is maintained during erection. No doubt some venous obstruction occurs when the thin-walled vascular spaces in the distended corpora are compressed against the thick fibrous membrane that surrounds the corpora cavernosa but not the corpus spongiosum. This no doubt contributes to tumescence, as evidenced by the fact that the spongiosum does not become as rigid as do the two corpora cavernosa.

Detumescence of the Penis   The return to the flaccid condition is termed detumescence. It too is the result of psychic or reflexogenic stimuli. It could occur either as the result of diminished cholinergic impulses to the polsters or of active vasoconstrictor impulses. At any rate, the hydraulics change so that more blood now leaves the corpora than enters. Detumescence generally begins slowly, after which the rate of collapse increases. This is probably due to release of the passive venous obstruction as the blood drains out of the corpora.

Inability to achieve or maintain an erection is termed impotence. Prolonged failure of an erection to subside in the absence of sexual stimulation is called priapism.

774

THE
ENDOCRINE
AND
REPRODUCTIVE
SYSTEMS

Higher-Center Control of Erection Although more than 90 percent of patients with complete spinal cord transections located above the sacral level are capable of erections, there is no doubt that in man cerebral impulses play a major role. There is ample evidence that the thought process can both elicit an erection and inhibit it. Interestingly, lesions or tumors in the anterior temporal lobe of the brain often are associated with impotence. The results of the current fad of taking large doses of vitamin E to combat impotence are probably more psychological than physiological. There is a saying that anything one thinks will improve an erection will do so.

Ejaculation Tactile stimulation of the penis not only causes it to become erect but ultimately evokes another reflex as well. Impulses propagated by sympathetic fibers bring about peristaltic contractions of the epididymis, the ductus deferens, the seminal vesicles, and the prostate gland. In addition, impulses are propagated by the pudendal nerves to the skeletal muscles at the base of the penis. The contraction of these muscles assists the expulsion of the sperm and the various secretions which make up the semen. The contractions of all these muscles and glands and the expulsion of semen constitute ejaculation. Ejaculation is usually accompanied by widespread sensations and movements. This total reaction, a response of intense excitement, is termed the orgasm. Thus, strictly speaking, an orgasm and an ejaculation are not the same.

Role of the Seminal Vesicles The seminal vesicles are misnamed in that they are not the repositories for the sperm. Actually, they are secretory glands which form a fructose-rich mucoid substance that is added to the semen during ejaculation. In addition to fructose, amino acids and ascorbic acid are present. The seminal vesicle secretion adds much of the normal volume of the semen and provides nutrient material for the sperm. The mucus probably also protects the sperm from the acid environment of the vagina.

Role of the Prostate Gland During ejaculation, the prostate gland is contracted by action of its smooth muscle capsule. As a result, a thin, alkaline, pale white fluid is added to the semen. This fluid contains calcium, citric acid, and acid phosphatase, and yet it has an alkaline pH. In all probability, the major role of the prostatic fluid is to help neutralize the vaginal secretions, which have a pH of 4 or even lower.

Sperm cannot survive long in an acid medium. Besides, their mobility appears to be a function of the pH; they are inactive at pH 5 or lower and optimally mobile at a pH of about 6.5. The ductus deferens secretes a small amount of a slightly acid fluid, and thus the sperm, while in the ductus deferens, are not mobile. But as a result of ejaculation, the prostatic fluid is added to the semen, the pH rises to at least 6.5, and the sperm become active. The alkalinity of the prostatic secretion and mucus of the seminal vesicle secretion combine to protect the sperm in the acid vagina.

The Semen Semen is the total material excreted during ejaculation. It is composed of a great number of sperm, some fluid from the prostate gland, and a greater volume from the seminal vesicles. In addition, the bulbourethral glands also contribute. The volume of semen varies from individual to individual, and also in the same male, depending upon many factors, including the elapsed time between ejaculations. The range is from about 1 ml to as much as 10 ml, the average being 3 to 4 ml. The pH is normally close to 6.5. Semen has a mucus consistency and appears milky because of the prostatic contribution.

The Female Role in Sexual Intercourse

In order for pregnancy to occur, the woman need take no active part in the sexual act. So long as the sperm are introduced into a favorable position at the proper time in the menstrual cycle, pregnancy can occur. In some animals, such as the rabbit, sexual excitement of the female is necessary to induce ovulation, but this is not true of women.

The clitoris, as stated previously, is the homologue of the corpora cavernosa penis. Like the penis, stimulation of the clitoris leads to its engorgement and erection, as well as to widespread sexual arousal in the person. Parasympathetic impulses are responsible for the engorgement. They also activate the greater vestibular glands, which, as a result, secrete a mucoid substance that assists in lubrication of the vagina during sexual intercourse.

There are rhythmic, reflex contractions at the zenith of sexual intercourse which correspond to the ejaculation in the male, but there is no actual production or expulsion of fluid. The rhythmic contractions do, however, facilitate the movement of the sperm toward the ovum.

FERTILIZATION

As the sperm penetrates the ovum, their nuclei unite, and at that instant fertilization has occurred. The resulting cell, called the zygote, now has the normal (46) number of chromosomes, half maternal and half paternal. All subsequent cell divisions result in cells with the full complement of chromosomes. Fertilization or conception usually occurs in the upper third of the uterine tube. As subsequent cell divisions in the zygote occur, the developing conceptus gradually moves through the tube into the cavity of the uterus.

Semen must contain a great number of spermatozoa in order for fertilization to take place. If the semen contains less than about 60 million spermatozoa per milliliter, fertilization rarely occurs. One wonders why this great number is required, since only one sperm actually penetrates the ovum. One answer to this question points out that the sperm, after being deposited in the vagina, move in all directions completely at random. If all the possible directions are taken into consideration and the distances involved calculated, it can be shown that purely on the basis of random walking, as the statistical phenomenon has been called, this huge number is required to assure that at least one will reach the ovum.

Another theory emphasizes the fact that the ovum is surrounded by a gel. In order for the spermatozoon to enter the ovum, the gelatinous envelope must be removed. The spermatozoa contain a proteolytic enzyme termed hyaluronidase. According to this theory, a large number of spermatozoa are required to provide adequate hyaluronidase to break down the gel and permit a spermatozoon to enter the ovum.

Multiple Fertilization

In view of the fact that there is such an abundance of spermatozoa, the number of ova fertilized at one time clearly depends on the number of ova present. In the great majority of instances, there is only one mature ovum available for fertilization; however, in some cases there are two, three, four, and in very rare cases even five. These multiple ova may represent separate and distinct germ cells that have matured simultaneously, or they may result from the division of a zygote into two or more cells. The former type of production of multiple ova accounts for fraternal multiple births, and the latter, for identical multiple births.

776

THE
ENDOCRINE
AND
REPRODUCTIVE
SYSTEMS

Time of Fertilization

Contrary to popular opinion, there is only a very short interval during the menstrual cycle when fertilization can occur. In the women with a normal menstrual cycle, ovulation takes place approximately 14 days before the onset of the menses. If the ovum is not fertilized in the uterine tubes, it continues on into the cavity of the uterus. At this time conception can still take place, but if it does not, the ovum undergoes rapid degeneration. From this time until the next ovulation, there can be no fertilization. The average woman ovulates but one time in each cycle, and it is only at this time that fertilization can occur. The length of time is somewhat extended by the fact that the spermatozoa may survive in the vagina, uterus, or uterine tubes for a day or two. Thus, if sexual relations take place just before ovulation, pregnancy may ensue. Taking all these possibilities into consideration, there is a period of about 4 days, in the middle of the menstrual cycle, during which fertilization can be successful.

The vagina is normally acid. Strangely, spermatozoa cannot survive in an acid medium. But the semen is alkaline and contains excellent buffers so that the spermatozoa are afforded some protection. Obviously, if the buffers of the semen are inadequate, the acidity of the vagina may not be neutralized, and in such a case the sperm may be killed before fertilization can occur. This is one of the causes of sterility.

Infertility

Fertilization requires the union of a normal spermatozoon and a viable ovum. Any physical condition that prevents this union will cause infertility, also referred to as sterility.

Male Infertility As already mentioned, in order for fertilization to occur, there must be an adequate sperm count. Normally each milliliter of semen contains over 100 million sperm; thus in 3 milliliters of semen there will be at least 300 million sperm. A count lower than about 60 million per ml may prove inadequate to cause fertilization, for reasons already considered.

Various diseases destroy the sperm-producing epithelium of the testes. These include mumps and typhus fever.

In some instances, the sperm are abnormal in that they have two heads, or multiple tails, or other malformations. Such sperm do not generally fertilize the ovum; thus if the great majority are abnormal, infertility may result.

Female Infertility In cases of infertile marriage where both parties have been examined, in two out of three instances, female infertility has been shown to be responsible. Female infertility is generally due either to a failure to ovulate, the ejection of an abnormal ovum, or a deformity of some component of the female sexual organs.

Contraception

Strictly speaking, contraception means prevention of fertilization of the ovum. However, the term today is used in a broader sense. For example, actual fertilization may occur, but some contraceptive procedures prevent implantation; thus pregnancy will not continue. Yet this is also referred to as contraception. Likewise, the newly fertilized and implanted ovum may be expelled by vigorous uterine contractions. Again, the method used is classified as contraceptive.

There is now a veritable armamentarium of contraceptive procedures, and yet millions of unwanted pregnancies occur. Clearly this is not a physiological problem.

Coitus Interruptus This is one of the oldest methods. It is mentioned in the Bible (Genesis 38). As the term indicates, coitus is interrupted before ejaculation so that the semen does not enter the vagina. Until recently it was the most widespread and frequently used procedure.

Condom A sheath, now most commonly made of rubber, is termed a condom. No one really knows the origin of this term, although fascinating stories abound. The condom not only prevents entrance of the semen; it also is valuable for protection against venereal disease and was first used for that purpose. The famous anatomist Fallopio is credited with first suggesting its use.

Vaginal Diaphragm About one hundred years ago, a German physician designed a diaphragm to cover the cervix. It quickly gained wide acceptance, and until more recent methods were developed, it was the method of choice among the more educated and affluent women.

Chemicals and Douches An amazing variety of substances have been introduced into the vagina either before or after intercourse to prevent conception. Very few are reliable and nearly all have disadvantages. Still, if prompt measures are taken to flush out the semen, conception can usually be prevented. Chemicals used today depend upon a foaming action to flush out the semen and a low pH to kill the sperm.

Rhythm Method The rhythm, or safe-period, method was first suggested over a hundred years ago. Unfortunately, until early in this century, knowledge of physiology was such that the authorities thought that ovulation occurred just prior to menses, and therefore they stated that midway through the cycle was the safe period. Now that this has been corrected, the method can be used effectively if the menstrual cycle is regular and if strict adherence is paid to the time of abstinence.

The "Pill" Development of a pill that is virtually foolproof in the prevention of pregnancy has revolutionized contraceptive practice. It is now the most widely accepted method, at least in relatively advanced countries. The pills are composed of a mixture of estrogen and progesterone. They are taken once a day for about 21 days. They act by inhibiting gonadotropin secretion, and therefore ovulation does not occur. After the pills are no longer taken at the end of 21 days, menstruation promptly follows. The method is reliable if the pills are taken as prescribed. Unfortunately, various undesirable side effects sometimes occur.

Intrauterine Devices Various shapes of loops, rings, and spirals have been devised which are placed in the uterine lumen. These so-called intrauterine devices (IUD) are left in place for long periods of time, in many cases more than a year, although a yearly examination is recommended. In some way they prevent pregnancy, probably by preventing implantation of the fertilized ovum. The menstrual cycle is unaffected and the results are good. Some women experience discomfort and others expel the device. But for those who can tolerate and retain it, the method proves extremely convenient and inexpensive.

Prostaglandins Quite recently the observation was made that prostaglandins cause vigorous uterine contractions, vigorous enough to prevent implantation or to expel a newly

778

THE
ENDOCRINE
AND
REPRODUCTIVE
SYSTEMS

implanted ovum. Accordingly, a procedure for contraception has been devised which involves placing a suppository containing prostaglandins in the vagina when menstruation is expected or if menstruation does not occur. Generally, within a day or two the menses appears. Since the substance needs to be used only once a month and only if there are reasons to believe that pregnancy could have occurred, the procedure may become popular. As yet, the material is not available for this purpose in the United States.

Sterilization A remarkable number of men and women are turning to sterilization to prevent pregnancy. In the male the relatively simple operation is termed vasectomy. The vasa deferentia are tied and severed, which prevents sperm from entering the ejaculate. Interestingly, an immune reaction may result so that if at a later date the tubes are sutured together, few if any viable sperm are produced, and the male remains sterile.

In the female the procedure is a major operation, but when properly done, that is, when the fallopian tubes are tied and severed and the ends buried in a tissue pocket, pregnancy cannot occur. Astonishingly, recent figures show that of all adults using some kind of contraceptives, 25 percent have been sterilized.

PREGNANCY

The first sign of pregnancy is a missed menstrual period, termed amenorrhea. But this may occur, of course, for other reasons. More definitive tests are based upon the fact that the placenta produces large quantities of gonadotropins which are excreted in the urine. The urine is administered to a female animal; some time later the ovaries are examined. If the woman is pregnant, the urine will contain the gonadotropins, and therefore the ovaries of the test animal will undergo significant maturation. Specifically, they will disclose corpora lutea and hemorrhagic points indicating ruptured follicles.

As soon as the sperm penetrates the ovum, the latter begins to divide and subdivide with ever-increasing frequency to undergo all the various transformations described in Chapter 4. The first mass of cells so produced is termed the "mulberry mass," or morula, and is covered by a membrane called the trophoblast. The trophoblast is thought to release a proteolytic enzyme which facilitates the implantation of the developing ovum into the mucosa of the uterine wall. The fertilized ovum takes a day or two to complete its journey down the fallopian tube and to enter the uterine cavity, where it remains an additional 3 or 4 days before implantation occurs. Thus, the conceptus is about 6 days old when it becomes implanted. The entire gestation period, that is, the time from fertilization to birth, is approximately 270 days.

By the time of implantation, the morula has become a blastocyst, a cell mass with a fluid-filled core (see page 74). The blastocyst settles deeply in the now thickened, highly vascular mucosal lining of the uterus, the endometrium. Because this modified endometrium is discarded at birth of the child, it is called the decidua, meaning "to fall off," or "to shed." The decidua deep to the blastocyst is termed the decidua basalis.

Some of the cells of the blastocyst become the embryo itself, and others form the membranes surrounding it. The inner embryonic membrane is the amnion. The space between this thin membrane and the embryo enlarges and fills with clear, watery fluid. Thus the embryo develops within this amniotic sac protected by a cushion of amniotic fluid. Another membrane, the chorion, thicker and tougher than the amnion, forms from blastocyst cells at the outer limit of the extraembryonic cavity. It develops fingerlike projections, or villi. The chorion adjacent to the decidua basalis is the chorion frondosum (see Figure 4-2).

Placenta

The placenta develops from the chorion of the embryo and the decidua basalis of the uterus. The villi of the chorion enlarge and are received into depressions of the decidua. Blood sinuses form around the villi. Waste materials, brought from the fetus by the umbilical arteries, diffuse out of the villi and across the sinuses into venules of the maternal circulation. Oxygen and nutrients transported in the maternal arterioles diffuse across the sinuses into villi and enter the umbilical vein.

Figure 33-2 shows a section through a full-grown placenta. Note that the umbilical artery ultimately gives rise to capillaries that lie in the villi. The villi are bathed by the maternal blood that flows in the sinuses. There is thus a large volume of maternal blood in intimate contact with the villi, which have a huge surface area. These physical relationships greatly facilitate the diffusion of nutrients and the respiratory gases.

The P_{O_2} in the maternal blood in the placental sinuses is about 40 mmHg; the pressure in the umbilical arteries is about 30 mmHg. There is thus a pressure gradient of 10 mmHg which drives the oxygen from the maternal to the fetal blood.

The P_{CO_2} in the fetal blood reaches about 50 mmHg. In the placental sinus blood it is about 45 mmHg. Thus there is only a 5-mmHg gradient, but this suffices.

Other substances, such as glucose, amino acids, fatty acids, and various ions, diffuse quite readily through the placental membranes. So do the end products of fetal metabolism. Active transport need not be postulated.

The placenta functions as an endocrine organ. The maternal gonadotropic and sex hormones are necessary for fertilization to occur and for the implantation and survival of the placenta and the prevention of abortion. But once the placenta has reached full development, the maternal ovaries and hypophysis may be removed in an experimental animal with fetal development continuing quite normally. This is so because the placenta secretes all the necessary hormones itself.

Gonadotropins   Fetal gonadotropins are differentiated from the ones secreted by the anterior lobe of the maternal hypophysis by being called chorionic gonadotropins. The term human chorionic gonadotropin (HCG) is commonly used. The amount of chorionic gonadotropin excreted in urine increases rapidly soon after fertilization, reaches a peak about the third month, and then rapidly declines (Figure 33-3). In short, in the absence of pregnancy the hypophyseal gonadotropins decline and the corpus luteum degenerates. If pregnancy occurs, the chorionic gonadotropin maintains the corpus luteum until the placenta can secrete its own female sex hormones.

Estrogen   As seen in Figure 33-3, the amount of estrogen in the urine begins to increase at about the third month of pregnancy and then progressively reaches higher levels until, just before birth, there may be 30 times as much estrogen in the urine as there is in the nonpregnant state. The blood level of estrogen only rises about tenfold, which simply means that the rate of excretion increases with the rate of production. Nonetheless, the blood level does increase sufficiently to support pregnancy and to further development.

Progesterone   At the same time that the placenta begins to secrete estrogens, it also produces progesterone. The rate of progesterone secretion progressively increases until the amount excreted in the urine becomes about 10 times that in the nonpregnant urine. These high levels of progesterone are also necessary for the maintenance of pregnancy, and the hormone is responsible for the development of the alveoli of the breasts in preparation for lactation.

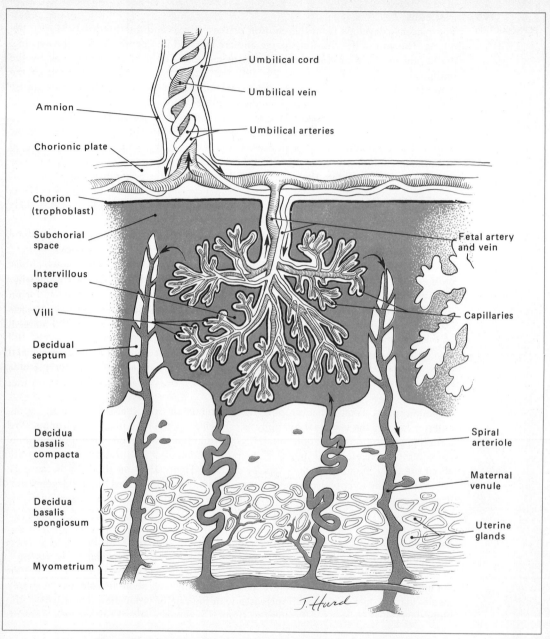

Figure 33-2. Section through the placenta. Nutrients, gases, and wastes are exchanged across intervillous space. Arrows indicate the direction of flow. Blood to the embryo moves from the placenta through umbilical veins and returns by way of umbilical arteries. Maternal blood is shown in color.

Labels (clockwise from top):
- Umbilical cord
- Umbilical vein
- Umbilical arteries
- Amnion
- Chorionic plate
- Chorion (trophoblast)
- Subchorial space
- Intervillous space
- Villi
- Decidual septum
- Decidua basalis compacta
- Decidua basalis spongiosum
- Myometrium
- Fetal artery and vein
- Capillaries
- Spiral arteriole
- Maternal venule
- Uterine glands

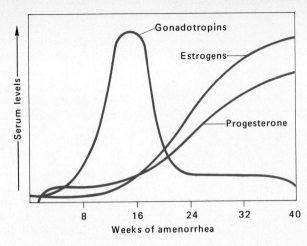

Figure 33-3. Hormones of pregnancy. The placenta secretes gonadotropins until pregnancy is well established. As gonadotropin secretion diminishes, secretion of estrogens and progesterone increases to maintain pregnancy through the normal gestation period.

Maternal Alterations during Pregnancy

During pregnancy there are significant alterations in almost all the physiologic mechanisms. The outstanding change, of course, is the appearance of the woman. The abdomen grows very large because of the development of the fetus within the uterus. The breasts enlarge and become engorged and often painful. The nipples become erect, enlarged, and deeply pigmented.

Weight varies. At first there may be a weight loss because of nausea, vomiting, and the inability to eat adequately. After this initial period there is then normally a progressive weight gain amounting to about 11 kg. A significantly greater weight gain indicates that the mother is overeating and depositing adipose tissue. The normal 11-kg weight gain is distributed as follows:

	kg
Fetus	3
Fluid and membranes	2
Uterine enlargement	1
Breast enlargement	1
Maternal fluid increase	4
Total	11

The increase in maternal body fluid, representing about 4 kg of the total weight gain, is probably a manifestation of the high concentration of steroid hormones, which act on the kidneys to cause fluid retention. After birth, the concentration of these hormones diminishes, and there is an accelerated production of urine, which serves to return body fluid to the normal volume.

In the maternal cardiovascular system there is a progressively increasing circulatory rate with a large blood volume. The blood volume is some 30 percent greater than normal at the end of pregnancy. The cardiac output is also 30 or 40 percent higher than normal.

The pregnant woman utilizes more oxygen and produces more carbon dioxide, due to the increase in her size, the increase in her metabolic rate, and the metabolism of the fetus. Thus, ventilation rises about 50 percent. In addition, the pressure of the

782

THE
ENDOCRINE
AND
REPRODUCTIVE
SYSTEMS

fetus on the diaphragm decreases vital capacity. There may thus be shortness of breath (dyspnea).

The maternal thyroid gland increases in size and secretes more hormone. During pregnancy, it functions for both the mother and fetus. The parathyroid glands also become hyperactive. As a result, more calcium is made available for the developing fetus. In pregnancy there is a positive balance, not only for calcium but for nitrogen, phosphorus, iron, and many other substances utilized by the developing fetus. Thus, the diet must be abundant in these materials to prevent undue drain in the mother.

PARTURITION

The term parturition means "a dividing." As used in this sense, it means the dividing of the fetus from the mother, that is, the act of giving birth.

At term, the uterine muscles begin to contract rhythmically and with ever-greater force. Estrogens stimulate uterine contractions; progesterone inhibits them. At the end of the gestation period there is apparently sufficient estrogen in the maternal blood to overcome the inhibition of progesterone, and therefore uterine contractions occur. In addition, oxytocin is liberated by the posterior lobe of the hypophysis. Oxytocin stimulates more forceful contractions of the uterus.

Stretch of the smooth muscle of the uterus in itself initiates a reflex contraction of that muscle. The greater the stretch, the greater the subsequent contractions. Thus, during pregnancy, as the fetus grows, the uterus undergoes contractions that are weak at first, then progressively stronger, until parturition occurs.

The uterine contractions occur in waves, similar to peristaltic waves, that begin at the top of the uterus and then spread downward. It is this downward wave that ultimately expels the fetus. The contractions at the beginning of labor occur about once every 30 min and last for about 1 min. They then become more frequent, finally occurring about every minute or two, and greatly increase in force.

Contractions of the uterine wall exert great pressure on the amniotic sac. Because the fluid is not very compressible, the sac bulges into the cervical canal. When this pressure so enlarges the opening that there is inadequate support for the sac, it bursts, releasing the amniotic fluid. The bursting of the sac marks the completion of the first stage of labor.

The cervix is now dilated and the head of the child begins to descend through it. Straining by the mother markedly increases intraabdominal pressure, which aids in moving the child through the vagina. Birth of the child is the end of the second stage of labor.

After birth, the uterus slowly contracts, taking up the huge slack left by the expulsion of the child. In a few minutes the uterus vigorously and rhythmically contracts once again, ridding itself of the placenta, or afterbirth. In addition, the contractions of the uterine muscle serve to constrict the various blood vessels which have been ruptured during parturition, reducing hemorrhage to a minimum.

NEONATAL PHYSIOLOGY

The extremely low mortality rate among newborn infants, especially in the more advanced countries, is apt to blind the student to the truly cataclysmic physiologic alterations at birth. While the fetus is carried in the womb, the placenta functions as both lungs and kidneys. At birth, this service is abruptly ended. Even more drastic, more urgent, are the alterations in circulation.

Circulation

Figure 33-4 illustrates the path of the fetal circulation. Note that the blood is oxygenated in the placenta and then passes through the umbilical vein to enter the ductus venosus, which bypasses the liver to enter the inferior vena cava. Thus, in the fetus, the blood which fills the right atrium has an oxygen content that is relatively higher than that in the adult right atrium, because the blood is oxygenated in the placenta.

Figure 33-4. Fetal circulation. Blood enters the fetus through the umbilical vein. The fetal structures allowing blood to bypass the liver and lungs are the ductus venosus, ductus arteriosus, and foramen ovale.

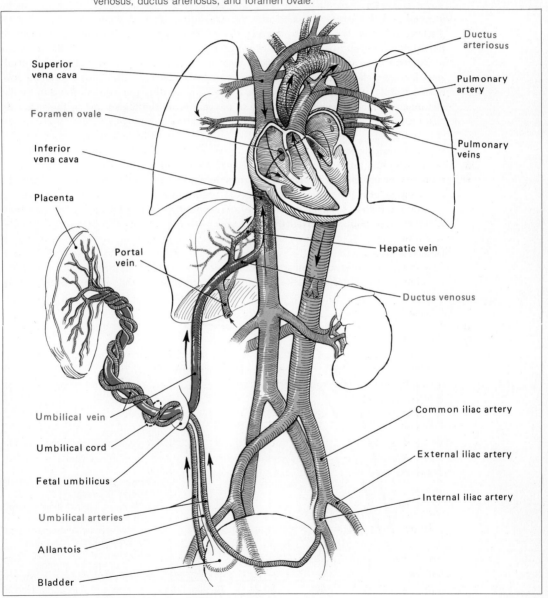

Superior vena cava

Foramen ovale

Inferior vena cava

Placenta

Portal vein

Umbilical vein

Umbilical cord

Fetal umbilicus

Umbilical arteries

Allantois

Bladder

Ductus arteriosus

Pulmonary artery

Pulmonary veins

Hepatic vein

Ductus venosus

Common iliac artery

External iliac artery

Internal iliac artery

784

THE
ENDOCRINE
AND
REPRODUCTIVE
SYSTEMS

Some of the blood entering the right atrium finds its way to the right ventricle and to the lungs, although they are not yet functioning. Most of the blood, however, bypasses the lungs through the foramen ovale, between the right and left atria, or the ductus arteriosus, between the pulmonary artery and aorta.

The P_{O_2} of blood in the umbilical arteries, as mentioned before, is only about 30 mmHg. Yet this suffices because the hemoglobin concentration in the fetal blood is twice that in adult blood and because fetal hemoglobin can carry about 25 percent more oxygen than can adult blood (Figure 33-5).

At birth, the child begins to breathe. With initiation of respiration the ductus arteriosus constricts vigorously, blocking this outlet to the aorta and forcing blood in the pulmonary trunk to flow instead into the expanding lungs. A greater volume of blood thus returns from the lungs to the left atrium. The increase in pressure this creates in the left atrium opposes the inflow through the foramen ovale. A flap of tissue begins to close the hole, and by the end of the first month, flow through the foramen has nearly ceased. By the end of the first year, it has ceased completely. Failure of the foramen ovale or the ductus arteriosus to close causes a mixing of venous and arterial blood in the left atrium and a blood supply inadequate in oxygen. From the bluish cast of their skin has come the term "blue baby" to describe such children.

In summary, the essential changes in fetal circulation include obliteration of (1) the foramen ovale, leaving the fossa ovalis to mark the site, (2) the ductus arteriosus, to form the ligamentum arteriosum, a solid cord of tissue, between the pulmonary trunk and the aorta, (3) the umbilical vein, which becomes the round ligament of the liver, (4) the ductus venosus, to form the ligamentum venosum on the inferior surface of the liver, and (5) the umbilical arteries, which become the lateral umbilical ligaments on the internal surface of the anterior abdominal wall (Figure 33-6).

Respiration

The fetus undertakes some breathing movements before birth. In the placental circulation, the carbon dioxide tension is kept within the normal range; thus there is little stimulus

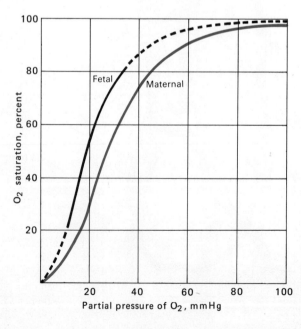

Figure 33-5. Comparison of the oxygen dissociation curve for fetal and maternal blood. Note that fetal hemoglobin has a greater affinity for oxygen, thus the fetal blood is adequately saturated even at the low O_2 partial pressures normally existent in the fetal circulation. (*Reproduced with permission of Dr. James Metcalfe.*)

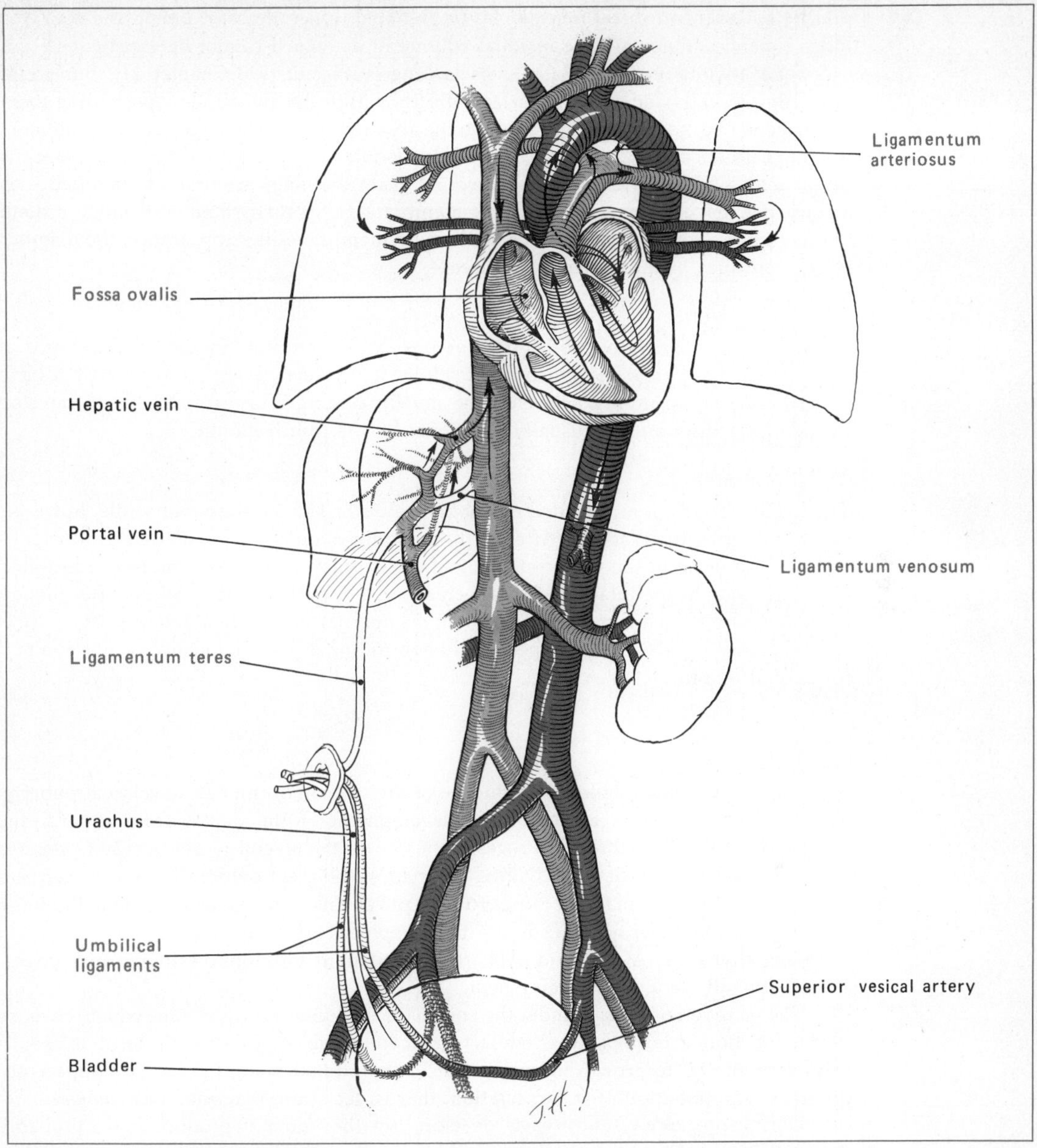

Ligamentum arteriosus
Fossa ovalis
Hepatic vein
Portal vein
Ligamentum venosum
Ligamentum teres
Urachus
Umbilical ligaments
Superior vesical artery
Bladder
J.H.

Figure 33-6. Changes in fetal circulation after birth. The umbilical vessels, ductus venosus, and ductus arteriosus become ligaments. The foramen ovale closes.

of the respiratory center. The breathing movements simply move amniotic fluid through the fetal lungs.

At birth the child is deprived of the placenta; thus the oxygenation of the blood and the removal of carbon dioxide cease. As a result, the respiratory center is strongly stimulated, and if the center has not been depressed by anesthetics administered to the

786

THE
ENDOCRINE
AND
REPRODUCTIVE
SYSTEMS

mother before birth, breathing quickly is initiated. Once the first breath is taken, the lungs expand and contain the residual volume of air which cannot be expelled.

Some infants are born with lungs lacking surfactant (see Chapter 21). Surfactant normally serves to decrease the surface tension within the alveoli, thus permitting them to open and expand during inspiration. In the absence of surfactant, respiration is difficult, the lungs do not expand properly, there is inadequate oxygenation of the blood, and the infant generally dies within a day or two. At death the lungs are found to be filled with a substance that appears to be a hyaline membrane but is really fluid with a high protein concentration. However, because of the hyaline membrane-like appearance, the disorder is referred to as hyaline membrane disease.

Nervous System

Many of the functions of the central nervous system are not completely developed at birth. The major pathways, which in the adult are myelinated, are unmyelinated at birth and take a year or more to develop the myelin coating. The various reflexes develop quickly after birth and are usually functional by the fourth month.

Renal Function

The kidneys function to a limited extent before birth. The main excretory role, however, is carried out by the placenta. After birth, the kidneys quickly assume their full responsibility. The urine in the newborn is found to become progressively more concentrated after birth and, as the intake of fluids increases, so does the urinary volume. But though the kidneys function adequately to maintain the internal environment fairly constant, their full capability in this respect, especially in controlling acid-base balance, is not attained for several months.

LACTATION

The mammary glands, under the influence of the ovarian hormones, develop at puberty. The estrogens are responsible for the development of the duct system and for the deposition of fat in the breasts. Progesterone causes the alveoli to develop and prepares them for the secretion of milk. During pregnancy, the great outpouring of estrogen and progesterone from the placenta causes further and extensive breast development. But these hormones have little or no effect unless the hypophysis is functioning. Apparently, growth hormone, thyroxine, ACTH, and LTH are all needed, in addition to estrogen and progesterone, for full mammary development.

The secretion of milk is under the control of the luteotropic hormone which, because of this function, is termed either the lactogenic hormone or prolactin. Prolactin is chemically very similar to growth hormone, but in the human being the two are apparently different. The initiation of milk secretion, that is, lactation, is termed lactogenesis.

During pregnancy, the breasts develop, but they do not produce milk, probably because the secretion of prolactin is inhibited by the estrogens and progesterone, which are in high concentration until loss of the placenta at parturition.

Other hormones apparently play roles in the secretion of milk. Thyroid activity increases during lactation in proportion to the milk yield. Further, the secretion of milk is seriously impaired by adrenalectomy.

Sympathetic nerve stimulation can reduce the blood flow through the mammary glands to zero, or close to it. When this occurs, milk secretion stops. It probably explains why emotional states can so drastically alter milk production.

The milk that is secreted by the alveoli cells does not simply pour through the ducts and out of the nipple. On the contrary, it accumulates, and if not ejected, will build up sufficient pressure to stop the secretion.

The ejection of milk is due to the contraction of the mammary myoepithelial cells which surround the alveoli and the small ducts. Impulses initiated by suckling are propagated to the hypothalamus. From there impulses travel to the posterior lobe of the hypophysis to effect the release of oxytocin. Oxytocin evokes contraction of the myoepithelial cells, and the milk is ejected. That this is a reflex is shown by the fact that suckling of one breast causes the ejection of milk from both breasts. Milk is ejected within a minute after suckling begins.

Suckling also reflexly stimulates the release of prolactin. Prolactin not only is responsible for the secretion of milk, but also delays mammary involution. Thus, if the child is not breast-fed, if there is no suckling, the secretion of milk will halt. If suckling does continue, milk production may continue for well over a year. Most children today are nursed for much shorter periods of time. The daily output is generally about 1 liter, but if the demand is greater, the output may increase to 2 or 3 liters.

The caloric value of human milk is about 67 Cal/100 ml. Most of the vitamins necessary for development are present in human milk with the exception of vitamin D. For this reason most diets for children have a high vitamin D supplement.

During the last month or two of pregnancy, the breasts are fully developed, and they generally secrete a few milliliters of fluid, termed colostrum. It differs from milk in the almost complete absence of fat. In addition, the volume secreted is very much less than that of milk. Following delivery, under the influence of prolactin, fat is added to the secretion, and the volume increases sharply.

QUESTIONS AND PROBLEMS

1   Describe gametogenesis of ova or spermatozoa.
2   How do primary and secondary spermatocytes differ?
3   What organs contribute to the total volume of semen? What function is served by the secretion of each organ?
4   Discuss the theory that proposes an enzymatic function for sperm during fertilization.
5   What is the genetic basis for fraternal and identical twins?
6   Name the chief factors in human infertility.
7   When during the menstrual cycle is fertilization most likely to occur? Why?
8   What function is served by the chorionic villi?
9   Describe the endocrine function of the placenta.
10   What hormones are involved in parturition, and what are their sources?

11   During pregnancy what changes occur in the maternal cardiovascular system? In the respiratory system? In the endocrine system?
12   Describe the initial uterine contractions and the subsequent events of the first stage of labor.
13   What event terminates the second stage of labor?
14   How does fetal circulation differ from the circulation of the newborn infant?
15   What is the major stimulus for the start of respiration in a newborn infant?
16   In what way is the autonomic nervous system thought to influence the secretion of milk in the lactating female?
17   Describe the relation of suckling to the ejection of milk.

RECOMMENDED READINGS

Gross Anatomy

Anson, B. J.: *Atlas of Human Anatomy,* 2d ed., W. B. Saunders Company, Philadelphia, 1963.

Christensen, J. B., and I. R. Telford: *Synopsis of Gross Anatomy,* 2d ed., Harper & Row Publishers, Inc., 1972.

Crouch, James E.: *Functional Human Anatomy,* 2d ed., Lea & Febiger, Philadelphia, 1972.

Gardner, Weston D., and William A. Osburn: *Structure of the Human Body,* W. B. Saunders Company, Philadelphia, 1967.

Gray, H.: in C. M. Goss (ed.), *Anatomy of the Human Body,* 29th ed., Lea & Febiger, Philadelphia, 1973.

Pansky, Ben, and Earl Lawrence House: *Review of Gross Anatomy,* 2d ed., The Macmillan Company, New York, 1969.

789

Biochemistry

Baker, J. J. W., and Garland E. Allen: *Matter, Energy and Life,* 2d ed., Addison-Wesley Publishing Company, Inc., Reading, Mass., 1970.

Binkley, Stephen B., and Max E. Rafelson, Jr.: *Basic Biochemistry,* 2d ed., The Macmillan Company, New York, 1968.

Conn, E. E., and P. K. Stumpf: *Outlines of Biochemistry,* 3d ed., John Wiley & Sons, Inc., New York, 1972.

Karlson, P.: *Introduction to Modern Biochemistry,* 3d ed., Academic Press, Inc., New York, 1968.

Mariella, Raymond P., and Rose Ann Blau: *Chemistry of Life Processes,* Harcourt, Brace & World, New York, 1968.

Sloane, Nathan H., and J. Lyndal York: *Review of Biochemistry*, The Macmillan Company, New York, 1969.

Weisz, P. B.: *The Science of Biology*, 4th ed., McGraw-Hill Book Company, New York, 1971.

————: *The Science of Zoology*, 2d ed., McGraw-Hill Book Company, New York, 1973.

Cellular Biology

Bronton, Daniel, and Roderick B. Park: *Biological Membrane Structure*, Little, Brown & Company, Boston, 1968.

De Robertis, E. D. P., W. W. Nowinski, and F. A. Saez: *Cell Biology*, 5th ed., W. B. Saunders Company, Philadelphia, 1970.

Du Praw, E. J.: *Cell and Molecular Biology*, Academic Press, Inc., New York, 1968.

Giese, Arthur: *Cell Physiology*, 3d ed., W. B. Saunders Company, Philadelphia, 1968.

Green, David E., and Robert F. Goldberger: *Molecular Insights into the Living Process*, Academic Press, Inc., New York, 1967.

Porter, Keith R., and Mary A. Bonneville: *Fine Structure of Cells and Tissues*, 4th ed., Lea & Febiger, Philadephia, 1973.

Readings from *Scientific American: From Cell to Organism* and *The Living Cell*, W. H. Freeman and Company, San Francisco, 1965.

Biology and Zoology

Curtis, Helena: *Biology*, Natural History Press, Doubleday & Company, Inc., 1968.

Hickman, Cleveland P.: *Integrated Principles of Zoology*, 4th ed., The C. V. Mosby Company, St. Louis, 1970.

Silverdale, Max N.: *Zoology*, The Macmillan Company, New York, 1969.

Simpson, George Gaylord, and William S. Beck: *Life: An Introduction to Biology*, 2d ed., Harcourt, Brace & World, Inc., New York, 1969.

Smallwood, W. L., and E. R. Green: *Biology*, Silver Burdett Company, Morristown, N.J., 1968.

Villee, C. A.: *Biology*, 6th ed., W. B. Saunders Company, Philadelphia, 1972.

————, Warren F. Walker, Jr., and F. E. Smith: *General Zoology*, 3d ed., W. B. Saunders Company, Philadelphia, 1968.

Embryology

Balinsky, B. I.: *An Introduction to Embryology*, 3d ed., W. B. Saunders Company, Philadelphia, 1970.

Ebert, James D.: *Interacting Systems in Development*, Holt, Rinehart and Winston, Inc., New York, 1970.

Haines, R. W., and A. Mohiuddin: *Handbook of Human Embryology*, 4th ed., The Williams & Wilkins Company, Baltimore, 1968.

Langman, J.: *Medical Embryology*, 2d ed., The Williams & Wilkins Company, Baltimore, 1969.

Endocrinology

Turner, C. Donnell, and Joseph T. Bagnara: *General Endocrinology*, 5th ed., W. B. Saunders Company, Philadelphia, 1971.

Williams, Robert H.: *Textbook of Endocrinology*, 4th ed., W. B. Saunders Company, Philadelphia, 1968.

Genetics

Beadle, George W., and Muriel Beadle: *The Language of Life*, Doubleday & Company, Inc., Garden City, N.Y., 1966.

Burns, George W.: *The Science of Genetics: An Introduction to Heredity*, The Macmillan Company, New York, 1969.

Gardner, Eldon J.: *Principles of Genetics*, 4th ed., John Wiley & Sons, Inc., New York, 1972.

Levine, Louis: *Biology of the Gene*, The C. V. Mosby Company, St. Louis, 1969.

Levine, R. P.: *Genetics*, 2d ed., Holt, Rinehart and Winston (Modern Biology Series), New York, 1968.

McKusick, Victor A.: *Human Genetics*, 2d ed., Prentice-Hall, Inc., Englewood Cliffs, N.J., 1969.

Moore, John A.: *Heredity and Development*, 2d ed., Oxford University Press, Fair Lawn, N.J., 1972.

Watson, James D.: *Molecular Biology of the Gene*, 2d ed., W. A. Benjamin, Inc., New York, 1970.

Winchester, A. M.: *Genetics*, 4th ed., Houghton Mifflin Company, Boston, Mass., 1972.

Histology

Bloom, W., and D. W. Fawcett: *A Textbook of Histology*, 9th ed., W. B. Saunders Company, Philadelphia, 1968.

Ham, A. W., *Histology*, 6th ed., J. B. Lippincott Company, Philadelphia, 1969.

Patt, Donald I., and Gail R. Patt: *Comparative Vertebrate Histology*, Harper & Row Publishers, Inc., New York, 1969.

Reith, Edward, and Michael H. Ross: *Atlas of Descriptive Histology*, 2d ed., Harper & Row Publishers, Inc., New York, 1970.

Windle, W. F.: *Textbook of Histology*, 4th ed., McGraw-Hill Book Company, New York, 1969.

Physiology

Ganong, William: *Review of Medical Physiology*, 5th ed., Lange Medical Publications, Los Altos, California, 1971.

Greisheimer, Esther, and Mary P. Wiedeman: *Physiology and Anatomy*, 9th ed., J. B. Lippincott Company, Philadelphia, 1972.

Guyton, A. C.: *Function of the Human Body*, 4th ed., W. B. Saunders Company, Philadelphia, 1969.

Langley, L. L., and E. Cheraskin: *The Physiology of Man*, 4th ed., Reinhold Publishing Corporation, New York, 1971.

Langley, L. L. (ed.): *Homeostasis: Origins of the Concept*, Dowden, Hutchinson & Ross, Inc., 1973.

Langley, L. L. (ed.): *Contraception*, Dowden, Hutchinson & Ross, Inc., 1973.

McNaught, A., and Robin Callander: *Illustrated Physiology*, 2d ed., The Williams & Wilkins Company, Baltimore, 1971.

Ruch, T. C., and Harry D. Patton (eds.): *Physiology and Biophysics*, 20th ed. (Howell-Fulton), W. B. Saunders Company, Philadelphia.

Shepro, David, and George P. Fulton (eds.): *Microcirculation as Related to Shock*, Academic Press, Inc., New York, 1968.

Tuttle, W. W., and Byron A. Schotelius: *Textbook of Physiology*, 16th ed., The C. V. Mosby Company, St. Louis, 1969.

SCIENTIFIC AMERICAN OFFPRINTS

Atkinson, Richard C., and Richard M. Shiffrin: The control of short-term memory, August 1971.

Davenport, Horace W.: Why the stomach does not digest itself, January 1972.

Dudrick, Stanley J., and Jonathan E. Rhoads: Total intravenous feeding, May 1972.

Edelman, Gerald M.: The structure and function of antibodies, August 1970.

Everhart, Thomas E., and Thomas L. Hayes: The scanning electron microscope, January 1972.

Fox, C. Fred: The structure of cell membranes, February 1972.

Frieden, Earl: The chemical elements of life, July 1972.

Gardner, Lytt I.: Deprivation dwarfism, July 1972.

Geschwind, Norman: Language and the brain, April 1972.

Gillie, R. Bruce: Endemic goiter, June 1971.

Gordon, Barbara: The superior colliculus of the brain, December 1972.

Guillemin, Roger, and Roger Burgus: The hormones of the hypothalamus, November 1972.

Ingram, Marylou, and Kendall Preston, Jr.: Automatic analysis of blood cells, November 1970.

Kandel, Eric R.: Nerve cells and behavior, July 1970.

Kretchmer, Norman: Lactose and lactase, October 1972.

McKusick, Victor A.: The mapping of human chromosomes, April 1971.

Margulis, Lynn: Symbiosis and evolution, August 1971.

Mayr, Otto: The origins of feedback control, October 1970.

Merton, P. A.: How we control the contraction of our muscles, May 1972.

Notkins, Abner L., and Hilary Koprowski: How the immune response to a virus can cause disease, January 1973.

Pengelley, Eric T., and Sally J. Asmundson: Annual biological clocks, April 1971.

Pryor, William A.: Free radicals in biological systems, August 1970.

Ptashne, Mark, and Walter Gilbert: Genetic repressors, June 1970.

Ratliff, Floyd: Contour and contrast, June 1972.

Reisfeld, Ralph A., and Barry D. Kahan: Markers of biological individuality, June 1972.

Solomon, Arthur K.: The state of water in red cells, February 1971.

Stent, Gunther S.: Cellular communication, September 1972.

Temin, Howard M.: RNA-directed DNA synthesis, January 1972.

Werblin, Frank S.: The control of sensitivity in the retina, January 1973.

Wessells, Norman K.: How living cells change shape, October 1971.

Wills, Christopher: Genetic load, March 1970.

Young, Richard W.: Visual cells, October 1970.

Young, Vernon R., and Nevin S. Scrimshaw: The physiology of starvation, October 1971.

Page numbers in **boldface** indicate illustrations.

793

Adenyl cyclase, 711, 712, 751
ADH (see Antidiuretic hormone)
Adiadochokinesis, 360
Adipose tissue, 62, 635–636
Aditus, **333,** 335
Adrenal cortex, 733, **734,** 750–751
Adrenal cortical hormones, 735–738
Adrenal glands, 732–739
Adrenal medulla, 733, **734**
Adrenal medullary hormones,
    738–739
Adrenergic fibers, 297
Adrenergic neurons, 476
Adrenocortical hormone, **721**
Adrenocorticotropic hormone
    (ACTH), 720–721, 751, 752
Adrenogenital syndrome, 738
Adrenoglomerulotropin, 736
After-discharge firing, **223**
Afterbirth, 782
Agglutinate, 394, **396**
Agglutinin, 394–396
Agglutinogen, 394–396
Agglutinogen D, 395
Agranulocytes, 381, 384
Agranulocytosis, 384
Air:
    movement of, in ventilation,
        536–543
    residual, 553
    tidal and alveolar, 541, 542
    (See also Alveolar air; Gases;
        Ventilation)
Air sacs, 525
    (See also Alveoli)
Air sinuses, 92, 527, 529
Air-velocity index, 544
Alanine cycle, 661–662
Alar cartilage, **526,** 527
Albumin, 385
Albuminuria, 686
Alcohol, 687
Aldosterone, 480, 494, **735–736, 737**
Aldosteronism, 738
Alimentary canal, 77, 588, 596
Alkaline reserve, 703
Alkalosis, 700, 704, 705
All-or-nothing response, 197–200,
    216–217
Allantois, **738**
Alleles, 47, **48**
Alpha cells, 600, 740
Alternating current, 21
Alveolar air, analysis of, 549–**550,**
    551
Alveolar arteries, **442**
Alveolar ducts, 525, **526**
Alveolar glands, 65–66
Alveolar nerves, **254,** 255
Alveolar process, 93, **95**
Alveoli, 525–**526, 552,** 574

Ambiguous nucleus, **236**
Ameba, 383
Ameboid movement, 383
Amelia, 111
Amenorrhea, 746
Amino acid sequence, **731**
Amino acids:
    absorption of, 626
    essential, 633
    metabolism of, 633–635
    nonessential, 633
    oxidation of, 633
Amino nitrogen, 633
Amitosis, 46
Ammonia, 633
Amnesic aphasia, 363
Amnion, 778, **780**
Amniotic cavity, **74**
Amniotic membrane, 77
Amniotic sac, **75, 76**
Amperes, 20
Amphiarthrosis, 113
Amplitude, 338, **340**
Ampulla:
    of ear, 256, **257, 336, 337,** 352, **353**
    of seminal vesicles, **761**
    of Vater, **589**
Amygdaloid nuclei, **231, 350, 364**
Amyl nitrite, 501
Amylase, 616
Anal canal, 160, 588, **595–**596
Anal columns, **595**
Anal fossa, 580
Anal sphincter, **158,** 159, **160, 595**
Anal valves, **595**
Analgesia, 277
Anaphase, **45,** 46
Anastomisis, 439–440
Anatomical position, **6**
    definition of, 7
Anatomy, 3–9
    (See also Body)
Anconeus, **168**
Androgen, 744
Anemia, 380, 384, 428, 557
    aplastic, 380
    pernicious, 380, 651
    sickle-cell, 381
Anesthesia, 22
Aneurysm, 476–**478**
Angina pectoris, 501
Angioblasts, 437
Angiotensin, 480, 736
Angiotensin II, 492, 494
Angiotensinogen, 492
Angular artery, **442**
Angular vein, **454**
Anion, 14, 387
Ankle, 107
Ankyloglossia, 583
Annular ligament, 106

Annular tendon, 129, **130**
Annulus fibrosus, 403
Annulospiral ending, **282**
Anode, 14
Anorectal hiatus, **157**
Anoxemia, 573
Anoxia, 573
Ansa cervicalis, 138, 242, **243, 263**
Antebrachial nerve, medial, **244,** 245
Antecubital nodes, 461
Anterior, as term, **6, 7**
Anthelix, **332**
Antibody, 382, 383, 394
Anticoagulants, 392
Antidiuretic hormone (ADH), 685,
    696, 697, 723, **724, 725**
Antigen, 382, 383, 394
Antigravity reflexes, 282
Antihemophilic factor (AHF), 393
Antipyretics, 302
Antithrombin, 392
Antitragus, **332**
Anus, **158,** 160, **595**
Aorta, 155, **399–401, 403,** 440
Aortic arch, **435, 441**
Aortic bodies, 436, 479
Aortic body reflexes, 566
Aortic insufficiency, 430
Aortic pressure, 419–**421**
Aortic sinus, 434, **435,** 478–479
Aortic sinus reflex, 566
Aortic stensois, 427–428, 430, 500
Aortic valve, **402, 417,** 427
Aphasia, 363
Apnea, 566, 569–571
Apneusis, 562
Apocrine glands, 66, **69**
Apoferritin, 638
Aponeurosis, 125, 153, **154**
Appendicitis, 277, **278**
Appendicular portion of body, **6,** 9
Appendicular skeleton, 102–116
Appendix, veriform, 592, **593**
Aqueduct of Sylvius, **237**
Aqueous humor, 312, 316, 327
Arachnoid, **238,** 239
Arachnoid space, **508**
Arachnoid villi, 455, **507, 508**
Arches:
    of foot, 111
    vertebral, **98, 100**
Arcuate artery, **670, 671**
Arcuate fasciculus, 363
Arcuate ligaments, **155**
Arcuate popliteal ligament, 114
Arcuate vein, 670, **671**
Area cribrosa, 313
Areflexia, 692
Areolar connective tissue, **60**
Argentaffin cells, 591
Argyll Robertson pupil, 327

Clotting, 383–385, 389–393
  extrinsic, **389**–390
  intrinsic, **390**
Clubfoot, 111
Coccygeal nerve, **248**
Coccygeus, **157**, 158
Coccyx, **96**, 97, **100**, **158**, 160
Cochlea, 81, 256, **257**, **331**, 334, **336**, **337**, **341**
Cochlear duct, 334, **336**, **337**, **341**
Cochlear nerve, **257**, 258, **337**, **338**
Cochlear nuclei, **251**, **339**
Coenzymes, 24, 38, 40
Cogan, D. G., 318
Coherin, 724
Coils, primary and secondary, 21
Coitus interruptus, 777
Cold, common, 344, 527
Colic arteries, **449**, 450
Collagenous fibers, 59, **60**
Collarbone, 103
Collateral circulation, 440
Collateral ligaments, 114, **115**
Collecting duct, 683
Collecting tubules, 669, **671**
Colliculi, superior and inferior, 233, **234**
Colloid-filled follicles, **726**
Colon, **449**, 592–593
Color blindness, 323, 324, 327
Color index, 377
Color vision, 323–**324**
Colostomy, 596
Colostrum, 787
Columnar epithelium, 56, **57**, **58**
Columns, ventral gray and dorsal sensory, 227
  (*See also* Spinal column)
Commissures, 229, **350**, 368
Common cold, 344, 527
Communicating arteries, **443**, 444
Concentration gradient, 31
Concentration of solution, 14–15
  hydrogen-ion, 15
Conception (*see* Fertilization)
Concha:
  ear, **332**
  nasal, 94, **95**, 527, **528**
Conditioned reflex, 284
Condom, 777
Conduction, heat loss through, 301
Conduction system of heart, **406**
Conductivity and stimulus, 28
Conductors, 20–21
Condylar process, 93
Condyles of femur, 109, **110**, **115**
Condyloid joint, 114
Cones and rods, 249, 309, **310**, 322, **323**, 327
Congenital anomalies, **82**–83

Congenital atelectasis, 524
Conjugate focus, **314**
Conjunctiva, **306**, **307**
Conjunctival fornix, **306**
Conjunctival sac, **305**
Conjunctivitis, 306, 527
Connective tissue, 59–62
  areolar, **60**
Connective-tissue envelopes, 123, 125
Constipation, 620
Constrictor muscles, **135**, **140**, **531**
Constrictor vaginae, 159
Contraception, 776–778
Contractility and stimulus, 28
Contraction, muscle, 190–**191**, 192–197, **198**–**200**, 204
  anaerobic, **194**
  cardiac, 203–**204**
  chemistry of, 193–195
  force of, in heart, 423, 427–431, **435**
  hunger, 602
  isotonic and isometric, 196
  of smooth muscle, 202
  of spleen, 514
  in stretch reflex, **282**
  (*See also* Uterus)
Contracture, muscle, 201
Contralateral nucleus, 342
Convection, heat loss through, 301
Convergence reflex, **320**
Convergent muscle, **123**
Converting enzyme, 492, 494
Coordination:
  centers of, 352–368
  muscle, 287
Copper, 658
Coracobrachialis, 162–**164**
Coracoid process, 103, **104**
Cords of brachial plexus, **244**, 245
Cori, C. F., 630
Cori, G. T., 630
Cori cycle, **194**, **630**, 631
Corium, 68
Cornea, **305**, **306**, 307, **308**, **309**, 316
Corniculate cartilage, 530, **531**
Corona radiata, **74**, **763**
Coronal plane, **8**
Coronary (*See also* Cardiac; Heart)
Coronary arteries, **404**–405, **458**
Coronary circulation 496–502
Coronary disease, diet in, 662
Coronary flow, 497–502
Coronary ligament, **115**, 596, **597**
Coronary occlusion, 501–502
Coronary plexus, 292
Coronary sinus, **403**, **404**, 405, 497
Coronary veins, **404**–405
Coronoid process, 93, **105**
Corpus albicans, 742, 764

Corpus callosum, 229, **233**, **364**, 368, **507**
Corpus cavernosum, **674**, 760–**762**, 773
Corpus hemorrhagicum, 742, 764
Corpus luteum, 742, 743, 745, 746, 764
Corpora quadrigemina, 227, 233, **234**
Corpus spongiosum, **160**, **674**, 761, **762**, 773
Corpus striatum, 231, 361, 362
Corpuscle of Ruffini, **213**, 214
Corresponding points in vision, 318
Corrugator muscle, 126–**128**, 129
Cortex:
  of cerebellum, 236, 356, 359
  of cerebrum, **229**–230, 274–**275**, 276, 287, 362–368, 571, **572**
  motor, 284, 359
  renal, 668, **670**, 671
  visual, lesions in, **328**–329
Cortical bone, **88**
Cortical sinus, 460, **461**
Corticobulbar pathway, 285
Corticoids, 737
Corticopontine tracts, **235**, **358**
Corticospinal nerves, **270**
Corticospinal pathway, 285
Corticospinal tract, 235, 285, **286**
Corticosteroids, 42
Corticosterone, **735**
Corticotropin-releasing factor (CRF), 720–721
Cortisol, **736**
Costal cartilage, **101**, **102**
Costal facet, upper, **98**
Costal groove of rib, **102**
Costal surface, 534
Costocentral articulation, **98**
Costocervical trunk, **442**, 444
Costochondral junction, **102**, **150**
Costodiaphragmatic recess, 534
Costotransverse articulation, **98**
Costotransverse facet, **98**
Coughing, 542
Countercurrent mechanism hypothesis of formation of urine, 681–684
Cowper's gland, **673**
Craniad, as term, 7
Cranial cavity, 12, 91–**93**
Cranial nerves, 248–263, 293, **294**, 295
Cranial root nerve, **262**
Craniosacral division, 291
Cranium, **90**–91
Cremaster muscle, **760**
Cretinism, 729
Cribriform plate, 91, **93**, 94, **95**, 252, 349, 527, **528**

Crick, Francis, 51, 52
Cricoarytenoid muscles, 140, **141**, 142
Cricoid cartilage, **141**, 530, **531**
Cricothyroid membrane, 530
Cricothyroid muscle, **141**, 142
Cristae, **29**, 37, **38**
Crista ampullaris, 353, 354
Crista galli, 91, **93, 95**
Crista terminalis, **403**
Cross bridges, 189, 191, **192**
Cross matching of blood, 396
Cross plane, **8**
Cruciate ligaments, **115**
Crura of diaphragm, 154, **155**
Crura of penis, 761, **762**
Crus of ear, **335**
Cryptorchism, 748
Crypts of Lieberkühn, 591
Cubital nodes, **462**
Cubital veins, 457, **460**
Cuboid bone, 107, 109, **111**
Cuboidal epithelium, 56, **57**
Cumulus oophorus, 763
Cuneiform bone, 107, 109, **111**
Currents:
  bioelectrical, 33
  direct and alternating, 21
  electrical, 20–21
Curvatures of stomach, **589**
Cushing reflex, 504, **505**
Cushing's syndrome, 724, 738
Cusps, 401–**402**
Cutaneous branches of brachial plexus, 245
Cutaneous nerves:
  of cervical plexus, 242, **243**
  of lumbar plexus, **246**, 247
  of sacral plexus, **248**, 249
Cyanosis, 573, 574
Cyclase, 322
Cyclic AMP, 322, 711–714
Cystic artery, **448**
Cystic duct, **589, 597**, 598
Cytochromes, 38
Cytology, 4
Cytopemphis, 32
Cytoplasm, 28, 75
Cytoplasmic granules, 381
Cytosine, **50**

Dale, Henry, 219
Dalton's law of partial pressure, 549
Danielli, James, 28
Dark adaptation, **323**
Davson, Hugh, 28
Dead space, anatomical and physiological, 539, **540–542**
Deamination, 633

1-Deamino-8-D-arginine vasopressin (DDAVP), **725**
Deafness, central and transmission, 344
Decerebrate rigidity, 284
Decibels, 338
Decidua basalis, 778, **780**
Decidual septum, **780**
Decortication, 287
Decussation of pyramids, **232**, 234–**236**, 285, **286**
Deep as term of reference, 7
Defecation, 620
Defibrinated blood, 392
Degeneration, 217–218
  retrograde, 218
  Wallerian, 217
Deglutition, 606–609
Deltoid branch of thoracoacromial trunk, **445**
Deltoid muscle, **144, 149**, 161, **163, 167**
Deltoid tuberosity, **105**
Dendrites, **210, 211, 219, 220**, 223
Dens, 99
Dense tissue, 61
Dentate ligament, **238**
Dentate nucleus, **358**, 359
Deoxyribonucleic acid (DNA), 44, **50, 51, 52**
Deoxyribose, 50
Depression, 362, 363
Depressor anguli oris, 127, **128**, 129
Depressor labii inferioris, 127, **128**, 129
Depressor septi, 127, 129
Dermal papillae, 66, **67, 69**
Dermal ridges, 68
Dermatome, **78**, 241
Dermis, **67**, 68
Descending colon, 592, **594**
Detrusor muscle, 672, 691
Deuteranopia, 327
Diabetes insipidus, 725
Diabetes mellitus, 600, 628, 637, 686, 718, 725, 741
Diabetogenic effect, 718
Dializing fluid, 687
Diapedesis, 383
Diaphragm:
  pelvic, **155–157**
  thoracic, 154–**155**
  urogenital, **158–159, 673, 674**, 754
Diaphragm, vaginal, 777
Diaphragma sellae, **238**
Diaphragmatic surface of heart, 400
Diaphysis, 85
Diarrhea, 620
Diarthrosis, 113–114
Diastasis, **417**, 419

Diastole, **402**, 416, **417**, 418, **421–422**, 426
Diastolic pressure, 469–470
Diastolic volume, end-, 425, **426, 428, 429**
Diathermy treatments, 21
Dicrotic notch, **417, 470**
Diencephalon, **226, 228, 232, 233**
Diet:
  normal, 658–664
  starvation, 662
  for weight loss, 661
Diffusion, **16**, 34, 548–553
Diffusion coefficient, 551
Digastric muscles, **136, 137**, 138
Digastric nerve, **254**
Digestion, 601–623
  biliary system in, 617–619
  of carbohydrates, 621–622
  deglutition in, 606–609
  of enzymes, 621
  of fats, 622–623
  food intake in, 601–603
  of hormones, 621
  large intestine in, 619–620
  lysosomal, **41**, 42
  mastication in, 603–604
  pancreas in, 615–616
  of proteins, 622
  salivation and, 604–606
  in small intestine, 614–615
  in stomach, 609–614
  summary of, 621–623
  (See also Absorption; Utilization)
Digestive system, 71–72
  anatomy of, 579–600
  embryology of, 579–**581**
  organs of, 581–596
    accessory, 596–600
Digestive vacuole, **41**, 42
Digital arteries, 445, **446**, 451, **453**
Dilator muscle, 308
Dilator nares, 127, 129
Dilution principle, 693–694
Diodrast, 689
Diopters, **315**, 316
Diphosphoglycerate (DPG), 556
Diphosphopyridine nucleotide (DPN), 40
Diploë, 89
Diploid number, 771, **772**
Direct current, 21
Disk:
  intercalated, **120**, 122, 203
  intervertebral, 99
  slipped, 99
Dissociation, 14
Dissociation curve, carbon dioxide, 558–**559**
Distal, as term, **6, 7**

Diuretics, 687
Dizziness, 570
DNA (deoxyribonucleic acid), 44, **50, 51,** 52
Dominance of cerebral hemispheres, 363, 368
Donnan equilibrium, 699, 700
Dopamine, 222–223
Dorsal, as term, 7
Dorsal digital vein, **459**
Dorsal root ganglion, **240,** 241, **242, 271,** 291, **293, 299**
Dorsal venous arch, **459**
Dorsal venous plexus, **459**
Dorsalis pedis artery, 451, **453**
Dorsalis pedis vein, **459**
Douches, 777
Down's syndrome, **82,** 83
Drugs and coronary flow, 501
Ductless glands, 709
Ducts, male genital, **673, 674,** 757–759
Ductus arteriosus, **783,** 784
Ductus deferens, **757,** 759, **760–762**
Ductus reuniens, 334, **336**
Ductus venosus, **783,** 784
Duodenum, **581, 589,** 591, **594**
Duplex theory, 342
Dura mater, **238, 311,** 313
Dural sinuses, 454
Dwarfism, 87, 719
Dye method in cardiac output, 424, **425**
Dysmenorrhea, 746
Dysmetria, 360
Dysphagia, 609
Dyspnea, 547, 572–573

Ear, 81, 256–258, 330–345
Ear infection, 344
Ear ossicles, **331,** 333, **335**
Eardrum, 332–333
Earlobe, **332**
Ectoderm, **74,** 77, **78,** 305
Ectopic systoles, 407
Edema, 19, 385, 483
Edinger-Westphal nerve, **251**
Einthoven's triangle, 412
Ejaculation, 774
Ejaculatory ducts, **673, 674,** 759, **761, 762**
Elastic cartilage, 61
Elastic fibers, 59, **60**
Elastic walls of blood vessels, 470
Electrical axis, 412–**413,** 414
Electrical current, 20–21
Electrical gradients, 19
Electrical potential, 19–21

Electrocardiogram (EKG), 408, 410–**411,** 412–414, **417**
abnormal, 414–**415**
twelve-lead, **409**
Electrocardiagraphy, 407–415
Electrochemical factors, 31
Electrodes, 14, 408–**409**
Electroencephalogram (EEG), 364
Electrolytes, 14, 387–388, 627
Electrolytic solution, 14
Electromyogram (EMG), 196, **197**
Electromyography, 196–197
Electron, **13**
Electrostatic solenoid theory, 193
Elements in nutrition, 654–658
Elephantiasis, 483
Emboliform nucleus, **358,** 359
Embolism, 393
pulmonary, 512
Embryogenesis, **74**
Embryo:
abnormal development of, 82–83
cross section of dorsal aspect of, **78**
development of, 77–81
epithelial cells of, 59
muscles of, 118–119
nervous system of, 225, **226–227**
23-day-old, somites of, **79**
28-day-old, **76**
31-day-old and 40-day-old, showing limb buds, **80**
40-day-old, **75**
Embryology, 5, 73–83
Emissary vein, **508**
Emphysema, 512, 574
Emulsification, 623
End arteries, 497
End bulbs of Krause, **213,** 214
End-diastolic volume, 425, **426, 428, 429**
End feet, 214, 218, **219**
Endings:
nerve, 213
visceromotor or secretory, 213
Endoabdominal fascia, **154**
Endocardium, 403
Endochondral bone formation, 85–86
Endocrine cells, **599**
Endocrine glands, **65,** 709–753
Endocrine system, 72
Endoderm, **74,** 77, **78**
Endolymph, 334, **336, 341,** 353–355
Endolymphatic duct, **336, 337**
Endometrium, **74, 79**
uterine, 766–767
Endomysium, 123
Endoneurium, **212**
Endoplasmic reticulum, **27,** 28, **29,** 43
Endothelial cell nucleus, 552

Endothelial tubes, 437, 438
Energy:
of activation, 24
calorimetry in, 641
mitochondria and, 37–40
of propagation, 217
utilization of, 40, 644
Enterogastric reflex, 611
Enterogastrone, 611
Enterokinase, 615
Enuresis, 692
Envelopes of connective tissue, 123
Enzymatic action, 22–24
Enzyme specificity, 24
Enzymes:
ATPase, 195
digestive, 621–622
of synapse, 220–221
table on principal, 656
Eosinophils, **375,** 381, 383
Ependymal layer, 227
Epicardium, 403
Epicondyles, **105,** 106, **110**
Epicranial aponeurosis, 126, **128**
Epidermis, 66–67, **69**
Epididymis, 757–759, **762**
Epigastric arteries, **447,** 448, **450, 452**
Epigastric region, **10,** 11
Epiglottic vallecula, 582
Epiglottis, **141, 346, 347,** 530, **531, 582**
Epilepsy, 368
brain wave showing, **365**
Epimysium, 123
Epinephrine, 711, 738, 739
Epineurium, **212**
Epiphyseal plate, **85,** 86–**88**
Epiphyses, 85, 86, 748
Epiploic appendages, **593**
Episiotomy, 767
Epispadias, 761
Epithalamus, 232
Epithelial membranes, 63–64
Epithelium, 56–59
ciliated, 524, **525**
columnar, 56, **57,** 58, 524, **525**
cuboidal, 56, **57**
germinal, 58
pseudostratified, **57,** 524, **525**
respiratory, 524, **525**
squamous, 56, **57,** 64
transitional, **57**
Epitympanic recess, **333**
Equatorial region, 283
Equilibrium, 81, 354, 355, 359, 360
Equivalents, 14
Erection, 773–774
Erector spinae muscles, 142–143, 147
Erythroblastosis fetalis, 395
Erythroblasts, 378

Filaments, actin and myosin, **121,**
   188–190, 191, **192,** 193
Filiform papillae, **346**
Filtration, 17, 638
Filtration fraction, 680
Filum terminale, 239, **240, 507**
Fimbriae, **764,** 765
Fingers:
   arteries of, 445, **446**
   bones of, **107**
   muscles of, 167, 170–**172**
Firing:
   continuous and repetitive, **223**
   rate of, 267
Fissure:
   of cerebrum, 228, **229,** 230
   of lung, 534
   of spinal cord, 239
Fixators, 125
Flaccid paralysis, 287–288
Flavin adenine dinucleotide (FAD),
   38
Flemming, Walther, 47
Flexion muscles, **124,** 125
Flexor carpi radialis, **164,** 165, **166,**
   167, 169
Flexor carpi ulnaris, 165, **166–168,**
   169
Flexor digiti minimi, 171, **172**
Flexor digiti minimi brevis, 183, 185,
   186
Flexor digitorum brevis, **183–186**
Flexor digitorum longus, 181, 183,
   184
Flexor digitorum superficialis and
   profuncus, **166–168,** 169
Flexor hallucis brevis, 183, 185,
   186
Flexor hallucis longus, 181, 183, **184**
Flexor pollicis brevis, 171, **172**
Flexor pollicis longus, 166, 169, 170
Flexor reflexes, 283–284
Flexor retinaculum, 165, **172**
Flexures:
   of brain, 226
   hepatic and splenic, 592, **594**
Flocculonodular lobe, 357, 359
Flower-spray ending, **282**
Flowmeters, 472
Fluid:
   body (see Body fluids)
   interstitial (see Interstitial fluid)
   movement of, through capillary
     wall, 482–483
Flutter, heart, 414
Focal distance, **314, 315**
Focal point, 317, 325
Focus and out of focus, **325,** 326
Folds of Kerckring, 592
Folia, 236
Folic acid, 650

Folicle-stimulating hormone (FSH),
   721, 743, **744,** 746–748, 750
Follicles:
   colloid-filled, **726**
   hair, 68, **69**
   ovarian, 742, 762–**763,** 764
Fontanel, 91
Food:
   calories in, table on, 660
   intake of, 300, 601–603
   preferences for, 348
   specific dynamic action of,
     644–645
   water in, 697
Foot:
   arteries of, 451, **453**
   bones of, 107, **111**
   muscles of, **182,** 183–186
   nerves of, 185
Foramen:
   obturator, of innominate bone, **108**
   of skull, **92, 94, 95**
   vertebral, **98,** 99, **100**
Foramen cecum, **346**
Foramen lacerum, **93**
Foramen of Luschka, 237, **507**
Foramen of Magendie, **237, 507**
Foramen magnum, 91, **93**
Foramen of Monro (interventricular),
   237
Foramen ovale, **93,** 403, **783,** 784
Foramen rotundum, **93**
Foramen spinosum, **93**
Forearm:
   arteries of, **446**
   bones of, 103, **105**
   muscles of, 164–170
   nerves of, 169, 170
Forebrain, **79,** 226
Foregut, 579–580
Forehead, 92
Formed elements, 374
Fornix, **233, 364**
Fossa:
   cranial, 91, **93**
   ileac, **108,** 109
   intercondylar, **110**
   of scapula, 103, **104**.
Fossa navicularis, 761, **762**
Fossa ovalis, **403**
Fovea centralis, **308,** 310, 318
Fractures, 112–113
Frenulum, **768**
Frequency, 338
Frontal bone, 89, **90, 92, 94**
Frontal lobe, 228, **229, 230**
Frontal nerve, **254,** 255
Frontal plane, **8**
Frontal sinus, **95, 528, 529**
Frontalis muscle, 126–128
Fructose, 27, 622, 626, 628, **629**

Fundic glands, **590**
Fundus:
   of stomach, **589**
   of uterus, **764**
Fungiform papillae, **346**
Funiculi of spinal cord, 268, **270**
Fusiform muscle, **123**

Gallbladder, **589,** 598–599, 617, 618
   embryonic, 580, **581**
Gallop rhythm, 422
Gallstones, 618
Galvanometer, string, 408
Gametogenesis, 771, **772**
Gamma aminobutyric acid (GABA),
   222
Gamma efferent, **282**
Gamma globulin, **385**
Gamma globulin antibodies, 382
Ganglia, 210
   of autonomic nervous system,
     290–292, **293**
   basal, **231,** 361–362
   celiac, **290,** 292
   cerebrospinal, 239
   cervical, **290,** 292
   ciliary, **253, 254, 290, 294**
   dorsal root, **240,** 241, **242, 271,**
     291, **293,** 299
   mesenteric, **290**
   otic, **254, 290, 294**
   parasympathetic, 292
   preaortic, 291
   pterygopalatine, **290, 294**
   sphenopalatine, 295
   submandibular, **290, 294, 295**
   sympathetic, 219–292
   trigeminal nerve and, **254**
   vertebral, 291
   vestibular and spiral, 256, **257**
   (See also specific names of
     ganglion)
Gangrene, 517, 518
Gas analyzer, Haldane, 549–**550**
Gases, respiratory, 548–560
Gastric (See also Stomach)
Gastric abnormalities, 613–614
Gastric arteries, **448,** 449
Gastric glands, **590,** 611
Gastric juice, 611–613
Gastric movements, 610–611
Gastric nerves, **260,** 261
Gastrin, 613
Gastritis, 614
Gastrocnemius, **115,** 175, 179,
   180–184
Gastrocolic ligament, 589
Gastroduodenal artery, **448,** 449
Gastroepiploic artery, 448, **458**

Iliac arteries, 448, **449, 450,** 451, **452**
  fetal, **783**
Iliac crest, **108,** 109
Iliac fossa, **108,** 109
Iliac nodes, 463
Iliac region, **10,** 11
Iliac spine, **108,** 109
Iliacus, 151, **155,** 156, **246**
Iliococcygeus, 156, **157,** 158
Iliocostalis muscles, 143, **146,** 147
Iliohypogastric nerve, **246,** 247
Ilioinguinal nerve, **246,** 247
Iliolumbar artery, **450**
Iliopsoas muscle, 151
Iliopsoas tendon, 154, 156
Iliotibial tract, 114
Illusions, 277, **279**
Ilium, **108,** 109
Image, formation of, 317–319
Immune reaction, 382, 383
Implantation, **74,** 76
Impotence, 773, 774
Impulses, retrograde auditory, 342
Incisive canal, **95**
Incus, 81, 89, **257, 333,** 334, **335**
Independent assortment, law of, 47,
  48
Inductorium, 21
Infant, newborn, 782–785
Infarct, 501, 502
Infarction, myocardial, **415**
Infection, resistance to, and white
  blood cells, 382–383
Inferior, as term, **6,** 7
Inferior colliculus, **339**
Inferior concha, 89, **307**
Infertility, 776
Inflammation:
  corticoids and, 737
  nasal, 527
Infrahyoid muscles, **137,** 138
Infraorbital artery, **442,** 443
Infraorbital foramen, **92**
Infraorbital groove, **94**
Infraorbital nerve, **254,** 255
Infraorbital vein, **454, 455**
Infraspinatus, 160–**162, 163**
Infratemporal fossa, 91, 443
Infundibular stalk, 714
Infundibulum, **95, 528,** 529
Inguinal canal, 152
Inguinal ligament, 151, **450, 761**
Inguinal nodes, **462,** 463
Inguinal region, **10,** 11
Inguinal ring, 152, 766
Inheritance, laws of, 47, 49
Inhibin, 748
Inhibition:
  autogenic, 283
  of reflex, 281

Inhibition:
  of synaptic transmission, 222–223
Inhibitory postsynaptic potential
  (IPSP), 222
Inner cell mass, **74,** 76
Innominate artery, 441
Innominate bone, 99, 107, **108,** 109,
  156
Inorganic compounds, 14
Inspiration, 536–538
Inspiratory capacity, 539, **540**
Inspiratory center, 565, **567**
Inspiratory level, **540**
Insufficiency, 423, 572
Insula, **229**
Insulin, 600, 628, 636, 718, 739–**741**
Intensity of sound, 338, **340,** 342,
  343
Intention tremor, 360
Interatrial septum, **403**
Intercalated disks, **120,** 122
  of cardiac muscle, 203
Intercavernous sinus, **455**
Intercondylar fossa, **110**
Intercostal arteries, **442,** 446, **447**
Intercostal muscles, 102, 150, **537**
Intercostal nerves, 150
Intercostal spaces of thoracic walls,
  143, 147, **150**
Intercostal veins, 456, **457**
Intercourse, sexual, 771, 773–775
Interferon, 382
Interlobar vessels of kidney, **670**
Interlobular veins, 456
Interlobular vessels of kidney, **670,**
  671
Intermediate gray column, 292
Intermuscular septa, 161
Intermedin, 722
Internal, as term, 7
Internal auditory meatus, **331**
Internal capsule, 231, **233**
Internal ear, 331, 334, **336,** 340
Internal nares, 530
International units, 647
Internuncials, **282**
Interossei muscles, 171, **172,** 185, **186**
Interosseous artery, 445, **446**
Interphase, **45,** 46
Interspinales, 143, 148
Interstitial cell-stimulating hormone
  (ICSH), 721, 747, 748
Interstitial cells of Leydig, 746
Interstitial cells of testis, **758**
Interstitial fluid, 482–485, 699, 700
  medullary, 683, **684**
Intertransversarii, 143, 148
Intertubercular line, **10,** 11
Interventricular foramen, **237, 507**
Interventricular septum, **401, 403,** 404
Interventricular sulci, 400, 404, 405

Intervertebral disk, 99
Intervillous space, **780**
Intestinal arteries, **449**
Intestinal flora, 619
Intestinal glands, 591
Intestinal juice, 614–615
Intestinal lining cells, 592
Intestines:
  in food intake, 602
  large, 592–593
    in digestion, 619–620
  secretions of, 620
  small, 591–**592**
    in digestion, 611, 613–615
Intracellular fluid, 698–700
Intracranial pressure, 503–**505**
Intrafusal fiber, **282,** 283
Intragastric pressure, 610
Intralobular veins, 456
Intramembranous bone formation,
  86–87
Intramural nerve plexuses, 608
Intramural pressure, 498–**499**
Intrapleural pressure, 538, **565**
Intrapulmonic pressure, 538
Intraocular pressure, **317,** 327
Intrathoracic pressure, 487–**488,**
  510
Intrauterine devices, 777
Intravesicular pressure, **691**
Intrinsic factor, 380, 627
Inulin, 699
Inulin clearance, 688, 689
Inversion muscles, **124,** 125
Iodide, 727
Iodine, 657, 729
Iodothyronines, 729
Ionization, 13–14
Ions:
  effect of, on arterioles, 480
  hydrogen (*see* Hydrogen-ion
    concentration)
  intra- and extracellular, table on,
    33
  movement, 31, 35
  plasma, 431
Iris, 305, 307, **308, 309, 316,** 317, 320,
  321
Irritability of cells, 21, 28
Iron:
  in blood, 378, 379
  in nutrition, 657
Iron metabolism, 638
Ischial tuberosities, **108,** 109, **158,**
  160
Ischiocavernosus, 159, **160**
Ischiococcygeus, 156
Ischiorectal fossa, 160
Ischium, **108,** 109
Islet cells, **599**

Islets of Langerhans, **599**, 600, 739, 741
Isometric contraction, 196
Isometric-contraction phase, **417,** 418
Isometric-relaxation phase, **417,** 419
Isotonic contraction, 196
Isotonic solution, 18
Isotopes, 27
Isovolumetric-contraction phase, 418
Isthmus of uterus, **764,** 765

Jaundice, 379, 617
Jaw bones (*see* Mandible; Maxilla)
Jejunum, 591
Joint(s), 113–115
   knee as, 114, **115**
Joint capsule, 113
Joint reflex, 568
Jugular foramen, **93**
Jugular notch, **101**
Jugular vein, **335**, 453, **454, 455,** 456, **459**
Juxtaglomerular cells, 492, 672, 678, 679
Juxtamedullary nephrons, 671, 683, 684

Katz, B., 219
Ketone bodies, 637
Ketonemia, 637, 686
Ketonuria, 637, 686
Ketosis, 637
Kety procedure, **497,** 502
Kidneys:
   anatomy of, 668–**669, 670**–672
   artificial, 686–687
   blood flow through, **492,** 494
   body fluid and, 695–697
   circulation in, 518–519
   transport through, 31
   (*See also* Renal)
Kinesiology, 4
Kinins, 480
Klinefelter's syndrome, **82,** 83
Knee, 107, 114–**115,** 181
Krause's end-bulb, **213,** 214
Krebs cycle, 631–**632**
Kupffer cells, 382, 513, 596, 638
Kyphosis, 97

L-dopa, 362
L-vasopressin, **725**
Labia, 767, **768**
Labium majus, 766
Labor in childbirth, 722, 723, 782
Lacrimal apparatus, 306–**307**
Lacrimal bone, 89, **94**

Lacrimal canaliculi, 306, **307**
Lacrimal ducts, 306, **307**
Lacrimal glands, **290, 294,** 295, 306, **307**
Lacrimal nerve, **254,** 255
Lacrimal papilla, **307**
Lacrimal reflex, 321
Lacrimal sac, 306, **307**
Lactase, 616
Lactation, 786–787
Lacteals, 591, 626
Lactic acid, **194,** 631, 632
Lactogenic factor, 719
Lactogenic hormone, 786
Lacunae, **61,** 86, 87
   Howship's, 87
Lamellae, bone, 87, **88**
Lamina, vertebral, **98,** 99, **100**
Lamina propria, **64, 349**
Laplace effect, 518
Laryngeal nerves, 142, **260,** 261
Laryngeal prominence, 530
Laryngitis, 533
Laryngopharynx, **582,** 587
Larynx, 530, **531, 532,** 533, 587
   muscles of, 139–**141**
   nerves of, 142
Lateral, as term, **6,** 7
Lateral lemniscus, **339**
Latissimus dorsi, 142, **144,** 145
Law(s):
   of independent assortment, 47, **48**
   of inheritance, 47, 49
   of Laplace, 476–**478,** 609
   of segregation, **47**
   of specific nerve energies, 266–267
Leads, electrocardiographic, 408–**409, 410**
Learning and memory, 367–368
Leg:
   arteries to, 451, **452, 453**
   bones of, 107
     (*See also* specific names of bones)
   muscles of, 177–179, **180–182,** 183, **184**
   nerves of, 178, 179, 183
   (*See also* Extremities)
Lemniscus, **236, 271,** 272, **273, 274**
Lengthening reaction, 283
Lens, **305,** 307, **308, 309,** 311–312, 316, **317–318**
   in accommodation, 317–318
   biconcave, **315, 325**
   biconvex, **314, 315, 325**
   disorders of, 325–327
   light refraction by, **314–315**
Lens placodes, 80, 81, **305**
Lens vesicle, **305**
Lenticular nucleus, **231, 233, 361**
Leukemia, 384

Leukocytes, 378
   (*See also* Blood cells, white)
Leukocytosis, 384
Leukopenia, 384
Levator anguli oris, 127, 129
Levator ani, 156, **157, 158,** 160, 595
Levator labii superioris, 127, **128,** 129
Levator palpebrae superioris, 131
Levator prostatae, **156,** 158
Levator scapulae, 138, 139, 142, **144,** 145
Levator veli palatini, 133–**135**
Ligament of Trietz, 591
Ligaments:
   knee, 114, **115**
   of liver, 596, **597**
   of ovaries, 761–762, **764**
   of uterus, 765–766
   visceral, 588, 589
Ligamentum teres, 596, **597**
Light:
   high and low intensity, 321–323
   pineal gland and, 751
   vision and, 313–327
Limb buds, **76,** 78, **80**
Limbic system, 363–**364**
Linea alba, 151, **152, 153**
Linea aspera, 109, **110**
Lines, muscle, 188–190
Lingual arteries, 442
Lingual branches, nerves, **259, 260,** 261
Lingual nerve, **254,** 255, 295
Lingual tonsil, 583
Lingula, **357**
Lining cells, intestinal, 591, 592
Linkage, 48, 49
   sex, 48, **49**
Lipase, 617
   gastric, 612, 622, 623
   pancreatic, 616
Lipid droplet, **29**
Lipid layer, **29**
Lipid metabolism, 635–637
Liquor folliculi, **763**
Lithium, **13**
Liver, 513–514, 596–**597, 598**
   cirrhosis of, 514, 618–619
   embryonic, **80,** 580, **581**
   function of, 637–638
   hepatic portal system and, 456
   (*See also* Hepatic)
Liver glycogen, 628, 629
Lobar branches, 440, 669
Lobes:
   of cerebellum, 356, **357,** 359
   hepatic, **597**
   of hypophysis, 714, 716
   of lungs, **441,** 534
   pulmonary, **532, 533**
Lobotomy, 363

Lobules:
  of cerebellum, 356, **357**
  of ear, **332**
  liver, 596–**598**
  of pancreas, 599
Loewi, Otto, 219
Longissimus muscles, 143, **146**, 147
Longitudinal arch, 111
Longitudinal muscle layer, 592
Longitudinal muscles of tongue, 133, **134**
Loop of Henle, **671**, 682–683
Loops of spinal nerves, **243**
Lordosis, 97
LRF, 722
Lumbar arteries, 448
Lumbar plexus, 246–247
Lumbar region, **10**, 11
Lumbar veins, **457**
Lumbar vertebra, 87, **98**, 99
Lumborum, 143, 147
Lumbosacral trunk, **246**, **248**
Lumbricales, 171, **172**, **184–186**
Lumirhodopsin, **322**
Lung(s), **526**, **532**, **533**
  circulation in, **441**, 509–512
  collapse of, 512, 546
  pneumonia, 574
  in respiration, **567**
  root of, 534
  (See also Pulmonary; Respiratory system)
Lung bud, **81**
Lung capacities, 539, **540**, 541
Lunate bone, 103, **106**
Luteinizing factor, 717
Luteinizing hormone (LH), 721–722, 743, **744**, 746, 747, 750
Luteotropic hormone (LTH), 719, 743, 747
Lymph, 71
  flow of, 484–485
  formation of, 484
Lymph follicles, 460
Lymph nodes, 460, **461**, 485
Lymph vessels, 437
Lymphatic duct, **462**, 463
Lymphatic system, 460–465, 484–485
Lymphatogogue, 485
Lymphocytes, 59, **60**, **375**, 381–384, 460
Lymphoid tissue, 378
Lysis, 383
  clot, 391–392
Lysosomes, 28, **29**, 40, **41**, 42
Lysozyme, 382

Macrophage, 59, **60**
Macrophage system, 62–63
Macula lutea, 310
Magnesium, 657–658, 732

Major petrosal nerve, **256**, 257
Malformations, 82–83
Malleolar arteries, **453**
Malleolus bones, 109, **110**
Malleus, 81, 89, **257**, **333**, 334, **335**
Malocclusion, 604
Malpighian corpuscles, 463, 672
  (See also Renal corpuscle)
Mammary gland, 66, 786
Mammary plexus, **462**
Mamillary body, **364**, **715**
Mandible, 89, **92**, 93, 131, **132**, **582**
Mandibular canal, 93
Mandibular foramen, 93
Mandibular nerves, **254**, 255, **256**, 257, 295
Manganese, 658
Manometer, 471
Mantle layer, 227
Mantoux test, 304–305
Manubrium, **101**, **335**
Marey's law of the heart, 434
Marginal layer, outer, 227
Marrow, **85**, 86, **88**, 101–102, 377–380, 381, 384
Marrow cavity, **85**, 86
Marrow spaces, primary, **85**
Masseter muscle, 131, **132**
Mast cells, 59
Mastectomy, 461, 463
Masters, William H., 747
Mastication, 603–604
  muscles of, 131–**132**, **254**
  nerves of, 132, **254**
Mastoid air cells, **333**
Mastoid process, **90**, 91, **92**, 135, 136
Mating and olfaction, 351
Maxilla, 89, **90**, **92**, **94**, 95
Maxillary arteries, **442**
Maxillary nerves, **254**, 255
Maxillary sinus, **95**, **528**, 529
Medial, as term, **6**, 7
Medial lemniscus, **236**, **271**, 272, **273**, **274**
Median eminence, 714, 716
Median nerve, 169, 171, **244–246**
Mediastinal nodes, **462**
Mediastinum, 399, **400**, 534
Medulla, **228**, **353**
  renal, 668, **670**, 671
  respiration and, 562
Medulla oblongata, 234–**235**, **236**
Medullary cords, 460, **461**
Medullary interstitial fluid, 683, **684**
Medullary pressor and depressor regions, 478
Medullated fibers, 211
Megakaryocytes, 385
Meiosis, 46, 771, **772**
Meissner's corpuscle, **213**, 214, **273**
Meissner's plexus, 295

Melanocyte-stimulating hormone (MSH), 722
Melanocytes, **69**
Melatonin, 749–**750**
Membrane:
  amniotic, 77
  basement, **58**, 64
  basilar, 337, **338**, 340, **341**
  cell, 16–**20**, 29–36
  epithelial, 63–64
  mitochondrial, 37, **38**, 40
  mucous, 63–**64**, 527, 529
  permeable, **16**, 551–552
  plasma, **27**, **28**, 29
  pleural, 534–535
  semipermeable, 17, **18**
  serous, 63
  of spinal cord, 237–239
  synovial, 113–115
  transfer through, 16–19
  tympanic, **257**, **294**, 331–333, 336, 338–339
  unit, **28**, 29
Membrane potential, 33, **34**, 35
Membranous labyrinth, 334, **336**
Memory, 367–368
Menarche, 762–763
Mendel, Gregor, 47–49
Meningeal veins, 454
Meninges, 237, **238**, 239
Meningocoele, 99
Menisci, 114, **115**
Menstrual cycle, **744**, **745**, 762–763, 766–767
Mental foramen, **92**
Mentalis, 127, **128**, 129
Mercurial compounds, 687
Meremyosin, 189
  light and heavy, 190, 191
Merocrine glands, 66
Mesencephalon, **226**, 227, 233–**234**, 235
Mesenchyme, 60, **305**
Mesenteric arteries, 448, **449**, **450**, **458**, **589**, 594
Mesenteric ganglis, **290**
Mesentery, 588, **593**, **594**
Mesoappendix, 593
Mesoderm, 77
Mesonephric ducts, 756
Mesonephros, 667–668
Mesothelium, 63
Metabolism, 28
  basal metabolic rate (BMR), 642, 644–645
  increasing, 300–302
  liver in, 638
Metacarpal bones, 103, **106**, 107
Metanephros, 668
Metaphase, **45**, 46
Metarhodopsin, 322

Metarteriole, **481**, 482
Metatarsal arteries, **453**
Metatarsal bones, 107, 109, **111**
Metencephalon, **226**, 227
Microcephaly, 91
Microglia, 215
Microscope, 4–5, 26, 28
Microsomes, 43
Microvilli, **58**, 592
Micturation, 690–692
Midbrain, 79, 226–**228**, 232–**234**, 235
Midclavicular lines, **10**, 11
Middle ear, **12**, 81, 344
Middle-ear cavities, 89, 91, **331**, **333**–334, **335**
Midgut, 579–580
Miescher, Friedrich, 50
Milk, 719, 723, 785–786
Milliequivalents, 15
Milligrams, 14, 15
Milliosmole, 15
Minerals in nutrition, 654–658
Mineralocorticoids, 735–736
Mitochondria, **27**, 28, **29**, 36–40, **58**, 193, 194, 210, **220**
Mitochondrial membranes and particles, 37, **38**, 40
Mitosis, **45**, 46, **58**, **74**, **772**
Mitral cell, **349**, **350**
Mitral insufficiency, 572
Mitral stenosis, 511
Mitral valve, **401**, **402**
Mixed nerves, **250**
Modiolus, 334, **337**
Molal solution, 14
Molar solution, 14
Molecular weights, 14
Moles, 14
Monocytes, **375**, 381–384
Monosaccharides, 31, 193–194, 625–626, 628–630
Monosomy, **82**
Mons pubis, 766
Morula, **74**, 75, 778
Motor areas of cortex, **230**
Motor end plate, **212**, 213
Motor nerves, **250**, 251
 (*See also* Nerves)
Motor neuron, **282**
 (*See also* Neurons)
Mouth, muscles of, 129
Mouth cavity, **81**, 581
Mouth-to-mouth resuscitation, 544–545
Movement, 277–288
 abnormalities of, 287–288, 360–362
 coordination and, 287, 356
 decomposition of, 361
 ionic and molecular, 16, 31, 35
 mass, of large intestine, 619
 postrotational errors of, 355

Movement:
 rate of, through membrane, 30–31
 reflex, 277–284
 semicircular canals and, 354
 stomach, 610–611
 terms used to describe, **124**–126
 voluntary, 284–288
Mucin, 612
Mucous alveolus, **586**
Mucous cells, 65
 of stomach, **590**, 591
Mucous membranes, 63–**64**, 527, 529
Mucus, 524, 527, 620
Müllerian duct, 756
Multifidus, 143, 148
Murmurs, 422–423
Muscle(s), 71, 118–187
 abdominal, 151–155
 action of, kinds of, 125–126
 adduction and abduction, **124**, 125
 antagonistic, 125
 appendicular, 160–186
 of arm, 161, **162–169**
 attachment of, 125–126
 of back, 140, 142–143
 blood supply of, 122–123
 branchiomeric, 119
 of chest, 143, **149**
 circumduction, **124**, 125
 contraction of, 190–200, 204, 282
 embryonic development of, 118–119
 of esophagus, 64
 of eye, 126, **128**–131, 253
 of face, 126–135
 fibers of (*see* Muscle tissue)
 fixating (stabilizing), 125
 flexion and extension, **124**, 125
 of forearm, 164–170
 gluteal, 172–**174**
 of hand, 164–167, 170–**172**
 histology of, 120–126
 intrinsic and extrinsic, 125
 inversion and eversion, **124**, 126
 knee, **115**
 of larynx, 139–**141**
 of leg, 177–184
 of mastication, 131–132, **254**
 migration of, 119
 morphogenesis of, 119
 of mouth, 129
 of neck, **128**, 135–140
 nerves of, 122–123
 nomenclature for, 126
 of nose, 129
 origins and insertions of, 125
 pectoral, 143, 149
 of pelvis, 155–159
 of perineum, 159–**160**
 of pharynx, **135**, 139, **140**
 physiology of, 188–205

Muscle(s):
 pronate and supinate, **124**, 125
 relaxation phase and twitch, 197
 response of, to stimulation, 197–200
 rotation, **124**, 125
 of scalp, 126
 of shoulder, 160–163
 of soft palate, 134–**135**
 synergistic, 125
 thigh, 173–178
 of thoracic diaphragm, 154–**155**
 of thoracic wall, 143, 147, **150**
 of tongue, 133, **134**
 of trunk, 140, 142–160
 types of architecture of, **123**
  table on, 122
 urogenital, **158**–159
 (*See also* specific names of muscles and organs)
Muscle bundle, 123
Muscle fatigue, 200–201
Muscle glycogen, 630–631
Muscle massage in venous circulation, 485, 488–489
Muscle nuclei, **120**, **121**, **213**
Muscle spindle, **213**, 214, **274**, **282**–283
Muscle support in venous circulation, 488
Muscle tissue (fibers), 118
 bands of, **121**–122
 cardiac, 118–**120**, 122, 203, **204**, 205
 patterns and orientation of, 123
  bipennate, 123
  convergent, 123
  fusiform, 123
  parallel, 123
  sphincter, 123
  unipennate, 123
 skeletal (voluntary or striated), 118–**121**, 122, 188–201, 518
 smooth (involuntary or visceral), 118–120, 122, 201–203
Muscle tonus, 200, 287, 360
Muscle process of larynx, 530
Muscularis mucosa, **64**, 588, **590**
Musculi pectinati, **403**
Musculocutaneous nerve, 163, **244**, 245
Musculophrenic artery, 446, **447**
Myasthenia gravis, 221
Myelencephalon, 226, **227**
Myelin sheath, 211, **212**
Myelinated fibers, 211
Myeloid tissue, 378
Myenteric plexus, 295, 608
Mylohyoid muscle, 136, **137**, 138, **254**, 255, **582**
Myoblasts, 119

Myocardial infraction, **415**
Myocardium, 403
Myofibrils, 119, **121**, 188
Myofilaments, 121
Myogenic spasm, 388
Myometrium, **766, 780**
Myopia, 324–**325**
Myosin, 189, **191, 192**
Myosin filament, **121**, 188, **189**–191, 193
Myotatic reflexes, 282
Myotomes, **78**, 119
Myxedema, 729

Nails, 69
Nasal bone, 89, **90**, 92–94, **95**, 525, **526**
Nasal cartilages, **95**, 525–**526**, 527
Nasal cavity, **12**, 89, 92–94, **95, 135**, 349, 527, **528, 582**
Nasal gland, **294**
Nasal meatuses, 527, **528**, 529
Nasal septum, 94, **95, 290**, 525, **526**, 527
Nasalis, 127, **128**, 129
Nasociliary nerve, **254**, 255
Nasolacrimal duct, 306, **307, 528**
Nasopharynx, 530, **582**
Navicular bone, 107, 109, **111**
Nearsightedness, 324–**325**
Neck:
    arteries to, 441–**442, 443**–444
    bones of, 89
    of humerus, **105**
    lymphatics of, 463
    muscles of, **128**, 135, **136–137**, 138–140
    nerves of, 138, 139
    veins of, 453–**454, 455**
Neonatal physiology, 782–785
Nephrons, 670–672
Nephrotome, 667, 668
Nernst equation, 34
Nerve(s), 212–213
    of alimentary canal, 596
    of arm, 163
    of back, 145, 147, 148
    cerebral, 503–504
    cervical, 241
    of coronary vessels, 500–501
    cranial, 248–263
    diencephalon and brainstem, **232**
    of eye (*see* Eye)
    facial (*see* Facial nerve)
    in food intake, 602–603
    of foot, 185
    of forearm, 169, 170
    gluteal, 174, **248**, 249
    of hand, 171

Nerve(s):
    of heart, **260**, 261, 405, 431–436
    intercostal, 150
    laryngeal, 142, **260**, 261
    of leg, 178, 179, 183
    of mastication, 132
    mixed, 212
    motor, 212, **250**, 251
    of muscles, 122–123
    of neck, 138, 139
    pectoral, 149
    pelvic, 156
    peripheral, **212**
    radial, **105**
    of renal blood vessels, 678
    sensory, 212
    of skin, 70
    spinal, 239, **240**, 241, **242**
    of thigh, 176, 177
    of thoracic wall, 150
    of tongue, 133
    (*See also* Cell body; Neuron; specific name of nerve and organ
Nerve bundle, **212**, 213
Nerve endings, **213**–215
Nerve energies, law of specific, 266–267
Nerve fiber, 210, **212**
Nerve of Hering, 259, **435**, 436
Nerve impulse, 215
Nerve pathways (*see* Pathways)
Nerve plexuses, 241–248, 292, 295, 608
    (*See also* specific name of plexus)
Nerve processes, 210
Nervi erigentes, 773
Nervous system, 71, 225–263
    autonomic, 289–302
    central, 227–239, 786
    peripheral, 239–263
    (*See also* Higher-center control)
Neural canal, **78**
Neural crest, 225
Neural crest cells, **78**
Neural groove, 78, 225
Neural regulation of ventilation, 571, **572**
Neural tube, **78**, 225, **226**, 227
Neurilemma, 211, **212**, 218
Neuroanatomy, 5
Neuroblasts, 226
Neurofibrils, 210
Neuroglia, **214**–215
Neurohumors, 710
Neurohypophysis, 714, 716, 722
Neuromuscular junction, 122
Neurons:
    afferent (sensory), 211–212, 279
    anatomy of, 209–215
    bipolar and unipolar, 210, **211**

Neurons:
    connection, association, or internuncial, 212
    continuous and repetitive firing of, **223**
    diameter of, 217
    efferent (motor), 211–212, 279–280
    first-order, second-order, and third-order, 270, **271**
    internuncial, 279, **280**
    motor, **280**, 285
    multipolar, 210, **211**
    myelinated and unmyelinated, 292, **293**
    physiology of, 215–218
    presynaptic and postsynaptic, 218, **219**
    prolonged activity of, 223
    in reflex, **280**
    sensory, **280**
    two-motor-, pathway, 291
    types of, **211**
    typical, **210**
Neuron cell body, synapsis upon, **219**
Neutrophils, **375, 381**, 383
Newborn, 556, 782–785
Niacin, 650
Nicotinamide adenine dinucleotide (NAD), 38, 40
Nicotinamide adenine dinucleotide phosphate (NADP), 40
Night blindness, 322
Nissl bodies, **210**
Nitrogen:
    nonprotein (NPN), in plasma, 387
    in respiration, 551
    in ventilation, 541
Nitrogen balance, 634–635
Nitroglycerin, 501
Nitrous oxide, 497–498
Nobel prize, 219, 355, 711, 738
Nociceptor, 283
Nodes:
    heart, **406**–407
    lymphatic, 460–**461, 462**, 485
    of Ranvier, 211, **212, 268**
Nodules of lymph nodes, **461**
Nodulus, **357**
Nomogram, **643**
Nonelectrolytes, 14
Nonnutrient vessels, 482
Nonprotein nitrogen, 387
Norepinephrine, 297, 517, 603, 738, 739
Normality, 14
Normoblast, 378
Nose, 79–80, 129, 525–530
Notochord, **78**
Nuclear bag region, **282**
Nuclear membrane, **29**

Nuclei of Luys, 361, 362
Nucleic acids, 50–51
Nucleolus, **29**, 44, **210**
Nucleoplasm, 29, 44
Nucleotide, 50
Nucleus, 13, **210**
  of brain, **231, 233, 235, 236**
  cell, **29**, 44
  of epithelial cell, **58**
  muscle fiber, **121**
  red, in midbrain, 233
Nucleus ambiguous, **251**
Nucleus cuneatus, **273**, 274
Nucleus gracilis, **273, 274**
Nucleus pulposus, 99
Nucleus solitarius, **347**
Nutrition, 646–663
Nystagmus, 354–355

Obesity, 659–662
Oblique ligaments, 114
Oblique muscles:
  of abdomen, 151, **152–154**
  of eye, 129–**130**, 131
Obturator artery, **450**, 451, **452**
Obturator canal, **157**
Obturator externus, 173, **174**, 177,
  248, 249
Obturator foramen, **108, 158, 761**
Obturator internus, 156, **157, 174,
  248**, 249
Obturator nerve, 177, **246**, 247
Occipital artery, **442**
Occipital bone, 89, **90**, 91
Occipital lobe, 228, **229**, 230
Occipital nerve, 242, **243**
Occipital nodes, **462**
Occipital sinus, **455**, 456
Occipitalis muscle, 126, 127
Occipitofrontalis muscle, 126
Ocular conjunctiva, **309**
Oculomotor nerve, 131, **232**, 250,
  251, **252–253**
Odontoid process, 99, **100**
Ohm, George S., 20
Ohms, 21
Olecranon process, 106, 165, **168**
Oleic acid, **627**
Olfaction, 349–351
Olfactory bulb, 249, 251, **252, 312,
  350**
Olfactory epithelium, 349, **350**, 527
Olfactory gland, **349**
Olfactory hairs, **349**
Olfactory nerve, 248–**250**, 251, **252,
  349**
Olfactory nerve filament, 349
Olfactory neuron, 349
Olfactory pits, 79–80

Olfactory placode, 349
Olfactory stria, **350**
Olfactory tract, 249, **252, 349, 350,
  364**
Olfactory trigone, 229, 249, **252**, 349,
  350
Oligodendroglia, 215
Olivary nucleus, **339**, 342
Olive, **232, 235**
  nucleus of, **236**
Olivocerebellar tract, 356–357
Omentum of stomach, **589**
O-methyl transferase, 297
Omohyoid muscle, **136, 137**, 138
Oogenesis, 771, **772**
Ophthalmic artery, 313, 443
Ophthalmic nerves, **254**, 255, 313
Opponens digiti minimi, 171
Oppenens pollicis, 171, 172
Opsins, 322
Optic bulb, 304
Optic chiasma, **252, 312**
Optic cup, 80, 81, 304, 305
Optic disk, 308, 310–**311**
Optic foramen, **94**
Optic nerve, **130**, 227, **232**, 249–**250**,
  251, **252**, 253, **304, 305**, 308, 309,
  311–313
Optic radiations, **312**, 313
Optic stalk, 80, 304
Optic tract, **233**, 250–252, **312**–313
Optic vesicles, 80, 227, 304, **305**
Optical system of eye, 316
Ora serrata, 307, **308, 309**
Oral cavity, **12**, 136, 581–**582**, 583
Oral fossa, 580
Oral pit, 79
Orbicularis oculi, 126–**128, 306**
Orbicularis oris, 127, **128**, 129
Orbital cavity, **12**, 89, **92, 94**
Orbital fissures, **93, 94**
Organ of Corti, 256, 334, 337
Organelles, **27, 28**, 29
Organic compounds, 14
Organology, 4
Orgasm, 774
Oropharynx, 530, **582**, 587
Orthopnea, 572–573
Os, internal and external, **764**
Oscilloscope, 408
Osmolarity, 15
Osmoreceptors, 696–697
Osmosis, 17, **18**, 19
Osmotic diuretics, 687
Osmotic pressure, 17, **18**, 385, 483,
  512
Osseus labyrinth, 334, **336**
Ossicles, ear, **331, 333, 335**
Ossification centers, **85**, 86
Osteoblasts, 86, 87, 731, 732
Osteoclasts, 87, 731, 732

Osteocytes, 84, 86, 87
Osteology, 4
Osteon, 87
Otic ganglion, **290, 294**, 295
Otic pits, 81, 331
Otic placode, 81, 331
Otitis media, acute, 344
Otocysts, 81, 331
Otolithic membrane, 353, 355
Oval window, 334, **335–337**, 340, **341**
Ovarian follicles, 721
Ovarian hormones, 742
Ovarian vessels, **764**
Ovary, 762, **763, 764**
  (*See also* Gonads)
Ovulation, 721, 744–745, 763–764
Ovum, 74, **763**, 765, **772**, 775
Oxalites, 392
Oxidation, 37
  beta, 637
  lipid, 637
Oxygen, **13**
  in blood, 377, 379–381, 424
  in carbon dioxide transport, 559
  in coronary flow, 500
  diffusion of, 550–553
  partial pressure of, 549–556, 559
  in physical solution, 556–557
  transport of, 553–557
  in ventilation, 564, 566–568
  (*See also* Ventilation)
Oxygen capacity, 553–554
Oxygen consumption in calorimetry,
  640–642
Oxygen debt in muscle contraction,
  195–**196**
Oxygen dissociation curve, **554**–555
Oxygen poisoning, 557
Oxyphil cells, **730**

P wave, 411
P-R interval, **411**, 412
Pacemaker, 204, **406**, 407, 423, 432,
  433
Pacinian corpuscle, **213**, 214, 267,
  268, 273
Pain, 270–272, 277, 278
Pain receptors, **212**, 567
Pain reflexes, **569**
Palate, 133–**135, 526**, 527, **528**, 581,
  582
Palatine bone, 89, **95**
Palatine nerves, 255
Palatine process, 94, **95**
Palatine tonsil, 81, **582**
Palatoglossal fold, **582**
Palatoglossus, 133, **134**, 135, 587
Palatopharyngeal fold, **582**
Palatopharyngeus, 133, **135**, 141, 587

Pallor, 517
Palmar, as term, 7
Palmar arches, 445, **446**
Palmar plexus, **459, 462**
Palmaris brevis, 171, **172**
Palmaris longus, 165–**167**, 169
Palpebrae superioris, **306**
Pancreas, 580–**581, 589, 594, 599–600**, 739–741
Pancreatic duct, 600
Pancreatic hormones, 739–741
Pancreatic juice, 615–616
Pancreaticoduodenal artery, **448**
Pancreatitis, 616
Pancreozymin, 616
Paneth cells, 591
Panhypopituitarism, 724
Pantothenic acid, 651
Papilla:
    dermal, 68
    filiform, fungiform, and vallate, of tongue, 583
    of kidney, 669
    of tongue, **345, 346**
Papillary layer, **69**
Papillary muscles, 402, **403**, 404, 420
Para-aminohipuuric acid (PAH), 689
Paraaortic nodes, **462**
Parallel muscle, **123**
Paralysis, 201
    flaccid, 287–288
    of respiratory muscles, 546
    spastic, 288
Paramesonephric ducts, 756
Paranasal sinuses, 527, **529**
Paraolfactory area, **350**
Parasympathetic ganglia, 292
Parasympathetic nerves, 291–296
Parathyroid hormone (PTH), 731–**733**
Parathyroids, **81**, 730–732
Paraventricular nucleus, **715**
Parietal bones, 89, **90, 92**
Parietal layer of membrane, 63
Parietal lobe, 228, **229, 230**
Parkinson's disease, 361
Parotid gland, **290, 294**, 584–586
Parotid nodes, **462**
Pars distalis, 714, **715**, 716
Pars intermedia, 714, **715**, 716
Pars nervosa, 714, **715**, 716
Pars tuberalis, 714, **715**, 716
Partial pressure, 549–556, 558–559
Particles, mitochondrial, 37, 38, 40
Partition coefficient, 30
Parturition, 782
Passages, conducting, respiratory, 524
Patella, 107, 109, **110**, 114, **115**, 174, **175**
Patellar ligament, 114, **115, 175**
Patellar tendon, 114

Pathways:
    afferent, 356–359
    auditory, **339**
    cerebellar, **358**
    corticobulbar, 285
    corticospinal, 285, **286**
    efferent, 359
    extrapyramidal, 285–287
    final common, 287
    motor, 284–287
    olfactory, **350**
    pyramidal, 285, **286**
    sensory, 268, 270–274
    taste, **347**
    vestibular, **353**
    visual, **312, 328**
Pectineus, 177
Pectoral branch of thoracoacromial trunk, **445**
Pectoral girdle, 102
Pectoral muscles, 143, 149
Pectoral nerves, 149, **244**, 245
Pectoralis major, 143, 149, **152, 153**, 167
Pectoralis minor, 143, 149
Pedicle, **98, 99**
Peduncles, cerebellar, **357, 358**
Pellagra, 650
Pelvic cavity, **10**, 11, 107, 156–158
Pelvic girdle, 102, 107
Pelvic muscles, 155–159
Pelvimetry, 107
Pelvis, 107
    arteries to, **450–451**
    lymphatics of, 463
    male, **157, 158, 674**
    muscles of, 155–159
    nerves of, 156
Penis, 159, **160, 674**, 760–762, 773
Pentoses, 50
Pepsin, 611, 613, 622
Perception, 265–277
Perforating branches, **452, 453**
Pericardiophrenic artery, 448
Pericardium, 63, 399–**400**, 403
Perichondrium, **61**, 86
Perilymph, 334, **336–338**, 340, **341**
Perimysium, 123
Perineal artery, **449, 450**
Perineal ligament, **158**
perineum, **768**
    arteries to, **450–451**
    central tendon of, **160**
    muscles of, 159–**160**
Perineurium, **212**
Perineus muscles, **158**, 159, **160**
Periodicity principle, 342
Periosteal bone collar, **85**
Periosteal bud, **85**
Periosteum, **85**, 86, 88
Peripheral, as term, 7

Peristalsis, 607–610, 620
Peritoneal cavity, 588
Peritoneum, 63, 588, **593**
Peritonitis, 592, 765
Perivascular feet, **214**, 215
Permeability, membrane, **16**, 551–552
Peroneal artery, 451, **453**
Peroneal nerve:
    common, 247–**248**, 249
    deep, 179, 185
    superficial, 179
Peroneal retinaculum, **180**
Peroneus brevis muscle, 178, 179, **180–182**
Peroneus longus muscle, 178, 179, **180–182**
Peroneus tertius muscle, 179, **180**, 182
Personality and prefrontal area, 362–363
Perspiration, 695, 696
    (See also Sweating)
Petrosal nerve, **259, 294**, 295, **335**
Petrosal sinus, **455**
Peyer's patches, 592
pH, 15
    of blood, 556, 563–**564**
    body fluid, 699–705
    of gastric juice, 611, 612
Phagocytes, 383
Phagocytosis, 33, **41**, 42, 382
Phagosome, 41
Phalanges, 103, **106, 107**, 109, **111**
Pharyngeal cavity, **582**
Pharyngeal nerves, **259, 260**, 261
Pharyngeal plexus, 133, 141
Pharyngeal pouches, **81**
Pharyngeal tonsil, **528**, 529
Pharyngotympanic tube, 134, **135**, **331, 333, 335**, 344–345, **528**, 530
Pharynx, **79, 81, 135**, 139, **140**, 587
Phases in cardiac cycle, 417–420
Phenylethanolamine-*N*-methyltransferase (PNMT), 738, **739**
Pheochromocytoma, 607, 739
Phlegm, 524
Phocomelia, 111
Phonation, 542
Phonocardiograph, 420
Phosphate, 686, 731–732
Phosphocreatine, 193, **194**
Phosphodiesterase, 711
Phospholipid, **389, 390**
Phosphorus, 657
Phosphorylase, 711
Phosphorylation, 300, 630
Photopic vision, 323
Phrenic arteries, **448**
Phrenic nerve, 242, **243**
Physiology, definition of, 4, 11–12
Pia mater, **238, 239, 311, 313**

Pseudopodia, 385
Pseudostratified epithelium, **57**
Psoas muscles, 151, 154, **155, 246**
Pterygoid canal, 295
Pterygoid hamulus, **95**
Pterygoid muscles, 131, **132**
Pterygoid plates, **95**
Pterygoid process, 94
Pterygomandibular raphe, **140**
Pterygomaxillary fissure, **90**
Pterygopalatine fossa, 443
Pterygopalatine ganglion, **290, 294**
Ptyalin, 586, 605, 609, 616
Pubic symphysis, **674**
Pubic tubercle, **108, 760**
Pubis, **108, 761**
Pubococcygeus, 156, **157,** 158
Puborectalis muscle, 156, **157,** 158
Pubovaginalis, 158
Pudendal artery, **449, 450,** 451, **452**
Pudendal nerve, **248**
Pulmonary (*See also* Lungs)
Pulmonary arteries, **399, 401,** 440,
    441, 512, **783**
Pulmonary capillaries, 512
Pulmonary circulation, 440, **441,**
    509–512
Pulmonary congestion, 573
Pulmonary edema, 572, 574
Pulmonary embolism, 512
Pulmonary fibrosis, 546–547
Pulmonary flow, 509–512
Pulmonary function, tests of,
    543–544
Pulmonary trunk, **400, 401, 403,** 440,
    441
Pulmonic valve, **402, 403**
Pulmonary veins, 440, **783**
Pulp, red and white, 463–**464**
Pulsatile, 470
Pulse pressure, 470, 509
Pump, sodium, 19, 35
Puncta lacrimalia, 306, **307**
Pupil, 307, **308,** 319–**321,** 327
Purines, 50
Purkinje fibers, 407
Purkinje system of heart, 405, 407
Pus, 383
Putamen, 231, 361
Putrefaction, 619
Pyelography, retrograde and
    intravenous, 675
Pyloric antrum, **589**
Pyloric arteries, **458**
Pyloric canal, **589,** 590
Pyloric glands, 591
Pyloric sphincter, **589,** 590
Pyloric stenosis, 591
Pylorus, **594**
Pyramidal cells, 210
Pyramidal pathways, 285, **286**

Pyramids:
    decussation of, **232,** 234–**236,** 285,
        286
    of medulla, 234–**236**
    of middle ear, 333, **335**
Pyramidalis muscle, 151, **152,** 153
Pyridoxine, 650–651
Pyrimidines, 50
Pyrogen, 302
Pyrroles, 378
Pyruvic acid, 631, 632

Q wave, 412, **415**
Q-T interval, **411,** 412
QRS complex, 411–**413,** 414, 415
Quadrate lobe, **597**
Quadratus femoris, 173, **174, 248,**
    249
Quadratus lumborum, 151, 154, **155**
Quadratus plantae, 184–**186**
Quadriceps femoris muscle, 109
Quadriceps tendon, 114, **115, 175**
Queckenstedt test, 509

R wave, 413
Radial artery, 444, **445, 446**
Radial nerve, **105,** 163, 170, **244**–246
Radialis indicis proprius artery, **446**
Radiation, heat loss through, 301,
    302
Radioimmunoassay, 714
Radiospiral nerve, 106
Radius, 103, **105, 106**
Rami communicantes, **240, 292, 293**
Ramus, 93, **108,** 109, **240,** 241, **242,**
    243
Raphe system, 366
Rapid-ejection phase, **417,** 418
Rapid-eye movement (REM) sleep,
    366
Rapid-inflow phase, **417,** 419
Ratchet theory, 193, 195
Raynaud's disease, 517
Reactive hyperemia, 474–**475**
Reactivity, thalamus and, 273–274
Rebound phenomenon, 360–361
Receptor endings in skin, 70
Receptors, **213**–214, 220, 266–**268,**
    269, 297–298, 340, 350–351,
    433–436
Rectal arteries, **449, 450,** 451
Rectal values, 594, **595**
Rectouterine pouch, 594, 765
Rectovesicular pouch, 594, 672, **674**
Rectum, 593–**594, 595**
Rectus abdominis muscle, 151,
    152–154
Rectus femoris muscle, **115,** 174, **175,**
    176

Rectus muscles of eye, 129, **130,** 131,
    308, 309
Rectus sheath, 153, **154**
Red nucleus, 233, 287, **358, 359**
Red pulp, 463–**464**
Reduced-ejection phase, **417,** 418
Reduction, 37
    open and closed, of bone,
        113
Reflexes, 281–284
    antigravity, 282
    aortic sinus, 566
    attenuation, 340
    Bainbridge, 435
    blood pressure and, 478, 479
    carotid body, 566
    conditioned, 284
    flexor, 283–284
    heart and, 435–436
    joint, 568
    pain and protective, 321, 569
    simple, **280**
    stretch, 282–283
    supporting, 284
    taste, 348
    of ventilation, 565–569
    visceral and somatic, 289
    visual, 319–321
Reflex arc, 279, 434
Reflex latent period, 281
Reflex movement, 277–288
Refraction, 313–**314, 315**–316
Refractive errors, 324–**325,** 326
Refractive media, 311–312
Refractory periods, 204, 216
Refractory state of muscle, 199
Regeneration:
    of muscle, 201, 202, 205
    of neuron, 218
Regurgitation, 423, 430, 609
Relaxin, 744
Releasing agents, 389
Renal (*see* Kidneys; Urinary system)
Renal arteries, **448,** 449, **669, 670**
Renal blood flow, 689
Renal capsule, **670**
Renal circulation, 518–519, 676–679
Renal clearance, concept of,
    687–690
Renal columns, 669
Renal corpuscle, 672
Renal function, 690, 703–704, 786
Renal hilus, 668, 669
Renal pelvis, 668, **669, 670**
Renal pyramids, 668, **670**
Renal tubule, 670–672
Renal vein, **669, 670**
Renal vessels, innervation of, 678
Renin, 492, 672, 678, 736
Rennin, 612
Replication, DNA, 52

Reproduction, 770–786
  cellular, 45–46
Reproductive cells, 46
Reproductive system, 72, 754–768
  female, 761–768
  male, 756–761
Reserve volume, 539, **540**
Residual air, 539, **540,** 553
Residual capacity, functional, 539, **540**
Residual volume, 539, **540**
Resistance to blood flow, 468–469, 475
Resistance vessels, 476
Respiration:
  artificial, 544, **545,** 546
  blood pressure and, **510**
  buffer systems and, 702–703
  control of, **567**
  fetal, 784–785
  internal and external, 423
  neural regulation of, 571
  (*See also* Ventilation)
Respirator, 546
Respiratory air volumes, 539, **540**
Respiratory bronchioles, 525, **526**
Respiratory center, 562, 571, **572, 573**
Respiratory chain, 37–**38,** 39
Respiratory gases, diffusion and transport of, 548–560
Respiratory infections and ear involvement, 344
Respiratory mechanism, **537**
Respiratory membrane, 551, **552**
Respiratory quotient, 559–560, 641–642
Respiratory system, 71, 523–525
Response:
  all-or-nothing, 216–217
  frequency of, in neuron, 216
Resting potential, 33–35
Rete testis, **757,** 758
Reticular fibers, 59, **60**
Reticular formation, 234, **235,** 355–356, **358,** 359
Reticular nucleus, **236**
Reticular tissue, **62**
Reticulocerebellar fibers, 357
Reticuloendothelial system, 62–63, 382
Reticulospinal tract, 359
Reticulum, 43, 382
Retina, 304, **305,** 308, 309–**310,** 312, 318, **320**
Retinacula, 114
Retinal artery and vein, **308, 311,** 313
Retinene, 322
Retrograde auditory impulses, 342
Retrolental fibroplasia, 557
Retropubic space, **674**
Rh factor, 395

Rhinitis, 527
Rhodopsin, 321–**322**
Rhombencephalon, 226
Rhomboideus muscles, 142, **144,** 145, **163**
Rhythm(s):
  biological, 751
  circadian, 366–**367,** 751
  (*See also* Heart rhythm)
Rhythm method of contraception, 777
Riboflavin, 649–650
Ribonucleic acid (RNA), 51, 52, 368
Ribose, 50
Ribosomes, **29,** 43, 52
Ribs, 98, **98, 101, 102**
Rickets, 111–112, 653
Rigor mortis, 201
Rima glottidis, 532
Rima palpebrarum, 305
Rigidity, decorticate and decerebrate, 287
Ring of spermatic cord, **760**
Ringer's solution, 431
Rinne test, 343
Risorius, 127, **128,** 129
RNA (ribonucleic acid), 51, 52, 368
Rods and cones, 249, 309, **310,** 321–**322, 323,** 327
Roentgenography, 112
Rootlets, **240, 262**
Roots, 238, 240–245, 293, 299
  (*See also* Dorsal root ganglion)
Rotameter, 471
Rotation muscles, **124,** 125
Rotatores, muscle, 143, 148
Rouleau formation, 375
Round window, 334, **335–337,** 340, **341**
Rubrospinal nerve, **270**
Rubrospinal tract, 286–287, **358**
Ruffini, corpuscle of, **213**
Rugae, 590

S-A node (*see* Sinoatrial node)
S-T segment, **411,** 412, 414, **415**
Saccule, 334, **336, 337,** 353, 354
Sacral arteries, **450,** 451
Sacral cornu, **100**
Sacral crest, **100**
Sacral hiatus, 99, **100**
Sacral plexus, 156, 247–**248**
Sacral promontory, 99, **100**
Sacral region, nerves of, 295
Sacral vertebra, **100**
Sacroiliac articulation, 99
Sacroiliac joint, 109
Sacrospinalis, **145**
Sacrum, **96,** 97, 99, **100, 108**

Sacs of serous membranes, 63
  (*See also* Air sacs)
Saddle joint, 114
Sagittal plane, **8**
Sagittal sinuses, 454–**455, 507, 508**
Saliva, 586
Salivary glands, 584–586
Salivation, 604–606
Salivatory nuclei, **251**
Saltatory propagation, 215
Salpingopharyngeal fold, **528**
Salpingopharyngeus, **135,** 139, 141
Saphenous veins, 458, **459**
Sarcolemma, **121**
Sarcomere, **121,** 189
Sarcoplasm, 120
Sarcoplasmic reticulum, 194
Sartorius, 174, **175,** 176
Satiety center, 602
Scala media, 334, 336
Scala tympani, 336, **337,** 340
Scala vestibuli, 336, **337,** 340
Scalene muscles, 138, 139
Scalp, muscles of, 126
Scaphoid bone, 103, **106**
Scapula, **101,** 103, **104, 105,** 142, 143
Scapular arteries, **442,** 444, **445**
Scapular nerve, dorsal, 145, **244,** 245
Scapular notch, **104**
Sciatic nerve, 177, 247–**248,** 249
Sciatic notches, **108,** 109
Sciatica, 99
Schally group, 717
Schwann cell, 211
Sclera, 81, **305,** 307, **308, 309**
Scleral spur, 308, **309**
Sclerocorneal junction, **309**
Sclerotome, **78**
Scoliosis, 97
Scotopic vision, 323
Scotopsin, 322
Scrotum, 756
Scurvy, 652
Sebaceous glands, 68, **69**
Sebum, 68
Second messenger, 711–713
Secretin, 615, 616
Secretions, mucous, 64, 65
Segmental branches, 440
Segmentation, 614
Segregation, law of, **47**
Sella turcica, 91
Semen, 713, 774, 775
Semicircular canals, 81, 256, **257, 331,** 334, **336,** 352, **353,** 354
Semicircular duct, **336**
Semicircular hiatus, **528**
Semilunar valves, 402
Semimembranosus muscle, **115,** 177, **178**
Seminal colleculus, **673**

Tubular cells, 31, **32**
Tubular glands, **65**, 66
Tubular mass, 690
Tubular reabsorption, 681
Tubular secretion, 689
Tubules:
    renal, 670–**671**, 681–683
    T and longitudinal, 194, 195
Tubuli recti, **757**, 758
Tunica adventitia, **438**, 439
Tunica albuginea, **757**, 762, **763**
Tunica intima, **438**
Tunica media, **438**, 439
Tunics:
    of alimentary canal, 588, **590**, 591
    of uterus, 766
Turbinate bones, 527
Turner's syndrome, **82**, 83
Twitch, muscle, 197
Tympanic cavity, **81**
Tympanic membrane, **257**, 294, **331**,
    332–**333**, 336, 338–339
Tympanic nerve, **259**, 294, 295
Tympanic plexus, **294**, 295, 334, **335**
Tympanum, 333–334

Ulcer, 613–614
Ulna, 103, **105, 106**
Ulnar arteries, **445, 446**
Ulnar nerve, 169, 171, **244**–246
Umbilical cord, **780**, **783**
Umbilical region, **10, 11**
Umbilical vessels, 779, **780, 783**, 784
Umbo, 333
Uncinate process, **589**
Unipennate muscle, **123**
Universal donor and recipients, 395
Urachus, 672, **761**
Uremia, 686
Ureteral orifice, **673**
Ureters, 670, 672, **673**, 691, **761, 762**
Urethra, **158**, 159, **160**, **673**–674
    female, **768**
    male, 758, 759, **761, 762**
Urethral hiatus, **157**
Urethral meatus, **674**
Urethral orifice, 672, **673**, 674
Urethral sphincter, **158**, 159
Urinary bladder, 298–**299**, 672, **673**,
    691 **761, 762**
    automatic, 692
Urinary system, 72, 667–692
    (*See also* Kidneys)
Urination, 690–692
Urine, **32**, 672, 679–685
Urobilin, 620
Urogenital diaphragm, **158**–159, **673**,
    674, 754
Uterine artery, 451
Uterine glands, **74**, 766, 767, **780**
Uterine tubes, 764–765

Uterovesicle pouch, 672, 765
Uterus, 722, 723, **764**, 765–**766**, 767,
    782
Utilization:
    carbohydrate, 631–632
    energy, table on, 644
Utilization coefficient, 555
Utricle, 334, **336**, **337**, **353**, 355
Utriculosaccular duct, **337**
Uvula, **133**–**135**, **357**, **582**

Vaccination, 382
Vacuoles, **41**–43
Vagal escape, 432
Vagal nerves, 565
Vegal tone, 432
Vagina, 156, **158**–**160**, **764**, 765, 767,
    768
Vaginal artery, 451
Vagus nerve, **232**, **236**, **250**, **251**, **258**,
    260, 261, 295, **347**, **406**, 431–**433**,
    435
Vagus nucleus, **236**
Valence, 14
Vallate papillae, 345, **346**
Vallecula, 587
Valves:
    aortic, **402**, **417**, 427–428, 430
    atrioventricular, **402**, 420–423, 489
    of heart, **401**–**402**, 420, 430
    of veins, 439
Vas deferens, **673**
Vasa recta, 670, **671**, 684
Vasa vasorum, 438
Vascular choroid, 81
Vascular spasm, 388
Vasconstriction, 475–**477**, 478, 487,
    518
Vasoconstrictors, 438
Vasodilation, 473, 475–**477**, 478, 490,
    518
Vasodilators, 438
Vasomotion, 482, 518
Vasomotor activity, 476–**477**,
    486–487, 510–511
Vasomotor center, 478
Vasomotor tone, 476, 478
Vasopressin, L-, **725**
Vastus lateralis muscle, **115**
Vastus medialis muscle, **115**
Vastus muscles, 174, **175**, 176
Veins, **438**, 439, **459**
    azygos, 456, **457**
    of extremities, superficial, 457–458
    of head and neck, 453–**454**
    of heart, **404**, 405
    hepatic, 456, **458**, 513, 597, **783**
    Thebesian, 405
    valves of, 439
    (*See also* specific vein; Venous)
Velocity of propagation, 216, 217

Vena vorticosa, **308**
Venae cavae, **155**, **399**–**401**, **403**, **435**,
    454, 456, **457**–**459**, **783**
Venae comitantes, 452
Venae cordis minimae, 404, 405
Venoconstriction, 476, 487
Venomotor changes, 487
Venous circulation, 485–489
Venous drainage, systemic, 452–460
Venous plexus, **516**
Venous pressure, 429, **471**, **472**
Venous return, 426–427
Venous sinuses, 454–**455**, 456
Ventilation, 536–547
    in exercise, 570–571
    reflexes of, 565–569
    regulation of, 561–574
    (*See also* Respiration)
Ventral, as term, 7
Ventral spinothalamic tract, 272,
    273
Ventricles:
    of brain, **233**, 234, **235**–**237**, 507
    fourth, of cerebellum, **357**
    of heart, 400, **401**, **403**, 486
    of larynx, **531**
    lateral, horns of, of eye, **312**
Ventricular pressure, **417**–419, **421**,
    432
Ventricular systole, 422
Venules, 439, **481**, 482, 779, **780**
Vermis, 236, 356
Vertebra(e), 89, **96**, 97, **98**
Vertebral artery, 99, **442**, **443**, 444
Vertebral canal, 239
Vertebral column, 96–101, 142, 143,
    148, 240
Vertebral foramen, **98**, 99, **100**
Vertebral ganglia, 291–292
Vertebral lamina, **98**, 99
Vertebral plexus of veins, 455
Vertebral vein, 99
Vertebrocostal triangle, **155**
Verticalis muscle, 133, **134**
Vesical arteries, **450**, 451
Vestibular apparatus, 352–355
Vestibular ganglion, **257**, 353
Vestibular glands, 767, 775
Vestibular ligament and fold of
    larynx, **531**
Vestibular nerve, **257**, 258
Vestibular nucleus, 236, **251**, 353
Vestibular pathways, **353**
Vestibules, 334, **337**, 527, **528**, 581
Vestibulocerebellar neurons, 357
Vestibulocerebellar tract, **358**
Vestibulocochlear nerve, **232**, **250**,
    256–**257**, 258
Vestibulospinal nerve, **270**
Vestibulospinal tracts, 286–287, **353**
Vibrations per second, 338, **340**
Vibrissae, 527, **528**